SPRINGER PROTOCOLS HANDBOOKS

For further volumes:
http://www.springer.com/series/8623

Animal Models of Acute Neurological Injuries II

Injury and Mechanistic Assessments, Volume 2

Edited by

Jun Chen

Department of Neurology and the Center for Cerebrovascular Disease Research, University of Pittsburgh School of Medicine, Pittsburgh, PA, USA

Xiao-Ming Xu

Departments of Neurological Surgery and Anatomy & Cell Biology, Spinal Cord and Brain Injury Research Group, Stark Neurosciences Research Institute, Indiana University School of Medicine, Indianapolis, IN, USA

Zao C. Xu

Department of Anatomy & Cell Biology, Indiana University School of Medicine, Indianapolis, IN, USA

John H. Zhang

Department of Neurosurgery, Loma Linda University School of Medicine, Loma Linda, CA, USA

Humana Press

Editors
Jun Chen
Department of Neurology
and the Center for Cerebrovascular
Disease Research
University of Pittsburgh School of Medicine
Pittsburgh, PA, USA

Zao C. Xu
Department of Anatomy & Cell Biology
Indiana University School of Medicine
Indianapolis, IN, USA

Xiao-Ming Xu
Departments of Neurological Surgery and
Anatomy & Cell Biology
Spinal Cord and Brain Injury Research Group
Stark Neurosciences Research Institute
Indiana University School of Medicine
Indianapolis, IN, USA

John H. Zhang
Department of Neurosurgery
Loma Linda University School of Medicine
Loma Linda, CA, USA

ISSN 1949-2448 ISSN 1949-2456 (eBook)
ISBN 978-1-4939-6052-1 ISBN 978-1-61779-782-8 (eBook)
DOI 10.1007/978-1-61779-782-8
Springer New York Dordrecht Heidelberg London

Softcover re-print of the Hardcover 1st edition 2012

Printed on acid-free paper

Humana Press is part of Springer Science+Business Media (www.springer.com)

Foreword

The second edition of this outstanding reference book is being published in very challenging times. Ischemic and traumatic CNS injuries remain a huge public health problem throughout the world. Stroke continues to be ranked as the third leading cause of death in all developed countries and arguably the major cause of long-term disability. Within the USA alone, estimates are currently at 1.2 million stroke incidents each year, leading to over 150,000 outright deaths. The picture for traumatic brain injury (TBI) and spinal cord injury (SCI) is equally daunting and impacts the young and the very old of our population disproportionately. In Europe, head-injured individuals living with significant disability number near seven million. In China, estimates indicate that over three million people each year endure significant brain injury. These facts emphasize the need for innovative approaches to CNS injury research and a reinvigoration of our mandate to translate data from bench-to-bedside.

Equally as challenging is the lack of safe pharmacological agents that consistently improve clinical CNS injury outcomes. Little new pharmacology has entered clinical practice for TBI in over three decades, and even investigational therapies, such as hypothermia remain uncertain in safety or efficacy for heterogeneous patient populations. At this writing, the stroke research community is thoughtfully questioning the achievability of therapeutic "neuroprotection" in humans. The latest large-scale clinical trial failure is still fresh in our minds, as we review summaries of the Stroke Acute Ischemic NXY-059 Treatment Trial (SAINT II). Bench *and* clinical researchers are looking for signs of hopeful progress, despite SAINT II's negative results. Many an international stroke conference opens its sessions with the reminder that recombinant tissue plasminogen activator (tPA) remains the only FDA approved agent for acute stroke treatment. While tPA does not repair post-ischemic cellular damage, its efficacy emphasizes the oldest research concept in our armamentarium, i.e., the importance of restoring perfusion. Studies of optimal recanalization remain crucial, fueled by the extension of tPA's therapeutic window beyond 3 h. The latter aspect may widen tPA's utility in acute stroke therapy in a variety of countries.

Despite this litany of challenges, basic and translational CNS injury researchers can point to many areas of progress and changing experimental paradigms that are vigorous and productive. Compelling directions include a focal shift from neuron-centric pharmacological agents into (1) integrated evaluations of the neurovascular unit, (2) increased investigational intensity into neuro-regeneration and the therapeutic boundary between molecular protection vs. recovery mechanisms, (3) a broader research stance that addresses stroke as a "systemic disease" that impacts non-CNS organs and the immune system, and (4) creative endovascular approaches to clot retrieval and manipulation.

Essential to all of these challenges and opportunities is the understanding, and precise use, of animal models in CNS injury research. In this second edition, Drs. Jun Chen, Xiao-Ming Xu, Zao Xu, and John Zhang provide a vital reference that will advance our field and educate researchers at all levels. These substantive volumes provide a manual for expert use of animal models of cerebral ischemia, intracerebral and subarachnoid hemorrhage, vasospasm, TBI, and SCI and emphasize mechanistic evaluations in all areas. An impressive array of internationally valued experts and leading investigators have authored chapters which are oriented around disease processes, culminating in a concise methodological reference that the reader will depend on as a guide to success. The volumes have been thoroughly updated and emphasize procedures in a way that captures underlying cellular and molecular mechanisms at work in injured tissue. Helpful material outlines procedural pitfalls and ways to avoid common failures that can confound in vivo studies.

For newcomers to CNS injury, Part I through V of Volume 1 are essential as they survey the principles and technical nuances of all current methodologies appropriate for comprehensive animal studies. Part VI–VIII focus on hemorrhage and vasospasm, a topic of great importance as patients with these diseases are commonly found in neuroscience critical care units around the world. Part I and II of Volume 2 contain the foundations of CNS ischemic models and offer broad coverage of morphological, physiological, molecular, and behavioral assessments, with new emphasis on neurogenesis and recovery parameters. Part III focuses on neonatal injury, a highly specialized and under-studied area of ischemic pathology that is frequently overlooked in reference compendia. Part IV and V complete the survey of injury models with key material on the assessment of TBI and SCI, emphasizing up-to-date imaging techniques that are elemental to the field.

The use of sophistical animal models remains the bedrock of CNS injury research. There is currently no substitute for the intact animal if our goal is to evaluate microcirculatory networks, the cells and units these networks serve, or the integrated functional output that is the basis of life. *Animal Models of Acute Neurological Injuries II* is an invaluable aid to junior investigators and more senior scientists who are forging new directions in their science or perhaps engaging translational hypotheses for the first time. This valuable collection of experience from the best in the field offers a thoughtful opportunity to all of us faced with challenging times.

Portland, OR, USA ***Patricia D. Hurn***

Preface

Our first book, entitled *Animal Models of Acute Neurological Injuries*, published by Humana Press/Springer (ISBN: 978-1-60327-184-4), was a great success. In that book, we constructed a benchwork manual for the most commonly used animal models of acute neurological injuries, including cerebral ischemia, hemorrhage, vasospasm, and traumatic brain, and spinal cord injuries. Since its publication, the book has been well received by both clinical and basic researchers due to its practicality.

Since the animal models for acute neurological injuries are in place, our next goal, i.e., the goal of these books, was to construct chapters on assessing these disorders from cells and molecules to behavior and imaging. These comprehensive assessments are the key for understanding disease mechanisms as well as developing novel therapeutic strategies to ameliorate or even prevent damages to the nervous system.

The layout of these books is disease/disorder oriented, which encapsulates several parts, including sections on global cerebral ischemia, focal cerebral ischemia, neonatal hypoxia–ischemia, subarachnoid hemorrhage, cerebral vasospasm, intracerebral hemorrhage, traumatic brain injury, traumatic spinal cord injury, and general assessments. Each part includes coverage of morphological, physiological, biochemical, neurobehavioral, and neuroimaging assessments. In general, each begins with an introductory discussion on the availability and selection of specific assessments for specific injury models as well as their pros and cons. In some chapters, the assessments are divided according to levels of importance into "recommended methods" that are well established and most extensively used, or "alternative methods" that depict less extensively used assessments that have their own utility.

These books are designed to provide both expert guidance and step-by-step procedures, along with multiple photographs and/or schematic drawings on assessments of acute neurological injuries. Throughout each chapter, the readers are aided in understanding what, why, when, where, and how a particular assessment is used. We hope that these books can be useful for trainees or beginners in their assessments of acute neurological injuries, for experienced scientists from other research fields who are interested in either switching fields or exploring new opportunities, and for established scientists within the field who wish to employ new assessments.

We are grateful to all chapter contributors for their time, effort, and dedication. It would be impossible to publish these books without their significant contributions. Finally, we hope that the volumes will be useful for students, postdoctoral fellows, and clinical and basic scientists who would like to study acute neurological disorders.

Pittsburg, PA, USA — *Jun Chen*
Indianapolis, IN, USA — *Xiao-Ming Xu*
Indianapolis, IN, USA — *Zao C. Xu*
Loma Linda, CA, USA — *John H. Zhang*

Contents

Contributors

DANIEL P. ANKENY • *Department of Neuroscience, Center for Brain and Spinal Cord Repair, Ohio State University Medical Center, Columbus, OH, USA*
COLEEN M. ATKINS • *The Miami Project to Cure Paralysis, Department of Neurological Surgery, Miller School of Medicine, University of Miami, Miami, FL, USA*
IRINA S. BALAN • *Department of Anesthesiology and Center for Shock, Trauma and Anesthesiology Research, University of Maryland School of Medicine, Baltimore, MD, USA*
JAMES W. BALES • *Department of Neurological Surgery, Brain Trauma Research Center, Safar Center for Resuscitation Research, Center for Neuroscience, School of Medicine, University of Pittsburgh, Pittsburgh, PA, USA*
JASON E. BEARE • *Department of Neurological Surgery, Kentucky Spinal Cord Injury Research Center, University of Louisville School of Medicine, Louisville, KY, USA*
RICHARD L. BENTON • *Departments of Neurological Surgery and Anatomical Sciences & Neurobiology, Kentucky Spinal Cord Injury Research Center, University of Louisville, Louisville, KY, USA*
MEHMET BILGEN • *Department of Biomedical Imaging, Faculty of Medicine, University of Malaya, Kuala Lumpur, Malaysia*
KLAS BLOMGREN • *Department of Pediatrics, Center for Brain Repair and Rehabilitation, Institute of Neuroscience and Physiology, The Queen Silvia Children's Hospital, University of Gothenburg, Gothenburg, Sweden*
JEFFREY M. BOSKEN • *Department of Anatomy and Neurobiology, Neurosurgery, Neurology and Physical Medicine and Rehabilitation, Lexington, KY, USA; Spinal Cord & Brain Injury Research Center (SCoBIRC), University of Kentucky Midical Center, Lexington, KY, USA*
EDWARD H. BROWN • *Department of Neurological Surgery, Kentucky Spinal Cord Injury Research Center, University of Louisville, Louisville, KY, USA*
DARLENE A. BURKE • *Department of Neurological Surgery, Kentucky Spinal Cord Injury Research Center, University of Louisville, Louisville, KY, USA*
PELIN CENGIZ • *Department of Pediatrics, University of Wisconsin-Madison, School of Medicine and Public Health, Madison, WI, USA*
JINHUI CHEN • *Department of Neurological Surgery, Stark Neuroscience Research Institute, Indiana University School of Medicine, Indianapolis, IN, USA*
JUN CHEN • *Department of Neurology and the Center for Cerebrovascular Disease Research, University of Pittsburgh School of Medicine, Pittsburgh, PA, USA*
XIANG YANG CHEN • *Laboratory of Neural Injury and Repair, Wadsworth Center, New York State Department of Health & State University of New York at Albany, Albany, NY, USA*

BEVERLY S. COLLEY • *Department of Anatomy and Neurobiology, Virginia Commonwealth University, Medical College of Virginia Campus, Richmond, VA, USA*
LINGXIAO DENG • *Spinal Cord and Brain Injury Research Group, Departments of Neurological Surgery, & Anatomy and Cell Biology, Stark Neurosciences Research Institute, Indiana University School of Medicine, Indianapolis, IN, USA*
DOUGLAS S. DEWITT • *Department of Anesthesiology, Moody Center for Traumatic Brain & Spinal Cord Injury Research/Mission Connect, The University of Texas Medical Branch, Galveston, TX, USA*
C. EDWARD DIXON • *Department of Neurological Surgery, University of Pittsburgh School of Medicine, Pittsburgh, PA, USA*
FUMIO ENDO • *Division of Neonatology, Perinatal Center, Kumamoto University Hospital, Kumamoto, Japan; Department of Pediatrics, Graduate School of Medical Science, Kumamoto University, Kumamoto, Japan*
JANELLE M. FASSBENDER • *Departments of Neurological Surgery and Anatomical Sciences & Neurobiology, M.D./Ph.D. Program, Kentucky Spinal Cord Injury Research Center, University of Louisville, Louisville, KY, USA*
YU GAN • *Department of Neurology, University of Pittsburgh School of Medicine, Pittsburgh, PA, USA*
XIANG GAO • *Department of Neurological Surgery, Stark Neuroscience Research Institute, Indiana University School of Medicine, Indianapolis, IN, USA*
YANQIN GAO • *State Key Laboratory of Department of Medical Neurobiology, Institutes of Brain Science, Fudan University, Shanghai, China*
STEVEN H. GRAHAM • *Department of Neurology and the Center for Cerebrovascular Disease Research, University of Pittsburgh School of Medicine, Pittsburgh, PA, USA*
THEO HAGG • *Departments of Neurological Surgery and Pharmacology & Toxicology, Kentucky Spinal Cord Injury Research Center, School of Medicine, University of Louisville, Louisville, KY, USA*
EDWARD D. HALL • *Department of Anatomy and Neurobiology, Neurosurgery, Neurology and Physical Medicine and Rehabilitation, Lexington, KY, USA; Spinal Cord & Brain Injury Research Center (SCoBIRC), University of Kentucky Medical Center, Lexington, KY, USA*
PHILIP J. HORNER • *Department of Neurological Surgery, Institute for Stem Cell and Regenerative Medicine, University of Washington School of Medicine, Seattle, WA, USA*
JOHN D. HOULÉ • *Department of Neurobiology and Anatomy, The Spinal Cord Research Center, Drexel University College of Medicine, Philadelphia, PA, USA*
BINGREN HU • *Department of Anesthesiology and Center for Shock, Trauma and Anesthesiology Research, University of Maryland School of Medicine, Baltimore, MD, USA*
XIAOMING HU • *Department of Neurology and the Center for Cerebrovascular Disease Research, University of Pittsburgh School of Medicine, Pittsburgh, PA, USA*
MASANORI IWAI • *Division of Neonatology, Perinatal Center, Kumamoto University Hospital, Kumamoto, Japan*
LYN B. JAKEMAN • *Department of Physiology and Cell Biology, Ohio State University College of Medicine, Columbus, OH, USA*

PINAR KENDIGELEN • *Department of Pediatrics, University of Wisconsin-Madison, School of Medicine and Public Health, Madison, WI, USA; Department of Anesthesiology, Istanbul University Cerrahpaşa Medical School, Istanbul, Turkey*
TIBOR KRISTIAN • *Department of Anesthesiology and Center for Shock, Trauma and Anesthesiology Research, University of Maryland School of Medicine, Baltimore, MD, USA*
CHIA-YI KUAN • *Division of Developmental Biology and Division of Neurology, Cincinnati Children's Hospital Medical Center, Cincinnati, OH, USA*
MICHELLE C. LAPLACA • *Neural Injury Biomechanics and Repair Laboratory, Parker H. Petit Institute for Bioengineering and Bioscience, Wallace H. Coulter Department of Biomedical Engineering, Georgia Institute of Technology/Emory University, Atlanta, GA, USA*
MARCUS CHRIS LESSING • *Neural Injury Biomechanics and Repair Laboratory, Parker H. Petit Institute for Bioengineering and Bioscience, Wallace H. Coulter Department of Biomedical Engineering, Georgia Institute of Technology/Emory University, Atlanta, GA, USA*
PEIYING LI • *Department of Neurology, University of Pittsburgh School of Medicine, Pittsburgh, PA, USA*
WEIMIN LIANG • *Department of Anesthesiology, Huashan Hospital, Fudan University, Shanghai, China*
JONATHAN LIFSHITZ • *Spinal Cord & Brain Injury Research Center, University of Kentucky Chandler Medical Center, Lexington, KY, USA*
CHUNLI LIU • *Department of Anesthesiology and Center for Shock, Trauma and Anesthesiology Research, University of Maryland School of Medicine, Baltimore, MD, USA*
JIALING LIU • *Department of Neurological Surgery, University of California, San Francisco, CA, USA*
WEI LIU • *Department of Neurological Surgery, University of California, San Francisco, CA, USA; Department of Neurological Surgery, Beijing Tiantan Hospital, Capital Medical University, Beijing, China*
KRISTIN MACFARLANE • *Department of Neurological Surgery, Brain Trauma Research Center, Safar Center for Resuscitation Research, Center for Neuroscience, School of Medicine, University of Pittsburgh, Pittsburgh, PA, USA*
DAVID S.K. MAGNUSON • *Department of Neurological Surgery, Kentucky Spinal Cord Injury Research Center, University of Louisville School of Medicine, Louisville, KY, USA*
SHIRO MATSUMOTO • *Division of Neonatology, Perinatal Center, Kumamoto University Hospital, Kumamoto, Japan*
HIROSHI MITSUBUCHI • *Division of Neonatology, Perinatal Center, Kumamoto University Hospital, Kumamoto, Japan*
JOHNNY R. MOREHOUSE • *Department of Neurological Surgery, Kentucky Spinal Cord Injury Research Center, University of Louisville School of Medicine, Louisville, KY, USA*
ANTHONY ANDREW OLIVA JR. • *Center for Research on Occupational and Environmental Toxicology, Oregon Health & Science University, Portland, OR, USA*

STEPHEN M. ONIFER • *Department of Anatomy and Neurobiology, College of Medicine, Spinal Cord and Brain Injury Research Center, University of Kentucky, Lexington, KY, USA*
MIGUEL A. PEREZ-PINZON • *Department of Neurology, Cerebral Vascular Disease Research Center, Miller School of Medicine, University of Miami, Miami, FL, USA*
PHILLIP G. POPOVICH • *Department of Neuroscience, Center for Brain and Spinal Cord Repair, Ohio State University Medical Center, Columbus, OH, USA*
DONALD S. PROUGH • *Department of Anesthesiology, The University of Texas Medical Branch, Galveston, TX, USA*
THOMAS M. REEVES • *Department of Anatomy and Neurobiology, Virginia Commonwealth University, Medical College of Virginia Campus, Richmond, VA, USA*
RACHEL K. ROWE • *Spinal Cord & Brain Injury Research Center, University of Kentucky Chandler Medical Center, Lexington, KY, USA*
YIWEN RUAN • *Department of Anatomy & Cell Biology, Spinal Cord and Brain Injury Research Group, Stark Neurosciences Research Institute, Indiana University School of Medicine, Indianapolis, IN, USA*
KATHRYN E. SAATMAN • *Spinal Cord & Brain Injury Research Center (SCoBIRC), University of Kentucky, Lexington, KY, USA*
LORI-ANN SACREY • *Department of Neuroscience, Canadian Centre for Behavioural Neuroscience, University of Lethbridge, Lethbridge, AB, Canada*
ANDREW J. SALADINO • *Department of Pathology, University of Maryland School of Medicine, and Pathology & Laboratory Medicine Service, Baltimore VA Medical Center, Baltimore, MD, USA*
LYANNE C. SCHLICHTER • *Toronto Western Research Institute, University Health Network, Toronto, ON, Canada; Department of Physiology, University of Toronto, Toronto, ON, Canada*
DREW L. SELLERS • *Department of Neurological Surgery, Institute for Stem Cell and Regenerative Medicine, University of Washington School of Medicine, Seattle, WA, USA*
HONG SHI • *Department of Anesthesiology, Huashan Hospital, Fudan University, Shanghai, China; State Key Laboratory of Department of Medical Neurobiology, Institutes of Brain Science, Fudan University, Shanghai, China*
YEJIE SHI • *Department of Neurological Surgery, Neuroscience Training Program, University of Wisconsin-Madison, Madison, WI, USA; Department of Neurological Surgery, Neuroscience Training Program University of Wisconsin-Madison Madison, WI, USA*
CHRISTOPHER B. SHIELDS • *Norton Neuroscience Institute, Louisville, KY, USA*
LISA B.E. SHIELDS • *Norton Neuroscience Institute, Louisville, KY, USA*
JED S. SHUMSKY • *Department of Neurobiology and Anatomy, The Spinal Cord Research Center, Drexel University College of Medicine, Philadelphia, PA, USA*
R. ANNE STETLER • *Geriatric Research, Educational and Clinical Center, Veterans Affairs Pittsburgh Health Care System, Pittsburgh, PA, USA; Center of Cerebrovascular Disease Research, University of Pittsburgh School of Medicine, Pittsburgh, PA, USA; State Key Laboratory of Medical Neurobiology, Fudan University, Shanghai, China*

DANDAN SUN • *Department of Neurology, University of Pittsburgh, Pittsburgh, PA, USA; Department of Neurological Surgery, Neuroscience Training Program University of Wisconsin-Madison Madison, WI, USA*
FENG-YAN SUN • *Department of Neurobiology and State Key Laboratory of Medical Neurobiology, Shanghai Medical College Fudan University, Shanghai, China; Institutes for Biomedical Sciences, Fudan University, Shanghai, China*
XIAO SUN • *Department of Neurobiology and State Key Laboratory of Medical Neurobiology, Shanghai Medical College Fudan University, Shanghai, People's Republic of China*
HIROKO TAJIRI • *Department of Pediatrics, Graduate School of Medical Science, Kumamoto University, Kumamoto, Japan*
THERESA CURRIER THOMAS • *Spinal Cord & Brain Injury Research Center, University of Kentucky Chandler Medical Center, Lexington, KY, USA*
DANIEL VAN BREGT • *Spinal Cord & Brain Injury Research Center, University of Kentucky Chandler Medical Center, Lexington, KY, USA*
CHANDLER L. WALKER • *Department of Anatomy & Cell Biology, Spinal Cord and Brain Injury Research Group, Stark Neurosciences Research Institute, Indiana University School of Medicine, Indianapolis, IN, USA*
JI-JIANG WANG • *Department of Neurobiology and State Key Laboratory of Medical Neurobiology, Shanghai Medical College, Shanghai, People's Republic of China*
JUAN A. WANG • *Department of Anatomy and Neurobiology, Neurosurgery, Neurology and Physical Medicine and Rehabilitation, Lexington, KY, USA; Spinal Cord & Brain Injury Research Center (SCoBIRC), University of Kentucky Medical Center, Lexington, KY, USA*
XIAOFEI WANG • *Department of Anatomy & Cell Biology, Spinal Cord and Brain Injury Research Group, Stark Neurosciences Research Institute, Indiana University School of Medicine, Indianapolis, IN, USA*
YONG-QUAN WANG • *Department of Neurobiology and State Key Laboratory of Medical Neurobiology, Shanghai Medical College, Shanghai, Shanghai, People's Republic of China*
IAN Q. WHISHAW • *Department of Neuroscience, Canadian Centre for Behavioural Neuroscience, University of Lethbridge, Lethbridge, AB, Canada*
SCOTT R. WHITTEMORE • *Departments of Neurological Surgery and Anatomical Sciences & Neurobiology, Kentucky Spinal Cord Injury Research Center, University of Louisville, Louisville, KY, USA*
JONATHAN R. WOLPAW • *Laboratory of Neural Injury and Repair, Wadsworth Center, New York State Department of Health & State University of New York at Albany, Albany, NY, USA*
XIAO-MING XU • *Departments of Neurological Surgery and Anatomy & Cell Biology, Spinal Cord and Brain Injury Research Group, Stark Neurosciences Research Institute, Indiana University School of Medicine, Indianapolis, IN, USA*
DIANER YANG • *Division of Developmental Biology and Division of Neurology, Cincinnati Children's Hospital Medical Center, Cincinnati, OH, USA*
FENG ZHANG • *Department of Neurology and the Center for Cerebrovascular Disease Research, University of Pittsburgh School of Medicine, Pittsburgh, PA, USA*

QIU-WAN ZHANG • *Department of Neurobiology and State Key Laboratory of Medical Neurobiology, Shanghai Medical College, Shanghai, People's Republic of China*
YI ZHANG • *Department of Neuroscience, Center for Brain and Spinal Cord Repair, Ohio State University Medical Center, Columbus, OH, USA*
YI PING ZHANG • *Norton Neuroscience Institute, Louisville, KY, USA*
CHANGLIAN ZHU • *Center for Brain Repair and Rehabilitation, Institute of Neuroscience and Physiology, University of Gothenburg, Gothenburg, Sweden; Department of Pediatrics, The Third Affiliated Hospital of Zhengzhou University, Zhengzhou, China*

Chapter 1

Introduction (Parts I–III)

R. Anne Stetler and Miguel A. Perez-Pinzon

Abstract

Cerebral ischemic models present highly complex scenarios affecting the interplay of multiple systems, cellular populations, signaling pathways, and molecular events. As the brief introduction to Parts I–III, Chap. 1 provides a general overview of the nature of the three primary cerebral ischemic models to be discussed: global cerebral ischemia (Part I), focal cerebral ischemia (Part II), and neonatal hypoxia/ischemia (Part III). Under each of these settings, the current morphological, biochemical, molecular, neurobehavioral, and neurogenic assessment strategies are discussed in the subsequent chapters. Parameters relevant to specific models have been identified and included in each section, such as the assessment of physiological alterations (e.g., cerebral blood flow), inflammation, and angiogenesis in the context of focal ischemic injury. The subsequent chapters provide a wide range of assessment strategies for investigating the mechanisms of cerebral ischemic injuries and potential therapeutic mechanisms.

Key words: Cerebral ischemia, Biochemical assessments, Morphology

The major rodent models of cerebral ischemia—global ischemia, focal ischemia, and neonatal hypoxia/ischemia—attempt to define and address a large variant of human ischemic conditions. These conditions are profoundly different in terms of vulnerable cellular populations, environmental conditions over time, and mode of cell death and survival. Thus, all three of these ischemic conditions pose distinct scenarios of cerebral ischemic injury and neurological outcomes.

The global ischemia model is intended to simulate what occurs after acute myocardial infarction (AMI). Full systemic models of AMI (e.g., cardiac arrest) are technically difficult and tend to add confounding factors to the pathophysiology that ensues after the insult. Thus, different models of global cerebral ischemia, including four-vessel occlusion or two-vessel occlusion/systemic hypotension, were developed with the goal of focusing on the effects of complete disruption of blood flow and resulting ischemic conditions in the brain. Important physiological parameters, such as

Jun Chen et al. (eds.), *Animal Models of Acute Neurological Injuries II*, Springer Protocols Handbooks,
DOI 10.1007/978-1-61779-782-8_1, © Springer Science+Business Media, LLC 2012

brain temperature regulation, have fine tuned this animal model to more accurately represent the neurological effects of AMI observed in humans.

Focal cerebral ischemia is perhaps the most replicable clinical model of human stroke. The model, typically induced by the occlusion of the middle cerebral artery (MCA) via an intraluminal thread leading from the common carotid artery to the juncture of the MCA, can be used as a transient ischemic episode model or a permanent occlusion. The occlusion targets the MCA territory, resulting in a core infarct within the striatum and extending a penumbral region through the adjacent cortical layers. Alternatively, thromboembolic ischemia can also be modeled by injection with coagulation factors or by induced photocoagulation, but the replication of the ischemic events suffers from variability and peripheral confounds.

The perinatal and neonatal periods pose particular risk for hypoxic brain conditions, as delayed heart or pulmonary development, anemia, and maternal complications resulting in reduced umbilical oxygenation can lead to asphyxia in the neonate. While hypoxia alone does not result in significant neural damage, transient induction of hypoxia can lower systemic oxygen tension and sensitize the brain to otherwise benign ischemic settings. Unlike adult brain, neonates are capable of maintaining cerebral blood flow (CBF) only over a fairly narrow window of systemic blood pressure. This decrease in CBF disrupts the compensation afforded by the circle of Willis in the presence of cerebral arterial blockage, which is modeled by unilateral MCA ligation. Thus, the MCA ligation and hypoxia are combined in rodent neonates, mimicking the human neonatal hypoxic/ischemic condition.

The first three parts of this book address these three major ischemic models (global ischemia, focal ischemia, and neonatal hypoxia/ischemia). As described in the chapters of Part I of this book, the different models of global cerebral ischemia are well suited to study and assess multiple aspects of the ischemia-induced pathology. Chapters 2 through 4 of Part I is dedicated to assessment strategies for macro- and micromorphological changes that occur following global cerebral ischemia, defining specifically techniques used to identify viable cells or cells in the process of degenerating. The final chapter of this section describes electron microscopy techniques used to define degeneration of neurons following global cerebral ischemia. The second section (Chapters 5 and 6) of Part I discusses several approaches to determine biochemical and molecular changes that occur following global cerebral ischemia. However, molecular and biochemical changes are insufficient for comprehensive neuroprotection assessments.

Neuroprotection must be assessed both morphologically and behaviorally to determine if therapeutic strategies are effective in ischemic models. Thus, Chapter 7 focuses on different neurobehavioral assessments following global cerebral ischemia. And finally, Chapter 8 draws methods from the important growing field of neurogenesis, which is likely to play a key role in recovery of function following ischemic insults in the near future.

Similar to Part I, Part II discusses the morphological, biochemical, molecular, neurobehavioral, and neurogenic assessments in the context of focal cerebral ischemia. While many of the methods described within these chapters may be used in other ischemic injury models, the relevance may be more appropriate for different ischemic conditions. In particular, an additional section (Chapter 11, Part II) is included to emphasize the importance of assessing physiological parameters over the course of focal ischemic injury, with the focus on blood flow. Accurate analysis of CBF alterations is critical to focal ischemic models to both establish the reproducibility of the insult itself and to monitor potential effects of therapeutics on blood flow vs. direct cellular effects. Furthermore, additional chapters focused on inflammation (Chapter 14, Part II) and angiogenesis Chapter 18, Part II) are included, as inflammatory responses represent an emerging focus in the focal ischemic model in mediating tissue response, and revascularization is critical for the recovery of surviving tissue.

Finally, Part III focuses on neonatal hypoxia/ischemia. The developing brain exhibits a highly distinct response to ischemic injury, thus necessitating unique assessment strategies. These are discussed throughout Part III in a similar manner as the adult models, but keeping in context the caveats and differences existing in neonatal cerebral ischemic models. A particular emphasis on white matter injury and recovery is made in Part III, Chapters 19–20 and 24, as this cellular population is highly sensitive to ischemic brain injury and leads to permanent impairment in neonates. Furthermore, morphological and neurobehavioral assessments following brain injury in neonates must be performed with regard to the normal course of neurological development. These are described in Part III, Chapters 19–20 and 23, respectively.

Taken together, these three initial parts should provide a wide range of assessment strategies in the investigation of cerebral ischemic injury and potential therapeutic mechanisms.

Part I

Global Cerebral Ischemia

Chapter 2

Morphological Assessments of Global Cerebral Ischemia: Viable Cells

Peiying Li

Abstract

Global cerebral ischemia (GCI) can cause selective neuronal damage in a variety of brain regions, including striatum, hippocampus, cortex, and so on. Hematoxylin and eosin and Nissl stains are commonly used histological methods for the morphological assessments of the GCI-viable cells. In addition, the immunohistochemical staining of neuronal nuclei or microtubule-associated protein 2 can be applied to identify the survived neurons after the ischemic insult in this disease model. This chapter introduces the sample preparations, staining procedures, and evaluation methods of these stains.

Key words: Global cerebral ischemia, Morphological assessment, H&E stain, Nissl stain, NeuN, MAP2

1. Introduction

Global cerebral ischemia (GCI) occurs in people who have a variety of clinical conditions, including cardiac arrest, shock, and asphyxia. The risk of GCI increases dramatically in patients undergoing complex cardiac surgery (1). Due to its catastrophic consequences, GCI has been extensively studied using experimental animal models.

Cessation of global cerebral perfusion for only a few minutes results in delayed and selective neuronal damage in a variety of brain regions, including the CA1 of the hippocampus, the striatum, and also the cortex (2). Hematoxylin and eosin (H&E) and Nissl stains are two commonly used methods for histological assessment of the neuronal damages. Immunohistochemical stain of neuronal nuclei (NeuN) or microtubule-associated protein 2 (MAP2) can also be applied to assess the surviving cells in this model.

Jun Chen et al. (eds.), *Animal Models of Acute Neurological Injuries II*, Springer Protocols Handbooks,
DOI 10.1007/978-1-61779-782-8_2,

As a simple and most frequently used staining method in histology, the double staining of hematoxylin and eosin has stood the test of time of more than 100 years since it was firstly introduced by Bhömer (3). Numerous formulations have appeared, among which the Harris', Gill's, Mayer's, and Weigert's have retained popularity. In this chapter, the Harris' formulation is introduced.

Hematoxylin is a natural dye extracted by boiling the wood of the South American and West Indian logwood tree. The oxidized product of hematoxylin, hematein, is the active dye serving as an indicator in the nuclear staining. It turns blue in aqueous alkaline conditions, and turns red and soluble in alcoholic acidic conditions. In acidic conditions, hematein can bind to lysine residues of nuclear histones via a metallic ion mordant. To ensure saturation of chemical binding sites, the Harris hematoxylin is applied longer than necessary, resulting in the overstaining of the tissues with much nonspecific background coloration. Differentiation in the alcoholic acidic solution can selectively remove the undesirable coloration. Differentiation is arrested by returning to an alkaline environment, whereupon the hematein takes on a blue hue—the process of "bluing up." The higher the pH of a bluing solution, the faster the rate of bluing will be. Longer immersion can loosen cell adhesion to glass and result in cell loss.

Counterstain of eosin confers sharp contrast of the blue nuclei. Eosinophilic structures of cells are generally composed of protein. Most of the cytoplasm is eosinophilic. Nucleoli also stain with eosin. There are three commonly used forms of eosin: eosin yellowish (eosin Y), eosin bluish (eosin B), and eosin alcohol soluble (ethyl eosin), all of which are similar in staining characteristics. Note that eosin Y is normally dissolved in 95% ethanol. Therefore, an eosin-stained slide is decolorized if it is left too long in 95% ethanol. Eosin is less soluble in 100% ethanol, but stain can still be lost slowly.

As a nonspecific method for neuronal cell viability assessment, the time course of the H&E stain in the GCI model is not well documented. Nevertheless, Dawson and Hallenbeck noted in their study that focal ischemic injury defined by H&E staining closely paralleled the loss of MAP2 (4).

Nissl stain is a widely used approach to examine the morphology of the brain. Neuronal cells contain Nissl substance, which is primarily composed of rough endoplasmic reticulum, with the amount, form, and distribution varying in different types of neurons. Nissl substance, with a large content of RNA, and DNA in the nuclei, can bind to basophilic dyes, cresyl violet for example, which yields a violet color.

Nissl stain exhibits cells with aberrant morphology as early as 2 h after reperfusion onset in the 2-h MCAO model. Unfortunately, it is not sensitive enough to detect subtle changes at very early reperfusion time following a mild ischemia or to identify perilesional affected tissue (5).

Despite their drawbacks in terms of the early assessment of ischemic injuries and precise identification of viable cells, the H&E and Nissl stains still gain popularity because of their easy application and excellent morphological performance. Even in current neuroscience research, they are still widely accepted as basic evaluation methods of cerebral ischemic injuries.

NeuN and MAP2 can be used for immunohistochemical detection of surviving neurons in the brain that underwent global ischemia. NeuN is a neuronal-specific nuclear protein first introduced by Mullen et al. (6). It is consistently detected by the corresponding monoclonal antibody in the nucleus of most types of neurons in a wide range of vertebrates, including mammals, birds, and amphibians (7). Immunohistochemically detectable NeuN protein first appears at the developmental time points which correspond with the withdrawal of the neuron from the cell cycle and/or with the initiation of terminal differentiation of the neuron (6). Therefore, anti-NeuN is widely used as a tool for detecting neurons when they become postmitotic and are initiating cellular and morphological differentiation.

As a neuronal-specific nuclear protein, NeuN primarily localizes in the nuclear, but the cytoplasm of neurons is also immunoreactive (Fig. 1). The loss of NeuN immunoreactivity can be used to assess the cell viability of the neurons following cerebral ischemia at the early phase of reperfusion. It was supportively reported

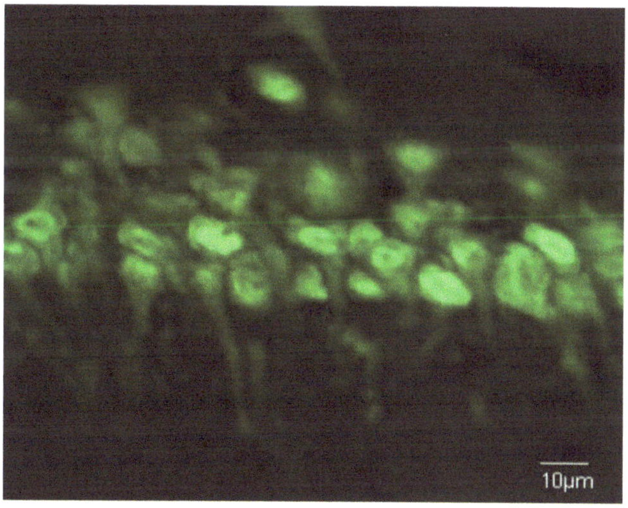

Fig. 1. NeuN immunofluorescent staining of CA1 neurons. Normal rat brain section was subjected to immunofluorescent staining of NeuN. The nuclei of the neurons show strong immunoreactivity (*green*, Alexa Fluor 488). The cytoplasm is also immunoreactive, although to a lesser extent. Scale bar = 10 μm.

that NeuN was greatly reduced after 1 h of reperfusion and declined further at 3 h in focal cerebral ischemia (8).

MAP2 is the major microtubule-associated protein of brain tissue. Cerebral ischemia and reperfusion will result in an extensive loss of MAP2 in the lesion area. In 1989, MAP2 immunohistochemistry was introduced as a means to investigate dendritic lesions in cerebral ischemia in the gerbil (9). Acute cytoskeletal rearrangements after focal ischemia were documented using immunostaining of MAP2 as early as 15 min after induction of ischemia (10). Such changes preceded neuronal death and eventually led to extensive MAP2 degradation by 24 h after ischemia. It was recently reported in the nonhuman primate model of stroke that neurons with reduced MAP2 immunostaining at early time points (8 h) after ischemia ultimately committed irreversible cell death (11). It is proposed that MAP2 was degraded by calpain, a calcium-dependent neutral proteinase, in response to ischemic injury, resulting in the loss of MAP2 immunoreactivity in neurons (10). Supportively, immunostaining of spectrin fragments, a cleaved product of calpain, increased in parallel with the loss of MAP2 immunostaining in the same brain regions. The MAP2 immunoreactivity is more intense in dendrites than in cell bodies (Fig. 2). Thus, it has the advantage over other neuronal markers in the staining of normal dendrites.

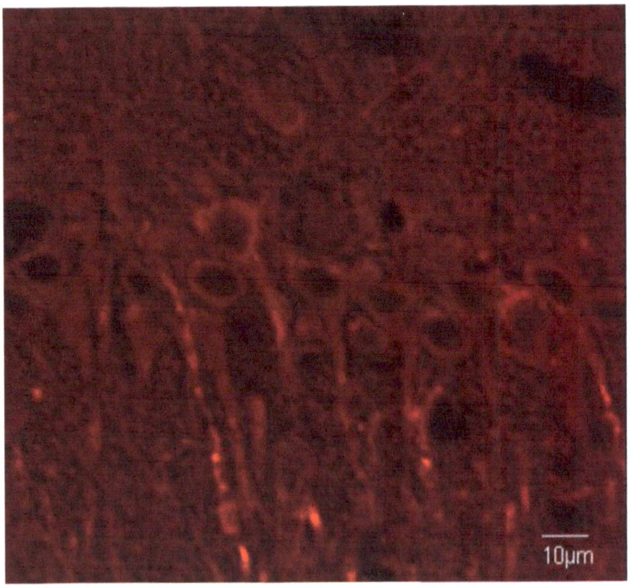

Fig. 2. MAP2 immunofluorescent staining of CA1 neurons on normal rat brain section. The dendrites and cell bodies of neurons in the CA1 region are highlighted by MAP2 immunoreactivity (*red*, Alexa Fluor 594). Scale bar = 10 μm.

2. Sample Preparations

1. Paraffin sections are recommended for histological staining because these sections have well-preserved morphology and are able to yield excellent stain.
2. Free floating sections are preferred for immunohistochemical or immunofluorescent staining because of their sufficient antibody binding properties.
3. Deparaffinize and rehydrate the paraffin sections as follows.
    (a) Immerse sections in two changes of xylene, 5 min each.
    (b) Hydrate in two changes of 100% ethanol for 3 min each, followed by 95 and 80% ethanol for 1 min each.
    (c) Rinse in running tap water for 5 min.
4. If floating sections are chosen for staining,
    (a) Mount sections on 2% gelatin-coated slides.
    (b) Air dry sections overnight or bake slides in slide warmer for at least 30 min.
    (c) Immerse sections in two changes of 100% ethanol for 3 min each, followed by 95 and 80% ethanol for 1 min each.
    (d) Rinse in running tap water for 5 min.
5. If fresh frozen sections are to be used,
    (a) Fix sections by immersion in 4% paraformaldehyde in PBS for 10 min at room temperature, followed by washing in PBS to remove traces of fixative.
    (b) Immerse sections in two changes of 100% ethanol for 3 min each, followed by 95 and 80% ethanol for 1 min each.
    (c) Rinse in running tap water for 5 min.

3. H&E Stain

3.1. Materials for H&E Stain

1. Harris Hematoxylin solution (Sigma, St. Louis, MO).
2. Differentiating solution: 0.25% HCl in 70% Alcohol (Sigma, St. Louis, MO).
3. Bluing solution: 0.1% NH_4OH (Sigma, St. Louis, MO).
4. Eosin Y solution (Sigma, St. Louis, MO).
5. Xylene (Sigma, St. Louis, MO).
6. Reagent alcohol (Sigma, St. Louis, MO).

7. Hydrochloric acid, concentrated (Sigma, St. Louis, MO).
8. Mounting medium (Thermo Fisher Scientific, Waltham, MA).
9. Microscope slides, coverslips, and staining dishes (Thermo Fisher Scientific, Waltham, MA).
10. MCID image analysis system (Imaging Research Inc., St. Catharines, ON, Canada) or NIH Image software (NIH, Bethesda, MD).

3.2. Staining Procedures

1. Stain in Harris Hematoxylin solution for 2–10 min. Empirical determination of the staining time is needed. Shorter staining time is required for thicker floating or frozen sections.
2. Rinse slides in running tap water.
3. Differentiate in acid alcohol (differentiation solution) for one to two dips. Check the stain under the microscope for a satisfactory effect.
4. Rinse slides in running tap water.
5. Differentiation can be repeated for best result, which can be checked microscopically.
6. Immerse slides in the bluing solution for 1 min.
7. Rinse in running tap water.
8. Immerse slides in 95% alcohol for 30 s.
9. Stain in Eosin Y solution for 30 s to 3 min.
10. Rinse in running tap water.
11. Dehydrate in 80% alcohol, 95% alcohol for 1 min each, and two changes of 100% alcohol, 3 min each.
12. Clear in two changes of 100% xylene for 5 min.
13. Mount with mounting medium.

4. Nissl Stain

4.1. Materials

1. 0.5% Cresyl violet solution (Sigma, St. Louis, MO).
2. Xylene (Sigma, St. Louis, MO).
3. Reagent alcohol (Sigma, St. Louis, MO).
4. Mounting medium (Thermo Fisher Scientific, Waltham, MA).
5. Microscope slides, coverslips, and staining dishes (Thermo Fisher Scientific, Waltham, MA).
6. MCID image analysis system (Imaging Research Inc., St. Catha-rines, ON, Canada) or NIH Image software (NIH, Bethesda, MD).

4.2. Staining Procedures

1. Stain in 0.5% cresyl violet solution for 2 min. For thicker sections, staining in warmed cresyl violet solution (warm up in 37.5 °C oven) can improve penetration and enhance even staining.
2. Rinse quickly in distilled water.
3. Differentiate in 95% alcohol for best result and check microscopically.
4. Dehydrate in two changes of 100% alcohol for 5 min.
5. Clear in two changes of xylene for 5 min.
6. Mount with mounting medium.

5. Immunohistochemistry/Immunofluorescent Staining of NeuN or MAP2

5.1. Materials

1. Primary antibodies: Anti-NeuN (Millipore, Billerica, MA) anti-MAP2 (Novus, Littleton, CO).
2. Secondary antibodies: Fluorescent-conjugated secondary antibody (Invitrogen, Carlsbad, CA; or Jackson Immunoresearch, West Grove, PA); biotinylated secondary antibody and avidin–biotin complex (ABC) kit (Vector, Burlingame, CA).
3. Mounting medium (Thermo Fisher Scientific, Waltham, MA).
4. Fluoromount-G or dapi Fluoromount-G mounting medium (SouthernBiotech, Birmingham, AL).
5. Microscope slides, coverslips, and staining dishes (Thermo Fisher Scientific, Waltham, MA).
6. MCID image analysis system (Imaging Research Inc., St. Catharines, ON, Canada) or NIH Image software (NIH, Bethesda, MD).

5.2. Pretreatment of Sections

1. Floating sections can be stained using a 24-well tissue culture plate and pretreated with PBS + 1% Triton X-100 for 20–30 min.
2. Frozen sections require fixation in 4% paraformaldehyde for about 10 min, followed by rinsing in distilled water to remove the residual paraformaldehyde.
3. Paraffin sections need to be deparaffinized as aforementioned. The frozen and paraffin sections can be stained on the microscope slides.

5.3. Staining Procedures

1. Rinse sections in PBST (PBS + 0.1% Tween-20) for 2 min ×2.
2. Immerse sections in 2% H_2O_2 in PBST for 10 min (only for immunohistochemistry staining).

3. Incubate sections with 5% normal serum for 30 min. The species of serum should be the same as that of the secondary antibody.
4. Incubate sections with primary antibody at appropriate dilution in antibody diluent for 1 h at room temperature or overnight at 4°C. If necessary, test and optimize antibody concentrations.
5. Rinse with PBST for 5 min ×3.
6. Incubate sections with the secondary antibody at appropriate dilution in PBST for 30 min at room temperature. If a fluorescent-conjugated secondary antibody is used, protect the sections from light, and rinse the sections in three changes of PBST for 5 min each, followed by mounting with Fluoromount-G or dapi Fluoromount-G mounting medium. If a biotinylated secondary antibody is applied, proceed to the next steps.
7. Rinse with PBST for 5 min ×3.
8. Incubate sections in ABC solution for 30 min.
9. Rinse in PBST for 5 min ×3.
10. Incubate sections in 0.05% 3,3 -diaminobenzidine (DAB) for up to 5 min.
11. Rinse in distilled water.
12. Immerse sections with 95% ethanol for 2 min, 100% ethanol for 3 min ×2, followed by two changes of xylene for 5 min each.
13. Mount with mounting medium.

6. Evaluations

6.1. Morphological Assessment of H&E and Nissl Stains

The ischemic-injured neurons can be demonstrated by the acidophilic/eosinophilic alteration of the cytoplasm and the shrunken nucleus in H&E and Nissl stains. The cells with aberrant morphology were triangular in shape and mostly exhibiting a dark staining due to the condensation of cytoplasm and karyoplasm. Thus, the degenerated cells can be inferred based on the morphological changes: neuronal shrinkage, vacuolation, and hyperchromatism.

In the GCI model, hippocampus, striatum, and cortex are the most susceptible areas to the ischemic injury. In Fig. 3, the white arrows show the degenerating cells in the CA1 region of hippocampus displaying the morphological changes as described above.

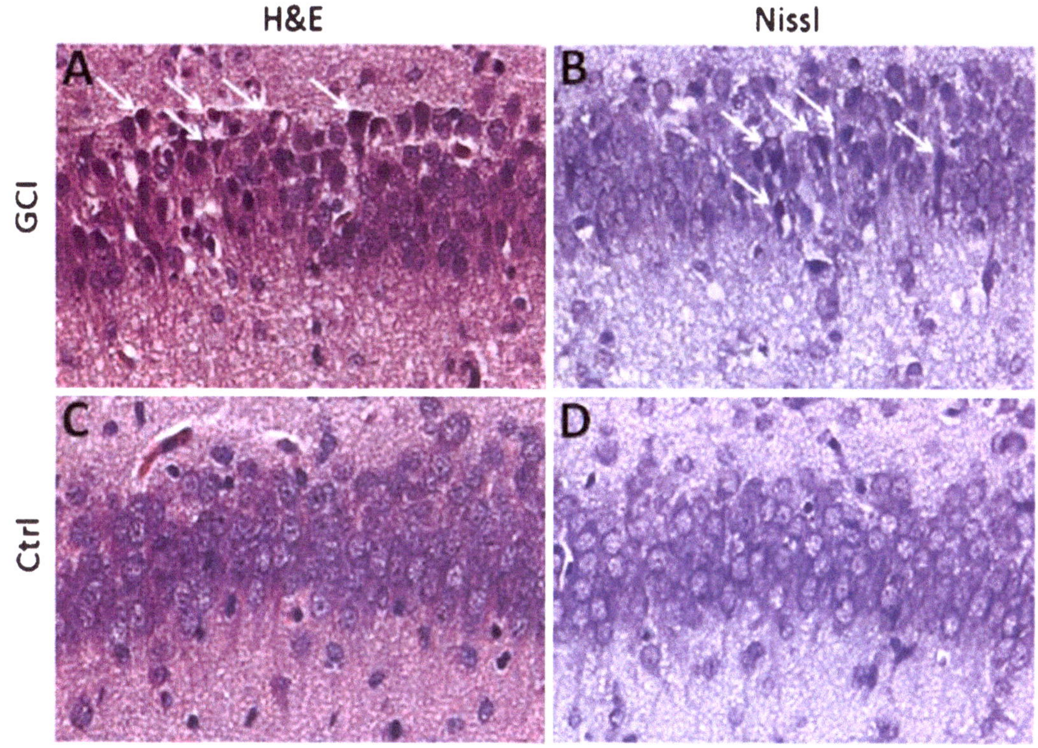

Fig. 3. H&E and Nissl staining. Brain sections from GCI (*upper*) and normal (*bottom*) rats were subjected to H&E (**a**, **c**) and Nissl (**b**, **d**) staining. *White arrows* in (**a**) and (**b**) show degenerating neurons in the CA1 region. Original magnification: ×400.

6.2. Quantitative Measurements of H&E and Nissl Stains

Counting of positive particles using ImageJ or counting of the degenerating cells manually can be used to quantify H&E and Nissl stains. Below are the procedures of counting degenerating cells with ImageJ in H&E- or Nissl-stained sections.

1. Take ×200 or ×400 pictures of the H&E- or Nissl-stained sections at designated regions of interest. Note that areas chosen for analysis should be consistent among different samples.
2. Launch the software, open the file for cell counting, and change the image type to an 8- or 16-bit black-and-white image.
3. Adjust the threshold until the positive particles best correlate with the degenerating cells, as shown in Fig. 4a.
4. Set the scale of the distance in pixels, and measure the areas of both the smallest and the largest positive particles in the picture obtained in the previous step.
5. Fill in the blanks of the size and circularity with the previously calculated or estimated values (Fig. 4b).

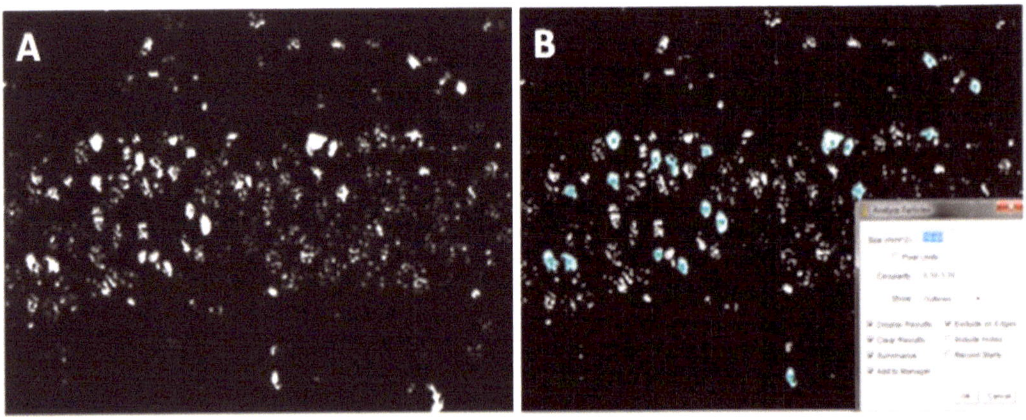

Fig. 4. Positive particle analysis for H&E stain. (**a**) Appropriate threshold to depict the degenerating neurons. (**b**) Cells marked with *blue circles* are taken as positive particles by the ImageJ software according to the corresponding parameters.

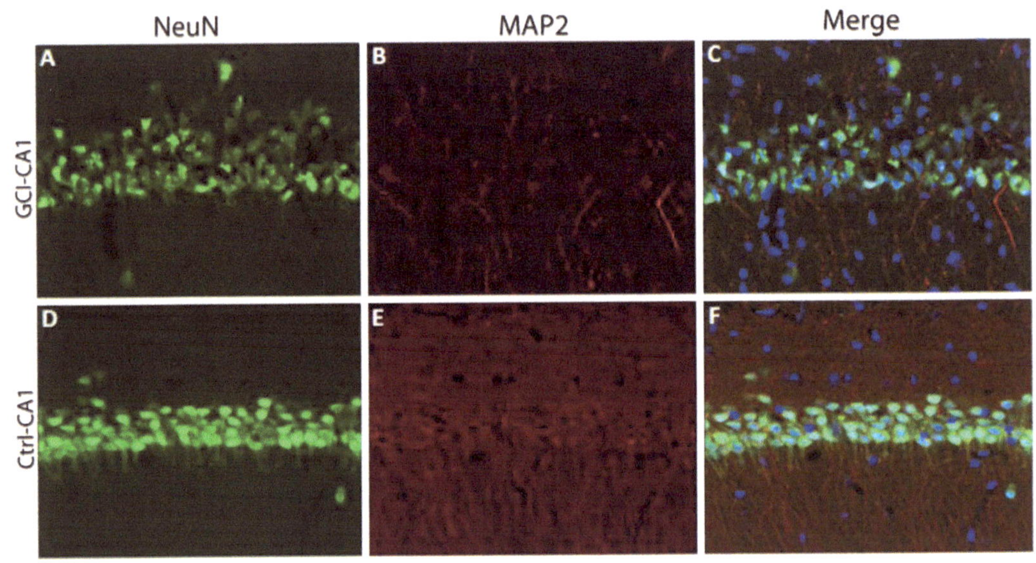

Fig. 5. Immunofluorescent staining of NeuN and MAP2. Immunofluorescent double staining of NeuN (*green*, Alexa Fluor 488) and MAP2 (*red*, Alexa Fluor 594) on hippocampus section from GCI rats (**a–c**) and normal rats (**d–f**). *Blue*, DAPI. Magnification: ×200.

6. The counted number is shown both in the result and summary window.
7. Set the same threshold and the same scale of estimated positive particles to count all the samples.

6.3. Evaluation of the Immunostaining of NeuN or MAP2

The typical immunofluorescent staining of NeuN and MAP2 is shown in Fig. 5. Immunohistochemistry staining can be semiquantitative by comparing the intensity of the stains or counting the number of positive particles. The stain intensity method applies best to NeuN or MAP2 staining. The detailed evaluation method is discussed in Chap. 13.

7. Summary

As a basic histological method which is especially valued by pathologists, H&E shows the shapes and sizes of the cell bodies through a simple rendition in two colors. Nissl stain provides detailed information about the perikaryon of neurons. However, these stains carry several drawbacks in terms of the early assessment of ischemic injuries and precise identification of viable cells, especially in cases of relatively mild injuries. NeuN and MAP2 are two specific neuronal markers, and immunostaining of these markers can be used to detect viable cells in the early phase after GCI.

References

1. Salazar JD et al (2001) Stroke after cardiac surgery: short- and long-term outcomes. Ann Thorac Surg 72(4):1195–1201; discussion 1201–1202
2. Pulsinelli WA (1985) Selective neuronal vulnerability: morphological and molecular characteristics. Prog Brain Res 63:29–37
3. Bhömer F (1865) Zur pathologischen, Anatomie der Meningitis cerebromedularis epidemia. Aerztl Intelligenzb (Manchen) 12:539–550
4. Dawson DA, Hallenbeck JM (1996) Acute focal ischemia-induced alterations in MAP2 immunostaining: description of temporal changes and utilization as a marker for volumetric assessment of acute brain injury. J Cereb Blood Flow Metab 16(1):170–174
5. Popp A et al (2009) Identification of ischemic regions in a rat model of stroke. PLoS One 4(3):e4764
6. Mullen RJ, Buck CR, Smith AM (1992) NeuN, a neuronal specific nuclear protein in vertebrates. Development 116(1):201–211
7. Weyer A, Schilling K (2003) Developmental and cell type-specific expression of the neuronal marker NeuN in the murine cerebellum. J Neurosci Res 73(3):400–409
8. Lee DR et al (2003) Losses of NG2 and NeuN immunoreactivity but not astrocytic markers during early reperfusion following severe focal cerebral ischemia. Brain Res 989(2): 221–230
9. Kitagawa K et al (1989) Microtubule-associated protein 2 as a sensitive marker for cerebral ischemic damage—immunohistochemical investigation of dendritic damage. Neuroscience 31(2): 401–411
10. Pettigrew LC et al (1996) Microtubular proteolysis in focal cerebral ischemia. J Cereb Blood Flow Metab 16(6):1189–1202
11. Kharlamov A et al (2009) MAP2 immunostaining in thick sections for early ischemic stroke infarct volume in non-human primate brain. J Neurosci Methods 182(2):205–210

Chapter 3

Morphological Assessments of Global Cerebral Ischemia: Degenerated Cells

Peiying Li

Abstract

Global cerebral ischemia (GCI) results in neurodegeneration in a variety of brain regions. Accumulated evidences suggest that DNA damages are involved in apoptotic cell death in cases of GCI. Klenow labeling assay, which detects single-strand DNA breaks, a relatively early event following the ischemic insult, serves as an effective approach to identify the degenerating cells. The terminal deoxynucleotidyl transferase-mediated dUTP nick-end labeling assay, which labels apoptotic DNA fragmentation or double-strand DNA breaks, is also utilized to determine the degenerated cells after the GCI injury. Fluoro Jade staining is a method recently proved to identify the degenerated cells with unknown mechanisms. It has soon become a frequently used method for the GCI injury assessment because of its convenience of staining and ease of evaluation. In this chapter, the principles, labeling procedures, and interpretation approaches regarding these three methods are discussed.

Key words: Global cerebral ischemia, Neurodegeneration, Klenow, TUNEL, Fluoro Jade

1. Introduction

Transient global cerebral ischemia (GCI) results in acute neurodegeneration in selective brain areas. Emerging studies suggest that delayed neuronal death in CA1 following GCI occurs via an apoptotic mechanism (1–5). Apoptotic DNA fragmentation is the best-studied active DNA damage, which is characterized by DNA double-strand breaks (DSBs). This fragmentation involves a cascade of cellular self-destruction, and usually occurs irreversibly at the late stage of cell injury. Single-strand DNA break (SSB) is a type of passive DNA damage which is an endonuclease-independent process. SSB is a reversible process that happens at a relatively early stage of cell injury and is involved in triggering apoptosis (6).

Jun Chen et al. (eds.), *Animal Models of Acute Neurological Injuries II*, Springer Protocols Handbooks,
DOI 10.1007/978-1-61779-782-8_3, © Springer Science+Business Media, LLC 2012

The Klenow labeling assay and the terminal deoxynucleotidyl transferase-mediated dUTP nick-end labeling (TUNEL) assay are two widely used methods to detect neuronal DNA damages following GCI. The Klenow fragment is derived from the enzymatic cleavage of intact *Escherichia coli* DNA polymerase I. It retains both the DNA polymerase activity and the 3′–5′ exonuclease activity of *E. coli* DNA polymerase I while lacks 5′–3′ exonuclease activity. Klenow fragment can bind at the site of cleavage in DNA SSBs with proficiency. With its 5′–3′ DNA polymerase activity, it uses the single-stranded DNA template as a primer, and fills in the gaps of SSBs with a free 3′-OH group (Fig. 1a). Besides SSBs, Klenow is also preferred for labeling the recessed 3′ termini of DSBs rather than protruding 3′ termini or blunt ends (Fig. 1b). Therefore, Klenow labeling assay is sensitive to detect the relatively early SSB, which represents the potentially repairable lesions. Klenow is not capable of nick translation, which is dependent on the 5′–3′ exonuclease activity.

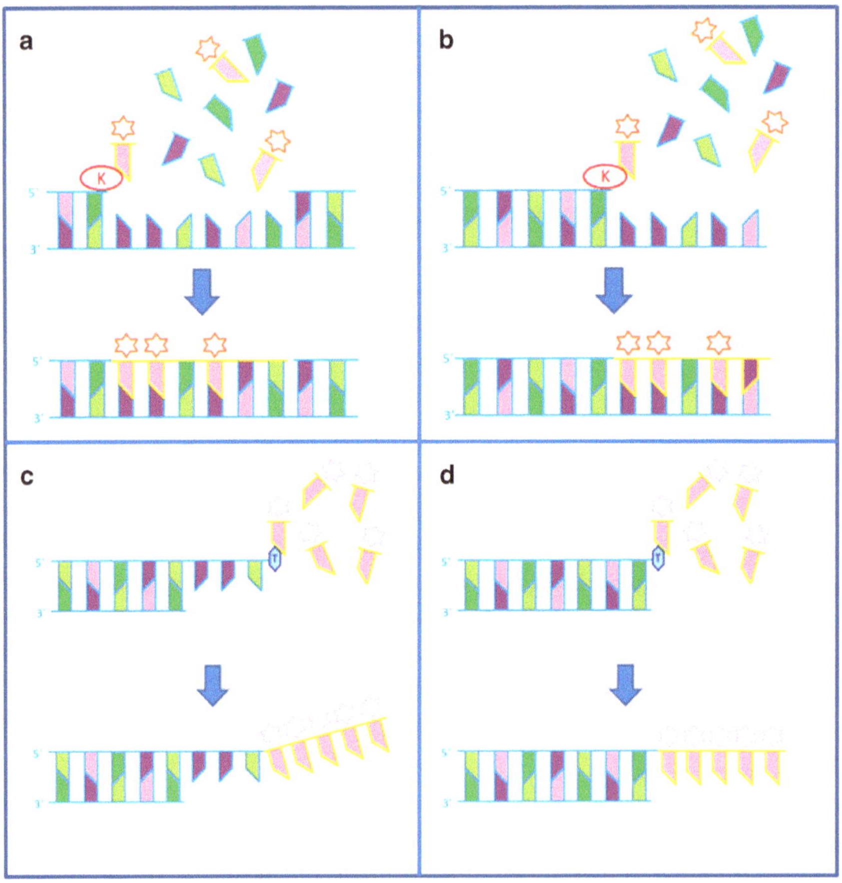

Fig. 1. Illustration of the Klenow and TUNEL labeling principle. Klenow labeling assay is sensitive to detect SSBs (**a**) and DSBs with recessed 3′ termini (**b**). TUNEL assay is sensitive to detect DSB with 3′ protruding termini (**c**) and blunt ends (**d**).

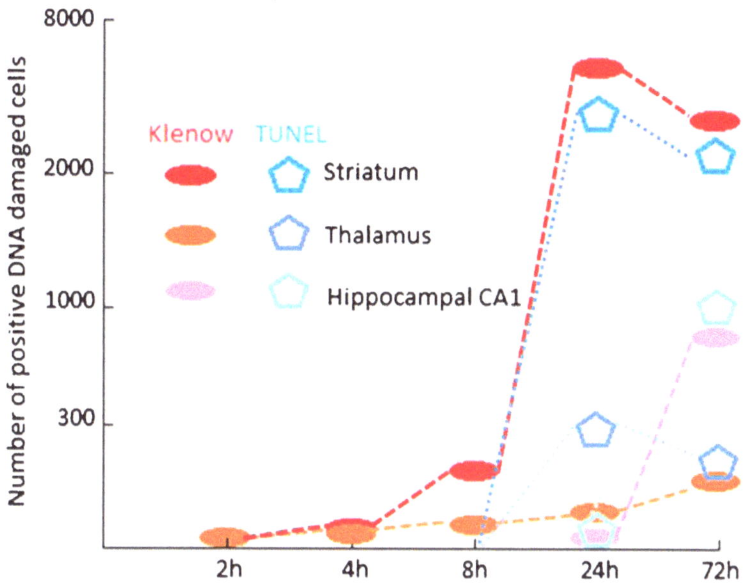

Fig. 2. Time course of detectable DNA damages by Klenow/TUNEL assay.

TUNEL assay is one form of the end labeling assay which utilizes terminal deoxynucleotidyl transferase (TdT), an exogenous terminal transferase, to label exposed DNA 3′-OH ends by incorporating modified nucleotides. TdT catalyzes the addition of nucleotides to 3′-OH termini of DNA molecules in a template-independent manner. The ideal substrates for TdT are 3′ protruding ends and blunt ends of double-stranded DNA. Therefore, TUNEL is preferred for labeling DSBs with protruding 3′ termini (Fig. 1c) or blunt ends (Fig. 1d) rather than recessed 3′ termini or SSB gaps.

There are evidences from Jin et al. that compared to the TUNEL technique the Klenow labeling assay could detect DNA damages at an earlier time point (within 2 h of reperfusion) and in the caudate putamen, reticular thalamus, thalamus, and cortex in the GCI model (Fig. 2) (7).

Fluoro Jade is an anionic fluorochrome capable of selectively staining degenerating neurons resulting from a variety of insults (8). It is faster and more reliable than other methods, suppressed silver for example, for the unequivocal qualitative detection and quantitative measurement of both gross and fine-scale neuronal degeneration (9). It is also more sensitive and definite than H&E or Nissl stains (10). The specific protein or lipid component labeled by Fluoro Jade, however, is not clarified.

Currently, there are two generations of commercial Fluoro Jade, Fluoro Jade® B and Fluoro Jade® C (Millipore). Different to its predecessors, Fluoro Jade® C exhibits the greatest signal-to-background ratio, as well as the highest resolution, making it ideal for localizing not only degenerating nerve cell bodies, but also distal dendrites, axons, and terminals. In this chapter, Fluoro Jade B is introduced as a routine dye for this staining.

2. Klenow Labeling

2.1. Sample Preparation

1. Klenow labeling assay may be used for in situ detection of DNA damages on most tissue section types, including fresh frozen, free floating, or paraffin-embedded sections.
2. Fresh frozen sections require little permeabilization, and therefore may be of benefit if immunohistochemistry is to be performed in addition to the detection of DNA damages. Prior to labeling, frozen sections need to be fixed with 4% paraformaldehyde/PBS for 30 min and then washed three times for 5 min each in PBS.
3. If free floating sections are to be labeled, sections can either be treated with PBS + 3% Triton X-100 for 30–60 min followed by washing three times for 5 min each in PBS or be permeabilized as follows.
   (a) Mount sections on 2% gelatin-coated slides.
   (b) Air dry sections overnight or bake slides in slide warmer for at least 30 min.
   (c) Immerse sections in 70% alcohol for 5 min and rinse in PBS to remove excess alcohol.
   (d) Lay slide flat and immediately pipette the working solution of Proteinase K on top of the samples and incubate for 5–30 min at 37°C in a humidified chamber. Empirical determination of incubation time is needed.
   (e) Rinse in PBS to terminate the reaction.

2.2. Materials for Klenow Labeling

1. 37-C humidified chamber.
2. Stocking buffer for Proteinase K: 100 mM Tris–HCl (pH 7.5), 200 mM EDTA 10% SDS, 1 mg/mL proteinase K (20 U/mg) in water. The working solution can be obtained from 1:50 dilution of the stocking buffer.
3. 3% Hydrogen peroxide (Sigma, St. Louis, MO) in 100% methanol (Sigma, St. Louis, MO).
4. Labeling buffer: 50 mM Tris (pH 7.5), 5 mM $MgCl_2$, 60 μM 2-mercaptoethanesulfonic acid (MESNA), 50 μg/mL BSA (Sigma, St. Louis, MO).
5. Klenow fragment (Invitrogen, Carlsbad, CA).

6. dNTP and biotinylated dATP (Invitrogen, Carlsbad, CA).
7. Biotin/streptavidin/peroxidase complex (Vectastain Elite ABC; Vector Laboratories, Burlingame, CA).
8. DAB detecting kit (Vector, Burlingame, CA).
9. DTAF (Jackson Immunoresearch, West Grove, PA).
10. Mounting medium (Thermo Fisher Scientific, Waltham, MA) or Fluoromount-G mounting medium (SouthernBiotech, Birmingham, AL).
11. Microscope slides, coverslips, and staining dishes (Thermo Fisher Scientific, Waltham, MA).

2.3. Labeling Procedures

1. Immerse sections in a freshly prepared solution of 3% hydrogen peroxide in methanol for 5 min at room temperature. If avidin-conjugated fluorescent dyes are to be used, omit this step.
2. Rinse slides briefly in PBS.
3. Equilibrate samples by immersing slides in Klenow labeling buffer for 2 min.
4. Prepare the enzyme labeling mix on ice: 10 μM dCTP, dGTP, and dTTP; biotinylated dATP; 40 U/mL *E. coli* Klenow fragment in 1× labeling buffer. Prepare immediately before use. Do not store.
5. Incubate the sections with the labeling mix in a moist air chamber at 37°C for 1 h.
6. Terminate the reaction by washing the slides twice in PBS.
7. To block nonspecific labeling, incubate the sections in PBST/BSA (5%) for 20 min.
8. Remove excess blocking buffer from around sample with a tissue.
9. Immediately pipette the biotin/streptavidin/peroxidase complex on the samples and incubate for 1 h at room temperature. Following 5 min in PBST (PBS + 0.1% Tween-20), the ABC complex was detected by incubating the sections with the DAB detecting kit. View periodically during development under microscope and stop the color development by washing in PBS. Dehydrate by passing through two changes of 95% ethanol and then 100% ethanol. Clarify by passing through two changes of xylene for 5 min each. Mount the sample and lay glass coverslip.
10. Alternatively, incubate the sections with the DTAF for 30 min at room temperature to detect the biotinylated dATP. Rinse in PBST for 5 min ×2. Air dry and mount the slides with Fluoromount-G mounting medium.

3. TUNEL Labeling Assay

3.1. Sample Preparation

Refer to Sect. 2.1.

3.2. Materials

1. 37°C humidified chamber.
2. 3% Hydrogen peroxide (Sigma, St. Louis, MO) in 100% methanol (Sigma, St. Louis, MO).
3. DNase I (New England Biolabs, Ipswich, MA).
4. TdT (New England Biolabs, Ipswich, MA).
5. 10× TdT buffer (New England Biolabs, Ipswich, MA).
6. Biotinylated dATP (Invitrogen, Carlsbad, CA).
7. Biotin/streptavidin/peroxidase complex (Vectastain Elite ABC; Vector Laboratories, Burlingame, CA).
8. DAB detecting kit (Vector, Burlingame, CA).
9. DTAF (Jackson Immunoresearch, West Grove, PA).
10. Mounting medium (Thermo Fisher Scientific, Waltham, MA) or Fluoromount-G mounting medium (SouthernBiotech, Birmingham, AL).
11. Microscope slides, coverslips, and staining dishes (Thermo Fisher Scientific, Waltham, MA).

3.3. Labeling Procedures

1. Quench sections with 3% hydrogen peroxide in 100% methanol for 15 min if the biotin/streptavidin/peroxidase method is to be used.
2. Rinse sections briefly in PBS.
3. Equilibrate sections by immersion in the labeling buffer for 2 min at room temperature.
4. Prepare the TUNEL assay labeling buffer from the TdT buffer at the dilution of 1:10. Add 20 μM biotin-16-dATP and 300 U/mL TdT.
5. Incubate the sections with the labeling mixture in a moist air chamber for 1 h at 37°C.
6. Detect the 3′-OH DNA end tailing of biotinylated dATP using the biotin/streptavidin/peroxidase method described in the Klenow labeling procedures.
7. Alternatively, detect the biotin-16-dATP incorporated into DNA with DTAF.
8. Obtain positive controls by pretreating normal nonischemic brain sections with DNase I followed by the identical TUNEL assay procedure.

4. Fluoro Jade B Staining

4.1. Sample Preparation

1. Fluoro Jade staining can be used on most tissue section types and thicknesses, including frozen, floating, or paraffin-embedded sections from 3 to 50 μm.
2. The sample preparation process of Fluoro Jade B staining is the same as that of H&E staining (Chap. 1).

4.2. Materials for Fluoro Jade B Staining

1. Fluoro Jade B stock solution: 10 mg of the dye powder (Millipore, Billerica, MA) dissolved in 100 mL of distilled water.
2. 5% Sodium hydroxide (Sigma, St. Louis, MO).
3. Potassium permanganate (Sigma, St. Louis, MO).
4. Reagent alcohol (Sigma, St. Louis, MO).
5. Xylene (Sigma, St. Louis, MO).
6. Acetic acid (Sigma, St. Louis, MO).
7. Mounting medium (Thermo Fisher Scientific, Waltham, MA).
8. Microscope slides, coverslips, and staining dishes (Thermo Fisher Scientific, Waltham, MA).

4.3. Staining Procedures

1. Air dry slides on a slide warmer at 50°C for at least half an hour.
2. Immerse slides in a solution containing 1% sodium hydroxide in 80% alcohol (20 mL of 5% sodium hydroxide added to 80 mL absolute alcohol) for 5 min.
3. Then, immerse slides in 70% alcohol for 2 min and then in distilled water for 2 min.
4. Transfer slides to a solution of 0.06% potassium permanganate for 10 min, preferably on a shaker table to insure consistent background suppression between sections.
5. Rinse slides in distilled water for 2 min.
6. Prepare the staining solution from a 0.01% stock solution for Fluoro Jade B. To make up 100 mL of staining solution, add 4 mL of the stock solution to 96 mL of 0.1% acetic acid vehicle. This results in a final dye concentration of 0.0004%. Prepare the staining solution within 10 min of use and do not reuse.
7. After 20 min in the staining solution, rinse the slides for 1 min in each of the three distilled water washes.
8. Remove excess water by briefly (about 15 s) draining the slides vertically on a paper towel.
9. Place the slides on a slide warmer, set at approximately 50°C, for 5–10 min until they are fully dry.
10. Clear the dry slides by immersion in xylene for 5 min before mounting with a nonfluorescent mounting media.

5. Evaluations

5.1. Klenow and TUNEL Labeling

Cells with DNA damage are labeled with incorporated biotinylated dATP, which can be visualized either with streptavidin-conjugated peroxidase and its substrate DAB or with streptavidin-conjugated fluorescence dye, as is shown in Fig. 3. Both Klenow and TUNEL assays are enzymatic reactions; therefore, it is crucial for the Klenow fragment, the dNTPs, and the damaged DNA to be simultaneously in the nuclear so that the reaction can proceed. Generating a positive control with DNase I on at least one sample helps to rule out false-negative labeling.

5.2. Fluoro Jade B Staining

The fluorochrome Fluoro Jade B has a similar excitation and emission profile as fluorescein (excitation: 495; emission: 521 nm), and thus can be visualized by the FITC filter. In the GCI model, Fluoro Jade B-positive cells predominantly localize in the hippocampus (Fig. 4), striatum, and cortex of the brain.

5.3. Quantification of the Positive Cells

Klenow, TUNEL, and Fluoro Jade B staining can be quantified with the same methods for the quantitation of immunofluorescence staining. The method of counting positive particles applies best to these three stains. Refer to Sect. 6.2 of Chap. 2 for the detailed procedures of counting the positive cells using ImageJ.

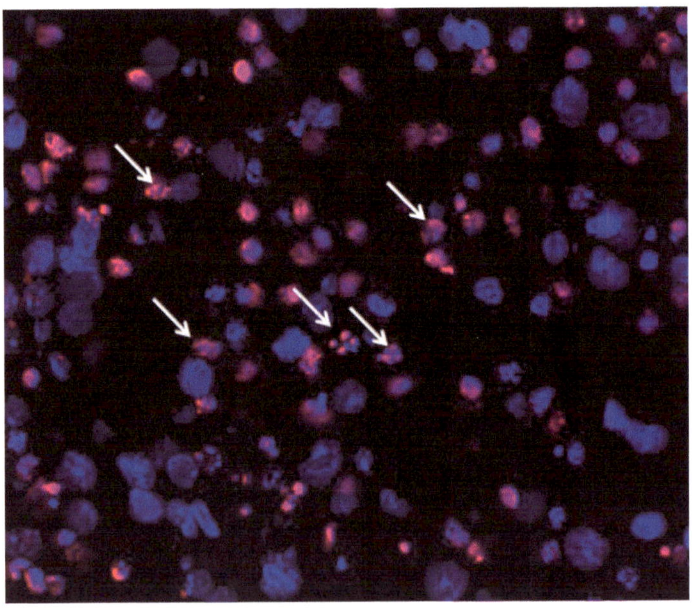

Fig. 3. DNA-damaged cells detected with the Klenow labeling assay. Klenow-positive cells were labeled by rhodamine (*red*). Some Klenow-positive cells displayed apoptotic nuclear morphological characteristics, including karyorrhexis and pyknosis (*white arrows*). *Blue*, DAPI. Magnification: ×400.

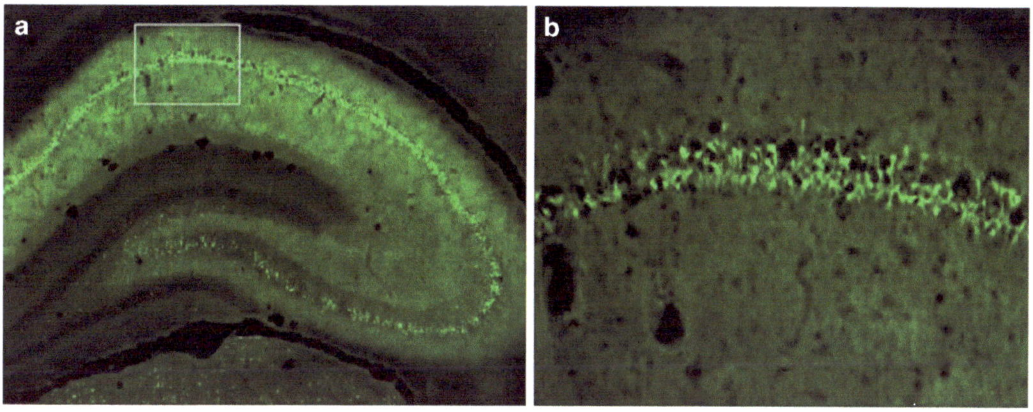

Fig. 4. Fluoro Jade B stained brain section of the hippocampus. (**a**) Free floating sections from GCI rat stained with Fluoro Jade B. The degenerating neurons in hippocampus were stained. The CA1 region (*boxed area*) in (**a**) is shown at a higher magnification in (**b**). Magnification: ×40 (**a**), ×200 (**b**).

6. Alternative Methods to Detect Degenerating Neurons in GCI Model

Immunostaining of active caspase protease, such as caspase-3 and caspase-9, can also be used to detect degenerating neurons because the caspase-dependent pathway is involved in the neural degeneration after GCI (1, 3). In addition, silver staining, a traditional method for detecting degenerated neurons, was used with success to map GCI-induced neuronal degeneration (11–13). However, the procedure of silver staining is time consuming and labor intensive, and the molecular basis of it is still largely unknown.

7. Summary

DNA damage, such as DNA strand break, is a feature of degenerating neurons following the global cerebral ischemic injury. TUNEL and Klenow assay are two well-developed methods to detect the DNA strand breaks, and thus can be used to characterize degenerating neuronal cells in the GCI model. Fluoro Jade B is a high-affinity fluorescent marker for the localization of neuronal degeneration in a variety of diseases. It is proved to be able to identify the neurodegeneration after GCI, and is gaining popularity because of the ease of staining, despite of its obscure mechanism.

References

1. Chen J et al (1998) Induction of caspase-3-like protease may mediate delayed neuronal death in the hippocampus after transient cerebral ischemia. J Neurosci 18(13): 4914–4928
2. Cao G et al (2001) Caspase-activated DNase/DNA fragmentation factor 40 mediates apoptotic DNA fragmentation in transient cerebral ischemia and in neuronal cultures. J Neurosci 21(13):4678–4690
3. Cao G et al (2002) Cloning and characterization of rat caspase-9: implications for a role in mediating caspase-3 activation and hippocampal cell death after transient cerebral ischemia. J Cereb Blood Flow Metab 22(5): 534–546
4. Ni B et al (1998) Transient global forebrain ischemia induces a prolonged expression of the caspase-3 mRNA in rat hippocampal CA1 pyramidal neurons. J Cereb Blood Flow Metab 18(3):248–256
5. Gillardon F et al (1999) Inhibition of caspases prevents cell death of hippocampal CA1 neurons, but not impairment of hippocampal long-term potentiation following global ischemia. Neuroscience 93(4):1219–1222
6. Li P et al (2011) Mechanistic insight into DNA damage and repair in ischemic stroke—exploiting the BER pathway as a model of neuroprotection. Antioxid Redox Signal 14(10): 1905–1918
7. Jin K et al (1999) In situ detection of neuronal DNA strand breaks using the Klenow fragment of DNA polymerase I reveals different mechanisms of neuron death after global cerebral ischemia. J Neurochem 72(3):1204–1214
8. Schmued LC, Hopkins KJ (2000) Fluoro Jade B: a high affinity fluorescent marker for the localization of neuronal degeneration. Brain Res 874(2):123–130
9. Fink RP, Heimer L (1967) Two methods for selective silver impregnation of degenerating axons and their synaptic endings in the central nervous system. Brain Res 4(4):369–374
10. Schmued LC, Albertson C, Slikker W Jr (1997) Fluoro Jade: a novel fluorochrome for the sensitive and reliable histochemical localization of neuronal degeneration. Brain Res 751(1): 37–46
11. Antonawich FJ, Federoff HJ, Davis JN (1999) BCL-2 transduction, using a herpes simplex virus amplicon, protects hippocampal neurons from transient global ischemia. Exp Neurol 156(1):130–137
12. Owen AJ et al (1997) Zonisamide as a neuroprotective agent in an adult gerbil model of global forebrain ischemia: a histological, in vivo microdialysis and behavioral study. Brain Res 770(1–2):115–122
13. Bartus RT et al (1998) Temporal ordering of pathogenic events following transient global ischemia. Brain Res 790(1–2):1–13

Chapter 4

Morphological Assessments of Global Cerebral Ischemia: Electron Microscopy

Irina S. Balan, Tibor Kristian, Chunli Liu, Andrew J. Saladino, and Bingren Hu

Abstract

Ischemic stroke in humans often results in acute and delayed neuronal death, as well as a wide range of chronic neurological deficits. In order to understand neuronal loss and neurological deficits after brain ischemia, several animal models have been established, including global and focal ischemic models. Transient cerebral ischemia leads to neuronal death that does not occur immediately, but takes place after 2–3 days of vascular reperfusion, the so-called delayed neuronal death. During this 2–3 days' delay period, all neurons destined to die appear normal under the light microscope. At the ultrastructural level, however, dramatic dissociation of polyribosomes, protein aggregation, and organelle damage takes place in postischemic neurons undergoing delayed neuronal death. Delayed neuronal death also occurs in the penumbral area after focal ischemia. The delay period provides a window of opportunity for understanding the underlying pathological processes and for developing therapies. This chapter describes the methods of transmission electron microscopy in the context of morphological studies of brain ischemia.

Key words: Brain ischemia, EM methods, Synapse, Apoptosis, Protein aggregation, Damaged organelles

1. Introduction

1.1. Ischemia-Induced Ultrastructural Changes in Brain

Brain ischemia triggers many pathological events in postischemic tissue before neurons undergo delayed cell death. These pathological events include mitochondrial dysfunction, acidosis, intracellular Ca^{2+} overload, reactive oxygen species (ROS) generation, protein aggregation, ER stress, Golgi fragmentation, and dysfunction of protein quality control systems. Most, if not all, major pathological changes in cellular morphology after brain ischemia can be studied with transmission electron microscopy (EM) (1–3).

Jun Chen et al. (eds.), *Animal Models of Acute Neurological Injuries II*, Springer Protocols Handbooks,
DOI 10.1007/978-1-61779-782-8_4,

Although established many decades ago, EM remains a powerful tool to study newly discovered pathological events after brain ischemia.

1.2. Protein Aggregation

Protein aggregation was discovered by EM methods as a major pathological event after brain ischemia (Fig. 1) (1–3). In an early EM study, increases in electron-dense deposits were observed in CA1 pyramidal neurons during the later periods of reperfusion in a rat transient forebrain ischemia model, and were termed

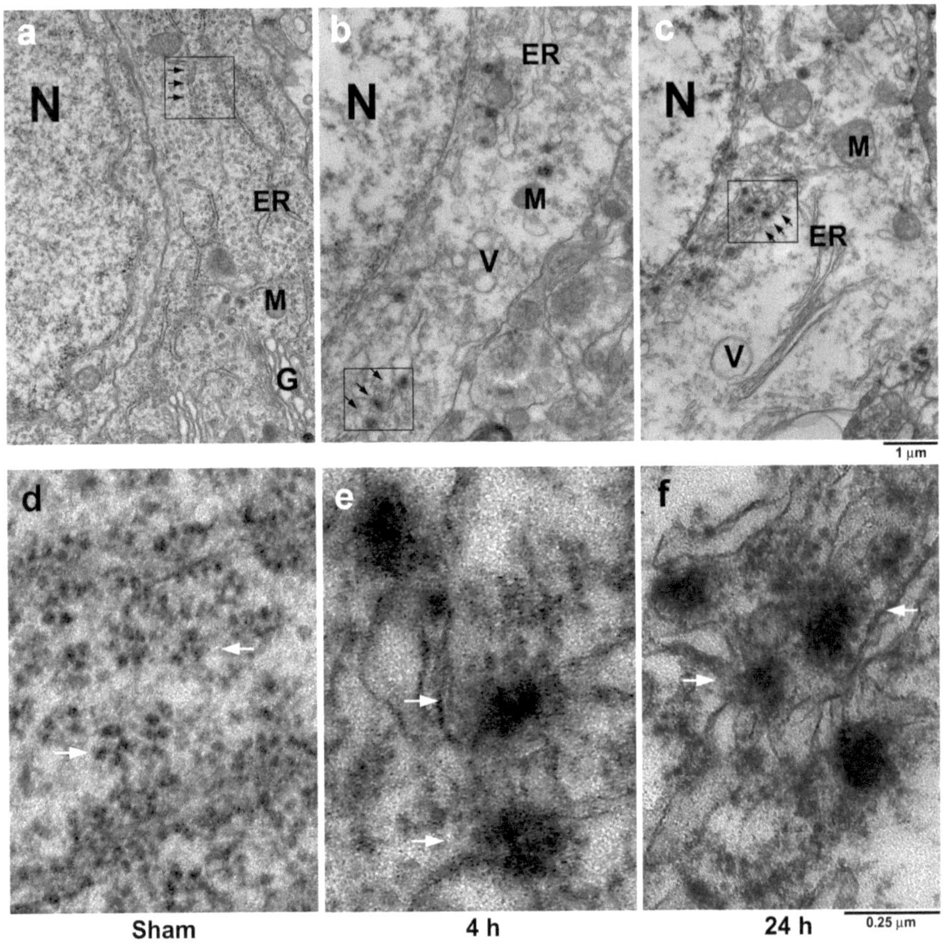

Fig. 1. Electron micrographs of osmium–uranium–lead-stained CA1 pyramidal neurons from a sham-operated control rat and rats subjected to 15 min of cerebral ischemia followed by 4 and 24 h of reperfusion. *Upper panel*: The rough endoplasmic reticulum (ER), mitochondria (M), nucleus (N), Golgi apparatus (G), and ribosomal rosettes (*arrows*) are normally distributed in sham-operated CA1 neurons (**a**, Sham, *upper panel*). At 4 and 24 h of reperfusion after ischemia, the ER and mitochondria (M) are dilated, membranous vesicles (V) are visible, and ribosomes (*arrows*) are clumped into large aggregates (**b**, **c**). *Lower panel*: Higher magnification of the ribosomal area indicated in the *upper panel* with three *black arrows* in sham (**a**), 4-h (**b**) and 24-h (**c**) micrographs. Ribosomal rosettes and ER-associated ribosomes are normally distributed in sham-operated control CA1 neurons (**d**, *arrows*). After ischemia, ribosomes are abnormally clumped into large aggregates (**e** and **f**, *arrows*). Scale bars = 1 μm in the *upper panel* and 0.25 μm in *lower panel* (adapted with permission and modification from ref. 7).

"dark substances" (4). This observation was recapitulated later and described as "electron-dense fluffy dark material" in an ultrastructural study of programmed cell death after brain ischemia (5). Hu and colleagues carried out a series of morphological, biochemical, and molecular studies demonstrating that these dark deposits are, in fact, protein aggregates made of abnormal proteins overproduced in postischemic neurons (1–3, 6–11). Furthermore, several molecular, biochemical, and confocal microscopic studies revealed that these protein aggregates are composed of at least four major categories of proteins: (1) components of the protein translation complex, including ribosomal proteins, protein synthesis initiation and elongation factors, molecular chaperones, and folding enzymes; (2) components of protein trafficking machinery; (3) components of ubiquitin proteasomal pathway; and (4) protein kinases, such as CaMKII and protein kinase C. Following ischemic insult, these proteins become unfolded and are ubiquitinated to be removed from cell. However, they are too numerous to be degraded by proteasomes that leads to their clumping into inactive protein aggregates (Fig. 1). These studies clearly demonstrated that combination of morphological EM studies with molecular and biochemical analysis provides a powerful tool to study the mechanisms of pathological events after brain ischemia.

1.3. Apoptotic and Necrotic Morphology Following Brain Ischemia

Apoptosis is a major type of cell death in neurons during brain development. However, the role of apoptosis in ischemic neuronal death has been controversial. Two types of neuronal necrotic morphologies characterized by EM have been described in the literature: one is the conventional cellular lysis with cell swelling followed by membrane rupture to release cell contents; the second type of necrosis is characterized by aggregation of nuclear contents and dilation of subcellular organelles followed by shrinkage of the entire cell. Neuronal cell lysis takes place in the infarct core area after focal brain ischemia (12) while the shrinkage type of necrosis occurs in a delayed manner in vulnerable brain regions after transient brain ischemia, in the penumbral regions after focal ischemia, as well as in excitotoxic neuronal death (2, 5, 13–16).

Morphological features of both neuronal apoptosis and necrosis have been repeatedly observed in neonatal hypoxia/ischemia (HI) models (9, 17–19). However, classical apoptotic morphology characterized by EM has not yet been found in adult postischemic neurons (3, 5, 9, 14, 15). Because each of these types of cell death may involve different molecular and biochemical events, understanding the nature of cell death after ischemia is important in order to determine effective therapeutic targets. EM evaluation remains the gold standard by which to characterize the structural aspects of apoptosis in neurons during development and after brain ischemia (2, 3). The studies by Hu and colleagues suggest that both classical apoptotic and necrotic morphologies were found in

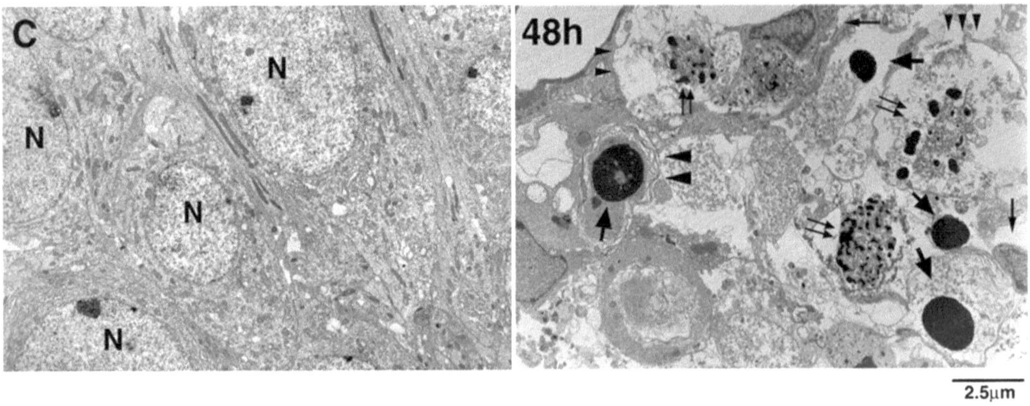

Fig. 2. Electron micrographs of postnatal 7-day (P7) DG neurons after hypoxia/ischemia (HI). P7 neonates were subjected to 1 h of HI followed by 48 h of recovery. Classical apoptotic cell death was not seen in adult neurons and neonatal CA1 neurons, but observed in neonatal DG neurons after HI. Ultrastructure of the contralateral (no ischemia) neurons was morphologically normal with normal nuclei (N) and cellular structures. The ipsilateral hypoxic/ischemic neurons were severely damaged after HI. Two types of nuclei were observed in HI-injured DG granule neurons: large/regular-shaped chromatin balls (*large arrows*) and small/irregular-shaped multichromatin masses (*double arrows*). The membrane became filament bundles (*arrowheads*). Phagocytes attached to the apoptotic neurons (*small arrows*). Scale bar = 2.5 μm (adapted with permission and modification from ref. 1).

postnatal 7-day-old (P7) dentate gyrus (DG) neurons after hypoxia/ischemia (Fig. 2), whereas necrotic morphology after HI was prominent in rat brains older than postnatal 15 days, consistent with several morphological studies (2, 5, 9, 14, 15). Therefore, the pathogenesis of HI-induced neuronal death shifts from apoptosis to necrosis during brain maturation.

1.4. Ultrastructural Characteristics of Mechanisms Removing Dysfunctional Intracellular Components

There are two major routes currently known for clearance of aberrant cellular components: (1) the ubiquitin-proteasomal pathway and (2) the autophagy-lysosomal pathway. The ubiquitin-proteasomal pathway is responsible for degradation of short-lived proteins and has been studied intensively (20), including following brain ischemia (3, 10). The autophagy pathway, originally described as a stress response to nutrient deprivation, is now emerging as the chief route for bulk degradation of aberrant organelles, protein aggregates, and invading foreign agents (21). There are three basic types of autophagy: macroautophagy, microautophagy, and chaperone-mediated autophagy. Bulk degradation of cytoplasmic organelles is mediated largely by macroautophagy, which is commonly referred to simply as autophagy (hereafter). Autophagy is a nonstop life-sustaining renewal process, which is active under normal conditions and is further enhanced in response to tissue injury (22, 23).

The autophagy pathway starts with formation of double-membraned cisternae that subsequently engulf cytoplasmic organelles to form double-membrane vacuoles, known as

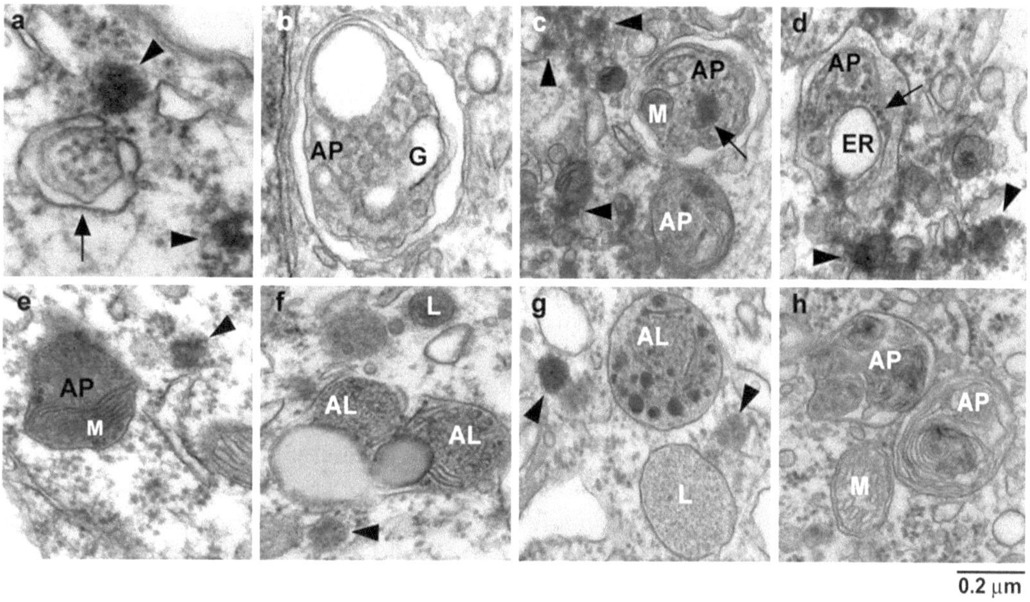

Fig. 3. Electron micrographs of autophagic ultrastructure in hippocampal neurons from rats subjected to 15 min of ischemia followed by 4 h of reperfusion. (**a**) A double-membraned cistern (*arrow*) and protein aggregates (*arrowheads*); (**b**) an AP containing dilated Golgi cistern (G) and its surrounding vesicles; (**c**) APs containing a damaged mitochondrion, a protein aggregate (*arrow*), and undigested membranous structures; (**d**) an AP containing dilated ER with ribosome; (**e**) an AP containing a shrunken mitochondrion (M) and a protein aggregate (*arrowhead*); (**f**) two autolysosomes (AL) with partially digested organelles, and bubble-like structures, as well as a lysosome (L); (**g**) an AL with partially digested membranous structures and protein aggregates (dark masses), as well as a lysosome; (**h**) two APs containing clusters of membrane whorls. Scale bar = 0.2 μm.

autophagosomes (APs). After maturation, APs merge with lysosomes for bulk degradation of the cargo contents (24).

Although labor intensive, EM remains an indispensable technique for evaluation of autophagy in situ and remains the most accurate method to detect induction of autophagy in tissue (25–28). The ultrastructural hallmarks for induction of autophagy are the manifestation of (1) double-membraned cistern structures; (2) autophagosomes containing cytoplasmic materials or aberrant organelles, such as ribosomes, ER, Golgi structures, and mitochondria; and (3) autolysosomes which contain partially digested subcellular structures, such as ER or mitochondrial remnants. Ultimately, autophagolysosome (AL) contents are digested and become homogeneous dense material. All these ultrastructural features of autophagy were clearly observed by transmission EM in postischemic neurons (Fig. 3).

1.5. Effect of Ischemia/Reperfusion on Intracellular Organelles' Morphology

Ischemic insult causes also significant morphological alterations of subcellular organelles. One consequence of ischemia-induced inhibition of cellular bioenergetic metabolism is cellular and mitochondrial calcium overload (29, 30). Mitochondria respond to calcium overload by opening of the permeability transition (PT) pore, an inner mitochondrial membrane megachannel that causes swelling of these organelles (31, 32). When mitochondria-isolated brain tissues are challenged with high levels of calcium, the swelling is observed only in a subpopulation of these organelles while the rest of the mitochondria do not show distinct characteristics of swelling (disrupted cristae and electron translucent matrix) (31, 33). Considering the numerous neural cell types from which brain mitochondria are isolated, this heterogeneity might reflect differences in the sensitivity of mitochondria to the permeability transition both within and among brain cells. In fact, cyclophilin D, a major regulator of the PT pore, has region- and cell type-specific distribution with highest expression in GABA-ergic interneurons (34).

Heterogeneous mitochondrial alterations have also been observed in animal models of cerebral ischemia. Ultrastructural studies of neuronal mitochondria after ischemic insult revealed several types of changes in mitochondrial morphology. Following focal ischemia and several hours of reperfusion, many mitochondria were moderately to severely swollen (35). Dispersed around the swollen mitochondria were organelles with poorly defined cristae and increased matrix density. Some mitochondria demonstrated dilated cristae (Fig. 4) (35). Similarly to swollen mitochondria,

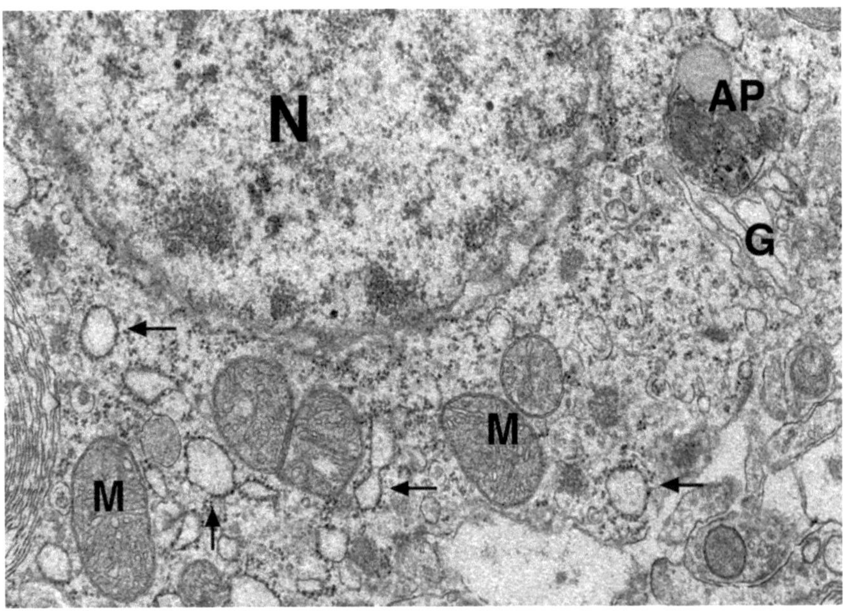

Fig. 4. Electron micrograph of hippocampal neuron following 30 min of global cerebral ischemia. One can clearly identify swollen endoplasmic reticulum (*arrows*) and Golgi apparatus (G). The mitochondria (M) also show mild swelling with dilated cristae. *N* nucleus; *M* mitochondria; *G* Golgi apparatus; *AP* apoptosomes containing lysosomal structure; *Arrows* denote swollen endoplasmic reticulum. Scale bar = 500 nm.

one can easily identify swollen endoplasmic reticulum and Golgi apparatus in the postischemic neurons (Fig. 4). These findings support the biochemical results suggesting inhibition of protein synthesis and bioenergetic failure of postischemic tissue.

1.6. Ischemia-Induced Alterations of Synaptic Ultrastructure

Synaptic ultrastructural changes can be studied by a preferential synaptic staining method at the EM level using ethanolic phosphotungstic acid (EPTA) (36). EPTA selectively stains the proteins in the postsynaptic density (PSD), the presynaptic grid, and some materials in the synaptic cleft. With the exception of nuclear chromatin, other structures are relatively unstained. This selective synaptic staining method has been utilized in a number of studies characterizing the structural organization and development of synapses in the CNS (37, 38). EPTA staining method has been employed to study synaptic ultrastructural changes in ischemic pathological conditions. These studies clearly demonstrate that synaptic ultrastructural and molecular composition changes are prominent events after brain ischemia (6, 39). These studies further suggest that the EPTA selective staining method is ideal for quantification of synaptic changes and analyses of synaptic morphology after brain injury. In contrast to the conventional osmium–uranium–lead staining method that cannot reveal clear changes in synaptic structure, robust synaptic ultrastructural alterations were observed in the EPTA-stained brain sections after ischemia (Fig. 5) (6, 39, 40). PSDs in the postischemic hippocampus appeared both thicker and more irregular than those from sham-operated controls (Fig. 5). These structural changes correlate very well with biochemical studies indicating that synaptic protein reorganization occurs in the postischemic brain (6).

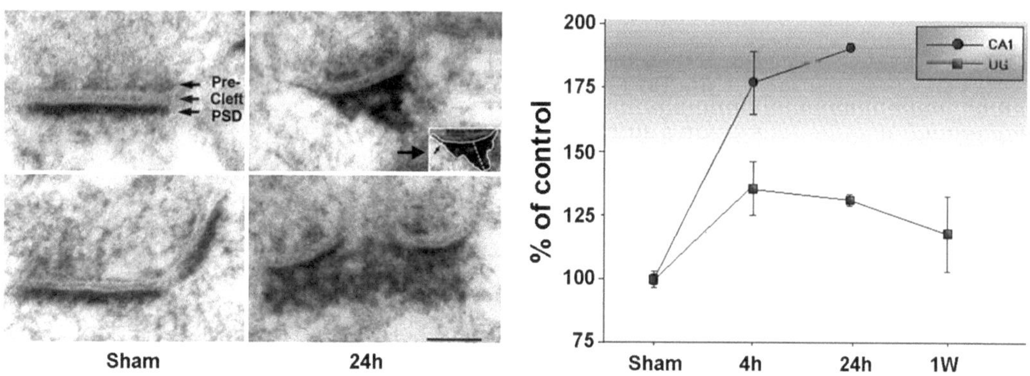

Fig. 5. (**a**) Electron micrographs of EPTA-stained synapses in CA1 regions from sham rats (sham) and rats subjected to a 15-min ischemia followed by 24 h of reperfusion. In comparison with control synapses (sham), postischemic synaptic PSDs were generally curved and surrounded by EPTA-stained material. Scale bar = 0.1 μm. Upper right insert (*arrow*): thickness measure used for PSD. *Pre* presynaptic specialization; *cleft* synaptic cleft; *PSD* postsynaptic density. (**b**) Maximum thickness of PSDs from control and postischemic brains after 4 h, 24 h, and 1 week of reperfusion, plotted as a percentage of controls.

The above studies indicate that EM methods remain a powerful tool to study most, if not all, pathological events after brain ischemia. The following are EM procedures used in the author's laboratories.

2. Materials

2.1. Animals

- Rodents (rats, mice, and gerbils) are the most commonly utilized animals for animal models of global and focal brain ischemia. After the animals are subjected to brain ischemia and designated time of recovery to apply electron microscopy technique, the brain tissue needs to be properly fixed.
- All protocols that use live animals must be reviewed and approved by an Institutional Animal Care and Use Committee (IACUC) prior to initiation of the study.

2.2. Equipment

- Amber glass bottles (Electron Microscopy Sciences, #64210-A, 64212-A, 64214-A).
- Analytical balance ED224S (VWR, #82028-322).
- Brushes (Electron Microscopy Sciences, #66105*).
- Capsules (Electron Microscopy Sciences, #70020).
- Centrifuge, benchtop (Fisher Scientific, #13-100-510).
- Decapicones (Fisher Scientific, #NC9745006).
- Desiccators (Fisher Scientific, #08-595).
- Duster w/nozzle (Electron Microscopy Sciences, #70705).
- Embedding capsule holder (Electron Microscopy Sciences, #70022).
- Grids, 200 mesh, copper (Electron Microscopy Sciences, #0200-Cu).
- Grid storage box for 50 grids (Electron Microscopy Sciences, #71152).
- Guillotine, small animals (VWR, #50000-450).
- Hotplate (Fisher Scientific, #11-496-5AQ).
- Lab oven (Fisher Scientific, #13-246-506GAQ).
- Latex gloves (Electron Microscopy Sciences, #71112).
- Lens tissue (Electron Microscopy Sciences, #71700).
- Marker pen, black (Electron Microscopy Sciences, #72120).
- Needles (VWR, #BD305198).
- Petri dishes, round (Electron Microscopy Sciences, #70680).

- Pipettes, 2 cc (Electron Microscopy Sciences, #70960-1).
- Precision balance ED2202S (VWR, #82028-360).
- Shell vials (Electron Microscopy Sciences, #72630, 72631).
- Razor blades (Electron Microscopy Sciences, #71960).
- Rotator/Rocker (Electron Microscopy Sciences, #71834).
- Sample vials (Electron Microscopy Sciences, #60992).
- Scissors (VWR, #82027).
- Serum Tubing Vials—Amber (Electron Microscopy Sciences, #60302).
- Spatula (Fisher Scientific, #14-373-25).
- Stirring Hotplate (Fisher Scientific, #11-520-49SH).
- Stirring Rods (Electron Microscopy Sciences, #66200-20).
- Syringes, 10 mL (Electron Microscopy Sciences, #72510).
- Syringes, 1 mL (Fisher Scientific, #NC9992493).
- Transmission Electron Microscope (Jeol 1200EX or equivalent).
- Tuberculin syringes (VWR, #BD-309624).
- Tubing for variable flow pump (VWR, #36934-128).
- Tweezers, #5 Dumont (Electron Microscopy Sciences, #72700-D).
- Ultramicrotome (Leica, EM UC7).
- Vacuum Nalgene Filter Units (VWR, #73520).
- Variable Flow Pump (VWR, #54856-075).
- Vial Racks (Electron Microscopy Sciences, #72641).
- Vibrating microtome (Electron Microscopy Sciences, EMS7000smz).
- Wood applicators (Electron Microscopy Sciences, #72300).
- Wooden picks (Electron Microscopy Sciences, #72340).

3. Procedures

3.1. Brain Tissue Fixation

There are two major models of chemical fixation: (1) vascular perfusion and (2) tissue immersion. Although each of these models has certain advantages and disadvantages, fixation by vascular perfusion yields better preservation of the fine structure of the brain.

A simple and reasonable requirement for satisfactory fixation for transmission electron microscopy is the uniformity of fixation throughout the tissue specimen. Uniformity in fixation is dependent primarily on the size of the specimen.

3.1.1. Brain Immersion Fixation

- For immersion fixation, the size of the brain tissue fragment should not exceed 1 mm^3. After brain removal, cut the brain tissue into small pieces (<1 mm^3) in the fixative solution consisting of a mixture of 4% paraformaldehyde and 1% glutaraldehyde with 0.1 M phosphate buffer, pH 7.4 (4F1G).
- Allow the tissue fragments to remain in the fixative for 24–48 h at 4°C. The cutting must be accomplished by one quick slashing motion; several back and forth movements of the razor blade result in mechanical damage.

Note: The choice of 4F1G (4 parts formaldehyde to 1 part glutaraldehyde) (41) is based on the following characteristics: (1) formaldehyde in solution is a rapidly penetrating fixative, but only partially cross-links proteins and nucleic acids and is not entirely permanent in solution or irreversible; (2) glutaraldehyde, on the other hand, is slower to penetrate tissues but cross-links more completely, thus better preserving ultrastructure. This results in the stabilizing of the intracellular system of labile microtubules, filaments, rough and smooth endoplasmic reticulum, mitotic spindles, platelets, and pinocytotic vesicles. However, the addition of glutaraldehyde fixation limits choices for additional histochemical and immunocytochemical staining studies. Therefore, the 4F1G combination fixative has proved to be the most effective for preserving CNS fine structure. Tissue specimens can be left in this fixative for days to weeks without apparent deterioration, and this flexibility of the pretreatment is one of its major assets. Both these aldehyde solutions are regarded as slightly to moderately toxic and noncorrosive to stainless steel.

3.1.2. Brain Removal Procedures

- Place the rat into the decapicone and hold the animal by closing the decapicone at the root of the tail. Place the animal head into the guillotine so that it is decapitated just behind the ears.
- To remove the brain, retract the scalp to reveal the skull. Make an incision in the bone along the midline by placing the blade of the bone scissors into the spinal canal and cut the bone towards the eyes. Make sure that the inserted blade does not damage the brain by leaning it against the inside of the skull while cutting.
- Make lateral horizontal incisions into the bone at the level of the ear on both sides by placing the blade of the bone scissors into the spinal canal and cutting the bone towards the eyes.
- Use a 12-in.-long nickel stainless spatula to apply a lateral force on the bone at the midline incision to reveal each hemisphere. Once the bones on both sides are retracted far enough exposing the whole brain, scoop out the brain from the skull by sliding the spatula between the surface of the brain and the skull, then carefully remove the brain, and place into 4F1G ideally within 30 s after decapitation.

3.1.3. Brain Fixation by Vascular Perfusion

- Anesthetize the animal by intraperitoneal injection of 100 mg/kg pentobarbital.
- Open chest cavity by making an inverted "T" incision of the skin over the chest and cutting into the peritoneal cavity to expose the diaphragm. Cut the diaphragm to expose the heart. Cut both sides of the rib cage and lift the sternum and ribs over the heart so that the entire heart is visible. Using a 1-mL syringe, inject 0.1 mL (100 U) of heparin into the heart apex. The heparin prevents blood clots from impairing flow of fixative to the brain tissue.
- Clamp descending aorta to eliminate whole-body perfusion.
- Insert a 15-G needle (connected to a tubing system of a peristaltic pump) into the heart apex.
- Cut a hole in the right atrium to allow fluids to drain from the circulatory system.
- The blood is flushed out with 0.9% NaCl for no longer than 1 min at 120 mm Hg.
- Immediately it is followed by perfusion with 4F1G for 10 min at 120 mm Hg. Approximately 200 ml of the fixative solution is used.
- Remove the brain from the skull, immerse in 4F1G, cut into small pieces (<1 mm^3), and leave the tissue fragments in 4F1G overnight at 4°C.

3.2. Subsequent Procedures with the Brain Tissue Fragments Fixed in 4F1G by the Immersion or Vascular Perfusion

3.2.1. Postfixation

- Rinse the brain tissue fragments in 0.1 M sodium cacodylate containing 0.2 M sucrose one time for 10 min.
- Postfix the brain fragments in 1% osmium tetroxide (OsO_4) with 0.1 M phosphate buffer, pH 7.4, for 1 h.
- En bloc stain with uranyl acetate for 1 h.

3.2.2. Dehydration

Dehydrate the brain fragments in 35% ethanol for 15 min; 70% ethanol for 15 min; two times in 95% ethanol, each time for 15 min; two times in 100% ethanol, each time for 15 min; and two times in 100% propylene oxide (Polysciences #441883), each time for 15 min.

3.2.3. Infiltration

- Drain the tissue fragments of most of the propylene oxide, leaving a little so that the tissue does not dry out.
- Incubate the tissue fragments with a 50:50 solution of propylene oxide to embedding medium for 1 h at room temperature.
- Incubate the tissue fragments with a 70:30 solution of embedding medium to propylene oxide at room temperature overnight in glass desiccators with cover off to allow any remaining propylene oxide to evaporate.

3.2.4. Embedding

- Make fresh embedding medium.
- Place predried embedding capsule (Electron Microscopy Sciences, #70020) in capsule board.
- Insert typed labels into each capsule.
- Fill capsule with well-mixed embedding medium avoiding unnecessary bubble formation.
- Transfer each piece of tissue to the top of the filled capsule using an applicator stick. Allow the tissue to sink to the bottom of the capsule. Gently center the tissue if necessary.
- Place capsules in 60-C oven for 48 h to polymerize.

Notes: All steps must be performed in a fume hood or cupboard and gloves should be worn throughout.

Osmium tetroxide, propylene oxide, and propylene oxide/embedding medium are all considered hazardous chemicals and should be collected in bottles for safe disposal.

For the steps before "Embedding," the processing vials should be on a rotating mixer.

3.3. Semithin Sectioning

- Cut semithin sections (0.5–1.0 μm) using glass knives and an ultramicrotome. Using a steel loop, transfer the sections to a drop of water (this flattens sections to avoid wrinkles) on glass slides.
- Dry the sections on a hot plate at 80°C.
- Stain the sections with toluidine blue solution for 1–2 min at 80°C.
- Rinse the stain off with distilled water, dry the sections, and cover with a glass coverslip using a synthetic mounting medium, such as D.P.X. (Sigma, #44581).
- Observe the sections under a light microscope for precise location to cut for ultrathin sections.

3.4. Ultrathin Sectioning for Electron Microscopy

- Cut ultrathin sections (around 100 nm) in the same way as for semithin sectioning but using a diamond knife.
- Pick the sections up onto 200-mesh (200 squares), thin-bar, copper grids.

3.4.1. Ultrathin Section Staining with Uranyl Acetate and Lead Citrate or Impregnation with Heavy Metals

- Dry the ultrathin sections overnight before staining.
- Stain grids with uranyl acetate for 30 min at room temperature.
- Stain grids with lead citrate for 4 min at room temperature.
- Observe under a transmission electron microscope.

3.5. EPTA Staining Procedure

3.5.1. Perfusion, Fixation, Dehydration, and Infiltration

- Perfuse an animal through the heart into ascending aorta with ice-cold saline (40 mL/rat at 20 mL/min) and then ice-cold fixative of 2% paraformaldehyde and 2.5% glutaraldehyde in 0.1 M sodium cacodylate buffer (300 mL/rat) with clamping of the descending aorta.
- Remove and dissect the brain through the levels of the dorsal hippocampus and striatum using a razor blade.
- Postfix the brain 1 h on ice on shaker in 4% glutaraldehyde in 0.1 M sodium cacodylate buffer.
- Take 150-μm coronal brain sections on a vibratome into a vial and fix the sections 30 min on ice on shaker in 4% glutaraldehyde in 0.1 M sodium cacodylate buffer.
- Wash for 10 min on ice on shaker in 0.1 M sodium cacodylate buffer.
- Dehydrate 1× for 10 min on ice on shaker 20, 50, 70, 90, and 100% ethanol.
- Stain 2× for 15 min at room temperature on shaker in 1% phosphotungstic acid (PTA) in ethanol.
- Dehydrate for 15 min under desiccant in 100% ethanol.
- Dehydrate 2× for 15 min under desiccant in 100% acetone.
- Infiltrate overnight in 50/50 mixture of acetone/durcupan at room temperature in a desiccant plastic box on shaker.

3.5.2. Infiltration and Embedding

- Infiltrate 2× for 1.5 h with 100% durcupan in a desiccant plastic box on shaker.
- Embed resin-coated sections between coated microscope slides.
- Clamp slides with binder clips.
- Polymerize at 60°C for 48 h.

4. Reagents and Solutions

4.1. Paraformaldehyde, 4%, Containing Glutaraldehyde, 1%, in 0.1 M Phosphate Buffer, pH 7.4 (4F1G)

- Dissolve 40 g of paraformaldehyde (Electron Microscopy Sciences, #19200) in 500 mL of distilled water. Stir and heat to 53–55°C (do not exceed 58°C).
- Add 12 N NaOH dropwise with stirring to clear.
- Cool to room temperature.
- Add 100 mL of 0.5 M sodium phosphate dibasic (Na_2HPO_4), 100 mL of 0.5 M sodium phosphate monobasic (NaH_2PO_4),

280 mL of distilled water, and 20 mL of 50% glutaraldehyde (Electron Microscopy Sciences, #16320).

- Filter through sterile disposable vacuum filter with 0.2-μm pore size.
- Check and adjust pH to 7.4 with either 12 N NaOH or 10 N HCl using a pH meter.

4.1.1. Sodium Phosphate Dibasic (Na_2HPO_4; Sigma-Aldrich, #S7907), 0.5 M

- Dissolve 70.98 g of sodium phosphate dibasic in 1,000 mL of distilled water.

4.1.2. Sodium Phosphate Monobasic (NaH_2PO_4; Sigma-Aldrich, #S8282), 0.5 M

- Dissolve 59.99 g of sodium phosphate monobasic in 1,000 mL of distilled water.

4.2. Sucrose, 0.2 M, in 0.1 M Sodium Cacodylate

- Add 21.4 g of sodium cacodylate trihydrate (Electron Microscopy Sciences, #12300) and 68.46 g of sucrose to 500 mL of distilled water.
- Stir the solution with magnetic stirring bar, and then add enough distilled water up to 1,000 mL. Keep the solution at 4°C.

Note: Cacodylate avoids the presence of extraneous phosphates that may interfere with cytochemical studies. It does not seem to decrease the nuclear dry mass, and there is little removal of acid-soluble proteins from nuclei. Sodium cacodylate contains arsenic and, thus, is a health hazard. If inhaled or absorbed through skin, it can cause dermatitis and liver and kidney inflammation. Hands should be protected by gloves and fume hoods should be used for handling.

4.3. Osmium Tetroxide, 1%, in 0.1 M Phosphate Buffer

- Add 5 mL of 2% osmium tetroxide stock solution to 5 mL of 0.2 M phosphate buffer, pH 7.4.

4.3.1. Osmium Tetroxide Stock Solution, 2%

- Transfer 1 g of dry, yellow osmium tetroxide crystal (Electron Microscopy Sciences #19100) into a clean brown glass bottle with 50-mL mark on it, add 49 mL of distilled water, and shake to suspend the crystal. This must be done under a fume hood.
- Leave it under a fume hood at room temperature; shake the bottle a couple of times a day. It takes a couple of days for the solution to be completed.
- Place this bottle in airtight plastic container and keep it in a refrigerator at 4°C.

4.3.2. Phosphate Buffer, 0.2 M, pH 7.4

- Dissolve 21.8 g of Na_2HPO_4 (Sigma-Aldrich, #S7907) and 6.4 g of NaH_2PO_4 (Sigma-Aldrich, #S8282) in 1,000 mL of distilled water.
- Check and adjust pH to 7.4 with either 10 N HCl or 12 N NaOH using a pH meter.

Note: Osmium tetroxide acts not only as a fixative complexing membrane lipids and lipid aggregates which are not well preserved with aldehyde fixatives, but it also acts as an electron dense stain, and this is its major advantage over most other known fixatives. Reduction of osmium imparts high contrast to the osmiophilic structures in the specimen. It also acts as a mordant. Osmium tetroxide-dependent enhancement of lead staining is well documented. The most important application of osmium tetroxide is its use as an aldehyde postfixative, for it preserves many lipids and imparts electron density to cell components. Osmium tetroxide is extremely dangerous. Its vapor can damage nasal mucosa and the cornea of the eye. Always handle in hood and always wear plastic gloves.

4.4. Uranyl Acetate Solution for En Bloc Stain

4.4.1. Prepare Stock Solution A

- Add 6.25 g of uranyl acetate (Electron Microscopy Sciences, #22400) to 500 mL of distilled water.
- Place on a slightly warm hot plate until it dissolves with continuous stirring, but do not overheat the solution.

4.4.2. Prepare Stock Solution B

- Add 4.86 g of sodium acetate (Sigma-Aldrich, #S2889) and 7.36 g of sodium barbital (Sigma-Aldrich, #B0500) to 250 mL of distilled water.

4.4.3. Prepare Stock Solution C

- Dissolve 8.5 g of sodium chloride (Sigma, #S3014) in 100 mL of distilled water.

4.4.4. Prepare Stock Solution D

- Prepare 0.1 N hydrochloric acid (HCl, 37%, Sigma-Aldrich, #320331) by mixing 9.854 mL of the hydrochloric acid with distilled water up to 1,000 mL.

4.4.5. Prepare Working Solution

First container:

- Place 500 mL of stock solution A in 2,000-mL flask.
- Add 400 mL stock solution D. *Note*: Solution should be pale yellow.

Second container:

- Place 80 mL of stock solution C in 500-mL flask.
- Add 250 mL of stock solution B.

Mixing:

- Add slowly solution of second container to first container with vigorous stirring. Note: Final working solution must be crystal yellow. If final working solution becomes milky with combination, it must be discarded. Final working solution must be stored in brown bottles and not be refrigerated.

4.5. Uranyl Acetate and Lead Citrate Solutions for Staining of Ultrathin Sections

4.5.1. Uranyl Acetate Solution, 2%

- Add 2 g of uranyl acetate (Electron Microscopy Sciences, #22400) to 100 mL of distilled water.
- Stir the solution and filter with Nalgene filter.

4.5.2. Lead Citrate Solution

- Prepare 10 N sodium hydroxide by adding 4 g of sodium hydroxide (Sigma, #S8045) to 10 mL of distilled water.
- Add 0.1 g of lead citrate (Electron Microscopy Sciences, #27800) and 0.5 mL of 10 N sodium hydroxide to 50 mL of distilled water and centrifuge the solution at 3,000 rpm for 15 min.

4.6. Toluidine Blue in Borax Solution

- Dissolve 5 g of sodium borate (Sigma-Aldrich, #S9640) in 500 mL of distilled water.
- Dissolve 4 g of toluidine blue (Sigma-Aldrich, #T3260) in 400 mL of the sodium borate solution.
- Dissolve 1 g pyronin Y (Sigma, #83200) in 100 mL of the sodium borate solution by boiling the solution on the hot plate with stirring.
- Mix the toluidine blue and pyronin Y solutions together.
- Heat the mixture to boiling, stirring occasionally.
- Cool to room temperature and then filter the solution. Store at room temperature and prior to use filter again.

4.7. Embedding Medium (Poly/Bed 812 Embedding Media/DMP-30 Kit, Polysciences #08792-1)

- Mix Poly/Bed 812, DDSA, and NMA using the following proportions presented in the table below in plastic beaker and stir very thoroughly with wood stick.
- Add DMP-30 just prior to infiltration or embedding. Use a tuberculin syringe to exactly measure the volume of DMP-30.

Notes: Measure or weigh all components under a fume hood. Prior to measuring and mixing, the resin and the anhydrite should be warmed to 60°C to reduce their viscosity. Inadequate mixing is one of the main causes of sectioning problems with epoxy resins. Although the mixture can be stored for up to 6 months at 4°C, it is highly recommended that freshly prepared embedding medium is always used. If you choose to store the mixture, you should warm it thoroughly prior to adding the accelerator.

Final volume (mL)	Poly/Bed 812 (mL)	DDSA (mL)	NMA (mL)	DMP-30 (~2%) (mL)
20.91	10	4.5	6	0.41
31.37	15	6.75	9	0.62
41.82	20	9	12	0.82
52.28	25	11.25	15	1.03

References

1. Hu BR, Janelidze S, Ginsberg MD, Busto R, Perez-Pinzon M, Sick T, Siesjo BK, Liu CL (2001) Protein aggregation after focal brain ischemia and reperfusion. J Cereb Blood Flow Metab 21:865–875
2. Hu BR, Liu CL, Ouyang Y, Blomgren K, Siesjo BK (2000) Involvement of caspase-3 in cell death after hypoxia-ischemia declines during brain maturation. J Cereb Blood Flow Metab 20:1294–1300
3. Hu BR, Martone ME, Jones YZ, Liu CL (2000) Protein aggregation after transient cerebral ischemia. J Neurosci 20:3191–3199
4. Kirino T, Tamura A, Sano K (1984) Delayed neuronal death in the rat hippocampus following transient forebrain ischemia. Acta Neuropathol 64:139–147
5. Deshpande J, Bergstedt K, Linden T, Kalimo H, Wieloch T (1992) Ultrastructural changes in the hippocampal CA1 region following transient cerebral ischemia: evidence against programmed cell death. Exp Brain Res 88:91–105
6. Hu BR, Park M, Martone ME, Fischer WH, Ellisman MH, Zivin JA (1998) Assembly of proteins to postsynaptic densities after transient cerebral ischemia. J Neurosci 18:625–633
7. Liu CL, Ge P, Zhang F, Hu BR (2005) Co-translational protein aggregation after transient cerebral ischemia. Neuroscience 134: 1273–1284
8. Liu CL, Martone ME, Hu BR (2004) Protein ubiquitination in postsynaptic densities after transient cerebral ischemia. J Cereb Blood Flow Metab 24:1219–1225
9. Liu CL, Siesjö BK, Hu BR (2004) Pathogenesis of hippocampal neuronal death after hypoxia-ischemia changes during brain development. Neuroscience 127:113–123
10. Ge P, Luo Y, Liu CL, Hu B (2007) Protein aggregation and proteasome dysfunction after brain ischemia. Stroke 38:3230–3236
11. Zhang F, Liu CL, Hu BR (2006) Irreversible aggregation of protein synthesis machinery after focal brain ischemia. J Neurochem 98: 102–112
12. Dodson RF, Aoyagi M, Hartmann A, Tagashira Y (1974) Acute cerebral infarction and hypotension: an ultrastructural study. J Neuropathol Exp Neurol 33:400–407
13. Siesjo BK (1985) Oxygen deficiency and brain damage: localization, evolution in time, and mechanisms of damage. J Toxicol Clin Toxicol 23:267–280
14. Fukuda T, Wang H, Nakanishi H, Yamamoto K, Kosaka T (1999) Novel non-apoptotic morphological changes in neurons of the mouse hippocampus following transient hypoxic-ischemia. Neurosci Res 33:49–55
15. Colbourne F, Sutherland GR, Auer RN (1999) Electron microscopic evidence against apoptosis as the mechanism of neuronal death in global ischemia. J Neurosci 19:4200–4210
16. Olney JW (2003) Excitotoxicity, apoptosis and neuropsychiatric disorders. Curr Opin Pharmacol 3:101–109
17. Mehmet H, Yue X, Squier MV, Lorek A, Cady E, Penrice J, Sarraf C, Wylezinska M, Kirkbride V, Cooper C, Brown GC, Wyatt JS, Reynolds EOR, Edwards AD (1994) Increased apoptosis in the cingulate sulcus of newborn piglets following transient hypoxia-ischaemia is related to the degree of high energy phosphate depletion during the insult. Neurosci Lett 181:121–125
18. Portera-Cailliau C, Price DL, Martin LJ (1997) Excitotoxic neuronal death in the immature brain is an apoptosis–necrosis morphological continuum. J Comp Neurol 378:70–87

19. Martin LJ, Al-Abdulla NA, Brambrink AM, Kirsch JR, Sieber FE, Portera-Cailliau C (1988) Neurodegeneration in excitotoxicity, global cerebral ischemia, and target deprivation: a perspective on the contributions of apoptosis and necrosis. Brain Res Bull 46:281–309
20. Ciechanover A (2006) The ubiquitin proteolytic system: from a vague idea, through basic mechanisms, and onto human diseases and drug targeting. Neurology 66:S7–S19
21. Nixon RA (2005) Endosome function and dysfunction in Alzheimer's disease and other neurodegenerative diseases. Neurobiol Aging 26:373–382
22. Klionsky DJ (2005) Autophagy. Curr Biol 15:R282–R283
23. Klionsky DJ (2006) Neurodegeneration: good riddance to bad rubbish. Nature 441:819–820
24. Yorimitsu T, Klionsky DJ (2005) Autophagy: molecular machinery for self-eating. Cell Death Differ 12:1542–1552
25. Eskelinen EL (2005) Macroautophagy in mammalian cells. In: Saftig P (ed) Lysosomes. Springer, New York, pp 166–177
26. Eskelinen EL (2005) Maturation of autophagic vacuoles in mammalian cells. Autophagy 1:1–10
27. Eskelinen EL (2006) Roles of LAMP-1 and LAMP-2 in lysosome biogenesis and autophagy. Mol Aspects Med 27:495–502
28. Ciechomska IA, Tolkovsky AM (2007) Non-autophagic GFP-LC3 puncta induced by saponin and other detergents. Autophagy 3: 586–590
29. Kristian T (2004) Metabolic stages, mitochondria and calcium in hypoxic/ischemic brain damage. Cell Calcium 36:221–233
30. Kristian T, Siesjo BK (1998) Calcium in ischemic cell death. Stroke 29:705–718
31. Kristian T, Gertsch J, Bates TE, Siesjo BK (2000) Characteristics of the calcium-triggered mitochondrial permeability transition in non-synaptic brain mitochondria: effect of cyclosporine A and ubiquinone 0. J Neurochem 74:1999–2009
32. Zoratti M, Szabo I (1995) The mitochondrial permeability transition. Biochim Biophys Acta 1241:139–176
33. Kristian T, Weatherby TM, Bates TE, Fiskum G (2002) Heterogeneity of the calcium-induced permeability transition in isolated non-synaptic brain mitochondria. J Neurochem 83: 1297–1308
34. Hazelton JL, Petrasheuskaya M, Fiskum G, Kristian T (2009) Cyclophilin D is expressed predominantly in mitochondria of gamma-aminobutyric acidergic interneurons. J Neurosci Res 87:1250–1259
35. Solenski NJ, diPierro CG, Trimmer PA, Kwan AL, Helms GA (2002) Ultrastructural changes of neuronal mitochondria after transient and permanent cerebral ischemia. Stroke 33: 816–824
36. Bloom FE, Aghajanian GK (1968) Fine structural and cytochemical analysis of staining of synaptic junctions with phosphotungstic acid. J Ultrastruct Res 22:361–375
37. Jones DG, Colangelo W (1985) Ultrastructural investigation into the influence of ethanol on synaptic maturation in rat neocortex. I. Qualitative assessment. Dev Neurosci 7: 94–106
38. Miyazaki M (1989) Electron microscopic study on the synaptic glomeruli of rat cerebellum: quantitative and qualitative analyses using the ethanol-phosphotungstic acid (EPTA) procedure. Neuropediatrics 20:73–78
39. Martone ME, Jones YZ, Young SJ, Ellisman MH, Zivin JA, Hu BR (1999) Modification of postsynaptic densities after transient cerebral ischemia: a quantitative and three-dimensional ultrastructural study. J Neurosci 19:1988–1997
40. Martone ME, Hu BR, Ellisman MH (2000) Alterations of hippocampal postsynaptic densities following transient ischemia. Hippocampus 10:610–616
41. McDowell EM, Trump BF (1976) Histologic fixatives suitable for diagnostic light and electron microscopy. Arch Pathol Lab Med 100: 405–414

Chapter 5

Biochemical and Molecular Biological Assessments of Global Cerebral Ischemia: mRNA

R. Anne Stetler and Yu Gan

Abstract

DNA microarrays have indicated that significant gene expression changes following global cerebral ischemia occur, and that neurotherapeutic strategies may specifically impact RNA expression in ischemic cells. Thus, the assessment of specific RNA species may provide an integral aspect to understanding the molecular effects of global cerebral ischemia and therapeutic strategies. We present here a brief discussion of RNA assessments following global ischemia, and describe two of the basic techniques, northern blotting and in situ hybridization, keeping in mind the specific context of global cerebral ischemic brain injury.

Key words: Global cerebral ischemia, RNA, Northern blot, In situ hybridization

1. Introduction

Rodent global cerebral ischemia models are developed to mimic human ischemic brain conditions that arise when total blood flow to the brain is arrested, such as during cardiac arrest. Following global cerebral ischemia, acute cell death occurs in the medium spiny neurons of the striatum, whereas a very specific subset of hippocampal neuronal populations (i.e., CA1 pyramidal neurons) undergo delayed cell death largely via apoptotic mechanisms. A large volume of literature detailing gene expression changes and the impact of therapeutic strategies on mRNA and protein alterations following global ischemia has steadily grown and expanded the understanding of molecular mechanisms of global cerebral ischemic injury and protection. Mass amounts of data garnered from DNA arrays following global ischemia provide the starting material for further investigation of

Jun Chen et al. (eds.), *Animal Models of Acute Neurological Injuries II*, Springer Protocols Handbooks,
DOI 10.1007/978-1-61779-782-8_5, © Springer Science+Business Media, LLC 2012

the alteration of gene transcription (1). However, microarrays are somewhat limited in that the exact cell population cannot be delineated. Thus, in situ assessments are often necessary to better characterize mRNA expression. For example, GRP78 has been found to decrease in the CA1, but transiently increases in the CA3 of the hippocampus following global ischemia (2). Likewise, caspase-3 mRNA was predominantly expressed in the degenerating CA1 following global ischemia, but showed limited expression in the surviving CA3 (3). Furthermore, activation or suppression of transcriptional signaling pathways that lead to modification of protein expression has been an active field of more recent research.

The most straightforward use of RNA assessments is in determining the specific upregulation or downregulation of mRNA products, and in comparing these results with the expression of the protein product. Often, this process is used simply to verify consequential changes in protein expression. However, mRNA expression can also be used as one end point of gene transactivation, and thus, in combination with other techniques, supports transcription factor activity (or promoter activity). For example, the CRE element is found in the promoter regions of many prosurvival factors (e.g., apurinic/apyrimidinic endonuclease-1 (APE1), brain-derived neurotrophic factor (BDNF), Bcl-2) (4, 5), and typically is bound by CREB or ATF2 leading to transactivation of downstream genes. Administration of neuroprotective agents often modulates large sets of gene products via activation of specific promoters; therefore, assessing the transcription (mRNA) of known downstream genes can verify the functional activity of promoter elements. Additional uses of mRNA gene expression analysis in global ischemic models center on confirmation of exogenous alteration of gene expression, such as in the application of knockout mouse models.

The primary basic techniques in RNA analysis are northern blot and in situ hybridization. Northern blot is used for the semiquantitative aspect, but is somewhat limited in the determination of cell-specific signal. On the other hand, in situ hybridization is a technique used primarily to determine the cellular expression of specific nucleotide sequences, particularly mRNA. The technique is being further applied to microRNA sequences, and has also been used for DNA genotyping (FISH). However, the semiquantification of in situ hybridization is less reliable. We present here a brief outline of both standard northern blotting and in situ hybridization protocols. The general flow of both techniques is presented in Fig. 1.

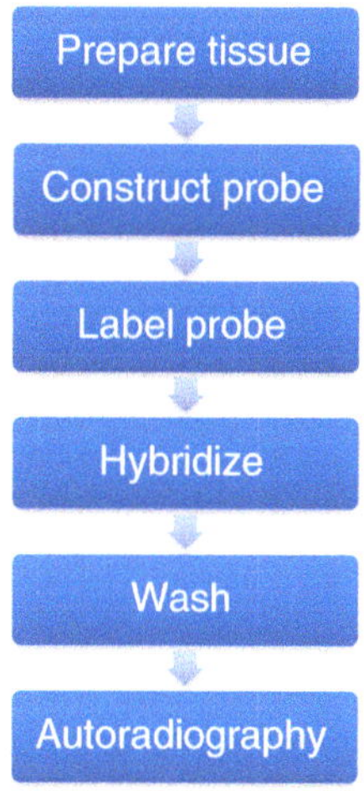

Fig. 1. General flowchart of northern and in situ RNA detection.

2. Northern Blot Analysis for mRNA

2.1. Materials

2.1.1. Reagents

1. RNeasy Lipid Tissue Mini Kit (Qiagen).
2. GE Nylon, Reprobing Charged Transfer Membrane (GE Life Science).
3. Rediprime Labeling System (GE Life science).
4. Illustra ProbeQuant G-50 Micro Columns (GE Life Science).

2.1.2. Solutions

5. 10× running buffer: 0.2 M MOPS, 0.05 M NaAc, and 0.005 M EDTA; adjust to pH 7.0 with NaOH.
6. Sample denaturation buffer (100 μL): 64.6 μL formamide, 22.6 μL formaldehyde, and 13 μL 10× running buffer.
7. Loading dye mix: 50% glycerol, 0.3% xylene cyanol, 0.3% bromophenol blue, and 1 mM EDTA.
8. 20× SSPE: 3 M NaCl, 0.25 M NaH_2PO_4, and 0.02 M EDTA; adjust to pH 7.4.
9. Prehybridization solution: 2.5 mL formamide, 1 mL 5× prehybridization stock, 1.4 mL H_2O, 0.292 g NaCl, and 0.1 mL sperm DNA.
10. 5× Prehybridization stock: 0.25 M Tris–HCl, pH 7.5, 0.5% sodium pyrophosphate, 1% polyvinylpyrolidone, 1% bovine serum albumin, 1% Ficoll, and 5% SDS.
11. Wash buffer: 1% SDS + 0.1% 20× SSPE.

2.2. Procedures

The protocol is divided into four sections: RNA isolation, electrophoresis, membrane transfer, and hybridization.

2.2.1. RNA Preparation

RNA isolation from brain is carried out with the RNeasy Lipid Tissue Mini Kit. The following procedure is based on Qiagen's protocol.

1. Place the tissue in a suitably sized vessel containing 1 mL QIAzol Lysis Reagent, and pipette up/down six times with 1-mL tip to break tissue into small pieces.
2. Pass six times through 3-mL syringe with twenty-two 1/2-gauge needle.
3. Place the tube containing the homogenate on the benchtop at room temperature for 5 min.
4. Add 200 μL chloroform. Securely cap the tube containing the homogenate, and shake it vigorously for 15 s.
5. Place the tube containing the homogenate on the benchtop at room temperature for 2–3 min.
6. Centrifuge at 12,000 × *g* for 15 min at 4°C.
7. Transfer the upper aqueous phase to a new tube. Add equal volume of 70% ethanol, and mix thoroughly by vortexing. Do not centrifuge. Proceed immediately to the next step.
8. Transfer the sample to an RNeasy Mini spin column placed in a 2-mL collection tube. Close the lid gently, and centrifuge for 15 s at 8,000 × *g* at room temperature. Discard the flow through.
9. Add 700 μL Buffer RW1 to the RNeasy spin column. Close the lid, and centrifuge for 15 s at 8,000 × *g* to wash the membrane. Discard the flow through.
10. Add 500 μL buffer RPE to the RNeasy spin column. Close the lid, and centrifuge for 15 s at 8,000 × *g* to wash the membrane. Discard the flow through.
11. Add 500 μL buffer RPE to the RNeasy spin column. Close the lid gently, and centrifuge for 2 min at 8,000 × *g* to wash the membrane.
12. Place the RNeasy spin column in a new 2-mL collection tube, and discard the old collection tube with the flow through. Close the lid gently, and centrifuge at full speed for 1 min.
13. Place the RNeasy spin column in a new 1.5-mL collection tube. Add 30–50 μL RNase-free water directly to the spin column membrane. Close the lid gently. Centrifuge for 1 min at 8,000 × *g* to elute the RNA.

2.2.2. Electrophoresis of RNA Samples Under Denaturing Conditions

1. Prepare gel:
   (a) Cast a 0.7–1.0% agarose gel, which requires at least 100 mL of agarose solution. Ethidium bromide is not added to the gel but rather to the sample buffer. Following is an example for casting a 1% agarose gel.
   (b) Dissolve 1 g agarose in 85 mL water by microwave and cool to 60°C in a water bath.
   (c) When the flask has cooled to 60°C, place in a fume hood and add 10 mL of 10× running buffer and 5 mL 35% formaldehyde.
   (d) Pour the gel and allow it to set. Remove the comb, place the gel in the gel tank, and add sufficient 1× running buffer to cover to a depth of ~1 mm.
2. Prepare RNA sample:
   Total RNA is isolated by using RNeasy Lipid Tissue Mini Kit as described in Sect. 2.2.1. RNA samples are prepared for loading as follows.
   (a) Adjust the volume of each RNA sample to 6 μL with water. Determine the amount of RNA to be loaded in each well (e.g., 2 μg of RNA).
   (b) Add 15 μL freshly prepared sample denaturation buffer and mix by vortexing.
   (c) Centrifuge briefly to collect liquid, and incubate for 5 min at 65°C.
   (d) Cool on ice for 2 min, and then add 2 μL loading dye mix.
3. Run gel:
   Run the gel in 1× running buffer at between 100 and 200 V for several hours, until the xylene cyanol dye front has migrated 3 4 cm into the gel and the bromophenol blue is about two-thirds down the gel. When the RNA has run an appropriate distance, photograph the gel.

2.2.3. Transfer of the mRNA from the Gel to a Nylon Membrane by Upward Capillary Transfer

4. Prepare gel for transfer:
   Place the gel in an RNase-free dish and rinse with sufficient water 4× for 20 min.
5. Transfer RNA from gel to membrane:
   (a) Fill the glass dish with enough 20× SSPE.
   (b) Cut two pieces of Whatman 1MM paper, place it on the glass plate, and wet it with 20× SSPE.
   (c) Place the gel on the filter paper and use a glass pipette to roll out any air bubbles.

(d) Cut four strips of plastic wrap and place over the edges of the gel.

(e) Cut one piece of nylon membrane to roughly the same dimensions of the gel and wet in water. Put the wet membrane on the top of the gel. Smooth out any bubbles between the gel and membrane.

(f) Wet the surface of the membrane with 20× SSPE. Cut five sheets of Whatman 3MM paper to the same size as membrane and place on top of the membrane.

(g) Put paper towels on top of the Whatman 3MM paper to a height of ~8 cm.

(h) Cover the stack with the gel casting tray and add weight to compress the stack.

(i) Leave overnight.

6. Prepare membrane for hybridization:
After overnight transfer, remove paper towels and filter papers and recover the membrane. Rinse the membrane in 5× SSPE to remove residual agarose. Bake the membrane at 80°C for 2 h.

2.2.4. Hybridization Analysis of the RNA of Interest Using a Labeled DNA Probe

1. Prepare labeled DNA primer: The probe labeled with Rediprime DNA labeling system (GE Life) based on the following random-primed labeling method:

(a) Dilute the interested cDNA to a concentration of 2.5–25 ng in sterile water.

(b) Place 25 ng of template DNA into a microcentrifuge tube and to it add 5 μL of primers and the appropriate volume of water to give a total volume of 50 μL in the final reaction.

(c) Denature by heating to 95–100°C for 5 min in a boiling water bath.

(d) Centrifuge briefly and keep the tube at room temperature.

(e) Add 10 μL of the labeling buffer followed by 5 μL of Redivue [^{32}P] dCTP and 2 μL enzyme. Mix by gently pipetting up and down and spin for a few seconds.

(f) Incubate at 37°C for 10 min. Stop the reaction by the addition of 5 μL of 0.2 M EDTA.

(g) Sephadex G-50 resin is resuspended and excess storage buffer is removed by centrifugation.

(h) Place the column in a 1.5-mL tube and slowly apply 50 μL of the sample to the top center of the resin.

(i) Spin the column at 3,000 × *g* for 2 min. The purified sample is collected in the bottom of the support tube.

2. Hybridization:
   (a) Wet the membrane in the 5× SSPE. And then, prehybridize membrane in a minimum volume of prehybridization solution at 42°C for 6 h.
   (b) Denature the labeled primer by heating to 95–100°C for 5 min, and then chill on ice.
   (c) Add the probe into the prehybridization solution (or remove prehybridization solution, and then add prehybridization solution and probe).
   (d) Continue to incubate with rotation overnight at 42°C.
3. Autoradiography:
   (a) Wash the membrane with wash buffer twice for 15 min at room temperature.
   (b) Wash the membrane with prewarmed wash buffer twice for 30 min at 65°C.
   (c) Remove final wash solution and rinse membrane in 5× SSPE at RT.
   (d) Cover membrane in UV-transparent plastic wrap. Do not allow membrane to dry out.
   (e) Expose blot to a film at −20°C.

2.3. Evaluation

Northern blot results can be semiquantitatively analyzed by band densitometry. To this end, the immunoblot films with the bands of the interested protein or internal control are digitized by using a flatbed scanner. All images should be acquired at the same setting, such as 256 gray shades, a resolution of 600 dpi, and no changes to contrast and brightness. The software ImageJ can be used to quantify the band density by using its "Gel Analysis" functions. The background correction is done using a "rolling ball" method as its default settings. The uncalibrated OD of the band of a given protein is corrected for that of internal control and the corrected ratios are then subjected to statistic analysis.

3. In Situ Hybridization

Several variations in in situ hybridization protocols exist, including choice of tissue preparation, type of probe, and choice of label. The advantages and disadvantages are summarized in Table 1. We present basic protocols using fresh frozen tissue and radioactively labeled oligonucleotide probes, which are the methods most commonly used in the literature. This section is divided into four parts: tissue preparation, probe construction and labeling, hybridization, and visualization.

Table 1
Options in tissue preparation, probe labeling, and probe type for in situ hybridization

	Advantages	Disadvantages
Tissue prep		
Fresh frozen	Less possibility of RNase contamination	Poor morphology
Paraffin	Better morphology	More likely to suffer RNase degradation
Label type		
Radioactive (^{32}S, ^{33}P, ^{3}H)	High sensitivity Ease of incorporation	Limited half-life (except ^{3}H) Radiation exposure Cannot perform multiple radioactive probe labeling
Digoxigenin (hapten) Biotin	Fairly sensitive Avoids radioactivity Can be used in combination for multiple probes	Amplified signal reduces ability to quantify results
Probe type		
Oligonucleotide	Simple construction Highly cell permeable Resistant to nuclease degradation	Less specific than longer probes Less stable hybridization
cRNA	Highly stable hybridization High specificity Higher sensitivity (amplified signal)	Complex construction Less cell permeable Probe sensitive to RNase degradation

3.1. Tissue Preparation

3.1.1. Materials

1. Charged glass slides (e.g., Fisher Superfrost or Probe-On).
2. 2-methylbutane.
3. Tissue Tek mounting medium.

3.1.2. Procedure

Typically, sections from the anterior half of the hippocampal formation, which clearly contain CA1, CA3, and the dentate gyrus formation, are used for global cerebral ischemic assessments. Frozen tissue offers the least contaminated option, and perhaps the easiest tissue preparation. Following cryosectioning, the brains can then be postfixed with RNase-free formalin. The major drawback to using fresh frozen tissue is the relatively poor cellular morphology (Table 1). Fixed tissue may also be used, but the animal should be perfused with RNase-free formalin, and all other solutions used (e.g., in the process of deparaffinization (see Chap. 13 under immunocytochemistry)) should be RNase free. If older sections are to be used that were fixed without regard to RNase-free treatment, the presence of RNA can be confirmed using in situ hybridization against a constitutively expressed housekeeping gene, such as actin. Additional alterations are recommended for paraffin-embedded tissue (6).

(a) Place 2-methylbutane in a beaker on dry ice using enough to be able to fully submerge the brain.

(b) Decapitate the animal and quickly remove the brain.

(c) Snap freeze the brain by quickly submerging it to the precooled 2-methylbutane for 15 s.

(d) Remove brain and quickly cover with embedding medium (e.g., Tissue Tek).

(e) Place in dry ice to freeze over the embedding medium, and repeat until a thick layer of frozen embedding medium *completely* surrounds the brain.

(f) Maintain tissue at –80°C until ready to section.

(g) Section tissue (16 μm thick) onto charged (Superfrost Plus) slides and store at –80°C until ready to use. For global ischemic brains, particular attention should be paid to the hippocampal sections, where both CA1 and CA3 are clearly present within the section. For focal ischemic brains, striatal and cortical levels are particularly important, covering both the ischemic core MCAO territory as well as the penumbral regions.

3.2. Probe Construction and Labeling

The choice of probe and label depends on a variety of factors, including the molecular expertise available in the lab, abundance of the RNA in the cells, and any intended further processing (Table 1). As we are limiting the discussion on targeting mRNA species, two major probe types are typically used—synthetic oligonucleotide and cRNA probes. In many cases, synthetic oligonucleotide probes are the most simple and reliable method, as long as the exact mRNA sequence is known. However, the longer length of a cRNA probe and higher stability of the cRNA/RNA hybrid allows for high-stringency posthybridization washes, thus greatly reducing background noise.

3.2.1. Materials

1. RNase-free Eppendorf tubes and pipette tips
   - All materials and solutions MUST be RNase free, particularly for steps prior to the posthybridization washes. Following hybridization, theoretically, the probe labeling survives exposure to RNases.
2. 3′ End labeling kit (Roche).
3. ^{35}S-dUTP (for oligonucleotide).

3.2.2. Solutions

1. RNase-free water.

3.2.3. Procedure

Probe Construction

1. Potential probes are screened against a known mRNA target using online search tools.
2. Avoid homology with other known RNAs.
3. Typically, to ensure stable hybridization, the probe length should be at least 40 nucleotides and contain ideally 55% guanines and cytosines (the G/C ratio). Sequences containing 40–60% G/C nucleotides are usable. However, the low-percentage probes may require decreased stringency in the posthybridization washes, and the high-percentage probes may be less specific and result in higher background.
4. If less G/C ratio is less than 55%, the posthybridization washes should be reduced in stringency.
5. Careful attention should be paid concerning the selected sequence region. Introns need to be avoided, as well as possible splice-variant exons.

End Labeling Using ^{35}S-dUTP

Note: For attempted semiquantification of signal, in situ hybridization should ideally use only single incorporation of the labeled nucleotide (e.g., ^{35}S-ddUTP or fluorescein-ddUTP). However, often, the signal strength is too weak to obtain detectable results. Therefore, extended tail labeling using dUTP alone or in combination with cold dATP is often used, but further limits the reliability of semiquantification.

1. Dilute oligonucleotide to 100 pmol with RNase-free water (supplied in most kits).
2. Most end-labeling reactions are performed using commercially available kits containing RNase-free reaction buffers, water (if necessary), cobalt chloride, and the enzyme terminal transferase (TdT).
3. Incubate labeling mixture with oligonucleotide at 37°C for 30 min (or according to kit instructions).
4. Stop the reaction by adding EDTA (pH 8.0) to a final concentration of 10 mM (can be skipped if performing probe purification mentioned below).
5. Probe purification
   - Although several kits state that probe purification is unnecessary, spin columns are often used to clean up the probe by removing the TdT enzyme, precluding the need for EDTA mentioned above. These columns are commonly available from commercial sources, but can also be made using Sephadex column protocols widely available. However, commercial columns save vast amounts of time and are more consistent.

6. Determination of labeling efficiency
   - Using a scintillation machine, measure specific activity of probe. Probes should measure at or above 10^6 cpm/μL.

3.3. Tissue Hybridization and Posthybridization Washes

1. Hybridization cocktail: 50% deionized formamide, 0.3 M NaCl, 20 mM Tris–HCl, pH 8, 5 mM EDTA, 10 mM $NaPO_4$, pH 8, 10% Dextran sulfate, 1× Denhardts, 0.5 mg/mL yeast tRNA, and salmon sperm DNA—*all solutions must be RNase free.*
2. 10× SSC (1,500 mmol/L NaCl, 150 mmol/L Na citrate, pH 7.4).

3.3.1. Materials

1. Hybridization oven or incubator, preheated to determined hybridization temperature (see below).
2. Humidified hybridization container or chamber, capable of being sealed and placed in oven.
3. Glass containers for post-hybridization washes.

3.3.2. Solutions

1. The only solution, other than RNase free water, is 0.2 M EDTA, pH 8.0, made with RNase-free water.

3.3.3. Procedure

1. Immediately prior to hybridization, remove frozen sections and thaw at room temperature (minimal time possible).
2. In some cases, postfixation and acetylation in RNase-free solutions may be performed, but is not always necessary. If desired, postfix in 3% RNase-free PFA for 5 min, followed by several washes with RNase-free PBS and alcohol dehydration (70, 90, and 100% for 2 min each). Acetylate 2× for 5 min in 750 μL acetic anhydride added immediately in 0.1 M triethanolamine-HCl (pH 7.5). Acetylation reduces nonspecific sticking of probe to charged groups in sections and slide.
3. Determine the hybridization temperature based on oligonucleotide sequence.
   - First, determine melting temperature of oligonucleotide

$$T_{\mathrm{m}} = 81.5 + 16.6\,(\log_{10}[\text{sodium ions in most stringent wash}]) + 0.41(\%\ \mathrm{G} + \mathrm{C}) - (500/\text{length of oligonucleotide})$$

   - Hybridization temperature is approximately 5–10°C below the melting temperature.
4. Dilute probe in hybridization cocktail (need 1×10^7 cpm/mL of probe).
5. Apply approximately 200 μL of the probe/cocktail solution to each slide and coverslip (parafilm), making sure to remove all bubbles.
6. Incubate overnight (18–20 h) at hybridization temperature (typically, around 42°C, based on a probe 40 nucleotides long

with a 55% G/C ratio) in a humidified chamber (layer paper towels soaked in 5× SSC and 50% formamide).

7. Posthybridization washes are done two times for 15 min each in 1× SSC, followed by three washes for 1 h each in 1× SSC.

3.4. Detection

3.4.1. Materials

1. Single sided emulsion film (Kodak).
2. Lightproof metal cassettes.

3.4.2. Procedure

1. Visualization on film: Tape slides to cardboard insert for film cassette, and expose to single-sided emulsion film in metal cassettes in darkroom.
2. Leave film on slides for 1 week for initial film, and reexpose to new film for 3 weeks for final film.
3. Develop (hand develop for best resolution—typically, 5 min in D-19 developer, 1 min in stop bath, 15 min in fixer, 30 min under running water).
4. Slide emulsion: Used for detection of label under light microscopy. The emulsion, once developed, leaves silver particles, where radioactivity was found. Emulsion is highly sensitive to light, so best if complete darkness can be attained during the entire process.
   - Warm liquid emulsion (available from Kodak) in 37°C water bath for at least 1 h prior to use.
   - Slowly and evenly dip slide into emulsion.
   - Air dry completely before storing—can wipe the back of slide, but do so thoroughly.
   - Store for 5 weeks in complete darkness.
   - Hand develop (4 min shaking in D-19 Kodak developer, 1 min shaking in stop bath, 5 min shaking in fixer solution (Kodak), 30 min under running water).
   - Silver deposits are visible under light microscopy.

3.5. Evaluation and Troubleshooting

1. In situ hybridization, while allowing for fairly precise detection of the targeted RNA to specific cellular localizations, is limited in several ways. First, qualitative observations of an increase or decrease of mRNA levels are not necessarily due to an increase or decrease in transcription—alterations in expression level could also be due to changes in degradation or stability of the target mRNA. Furthermore, the resulting films are semiquantitative at best via optical density measurements.
2. In the case of high background noise, the following strategies can be employed:
   - Redesigning the probe (e.g., altering the G/C ratio, or targeting a different location within targeted mRNA)

- Increasing the salt content in the washes
- Increasing the temperature of the posthybridization washes

4. Summary

Evaluation of RNA expression following global cerebral ischemia has and continues to be an important aspect in support for evidence of de novo gene transcription or suppression, of promoter function or silencing, and of confirmation of targeted gene knockdown. Recent studies investigating noncoding RNAs are emerging, and will certainly use procedures based on the basic ones presented in this chapter.

References

1. Büttner F, Cordes C, Gerlach F, Heimann A, Alessandri B, Luxemburger U, Türeci O, Hankeln T, Kempski O, Burmester T (2009) Genomic response of the rat brain to global ischemia and reperfusion. Brain Res 1252: 1–14
2. Roberts GG, Di Loreto MJ, Marshall M, Wang J, DeGracia DJ (2007) Hippocampal cellular stress responses after global brain ischemia and reperfusion. Antioxid Redox Signal 9: 2265–2275
3. Chen J, Nagayama T, Jin K, Stetler RA, Zhu RL, Graham SH, Simon RP (1998) Induction of caspase-3-like protease may mediate delayed neuronal death in the hippocampus after transient cerebral ischemia. J Neurosci 18: 4914–4928
4. Stetler RA, Gao Y, Zukin RS, Vosler PS, Zhang L, Zhang F, Cao G, Bennett MV, Chen J (2010) Apurinic/apyrimidinic endonuclease APE1 is required for PACAP-induced neuroprotection against global cerebral ischemia. Proc Natl Acad Sci U S A 107:3204–3209
5. Mayr B, Montminy M (2001) Transcriptional regulation by the phosphorylation-dependent factor CREB. Nat Rev Mol Cell Biol 2: 599–609
6. Liu CQ, Shan L, Balesar R, Luchetti S, Van Heerikhuize JJ, Luo JH, Swaab DF, Bao AM (2010) A quantitative in situ hybridization protocol for formalin-fixed paraffin-embedded archival post-mortem human brain tissue. Methods 52(4):359–366

Chapter 6

Biochemical and Molecular Biological Assessments of Global Cerebral Ischemia: Protein

Yu Gan and R. Anne Stetler

Abstract

The evaluation of protein expression and modifications following global cerebral ischemia has led to better characterization of selectively vulnerable neuronal populations and exploration of potential targets for neuroprotection. The absence or decrease of protein product must also be validated for knockdown or knockout models in order to ensure efficacy of gene silencing. Thus, two basic approaches for protein assessments (western blotting and immunohistochemistry) used in global cerebral ischemic models are presented here to serve as the basis of further experimental designs.

Key words: Global cerebral ischemia, Western blot, Immunohistochemistry, Protein

1. Introduction

Rodent global cerebral ischemia models are developed to mimic human ischemic brain conditions that arise when total blood flow to the brain is arrested. After global cerebral ischemia, a series of events occur from minutes to days post reperfusion. Among these events, neural cell death affects a highly specific neuronal subpopulation (CA1 pyramidal neurons) while surrounding cells (e.g., CA3 pyramidal neurons) survive (1). The underlying basis of cell survival and death following global cerebral ischemia is still poorly understood, but appears to be highly influenced and controlled—at least in part—by alterations in protein expression and function. Thus, basic techniques of assessing protein expression and post-translational alterations provide groundwork for further investigation in the mechanisms of global cerebral ischemic injury.

Jun Chen et al. (eds.), *Animal Models of Acute Neurological Injuries II*, Springer Protocols Handbooks,
DOI 10.1007/978-1-61779-782-8_6, © Springer Science+Business Media, LLC 2012

In addition to assessment of mechanisms underlying the global ischemic state, gene targeting by knockout, antisense, or siRNA knockdown requires confirmation that the targeted protein gene product is successfully decreased. Evaluation of mRNA expression alone is insufficient for transient gene knockdown, as the preexisting proteins may have fairly extended half-lives and thus be present at levels sufficient to maintain function. Therefore, assessment of protein expression level decrease is mandatory for ensuring that the observed effects are truly correlated with reduced protein expression.

Finally, immunochemical reactions are often used as tools for the visualization of tagged labels. For example, the visualization of DNA-damaged or proliferating cells following global ischemia often relies on incorporation of bromodeoxyuridine (BrdU) into the damaged strand (2). Following incorporation, BrdU is visualized using immunohistochemical detection with an anti-BrdU antibody. Likewise, fluorescent-based in situ hybridization (discussed in Chap. 12) relies on the immunohistochemical detection of the digoxigenin (DIG) label incorporated into the probe with an anti-DIG antibody (3). Thus, protein detection not only serves to monitor endogenous proteins, but also can be exploited in the use of proteins as molecular tools.

The two common methods to detect protein expression are western blot and immunohistochemistry (IHC). Both of these techniques are based on the specific binding between antigen and antibody. Western blot is used to examine specific protein in an SDS-PAGE gel-separated sample on the membrane. It is an effective method for semiquantitative analysis of protein expression levels. Also, western blot has the advantage to provide the information of protein size. This is helpful to determine the activation of some proteases, such as caspase-3, which is initiated by a proteolytic cleavage to produce an activated form with different size. The detection of activated caspase-3 has led to the investigation into its important role in the execution of neuronal cell death after global ischemia (4). IHC is used to examine protein expression on tissue sections. Compared to western blot, IHC is weaker in quantitative ability. However, IHC has the advantage of being able to show the distribution and location of a given protein, and can be used to identify expression of several different antigens within the same sample (to be discussed in later chapters). Since each technique has its limitation, the combination of western blot and IHC is preferred for evaluating protein expression after global cerebral ischemia.

While many protocols exist now for detailed exploration into protein expression and biochemical function, most of the protocols are based on either western blotting or IHC. Therefore, in this chapter, we present these two techniques as foundations for further experimental design, some of which are discussed in other sections of the book.

2. Western Blot Analysis

2.1. Material

2.1.1. Reagent

1. CelLytic MT reagent (Sigma, St. Louis, MO).
2. BenchMark Pre-stained Protein Ladder (Invitrogen, Grand Island, NY).
3. Precast Gel (BioRad, Hercules, CA).
4. LumiGLO Peroxidase Chemiluminescent Substrate Kit (KPL, Gaithersburg, MD).

2.1.2. Solution

1. 10× running buffer: 30.28 g Tris, 144.13 g glycine, 10 g SDS (or 10 mL 10% SDS); add ddH_2O to 1 L.
2. 4× loading dye: 0.25 M Tris–HCl, pH 6.8, 8% SDS, 30% glycerol, 0.02% bromophenol blue, 5% B-ME.
3. 1× transfer buffer: 6.04 g Tris, 28.8 g glycine, 200 mL methanol; add ddH_2O to 2 L.
4. 10× TBS-Tween: 12.2 g Tris, 87.65 g NaCl, 5 mL Tween-20; bring up pH to 8.0, and add ddH_2O to 1 L.
5. Blocking buffer: 5% fat-free dry milk in 1× TBS-Tween.

2.2. Procedure

2.2.1. Preparation of Protein Sample from Brain Tissue

1. Brain tissue is extracted by separating CA1 from CA3 regions, and protein extraction is performed by using Sigma CelLytic MT reagent.
2. Add cold CelLytic MT reagent to the tissue sample. The ratio of tissue to the extraction reagent is about 1:20 (1 g of tissue/20 mL of reagent).
3. Homogenize the tissue with a prechilled microhomogenizer.
4. Put the lysed sample on ice for 20 min.
5. Centrifuge at 13,000 × *g* for 10 min at 4°C.
6. Transfer the protein-containing supernatant to a new microtube.
7. Determine the protein concentration with Bio-Rad protein assay according to the manufacturer's protocol.
8. Adjust the protein concentration with PBS, and then add 1/3 volume 4× loading buffer.
9. Boil the sample for 5 min. Then, it is ready for electrophoresis.

2.2.2. Protein Electrophoresis

1. Choose appreciate precast SDS-PAGE gel (5–18%, or gradients gel) for electrophoresis depending on the molecular weight of interested protein.
2. Assemble gel in gen tank, and fill with 1× running buffer.
3. Load prestained protein ladder and protein sample. The amount of sample varies with different proteins (from 10 to 100 μg).
4. Run at 100 V (constant voltage) for 1.5–2 h.

2.2.3. Membrane Transfer

1. PVDF membranes should be presoaked in methanol for 30 s, and then put it in 1× transfer buffer until use.
2. Wet the filter papers and sponges in 1× transfer buffer.
3. Assemble transfer "sandwich" and put it in the tank filled with prechilled 1× transfer buffer.
4. Transfer at constant current of 40 mA overnight.
5. After transfer, wash the membrane briefly with 1× TBS-Tween.
6. Block the membrane in blocking buffer for 1 h at room temperature (RT).

2.2.4. Antibody Detection

1. Incubate with primary antibody diluted in blocking buffer for 1–2 h at RT or overnight at 4°C.
2. Wash blot 3× for 10 min with 1× TBS-Tween.
3. Incubate blot with HRP-labeled secondary antibody diluted in blocking buffer for 60 min at RT.
4. Wash blot 3× for 10 min with 1× TBS-Tween.
5. Detect with ECL detection kits.

2.3. Evaluation

Western blot results can be semiquantitatively analyzed by band densitometry. To this end, the immunoblot films with the bands of the interested protein or internal control are digitized by using a flatbed scanner. All images should be acquired at the same setting, such as 256 gray shades, a resolution of 600 dpi, and no changes to contrast and brightness. The software ImageJ is applied to quantify the band density by using its "Gel Analysis" functions. The background correction is done using a "rolling ball" method as its default settings. The uncalibrated OD of the band of a given protein is corrected for that of internal control and the corrected ratios are then subjected to statistic analysis.

3. Immunohistochemistry

3.1. Solutions

1. 10× Tris-buffered saline (TBS): 500 mM Tris–HCl, pH 7.4, and 1,500 mM NaCl.

3.2. Procedure

3.2.1. Tissue Preparation

1. Tissue should be processed with particular focus on mid-hippocampal regions, where both the CA1 and CA3 subfields are present in the hippocampal formation.
2. If using paraffin-embedded sections, tissue should be mounted on slides and deparaffinized by immersion in four sequential baths of histology-grade xylene for 2 min each, followed by rehydration in a series of alcohol gradients (1 min in absolute

ETOH, 1 min in a second bath of ETOH, 30 s in 95% ETOH, 45 s in 70% ETOH) finalized by a 1-min wash in water.

3. Free floating formalin-fixed tissue can be processed completely in wells, wherein mounting on the slides constitutes the final step after IHC is completed. The brains are removed from animals perfused with 10% formalin, postfixed at 4°C for 4 h, then submerged in 30% sucrose in PBS (as a cyroprotectant) in a 50-mL centrifuge tube, and kept at 4°C until the brain drops to the bottom of the tube. Once the brain drops, it can then be cut on a frozen microtome in 20-μm-thick sections.
4. Fresh frozen fixed tissues cut onto glass slides may also be used (obtained as described in the previous chapter under in situ hybridization), but are not optimal for maintaining morphology. However, in the case of colabeling for protein and mRNA, this technique may be desired. In this case, the tissue should be postfixed for 10 min in 4% paraformaldehyde prior to IHC, followed by several washes in PBS.

3.2.2. Permeabilization and Blocking

1. Permeabilization: Typically, incubation of sections with 0.3% Triton X-100 in TBS for 30-min shaking is sufficient. However, for membrane proteins, 0.3% saponin may give better results.
2. Blocking: Incubate sections in 3% BSA/0.3% Triton X-100 in TBS for 1 h at room temperature. Some antibodies may have cleaner results if blocking is performed using the serum of the host of the secondary antibody.

3.2.3. Immunochemical Detection

1. Following blocking, incubate tissue with primary antibody solution at 4°C overnight (the antibody should be diluted in 3% BSA/0.3% Triton X-100 in TBS). The concentration of the antibody to be used is highly variable; references to previously published results with the particular antibody are most useful. If difficulty arises in obtaining a signal, a dose range can be set up to evaluate the optimal antibody concentration. Although this is somewhat costly, it can be valuable in the long run for antibodies commonly used in the lab.
2. Wash in TBS three times for 10 min each.
3. Incubate tissue with a fluorophore-conjugated secondary antibody diluted in 3% BSA/0.3% Triton X-100 in TBS for 1 h at room temperature.
    - The secondary is anti-IgG raised toward the host of the primary antibody. Thus, if you use mouse anti-caspase-3, then the secondary antibody would be anti-mouse IgG.
    - The concentrations of secondary antibodies are typically 1:1,000–2,000.
4. Wash in TBS three times for 10 min each.

5. Perform Hoescht stain for nuclei (incubate tissue in 1 μg/mL of Hoescht 33342 for 10 min, followed by several washes in TBS).
6. Mount tissue onto Superfrost slides (Fisher, Pittsburgh, PA), and dry thoroughly.
7. Coverslip using fluorescent-specific mounting medium.
8. Evaluate fluorescent signal under microscopy.

4. Other Method of Evaluating Protein Expression

Enzyme-linked immunosorbent assay (ELISA) is another commonly used method for protein detection. It provides a powerful technique for protein quantitative assessment, and determining the protein concentration with a standard curve. ELISA also has the advantages of high specificity, reliability, relatively low cost, and ease of use, which make it a good option for high-throughput detection of protein levels. Since ELISA is usually used for cytokine assessment, the detailed procedure is discussed in Chap. 14.

5. Summary

Basic protein assessments relevant to global cerebral ischemia, including western blotting and IHC, provide a basic framework from which more intricate experimentation can be designed. Sample manipulation—such as subcellular fractionation, coimmunoprecipitation, or in vitro activity assays—all require western blotting as the final detection step. Similarly, IHC allows localization of protein expression to specific populations, and further analysis can be performed to visualize translocation or posttranslational modification events in cells that are selectively vulnerable to global ischemic injury.

References

1. Lipton P (1999) Ischemic cell death in brain neurons. Physiol Rev 79:1431–1568
2. Kawai T, Takagi N, Miyake-Takagi K, Okuyama N, Mochizuki N, Takeo S (2004) Characterization of BrdU-positive neurons induced by transient global ischemia in adult hippocampus. J Cereb Blood Flow Metab 24:548–555
3. Arnold N, Seibl R, Kessler C, Wienberg J (1992) Nonradioactive in situ hybridization with digoxigenin labeled DNA probes. Biotech Histochem 67:59–67
4. Chen J, Nagayama T, Jin K, Stetler RA, Zhu RL, Graham SH, Simon RP (1998) Induction of caspase-3-like protease may mediate delayed neuronal death in the hippocampus after transient cerebral ischemia. J Neurosci 18: 4914–4928

Chapter 7

Neurobehavioral Assessments of Global Cerebral Ischemia

Pinar Kendigelen, Dandan Sun, and Pelin Cengiz

Abstract

Global cerebral ischemia (GCI) usually occurs after cardiac arrest and severe shock due to lack of blood flow and oxygen delivery to the brain. Some areas of the brain, such as hippocampus, are more susceptible to injury than the others following GCI. Several animal models have been developed to study the pathophysiology of GCI. In this chapter, we describe the most widely used protocols in assessing the neurological function and hippocampus-dependent learning and memory deficit following GCI.

Key words: Global cerebral ischemia, Morris water maze, Open field, T-maze, Radial arm maze

1. Introduction

Global cerebral ischemia (GCI) usually occurs after cardiac arrest and severe shock due to lack of blood flow and oxygen delivery to the brain. The interruption of cerebral blood flow can damage many brain regions, particularly hilar cells in the dentate gyrus (DG) and pyramidal cells in the hippocampal cornu ammonis field 1 (CA1) (1, 2). These cells are highly vulnerable to reduced blood supply and oxygen. The hippocampus belongs to limbic system and plays a very important role in the process of learning, memory, and spatial navigation. Therefore, hippocampal CA1 cell loss following GCI can result in cognitive deficits (3).

The most commonly used rodent GCI models are the four-vessel occlusion (4-VO) and the two-vessel occlusion plus hypotension (2-VO + H) models. The 4-VO model is a two-stage procedure which includes permanent coagulation of two vertebral arteries and temporary ligation of two common carotid arteries for 10–30 min (4). On the other hand, the 2-VO + H model is produced by transient occlusion of both common carotid arteries and systemic

Jun Chen et al. (eds.), *Animal Models of Acute Neurological Injuries II*, Springer Protocols Handbooks, DOI 10.1007/978-1-61779-782-8_7, © Springer Science+Business Media, LLC 2012

hypotension. Systemic hypotension is induced by phlebotomy or administration of peripheral vasodilators (5). These models are routinely used to investigate mechanisms underlying ischemic damage as well as neuroprotection.

2. Behavioral Assessment

Neurological functional changes following GCI can be assessed with various neurobehavioral tests. In this section, we describe the neurobehavioral tests that are commonly used in the global ischemia models in adult rodents. These include the Morris water maze (MWM) task, open field test, T-maze test, and radial arm maze test.

2.1. Morris Water Maze Task

The MWM task is the most frequently used test to investigate spatial learning and memory in rodents. It was developed by Richard G.M. Morris 30 years ago for studying the mechanism of spatial localization in rats. Several studies demonstrated spatial learning and memory deficits, which likely resulted from hippocampal CA1 damage in GCI models using the MWM task (5, 6).

There are two disadvantages associated with MWM tests. Immersion of the animal into the water may cause significant stress and endocrinological changes that may interfere with the purpose of the study. These problems are usually resolved with continued exposure of the animal to the water. Moreover, this stress may be sufficient to induce cardiovascular system collapse and lead to death or stroke for older animals. Secondly, if powdered milk is added to water to make it opaque, the pool must be drained every day to avoid bacterial contamination and odor. If powdered white paint is used, it must be ensured that it is not toxic to the animal (7). The major advantage of the MWM task over the other tasks is that the animals do not need to be food deprived and that they are quite motivated to escape from the water (7).

In this section, different types of MWM tasks are described. These include:

(a) Hidden Platform MWM task.

(b) Probe trial of MWM task.

(c) Cued platform MWM task.

2.1.1. Materials and Tools

1. Circular white plastic pool
   Different sizes of pools are used, approximately 1–2 m (75–150 cm for mice, 150–214 cm for rats) in diameter. The size of the circular pool may or may not affect the performance of animal depending on strain (8). The water temperature should be 21–25°C (room temperature).

2. White, nontoxic paint
   White color paint (550 g) is added until the water is opaque enough.
3. Escape platform
   A clear acrylic hidden escape platform with a circular top (5–20-cm diameter) is hidden 1–2 cm below the surface of opaque water (9).
4. Video camera and software
   A video camera [Panasonic Digital CCTV (Model WV-BP334)] is mounted to the ceiling, directly above the center of the pool (10). It can be connected to a personal computer via coaxial cable to a Piccolo Diligent video capture board. Experimental events and data can be controlled and recorded with Noldus Instruments EthoVison XT software (Version 5.00) (10).
5. Chronometer and remote video control
6. Visual cues
   Visual cues are placed on the walls of the testing room so that the animals can use these visual cues to find the platform location. These cues need to be visible consisting of colorful geometric images (squares, triangles, circles, etc.) (Fig. 1a).
7. Visible platform
8. Long bench to rest the individual cages

2.1.2. Hidden Platform MWM Task

1. Divide the pool virtually into four equal quadrants as described before using the special software designed for MWM Task [Noldus Instruments EthoVison XT software (Version 5.00)] (10).
2. Label the four points consisting of west, north, east, and south as A, B, C, and D.
3. Place the escape platform in the center of northwest (NW) quadrant and 45° diagonally to the center.
4. Transport rodents from the housing room to the testing room and place them into the individual cages, identical to their housing cages. Place an absorbent paper towel in those individual holding cages and put them on the benches that are 30 cm above the ground along one of the walls.
5. Allow rodents to acclimate to the room for 30 min prior to training or testing.
6. Test one rodent at a time. First, ensure that rodents can swim and climb onto the platform.
7. Keep the position of experimenter at a constant location. Experimenter is not supposed to wear any perfume, lotion, or scented cream to minimize the noxious stimuli to the rodent.

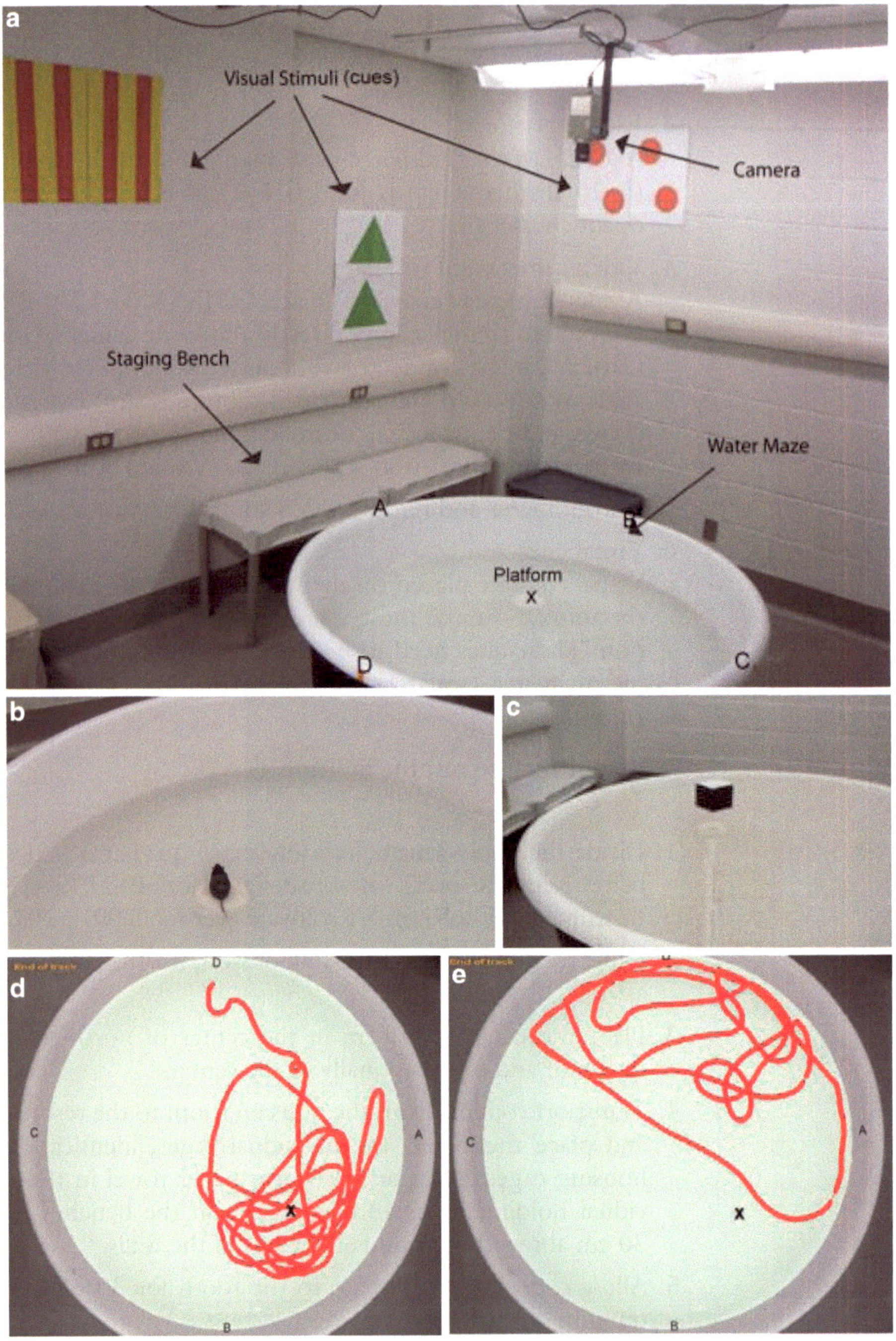

Fig. 1. (**a**) Morris water maze task testing room. (**b**) Rodents rest on the platform and survey the surroundings for 10 s. (**c**) Visible platform (place a flag or box on the platform). (**d**) A probe trial trace for a sham animal, showing spending more time in the escape platform area and good spatial memory. (**e**) Representative probe trial trace for an animal after hypoxia–ischemia injury. HI animal did not spend time in the escape platform area (indicated as x), which implies poor spatial memory (4).

8. Start the testing at the same time of the day.
9. Remove an individual rodent from the holding cage.
10. Place the rodent through the edge of the pool from one of the four starting points: A, B, C, or D (choose randomly). Press the record button on the remote control. Trials last for 1–2 min (usually, 2 min for rats, 1 min for mice) or until the rodent finds the platform (Fig. 1a).
11. If the rodent *finds* the platform, allow it to rest on the platform for about 10 s (Fig. 1b).
12. If the rodent *does not find* the platform within 1–2 min, gently put it onto the platform and allow it to survey the surroundings for 10 s. In the Hidden Platform MWM task, the primary dependent variable is the "escape latency." If a rodent does not find the platform on a given trial, escape latency is set at 60 or 120 s for a mouse or a rat, respectively.
13. Randomly choose the second, third, and fourth trial starting point and repeat steps 9–12.
14. At the end of the fourth trial of any given block (1 block = 4 trials), the animal is returned to the cage and placed on a heating pad to help animal dry.
15. Give 1 block for 2 times per day for 4 consecutive days (total of 32 trials) (10).
16. Experimenter makes sure that there is at least a 1-h interval between the consecutive blocks.
17. Create a graph of the performance measures that compare consecutive trials or blocks for each experimental group to assess the spatial learning.
18. Use repeated measures, two-way analysis of variance (ANOVA), to test the differences between each experimental group.

2.1.3. Morris Water Maze Probe Trial

After all of the 32 trials are completed, on the fourth day of Hidden Platform MWM task, the platform is removed from the pool and one additional (probe trial) 1-min trial is run. Probe trial can also be run 24 h after the last acquisition day. The time spent in the target (training) quadrant that previously contained the escape platform is primary dependent measure. It is important that the animals know the location of hidden platform before beginning this test. This knowledge is demonstrated by the animal swimming quickly and directly to where the hidden platform was (Fig. 1d, e). The animal learns the spatial location of the platform using the external cues (7).

1. Place the animal 180° from the original platform position (11). We use the point D (east) when our platform quadrant is northwest (10).

2. Press the record button on the remote control.
3. Trial lasts for 1 min and then stop recording.
4. The time spent in four individual quadrants is measured and compared using ANOVA (10). Spending more time in exploring the training quadrant reflects better learning and memory.

2.1.4. Cued Platform Morris Water Maze Task (Visible Platform Test)

The visible platform or a flag is used to assess sensorimotor (swimming ability, vision) function and/or motivational deficits that could affect performance during the Hidden Platform MWM task. The procedure is nearly identical to the standard Hidden Platform MWM task. The only difference is replacing the hidden platform with the visible platform (box or flag on the hidden platform) that mouse can see and placing the visible platform through the each quadrant trial (choose randomly) (Fig. 1c). Give 4 trials (1 block) for 2 times per day for 2 consecutive days (total of 16 trials) (10).

2.2. Open Field Test

The open field (OF) test is the most commonly used neurobehavioral assessment tool to assess locomotor function, exploratory behavior, and anxiety in laboratory animals (12, 13). Animals are observed for different behaviors of concern, such as exploration, walking, spinning, turning, self-grooming, sitting, sleeping, stretch attending, returning, rearing, and jumping (10). Locomotor activity is measured by total distance traveled, including walking and exploring in OF arena. Total distance traveled and the proportion of the distance traveled in the center four squares (center ratio) serve as the primary dependent measures for motor and exploratory function. The distance traveled in the 12 squares near the walls (the outside of the grid) is compared with the distance traveled in the center 4 squares serve as primary dependent measures of anxiety (center ratio) (10).

2.2.1. Materials and Tools

1. The OF arena
   It is an empty and bright square (or rectangular, circular) arena, surrounded by walls to prevent animal from escaping. Size of OF arena is 54 × 54 cm (Fig. 2a). The OF arena is divided virtually into 16 equal squares, via a 4 × 4 grid to assist with data analysis. The performance in OF arena is evaluated by looking at the four different areas: corners, walls, outside of the grid (12 squares near the walls), and center square (4 squares), in which the various behaviors are recorded (Fig. 2b).
2. Video camera and software
   A video camera is mounted on the ceiling above the center of the OF arena to record and records different behaviors of animals. The animal is placed in the center of the OF arena and its behavior recorded during the testing period (1–30 min).

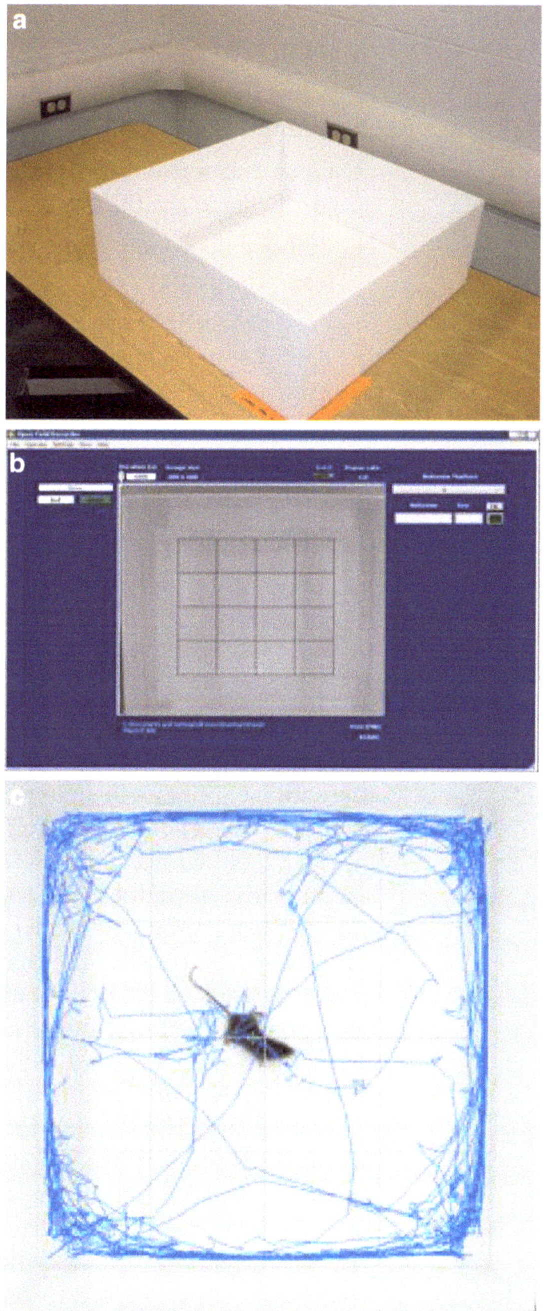

Fig. 2. (**a**) Open field arena. (**b**) Open field task software program divides the OF arena into 16 equal squares. (**c**) Open field task software program records the various behaviors.

A digital camera is attached to a National Instruments® video capture board mounted directly above the OF arena. Video and data are recorded using Coulbourn Instruments LimeLight2 software (Whitehall, PA).

2.2.2. Procedures

1. Transport the rodents from the housing room to the testing room (10).
2. Allow the rodents to acclimate to the room for 30 min prior to testing.
3. Open the software program and be sure that the digital grid (4×4) fits to the OF arena (place OF arena in correct position).
4. Remove the rodent from its home cage and place it in the center of the arena.
5. Press the record button on the remote control. Sessions last for 15–30 min; each rodent receives one session.
6. After placement, leave quickly and quietly from the OF test room for the whole session.
7. Return to the OF test room at the completion of 15–30 min, remove the rodent from the arena, and place it back into the home cage.
8. Remove any fecal boli from OF arena and wipe up any urine spots (record the number of boli and spots if you like).
9. Clean OF arena with 70% ethanol.

2.3. T-Maze Test

The T-maze test is used in various ways to assess the cognitive function, including spatial learning, memory, and spontaneous alternation. Spontaneous alternation is a measure of exploratory behavior (14). The T-maze is an enclosed apparatus in the form of a capital T placed horizontally (Fig. 3). The side arms can contain discriminative stimuli (cues), such as patterns or objects that the animal must respond to, generally to obtain a reward (14). The task is based on the premise that animals have evolved an optimal strategy to explore their environment and obtain food with a minimum amount of effort. In the absence of food deprivation, a simple T-maze can be used as described to assess spontaneous alternation. The principle of alternation is based on the fact that the animals prefer to visit less recently visited arm, thus implicating that it needs to recall the last arm visited (15). These tests can be performed either using the free trial or forced trial methods. In the forced trial method, one of the mazes is blocked during the first trial. However, with the free trial method, the animal can choose which arm to go. In the forced trial method, the possibility of introducing the spatial bias diminishes. On the contrary, in the free trial method, effects of lesions on spontaneous alternation are better evaluated.

The most important advantage of using a T-maze is that it is the simplest device to assess spatial working memory. There is no need for a video recording system. T-maze test only requires constant observation by the experimenter (15). The results are highly reproducible.

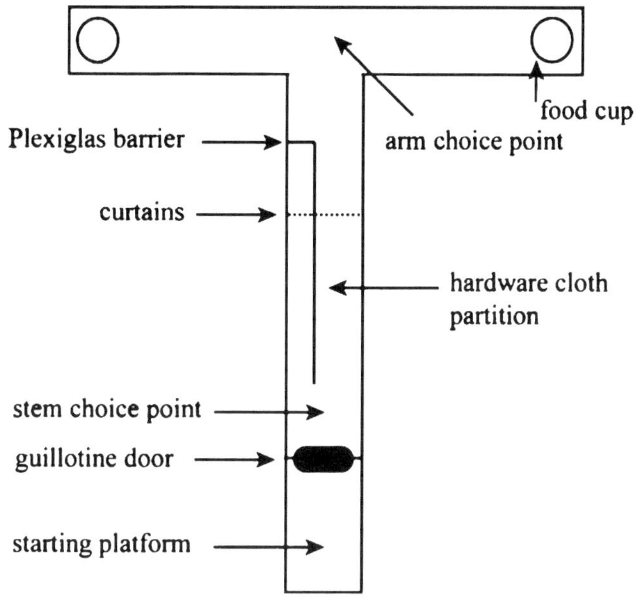

Fig. 3. Split-stem T-maze.

One of the disadvantages of this test is that there are only two alternatives for the rodents to go. This increases the possibility of success by giving the animal 50% chance to choose the correct arm. Rodent may also use a strategy other than spatial to find the correct arm (16).

2.3.1. Materials and Tools

1. T-maze
   The T-maze is shaped like a capital T that is made of opaque Plexiglas which is handmade or commercial (Coulbourn or Columbus Instruments) with three enclosed sides and a stem length of 35 cm for mice, 50 cm for rats, an arm length of 28 cm for mice, 40 cm for rats (16).
2. Food reward
   For example, 1-mg piece of normal chow, flavored (chocolate) or sweetened breakfast cereal, chocolate milk, or water.

2.3.2. Procedures

1. Habituate the animals before T-maze training; unless they are well habituated and relaxed, they will not learn well (16).
2. During habituation days, weigh each animal daily and restrict the food available to the animals so that their body weights attain 85% of their baseline weight prior to experiment.
3. Give food reward in the home cage a few days prior to experiment in order to acclimate the animal to the reward.

2.3.3. Training Protocol

1. Place the animals (two animals on the maze at the same time will make the animals more comfortable during the initial phase of exploration and will reduce the time of acclimation) at the base of the T-maze task and allow exploring and choosing freely one of the side arms without reward in either arm. Determine the side preferences (16).
2. Allow animals to acclimate to the T-maze for 1 or 2 days prior to testing.
3. Replace the consumed food rewards on the T-maze during acclimation.
4. On day 1, place food rewards only on the arms of T-maze and on day 2, place food rewards only at the ends of the arms.
5. Place one animal on the T-maze that has food only in the food cup at the end of the arms. When the animal is comfortable with the experimenter and has acclimated to the T-maze, it is ready for testing.
6. Clean the T-maze apparatus with 70% ethanol between animals to eliminate odors and prevent the use of olfactory cues.

2.3.4. Basic Testing Protocol

1. Place a food reward at the end of an open arm (16).
2. Place a barrier on one of the arms of the T-maze to prevent the animal from entering the other arm.
3. Place the animal at the base of the T-maze behind the closed guillotine door.
4. Raise the guillotine door and allow the animal to enter the stem of T-maze. Close the door.
5. Allow the animal to enter to the open arm and eat the food reward. Return the animal to the starting point, do not allow the animal to explore the T-maze task.
6. Remove the barrier from the unbaited arm and place a food reward on this arm.
7. Raise the guillotine door and allow the animal to enter the stem and either arm of the T-maze. Close the guillotine door.
8. Allow the animal to eat food reward and return to the starting point. If the animal chooses the unbaited arm, allow it to reach and explore the unbaited arm.
9. Repeat steps 7 and 8 one to nine times. Place the food reward on the arm that the animal did not enter on the initial trial. Allow the animal to choose between two arms. If the animals choose the baited arm, consider that as a correct response. If the animals choose unbaited arm, record it as an error and place the animal back at the start position without food. (The food reward remains until the animal finds and eats it).

10. Record the number of correct entries into baited arms and number of reentries into the unbaited arms. Repeated visitation to an unbaited arm is considered as a working memory error.
11. When choice accuracy is high, begin testing the experimental animals that are exposed to GCI.
12. Express performance as the percentage of the correct choices made in each test session.
13. Clean the T-maze apparatus with 70% ethanol after each test to eliminate odors and prevent the use of olfactory cues.

2.3.5. Alternate Testing Protocol

Alternate T-maze protocol is achieved by changing the external cues, for example moving the T-maze to a new location or using an additional partition and barrier in the T-maze stem. This alternate protocol assesses the reference memory. Additional materials include hardware cloth partition, Plexiglas barrier, and curtain for split-stem T-maze (Fig. 3).

1. Repeat steps described in Sects. 2.3.2 and 2.3.3.
2. Move the T-maze to a different location.
3. Place the hardware cloth partition, Plexiglas barrier, and curtain on the T-maze.
4. Repeat Sect. 2.3.4 (steps 1–4). If the animal enters the side of the stem that is blocked by the Plexiglas barrier, allow the animal to make a correction and enter the open side of stem.
5. Complete Sect. 2.3.4 (steps 5–9).
6. Record the number of correct entries into open side of stem, baited arms, and blocked side of stem (reference memory errors) and the number of reentries into unbaited arms (working memory errors).When choice accuracy is high, begin testing the experimental animals that are exposed to GCI.
7. Repeat Sect. 2.3.4 (steps 12–13).

2.4. Radial Arm Maze Test

The radial arm maze task has also been used to investigate spatial working and reference memory (7, 15). The principle of the radial arm maze test is similar to that of the T-maze test. An animal who has been food deprived is placed in the central chamber with all guillotine arm doors closed and allowed to collect food pellets from the end of the each arm (Fig. 4).

The disadvantage of radial arm maze is that an animal can use other strategies going from one arm to another arm instead of using spatial cues, such as olfactory cues (15). Excessive levels of fear (maze or experimenter) prevent the animal from performing. Animals may remain frozen in one place and not explore. A lack of motivation produces similar results. Motivation can be increased by

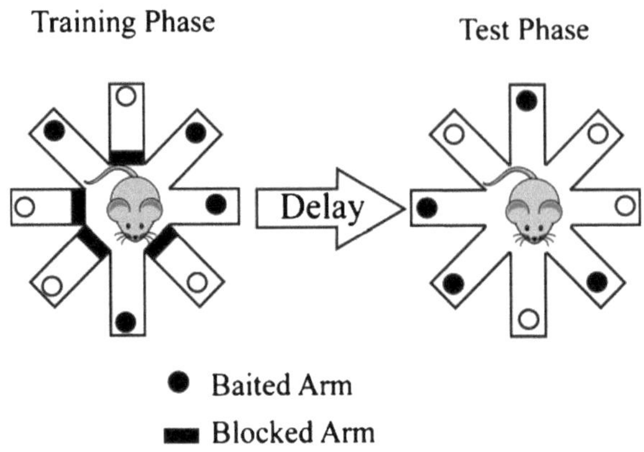

Fig. 4. Baited arm test.

restriction of food intake. The animal's weight and general health must be carefully monitored during food restriction (7).

2.4.1. Materials and Tools

1. Radial arm maze
   The radial arm maze usually is used for rats but if it is scaled down in size (75%), it could be used for mice. The central platform should be ≥45 cm (20 cm for mice) in diameter to allow the rat to turn easily between arms. A Plexiglas wall (25 cm) surrounds the central platform. The arms are 87 cm long and 10 cm wide (35 × 5 cm for mice), radiating from the central platform at equal angles. Each arm has food cup. The maze is set up 1 m above the floor. Most common radial arm maze consists of eight arms connected to a central chamber. A Plexiglas barrier surrounds the central chamber and contains eight guillotine doors that can be opened and closed individually or simultaneously separating the central platform from the arms (7). This maze is handmade or fully automated (Coulbourn or Columbus Instruments).
2. Visual cues
3. Food reward

2.4.2. Procedures

Repeat steps in Sect. 2.3.2.

2.4.3. Training Protocol

1. Repeat Sect. 2.3.3, step 1.
2. Spread food rewards around the radial arm maze (acclimation requires only 1 or 2 days).
3. On subsequent days, place food rewards only on the arms, and then only at the ends of arms (in food cups).
4. Place the animal alone in the maze and place food reward only in the food cup that is at the end of arms.

5. Start testing when the animal is comfortable with the experimenter and the maze (the animal explores without hesitation and excessive defecation or urination).

2.4.4. Basic Testing Protocol

1. Place food rewards in the each food cup (end of the eight arms) before staring the testing session.
2. Place the animal on central platform with all doors closed.
3. Open all doors and allow the animal to enter in an arm. After the animal enters an arm, close doors to all other arms.
4. Allow time to the animal to collect food and return to central platform.
5. Close the door to that arm and confine the animal to the central platform for a set time (from 0 s to minutes; 5 s is ideal to begin). Longer waits make the task more difficult to solve.
6. Repeat steps 3–5 until all food pellets are collected or until a predetermined length of time elapsed between the beginning of the training and the animal's obtaining of all eight food rewards.
7. The experimenter always sits in the same corner of the room.
8. Record the following: 1. the arms the animal entered each time and whether it received a food reward, 2. number of correct (the animal chose it the first time) and incorrect arm choices (the animal visited to the same arm more than once during single test session), and 3. the time elapsed between the beginning of the test and the animal's obtaining of all eight food rewards. Time elapsed shows how fast the animal is making choices and finding the food rewards. This is an indirect indication of motivation. Reentry into a previously chosen arm is considered as a working memory error (7).
9. Analyze the percentage of correct choices made in each test session in relation to the total number of arms entered, the absolute number of correct choices made in the first 8–12 choices of each session, or the percentage of correct choices made in relation to the number of incorrect choices.
10. Repeat Sect. 2.3.4, steps 11–13.

References

1. Nelson A et al (1997) Comparison of effects of global cerebral ischaemia on spatial learning in the standard and radial water maze: relationship of hippocampal damage to performance. Behav Brain Res 85(1):93–115
2. Gobbo OL, O'Mara SM (2004) Impact of enriched-environment housing on brain-derived neurotrophic factor and on cognitive performance after a transient global ischemia. Behav Brain Res 152(2):231–241
3. Yan XB et al (2007) Lithium improves the behavioral disorder in rats subjected to transient global cerebral ischemia. Behav Brain Res 177(2):282–289
4. Hartman RE et al (2005) Characterizing learning deficits and hippocampal neuron loss following

transient global cerebral ischemia in rats. Brain Res 1043(1–2):48–56
5. Block F, Schwarz M (1998) Global ischemic neuronal damage relates to behavioural deficits: a pharmacological approach. Neuroscience 82(3):791–803
6. Nunn JA et al (1994) Global ischaemia: hippocampal pathology and spatial deficits in the water maze. Behav Brain Res 62(1):41–54
7. Wenk GL (2004) Assessment of spatial memory using the radial arm maze and Morris water maze. Curr Protoc Neurosci Chapter 8:Unit 85A
8. Van Dam D, Lenders G, De Deyn PP (2006) Effect of Morris water maze diameter on visual-spatial learning in different mouse strains. Neurobiol Learn Mem 85(2):164–172
9. Vorhees CV, Williams MT (2006) Morris water maze: procedures for assessing spatial and related forms of learning and memory. Nat Protoc 1(2):848–858
10. Cengiz P et al (2011) Inhibition of Na+/H+ exchanger isoform 1 is neuroprotective in neonatal hypoxic ischemic brain injury. Antioxid Redox Signal 14(10):1803–1813
11. Venault P et al (2006) Mouse lines selected for difference in sensitivity to beta-CCM also differ in memory processes. Behav Brain Res 173(2):282–287
12. Choleris E et al (2001) A detailed ethological analysis of the mouse open field test: effects of diazepam, chlordiazepoxide and an extremely low frequency pulsed magnetic field. Neurosci Biobehav Rev 25(3):235–260
13. Walsh RN, Cummins RA (1976) The open-field test: a critical review. Psychol Bull 83(3):482–504
14. Deacon RM, Rawlins JN (2006) T-maze alternation in the rodent. Nat Protoc 1(1):7–12
15. Sharma S, Rakoczy S, Brown-Borg H (2010) Assessment of spatial memory in mice. Life Sci 87(17–18):521–536
16. Wenk GL (2001) Assessment of spatial memory using the T maze. Curr Protoc Neurosci Chapter 8:Unit 85B

Chapter 8

Assessment of Neurogenesis in the Dentate Gyrus After Global or Focal Ischemia

Wei Liu and Jialing Liu

Abstract

Neurogenesis in the mammalian brain continues throughout adulthood. Emerging evidence indicates that cerebral ischemia-induced increase in neurogenesis in the dentate gyrus may be an adaptive response to hypoxia or ischemia-associated injury, and a process involved in the recovery of function. This chapter introduces the classic method for studying dentate gyrus neurogenesis in models of global and focal ischemia by immunohistochemistry, confocal microscopy, and unbiased stereology. Details in tissue preparation, staining procedures, and methods of detection and evaluation are also discussed.

Key words: BrdU, DCX, GFAP, NeuN, Oligo, Confocal microscopy, Unbiased stereology

1. Introduction

Neural plasticity is thought to be one of the major mechanisms underlying functional recovery after central nervous system (CNS) injury (1–4). In addition to axonal sprouting, neurogenesis constitutes other major cellular pathways involved in the ischemia-induced neuroplasticity change in peri-infarct and connected cortical regions. Current available data suggest that there is a strong association between post-brain ischemia functional recovery and regeneration events, such as neurogenesis (5), occurring in discrete brain regions of the adult mammals including humans. During the recovery of stroke, neurons are produced persistently from stem cells (6, 7) in the forebrain subventricular (SVZ) and hippocampal dentate gyrus (DG) germinal zones of the adult brains. Post-stroke newly generated neurons in the peri-infarct cortex are found in a neurovascular niche, where neurogenesis is

Jun Chen et al. (eds.), *Animal Models of Acute Neurological Injuries II*, Springer Protocols Handbooks,
DOI 10.1007/978-1-61779-782-8_8, © Springer Science+Business Media, LLC 2012

causally linked to angiogenesis through the vascular factors SDF-1 and angiopoietin-1 (8, 9). The adult-born immature neurons in the normal brain are functionally integrated into existing neuronal circuitries (10, 11); have distinct electrophysiological properties from old neurons; express proteins related to synaptic stability and activity circuitries (10, 11); and are likely to play important roles in a variety of functions, including olfaction, learning, and mood regulation (12). Specifically, hippocampal neurogenesis enhances long-term potentiation in the medial perforant path/granule cell synapse of the dentate gyrus (13, 14). Similar to the new neurons in the normal brain with regard to functional integration, post-stroke newly generated GABAergic or cholinergic neurons in the striatum could fire action potentials and display spontaneous post-synaptic currents (15).

Immunohistochemistry is the most established method to detect and quantify neurogenesis. A breakthrough in the identification of S-phase cells in the brain was made since the development of BrdU immunohistochemistry (16). As a thymidine analog, BrdU is used to mark cells that have undergone DNA synthesis. When used in conjunction with other cell type-specific markers in multiple-fluorochrome labeling experiment, BrdU can reveal the phenotypes of the progenies derived from the neuroprogenitor cells. During specific time frames of adult neurogenesis, various markers are expressed that correlate with the differentiation steps along the pathway from early progenitor cells to newly generated postmitotic neurons within the DG (Fig. 1) (17).

The aforementioned cell type-specific markers that are widely used for the investigation of adult hippocampal neurogenesis and gliogenesis include doublecortin (DCX), neuronal nuclear protein (NeuN), glial fibrillary acidic protein (GFAP), vimentin, Calbindin, the progenitor cell marker NG2, as well as Olig1 and Olig2 of oligodendrocytic origin. A few examples of confocal images showing colocalization between BrdU and markers are shown in Fig. 2.

Vimentin is an early-stage neuronal marker. Vimentin-immunoreactive radial glia are observed in the dentate gyrus with cell bodies in the subgranular zone and processes extending into the granule cell layer (18). GFAP is widely known as a marker for mature astrocytes in the adult brain. However, a large proportion of the newborn cells in the SGZ of the hippocampal region have also been found to be GFAP positive. During the proliferation phase, i.e., the first stage of adult hippocampal neurogenesis, the newly generated cells express markers GFAP and nestin (19). However, the use of GFAP as a marker for neurogenesis is hampered by the finding that the glial cell lineage and mature astrocytes are also labeled. To confirm the astrocytic nature of the progeny, a second astroglia marker S100b should be used.

DCX, a protein that promotes microtubule polymerization, is present in migrating neuroblasts and young neurons (20), and can

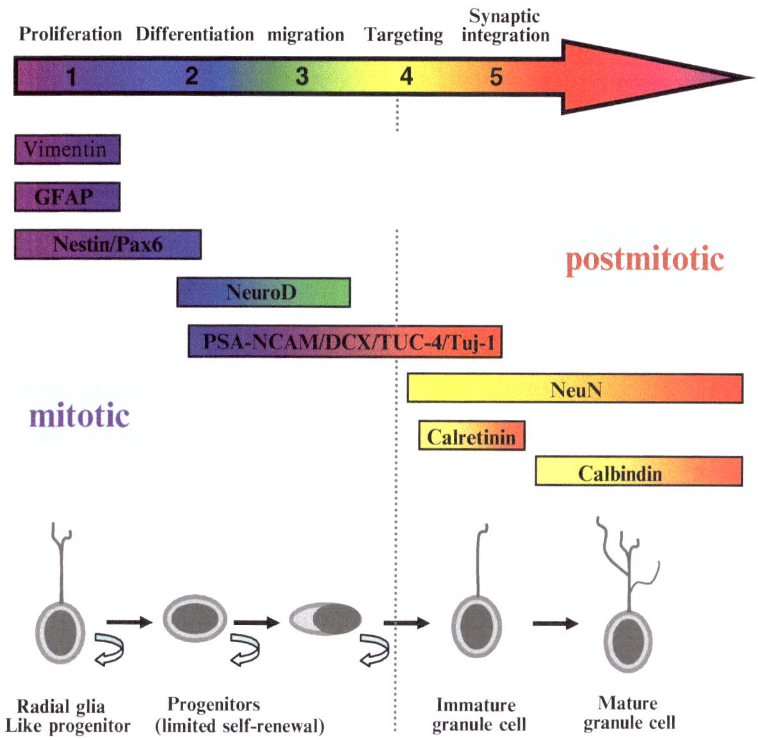

Fig. 1. Stages of adult neurogenesis in the DG and expression pattern of specific markers. Generation of new neurons within the granular layer of the DG can be subdivided into five developmental stages (adapted and modified from ref. 17).

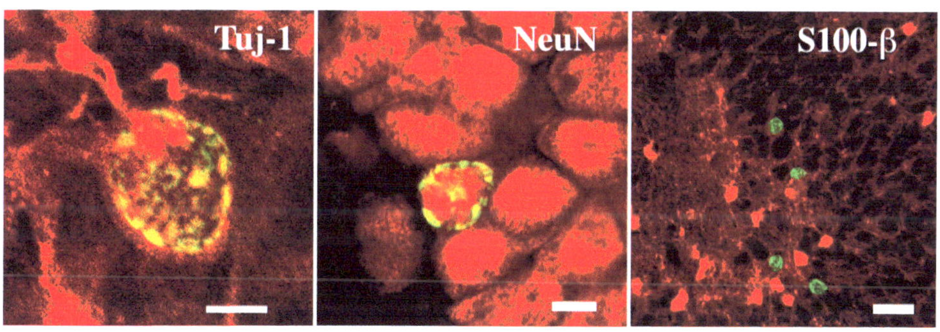

Fig. 2. Confocal images showing the neuronal identity of newly divided cells in the granule cell layer by BrdU in green, and neuronal/glial markers in red for Tuj1, NeuN, and S100-β. BrdU/Tuj1 and BrdU/NeuN colocalized cells are labeled in *yellow* as shown in merged images. Lack of colocalization between BrdU and S100-β suggests that the newborn granule cells do not differentiate into astrocytes. Scale bars: 5, 5, and 25 μm.

serve as a marker of adult neurogenesis in the hippocampus. DCX labels late mitotic neuronal precursors and early postmitotic neurons (21). PSA-NCAM is another widely used marker for immature postmitotic neurons. Although sometimes described as an immature neuronal marker, Tuj-1 is expressed mostly in mature postmitotic neurons in many parts of the brain in addition to the granule cell layers.

NeuN (neuronal nuclear protein) is expressed in relatively mature neurons in the hippocampus, compared to other early neuronal markers, such as PSA-NCAM or DCX. As another late-stage neuronal marker, Calbindin is expressed in all mature granule cells in the dentate gyrus, and in a large proportion of CA1 and CA2 pyramidal neurons (17, 22).

So far, NG2-positive cells are thought to be mainly oligodendrocyte precursor cells. Several findings indicated that NG2 is one of the neural precursor markers for hippocampal progenitor cell proliferation (23). NG2-positive cells, which are located in or adjacent to the subgranular layer of the dentate gyrus, may represent multipotent neural progenitor cells in the hippocampus. Also, NG2 is capable of modulating proliferation of several cell types (11) and may play an important role in presenting growth factors.

The oligodendrocyte lineage genes *Oligo1* and *Oligo2* encode transcription factor-related proteins that are coexpressed in neural progenitors (24). They are expressed in mature oligodendrocytes while expression is weak to moderate in astrocytomas. *Oligo1* has roles in development and maturation of oligodendrocytes, especially within the brain. *Oligo2* is required for oligodendrocyte and motor neuron specification in the spinal cord. Both *Oligo* genes contribute to neural pattern formation.

2. Materials and Instruments

2.1. BrdU Injection

- BrdU can be easily administered by intraperitoneal injection. Depending on the application mode, the injection paradigms and the numbers of labeled cells within the brain can vary. In general, the longer the survival time after the labeling, the greater number of injections is needed to guarantee a good staining result. In order to avoid potential DNA repair-induced BrdU incorporation, do not exceed the dose of a single BrdU injection for more than 50 mg/kg.
- Store BrdU (50 mg/kg; Sigma, St. Louis, MO, USA) powder at –20°C. Equilibrate to room temperature (RT) before opening. Seal with paraffin after use.
- Prepare BrdU solution just before use to ensure the quality. The maximal solubility of BrdU solution in saline does not exceed 15 mg/mL. The maximal volume of BrdU solution tolerated by a rat and a mouse is about 2 and 1 mL, respectively. Warm the solution to 37°C prior to injection.
- BrdU has a clearance rate of approximately 4 h; thus, a daily two-dose (with 8-h interval) or three-dose (with 4-h interval) paradigm can be easily implemented. Following focal ischemia, the optimal time to capture most of the proliferating progenitor

cells is between 3 and 8 days. However, the peak of proliferation at the dentate gyrus after global ischemia shifts towards 7–12 days after the insult.

- Warning signs of BrdU being potential mutagen are posted on the doors of rooms and on the cages of animals receiving BrdU injections. All personnel involved should be trained in chemical hygiene plan and should dispose of bedding as hazardous chemical waste. Gloves and safety goggles should be worn when handling the solution or waste.

2.2. Tissue Preparation

- Animals are anesthetized with ketamine (80 mg/kg; Parke-Davis, MorrisPlains, NJ) and xylazine (20 mg/kg; Butler, Columbus, OH).
- Followed by exsanguination and perfusion with cold 4% paraformaldehyde (PFA) in 0.1 M pH 7.4 phosphate buffer (PB).
- Remove brains and postfix overnight in 4% paraformaldehyde.
- Immerse brains in 20% sucrose in PB for 24–48 h (mouse) or 48–72 h (rat) or until the brains sink.
- 40-μm coronal sections were cut and collected serially by a cryostat for frozen sections and store the sections at −70°C; by a freezing microtome and store the sections at 4°C in 0.1 M PB containing 0.02% sodium azide; or by a microtome following paraffin embedding and store the sections at room temperature.

2.3. Pretreatments of Sections

Frozen or free-floating sections do not require pretreatment. Paraffin sections need to be deparaffinized as follows:

- Xylene: 3 min ×2.
- Xylene 1:1 with 100% ethanol: 3 min.
- 100% ethanol: 2× 3 min.
- 95% ethanol: 3 min.
- 70% ethanol: 3 min.
- 50% ethanol: 3 min.
- Running cold tap water to rinse.
- For optimal staining: Perform antigen retrieval.

Keep the slides in the tap water until ready to perform antigen retrieval. At no time from this point onwards should the slides be allowed to dry. Drying out causes nonspecific antibody binding and, therefore, high background staining.

2.4. Immunohistochemistry Using DAB

Free-floating sections are immunostained as described previously (22, 25, 26) using the following reagents (see Table 1) and multiwell cell culture plates (6-well plate for rat sections and 12-well

Table 1
Common primary antibody in IHC

Marker	Primary antibody
BrdU	Rat-anti-BrdU (Clone BU1/75, ICR1) (2 μg/mL) (Accurate Chemicals & Scientific, Westbury, NY)
NeuN	Mouse anti-neuronal nuclei (NeuN) biotin conjugated monoclonal antibody (MAB377B) (1 μg/mL) (Chemicon, Temecula, CA)
GFAP	Rabbit anti-glial fibrillary acidic protein-δ (GFAP-δ) polyclonal antibody (AB9598) (1:500) (Chemicon, Temecula, CA)
Vimentin	Mouse anti-vimentin monoclonal antibody (MAB3400) (5–10 μg/mL) (Chemicon, Temecula, CA)
NG2	Rabbit anti-NG2 chondroitin sulfate proteoglycan polyclonal antibody (AB5320) (2 μg/mL) (Millipore, MA)
Oligo	Rabbit anti-olig-1 polyclonal antibody (AB15620) (1:1,000–1:5,000) (Chemicon, Temecula, CA)
Oligo2	Rabbit anti-olig-2 polyclonal antibody (AB9610) (1:500–1:1,000) (Chemicon, Temecula, CA)
DCX	Anti-doublecortin (1:1,000–1:15,000) (AB2253) (Millipore, MA)
Calbindin	Rabbit anti-calbindin D-28K polyclonal antibody (P05937) (Millipore, MA)

plate for mouse sections). For sections on the slides (frozen or paraffin sections), use Pap pen (Ted Pella Inc., Product no. 22311) to draw an enclosed circle around the brain sections to keep the solutions.

- If perfusion is not satisfactory and one does not want the risk of sections falling apart during staining or mounting, gently fix the brain sections.
  - Immerse sections in PFA for 5 min.
  - Rinse sections in PB for 5 min ×2.
  - Incubate at 65–75°C for 1 h.
- BrdU pretreatment:
  - Wash sections twice with 0.1 M PB.
  - 50% formamide/PB at 65°C for 2 h.
  - Wash with PB twice.
  - 2 N HCl 37°C for 30 min.

- Wash with PB twice.
- 0.5 M boric acid for 10 min.
- Wash with 0.1 M PB twice.

- Incubate sections with blocking serum (2% serum, 1% BSA, 0.3% Triton in 0.1 M PB) for 30 min. The species of serum should be the same as that of the secondary antibody.
- Incubate sections with primary antibody at appropriate dilution in blocking serum overnight at room temperature. For best results, optimize antibody concentrations by serial dilutions.
- Rinse with PB for 5–10 min ×2–3.
- Incubate sections with an appropriate secondary antibody (biotinylated sheep-anti-mouse, anti-rabbit, anti-goat, or anti-rat secondary antibodies, 1:2,000, Jackson Immunoresearch) in blocking serum for 2 h at room temperature.
- Rinse with PB for 5–10 min ×2–3.
- Incubate sections in avidin–biotin complex (ABC) (ABC kit, Vector laboratories, Burlingame, CA) solution for 2 h at room temperature. Premix solutions A and B for 30 min prior to use.
- Rinse with PB for 5–10 min ×2–3.
- Incubate sections in 0.05% diaminobenzine (DAB)-tetrachloride (one pair of tablets in 15 mL 0.1 M PB; DABFast; Sigma, St. Louis, MO) and 0.04% $NiCl_2$ for up to 5 min. DAB solution should be made immediately prior to use for best results.
- Rinse in distilled water and mount on slide. Warning: DAB is a carcinogen; thus, all waste containing DAB should be treated as hazardous materials.
- Dehydrate with 95% ethanol for 2 min, and 100% ethanol for 3 min ×2.
- Coverslip with mounting medium.

2.5. Double Immunofluorescence

- Same as "Immunohistochemistry staining" before incubation with secondary antibody.
- Rinse with PB for 5–10 min ×2–3.
- Incubate sections with secondary antibody at appropriate dilution in PB for 2 h at room temperature. If a fluorescent-conjugated secondary antibody is used, dilute it in the blocking serum and protect the sections from light. Alexa Fluor 488 or Cy2 is a good green dye. Alexa Fluor 594, Texas red, Cy3, or Cy3.5 is a good red dye. Dye light are a new series of fluorochromes that are conjugated to various kinds of secondary antibodies (can be purchased from Jackson ImmunoResearch).

- Rinse with PB for 5–10 min ×2.
- Rinse once with blocking serum for 5 min in dark.
- Incubate sections with a second primary antibody in blocking serum overnight at room temperature.
- Following two washes in PB, incubate sections with appropriate secondary antibody conjugated to a different fluorochrome dye in proper dilution in blocking serum for 2 h at room temperature. Protect the sections from light.
- Rinse with PB for 5–10 min ×2.
- Rinse once with blocking serum for 5 min in dark.
- Mount sections on slides in dim light.
- Drip a small quantity of mounting solution with anti-fade (Molecular probes) before coverslipping. Seal the coverslip under microscope in dark with nail polish to prevent drying and movement.
- Store in dark at 4°C.

2.6. Photography and Image Processing Tools

2.6.1. Microscope (Bright Field or Fluorescent)

Following double immunofluorescence, fluorescence signals are detected by using the Zeiss LSM 510 confocal image system (Zeiss) with a sequential scanning model for Alexa 488 and 594 or other combinations. Stacks of images (1,024 × 1,024 pixels) from consecutive slices of 0.66–0.7 μm in thickness can be obtained by averaging four scans per slice. The desired images can be exported from the LSM browser during off-line analysis and processed with Adobe Photoshop (Adobe System, Mountain View, CA).

2.7. Evaluation

2.7.1. Cell Count and Semiquantitative Measurement of Optical Density

If the confocal imager is equipped with a stereology system (motorized stage plus stereology software), the number of colocalized cells can be estimated directly off-line following online acquisition of images. Alternatively, the number of BrdU-labeled cells in the dentate gyrus can be estimated by unbiased stereology with a Zeiss Axioskope 2 Plus epifluorescent microscope equipped with stereology software (StereoInvestigator, MicroBrightField, Colchester, VT). The number of BrdU/NeuN or other double-labeled cells can be estimated by multiplying the percentages of colocalization (determined by confocal microscopy in at least 50 double-labeled cells) to the total number of BrdU-positive cells (determined by Zeiss Axioskope 2 Plus epifluorescent microscope) (26–28).

To improve the precision of quantitative analysis and reduce the variability, the following are necessary.

- Carry out all immunohistochemistry with one antibody on the same day, with the same reagent.
- Capture all images with the same pinhole size, intensity, and contrast settings, and use short exposure times to prevent fading.

- Test for fading (decreased intensity) with several consecutive images taken from the same location.

2.7.2. Stereology

The use of unbiased stereology has become the mainstream method for estimating numbers of newborn neurons, total granule cells, or volume of the granule cell layer, all of which are related to neurogenesis. The random and systematic sampling methods of stereology may yield more accurate and unbiased results than the profile counting method, and therefore are preferred in getting more consistent results for a defined brain region, such as the dentate gyrus. To achieve a coefficient of error (CE) of less than 0.10, counting frames of $15 \times 15 \times 20$ μm are used in a 100×100-μm matrix (for estimating granule cells) and in a 40×40-μm matrix (for estimating BrdU and DCX) that are randomly superimposed onto the region of interest by the program. The numbers of BrdU nuclei or DCX-labeled cell bodies are determined using a ×63 oil objective in serial coronal sections spanning the entire hippocampus with the optical fractionator probe (29) (Stereo Investigator, MicroBrightField, Colchester, VT). Hoechest 33342 (50 μg/mL; Molecular Probes) staining is used for the counting of granule cells. The volume of the DG can be determined using the Cavalieri's principle (30).

3. Summary

In summary, the current research methods for studying neurogenesis after global and focal cerebral ischemia are primarily quantitative morphometry-based techniques. The role of adult neurogenesis in hippocampal function and disease etiology will be better understood as more selective, inducible, and reversible manipulations of in vivo neurogenesis are developed. The discovery of novel therapeutic compounds for various diseases may involve mechanisms that induce a superior regulation of adult hippocampal neurogenesis.

References

1. Goldberger M (1981) The role sprouting might play during the recovery of motor function. Springer, New York
2. Stroemer RP, Kent TA, Hulsebosch CE (1998) Enhanced neocortical neural sprouting, synaptogenesis, and behavioral recovery with d-amphetamine therapy after neocortical infarction in rats. Stroke 29:2381–2393; discussion 2385–2393
3. Liepert J, Hamzei F, Weiller C (2004) Lesion-induced and training-induced brain reorganization. Restor Neurol Neurosci 22:269–277
4. Barbay S, Zoubina EV, Dancause N, Frost SB, Eisner-Janowicz I, Stowe AM, Plautz EJ, Nudo RJ (2006) A single injection of d-amphetamine facilitates improvements in motor training following a focal cortical infarct in squirrel monkeys. Neurorehabil Neural Repair 20:455–458

5. Carmichael ST (2008) Themes and strategies for studying the biology of stroke recovery in the poststroke epoch. Stroke 39:1380–1388
6. Thored P, Wood J, Arvidsson A, Cammenga J, Kokaia Z, Lindvall O (2007) Long-term neuroblast migration along blood vessels in an area with transient angiogenesis and increased vascularization after stroke. Stroke 38:3032–3039
7. Thored P, Arvidsson A, Cacci E, Ahlenius H, Kallur T, Darsalia V, Ekdahl CT, Kokaia Z, Lindvall O (2006) Persistent production of neurons from adult brain stem cells during recovery after stroke. Stem Cells 24:739–747
8. Ohab JJ, Fleming S, Blesch A, Carmichael ST (2006) A neurovascular niche for neurogenesis after stroke. J Neurosci 26:13007–13016
9. Ohab JJ, Carmichael ST (2008) Poststroke neurogenesis: emerging principles of migration and localization of immature neurons. Neuroscientist 14:369–380
10. Song HJ, Stevens CF, Gage FH (2002) Neural stem cells from adult hippocampus develop essential properties of functional CNS neurons. Nat Neurosci 5:438–445
11. van Praag H, Schinder AF, Christie BR, Toni N, Palmer TD, Gage FH (2002) Functional neurogenesis in the adult hippocampus. Nature 415:1030–1034
12. Doetsch F, Hen R (2005) Young and excitable: the function of new neurons in the adult mammalian brain. Curr Opin Neurobiol 15: 121–128
13. Snyder JS, Kee N, Wojtowicz JM (2001) Effects of adult neurogenesis on synaptic plasticity in the rat dentate gyrus. J Neurophysiol 85:2423–2431
14. Son H, Yu IT, Hwang SJ, Kim JS, Lee SH, Lee YS, Kaang BK (2003) Lithium enhances long-term potentiation independently of hippocampal neurogenesis in the rat dentate gyrus. J Neurochem 85:872–881
15. Hou SW, Wang YQ, Xu M, Shen DH, Wang JJ, Huang F, Yu Z, Sun FY (2008) Functional integration of newly generated neurons into striatum after cerebral ischemia in the adult rat brain. Stroke 39:2837–2844
16. Miller MW, Nowakowski RS (1988) Use of bromodeoxyuridine-immunohistochemistry to examine the proliferation, migration and time of origin of cells in the central nervous system. Brain Res 457:44–52
17. von Bohlen Und Halbach O (2007) Immunohistological markers for staging neurogenesis in adult hippocampus. Cell Tissue Res 329:409–420
18. Encinas JM, Vaahtokari A, Enikolopov G (2006) Fluoxetine targets early progenitor cells in the adult brain. Proc Natl Acad Sci U S A 103:8233–8238
19. Maslov AY, Barone TA, Plunkett RJ, Pruitt SC (2004) Neural stem cell detection, characterization, and age-related changes in the subventricular zone of mice. J Neurosci 24:1726–1733
20. Gleeson JG, Lin PT, Flanagan LA, Walsh CA (1999) Doublecortin is a microtubule-associated protein and is expressed widely by migrating neurons. Neuron 23:257–271
21. Brown JP, Couillard-Despres S, Cooper-Kuhn CM, Winkler J, Aigner L, Kuhn HG (2003) Transient expression of doublecortin during adult neurogenesis. J Comp Neurol 467:1–10
22. Liu J, Solway K, Messing RO, Sharp FR (1998) Increased neurogenesis in the dentate gyrus after transient global ischemia in gerbils. J Neurosci 18:7768–7778
23. Thallmair M, Ray J, Stallcup WB, Gage FH (2006) Functional and morphological effects of ng2 proteoglycan deletion on hippocampal neurogenesis. Exp Neurol 202:167–178
24. Lu QR, Sun T, Zhu Z, Ma N, Garcia M, Stiles CD, Rowitch DH (2002) Common developmental requirement for olig function indicates a motor neuron/oligodendrocyte connection. Cell 109:75–86
25. Matsumori Y, Hong SM, Fan Y, Kayama T, Hsu CY, Weinstein PR, Liu J (2006) Enriched environment and spatial learning enhance hippocampal neurogenesis and salvages ischemic penumbra after focal cerebral ischemia. Neurobiol Dis 22:187–198
26. Liu Z, Fan Y, Won SJ, Neumann M, Hu D, Zhou L, Weinstein PR, Liu J (2007) Chronic treatment with minocycline preserves adult new neurons and reduces functional impairment after focal cerebral ischemia. Stroke 38:146–152
27. Raber J, Fan Y, Matsumori Y, Liu Z, Weinstein PR, Fike JR, Liu J (2004) Irradiation attenuates neurogenesis and exacerbates ischemia-induced deficits. Ann Neurol 55:381–389
28. Hong SM, Liu Z, Fan Y, Neumann M, Won SJ, Lac D, Lum X, Weinstein PR, Liu J (2007) Reduced hippocampal neurogenesis and skill reaching performance in adult emx1 mutant mice. Exp Neurol 206:24–32
29. West MJ, Slomianka L, Gundersen HJ (1991) Unbiased stereological estimation of the total number of neurons in the subdivisions of the rat hippocampus using the optical fractionator. Anat Rec 231:482–497
30. Geinisman Y, Gundersen HJ, van der Zee E, West MJ (1996) Unbiased stereological estimation of the total number of synapses in a brain region. J Neurocytol 25:805–819

Part II

Focal Cerebral Ischemia

Chapter 9

Infarct Measurement in Focal Cerebral Ischemia: TTC Staining

Feng Zhang and Jun Chen

Abstract

Focal cerebral ischemia is the commonest type of human strokes. Several animal models of focal brain ischemia have been developed to resemble human strokes closely. Infarct volume is an essential indicator of how severe the ischemic damage is 2,3,5-triphenyltetrazolium chloride (TTC) staining is a frequently used method to determine the infarct induced by focal cerebral ischemia in rat or mouse. TTC is white salt and substrate of dehydrogenases. TTC metabolite is a red dye, which turns the cell red. As a result, viable tissue is stained red while dead tissue is left unstained. This chapter introduces TTC stain's brief history, required material and tools, and typical procedure and its alternative approaches.

Key words: Focal, Formazan, Injury assessment, Rat, Stroke, Tetrazolium

1. Introduction

Focal cerebral ischemia in animals mimics the thrombotic and embolic strokes in human. It causes nonselective cell death in the territory of occluded artery, forming the so-called infarct. The standard and essential approach to determine the severity of ischemic damage is to measure the size of the infarct.

Common method to measure infarct is to stain brain slices or sections using 2,3,5-triphenyltetrazolium chloride (TTC), Hematoxylin & Eosin (H&E), or crystal violet solutions. Among them, TTC staining is simple and accurate; therefore, it is the focus of this chapter.

TTC staining was first introduced by two independent groups in 1984 to detect infarct after focal cerebral ischemia in cats, gerbils, and rats (1, 2). It was further characterized by Bederson et al. (3). TTC itself is water-soluble tetrazolium salt in a white crystalline

Jun Chen et al. (eds.), *Animal Models of Acute Neurological Injuries II*, Springer Protocols Handbooks,
DOI 10.1007/978-1-61779-782-8_9, © Springer Science+Business Media, LLC 2012

powder appearance. Typically, TTC is the substrate of cellular dehydrogenases that are most plentiful in mitochondria (4). Once inside the cells, TTC may be reduced to formazan by those enzymes. In contrast to TTC, formazan is water-insoluble dye and is deep red in color. Formazan, therefore, stains the tissues red. Thus, with TTC staining, viable tissues will be red and dead tissues will stay in their original color—white.

The advantages of TTC staining are low cost, simplicity, and accuracy. It is easy to perform without sophisticated equipment and skills. Most importantly, TTC staining accurately delineates the cerebral infarct. Infarct volume measured with TTC staining correlates well with the infarct measured with H&E (3, 5) or crystal violet staining (6). On the other hand, TTC staining has a couple of disadvantages. It is a single-purpose stain, only for macroscopic measurement of cerebral infarct. TTC staining has relatively narrow time window to perform.

2. Materials and Instruments

2.1. TTC Solution

1. Normal saline.
2. TTC powder (Sigma, St. Louis, MO).
3. Formalin solution.

2.2. Brain Sectioning

1. Rat or mouse brain slicer matrix, with 1.0-mm coronal slice intervals (Fig. 1).
2. Thin and sharp blades (Leica® Disposable Microtome Blades, Cat. 63065).

2.3. Staining

1. Small beaker or Petri dish.
2. Water bath.

2.4. Photography and Processing Tools

1. Digital camera.
2. Ruler.
3. Blue plastic that has good contrast to red and white.
4. MCID image analysis system (Imaging Research Inc., St. Catharines, ON, Canada) or NIH image (NIH, Bethesda, MD) software.

3. Staining Procedure

3.1. Preparation of TTC Solution

1. Dissolve TTC in normal saline at a concentration of 2%.
2. Pour 25 mL 2% TTC solution into a Petri dish or a small beaker.
3. Prewarm TTC solution to 37°C in water bath.

Fig. 1. A mouse brain matrix (*left*) and a microtome blade (*right*).

3.2. Brain Sectioning

1. Remove whole brain from the skull at a certain time point after ischemia.
2. Place brain into the matrix with ventral side up.
3. Line up brain longitudinally with matrix.
4. Slice brain with wet blades, with intervals of 2 mm in rat or 1 mm in mouse.
5. Lift brain slices and blades together off the matrix. Six or seven slices are expected.

3.3. TTC Staining

1. Gently brush brain slices off the blade one by one, and carefully drop the slices into TTC solution.
2. Spread the slices flat on the bottom, making sure that they are totally immersed in the solution.
3. Keep the brain slices at 37°C for 30 min in dark.
4. Carefully transfer the slices into formalin and fix for 1–24 h.
5. Transfer the slices into saline.

3.4. Photography

1. On a piece of blue plastic, line up the slices from anterior to posterior, with left hemisphere on left.
2. Remove excessive water on and around the slices with paper towel.
3. Place a ruler along the slices.
4. Shoot pictures with digital camera (Fig. 2). Do not flash, but use macro close-up mode.
5. Transfer pictures to computer, and save them in tif format.

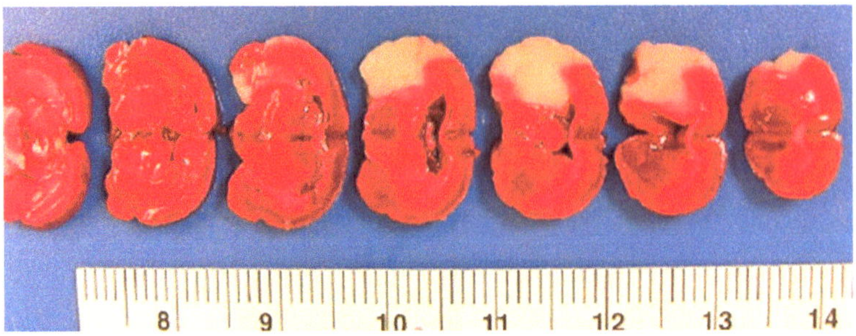

Fig. 2. Representative picture of TTC staining of ischemic rat brain. The rat was subjected to 90-min focal cerebral ischemia induced by left middle cerebral artery occlusion. The brain slices were stained with 2% TTC for 30 min and then photographed.

3.5. Measuring Infarct

1. Launch software (MCID or NIH image), and load images.
2. Calibrate the distance using the ruler as a reference.
3. Select the regions of interest (ROIs) manually or automatically with threshold setting (7). Typical ROIs are the whole area of contralateral hemisphere and the healthy tissue of ipsilateral hemisphere. Both of them are red. Another ROI is the infarct in white.
4. Calculate infarct area and volume on each slice using the following equations (8):
   - Infarct area:
     - Contralateral hemisphere area – healthy area of ipsilateral hemisphere.
   - Infarct volume:
     - Infarct area × thickness of slice.

3.6. Presentation of Infarct

There are three ways to present as follows:

- Infarct area of each slice.
- Infarct ratio of each slice: Infarct area divided by the area of contralateral hemisphere.
- Total infarct volume: By adding up infarct volume of each slice.

4. Variation in TTC Staining

4.1. When to Perform TTC Staining After Ischemia?

TTC staining is only applicable within certain duration following cerebral ischemia. Studies have demonstrated that TTC staining is applicable as early as 4 h and as late as 7 days after ischemia (4, 9); however, the best time points are 24 or 48 h after ischemia. When the duration is less than 24 h, the contrast between healthy and dead tissues is not clear enough as infarct takes times to mature.

After 48 h, macrophages increasingly infiltrate the infarct, and smudge infarct border and interfere the measurement of infarct volume.

4.2. When to Perform TTC Staining After Animal Death?

TTC staining is commonly performed right after the animals were sacrificed, which generally yields satisfied results. Due to the fluidity of brain, it may not be easy to perform the staining for beginners; therefore, it is acceptable to freeze whole brain for a short duration, for instance 15–20 min, before sectioning. Interestingly, a study demonstrated that the delayed staining 8 h after animal death could still reliably assess the infarct volume in a rat (10). However, this is not recommended as a routine practice.

4.3. Concentration of TTC Solution

TTC concentrations vary from 0.05 to 4% for measuring infarct of TTC (11). However, majority of studies found that 2% is an appropriate concentration (3, 6, 9, 11).

4.4. Immersion or Perfusion

An alternative way to perform TTC staining is perfuse animals with TTC solution transcardially (4, 12, 13). This method is not preferred as it is more complicated but is not as accurate as immersion method (4).

4.5. Alternatives for Infarct Measurement

In addition to TTC staining, other commonly used methods include H&E, CV, and Fluoro-Jade staining. These methods are discussed in related chapters.

References

1. Bose B, Osterholm JL, Berry R (1984) A reproducible experimental model of focal cerebral ischemia in the cat. Brain Res 311:385–391
2. Liszczak TM, Hedley-Whyte ET, Adams JF, Han DH, Kolluri VS, Vacanti FX, Heros RC, Zervas NT (1984) Limitations of tetrazolium salts in delineating infarcted brain. Acta Neuropathol 65:150–157
3. Bederson JB, Pitts LH, Germano SM, Nishimura MC, Davis RL, Bartkowski HM (1986) Evaluation of 2,3,5-triphenyltetrazolium chloride as a stain for detection and quantification of experimental cerebral infarction in rats. Stroke 17:1304–1308
4. Benedek A, Moricz K, Juranyi Z, Gigler G, Levay G, Harsing LG Jr, Matyus P, Szenasi G, Albert M (2006) Use of TTC staining for the evaluation of tissue injury in the early phases of reperfusion after focal cerebral ischemia in rats. Brain Res 1116:159–165
5. Isayama K, Pitts LH, Nishimura MC (1991) Evaluation of 2,3,5-triphenyltetrazolium chloride staining to delineate rat brain infarcts. Stroke 22:1394–1398
6. Tureyen K, Vemuganti R, Sailor KA, Dempsey RJ (2004) Infarct volume quantification in mouse focal cerebral ischemia: a comparison of triphenyltetrazolium chloride and cresyl violet staining techniques. J Neurosci Methods 139: 203–207
7. Goldlust EJ, Paczynski RP, He YY, Hsu CY, Goldberg MP (1996) Automated measurement of infarct size with scanned images of triphenyltetrazolium chloride-stained rat brains. Stroke 27:1657–1662
8. Swanson RA, Sharp FR (1994) Infarct measurement methodology. J Cereb Blood Flow Metab 14:697–698
9. Liu F, Schafer DP, McCullough LD (2009) TTC, fluoro-Jade B and NeuN staining confirm evolving phases of infarction induced by middle cerebral artery occlusion. J Neurosci Methods 179:1–8
10. Li F, Irie K, Anwer MS, Fisher M (1997) Delayed triphenyltetrazolium chloride staining remains useful for evaluating cerebral infarct volume in a rat stroke model. J Cereb Blood Flow Metab 17:1132–1135

11. Joshi CN, Jain SK, Murthy PS (2004) An optimized triphenyltetrazolium chloride method for identification of cerebral infarcts. Brain Res Brain Res Protoc 13:11–17
12. Dettmers C, Hartmann A, Rommel T, Kramer S, Pappata S, Young A, Hartmann S, Zierz S, MacKenzie ET, Baron JC (1994) Immersion and perfusion staining with 2,3,5-triphenyltetrazolium chloride (TTC) compared to mitochondrial enzymes 6 hours after MCA-occlusion in primates. Neurol Res 16: 205–208
13. Park CK, Mendelow AD, Graham DI, McCulloch J, Teasdale GM (1988) Correlation of triphenyltetrazolium chloride perfusion staining with conventional neurohistology in the detection of early brain ischaemia. Neuropathol Appl Neurobiol 14:289–298

Chapter 10

Morphological Assessments of Focal Cerebral Ischemia: White Matter Injury

Feng Zhang and Lyanne C. Schlichter

Abstract

A common type of ischemic stroke, focal cerebral ischemia causes nonselective tissue damage in the ischemic area, including damage to both gray and white matter. White matter is made up of neuronal axons and their surrounding oligodendrocytic myelin sheath, and is susceptible to ischemic injury. The contribution of white matter injury to neurological dysfunction was largely ignored in the past, but its significance has been increasingly noticed in recent years. This chapter introduces the classic methods for studying white matter injury after brain ischemia, including immunohistochemical staining of amyloid precursor protein and myelin basic protein. Tissue preparation, staining procedures, evaluation methods, and alternative methods are discussed.

Key words: White matter injury, Brain ischemia, APP, MBP, dMBP, Luxol fast blue, Silver stain

1. Introduction

Brain tissues are composed of gray matter and white matter. Gray matter refers to cerebral cortex and subcortical ganglia, which are mainly composed of cell bodies and neuronal dendrites, while white matter mainly refers to the corpus callosum, internal capsule, and anterior commissure (Fig. 1). In addition, there are small bundles of white matter in the striatum. Microstructurally, white matter is composed of bundles of axonal fibers and their surrounding myelin sheath, and oligodendrocytes. The function of white matter is to conduct electrical impulses between neurons: the corpus callosum and commissure connect the two cerebral hemispheres, and the internal capsule connects cortex with subcortical ganglia, brain stem nuclei, and spinal cord.

Jun Chen et al. (eds.), *Animal Models of Acute Neurological Injuries II*, Springer Protocols Handbooks, DOI 10.1007/978-1-61779-782-8_10, © Springer Science+Business Media, LLC 2012

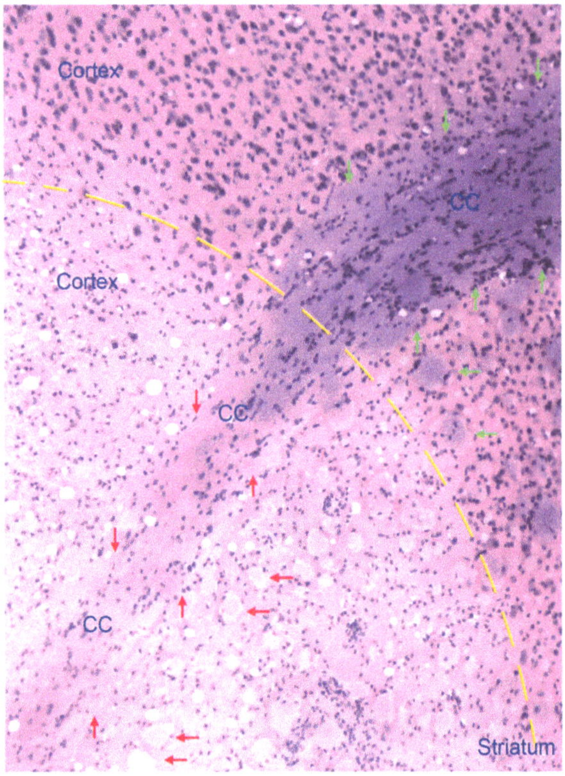

Fig. 1. Brain white matter. The mouse was subjected to a 1-h focal cerebral ischemia (MCAO on left) and a 48-h reperfusion. A fresh-frozen brain section (20 M) was stained with Hematoxylin and Eosin. The *yellow dashed line* indicates the boundary between normal and damaged brain tissues, with the infarct on the bottom left part. *Green arrows* point to normal white matter, and *red arrows* point to damaged white matter. *cc* corpus callosum, ×100.

Following focal cerebral ischemia, all tissues within the ischemic area are damaged, including both gray matter and white matter. In the experimental model of middle cerebral artery occlusion, the injured white matter includes corpus callosum, striatal white matter, and possibly internal capsule, depending on the severity of ischemia. If white matter injury occurs after focal cerebral ischemia, corpus callosum will be the primary site of injury, followed by striatum. To produce focal ischemia that is localized to the striatum, stereotaxic injection of the vasoconstrictor peptide, endothelin-1, can be used (1).

The preferred approaches to evaluating white matter injury are to monitor both axonal injury and myelin sheath injury. The common method is to use immunohistochemistry to detect the levels of a number of relevant proteins. These proteins include amyloid precursor protein (APP) of neuronal origin, as well as myelin basic protein (MBP) or degraded MBP (dMBP) of oligodendrocytic origin. APP is a membrane protein, mainly located at neuronal synapses. Its function is not clear, although a relation to synaptic structure is suggested. APP is barely detectable in normal axons because neuronal anterograde transport keeps it at a low level (1–3). When axonal injury occurs, however, the axonal transport stops and APP accumulates to a measurable level in axonal membranes.

MBP is one of the most abundant proteins in the myelin sheath, comprising about 30% of the total protein fraction. MBP is indispensable for myelin biogenesis and maintenance, as it is a key molecule for myelin assembly and functions as a "glue," mediating the static interaction between the leaflets of oligodendrocytic membranes (4). When white matter is injured, the MBP level decreases, and this can be used as an indicator of white matter injury (1, 5). A specific antibody recognizing only dMBP (6) is also valuable and has been used to detect white matter damage (1, 7).

2. Materials

2.1. Primary Antibodies

1. APP (Zymed, South San Francisco, CA; or Chemicon, Temecula, CA).
2. MBP (Abcam, Cambridge, MA; or Sigma, St. Louis, MO).
3. dMBP (Chemicon, Temecula, CA).

2.2. Secondary Antibodies

1. Fluorescent-conjugated secondary antibody (Jackson Immunoresearch, West Grove, PA) or biotinylated secondary antibody and avidin–biotin complex (ABC) kit (Vector, Burlingame, CA).

2.3. Photography and Image Processing Tools

1. Microscope (light or fluorescent).
2. MCID image analysis system (Imaging Research Inc., St. Catharines, ON, Canada) or NIH Image software (NIH, Bethesda, MD).

3. Staining Procedures

3.1. Tissue Preparation

1. Perform transcardial perfusion on the animal using cold phosphate-buffered saline (PBS) and 4% paraformaldehyde.
2. Remove brains and postfix overnight in 4% paraformaldehyde.
3. Immerse brains in 15% sucrose in PBS for 24 h or until the brains sink.
4. Immerse brains in 30% sucrose in PBS until the brains sink.
5. Cut the brains coronally using a cryostat for frozen sections and store the sections at –20°C until used; or embed brains in paraffin blocks and then cut them for paraffin sections.

3.2. Pretreatments of Sections

Frozen sections do not require pretreatment. Paraffin sections need to be deparaffinized as follows.

1. Immerse sections in xylene for 5 min ×2.
2. Immerse sections in 100% ethanol for 3 min ×2.
3. Hydrate with 95% ethanol for 2 min.
4. Rehydrate with 70% ethanol for 2 min.
5. Rinse in distilled water.
6. Optional: Perform antigen retrieval.

3.3. Staining

1. Rinse sections in PBS + Tween 20 (PBST) for 2 min ×2.
2. Immerse sections in 2% H_2O_2 in PBST for 10 min.
3. Incubate sections with 5% normal serum for 30 min. The species of serum should be the same as that of the secondary antibody.
4. Incubate sections with primary antibody at appropriate dilution in antibody diluent for 1 h at room temperature or overnight at 4°C. If necessary, test and optimize antibody concentrations.
5. Rinse with PBST for 5 min ×3.
6. Incubate sections with the secondary antibody at appropriate dilution in PBS for 30 min at room temperature. If a fluorescent-conjugated secondary antibody is used, protect the sections from light and skip to the water-rinsing step (below).
7. Rinse with PBST for 5 min ×3.
8. Incubate sections in ABC solution for 30 min.
9. Rinse in PBST for 5 min ×3.
10. Incubate sections in 0.05% 3,3 -diaminobenzidine (DAB) for up to 5 min.
11. Rinse in distilled water.
12. Dehydrate with 95% ethanol for 2 min, and 100% ethanol for 3 min ×2.
13. Coverslip with mounting medium.

3.4. Evaluation

Without a reference standard, immunohistochemical staining can only be analyzed semiquantitatively. Common methods include comparison of stain intensity, counting the number of positive particles, or both (8). The stain intensity method applies best to MBP and dMBP staining (Fig. 2) while the positive particle method applies best to APP staining (Fig. 3) (1, 9, 10).

3.4.1. Semiquantitative Measurement of Optical Density

Several procedures reduce the variability and improve the semi-quantitative analysis. Carry out all immunohistochemistry with one antibody on the same day, with the same reagents. For analysis, capture all images with the same pinhole size, intensity, and contrast

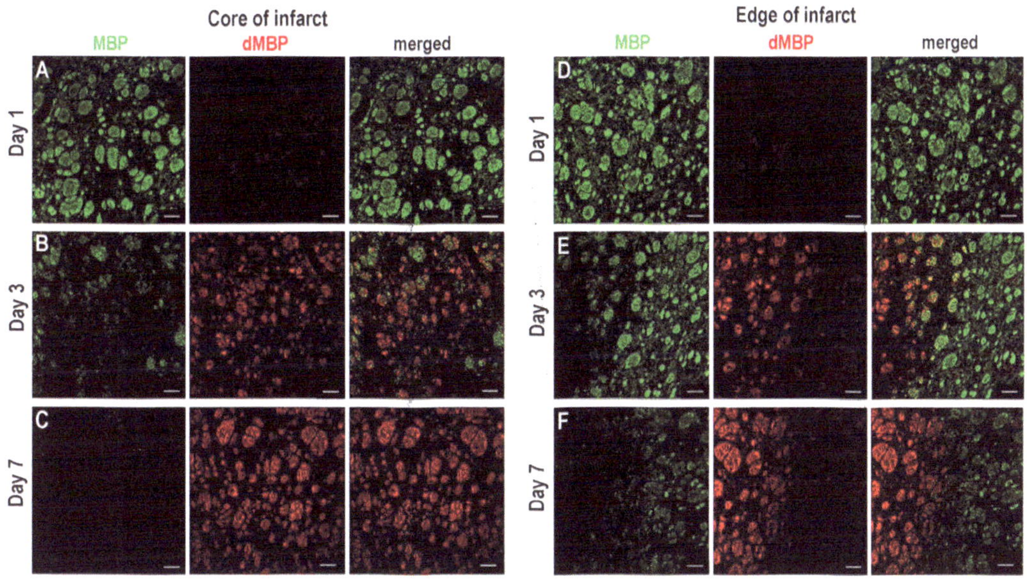

Fig. 2. MBP staining to show the temporal and spatial changes in normal and damaged myelin after focal cerebral ischemia. (**a**–**c**) Core of the infarct. Representative confocal images of sections double labeled for normal myelin (antimyelin basic protein [MBP], *green*) and damaged myelin ([dMBP] *red*) at 1, 3, and 7 days after stroke induction. (**d**–**f**) Edge of the infarct showing representative confocal images of double-labeled sections, as in (**a**–**c**). The core is out of view to the left. Scale bars = 100 μm (adapted with permission and modification from ref. 1).

settings, and use short exposure times to prevent fading. As a further validation, test for fading (decreased intensity) with several consecutive images taken from the same location.

Use ImageJ to measure stain intensity or to count positive particles in several sampling areas from each section and take the average. Regional differences (e.g., core of infarct, penumbra, surrounding tissue) can also be analyzed in this way. For most stroke models, the staining intensity or density of positive particles in the ipsilateral (damaged) hemisphere can be normalized to the contralateral hemisphere of each animal. The undamaged hemisphere yields control values (MBP) or background levels (dMBP, APP). In addition, background fluorescence can be subtracted if there are blank areas within the tissue.

3.4.2. Lesion Area Measurement

The lesion area of white matter can be measured as loss of the MBP signal (5, 11) or as areas of dMBP staining (1), as follows.

1. Take pictures of immunohistochemical staining and save as .tif files.
2. Launch software (MCID or NIH ImageJ), and load the microscopic image files.
3. Perform spatial calibration.
4. Select the regions of interest (ROIs) manually or automatically using the threshold setting (12). Typical ROIs are the areas of the damaged white matter.

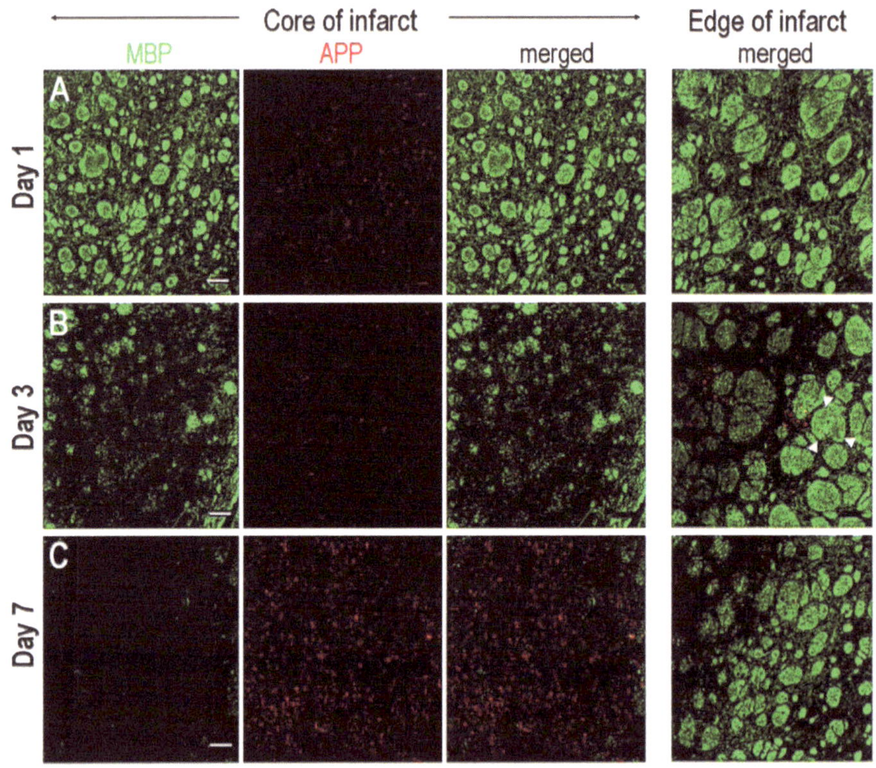

Fig. 3. APP staining to show the temporal and spatial analysis of axonal damage after focal cerebral ischemia. Accumulation of amyloid precursor protein (APP) was monitored with a rabbit polyclonal anti-APP antibody (*red*) and compared with staining for normal myelin using a mouse monoclonal anti-MBP (*green*). (**a–c**) Representative confocal images of sections double labeled for MBP and APP in the core of the infarct (*left*). The center is out of view to the left. In the merged images from the edge of the infarct (*right*), *arrowheads* surround a myelinated bundle containing puncta of APP staining. Scale bars = 100 μm (adapted with permission and modification from ref. 1).

5. Measure the lesion area on each section. White matter damage can be presented as lesion areas in selected brain levels or as lesion volume, by integrating lesion areas into volumes.

4. Other Methods of Evaluating White Matter Injury

4.1. Axons

Jackson Laboratories has a line of YFPH mice (stock number: 003782; strain name: B6.Cg-Tg (Thy1-YFPH)2Jrs/J) that express soluble yellow fluorescent protein in the layer-5 pyramidal neurons of cerebral cortex. Both neuronal bodies and axons are visible, which makes this strain valuable in studying axonal injury (11). Following brain ischemia, white matter injury presents as a loss of axons and as the presence of beaded axons and retraction bulbs (11). Alternatively, Bielschowsky's silver stain or tau-1 immunostaining can be used to show axons (5, 13).

4.2. Myelin

Luxol fast blue-periodic acid-Schiff stain for myelin is the traditional method for staining myelin sheath (5, 13). Luxol fast blue is a copper phthalocyanine dye that reacts with the lipoprotein of myelin sheath and stains myelin blue. Thus, reduced staining reflects loss of, or damage to, myelin.

5. Summary

In summary, the current research methods for studying white matter injury are morphological staining techniques. They usually demonstrate white matter injuries no earlier than 1–3 days after brain ischemia. The lack of earlier timer points reflects either the nature of white matter injury or the possible lack of sensitivity of these methods for detecting early events in white matter after ischemia. The future development of functional methods, such as MRI, will be helpful (14).

References

1. Moxon-Emre I, Schlichter LC (2010) Evolution of inflammation and white matter injury in a model of transient focal ischemia. J Neuropathol Exp Neurol 69:1–15
2. Gentleman SM, Nash MJ, Sweeting CJ, Graham DI, Roberts GW (1993) Beta-amyloid precursor protein (beta APP) as a marker for axonal injury after head injury. Neurosci Lett 160: 139–144
3. Thinakaran G, Koo EH (2008) Amyloid precursor protein trafficking, processing, and function. J Biol Chem 283:29615–29619
4. Baron W, Hoekstra D (2010) On the biogenesis of myelin membranes: sorting, trafficking and cell polarity. FEBS Lett 584:1760–1770
5. Irving EA, Bentley DL, Parsons AA (2001) Assessment of white matter injury following prolonged focal cerebral ischaemia in the rat. Acta Neuropathol 102:627–635
6. Matsuo A, Lee GC, Terai K, Takami K, Hickey WF, McGeer EG, McGeer PL (1997) Unmasking of an unusual myelin basic protein epitope during the process of myelin degeneration in humans: a potential mechanism for the generation of autoantigens. Am J Pathol 150:1253–1266
7. Wakita H, Tomimoto H, Akiguchi I, Matsuo A, Lin JX, Ihara M, McGeer PL (2002) Axonal damage and demyelination in the white matter after chronic cerebral hypoperfusion in the rat. Brain Res 924:63–70
8. Taylor CR, Levenson RM (2006) Quantification of immunohistochemistry—issues concerning methods, utility and semiquantitative assessment II. Histopathology 49:411–424
9. Arai K, Lo EH (2009) Experimental models for analysis of oligodendrocyte pathophysiology in stroke. Exp Transl Stroke Med 1:6
10. Hagberg H, Peebles D, Mallard C (2002) Models of white matter injury: comparison of infectious, hypoxic-ischemic, and excitotoxic insults. Ment Retard Dev Disabil Res Rev 8: 30–38
11. Sozmen EG, Kolekar A, Havton LA, Carmichael ST (2009) A white matter stroke model in the mouse: axonal damage, progenitor responses and MRI correlates. J Neurosci Methods 180: 261–272
12. Goldlust EJ, Paczynski RP, He YY, Hsu CY, Goldberg MP (1996) Automated measurement of infarct size with scanned images of triphenyltetrazolium chloride-stained rat brains. Stroke 27:1657–1662
13. Schabitz WR, Li F, Fisher M (2000) The N-methyl-D-aspartate antagonist CNS 1102 protects cerebral gray and white matter from ischemic injury following temporary focal ischemia in rats. Stroke 31:1709–1714
14. Yang D, Nemkul N, Shereen A, Jone A, Dunn RS, Lawrence DA, Lindquist D, Kuan C-Y (2009) Therapeutic administration of plasminogen activator inhibitor-1 prevents hypoxic-ischemic brain injury in newborns. J Neurosci 29:8669–8674

Chapter 11

Blood Flow Reduction: Laser Doppler, ^{14}C-IAP

Feng Zhang, Steven H. Graham, and Jun Chen

Abstract

Ischemic strokes are caused by a sharp fall of the cerebral blood flow (CBF) to certain regions of brain tissues. In experimental research of cerebral ischemia, the animal models are produced by a variety of methods to reduce or stop the blood supply to the brain, such as using suture, embolus, or direct cut of artery. The reduction in CBF is the key for making ischemic models; therefore, the monitoring of regional CBF is an important experimentation in the field of stroke research. This chapter introduces the common methods for monitoring CBF, with focus on the laser Doppler flowmetry, Carbon-14 iodoantipyrine autoradiography, and laser speckle flowmetry.

Key words: Cerebral ischemia, rCBF, Laser Doppler flowmetry, C-14 IAP autoradiography, Laser speckle flowmetry

1. Introduction

By definition, cerebral ischemia is a disorder in which the cerebral blood flow (CBF) to certain brain tissue is reduced so that cerebral metabolism cannot be carried out efficiently and the brain tissue may die. CBF change is a key factor in the occurrence and intensity of strokes. Several methods have been developed to monitor CBF before, during, and after strokes. In the setting of experimental researches, laser Doppler flowmetry, Carbon-14 iodoantipyrine (^{14}C-IAP) autoradiography, and laser speckle imaging are the common methods to monitor regional CBF (rCBF). The measurement of rCBF is a valuable parameter in verifying the success of the occlusion of cerebral middle artery and evaluating vascular dynamics of certain drugs.

In laser Doppler flowmetry, low-power laser light is applied to brain surface via fiber optics. The laser light hits moving blood cells

Jun Chen et al. (eds.), *Animal Models of Acute Neurological Injuries II*, Springer Protocols Handbooks,
DOI 10.1007/978-1-61779-782-8_11, © Springer Science+Business Media, LLC 2012

in the microcirculation, which scatters the light and causes the laser to change its frequency. The change of light frequency depends on the number of blood cells within the illuminated region and their moving speed. A quantity of the laser light is returned to a photodetector in the system. The value form the detector is processed and displayed as perfusion unit. As such, the number of moving cells and the moving speed determine the local blood perfusion. This is the most frequently used method in the research of cerebral ischemia (1–5) because it is easy to perform, can be applied repeatedly or continuously during the entire procedure of focal cerebral ischemia (4, 6), and is not a terminal study.

^{14}C-IAP autoradiography is a quantitative method to determine the rate of blood flow in a given region of brain tissues (7–9). ^{14}C-IAP itself is an isotope-labeled diffusible tracer with permeability to blood–brain barrier (8). The amount of ^{14}C-IAP in brain is in proportion with the blood flow and the duration after the injection of the tracer. After exposing sections of brain to film, rCBF can be calculated in reference to the signal strength and the radiation in the blood samples. This is an accurate and quantitative method, but it is a terminal study.

Along with laser Doppler flowmetry, laser speckle flowmetry or laser speckle imaging belongs to the category of laser Doppler velocimetry (10–13). Laser speckle itself refers to an optical interference effect that gives a grainy appearance to the objects illuminated by laser light. If the object consists of moving particles, such as blood cells, the pattern and intensity of the speckle fluctuate (13). In general, the fast-moving particles have lower speckle visibility due to the stronger blurring effect (13, 14). Thus, these fluctuations provide information about the velocity distribution of the particles. Laser Doppler flowmetry provides a two-dimensional (spatial and temporal) measurement of rCBF with high resolution.

In this chapter, we give step-by-step operational procedures for rCBF measurement in the model of middle cerebral artery (MCA) occlusion.

2. Materials and Instruments

2.1. Laser Doppler Flowmetry

1. Stereotactic frame (David Kopf instruments, Tujunga, CA, USA).
2. Dental drill.
3. Dental cement.
4. PeriFlux system 5000 (Perimed, Stockholm, Sweden).
5. PeriSoft software (Perimed, Stockholm, Sweden).

2.2. ^{14}C-IAP Autoradiography

1. ^{14}C-IAP solution, 250 μCi in 2.5 mL ethanol (PerkinElmer, Waltham, MA, USA).
2. Liquid scintillation counter, LS5600 (Beckman, Fullerton, CA, USA).
3. Carbon-14 standard microscale (Amersham, San Diego, CA, USA).
4. X-ray films (Kodak, Rochester, NY, USA).
5. 30% hydrogen peroxide.
6. Soluene-350.
7. Ethanol.
8. Cryostat for brain sectioning.
9. MCID image analysis system (Imaging Research Inc., St. Catharines, ON, Canada) or NIH image (NIH, Bethesda, MD) software.

2.3. Laser Speckle Imaging

1. Stereotactic frame (David Kopf instruments, Tujunga, CA, USA)
2. Dental drill.
3. Charge coupled device (CCD) camera (Cohu, San Diego, CA, USA).
4. Collimated 785 nm laser diode (Hitachi, Thorlabs, Newton, NJ, USA).
5. Imaging software (StreamPix, NorPix, Montreal, QC, Canada).

3. Procedures

3.1. Laser Doppler Flowmetry in Rats

3.1.1. Surgical Procedure

1. Anesthetize animal, and fix head to the stereotactic frame.
2. Cut the scalp along the midline, and expose the skull. Stop bleeding using hydrogen peroxide when necessary.
3. Drill burr holes on the skull using dental drill and expose the dura underneath. Use Table 1 as reference for coordinations.
4. Position the probe into the burr hole. Make sure that the probe tip touches the dura without pressure.

3.1.2. Measurement of CBF

1. Calibrate the probe using the motility standard.
2. Measure rCBF before the occlusion of MCA as baseline.
3. Insert the suture to occlude MCA to induce focal ischemia.
4. Measure the rCBF during ischemia. Multiple measurements can be made when necessary.
5. Withdraw the suture to restore blood flow.
6. Measure the rCBF after reperfusion.

Table 1
The coordinations for rCBF measurement

	Bregma (mm)	Lateral (mm)
Core	1.0	3.0
Penumbra	–3.0	4.0

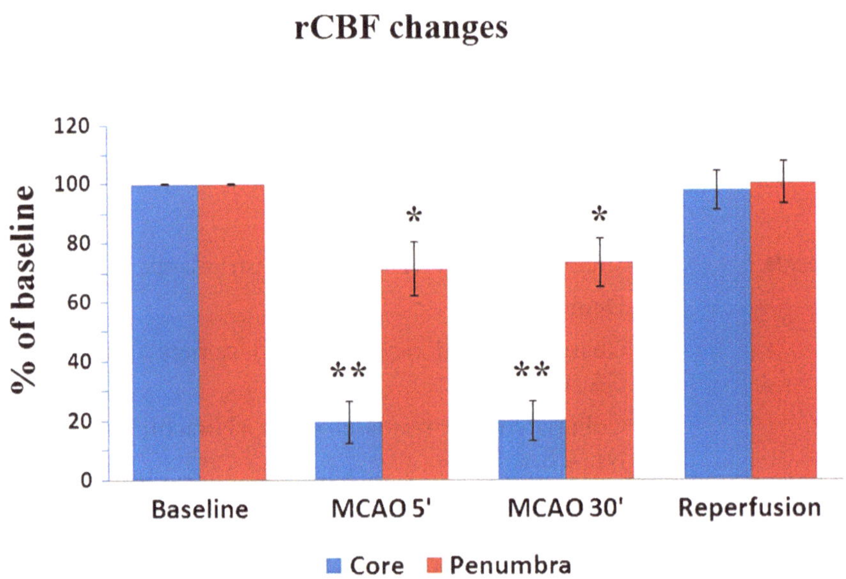

Fig. 1. The rCBF changes in the ischemic core and penumbra in rats after MCAO. The data were with laser Doppler flowmetry and presented as percentage of basal levels of the perfusion units. The baseline level was set as 100%. *$P<0.05$, **$P<0.01$ compared with baseline in the same region.

3.1.3. Presentation of CBF

1. Collect readings of CBF in perfusion unit.
2. The data can be presented with perfusion units, commonly for continuous monitoring.
3. The data can also be presented as percentage of baseline perfusion unit, dividing each reading by the baseline unit (Fig. 1). This is typical for intermittent measurements.

3.2. ^{14}C-IAP Autoradiography in Mouse

3.2.1. Preparation of ^{14}C-IAP Aliquots

1. Each animal is injected 1.5 μCi of ^{14}C-IAP in 150 μL normal saline; however, 2.0 μCi of ^{14}C-IAP in 200 μL normal saline is prepared.
2. On the day before the injection, transfer 20 μL of IAP form the stock vial to a clean and sterile Eppendorf tube for each mouse.
3. On the day of the injection, dissolve each dose of IAP into 200 μL of sterile normal saline.

3.2.2. Injection of ^{14}C-IAP Solution

1. Prepare a radiation waste bag and a sharp container for C-14.
2. Grind half ice bucket of dry ice, and precool 75–100 mL isopentane (also known as 2-methylbutane) in a baker sitting in dry ice.
3. Perform animal surgery to occlude MCA.
4. Thirty minutes after the arterial occlusion, with the suture in place and animal under anesthesia, intraperitoneally inject 150 μL of ^{14}C-IAP in saline using a insulin syringe.

3.2.3. Collection of Arterial Blood and the Brain

1. At about 90 s after IAP injection, expose the heart.
2. At exactly 120 s after IAP injection, insert a large 22-gauge needle connected to a 1-mL syringe into the left ventricle, and remove as much blood as possible.
3. Transfer the blood into an Eppendorf tube, freeze the tube in dry ice power, and then store the tube in –80-C freezer.
4. Also at exactly 120 s after IAP injection, another person decapitates the mouse and harvests the brain.
5. Drop the brain into precooled isopentane for 10–15 s, wrap the brain with foil, and store the brain at –80°C.

3.2.4. Measurement of the Blood Radiation Level

1. Thaw the blood sample, and centrifuge for 10 s at 14,000 rpm. Repeat two times.
2. Transfer 20 μL of lysed blood into scintillation tube in duplicate.
3. Bleach the blood samples using 350 μL of mixture of Soluene-350 and ethyl alcohol, and then adding 150 μL of 30% hydrogen peroxide per sample. Wait for 30 min for complete reaction.
4. Seal the vials and incubate the vials at 60°C for 30 min.
5. Cool to room temperature and add 5.25 mL scintillation fluid.
6. Count the radiation activities using scintillation counter.
7. Divide the DPMs by the number of μg of blood.

3.2.5. Brain Section Autoradiogram

1. Cut the brains at 20 μm thick.
2. Choose five sections every 1 mm throughout the length of the brains.
3. Arrange the sections in order in the developing cassette, along with Carbon-14 standard microscale.
4. In dark room, place film on the top of the sections. Make sure that the emulsion faces the slides.
5. Close the cassette. Expose for 4–6 weeks.
6. Develop the film. Scan the films and save them as TIFF file (Fig. 2).

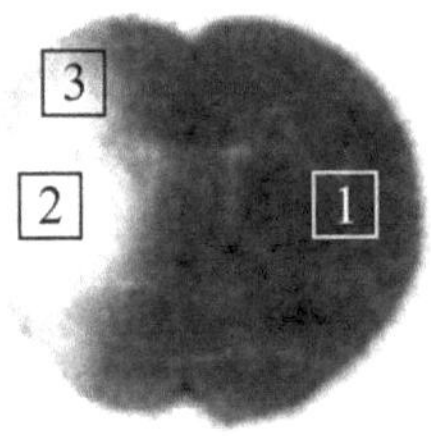

Fig. 2. A representative C-14 IAP autoradiogram and rCBF values in mouse after MCAO. Region 1 refers to the contralateral hemisphere while region 2 refers to the ischemic core and region 3 the penumbra.

3.2.6. Calculate the Rate of Cerebral Blood Flow

1. Open saved files in MCID software.
2. Measure the optical density of each panel on the C-14 microscale. Establish a standard curve.
3. Measure the optical density of the regions of interest on brain autoradiogram. Convert the optical densities to radiation activities using the standard curve.
4. Calculate the rate of CBF of interested areas using blood radiation activity as a reference (7, 9). The rCBF is present as milliliters of blood flow per 100 g of brain tissue per minute.

3.3. Laser Speckle Imaging in Mouse

3.3.1. Surgical Procedure

1. Anesthetize mouse, and place the animal into the stereotactic frame.
2. Cut the scalp along the midline, and expose the skull. Stop bleeding using hydrogen peroxide when necessary. Prevent the skull from drying with a thin layer of mineral oil.
3. Set the laser diode aside the mouse, and illuminate the skull in about 40° angle.
4. Position the CCD camera directly above the skull to capture images.

3.3.2. Speckle Contrast Calculation

1. Capture ten consecutive raw speckle images at 15 Hz using a window of 7×7 pixels.
2. Calculate the speckle contrast using the equation in refs. (11, 15). Speckle contrast is a parameter showing the speckle visibility related to the velocity of the scattering particles. It is calculated by dividing the mean pixel intensity in a small region of the image by the standard deviation of pixel intensities.

3.3.3. Calculation of CBF Arbitrary Unit

1. Extract τ value form the speckle contrast.
2. Calculate the value of $1/\tau$, which is the index of CBF value.

4. Variation of rCBF Measurement

In laser Doppler flowmetry, the laser light is susceptible to interference. To get consistent readings in the repeated measurements of rCBF, the probe should be vertical to the subject surface and it should be closely in contact with the surface. For continuous measurement of rCBF with laser Doppler flowmetry, the probe needs to be fixed in place using dental cement. The animal should be placed on a bed that is lifted from the table and has a hole under the head so that the probe can go through the hole freely.

For ^{14}C-IAP autoradiography, the accuracy of rCBF values can be compromised by the postmortem diffusion of ^{14}C-IAP in brain tissue (8, 15). To prevent this, it is critical to harvest brain tissues at exact 120 s after the injection, to freeze the brain immediately after harvest, and to cut and expose the brain early.

In addition to these three methods for CBF measurement, magnetic resonance image (MRI) is also a valuable approach for assessing rCBF changes during focal cerebral ischemia (16, 17).

References

1. Dirnagl U, Kaplan B, Jacewicz M, Pulsinelli W (1989) Continuous measurement of cerebral cortical blood flow by laser-Doppler flowmetry in a rat stroke model. J Cereb Blood Flow Metab 9:589–596
2. Haberl RL, Heizer ML, Ellis EF (1989) Laser-Doppler assessment of brain microcirculation: effect of local alterations. Am J Physiol 256: H1255–H1260
3. Haberl RL, Heizer ML, Marmarou A, Ellis EF (1989) Laser-Doppler assessment of brain microcirculation: effect of systemic alterations. Am J Physiol 256:H1247–H1254
4. Soriano MA, Sanz O, Ferrer I, Planas AM (1997) Cortical infarct volume is dependent on the ischemic reduction of perifocal cerebral blood flow in a three-vessel intraluminal MCA occlusion/reperfusion model in the rat. Brain Res 747:273–278
5. Zhang F, Wang S, Luo Y, Ji X, Nemoto EM, Chen J (2009) When hypothermia meets hypotension and hyperglycemia: the diverse effects of adenosine 5'-monophosphate on cerebral ischemia in rats. J Cereb Blood Flow Metab 29:1022–1034
6. Reith W, Forsting M, Weber J, Stingele R, Hacke W, Sartor K (1994) Laser Doppler flowmetry of focal ischaemia and reperfusion in deep brain structures in rats. Acta Neurochir (Wien) 131:151–156
7. Van Uitert RL, Sage JI, Levy DE, Duffy TE (1981) Comparison of radio-labeled butanol and iodoantipyrine as cerebral blood flow markers. Brain Res 222:365–372
8. Sakurada O, Kennedy C, Jehle J, Brown JD, Carbin GL, Sokoloff L (1978) Measurement of local cerebral blood flow with iodo [14C] antipyrine. Am J Physiol Heart Circ Physiol 234:H59–H66
9. Chen J, Graham SH, Zhu RL, Simon RP (1996) Stress proteins and tolerance to focal cerebral ischemia. J Cereb Blood Flow Metab 16:566–577
10. Dunn AK, Bolay H, Moskowitz MA, Boas DA (2001) Dynamic imaging of cerebral blood flow using laser speckle. J Cereb Blood Flow Metab 21:195–201
11. Ayata C, Dunn AK, Gursoy OY, Huang Z, Boas DA, Moskowitz MA (2004) Laser speckle flowmetry for the study of cerebrovascular physiology in normal and ischemic mouse cortex. J Cereb Blood Flow Metab 24:744–755
12. Atochin DN et al (2007) The phosphorylation state of eNOS modulates vascular reactivity and outcome of cerebral ischemia in vivo. J Clin Invest 117:1961–1967
13. Briers JD (2001) Laser Doppler, speckle and related techniques for blood perfusion mapping and imaging. Physiol Meas 22:R35–R66

14. Zhou C et al (2008) Acute functional recovery of cerebral blood flow after forebrain ischemia in rat. J Cereb Blood Flow Metab 28: 1275–1284
15. Zhao L, Nowak TS (2006) CBF changes associated with focal ischemic preconditioning in the spontaneously hypertensive rat. J Cereb Blood Flow Metab 26:1128–1140
16. van Dorsten FA, Olah L, Schwindt W, Grune M, Uhlenkuken U, Pillekamp F, Hossmann KA, Hoehn M (2002) Dynamic changes of ADC, perfusion, and NMR relaxation parameters in transient focal ischemia of rat brain. Magn Reson Med 47:97–104
17. Spaanderman ME, Aardenburg R, Ekhart TH, van Eyndhoven HW, van der Heijden OW, van Eyck J, de Leeuw PW, Peeters LL (2001) Non-pregnant circulatory volume status predicts subsequent pregnancy outcome in normotensive thrombophilic formerly preeclamptic women. Eur J Obstet Gynecol Reprod Biol 95: 218–221

Chapter 12

Biochemical and Molecular Biological Assessments of Focal Cerebral Ischemia: mRNA and MicroRNA

Yu Gan and R. Anne Stetler

Abstract

Evaluation of messenger RNA (mRNA) expression has long been used as evidence demonstrating alterations in gene expression at the transcriptional level following global ischemia. Given the complexity of focal cerebral ischemic injury, in situ analysis of the localization and coexpression of specific RNAs is necessary to tease apart differentially susceptible populations of neurons to ischemic injury. Historically, RNA has been long characterized by its role as the "messenger in the middle" between DNA and proteins, where it was seen as primarily a stopping ground and potential added layer of control of protein expression. However, in recent years, RNA has begun to be understood as much more than simply a messenger. The discovery of microRNA (miRNA) and other small noncoding RNA species has led to a reemergence of RNA-based applications, particularly in the investigation of pathological conditions, such as focal cerebral ischemia. One of the strongest of these techniques lies in the modification of polymerase chain reaction (PCR). In this chapter, we present the basic techniques of reverse transcription-PCR for mRNA and miRNA detection and DIG-labeled oligonucleotide in situ hybridization in the context of focal cerebral injury.

Key words: mRNA, RT-PCR, miRNA, DIG, Focal ischemia

1. Introduction to RNA Assessments

The intricate and precise control of molecular mechanisms contributing to the pathophysiology of focal cerebral ischemia has led to the application of a variety of tools to monitor the presence of particular gene products. While RNA assessments have been lagging, new evidence to support a more functional role of RNA in cell death and survival has sparked new interest in the evaluation of RNA species. Currently, the procedures themselves are fairly straightforward, and their application can be used to address a variety of questions.

Jun Chen et al. (eds.), *Animal Models of Acute Neurological Injuries II*, Springer Protocols Handbooks,
DOI 10.1007/978-1-61779-782-8_12,

Focal cerebral ischemia specifically induces a wide variety of physiological states over distinct time periods following reperfusion. These physiological states are reflected by and indeed controlled by cellular molecular responses that underlie cell survival or demise. Excitotoxicity, inflammation, apoptosis, and survival signaling are all targets of specific molecular questions and possible therapeutics in the treatment of stroke. It has widely been established that stroke is a potent regulator of gene expression. Protein expression can be regulated at the level of messenger RNA (mRNA) transcription after minutes to days of focal cerebral ischemia. For example, the expression of mRNAs encoding immediate early genes, mainly including various transcription factors, is induced rapidly (1), whereas delayed alterations are observed for other mRNAs, such as those encoding proteins related to adhesion molecules (ICAM-1 and ELAM-1), apoptosis-associated genes (p53, Bax, and caspase-3), DNA damage-inducible and DNA repair genes (GADD45), growth factors and their receptors (VEGF, bFGF, BDNF, and VEGFR), and heat-shock proteins (HSP70 and HSP27) (2, 3). The alteration in mRNA expression can extend well beyond the ischemic zone, reaching even to the ipsilateral hemisphere. In addition, the same mRNA induced in different regions or different cells may have different implications for pathophysiological process at the same time period (2). Thus, it is important to study the spatial and cellular basis for the changes of mRNA expression in the context of the ischemic injury state. The study of any of these molecules necessitates reliable molecular assessments. In this chapter, we discuss the standard assessment of RNA, primarily in the context of mRNA.

Northern blotting is considered the "gold standard" of RNA detection (4). Gel-separated RNA fragments are transferred to a membrane, and then detected using a hybridization probe with a base sequence complementary to the target mRNA. Northern blotting has an advantage in that mRNA size and small changes in gene expression can be detected with fairly high accuracy. However, it requires a long period of time and large amounts of RNA samples. Compared to Northern blotting, real-time reverse transcription (RT)-PCR is a more sensitive method for mRNA detection (5). Traditionally, real-time RT-PCR involves two steps. RNA is first reverse transcribed into cDNA using a reverse transcriptase. Subsequently, the resulting cDNA serves as templates for real-time polymerase chain reaction (PCR) amplification by addition of specific primers. The products in real-time PCR can be detected by nonspecific fluorescent dyes, such as SYBR Green, that intercalate with any double-stranded DNA. Another common method is using fluorescent reporter-labeled specific DNA probes, such as TaqMan probes. The reporter dye allows detection only after the probe specifically anneals with target sequence of the PCR product. Application of these probes increases the specificity of real-time RT-PCR detection.

Although much of the focus in the past has been on mRNA, noncoding RNAs make up a major percentage of total cellular RNA. Recently, microRNAs (miRNAs) have been discovered as a novel family of noncoding short RNA molecules. These short noncoding RNA can hybridize to partially complementary binding sequences within the 3′ untranslated regions (3′ UTRs) of target mRNAs, resulting in negative regulation of protein expression by either cleavage or translational repression of mRNA in a sequence-specific manner (6). It is now evident that miRNAs are able to regulate at least one-third of the human genes (7) and play a critical role in a variety of normal physiological and pathological processes. Studies have demonstrated that cerebral miRNA expression pattern was altered after focal cerebral ischemia (8, 9). Another study identified an individual miRNA, miR-497, as a critical player in the regulation of ischemic neuronal death in the focal cerebral ischemia model (10). These findings suggest miRNAs as potential candidates for possible biomarkers or therapeutic targets. miRNA expression is often investigated with the use of real-time RT-PCR. Unlike mRNA, miRNAs are only ~21 nucleotides long. Direct application of conventional RT-PCR protocols to miRNA detection is limited by the similar sizes of mature miRNAs and standard PCR primers. Two strategies developed to solve this problem are (1) application of stem-loop primers and (2) application of poly(A) polymerase for template-independent addition of adenosine monophosphate to the 3′ end of RNA. Current real-time RT-PCR technique provides a highly sensitive and specific method not only for individual miRNA detection, but also for high-throughput miRNA expression profiling.

Besides northern blot and real-time RT-PCR, in situ hybridization is another widely used method for mRNA detection (11). The basic technique utilizes the fact that DNA and RNA are hydrogen bonded to mRNA containing complementary sequences. By using labeled DNA or RNA probes of sufficient length (approximately 50–300 bp), detection of mRNA sequences in tissue slides is possible. Although it is not as sensitive as RT-PCR, in situ hybridization is a powerful technique that allows one to study the macroscopic distribution and cellular localization of mRNA sequences in a heterogeneous cell population. This is particularly useful in studying the spatial, temporal, and cellular patterns of mRNA expression after focal ischemia in brain.

In this chapter, we outline protocols geared toward real-time RT-PCR for mRNA detection, SYBR Green-based real-time RT-PCR assessment for individual miRNA expression, and DIG-labeled oligonucleotide in situ hybridization for RNA detection within tissue sections.

2. Real-Time RT-PCR for mRNA Detection

2.1. Material

2.1.1. Reagent

1. RNeasy Lipid Tissue Mini Kit (Qiagen, Valencia, CA).
2. SuperScriptIII First-Strand Synthesis System for RT-PCR (Invitrogen, Grand Island, NY).
3. SYBR® Green PCR Master Mix (Applied Biosystems/Life Technologies, Carlsbad, CA).

2.1.2. Tools

1. BioRad Real-time PCR machine.

2.2. Procedures

2.2.1. RNA Preparation

RNA isolation from brain is carried out with the RNeasy Lipid Tissue Mini Kit. The following procedure is based on Qiagen's protocol.

1. Place the tissue in a suitably sized vessel containing 1 mL QIAzol Lysis Reagent, and pipette up/down six times with 1-mL tip to break tissue into small pieces.
2. Pass six times through 3-mL syringe with twenty-two 1/2-gauge needle.
3. Place the tube containing the homogenate on the benchtop at room temperature for 5 min.
4. Add 200 μL chloroform. Securely cap the tube containing the homogenate, and shake it vigorously for 15 s.
5. Place the tube containing the homogenate on the benchtop at room temperature for 2–3 min.
6. Centrifuge at 12,000 × *g* for 15 min at 4°C.
7. Transfer the upper aqueous phase to a new tube. Add equal volume of 70% ethanol, and mix thoroughly by vortexing. Do not centrifuge. Proceed immediately to the next step.
8. Transfer the sample to an RNeasy Mini spin column placed in a 2-mL collection tube. Close the lid gently, and centrifuge for 15 s at 8,000 × *g* at room temperature. Discard the flow through.
9. Add 700 μL Buffer RW1 to the RNeasy spin column. Close the lid, and centrifuge for 15 s at 8,000 × *g* to wash the membrane. Discard the flow through.
10. Add 500 μL Buffer RPE to the RNeasy spin column. Close the lid, and centrifuge for 15 s at 8,000 × *g* to wash the membrane. Discard the flow through.
11. Add 500 μL Buffer RPE to the RNeasy spin column. Close the lid gently, and centrifuge for 2 min at 8,000 × *g* to wash the membrane.
12. Place the RNeasy spin column in a new 2-mL collection tube, and discard the old collection tube with the flow through. Close the lid gently, and centrifuge at full speed for 1 min.

13. Place the RNeasy spin column in a new 1.5-mL collection tube. Add 30–50 μL RNase-free water directly to the spin column membrane. Close the lid gently. Centrifuge for 1 min at 8,000 × *g* to elute the RNA.

2.2.2. Reverse Transcription

Reverse transcription is carried out with the SuperScriptIII First-Strand Synthesis System for RT-PCR. The following procedure is based on Invitrogen's protocol.

1. Prepare the following RNA/primer mixture in each tube:

Total RNA	5 μg
Oligo(dT)20 (50 μM)	1 μL
10 mM dNTP mix	1 μL
DEPC H_2O	To 10 μL

2. Incubate the samples at 65°C for 5 min and then on ice for at least 1 min.
3. Prepare reaction master mixture. For each reaction combine:

10× RT buffer	2 μL
25 mM $MgCl_2$	4 μL
0.1 M DTT	2 μL
RNAaseOUT	1 μL
SuperScriptIII RT	1 μL

4. Add the reaction mixture to the RNA/primer mixture, mix gently, and then incubate at 50°C for 50 min.
5. Terminate the reaction at 85°C for 5 min, and then chill on ice.
6. Add 1 μL RNase H and incubate at 37°C for 20 min.
7. Store the first-strand cDNA at −20°C until use for real-time PCR.

2.2.3. Real-Time PCR

1. Normalize the primer concentrations and mix gene-specific forward and reverse primer pair. Each primer (forward or reverse) concentration in the mixture is 5 pmol/μL.
2. Set up the experiment and the following PCR program on BioRad. Do three duplicated reactions for each diluted templates. Do not click on the dissociation protocol if you want to check the PCR result by agarose gel firstly.

50°C	2 min	1 cycle
95°C	10 min	1 cycle
95°C	15 s	40 cycles

60°C	30 s	40 cycles
72°C	30 s	40 cycles
72°C	10 min	1 cycle

3. Prepare the real-time PCR reaction mixture as follows:

SYBR® Green PCR Master Mix (2×)	25 μL
cDNA	1 μL
Primer pair mix	2 μL
H_2O	22 μL

4. Run RT-PCR as soon as possible.
5. After PCR is finished, remove the tubes from the machine. The PCR specificity is examined by 3% agarose gel using 5 μL from each reaction.
6. Put the tubes back in PCR machine and perform dissociation curve analysis.
7. Analyze the real-time PCR result with the software.
   (a) Check dissociation curve to make sure that every reaction just has a single peak.
   (b) Set up proper threshold to get the cycle threshold (C_t) values in the range of exponential amplification (usually, 0.3 works fine).
   (c) Export (C_t) values to Excel file.
   (d) Calculate the fold change of each gene for the two samples and normalize them by using the spiking gene as reference.

3. Northern Blot Analysis for mRNA

3.1. Material

3.1.1. Reagents

1. RNeasy Lipid Tissue Mini Kit (Qiagen).
2. GE Nylon, Reprobing Charged Transfer Membrane (GE Life Science).
3. Megaprime DNA Labeling System (GE Life Science).
4. Illustra ProbeQuant G-50 Micro Columns (GE Life Science).

3.1.2. Solutions

1. 10× running buffer: 0.2 M MOPS, 0.05 M NaAc, and 0.005 M EDTA; adjust to pH 7.0 with NaOH.
2. Sample denaturation buffer (100 μL): 64.6 μL formamide, 22.6 μL formaldehyde, and 13 μL 10× running buffer.
3. Loading dye mix: 50% glycerol, 0.3% xylene cyanol, 0.3% bromophenol blue, and 1 mM EDTA.

4. 20× SSPE: 3 M NaCl, 0.25 M NaH_2PO_4, and 0.02 M EDTA; adjust to pH 7.4.
5. Prehybridization solution: 2.5 mL formamide, 1 mL 5× prehybridization stock, 1.4 mL H_2O, 0.292 g NaCl, and 0.1 mL sperm DNA.
6. 5× Prehybridization stock: 0.25 M Tris–HCl, pH 7.5, 0.5% sodium pyrophosphate, 1% polyvinylpyrolidone, 1% bovine serum albumin, 1% Ficoll, and 5% SDS.
7. Wash buffer: 1% SDS + 0.1% 20× SSPE.

3.2. Procedures

The protocol is divided into three sections.

3.2.1. Electrophoresis of RNA Samples Under Denaturing Conditions

1. Prepare gel:
   (a) Cast a 0.7–1.0% agarose gel, which requires at least 100 mL of agarose solution. Ethidium bromide is not added to the gel but rather to the sample buffer. Following is an example for casting a 1% agarose gel.
   (b) Dissolve 1 g agarose in 85 mL water by microwave and cool to 60°C in a water bath.
   (c) When the flask has cooled to 60°C, place in a fume hood and add 10 mL of 10× running buffer and 5 mL 35% formaldehyde.
   (d) Pour the gel and allow it to set. Remove the comb, place the gel in the gel tank, and add sufficient 1× running buffer to cover to a depth of ~1 mm.
2. Prepare RNA sample:

   Total RNA is isolated by using RNeasy Lipid Tissue Mini Kit as described in Sect. 2.2.1. RNA samples are prepared for loading as follows.
   (a) Adjust the volume of each RNA sample to 6 μL with water. Determine the amount of RNA to be loaded in each well (e.g., 2 μg of RNA).
   (b) Add 15 μL freshly prepared sample denaturation buffer and mix by vortexing.
   (c) Centrifuge briefly to collect liquid, and incubate for 5 min at 65°C.
   (d) Cool on ice for 2 min, and then add 2 μL loading dye mix.
3. Run gel:

   Run the gel in 1× running buffer at between 100 and 200 V for several hours, until the xylene cyanol dye front has migrated 3–4 cm into the gel and the bromophenol blue is about two-thirds down the gel. When the RNA has run an appropriate distance, photograph the gel.

3.2.2. Transfer of the mRNA from the Gel to a Nylon Membrane by Upward Capillary Transfer

1. Prepare gel for transfer:

   Place the gel in an RNase-free dish and rinse with sufficient water 4× for 20 min.

2. Transfer RNA from gel to membrane:
   (a) Fill the glass dish with enough 20× SSPE.
   (b) Cut two pieces of Whatman 1MM paper, place it on the glass plate, and wet it with 20× SSPE.
   (c) Place the gel on the filter paper and use a glass pipette to roll out any air bubbles.
   (d) Cut four strips of plastic wrap and place over the edges of the gel.
   (e) Cut one piece of nylon membrane to roughly the same dimensions of the gel and wetted in water. Put the wet membrane on the top of the gel. Smooth out any bubbles between the gel and membrane.
   (f) Wet the surface of the membrane with 20× SSPE. Cut five sheets of Whatman 3MM paper to the same size as membrane and place on top of the membrane.
   (g) Put paper towels on top of the Whatman 3MM paper to a height of ~8 cm.
   (h) Cover the stack with the gel casting tray and add weight to compress the stack.
   (i) Leave overnight.

3. Prepare membrane for hybridization:

   After overnight transfer, remove paper towels and filter papers and recover the membrane. Rinse the membrane in 5× SSPE to remove residual agarose. Bake the membrane at 80°C for 2 h.

3.2.3. Hybridization Analysis of the RNA of Interest Using a Labeled DNA Probe

1. Prepare labeled DNA primer:

   The probe labeled with Ridiprimer DNA labeling system based on the random-primed labeling method:
   (a) Dilute the interested cDNA to a concentration of 2.5–25 ng in sterile water.
   (b) Place 25 ng of template DNA into a microcentrifuge tube and to it add 5 μL of primers and the appropriate volume of water to give a total volume of 50 μL in the final reaction.
   (c) Denature by heating to 95–100°C for 5 min in a boiling water bath.
   (d) Centrifuge briefly and keep the tube at room temperature.
   (e) Add 10 μL of the labeling buffer followed by 5 μL of Redivue [^{32}P] dCTP and 2 μL enzyme. Mix by gently pipetting up and down and spin for a few seconds.

(f) Incubate at 37°C for 10 min. Stop the reaction by the addition of 5 μL of 0.2 M EDTA.

(g) Sephadex G-50 resin is resuspended and excess storage buffer is removed by centrifugation.

(h) Place the column in a 1.5-mL tube and slowly apply 50 μL of the sample to the top center of the resin.

(i) Spin the column at 3,000 × *g* for 2 min. The purified sample is collected in the bottom of the support tube.

2. Hybridization:

(a) Wet the membrane in the 5× SSPE. And then, prehybridize membrane in a minimum volume of prehybridization solution at 42°C for 6 h.

(b) Denature the labeled primer by heating to 95–100°C for 5 min, and then chill on ice.

(c) Add the probe into the prehybridization solution (or remove prehybridization solution, and then add prehybridization solution and probe).

(d) Continue to incubate with rotation overnight at 42°C.

3. Autoradiography:

(a) Wash the membrane with wash buffer twice for 15 min at room temperature.

(b) Wash the membrane with prewarmed wash buffer twice for 30 min at 65°C.

(c) Remove final wash solution and rinse membrane in 5× SSPE at RT.

(d) Cover membrane in UV-transparent plastic wrap. Do not allow membrane to dry out.

(e) Expose blot to a film at −20°C.

4. Real-Time RT-PCR for miRNA Detection

4.1. Material

4.1.1. Reagents

1. miRNeasy Mini Kit (Qiagen).
2. RNeasy MinElute Cleanup Kit (Qiagen).
3. miScript Reverse Transcription Kit (Qiagen).
4. miScript SYBR Green PCR Kit (Qiagen).
5. miScript Primer Assay for target RNA and endogenous reference (Qiagen).

4.1.2. Tool

1. BioRad Real-time PCR machine.

4.2. Procedure

SYBR Green-based real-time RT-PCR for miRNA analysis is performed by using Qiagen's miScript PCR system. The protocol is similar to that of mRNA RT-PCR detection but with different kits.

4.2.1. RNA Preparation

RNA isolation from brain is carried out with the miRNeasy Mini Kit combined with RNeasy MinElute Cleanup Kit for further purification of the miRNA-enriched fraction. The following procedure is based on Qiagen's protocol.

1. Remove the brains and isolate the tissue(s) of interest using sterile RNase-free equipment and instruments. Determine the amount of tissue (no more than 50 mg) and place it immediately into 700 μL QIAzol Lysis Reagent.
2. Pipette up/down 5× with 1-mL tip to break tissue into small pieces. Then, pass through 3-mL syringe with twenty-two 1/2-gauge needle for five times.
3. Place the tube on the benchtop at RT for 5 min.
4. Add 140 μL chloroform, securely cap the tube, and shake it vigorously for 15 s.
5. Place the tube at RT for another 2–3 min.
6. Centrifuge at 12,000 × *g* for 15 min at 4°C.
7. Transfer the upper aqueous phase to a new tube. Add one volume of 70% ethanol, and mix thoroughly by vortexing. Proceed immediately to the next step without centrifuge.
8. Transfer 700 μL sample to an RNeasy Mini spin column and centrifuge for 15 s at 8,000 × *g* at RT. Pipette the flow through into a new tube.
9. Add 0.65 volumes of 100% ethanol to the flow through and mix thoroughly by vortexing. Proceed immediately to the next step without centrifuge.
10. Add 700 μL of the sample to the RNeasy MinElute spin column. Centrifuge for 15 s at 8,000 × *g* at RT. Discard the flow through.
11. Add 500 μL buffer RPE into the RNeasy MinElute spin column. Centrifuge for 15 s at 8,000 × *g*.
12. Add 500 μL of 80% ethanol to the RNeasy MinElute spin column. Centrifuge for 2 min at 8,000 × *g* to dry the spin column membrane.
13. Put the RNeasy MinElute spin column in a new 2-mL tube, and centrifuge at 8,000 × *g* for 5 min.
14. Place the RNeasy MinElute spin column in a new 1.5-mL tube. Add 14 μL RNase-free water to the spin column membrane. Close the lid gently. Centrifuge for 1 min at 8,000 × *g* to elute the miRNA-enriched fraction.

4.2.2. Reverse Transcription

Reverse transcription is carried out with the miScript Reverse Transcription Kit containing a mix of poly(A) polymerase and reverse transcriptase. During this procedure, miRNAs are polyadenylated firstly, and then converted to cDNA using oligo-dT in the same reaction. The following procedure is based on Qiagen's protocol.

1. Prepare the following mix on ice in the tube:

Template RNA	1 μg
miScript RT buffer	4 μL
miScript reverse transcriptase mix	1 μL
DEPC H_2O	To 20 μL

2. Incubate the samples at 37°C for 60 min.
3. Then, incubate at 95°C for 5 min, and chill on ice immediately.
4. Store the cDNA at −20°C or keep on ice for continued real-time PCR analysis.

4.2.3. Real-Time PCR

The real-time PCR is performed by using miScript SYBR Green PCR Kit, including the miScript Universal Primer in combination with miScript Primer Assay which contains a specific primer for the target miRNA. It is necessary to choose another specific primer for amplifying an appropriate endogenous reference RNA, such as RNU6B. The following procedure is based on the Qiagen's protocol.

1. Set up the experiment and the following PCR program on BioRad. Do three duplicated reactions for each diluted templates. Do not click on the dissociation protocol if you want to check the PCR result by agarose gel firstly.

95°C	15 min	1 cycle
94°C	15 s	40 cycles
55°C	30 s	40 cycles
72°C	30 s	40 cycles

2. Prepare the real-time PCR reaction mixture as follows:

SYBR Green PCR Master Mix (2×)	25 μL
miScript Primer Assay	5 μL
miScript Universal Primer	5 μL
Template cDNA	2 μL
DEPC H_2O	18 μL

3. Run RT-PCR as soon as possible.
4. Perform dissociation curve analysis.
5. Analyze the real-time PCR result with the software.
   (a) Check dissociation curve to make sure that every reaction has a single peak.
   (b) Set up proper threshold to get the C_t values in the range of exponential amplification (usually, 0.3 works fine).
   (c) Export C_t values to Excel file.

4.3. Evaluation

A comparative method to calculate the real-time result is termed "delta-delta C_t method." This simple method is used to compare the C_t values of samples from experimental group with that from control group. Both C_t values are normalized to an appropriate endogenous reference. The fold change in the target RNA can be calculated by the following equation:

$$\text{Fold change of target RNA} = 2^{-\Delta\Delta C_t}$$

Here,

$$\Delta\Delta C_t = (C_{t,\text{sample}} - C_{t,\text{reference}})_{\text{experiment}} - (C_{t,\text{sample}} - C_{t,\text{reference}})_{\text{control}}$$

For this calculation to be valid, the amplification efficiencies of the target RNA and the endogenous reference must be approximately equal. And the experiment conditions, such as the amplicon length and the Mg^{2+} concentration, should be optimized to make the amplification efficiency close to 100%.

5. In Situ Hybridization (DIG-Fluorescent Tag)

5.1. Materials

5.1.1. Reagents

1. DIG 3′ end labeling kit (Roche).
2. Sephadex spin column.
3. DIG antibody conjugated with fluorophore.

5.1.2. Solutions

1. 20× SSC (3 M NaCl, 0.3 M sodium citrate).
2. 50× Denhardt's solution (1% polyvinylchloride, 1% pyrroldone, 2% BSA).
3. Hybridization solution (50% deionized formamide, 0.3 M NaCl, 20 mM Tris–HCl, pH 8, 5 mM EDTA, 10 mM $NaPO_4$, pH 8, 10% Dextran sulfate, 1× Denhardts, 0.5 mg/mL yeast tRNA, and salmon sperm DNA).
4. Wash buffer: 1× SSC (150 mmol/L NaCl, 15 mmol/L Na citrate, pH 7.4).

5.2. Procedure

The protocol is divided into two sections: probe preparation and hybridization.

5.2.1. Probe Preparation (Oligonucleotide)

Potential Probes Are Screened Against a Known mRNA Target Using Online Search Tools (Avoid Homology with Other Known RNAs)

(a) Typically, to ensure stable hybridization, the probe length should be at least 40 nucleotides and contain ideally 55% guanines and cytosines (the G/C ratio).

(b) Sequences containing 40–60% G/C nucleotides are usable. However, the low-percentage probes may require decreased stringency in the posthybridization washes, and the high-percentage probes may be less specific and result in higher background.

(c) If G/C ratio is less than 55%, the posthybridization washes should be reduced in stringency.

(d) Careful attention should be paid concerning the selected sequence region. Introns need to be avoided, as well as possible splice-variant exons.

End Labeling Using Either DIG-dUTP for Short or Extended Tail Labeling or Tagged DIG-ddUTP for Single-End Labeling

(a) For attempted semiquantification of signal, in situ hybridization should ideally use only single incorporation of the labeled nucleotide (e.g., ^{35}S-ddUTP or DIG-ddUTP). However, often, the signal strength is too weak to obtain detectable results. Therefore, extended tail labeling using dUTP alone or in combination with cold dATP is often used, but further limits the reliability of semiquantification.

(b) Most end-labeling reactions are performed using commercially available kits, containing RNase-free reaction buffers, water (if necessary), cobalt chloride, and the enzyme terminal transferase (TdT).

(c) All steps should be carried out under RNase-free conditions using RNase-free pipette tips and Eppendorf tubes, as well as appropriate safety shields in the case of radioactivity.

(d) Dilute oligonucleotide to 100 pmol with RNase-free water (supplied in most kits).

(e) Add kit components as per manufacturer's instructions (usually, first combine 100–200 pmol oligo, reaction buffer, cobalt chloride, and DIG-labeled UTP. Mix well, and then add TdT).

(f) Incubate at 37°C for 30 min (or according to kit instructions).

(g) Stop the reaction by adding EDTA (pH 8.0) to a final concentration of 10 mM (can be skipped if performing probe purification below).

Probe Purification and Labeling Efficiency

(a) Although several kits state that probe purification is unnecessary, spin columns are often used to clean up the probe by removing the TdT enzyme, precluding the need for EDTA mentioned above. These columns are commonly available from commercial sources, but can also be made using Sephadex column protocols widely available. However, commercial columns save vast amounts of time and are more consistent.

(b) Using a positively charged nylon membrane, spot 1 μL of each of a series of the diluted probe (100, 31, and 10 fmol/μL), as well as a labeled control oligo (supplied in kit), cross-link the membrane, and detect using direct immunofluorescence methods below. This will yield the labeling efficiency of the reaction.

5.2.2. Hybridization

1. Immediately prior to hybridization, remove frozen sections and thaw at room temperature (minimal time possible).
2. In some cases, postfixation and acetylation in RNase-free solutions may be done, but it is not always necessary.
   (a) Postfix in 3% PFA for 5 min, followed by several washes with RNase-free PBS and alcohol dehydration (70, 90, and 100% for 2 min each).
   (b) Acetylate 2× for 5 min in 750 μL acetic anhydride added immediately in 0.1 M triethanolamine-HCl (pH 7.5).
   - This step reduces nonspecific sticking of probe to charged groups in sections and slide
3. Determine the hybridization temperature based on oligonucleotide sequence.
   (a) First, determine melting temperature of oligonucleotide:

$$T_m = 81.5 + 16.6(\log_{10}[\text{sodium ions in most stringent wash}]) + 0.41(\%G + C) - (500/\text{length of oligonucleotide})$$

   (b) Hybridization temperature is approximately 5–10°C below the melting temperature.
4. Dilute probe in hybridization cocktail.
5. Apply approximately 200 μL of the probe/cocktail solution to each slide and coverslip (parafilm), making sure to remove all bubbles.
6. Incubate overnight (18–20 h) at hybridization temperature (typically, around 42°C, based on a probe 40 nucleotides long with a 55% G/C ratio).
   (c) Place slides in a humidified chamber (layer paper towels soaked in 5× SSC and 50% formamide).

5.2.3. Posthybridization Washes

1. Wash twice for 5 min in 2× SSC and 0.1% SDS at 20°C with shaking.
2. Wash twice for 15 min in 0.5× SSC+0.1% SDS at hybridization temperature with shaking.

5.2.4. Visualization

1. Wash with PBS three times for 10 min.
2. Block tissue with 3% BSA/1% Triton X-100/PBS for 1 h at room temperature.
3. Incubate with anti-DIG conjugated with the desired fluorophore overnight at 4°C.
4. Wash slides with PBS three times for 10 min.
5. Visualize fluorescence under microscope.

6. Summary

Evaluation of RNA following focal cerebral ischemia is a still-emerging field, requiring the alteration of standard RNA techniques. Recent literature suggests a previously unrecognized function of RNA in the control of cell fate and response to injury, thus mandating the incorporation of the above-described protocols in new and interesting fashions to address the current questions.

References

1. Akins PT, Liu PK, Hsu CY (1996) Immediate early gene expression in response to cerebral ischemia. Friend or foe? Stroke 27:1682–1687
2. Sharp FR, Lu A, Tang Y, Millhorn DE (2000) Multiple molecular penumbras after focal cerebral ischemia. J Cereb Blood Flow Metab 19:1247–1255
3. Shin YJ, Choi JS, Choi JY, Cha JH, Chun MH, Lee MY (2010) Enhanced expression of vascular endothelial growth factor receptor-3 in the subventricular zone of stroke-lesioned rats. Neurosci Lett 469:194–198
4. Ding Y, Xu L, Jovanovic BD, Helenowski IB, Kelly DL, Catalona WJ, Yang XJ, Pins M, Bergan RC (2007) The methodology use to measure differential gene expression affects the outcome. J Biomol Tech 18:321–330
5. Murphy J, Bustin SA (2009) Reliability of real-time reverse-transcription PCR in clinical diagnostics: gold standard or substandard? Expert Rev Mol Diagn 9:187–197
6. Meister G (2007) miRNAs get an early start on translational silencing. Cell 131:25–28
7. Lewis BP, Burge CB, Bartel DP (2005) Conserved seed pairing, often flanked by adenosines, indicates that thousands of human genes are microRNA targets. Cell 120:15–20
8. Jeyaseelan K, Lim KY, Armugam A (2008) MicroRNA expression in the blood and brain of rats subjected to transient focal ischemia by middle cerebral artery occlusion. Stroke 39:959–966
9. Dharap A, Bowen K, Place R, Li LC, Vemuganti R (2009) Transient focal ischemia induces extensive temporal changes in rat cerebral microRNAome. J Cereb Blood Flow Metab 29:675–687
10. Yin KJ, Deng Z, Huang H, Hamblin M, Xie C, Zhang J, Chen YE (2010) miR-497 regulates neuronal death in mouse brain after transient focal cerebral ischemia. Neurobiol Dis 38:17–26
11. Kiyama H, Emson PC, Tohyama M (1990) Recent progress in the use of the technique of non-radioactive in situ hybridization histochemistry: new tools for molecular neurobiology. Neurosci Res 9:1–21

Chapter 13

Biochemical and Molecular Biological Assessments of Focal Cerebral Ischemia: Protein

R. Anne Stetler and Yu Gan

Abstract

Focal cerebral ischemia induces a complex array of cellular states, which significantly influence and ultimately are controlled by molecular changes. While certainly some of these changes occur at the nucleotide or lipid level, many processes require significant modification of protein expression and/or function. The detection of protein alterations following focal cerebral ischemia has long been studied, but—given the complexity of cell type and protein signaling—recent studies have focused more on modulated protein detection. In this chapter, we discuss the nature of focal cerebral ischemia that impacts protein detection, as well as discuss some further techniques that are often used in the field.

Key words: Protein, Subcellular fractionation, Cell-type markers, Immunohistochemistry

1. Introduction

Rodent focal cerebral ischemia models are developed to mimic human ischemic stroke. After focal cerebral ischemia, a series of cerebral events, including neural cell death, inflammation, neurogenesis, and angiogenesis, are triggered (1). These pathophysiologic processes are considered to be relevant to human ischemic stroke, and associated with the alteration of gene expression. Although mRNA is frequently studied to estimate gene regulation at the transcriptional level, it is essential to examine protein expression and modifications that typically serve as the main mediators of cellular events after focal cerebral ischemia. Evaluations of these protein alterations are helpful for understanding the molecular pathways underlying postischemic cellular events and identifying novel targets for treatment.

Jun Chen et al. (eds.), *Animal Models of Acute Neurological Injuries II*, Springer Protocols Handbooks,
DOI 10.1007/978-1-61779-782-8_13, © Springer Science+Business Media, LLC 2012

Similar to the discussion of protein expression following global cerebral ischemia, protein expression may be investigated for induction of cell death or survival programs, translocation of proteins to different subcellular compartments, or posttranslational modifications of specific proteins. Following focal cerebral ischemia, a large-scale suppression of protein synthesis in the infarct core was long described as one of the first biochemical alterations (2). However, specific protein expression has been found over the course of reperfusion—both acute and delayed—and may occur to a large extent outside of the ischemic core, extending even into the contralateral hemisphere (3). Hence, it is important to detect the protein expression over a wide time frame and a wide neural region (i.e., in the ischemic core, at the margins, and even at some distance from the ischemic region). As an example, immunoreactivity of c-Fos, an immediate early gene, was increased starting at 3 h in the affected cortex, but was rarely detected by 24 h (4, 5). In addition, it was strongly expressed in the undamaged cortex of the ipsilateral brain side (5). Thus, protein expression following focal ischemic injury cannot be studied in isolation from a spatial and temporal basis.

Many cellular events, including mitochondrial dysfunction and apoptotic neural death, are predominately associated with the subcellular translocation of certain proteins in the absence of any increased induction of gene expression. Some proteins are found to execute a different function when they translocate from one subcellular structure to another. For example, apoptosis-inducing factor (AIF) is a conserved mitochondrial protein and is targeted to the mitochondrial intermembrane space under normal condition (6). Once migrating to the nucleus, AIF is demonstrated to induce large-scale DNA fragment and apoptosis by a caspase-independent mechanism (6). Following focal cerebral ischemia, the nuclear translocation of AIF from dysfunctional mitochondria is observed in ischemic neurons, which subsequently mediates apoptotic neuron death (7). Hence, subcellular distribution should also be studied in some cases of protein assessment after focal ischemia. Subcellular fractionation followed by western blot analysis is one of the common methods for detecting the redistribution of proteins. Since mitochondria plays a critical role in mediating postischemic subcellular events (8), we describe a method for crude subcellular fractionation followed by a procedure of percoll gradient centrifuge-based mitochondria isolation. Furthermore, confocal analysis of colabeled immunohistochemistry (IHC) directed at a specific protein and an organelle-specific marker is also a useful technique in demonstrating translocation under specific cellular contexts following focal ischemia (9).

While detection of specific protein activation and expression patterns has long been the focus of IHC, a critical use of IHC in focal cerebral ischemic models lies in the colabeling of cells for

Table 1
Commonly used markers for cell phenotypes relevant to focal cerebral ischemia

Antigen	Target protein	Associated phenotype	Caveats
NeuN	Neuronal nuclear protein	Neurons	Not present in Purkinje, photoreceptor, and some retinal neurons
GFAP	Glial fibrillary acidic protein	Astrocytes	Labels glia as well as immature neural precursors
OX42	CD11b	Microglia	
Iba1	Ionized calcium-binding adaptor protein-1	Activated microglia	
MAP2	Microtubule-associated protein 2	Neurons	Cytoplasmic
NF	Neurofilament	Neurons	Particularly dense in axons
MBP	Myelin basic protein	Oligodendrocytes	
DARPP-32	Protein phosphatase 1 regulatory subunit 1B	GABAergic medium spiny neurons	
NSE	Neuron-specific enolase	Neuronal lineage cells	No longer used—found in non-neuronal lineage

identification of specific cell phenotype. The region affected by focal ischemic injury includes striatal, cortical, and hippocampal regions (as well as other areas) that are interspersed with a wide range of neurons and glia. Immunohistochemical labeling of cell-specific markers has now become standard fare when investigating cellular morphology or expression of proteins of interest. We have presented some of the more commonly used cell-specific markers in IHC in Table 1. As mentioned above, the same rationale can be used to colabel subcellular organelles to demonstrate the presence of a particular protein in specific regions of the cells. Both of these techniques should be visualized and evaluated under confocal microscopy.

As described in the previous chapter on protein detection following global ischemia (Part I, Chapters 5 and 6), the two most commonly available methods to detect protein expression are western blot and IHC. The techniques outlined in the previous chapter are also applicable under focal ischemic settings, but applied keeping in mind the differential injury distribution in focal vs. global ischemic settings. Given the complexity of focal ischemic injury, we take the opportunity here to focus on some commonly used modifications, including cell fractionation and cell-specific colabeling. In combination with the western blot and

IHC protocols outlined in Part I, Chapter 6, the following procedures will hopefully give a basic illustration on the flexible application of the basic techniques of protein detection.

2. Subcellular Fractionation

2.1. Solutions

1. MSE buffer: 225 mM mannitol, 75 mM sucrose, 1 mM EGTA, 10 mM HEPES-KOH, pH 7.4, 1 mg/mL BSA (Fraction V, Fatty-acid free).
2. 80% Percoll-MSE (200 mL): 160 mL Percoll, 225 mM mannitol, 75 mM sucrose, 1 mM EGTA, 10 mM HEPES-KOH, pH 7.4, 1 mg/mL BSA (Fraction V, Fatty-acid free).
3. 40% Percoll-MSE (200 mL): 100 mL 80% Percoll-MSE, 100 mL MSE.
4. 24% Percoll-MSE (200 mL): 60 mL 80% Percoll-MSE, 140 mL MSE.
5. 15% Percoll-MSE (200 mL): 37.5 mL 80% Percoll-MSE, 162.5 mL MSE.
6. RIPA lysis buffer: 50 mm Tris–HCl, 150 mm NaCl, 1% NP40, 0.5% deoxycholate, and 0.1% SDS.

2.2. Procedure

1. Place the fresh brain tissue in 10 mL ice-cold MSE with protein inhibitor cocktail.
2. Homogenize by hand using ten up and down strokes, centrifuge at 1,200 × *g* for 8 min, and transfer the supernatant to a fresh tube.
3. Resuspend the pellet in 10 mL MSE and repeat the above step.
4. The pellet is resuspended in RIPA buffer as crude nuclear fraction.
5. Combine the supernatants, centrifuge at 1,200 × *g* for 8 min, and transfer the supernatant to a fresh tube.
6. Centrifuge at 1,200 × *g* for 8 min to further spin down intact cells and nuclei.
7. Centrifuge at 12,500 × *g* for 10 min. The pellet is the crude heavy membrane fraction.
8. Transfer the supernatant to a fresh ultracentrifuge tube and centrifuge at 100,000 × *g* for 60 min to yield the light membrane fraction (pellet) and cytosolic fraction (supernatant).

9. Resuspend crude heavy membrane fraction in 15% Percoll (7.0 mL).
10. Add 3.5 mL 24% Percoll-MSE in two 11-mL-ultra-clear centrifuge tubes (Beckman 344059), draw 2 mL of 40% Percoll-MSE using 3-mL syringe with 19G 11/2″ needle, then insert the needle to the bottom of the tube, and add *very slowly*! Do not mix it with 24% Percoll-MSE.
11. Add the mitochondria (3.5 mL each tube) obtained from step 6 using 3-mL syringe with 19G 1″ needle on the top of 24% Percoll-MSE *very slowly*. Do not mix it with 24% Percoll-MSE.
12. Centrifuge at 50,000 × *g* for 20 min (16,600 rpm, SW41 Rotor).
13. Draw the mitochondrial fraction located at the interface between 24 and 40% layer.
14. Dilute at 1:4–5 in MSE and centrifuge at 25,000 × *g* (14,000 rpm, SW41 Rotor) for 15 min. Remove the supernatant *carefully*! Pellet is very loosely formed.
15. Resuspend the pellet in 1.5 mL MSE and centrifuge at 12,000 × *g* for 10 min. Resuspend the mitochondrial pellet in MSE buffer without EGTA, but with BSA (1 mg/mL) for mitochondrial function study, or resuspend in lysis buffer for protein extraction.

2.3. Subcellular Fraction Markers

Subcellular fractions are then subject to western blot and semiquantitative analysis for determining the subcellular distribution and translocation of a given protein. For these experiments, various subcellular markers should be used as loading control and to confirm the identity of fractions. Cytoskeletal proteins actin and tubulin are commonly used as cell cytosol markers. Chromosome structural protein histone and basal transcription factor TFIID are used for nuclear fraction while cytochrome c oxidase IV (COX IV) and manganese superoxide dismutase (MnSOD) are usually used as markers of mitochondria.

2.4. Evaluation

Western blot results can be semiquantitatively analyzed by bands densitometry. To this end, the immunoblot films with the bands of the interested protein or internal control are digitized by using a flatbed scanner. All images should be acquired at the same setting, such as 256 gray shades, a resolution of 600 dpi, and no changes to contrast and brightness. The public software ImageJ is applied to quantify the band density by using its “Gel Analysis” functions. The background correction is done using a “rolling ball” method as its default settings. The uncalibrated OD of the band of a given protein is corrected for that of internal control and the corrected ratios are then subjected to statistic analysis.

3. Colabeling Immunohistochemistry

Immunohistochemical protocols can vary significantly based on both the particular antibody as well as the target antigen. While the protocol listed below is a basic starting point in IHC, modifications that include pretreatments for antigen retrieval, different blocking solution (serum of secondary antibody host vs. BSA), and alteration of histological tissue preparation may be necessary for optimal staining. If possible, follow previously published protocols specific for the antibody or antigen. (Note: Secondary antibodies *must be labeled with different fluorophores.*)

3.1. Solutions

1. 10× Tris-buffered saline (TBS): 500 mM Tris–HCl, pH 7.4, and 1,500 mM NaCl.

3.2. Procedure

3.2.1. Tissue Preparation

1. Tissue should be processed with particular focus on frontal striatal and hippocampal regions, which should encompass both the ischemic core and surrounding penumbral regions (e.g., cortex and hippocampus).
2. If using paraffin-embedded sections, tissue should be mounted on slides and deparaffinized by immersion in four sequential baths of histology-grade xylene for 2 min each, followed by rehydration in a series of alcohol gradients (1 min in absolute ETOH, 1 min in a second bath of ETOH, 30 s in 95% ETOH, 45 s in 70% ETOH) finalized by a 1-min wash in water.
3. Free floating formalin-fixed tissue can be processed completely in wells, wherein mounting on the slides constitutes the final step after IHC is completed.
4. Fresh frozen fixed tissues cut onto glass slides may also be used, but are not optimal for maintaining morphology. However, in the case of colabeling for protein and mRNA, this technique may be desired. In this case, the tissue should be postfixed for 10 min in 4% paraformaldehyde prior to IHC, followed by several washes in PBS.

3.2.2. Permeabilization and Blocking

1. Permeabilization: Typically, incubation of sections with 0.3% Triton X-100 in TBS for 30-min shaking is sufficient. However, for membrane proteins, 0.3% saponin may give better results.
2. Blocking: Incubate sections in 3% BSA/0.3% Triton X-100 in TBS for 1 h at room temperature. Some antibodies may have cleaner results if blocking is performed using the serum of the host of the secondary antibody.

3.2.3. Coimmunochemical Detection

1. Two different methods can be used for colabeling two or more antigens: sequential or combinatorial. Sequential labeling requires a separate incubation for the first primary antibody (diluted in 1% BSA/0.3% Triton X-100 in TBS at 4°C overnight), followed by incubation with its secondary antibody (diluted in 1% BSA/0.3% Triton X-100 in TBS for 1 h at room temperature). The sections are then washed several times in 1% BSA/0.3% Triton X-100 in TBS, reblocked, incubated with the second primary antibody (diluted in 1% BSA/0.3% Triton X-100 in TBS at 4°C overnight), followed by incubation with its secondary antibody (diluted in 1% BSA/0.3% Triton X-100 in TBS for 1 h at room temperature).
2. Combinatorial IHC combines both primary antibodies diluted together in 1% BSA/0.3% Triton X-100 in TBS, incubated at 4°C overnight, followed by incubation with both secondary antibodies diluted together in 1% BSA/0.3% Triton X-100 in TBS for 1 h at room temperature. This method can be used particularly when the secondary antibodies are raised in different species.
3. Following the secondary antibody incubation, sections are rinsed several times with 1% BSA/0.3% Triton X-100 in TBS, mounted onto glass slides, and coverslipped using fluorescent mounting medium.

3.2.4. Evaluation

1. Immunofluorescence can be visualized under microscopes outfitted with the appropriate fluorescent cubes. Care should be given to the band pass to avoid bleeding of signal. For colabeling IHC, confocal microscopy should be used in order to accurately identify colocalization of signal.

4. Summary

Basic protein detection following focal cerebral ischemia is a highly flexible and widely applicable procedure to study an array of cellular processes and responses. More recently, greater attention has been paid to establishing coexpression of various proteins within the same cell, tracking the intracellular location and dynamic translocation of effector molecules, and investigating protein–protein interactions. Other more complicated permutations centered on the presented protocols in this chapter and in Part I, Sect. 2 are quickly advancing our understanding of cerebral ischemic injury at the molecular level.

References

1. Carmichael ST (2003) Gene expression changes after focal stroke, traumatic brain and spinal cord injuries. Curr Opin Neurol 16:699–704
2. Sharp FR, Lu A, Tang Y, Millhorn DE (2000) Multiple molecular penumbras after focal cerebral ischemia. J Cereb Blood Flow Metab 19:1247–1255
3. Föcking M, Besselmann M, Trapp T (2006) Proteomics of experimental stroke in mice. Acta Neurobiol Exp (Wars) 66:273–278
4. Akaji K, Suga S, Fujino T, Mayanagi K, Inamasu J, Horiguchi T, Sato S, Kawase T (2003) Effect of intra-ischemic hypothermia on the expression of c-Fos and c-Jun, and DNA binding activity of AP-1 after focal cerebral ischemia in rat brain. Brain Res 975:149–157
5. Sanz O, Estrada A, Ferrer I, Planas AM (1997) Differential cellular distribution and dynamics of Hsp70, cyclooxygenase-2, and c-Fos in the rat brain after transient focal ischemia or kainic acid. Neuroscience 80:221–232
6. Susin SA, Lorenzo HK, Zamzami N, Marzo I, Snow BE, Brothers GM, Mangion J, Jacotot E, Costantini P, Loeffler M, Larochette N, Goodlett DR, Aebersold R, Siderovski DP, Penninger JM, Kroemer G (1999) Molecular characterization of mitochondrial apoptosis-inducing factor. Nature 397:441–446
7. Culmsee C, Zhu C, Landshamer S, Becattini B, Wagner E, Pellecchia M, Blomgren K, Plesnila N (2005) Apoptosis-inducing factor triggered by poly(ADP-ribose) polymerase and Bid mediates neuronal cell death after oxygen-glucose deprivation and focal cerebral ischemia. J Neurosci 25:10262–10272
8. Fiskum G, Murphy AN, Beal MF (1999) Mitochondria in neurodegeneration: acute ischemia and chronic neurodegenerative diseases. J Cereb Blood Flow Metab 19: 351–369
9. Zhao H, Yenari MA, Cheng D, Barreto-Chang OL, Sapolsky RM, Steinberg GK (2004) Bcl-2 transfection via herpes simplex virus blocks apoptosis-inducing factor translocation after focal ischemia in the rat. J Cereb Blood Flow Metab 24:681–692

Chapter 14

Assessments of Inflammation After Focal Cerebral Ischemia

Xiaoming Hu and Jun Chen

Abstract

Inflammation is an important factor in the development of ischemic neuronal damage. Assessment of CNS and peripheral inflammation after focal ischemia is important for both mechanistic study and therapeutic strategy development. One preferred approach to evaluate inflammation is to measure the expression of inflammatory mediators in the blood or organs of interest. Quantitative polymerase chain reaction and enzyme-linked immunosorbent assay (ELISA) are commonly used to quantify the inflammatory mediators at mRNA and protein levels, respectively. Another strategy to assess cerebral and peripheral inflammation is to analyze the number and morphology of inflammatory cells using different approaches, including immunohistochemistry staining, Western blot, and flow cytometry. This chapter focuses on the procedures of ELISA and flow cytometry. In addition, the commonly measured inflammatory mediators in focal ischemia model and specific markers for different inflammatory cell types are listed.

Key words: Inflammation, Cerebral ischemia, Enzyme-linked immunosorbent assay, Flow cytometry

1. Introduction

Inflammation is an important factor in the development of ischemic neuronal damage. A large number of studies have revealed that focal cerebral ischemia/reperfusion elicits robust and prolonged cerebral inflammation, which is characterized by the accumulation of inflammatory cells (1, 2) and subsequent production of inflammatory mediators (3, 4), including interleukin (IL)-1β, tumor necrosis factor (TNF)-α, IL-6, macrophage inflammatory protein (MIP)-1, monocyte chemotactic proteins (MCP), and reactive oxygen species (ROS). Neutrophils are among the first inflammatory cells that infiltrate into injured brain tissue. Macrophages move into the brain over 1–5 days and significantly accumulate in

Jun Chen et al. (eds.), *Animal Models of Acute Neurological Injuries II*, Springer Protocols Handbooks,
DOI 10.1007/978-1-61779-782-8_14, © Springer Science+Business Media, LLC 2012

the infarcted brain. T lymphocytes also migrate and accumulate to the ischemic boundary zone within 24 h after reperfusion. In addition to infiltrated inflammatory cells, the activation of residential microglia and astrocytes could be observed since 1–3 days after ischemia and last for several weeks post injury (5). Infiltrated leukocytes and activated glial cells in the brain could function in a neuroprotective manner by removing dying cells and cellular debris and stimulating repair processes after brain damage (6, 7). Prolonged activation of these inflammatory cells and excessive production of inflammatory mediators, however, exacerbate brain injury, leading to secondary expansion of ischemic infarction and deterioration of neurological outcomes (8, 9).

Systemic inflammatory responses and immune disturbance after focal ischemia have also been reported and shown to be associated with poor outcomes (10). Higher concentrations of plasma proinflammatory markers, including IL-6, TNF-α, and ICAM-1, have been observed to be associated with early neurological deterioration and poor functional outcomes in stroke patients (11, 12). The mice model of focal cerebral ischemia has been shown to result in dynamic and widespread activation of inflammatory cytokines, chemokines, and chemokine receptors in the peripheral immune system (13). Other studies demonstrated that focal cerebral ischemia induces immunodeficiency characterized by the loss of lymphocytes and a shift from T helper cell (Th)1 to Th2 cytokine production (14) in the experimental animals. Therefore, it becomes more and more important to evaluate peripheral inflammatory response in the studies of focal cerebral ischemia.

One preferred approach to evaluate inflammation is to measure the expression of inflammatory mediators in the blood or organs of interest. Quantitative polymerase chain reaction (PCR) and enzyme-linked immunosorbent assay (ELISA) are commonly used to quantify the inflammatory mediators at mRNA and protein levels, respectively. Another strategy to assess cerebral and peripheral inflammation is to analyze the number and morphology of inflammatory cells using different approaches, including immunohistochemistry staining, Western blot, and flow cytometry.

Since the techniques of PCR, Western blot, and immunochemistry staining have been described in other chapters, they are not discussed in detail here. This chapter focuses on the procedures of ELISA and flow cytometry. In addition, the commonly measured inflammatory mediators in focal ischemia model and specific markers for different inflammatory cell types are listed.

2. Enzyme-Linked Immunosorbent Assay for Inflammatory Mediator Assessment

ELISA is a plate-based assay which could be used to quantify the protein expression of inflammatory mediators in blood and tissues. These mediators include proinflammatory cytokines and chemokines. Some anti-inflammatory cytokines could also be measured. Table 1 lists inflammatory mediators which are frequently detected after focal cerebral ischemia. ELISA kits for commonly measured cytokines are commercially available.

Different from mRNA transcription, protein synthesis and accumulation after stimuli usually take several hours or days. Therefore, the samples (blood or brain tissues) for ELISA analysis of inflammatory mediators are usually collected after 6–12 h post ischemia, whereas the changes in mRNA expression of many cytokines can be detected by quantitative PCR as early as 1 h post ischemia. Since inflammatory response after ischemia happens early after injury and lasts for several days, samples for ELISAs are usually collected at multiple time points to depict the temporal profile of postischemic inflammation.

In an ELISA, the antigen of interest is immobilized to a solid surface and then bound to an enzyme-conjugated antibody. The enzyme activity is then assessed by incubation with a substrate to produce a measurable product. ELISAs can be performed with different modifications on how the antigen of interest is immobilized and detected. The most sensitive and commonly used ELISA is the so-called sandwich ELISA because the analyte is bound between two primary antibodies: the capture antibody and the detection antibody. A protocol for sandwich ELISA is described below.

2.1. Materials and Instruments

1. Standard.
2. Capture antibody.

Table 1
Inflammatory mediators frequently detected after focal cerebral ischemia

Class	Inflammatory mediators
Inflammatory cytokines	IL-1 α, β, IL-6, TNF-α, β
Chemokines	IL-8, MIP1 α, β, MCP 1,2,3, and 4, platelet factor 4, regulated on activation, normal T-expressed, and secreted (RANTES)
Anti-inflammatory cytokines	IL-4, IL-10

3. Detection antibody: Usually generated in a different host species than the capture antibody or biotinylated.
4. Secondary antibody horseradish peroxidase (HRP) or streptavidin–HRP.
5. Substrate: 3,3′,5,5′-tetramethylbenzidine (TMB).
6. Stop solution: 2 N H_2SO_4.
7. Standard diluent buffer: 1% BSA in PBS.
8. Washing buffer: 0.05% Tween 20 in PBS.
9. 96-well frame.
10. ELISA plate reader.

2.2. Procedure

2.2.1. Sample Preparation

1. Serum preparation: The whole blood is collected from animals and left undisturbed at room temperature until clot. This usually takes 15–30 min. Remove the clot by centrifuging at 1,000–2,000 × *g* for 10 min in a refrigerated centrifuge. The resulting supernatant is designated serum. Immediately aliquot the serum into tubes and store at −20°C until use.
2. Brain tissue preparation (15, 16): Animals are perfused with saline solution. Appropriate brain regions (cerebral cortex and striatum) are dissected and homogenized using Dounce homogenizer and sonicated for 10 s at 10 mV in 100 mg tissue/mL ice-cold lysis buffer containing 20 mM Tris, 0.25 M sucrose, 2 mM EDTA, 10 mM EGTA, 1% Triton X-100, 10 μg/mL pepstatin, aprotinin, and leupeptin, and 1 mM PMSF. The homogenates are centrifuged at 100,000 × *g* for 40 min and stored at −80°C.

2.2.2. Assay Procedure

1. Coat the 96-well plate with 100 μL capture antibody at room temperature overnight.
2. Wash the plate three times with washing buffer.
3. Block the plate with 300 μL standard diluent buffer. Incubate for 1 h at room temperature.
4. Wash the plate three times with washing buffer.
5. Samples are diluted in standard diluent buffer if necessary.
6. Standards are prepared using twofold serial dilution in standard diluent buffer.
7. Add 100 μL of standard, samples, or controls into the wells in duplicate. The side of the frame is tapped for gentle mixture of solutions.
8. Wells are covered and left stationary at room temperature for 2 h.
9. Wash the plate three times with washing buffer.
10. Add 100 μL of detection antibody into each well. Mix wells by gently tapping.

11. Wells are covered with plate covers and left stationary at room temperature for 2 h. Plate is washed as described above.
12. Add 100 μL of HRP-conjugated secondary antibody (or streptavidin–HRP if biotinylated antibody was used in step 10) into each well.
13. Wells are covered and left stationary at room temperature for 20 min. Plate is washed as described above.
14. Add 100 μL of substrate into each well. Incubate for 20 min at room temperature. Avoid direct light. A change in color from clear to blue will be observed.
15. Add 50 μL of stop solution into each well. Gently mix wells. A change in color from blue to yellow will be observed.
16. Determine the optical density immediately using a microplate reader set to 450 nm.

2.2.3. Data Analysis

1. Standard curve: Construct the standard curve by plotting the mean absorbance for standards on the *x*-axis and their concentrations on the *y*-axis. A best-fit curve could be drawn through the points on the graph. The data could be linearized by plotting the log of the concentrations vs. the log of the absorbance values. The best-fit line can be determined by regression analysis.
2. Concentration calculation: The concentration of the unknown samples could be determined from the standard curve. For diluted samples, the concentration read from the standard curve must be multiplied by the dilution factor.

2.3. Alternative Methods

Quantitative PCR is commonly used to measure the amount of inflammatory mediators at mRNA level. This method is discussed in related chapters.

3. Flow Cytometry for Inflammatory Cell Assessment

Flow cytometry is a technique that allows the measurement of cell properties in a flowing suspension of cells. The most common detection system in flow cytometry uses fluorescent molecules that are attached to the cells of interest. The key advantage of flow cytometry is rapid simultaneous multiparameter analysis. Moreover, the acquired data could be displayed and reanalyzed in different ways.

Flow cytometry is originally developed to study blood cells. Since only cell suspension solutions are suitable for flow cytometry analysis, cells from solid tissues have to be disaggregated. The main difficulties encountered in the application of flow cytometry to study cerebral ischemia reside in obtaining a cell suspension, by enzymatic or mechanical dissociation followed by density gradient separation, without harming antigen determinants. Nevertheless, the remarkable quantitative potential of flow cytometry technique encouraged researchers to develop reliable methods to overcome this incumbrance and extrapolate this technique to study cells from solid tissues, including brain (17).

3.1. Cell-Specific Markers for Inflammatory Cells Frequently Detected by Flow Cytometry

The activation of CNS immune cells and the infiltration of peripheral inflammatory cells are dynamic processes that start almost immediately after ischemic injury and last till several weeks after. A comprehensive characterization of immune cells in the ischemia–reperfusion stroke model could be achieved by flow cytometric analysis or immunohistochemistry staining using cell-specific markers. Table 2 lists the commonly used immune cell markers for flow cytometry studies with ischemic stroke models. Checking the dynamics of postischemic inflammatory cells is important to evaluate therapeutic intervention, especially the anti-inflammatory strategies. Figure 1 summarizes the temporal profile of the immune cells after cerebral ischemia (5, 8).

3.2. Materials and Instruments

1. Fluorochrome-labeled antibodies.
2. Percoll solution.
3. Fix solution: 2% paraformaldehyde.
4. Staining buffer: HBSS with 1% fetal bovine serum (FBS).
5. 0.5% saponin.
6. Flow cytometer.

Table 2
Cell-specific markers for immune cells frequently detected after focal cerebral ischemia

Cell	Specific marker
T lymphocyte	CD3
T helper cells	CD4
Th1	IFNγ[a]
Th2	IL-4[a]
Th17	IL17[a]
T cytotoxic cells	CD8
T regulatory cells	CD4, CD25, Foxp3[a]
Myeloid cells	
Neutrophil	CD15, CD16
Monocyte/macrophage	CD14, CD11b, CD45 (highly positive)
Microglia	CD11b, CD45 (intermediate positive)
Dendritic cell	CD11c

[a]Indicates intracellular marker

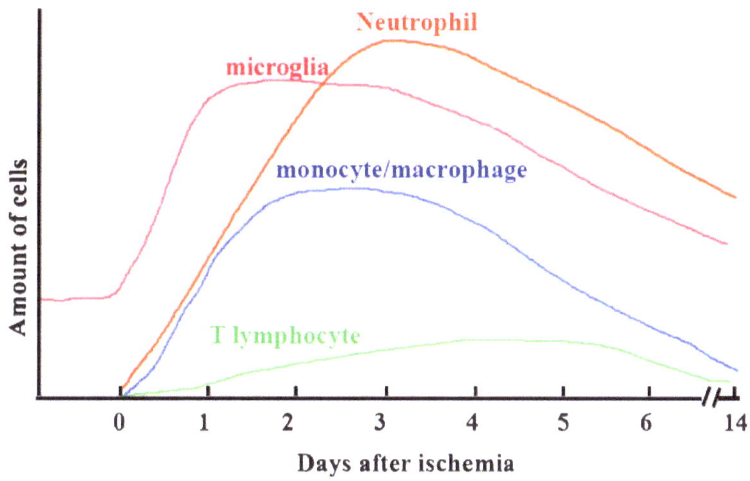

Fig. 1. Temporal profile of the immune cells after cerebral ischemia.

3.3. Procedures

3.3.1. Sample Preparation (17)

Appropriate brain regions (cerebral cortex and striatum) are dissected and divided into ipsilesional and contralesional sides. Samples are disrupted in a glass homogenizer and pressed through a 40-μm nylon cell strainer. The suspension is centrifuged at 400 × *g* for 10 min at room temperature. The pellet is resuspended in 4 mL of 30% Percoll and overlaid on the top of a gradient containing 3.5 mL of 37% and 3.5 mL of 70% Percoll solution. The gradient is centrifuged at 500 × *g* for 20 min at room temperature. Cells in the 37–70% interface (approximately 5 mL) are collected, washed once with HBSS containing 1% FBS, and resuspended for further staining procedure.

3.3.2. Cell Surface Marker Staining

Cell Marker Staining

1. Adjust cell suspension to a concentration of 1×10^6 cells/mL in ice-cold HBSS with 1% FBS.
2. Add adequate amount of fluorochrome-conjugated primary antibodies (usually, 1 μg/1×10^6 cells). Antibodies should be diluted in 1% FBS/HBSS.

   *When attempting to make simultaneous immunofluorescence measurements from several cell-bound antibodies, compensation is necessary to correct the overlap of emission spectra from antibody fluorescent labels. Choosing the fluorochromes which have the least spectral overlap reduces the need for compensation and enhances the quality of results. Generally, the further apart the maximal emission, the less spectral overlap will occur.
3. Incubate for at least 30 min at room temperature or 4°C in dark. Lower temperature needs longer incubation time.
4. Wash the cells two times by centrifugation at 400 × *g* for 5 min and resuspend them in 200 μL to 1 mL of ice-cold HBSS

with 1% FBS. For extended storage, cells could be resuspended in 2% paraformaldehyde.

5. Cells are transferred in polystyrene round-bottom falcon tubes. Keep the cells in the dark at 4°C before acquisition on the flow cytometer.

Intracellular Marker Staining

1. Adjust cell suspension to a concentration of 1×10^6 cells/mL in ice-cold HBSS with 1% FBS.
2. Stain cell surface antigen(s) as described above.
3. After staining, spin down the cells and suck off the supernatant.
4. Fix cells with 200 μL 2% paraformaldehyde for 20 min at room temperature in dark.
5. Wash cells with HBSS PBS twice. Cells can be put away for 2–3 days at 4°C in dark at this step.
6. Wash cells with 0.5% saponin. This step helps to punch holes on cell membrane.
7. Prepare antibody for intracellular marker in 0.5% saponin at optimal concentration. Incubate cells with the antibody solution at room temperature for 30 min in dark.
8. Wash cells with 0.5% saponin once. Then, wash cells with PBS once.
9. Resuspend cells in 200 μL to 1 mL of ice-cold HBSS with 1% FBS.
10. Cells are transferred into polystyrene round-bottom falcon tubes. Keep the cells in the dark at 4°C before acquisition on the flow cytometer.

3.4. Data Analysis

The flow cytometer measures the amount of fluorescent dye on cells. Other physical characteristics of cells independently from the fluorescence are also measured: (1) forward-angle light scatter (FSC) is roughly equivalent to particle size and (2) side-angle light scatter (SSC) is proportional to granularity.

Flow cytometry data can be displayed in many forms. Two most commonly used forms are (1) one-parameter histogram (Fig. 2a), which allows viewing a single parameter (fluorescence signal intensity or FSC on horizontal axis) against the number of events (on vertical axis), and (2) two-dimensional dot plots, whereby the relationship between two parameters is displayed. It could be two fluorochromes or one fluorochrome vs. a light scatter characteristic. As shown in Fig. 2b, each axis displays the signal intensity of one fluorescence. This information allows researchers to quantitatively characterize various subpopulations of cells.

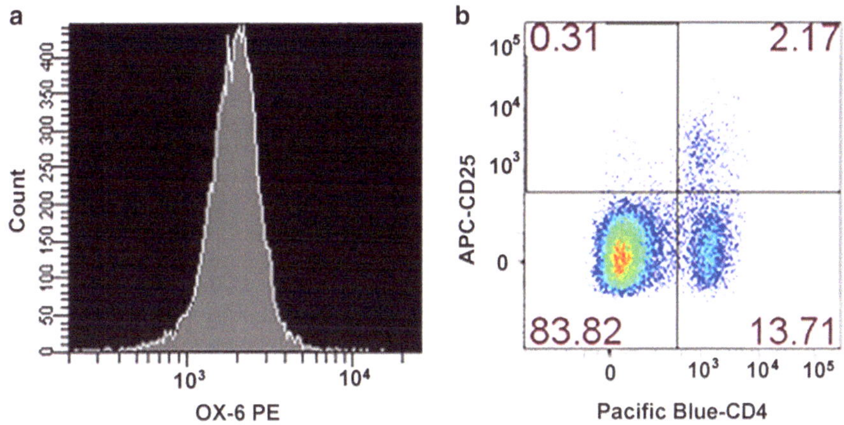

Fig. 2. Data presentation in flow cytometry: (**a**) One-parameter histogram of OX-6 PE staining. (**b**) Two-dimensional dot plot of pacific blue-CD4 (*x*-axis) and APC-CD25 (*y*-axis) staining.

For example, the upper right quadrant in Fig. 2b represents $CD4^{+}CD25^{+}$ population, which is 2.17% of total splenocytes obtained from adult mouse spleen.

3.5. Alternative Methods

1. Immunohistochemistry staining: Majority of inflammatory cells, especially the CNS residential glial cells, undergo morphological alteration upon activation, displaying intensified marker staining, enlarged cell size, cell spreading, or polarization. Immunohistochemistry staining of cell-specific markers is a commonly used method to reveal the morphology of inflammatory cells. The images of immunohistochemistry staining can be further analyzed semiquantitatively by counting the number of positively stained cells in several sampling areas. Cell quantification in immunohistochemical sections, however, introduces a number of artifacts into the experiments and could lead to biased conclusions. This problem can be at least partly overcome by manual counting in a double-blinded manner or better eradicated with the use of unbiased stereological methods (18).
2. Western blot: Western blot using antibodies against cell-specific markers followed by densitometry analysis is another semiquantitative method to quantify specific inflammatory cells. It is notable that both the cell number and the cell activation status affect the expression of cell-specific markers, which may confound the interpretation of the Western blot data.

   The technique details of immunohistochemistry staining and Western blot are discussed in related chapters.

References

1. Schwab JM, Nguyen TD, Postler E, Meyermann R, Schluesener HJ (2000) Selective accumulation of cyclooxygenase-1-expressing microglial cells/macrophages in lesions of human focal cerebral ischemia. Acta Neuropathol 99: 609–614
2. Mabuchi T, Kitagawa K, Ohtsuki T, Kuwabara K, Yagita Y, Yanagihara T, Hori M, Matsumoto M (2000) Contribution of microglia/macrophages to expansion of infarction and response of oligodendrocytes after focal cerebral ischemia in rats. Stroke 31:1735–1743
3. Nakase T, Yamazaki T, Ogura N, Suzuki A, Nagata K (2008) The impact of inflammation on the pathogenesis and prognosis of ischemic stroke. J Neurol Sci 271:104–109
4. Kunz A, Abe T, Hochrainer K, Shimamura M, Anrather J, Racchumi G, Zhou P, Iadecola C (2008) Nuclear factor-kappaB activation and postischemic inflammation are suppressed in CD36-null mice after middle cerebral artery occlusion. J Neurosci 28:1649–1658
5. Gelderblom M, Leypoldt F, Steinbach K, Behrens D, Choe CU, Siler DA, Arumugam TV, Orthey E, Gerloff C, Tolosa E et al (2009) Temporal and spatial dynamics of cerebral immune cell accumulation in stroke. Stroke 40:1849–1857
6. Lalancette-Hebert M, Gowing G, Simard A, Weng YC, Kriz J (2007) Selective ablation of proliferating microglial cells exacerbates ischemic injury in the brain. J Neurosci 27:2596–2605
7. Hanisch UK, Kettenmann H (2007) Microglia: active sensor and versatile effector cells in the normal and pathologic brain. Nat Neurosci 10:1387–1394
8. Barone FC, Feuerstein GZ (1999) Inflammatory mediators and stroke: new opportunities for novel therapeutics. J Cereb Blood Flow Metab 19:819–834
9. Dirnagl U, Iadecola C, Moskowitz MA (1999) Pathobiology of ischaemic stroke: an integrated view. Trends Neurosci 22:391–397
10. Whiteley W, Jackson C, Lewis S, Lowe G, Rumley A, Sandercock P, Wardlaw J, Dennis M, Sudlow C (2009) Inflammatory markers and poor outcome after stroke: a prospective cohort study and systematic review of interleukin-6. PLoS Med 6:e1000145
11. Castellanos M, Castillo J, Garcia MM, Leira R, Serena J, Chamorro A, Davalos A (2002) Inflammation-mediated damage in progressing lacunar infarctions: a potential therapeutic target. Stroke 33:982–987
12. Shenhar-Tsarfaty S, Ben Assayag E, Bova I, Shopin L, Fried M, Berliner S, Shapira I, Bornstein NM (2010) Interleukin-6 as an early predictor for one-year survival following an ischaemic stroke/transient ischaemic attack. Int J Stroke 5:16–20
13. Offner H, Subramanian S, Parker SM, Afentoulis ME, Vandenbark AA, Hurn PD (2006) Experimental stroke induces massive, rapid activation of the peripheral immune system. J Cereb Blood Flow Metab 26:654–665
14. Prass K, Meisel C, Hoflich C, Braun J, Halle E, Wolf T, Ruscher K, Victorov IV, Priller J, Dirnagl U et al (2003) Stroke-induced immunodeficiency promotes spontaneous bacterial infections and is mediated by sympathetic activation reversal by poststroke T helper cell type 1-like immunostimulation. J Exp Med 198: 725–736
15. Qin L, Wu X, Block ML, Liu Y, Breese GR, Hong JS, Knapp DJ, Crews FT (2007) Systemic LPS causes chronic neuroinflammation and progressive neurodegeneration. Glia 55:453–462
16. Urakubo A, Jarskog LF, Lieberman JA, Gilmore JH (2001) Prenatal exposure to maternal infection alters cytokine expression in the placenta, amniotic fluid, and fetal brain. Schizophr Res 47:27–36
17. Campanella M, Sciorati C, Tarozzo G, Beltramo M (2002) Flow cytometric analysis of inflammatory cells in ischemic rat brain. Stroke 33:586–592
18. Coggeshall RE, Lekan HA (1996) Methods for determining numbers of cells and synapses: a case for more uniform standards of review. J Comp Neurol 364:6–15

Chapter 15

Neurobehavioral Assessments of Focal Cerebral Ischemia: Sensorimotor Deficit

Yejie Shi and Dandan Sun

Abstract

Stroke induces long-term sensorimotor deficits that severely compromise the daily activity of patients. Experimental focal cerebral ischemia (FCI) models in rodents can exhibit certain aspects of clinical deficits in sensorimotor function following cerebral ischemia. Therefore, both histological changes and sensorimotor function can be evaluated in the experimental studies using FCI models. FCI is a common type of ischemic stroke. There are several rodent models for focal ischemia, including the ones with reperfusion (transient FCI) and those without reperfusion (permanent FCI), including endovascular filament middle cerebral artery occlusion (MCAO) model, transcranial MCAO model, and cerebrocortical photothrombosis model.

Various tests can be used to assess sensorimotor function in rodents. The rotarod test, which evaluates both the motor function and motor learning of the animals, is the most commonly used one. Other tests, including the cylinder test, adhesive removal test, and corner test, are important in investigating the asymmetric deficits caused by FCI. This chapter provides a brief introduction of each neurobehavioral assessment test on sensorimotor function and its procedures.

Key words: Focal cerebral ischemia, Sensorimotor deficits, Rotarod, Cylinder test, Corner test, Adhesive removal test

1. Introduction

Stroke is the leading cause of long-term disabilities affecting sensorimotor, cognitive, and emotional functions. In stroke patients, sensorimotor deficits can persist for a long period of time and severely compromise daily activity. Therefore, improvement of functional impairment and enhancement of neurological recovery following therapeutic intervention are important in evaluating efficiency for new treatments and rehabilitative strategies for stroke.

Focal cerebral ischemia (FCI) can mimic certain aspects of pathophysiology seen in clinical stroke. Transient FCI using unilateral MCA occlusion followed by reperfusion induces significant

Jun Chen et al. (eds.), *Animal Models of Acute Neurological Injuries II*, Springer Protocols Handbooks,
DOI 10.1007/978-1-61779-782-8_15, © Springer Science+Business Media, LLC 2012

sensorimotor deficits in mice (1). Several behavioral tests have been developed to examine changes in sensorimotor function in rodents. The rotarod test is the most commonly and widely used one in studying neurological behavior changes in rodents. By assessing the performance of the animals on the accelerating rotarod, researchers can evaluate both the motor function and motor learning (2). Other tests have also been developed to evaluate asymmetric limb usage, such as cylinder test, adhesive removal test, and corner test (3–5). These tests are of great importance for the ischemic model, since the deficits caused by cerebral ischemia are mainly asymmetric and can thus be readily highlighted by these tests.

This chapter describes the procedures of these four tests and lists the required materials and apparatus. All tests can be applied for both rats and mice, which are referred as "animals" throughout the chapter.

2. Rotarod Test

2.1. Introduction

The rotarod test was first developed in 1960s and has been successfully used to test motor function and motor learning in rodents (2). Briefly, animals are required to walk on a rotating cylinder. This performance requires good coordination, balance, and sensorimotor function. The animals following FCI performed significantly worse, compared to control animals (1, 6). This test can be used to assess motor function deficits in animals following FCI or evaluate neuroprotection mediated by new experimental interventions.

2.2. Materials and Apparatus

Rotarod instrument (Fig. 1, Ugo Basile, Collegeville, PA)

2.3. Procedure

1. Animals are first habituated to a stationary rod for 2 min.
2. The animals are exposed to a rotating rod. The rod is started at 2 rpm and accelerates linearly to 20 rpm within 300 s.
3. Animals are required to stay on the accelerating rod for a minimum of 30 s on two separate trials. If animals are unable to reach this criterion, the trial is repeated for a maximum of five times at 1-min intervals.
4. Animals are acclimatized to the rotarod for five trials each day for 2 days before surgery.
5. After surgery, each animal undergoes testing every day. On each testing day, animals undergo five trials on the rotarod, with an interval of 1 min. The motor function can be monitored for 1 day up to 1 month.

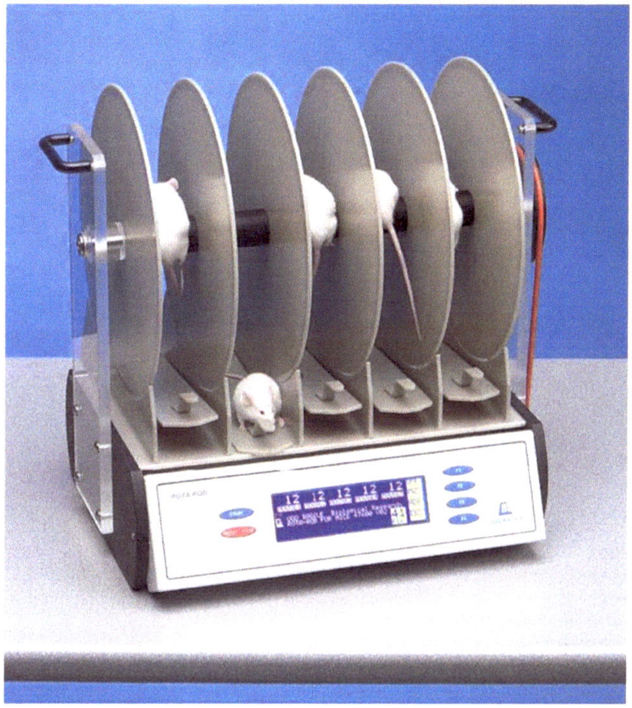

Fig. 1. Rotarod instrument for mice (Cat. # 47600, Ugo Basile, Collegeville, PA, http://www.ugobasileusa.com/Behaviour_Motory.html).

2.4. Data Analysis

The latency time(s) to fall from the rod is measured automatically with stop watches connected to detectors. The two longest fall latency values an animal could achieve during this procedure are averaged and used for data analysis. The performance of animals is compared among different groups (e.g., sham, ischemic, or drug-treated groups). Longer time spent on the rod suggests better sensorimotor function.

3. Cylinder Test

3.1. Introduction

The cylinder test was developed 10 years ago and is used to assess forelimb usage deficit and rotation asymmetry (5). After injury to the motor system, the animals tend to frequently use the forelimb ipsilaterally to the lesion hemisphere. It is usually performed in rats, although the application on mice has also been reported recently (7).

3.2. Materials and Apparatus

1. A clean, transparent plexiglass cylinder (20 cm in diameter for rats).

Fig. 2. Cylinder test. The animal is placed in a transparent cylinder, and forepaw exploration is assessed by the number of instances the animal uses each paw to touch the cylinder wall.

2. Video camera.
3. Light source.
4. Mirrors.

3.3. Procedure

1. Acclimatize animals to the testing room for at least 30 min prior to sessions.
2. Set up light source below the cylinder. Check to make sure that there is no glare on the cylinder from the light.
3. Set up the video camera. Mirrors are placed behind the cylinder to view the rearing movements. Ensure that the entire cylinder is in view before recording.
4. Place the animal in the cylinder and record movements for 5 min (Fig. 2).
5. Return the animal to home cage and clean the cylinder.
6. Care must be taken to avoid overuse of this test since habituation reduces exploratory behavior.
7. Animals are excluded if they make fewer than eight independent touches on the wall (fewer touches may not give an accurate estimate of asymmetry limb usage).

3.4. Data Analysis

1. Three categories of placements are recorded: independent ipsilateral limb usage, independent contralateral limb usage, and usage of both forelimbs (simultaneous movement).

2. The first forelimb to contact the wall during a full rear is recorded as an independent wall placement for that limb.
3. A simultaneous (both forelimbs) movement is recorded under the following conditions.
   - The first limb maintained its position and the other forelimb is placed on the wall.
   - Both forelimbs are placed simultaneously on the wall.
   - One forelimb is stationary and the other makes several movements.
   - The animal is exploring the wall laterally, alternating both limbs.
4. To repeat the test, both forelimbs of the animal have to be removed from the surface of the cylinder.
5. The numbers of quarter turns in either direction during rearing and ground are also counted from the videotapes to measure turning preference.
6. The paw preference is determined according to the formula (7):

$$[\text{contralateral}/(\text{contralateral} + \text{ipsilateral} + \text{both})] \times 100$$

4. Adhesive Removal Test

4.1. Introduction

The adhesive removal test was developed by Schallert et al. (8) to evaluate both sensorimotor and motor function in rats and later adapted to mice (3). Briefly, this test consists of placing adhesive tape strips on both sides of different parts of the animal's body. Animal performance is assessed by measuring the time needed to sense and remove the adhesives.

4.2. Materials and Apparatus

1. Transparent Perspex box (15 × 25 cm for mice).
2. Small adhesive tape strips (0.3 × 0.4 cm).
3. Timers.

4.3. Procedure

1. Acclimate the animals to the testing box for 1 min.
2. Move the animal from the testing box and apply two adhesive tape strips on the hairless parts of the forepaws with equal pressure (Fig. 3).
3. Replace the animal in the testing box and start timing.
4. Observe the animal's behavior. The animal may react to the presence of the adhesive tape strips by either shaking its paws

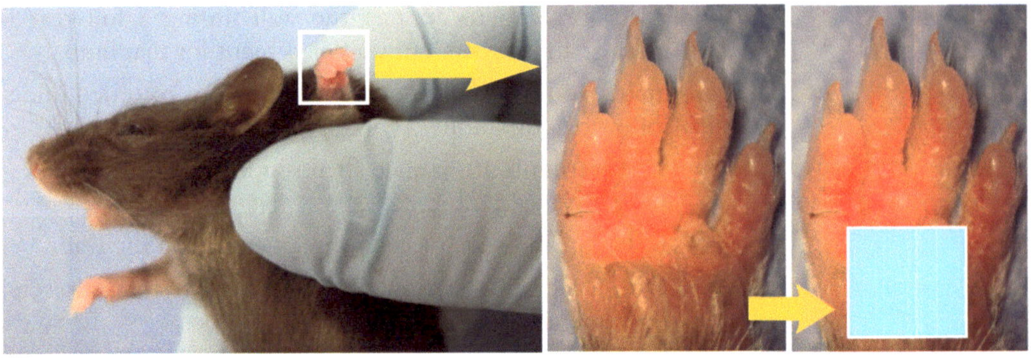

Fig. 3. Mice handling during the adhesive removal test. Grasp the tail with one hand and then handle the mouse by the scruff of its neck with the other hand. In order to immobilize the head of the mouse, all loose skin around the back of the neck must be gripped. To get all this skin in the grip, start low on the shoulders of the mouse with a thumb on one side and index finger on the other. This figure represents a zooming sequence of where to stick the adhesive tape onto the animal paw.

or directly bringing its paws to its mouth. Both behaviors represent the end of the time to contact.

5. The contact time is defined as the point that the animal reacts to the adhesive tape strips. Record the time when the animal feels the first tape as well as the time when the animal feels the second tape.
6. The removal time is defined as the point that the animal removes the tape. Record the time when the animal removes the first and second tapes.
7. After both tapes are removed, return the mouse to its cage. Repeat the procedure if needed.

4.4. Data Analysis

For each animal and trial, four parameters are to be collected: left and right time to contact and left and right time to remove the adhesive tapes. Statistical analysis can be carried out on these parameters depending on the purpose of the study.

5. Corner Test

5.1. Introduction

Corner test was initially designed by Schallert et al. (8) for rats and later adapted by Zhang et al. (4) for mice. The test is used to detect sensorimotor and postural asymmetries. The following uses mice as an example (4).

5.2. Materials and Apparatus

1. Two boards to make the corner, each with dimension of $30 \times 20 \times 1$ cm^3.

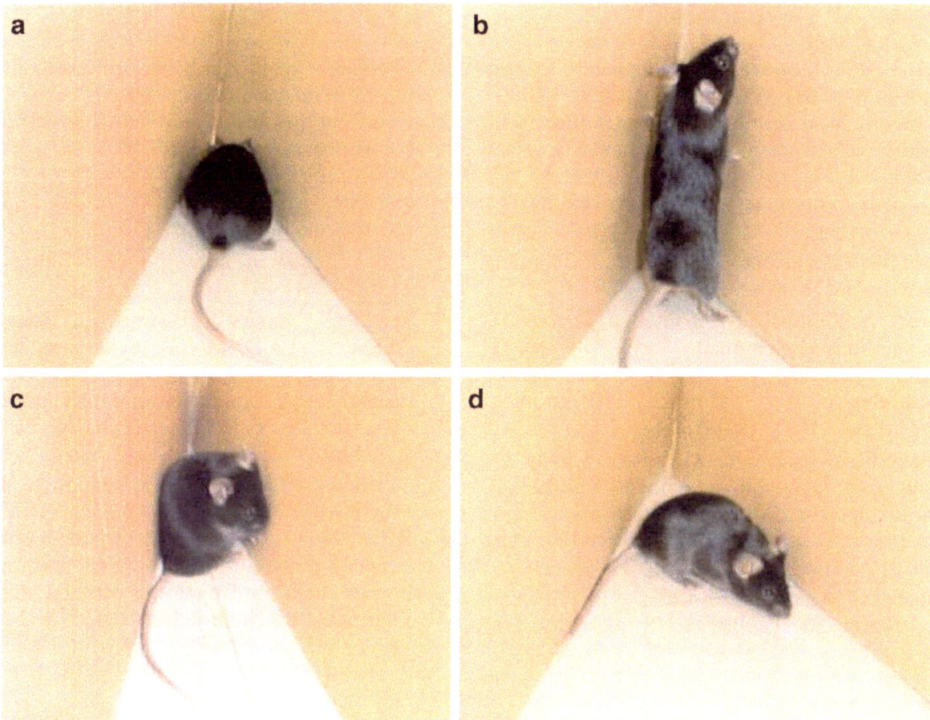

Fig. 4. The front view of the corner test from a representative mouse 90 days after MCA occlusion. The mouse walked forward until it reached the corner (**a**). The mouse then reared forward and upward, and then turned back along the right-side wall to face the open end (**b**–**d**). Figure from ref. (4) with permission.

2. The edges of the two boards are attached at a 30° angle with a small opening along the joint between the two boards to encourage entry into the corner.

5.3. Procedure

1. In the home cage, an animal is placed between two boards (Fig. 4).
2. The animal is placed between the two angled boards facing the corner and half way to the corner.
3. When entering deep into the corner, both sides of the vibrissae are stimulated together. The animal then rears forward and upward, and then turns back to face the open end.
4. The turns in one vs. the other direction are recorded from ten trials for each test.
5. Turning movements that are not part of a rearing movement are not scored.

5.4. Data Analysis

During the turning-back movement, the nonischemic mouse turns either left or right, but the ischemic mouse preferentially turns toward the nonimpaired ipsilateral side.

References

1. Gerlai R, Thibodeaux H, Palmer JT, van Lookeren CM, Van Bruggen N (2000) Transient focal cerebral ischemia induces sensorimotor deficits in mice. Behav Brain Res 108(1):63–71
2. Gerlai R, Millen KJ, Herrup K, Fabien K, Joyner AL, Roder J (1996) Impaired motor learning performance in cerebellar En-2 mutant mice. Behav Neurosci 110:126–133
3. Bouet V, Boulouard M, Toutain J, Divoux D, Bernaudin M, Schumann-Bard P, Freret T (2009) The adhesive removal test: a sensitive method to assess sensorimotor deficits in mice. Nat Protoc 4(10):1560–1564
4. Zhang L, Schallert T, Zhang ZG, Jiang Q, Arniego P, Li Q, Lu M, Chopp M (2002) A test for detecting long-term sensorimotor dysfunction in the mouse after focal cerebral ischemia. J Neurosci Methods 117(2):207–214
5. Schallert T, Fleming SM, Leasure JL, Tillerson JL, Bland ST (2000) CNS plasticity and assessment of forelimb sensorimotor outcome in unilateral rat models of stroke, cortical ablation, parkinsonism and spinal cord injury. Neuropharmacology 39(5):777–787
6. Gibson CL, Bath PM, Murphy SP (2005) G-CSF reduces infarct volume and improves functional outcome after transient focal cerebral ischemia in mice. J Cereb Blood Flow Metab 25(4):431–439
7. Freret T, Bouet V, Leconte C, Roussel S, Chazalviel L, Divoux D, Schumann-Bard P, Boulouard M (2009) Behavioral deficits after distal focal cerebral ischemia in mice: usefulness of adhesive removal test. Behav Neurosci 123(1):224–230
8. Schallert T, Upchurch M, Lobaugh N, Farrar SB, Spirduso WW, Gilliam P, Vaughn D, Wilcox RE (1982) Tactile extinction: distinguishing between sensorimotor and motor asymmetries in rats with unilateral nigrostriatal damage. Pharmacol Biochem Behav 16:455–462

Chapter 16

Neurobehavioral Assessments of Focal Cerebral Ischemia: Cognitive Deficit

Yejie Shi and Dandan Sun

Abstract

Focal cerebral ischemia (FCI) is a commonly used model for ischemic stroke. It induces rapid neuronal death in the ischemic core and causes a wide range of brain injury which affects cerebral cortex, striatum, hippocampus, and thalamus. In addition to the sensorimotor function deficits, FCI can also alter the cognitive function of animals. Several behavioral tests have been used to evaluate the learning and memory deficits in rodents following focal ischemia, including the Morris water maze, the passive avoidance test, and the elevated plus maze. This chapter introduces a brief history of each neurobehavioral assessment in animals following FCI, and the procedures of the tests.

Key words: Focal cerebral ischemia, Cognitive deficits, Learning and memory, Morris water maze, Passive avoidance, Elevated plus maze

1. Introduction

Focal cerebral ischemia (FCI) causes nonselective tissue damage in the ischemic area, including cerebral cortex, striatum, hippocampus, and thalamus. Thus, both sensorimotor and cognitive function could be impaired after FCI. In addition to the classical histological study of brain damage, the assessment of long-lasting functional outcome is highly recommended to evaluate ischemia-induced deficit and efficacy of certain new experimental therapeutic strategies.

Morris water maze (MWM) is the most frequently used test for assessing learning and memory deficit in rodents (1). MWM was described nearly 30 years ago (2) investigating spatial learning and memory in laboratory rats. It is a task used to train animals to locate a hidden platform within a pool of opaque water using spatial cues around the pool as a guide. During a number of trials, animals learn to find the platform and escape from the pool.

Jun Chen et al. (eds.), *Animal Models of Acute Neurological Injuries II*, Springer Protocols Handbooks, DOI 10.1007/978-1-61779-782-8_16, © Springer Science+Business Media, LLC 2012

It is well established that the hippocampal formation is essential for spatial learning; however, research also shows that MWM tends to depend on coordinated actions of different brain regions to produce a functionally integrated neural network (3). Several characteristics of the experimental animals may influence the experimental results. Factors, such as body weight, physical development, age, and sex, have been reported to have effects on MWM performance (1). Males tend to perform better than females, and there is significant difference in the performance in animals of different species and strains (4). One major advantage of the MWM is its relative simplicity compared to other tests. Another advantage of swimming maze over dry mazes is the reduced likelihood of odor trail interference, which can trigger the animals to employ a nonspatial navigation strategy.

Other tests used to assess the cognitive deficit following FCI include the passive avoidance test and the elevated plus maze. The passive avoidance behavior is known to require the integrity of the basal forebrain, cortico-hippocampal connections, frontal cortex, and striatum (5). Among these structures, the striatum and parietal cortex are massively affected by focal ischemia. Learning deficits in passive avoidance test have been reported recently in mice undergoing MCAO (5). Another commonly used test is the elevated plus maze. It was originally introduced as a model for studying anxiolytic agents. Later on, it was found that acquisition and retention processes of memory in ischemic brains could also be studied using the elevated plus maze. Hattori et al. and Rehni et al. show acquisition and retention deficit in mice during the elevated plus maze following FCI (6, 7).

2. Morris Water Maze

MWM can be used to assess the spatial reference memory of both mice and rats. It is usually carried out 14–20 days after FCI.

2.1. Materials and Apparatus

1. A large circular tank (usually, 130–150 cm in diameter) is filled with opacified water at 21°C (±1°C) to the depth of 33 cm.
2. A square Plexiglass platform (10×10 cm^2) is submerged 1.2 cm below water level. The platform is left in the same location (north, east, south, or west, 31 cm from the edge) for the duration of each trial (Fig. 1).
3. Extravisual cues are placed around the room and the lighting is kept constant during testing.

2.2. Procedure

1. Each animal is randomly assigned a platform location, which remains constant throughout the duration of testing.

Fig. 1. Testing tank for the Morris water maze (MWM). It is filled with opacified water with a platform submerged below water level.

2. Animals are placed into the maze and undergo two acclimatization trials, whereby they are allowed to swim freely in the maze, without platform, for 90 s.
3. After the acclimatization trials, animals are tested each day for five consecutive days. Each day consists of two testing sessions, which are made up of four trials. For each of the four trials, animals are placed in the pool facing the wall at one of the four starting locations (northeast, southeast, southwest, northwest) and allowed to swim for a maximum of 90 s.
4. If the animals find the platform, they are allowed to remain on it for 30 s; animals not finding the platform are placed on it and allowed to remain there for 30 s.
5. Intertrial intervals are kept constant at 30 min and a 2-h interval is permitted between each session.
6. Each trial is videotaped via a ceiling-mounted video camera and the animal's movement tracked using Ethovision software, which allows the calculation of various measures (Fig. 2).
7. On the last session, animals undergo the "probe trial," whereby the platform is removed and the mouse placed into the center of the pool, where it is allowed to swim freely for 30 s.
8. To prevent hypothermia, the animals are gently dried with a cotton towel between and after the trials.

2.3. Parameters Recorded

1. Latency before reaching the target quadrant.
2. Time spent in the target quadrant.
3. Latency before finding the previous platform position.

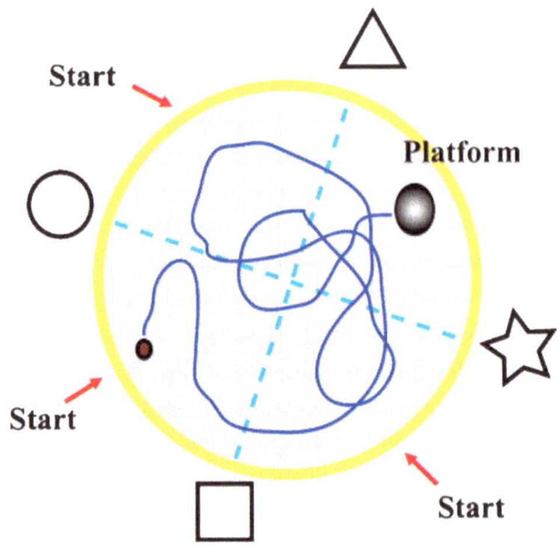

Fig. 2. Test procedure of the MWM. Extravisual cues (different shapes) are placed around the room to assist the animal to identify the direction. *Blue line* is a representative swimming trace recorded.

4. The number of crossing of the previous place of the platform.
5. Swim speed.

The parameters are compared among experimental groups (e.g., sham, ischemic, or drug-treated groups). Shorter latency in finding the platform position and longer time spent in the training quadrant suggest better learning and memory.

3. Passive Avoidance Test

Passive avoidance test is used to evaluate the long-term memory of rodents. It is usually carried out 8–14 days after cerebral ischemia.

3.1. Materials and Apparatus

1. A step-through-type passive avoidance box (Fig. 3a, Bioseb, France), divided into a wide, white, and illuminated compartment ($22 \times 21 \times 30$ cm^3) and a small, black, and dark compartment ($7.3 \times 7.5 \times 14$ cm^3).
2. A weight sensor under the floor covering both compartments can automatically detect the animal.
3. The grid of the dark compartment was designed to deliver an electric foot shock.
4. A guillotine door, automatically controlled, separates the two compartments.

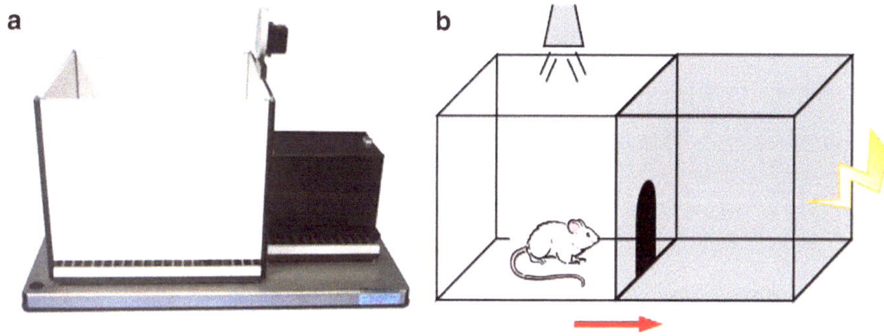

Fig. 3. (**a**) Step-through passive avoidance box. (**b**) Illustration of the experimental procedure. Animal goes from light to dark chamber and receives an electric foot shock.

3.2. Procedure

1. Acquisition: The animal is placed in the illuminated compartment furthest away from the closed door. After 30 s, the door is opened. As soon as the animal has entered the dark compartment (the step-through maximal latency was fixed at 50 s; usually, all animals enter the dark compartment within this latency), the door closes and the animal receives a single inescapable electric foot shock (0.4 mA, 2 s) before being placed back in its own cage (Fig. 3b).
2. Retention: 24 h and 7 days later, the same animal is tested twice for retention.
3. Testing: To this end, the animal is placed again in the white compartment. No electric shock is given during these retention sessions. In the testing experiment, the time before entering the dark compartment is recorded. The step-through maximal latency is fixed at 300 s, and it is the value given to the animals that do not go in the dark compartment. If the animals remember that the dark compartment is aversive, then it will avoid entering it and exhibit longer step-through latency. No changes in the step-through latency indicate learning deficits.

4. Elevated Plus Maze

The elevated plus maze is used to assess cognitive behavior in rodents. It was originally introduced as a model for studying anxiolytic agents. Later on, it was found that acquisition and retention processes of memory could also be studied in ischemic brains. The test is performed 6 days after ischemia.

4.1. Materials and Tools

The elevated plus maze apparatus consists of a central platform (10 × 10 cm) connected to two open arms (16 × 5 cm) and two

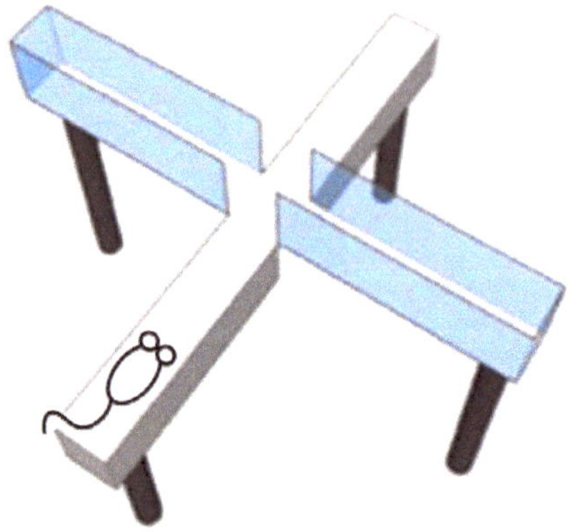

Fig. 4. Apparatus for the elevated plus maze. Each animal is placed at the end of an open arm.

covered (enclosed) arms (16 × 5 × 12 cm). The maze was elevated to a height of 25 cm from the floor (Fig. 4).

4.2. Procedure

1. Each animal is placed at the end of an open arm facing away from the central platform.
2. The animal is allowed to explore the maze for 10 s and then returns to its home cage.
3. The retention of this learned task is examined at 24 h after the initial exposure. Care should be taken for not changing the relative location of plus maze with respect to any object serving as visual clue in laboratory.

4.3. Data Analysis

The transfer latency time (TLT), defined as the time in seconds taken by the animal to move into one of the enclosed arms with all its four legs, is a parameter of memory. Longer TLT indicates memory deficits.

References

1. D'Hooge R, De Deyn PP (2001) Applications of the Morris water maze in the study of learning and memory. Brain Res Brain Res Rev 36(1):60–90
2. Morris R (1984) Developments of a water-maze procedure for studying spatial learning in the rat. J Neurosci Methods 11:47–60
3. Gibson CL, Bath PM, Murphy SP (2005) G-CSF reduces infarct volume and improves functional outcome after transient focal cerebral ischemia in mice. J Cereb Blood Flow Metab 25(4):431–439
4. Crawley JN (1996) Unusual behavioral phenotypes of inbred mouse strains. Trends Neurosci 19:181–182
5. Bouet V, Freret T, Toutain J, Divoux D, Boulouard M, Schumann-Bard P (2007) Sensorimotor and cognitive deficits after transient middle cerebral artery occlusion in the mouse. Exp Neurol 203(2):555–567
6. Rehni AK, Singh N, Jaggi AS, Singh M (2007) Amniotic fluid derived stem cells ameliorate focal cerebral ischaemia-reperfusion injury induced behavioural deficits in mice. Behav Brain Res 183(1):95–100
7. Hattori K, Lee H, Hurn PD, Crain BJ, Traystman RJ, DeVries AC (2000) Cognitive deficits after focal cerebral ischemia in mice. Stroke 31(8):1939–1944

Chapter 17

Assessment of Neurogenesis in Models of Focal Cerebral Ischemia

Feng-Yan Sun, Xiao Sun, Ji-Jiang Wang, Qiu-Wan Zhang, and Yong-Quan Wang

Abstract

Neurogenesis is the generation of new neurons. In hippocampal dentate gyrus and olfactory bulb, which usually are referred to as the neurogenic region in the mammalian brain, new neurons are developed daily throughout their lifetime. In the non-neurogenic region of the brain, neurogenesis occurs only under certain pathological conditions. Focal cerebral ischemia is a common neurological disease that stimulates neurogenesis in both neurogenic and non-neurogenic regions in adult brains. The integration between the preexisting and the newly developed neurons in the ischemic regions happens after the new ones matured morphologically and functionally. This chapter introduces the methods of studying neurogenesis using focal cerebral ischemia models, including morphological analysis by performing immunohistochemical and immunofluorescent multiple staining with confocal scanning and immunoelectron microscopy, and functional determination of neurons by using electrophysiological recording and FM dye synaptic vesicle staining in animal brain slices containing retroviral green fluorescent protein-labeled new neurons.

Key words: Adult neurogenesis, Morphological assessment, Functional assessment

1. Introduction

Neurogenesis is the generation of new neurons. Regardless of the physiological and pathological conditions, the process of generating new neurons exists in the adult mammalian brain throughout the life span. From the research in the past few decades, we find that the concept of adult neurogenesis was defined upon the development of innovated techniques. In the 1960s, Altman first used the combination of [^{3}H]-thymidine ([^{3}H]-dT) injection and autoradiography in an attempt to study neurogenesis in the hippocampus and olfactory bulb (OB) of adult mammalian brains (1–3). Although [^{3}H]-dT is incorporated into the cell during DNA

Jun Chen et al. (eds.), *Animal Models of Acute Neurological Injuries II*, Springer Protocols Handbooks,
DOI 10.1007/978-1-61779-782-8_17,

synthesis at the S phase of the cell cycle, in vivo [^{3}H]-dT labeling does not distinguish new neurons from other cell types. This limitation prevents us from providing convincing evidence for adult neurogenesis. In the 1980s, Nowakoski used bromodeoxyuridine (BrdU), an immunohistochemical staining that is analogous to thymidine, to study the length of the cell cycle and DNA synthetic phase for an anatomically defined population (4). Multiple immunohistochemical staining of BrdU with specific makers for different lineages of neurons is obtained to phenotype the newly generated cells (5). Besides, BrdU labeling is easier and faster to perform. Labeled cells in the slices can be used for stereological counting. Using these techniques, biologists successfully found the existence of neurogenesis in adult mammalian brains (6, 7). Generally, BrdU-positive signals in the newborn neurons should mainly locate in the nucleus of cells, but not in the cytoplasm. Green fluorescent protein (GFP)-targeted retroviral vectors replicate and express GFP in the dividing cells. The expressed GFP distributes fully around the cell body and neurite, and can be observed directly from fresh or fixed animal brain slices under microscope (8). Therefore, retroviral tracing technique is introduced to use for the detection of neurite development and synaptic formation, as well as functional analysis of newborn neurons (8–11). However, following in vivo injection, retroviral infection is not controlled and labeling is quite variable. Thus, retroviral tracing is unsuitable for the quantitative analysis between different experimental groups (12).

New neurons in the OB are generated from neural progenitors/precursors in the anterior part of the subventricular zone and dentate gyrus (DG) from the subgranular zone of the hippocampus in the adult mammalian brains (12). Both the OB and hippocampal DG, referring to as the constitutive neurogenic region in adult brains, produce new neurons under physiological conditions. However, other brain regions, such as the striatum, cerebral cortex, and hippocampal CA1 and CA3 areas, which are usually referred to as the non-neurogenic regions, undergo neurogenesis only under certain pathological conditions, such as ischemic brain injury (13–17). New neurons in both neurogenic (18, 19) and non-neurogenic (9, 18) regions can mature morphologically, and form functional synapses with the preexisting neurons. These findings have drawn increased interest in treating central nervous system (CNS) diseases with cellular replacement strategies, and improving functional recovery of neurons (17, 20). Since the number of newborn neurons is very limited in the adult brain, new methods to enhance neurogenesis are in need to be developed. In our case, how to assess neurogenesis in the adult brain with diseased models becomes crucial.

Adult neurogenesis incorporates all events of neuronal development, including morphological maturation and functional integration of neurons generated during animals' adulthood. These processes involve the proliferation of progenitors and their migration into damaged regions, further differentiation into region-proper mature neurons, formation of neurites, and functional integration with neuronal networks. In this chapter, we introduce methodology for morphological and functional assessment of adult neurogenesis of rats using models of focal cerebral ischemic injury.

2. Morphological Assessment of Neurogenesis in Focal Cerebral Ischemia Models

The most important steps for evaluating neurogenesis involve labeling proliferating cells and phenotyping newborn cells. If BrdU is incorporated into DNA synthesis during the S phase of the cell cycle, its labeling signals can be revealed by immunohistochemical staining. Maturation of newborn cells can be identified by multiple immunohistochemical staining of BrdU with neuronal markers for different lineage. Thus, to evaluate neurogenesis in the brain, the selection of specific markers for multiple immunostaining becomes significant. Many publications have extensively reviewed the availability of current specific markers for different stages of developing neurons (12, 21, 22). We summarize the stage and markers for neural development of newly generated neurons in the striatum of rat brain after a transient focal ischemic stroke as illustrated in Fig. 1. To simplify the illustration of the methodology, we selectively choose the following specific neuronal markers, which are widely used to phenotype newly generated cells in the adult brain (12, 17, 18, 22–24).

2.1. Materials and Tools

2.1.1. Primary Antibodies

1. Proliferation marker: Anti-BrdU mouse monoclone (Roche, Mannheim, Germany) or rabbit polyclonal antibody (Megabase research Products, Nebraska, USA).
2. Stem cell marker: Anti-nestin mouse monoclonal antibody (San Jose, CA or BD PharMingen, USA).
3. Neuroprogenitor marker: Anti-doublecortin (DCX) goat monoclonal antibody (Chemicon, Temecula, CA, USA).
4. Immature neuronal markers: Anti-βIII tubulin (Tuj-1) mouse monoclonal antibody (Chemicon, Temecula, CA, USA), anti-Tuj-1 rabbit polyclonal antibody (Covance, California, USA), anti-CRMP-4 (also called TUC-4 protein) rabbit polyclonal antibody (Chemicon, Temecula, CA, USA).
5. Mature neuronal markers: Anti-microtubule-associated protein-2 (MAP2) rabbit polyclonal antibody (Chemicon, Temecula, CA, USA), anti-neuron-specific endolase (NSE) mouse monoclonal antibody (Dako, Denmark), anti-neuronal nuclei (NeuN) mouse monoclone antibody (Abcam, Cambridge, UK).
6. Cholinergic neuronal marker: Anti-choline acetyltransferase mouse monoclonal antibody (ChAT, Chemicon, Temecula, CA, USA).

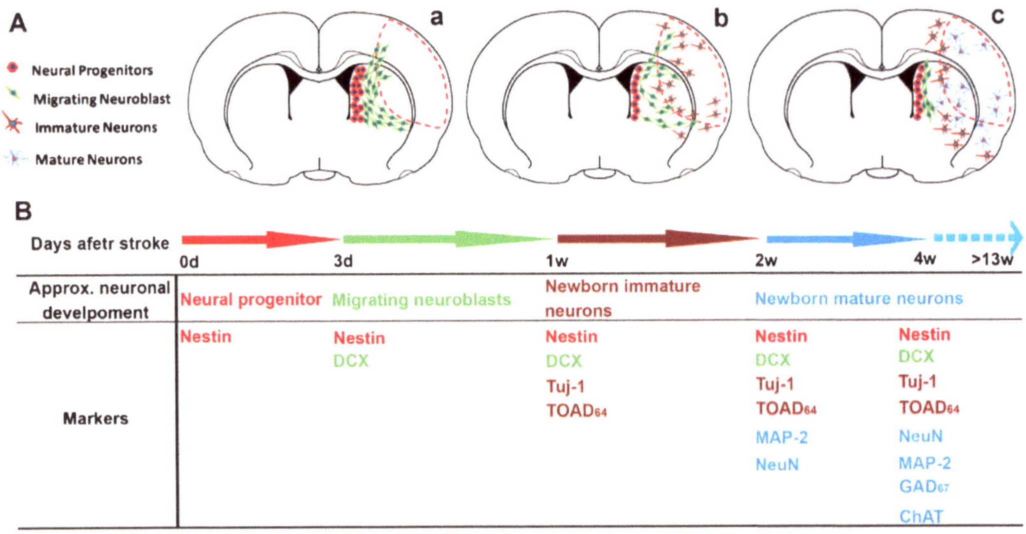

Fig. 1. Stage and markers of development of newborn neurons in the striatum and cortical non-neurogenic regions of rat brains after a transient middle cerebral artery occlusion (MCAO). (**a**) Depicts that ischemic injury induces activation of neural stem cells/progenitors (**Aa**), stimulates the activated neural progenitors migrating toward the ischemic injured areas (**Ab**), and differentiates new neurons including immature and mature ones after then (**Ac**). (**b**) Illustrates that the stage of neural development corresponds to the proximal days after stroke and markers are mostly used for revealing newly generated neurons at different lineage as indicated corresponding colors. Neurogenesis exists through the life span in the SVZ, a neurogenic region. Therefore, nestin and doublecortin (DCX) protein can be detected near the SVZ whole periods we studied from 3 days to 13 weeks after MCAO. Migrating neural progenitors (BrdU and DCX double-staining cells) differentiate into newly generated neurons, which further develop into immature neurons (BrdU and Tuj-1/TOAD$_{64}$ double-staining cells) from 7 days after MCAO and become mature ones (BrdU and MAP-2/NeuN double staining cells) from 2 weeks. Those new mature neurons develop into region-appropriate neurons (BrdU and GAD$_{67}$/ChAT double-staining cells) about 4 weeks late.

7. GABAergic neuronal marker: Anti-glutamate decaboxylases of 67 kDa rabbit polyclonal antibody (GAD-67, Santa Cruz Biotechnology, Inc., Santa Cruz, CA, USA).
8. Astrocyte glial marker: Anti-GFAP rabbit polyclonal antibody (Chemicon, Temecula, CA, USA).
9. Anti-GFP goat polyclonal antibody (Abcam, Cambridge, UK).

2.1.2. Secondary Antibodies

1. Fluorescein isothiocyanate (FITC)-conjugated anti-rabbit (mouse) IgG secondary antibody (Santa Cruz Biotechnology, Santa Cruz, CA, USA).
2. Rhodamine-conjugated anti-rabbit (mouse) IgG secondary antibody (Santa Cruz Biotechnology, Santa Cruz, CA, USA).
3. Biotinylated anti-goat (rabbit or mouse) IgG secondary antibodies (Vector Laboratories, Inc. Burlingame, CA, USA).

2.1.3. ABC Kits

1. ABC–alkaline phosphatase Kits (Anti-mouse, rabbit and goat IgG) (Vector Laboratories, Inc. Burlingame, CA, USA).
2. ABC–peroxidase Kits (Anti-mouse, rabbit and goat IgG) (Vector Laboratories, Inc. Burlingame, CA, USA).
3. Vector Blue (Vector Laboratories, Inc. Burlingame, CA, USA).

2.1.4. Reagents

1. pIB-hrGFP retroviral vector supernatant (1×10^9 units/mL of titers, Stratagene, USA).
2. 5′-Bromodeoxyuridine (Roche applied science, Mannheim, Germany).
3. Picrotoxin and 6-cyano-7-nitroquinoxaline-2,3-dione (CNQX, Sigma, St. Louis, MO, USA).
4. Diaminobenzidine (DAB, Sigma, St. Louis, MO, USA).

2.1.5. Equipment

1. Microscope with Stereo Investigator Software (Micro-Bright Field, Inc., Williston, VT).
2. Confocal laser scan microscope (TCS, SP2, Leica, Germany).
3. Electron microscope (JEM1200, Japan).
4. Ultramicrotome (Reichert Ultracut E, Leica, Germany).

2.2. Procedure

2.2.1. BrdU Labeling

During the process of DNA synthesis and repair, BrdU can be incorporated into the S phase of cell cycle and lesion sites, respectively. After stroke, DNA repair usually occurs within hours, and cell proliferation within days. Therefore, it has been recommended that, to label proliferating cells, fresh prepared BrdU should be given to rats 4 days after the middle cerebral artery occlusion (MCAO) (17, 25). Because of the capable role that BrdU could play in DNA synthesis, high dose of BrdU injection (150 mg/kg for 5 days, ip) will induce a neurotoxic effect (26). However, low dose or a single injection of commonly used dose of BrdU will result in a false negative outcome due to the partially labeled proliferating cells (12). Research from the Gage laboratory has suggested that repeating BrdU injections with doses of 50 mg/kg/day (ip) should be used to label proliferating cells in adult rat brains (27–29). Lindvall's laboratory suggested that BrdU could be given to the MCAO rat model intraperitoneally at 50 mg/kg/day for 3–5 days, starting from the fourth day after ischemic operation (17, 25). In the past few years, we have used this protocol to successfully detect adult neurogenesis in the rat brains after the transient MCAO.

BrdU Injection

Rats were returned to their cages after recovering from the MCAO operation and given intraperitoneal injection of BrdU (50 mg/kg body weight once daily, freshly prepared) 4–6 days after the operation.

BrdU Immunostaining

Brain Section Preparation

1. Anesthetize the rats with an overdose of 10% chloral hydrate.
2. Perfuse transcardially with 0.9% saline solution followed by 4% ice-cold phosphate-buffered paraformaldehyde (PFA).
3. Remove brain and fix in 4% PFA for 4–6 h.
4. Immerse brains sequentially in 20 and 30% sucrose solutions in 0.1 M phosphate buffer (pH 7.4) until sank.

5. Cut brain sections to thickness of 30–50 μm on a freezing microtome (Jung Histocut, Model 820-II, Leica, Germany).
6. Store sections at −20°C in cryoprotectant solution for further immunostaining.

Single Staining of BrdU

Free-floating sections were used for immunostaining as follows. On the first day:

1. Incubate sections with 50% formamide/2× SSC (0.3 M NaCl, 0.03 M sodium citrate) at 65°C for 2 h, and 2 N HCl at 37°C for 30 min.
2. Rinse sections with Tris buffer (0.1 M, pH 7.6).
3. Treat sections with 1% H_2O_2 for 10 min.
4. Wash sections with buffer A (0.1 M Tris buffer/0.1% Triton X-100) and then buffer B (0.1 M Tris buffer/0.1% Triton X-100, 0.05% BSA) for 15 min each.
5. Incubate sections with 10% goat serum for 1 h at 37°C.
6. Incubate sections overnight with mouse monoclonal anti-BrdU antibody at 4°C.

On the second day:

7. Wash sections with buffer A and B for 15 min each on the shaker at room temperature (RT).
8. Incubate sections with biotinylated secondary antibody for 30 min at 37°C.
9. Wash sections with buffer A and B three times for 15 min each on the shaker at RT.
10. Incubate sections with avidin–biotin–peroxidase complex for 45 min at 37°C.
11. Rinse sections three times with Tris buffer (0.1 M, pH 7.6) on the shaker at RT.
12. Develop immunoreactive signals with 0.05% DAB in Tris–HCl buffer (0.1 M, pH 7.6) containing 0.03% H_2O_2.
13. Mount sections on glass slides.
14. Dry slides on hot plate (HI1220, Leica) at 37°C.
15. Dehydrate slides in successively increasing concentrations of alcohol, optimally 4–5 times until the concentration reaches 100%.
16. Rinse slides in 3–4 times in xylene and coverslip.
17. Observe BrdU immunopositive ($BrdU^+$) signals under light microscope.

Multiple Staining of BrdU with Neuronal Markers

Multiple staining of BrdU with neuronal markers was used to phenotype newly generated neurons. Immunoreactive signals for the primary antibody in the sections can be revealed by immunohistochemical staining or immunofluorescent staining. The signals revealed by immunohistochemical staining usually remain longer in the room temperature and can be observed under light microscope. Therefore, cell counting is allowed to be carried out easily. However, double-staining signals revealed by immunohistochemical staining are sometimes difficult to be identified because it is difficult to distinguish whether two signals in the same cell are specific cellular colocalization or nonspecific overlapping. To solve this problem, fluorescent double staining combined with confocal laser scanning microscopy should be used in further experiments. Thus, the following content is divided into histochemical double staining and fluorescent double staining.

Histochemical Double Staining

On the first day:

1. Immunoreact sections with mouse monoclonal anti-BrdU antibody (1:200) overnight at 4°C, as described in steps 1–6 in section "Single Staining of BrdU."

On the second day:

2. Wash sections three times with buffer A and B for 15 min each on the shaker at RT.
3. Incubate sections with biotinylated anti-mouse IgG (1:200).
4. Wash sections three times with buffer A and B for 15 min each on the shaker at RT.
5. Incubate sections with avidin–biotin–alkaline phosphatase complex (1:200) for 1 h at 37°C.
6. Develop BrdU immunoreactive signals with Vector Blue (blue color).
7. Wash sections three times with Tris buffer (0.1 M, pH 7.6) for 15 min on the shaker at RT.
8. Incubate sections overnight at 4°C with one of the primary antibodies against rabbit polyclonal anti-MAP-2 (1:200), Tuj-1 (1:200), GFAP (1:200), or GAD67 (1:200).

On the third day:

9. Rinse sections three times with PBS for 15 min on the shaker at RT.
10. Incubate sections with corresponding biotinylated anti-rabbit (goat) secondary antibodies (1:200).
11. Rinse sections three times with PBS for 15 min on the shaker at RT.
12. Incubate sections with avidin–biotin–peroxidase complex (1:200) for 45 min at 37°C.

13. Rinse sections three times with PBS for 15 min on the shaker at RT.
14. Develop immunoreactive signals with 0.05% DAB (brown color) in Tris–HCl buffer (0.1 M, pH 7.6) containing 0.03% H_2O_2.
15. Mount sections on glass slides.
16. Dry slides on hot plate at 37°C for 1 h.
17. Dehydrate slides in successively increasing concentrations of alcohol, optimally 4–5 times until the concentration reaches 100%.
18. Rinse slides 3–4 times in xylene and coverslip.
19. Observe double-stained cells under light microscope as shown in examples (Fig. 2a–d).
20. Capture double-staining cells in the region of interest based on the instruction of Stereo Investigator Software.

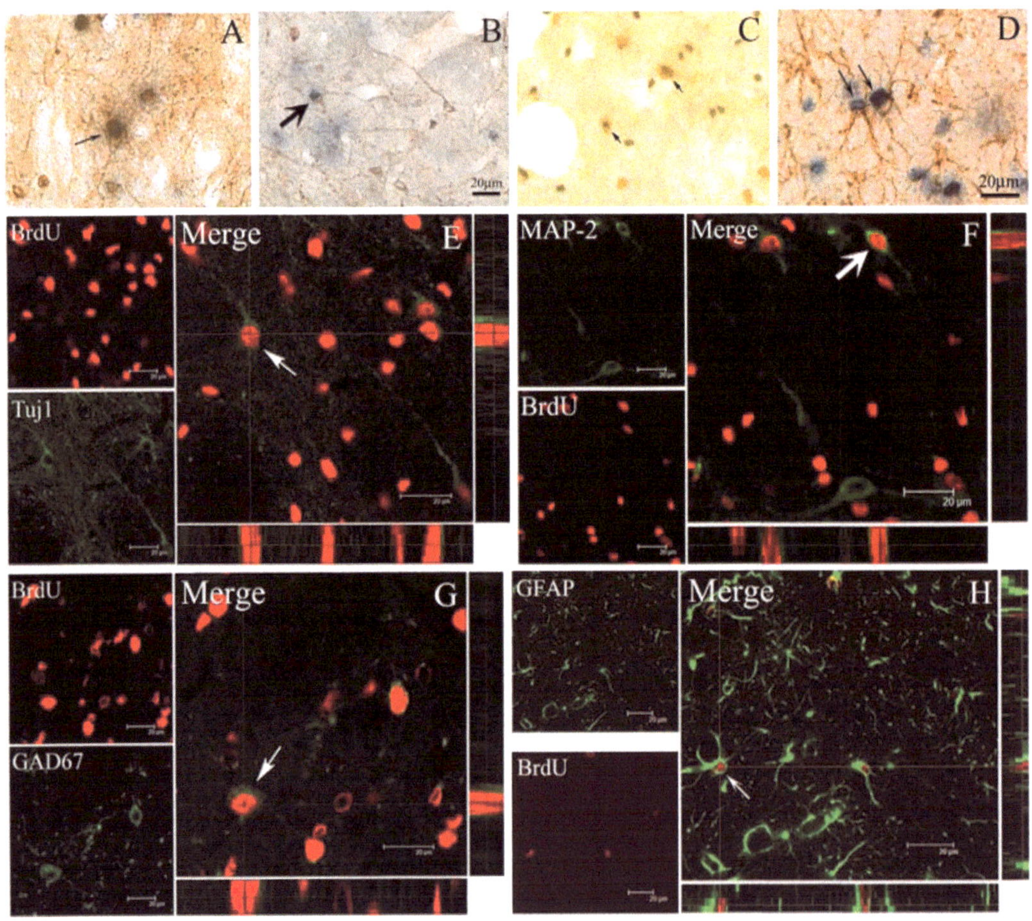

Fig. 2. Double staining of immunohistochemistry (**a**–**d**) and immunofluorescence (**e**–**h**) of BrdU with Tuj-1 (**a**, **e**), MAP-2 (**b**, **f**), GAD67 (**c**, **g**), and GFAP (**d**, **h**) in the striatum of rat brains at the fourth week after ischemic injury. (**a**–**d**) *Blue* and *brown colors* indicate positive BrdU staining and neuronal markers, respectively. Photographs show BrdU (*blue* in **a**–**d**, *red* in **e**–**h**) in the nuclei of newly generated immature (**a**, **e**), mature (**b**, **f**), and GABA neurons (**c**, **g**) and astrocytes (**d**, **h**).

21. Count the number of double-staining cells in each brain section with Stereo Investigator Software.

Fluorescent Double Staining and Confocal Laser Scanning Microscopy

On the first day:

1. Treat section as described in steps 1–5 in the "Single Staining of BrdU" section.
2. Incubate sections overnight at 4°C with a mixture of two primary antibodies raised in separate species: a mouse monoclonal antibody against BrdU and a rabbit polyclonal antibody against Tuj-1/MAP-2/GAD67/GFAP.

On the second day:

3. Wash sections three times with PBS (0.01 M, pH 7.2) for 15 min on the shaker at RT.
4. Incubate sections for 1 h at 37°C in the combination of two secondary antibodies: an anti-mouse IgG–rhodamine (1:50–100) and an anti-rabbit IgG–FITC (1:50–100).
5. Wash sections three times with PBS (0.01 M, pH 7.2) for 15 min on the shaker at RT.
6. Mount sections on the glass slide and coverslip using fluorescence mounting medium (Vector Laboratories).
7. Detect and capture double signals at excitation 535 nm and emission 565 nm (rhodamine) and 492 and 520 nm (FITC) by confocal laser scanning microscopy as shown in Fig. 2e–h.
8. Perform cell counting in the region of interest in the brain sections.

2.2.2. Retroviral Labeling

A retroviral vector targeted with GFP can be used to specifically label proliferating cells and successfully identify morphological and functional maturation of newly generated neurons in adult rodent brains (10, 30). In general, mature neurons consist of cell body and multiple neurites, which include an axon and multiple dendrites. New neuron-replicated viral GFP shows fluorescence throughout the entire cell body and neurites. Although GFP-expressed new neurons in the brain sections can be directly observed under fluorescent or confocal microscope (Fig. 3a), GFP immunohistochemical staining is also a recommended detecting method since it gives more details (Fig. 3b). With GFP immunostaining, the detection of positive staining signals can be easily made under a light microscope. Besides, the staining signals can be well maintained for a longer period of time for further quantification, such as measuring the soma size and counting the dendritic length and the number of its branches (9, 30).

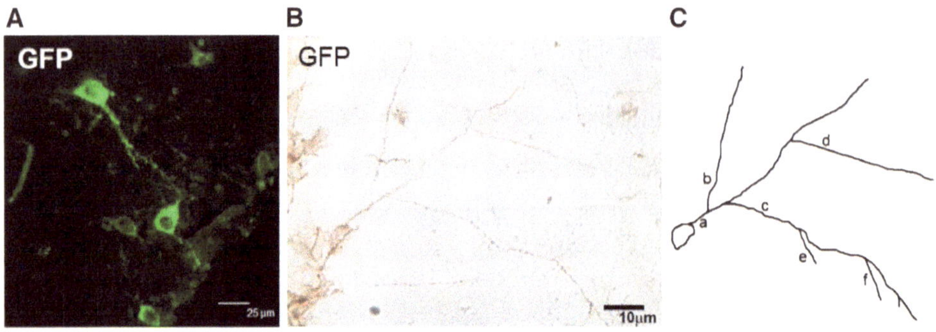

Fig. 3. Fluorescent and immunohistochemical signals of GFP in the striatum of rat brain at 13th week after a transient MCAO. (**a**) GFP-expressed neurons were detected under confocal microscope. (**b**) GFP immunohistochemical signals show long process and several branches of neurite of newborn neuron. (**c**) A mimic figure of (**b**) represents the calculation of the dentritic branch of GFP+ newborn neuron.

Retroviral Injection

To mark the ependymal cells, which line the walls of lateral ventricle, GFP gene-bearing retrovirus supernatant (5×10^6 infectious units/5 μL) is stereotaxically injected into the lateral ventricle contralaterally (coordinates around bregma: AP −0.8 mm, ML 1.4 mm, DV −3.6 mm from the pial surface) 24 h before the operation.

To label cells in specific brain nuclei, GFP gene-bearing retrovirus supernatant ($0.5–1 \times 10^6$ infectious units/0.5 μL) was stereotaxically injected into the desired nuclei on the same or third day after the MCAO operation.

Viral-Replicated Signal Detection

GFP Fluorescent Detection

1. The animals preinjected with GFP-targeted retroviral vectors are sacrificed at different days based on your research purpose.
2. Cut brain sections at the thickness of 30 μm as described above.
3. Observe GFP fluorescence of new neurons in the brain sections directly under microscope or confocal laser scanning microscope at excitation and emission wavelengths of 480 and 500 nm, respectively (Fig. 3a).

Detection of Neurite and Synapses' Formation

GFP Immunostaining

1. Wash brain slices, which were prepared for GFP fluorescent detection as mentioned above, three times at RT.
2. Incubate brain sections with anti-GFP antibody (1:1,000) at 4°C overnight. GFP^+ signals are visualized by histochemical reaction with ABC kit and 0.05% DAB.
3. The procedure for GFP immunohistochemical staining can be referred to steps 9–18 in the "Histochemical Double Staining" section.

Neurite Detection

1. Observe GFP^+ signals under a light microscope (×40 objective).

2. Capture cells expressed with GFP^+ signals by a computer-assisted imaging system with a cooled CCD camera (Fig. 3b).
3. Measure the length of neurite branches of GFP^+ neuron with Stereo Investigator Software (Micro-Bright Field, Inc., Williston, VT).
4. Count the number of neurite branches of GFP^+ neuron manually as demonstrated in Fig. 3c.

Detection for Synapses with Immunoelectron Microscopy

1. Anesthetize rats preinjected with GFP retroviral vector at 8–10 weeks after the MCAO.
2. Perfuse transcardially with 4% PFA containing 0.25% glutaraldehyde.
3. Cut coronal sections at a thickness of 50 μm.
4. Perform GFP immunohistochemical staining as described above.
5. Observe GFP^+ neurons in the sections under light microscope.
6. Cut the sample containing GFP^+ neurons (1 × 3 mm) from sections mentioned in step 4.
7. Fix GFP^+ neuronal sample in 2.5% glutaraldehyde for 2 h at 4°C.
8. Wash the sample with PBS (0.1 M, pH 7.2).
9. Dehydrate the sample in successively increasing concentrations of alcohol, optimally four to five times until the concentration reaches 100%.
10. Rinse the sample in acetone.
11. Embed the sample with Epon812 resin.
12. Cut the embedded sample at the thickness of 70 nm with an ultramicrotome (Reichert Ultracut E, Leica, Germany).
13. Visualize GFP^+ signals with lead acetate and uranyl staining.
14. Observe the sample and take pictures with an electron microscope (JEM1200, Japan).

3. Functional Assessment of Neurogenesis in Models of Focal Cerebral Ischemia

According to Kandel, the author of the book *Principles of Neural Science*, mature neurons of the CNS fire action potentials and, most importantly, communicate with other neurons when they become morphologically matured. To analyze synaptic functions of the neurons, use freshly prepared brain slices combined with electrophysiological recording and/or FM dye staining to observe how GFP-expressed new neurons work.

3.1. Materials and Equipment

3.1.1. Reagents

1. pIB-hrGFP retroviral vector supernatant (1×10^9 units/mL of titers, Stratagene, USA).
2. 5′-bromodeoxyuridine (Roche applied science, Mannheim, Germany).
3. Picrotoxin and CNQX (Sigma, St. Louis, MO, USA).
4. 4-Aminopyridine (4-AP, Sigma, St. Louis, MO, USA).
5. Artificial cerebrospinal fluid (CSF, in mM, pH at 7.4): 130 NaCl, 3 KCl, 2 $CaCl_2$, 2 $MgCl_2$, 1.25 NaH_2PO_4, 26 $NaHCO_3$, and 10 glucose.
6. Patch pipette prefilled with solution (in mM, pH 7.25) containing the following: 135 KCl, 2 $MgSO_4{\cdot}7H_2O$, 0.1 $CaCl_2$, 1.1 EGTA, 2 K_2ATP, 0.1 Na_3GTP, 10 HEPES (3–5 mol/I resistance).

3.1.2. Equipment

1. Vibrating-blade microtome (Leica VT1000s, Germany).
2. Confocal microscope (TCS, SP2, Leica, Germany).
3. Patch clamp recorder (Axon Instruments/Molecular Devices Corp. Foster City, California, USA).

3.2. Procedure

3.2.1. Retroviral Injection and Brain Slice Preparation

1. Inject retroviral vector-targeted GFP into MCAO-operated rats as mentioned above.
2. Anesthetize rats with overdose of 10% chloride hydrate at 6–8 weeks (or at the desired time) after the MCAO operation.
3. Quickly remove the brains.
4. Immerse tissues in cooled CSF constantly bubbled with 95% O_2 to 5% CO_2.
5. Cut coronal striatal (or the interesting regions) slices at the thickness of 250–300 μm using a vibratome (Fig. 4a).
6. Incubate brain slices between 22 and 24°C for 1 h for patch-clamp recording or FM4-46 dye staining.

3.2.2. Electrophysiological Recording

Whole-cell patch clamp recording is performed as previously reported (9, 18). Main steps are as follows.

1. Transfer slices into a recording chamber perfused continuously with oxygenated CSF between 22 and 24°C (Fig. 4b).
2. Observe under fluorescent microscope for GFP-fluorescent neurons in the slices.
3. Select GFP-labeled neurons for electrophysiological recording (Fig. 4c).
4. Patch GFP-labeled cells with patch pipette (3–5 MΩ resistance).
5. Measure the resting membrane potential and membrane capacitance immediately after rupture of the cell membrane.

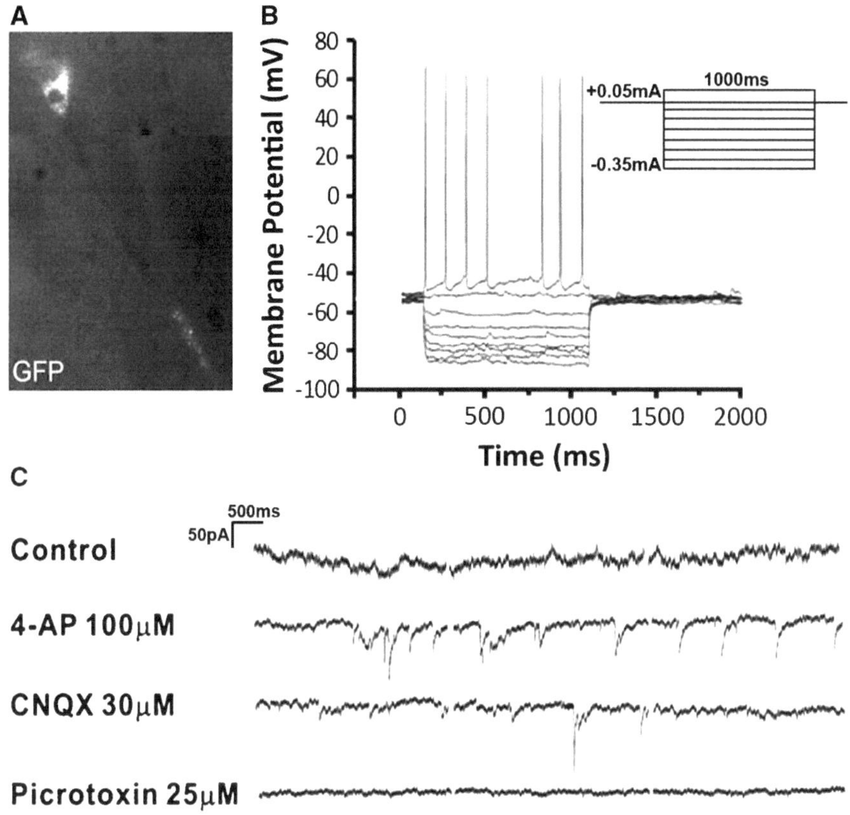

Fig. 4. Electrophysiological recording of newly generated striatal neuron in the freshly prepared brain section of rat at 8 weeks after a transient MCAO. Newborn neuron was indicated by GFP fluorescence (**a**). Electrophysiological recording shows membrane potential, frequency, and amplitude of action potentials (**b**), and postsynaptic currents (**c**). Photographs show that 4-AP-elicited postsynaptic currents can be partially blocked by CNQX, an antagonist of NMDA receptor, and abolished buy picrotoxin, a GABA receptor blocker. Thus, newly generated striatal neurons are capable of firing action potentials and receiving excitatory and inhibitory synaptic inputs.

6. Calculate the input resistance by measuring the current amplitude evoked by a steady-state hyperpolarizing pulse of 2 mV for 50 ms, under a holding potential of –70 mV.
7. Measure the threshold of action potentials and spiking frequency of neurons by injecting a hyperpolarizing-depolarizing current from –500 to +50 pA at a step of 50 pA, under the current clamp configuration (Fig. 4b).
8. Record spontaneous postsynaptic currents (SPSCs) under the voltage patch clamp configuration at a holding potential of –70 mV to identify whether newborn neurons received synaptic input (control in Fig. 4c). If the SPSC is too small to detect, use 4-AP to excite the whole slice (4-AP 100 μM in Fig. 4c).
9. Record the changes of the SPSC in the presence of glutamate receptor antagonist and/or GABA receptor antagonist to identify whether newborn neurons received excitatory or inhibitory synaptic inputs (CNQX and Picrotoxin in Fig. 4c).

3.2.3. FM Dye Staining and Release Analysis

FM 4-46 and its analogous are nontoxic and water soluble. They can be inserted into the outer leaflet of surface membrane. The FM dye has two different forms: membrane binding form and soluble form. The membrane binding form of the FM dye gives out fluorescent, but the soluble form does not. When the neurons actively release neurotransmitters, the membrane-bounded FM dye moves inside the recycled synaptic vesicles (31, 32). As a result, the nerve terminals become brightly stained. The fluorescent signals in the vesicles can then be observed under confocal microscope using the proper absorption/emission wavelength. The nonspecific staining on the cell surface membrane can be easily washed off prior to viewing. Therefore, the FM dye staining test can be used to identify active neuron firings and investigate the mechanism of activity-dependent vesicle recycling in cultured neurons and animal brain slices (31, 32). Thus, to study whether GFP-expressed newborn neurons actively release neurotransmitters or regulate synaptic vesicle recycling, the use of confocal imaging of FM dye staining on freshly prepared animal brain slices is recommended.

1. Prepare animal brain slices as mentioned above.
2. Incubate slices in oxygenated CSF containing 3 mM potassium chloride (KCl) at between 22 and 24°C for 10 min.
3. Stain with 10 μM FM4-64 dye for 2 min in oxygenated CSF containing 3 mM KCl (normal CSF) at 22–24°C.
4. Wash with normal CSF three times for 2 min to remove non-specific membrane-binding dye.
5. Transfer slices into a recording chamber on the stage of confocal microscope continuously perfused with oxygenated normal CSF.
6. Observe under confocal microscope for GFP-expressed neurons and neurites at excitation and emission wavelength of 488 and 525 nm, respectively (Fig. 5a).
7. Observe under confocal microscope for FM4-64 fluorescent signals near GFP-expressed dendrites at excitation and emission wavelength of 515 and 640 nm, respectively (Fig. 5b). Capture GFP and FM4-46 fluorescent signals simultaneously in real time using a confocal microscopic scanning for 120 s (Fig. 5c).
8. Perfuse with CSF containing 120 mM KCl (high KCl) to induce membrane depolarization and synaptic release (Fig. 5d).
9. Continue to capture GFP and FM4-46 fluorescent signals for another 120 s under the condition mentioned above.
10. Measure the density of each punctum located along (arrows) and away from (arrowhead) GFP-expressed neuritis with imaging analysis software.

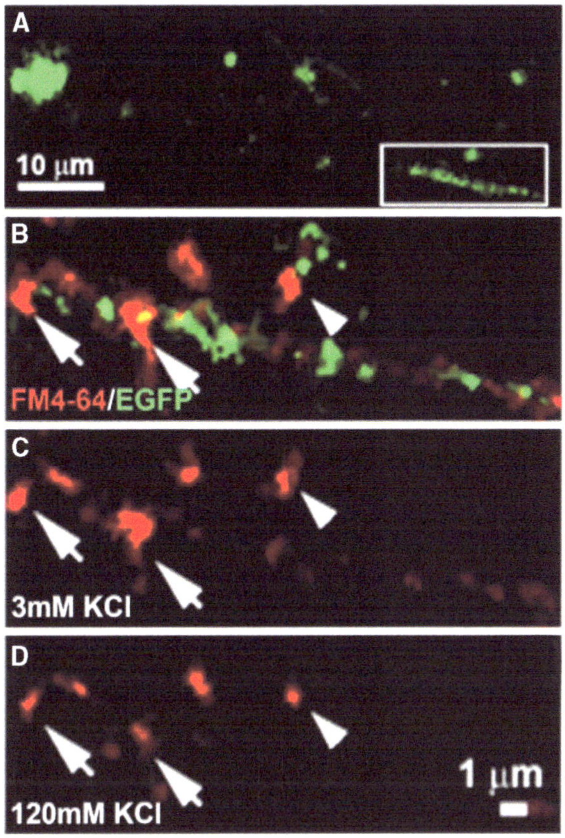

Fig. 5. Confocal imaging of functional vesicle release from GFP-expressed neuronal dendrite in the striatal brain slices of rat at 8 weeks after MCAO. (**a**) Indicated newborn neuron and dendrite expressed with GFP fluorescence. (**b**) is the magnification corresponding to the square region in (**a**), which showed FM4-64 dye-stained synaptic vesicle (*red*) near dentrite of newborn neuron (*green*). The signals were scanned in real time at 1 s (**b**) and 120 s (**c**) after perfusion with CSF containing 3 mM KCl, and at 120 s after perfused with CSF containing 120 mM KCl (**d**). The photographs show that the synaptic vesicles of newly developed neurite (*arrows* in **b**–**d**) possess the capability to uptake and release transmitters as preexisted neurons (*arrowhead* in **b**–**d**).

Acknowledgments

This work was supported by grants from National Basic Research Program of China (2006CB943702 and 2006CB504100) and National Nature Science Foundation of China (81030020 and 30770660).

References

1. Altman J, Chorover SL (1963) Autoradiographic investigation of the distribution and utilization of intraventricularly injected adenine-3h, uracil-3h and thymidine-3h in the brains of cats. J Physiol 169:770–779
2. Altman J, Das GD (1965) Autoradiographic and histological evidence of postnatal hippocampal neurogenesis in rats. J Comp Neurol 124(3):319–335
3. Altman J (1969) Autoradiographic and histological studies of postnatal neurogenesis. 3. Dating the time of production and onset of differentiation of cerebellar microneurons in rats. J Comp Neurol 136(3):269–293
4. Nowakowski RS, Lewin SB, Miller MW (1989) Bromodeoxyuridine immunohistochemical determination of the lengths of the cell cycle and the DNA-synthetic phase for an anatomically defined population. J Neurocytol 18(3): 311–318
5. Nowakowski RS, Hayes NL (2000) New neurons: extraordinary evidence or extraordinary conclusion? Science 288(5467):771
6. Reynolds BA, Tetzlaff W, Weiss S (1992) A multipotent EGF-responsive striatal embryonic progenitor cell produces neurons and astrocytes. J Neurosci 12(11):4565–4574
7. Sohur US et al (2006) Adult neurogenesis and cellular brain repair with neural progenitors, precursors and stem cells. Philos Trans R Soc Lond B Biol Sci 361(1473):1477–1497
8. Overstreet-Wadiche LS, Westbrook GL (2006) Functional maturation of adult-generated granule cells. Hippocampus 16(3):208–215
9. Hou SW et al (2008) Functional integration of newly generated neurons into striatum after cerebral ischemia in the adult rat brain. Stroke 39(10):2837–2844
10. van Praag H et al (2002) Functional neurogenesis in the adult hippocampus. Nature 415(6875):1030–1034
11. Tonnesen J et al (2010) Functional properties of the human ventral mesencephalic neural stem cell line hVM1. Exp Neurol 223(2):653–656
12. Abrous DN, Koehl M, Le Moal M (2005) Adult neurogenesis: from precursors to network and physiology. Physiol Rev 85(2):523–569
13. Parent JM et al (2002) Rat forebrain neurogenesis and striatal neuron replacement after focal stroke. Ann Neurol 52(6):802–813
14. Arvidsson A, Kokaia Z, Lindvall O (2001) N-methyl-D-aspartate receptor-mediated increase of neurogenesis in adult rat dentate gyrus following stroke. Eur J Neurosci 14(1):10–18
15. Zhang R et al (2004) Activated neural stem cells contribute to stroke-induced neurogenesis and neuroblast migration toward the infarct boundary in adult rats. J Cereb Blood Flow Metab 24(4):441–448
16. Yang ZJ et al (2005) Electroacupuncture enhances striatal neurogenesis in adult rat brains after a transient cerebral middle artery occlusion. Acupunct Electrother Res 30(3–4): 185–199
17. Arvidsson A et al (2002) Neuronal replacement from endogenous precursors in the adult brain after stroke. Nat Med 8(9):963–970
18. Song HJ, Stevens CF, Gage FH (2002) Neural stem cells from adult hippocampus develop essential properties of functional CNS neurons. Nat Neurosci 5(5):438–445
19. Doetsch F, Hen R (2005) Young and excitable: the function of new neurons in the adult mammalian brain. Curr Opin Neurobiol 15(1): 121–128
20. Kokaia Z, Lindvall O (2003) Neurogenesis after ischaemic brain insults. Curr Opin Neurobiol 13(1):127–132
21. Sharp FR, Liu J, Bernabeu R (2002) Neurogenesis following brain ischemia. Brain Res Dev Brain Res 134(1–2):23–30
22. Cayre M, Canoll P, Goldman JE (2009) Cell migration in the normal and pathological postnatal mammalian brain. Prog Neurobiol 88(1):41–63
23. Jin K et al (2003) Directed migration of neuronal precursors into the ischemic cerebral cortex and striatum. Mol Cell Neurosci 24(1): 171–189
24. Wang YQ et al (2009) VEGF enhance cortical newborn neurons and their neurite development in adult rat brain after cerebral ischemia. Neurochem Int 55(7):629–636
25. Zhang R et al (2006) Bcl-2 enhances neurogenesis and inhibits apoptosis of newborn neurons in adult rat brain following a transient middle cerebral artery occlusion. Neurobiol Dis 24(2):345–356
26. Drapeau E et al (2003) Spatial memory performances of aged rats in the water maze predict levels of hippocampal neurogenesis. Proc Natl Acad Sci U S A 100(24):14385–14390
27. Kempermann G, Kuhn HG, Gage FH (1997) More hippocampal neurons in adult mice living in an enriched environment. Nature 386(6624): 493–495
28. Kempermann G, Kuhn HG, Gage FH (1997) Genetic influence on neurogenesis in the dentate gyrus of adult mice. Proc Natl Acad Sci U S A 94(19):10409–10414

29. Kempermann G, Kuhn HG, Gage FH (1998) Experience-induced neurogenesis in the senescent dentate gyrus. J Neurosci 18(9): 3206–3212
30. Ge S et al (2006) GABA regulates synaptic integration of newly generated neurons in the adult brain. Nature 439(7076):589–593
31. Shen W et al (2006) Activity-induced rapid synaptic maturation mediated by presynaptic cdc42 signaling. Neuron 50(3):401–414
32. Pan PY et al (2005) SNAP-29-mediated modulation of synaptic transmission in cultured hippocampal neurons. J Biol Chem 280(27): 25769–25779

Chapter 18

Assessment of Angiogenesis in Models of Focal Cerebral Ischemia

Hong Shi, Yanqin Gao, and Weimin Liang

Abstract

Angiogenesis could be interpreted as a natural defense mechanism, helping to restore oxygen and nutrient supplies to the affected brain tissue. It has been demonstrated that angiogenesis is involved in functional recovery after ischemic stroke. This chapter introduces the classic methods for assessing angiogenesis after brain ischemia. Tissue preparation, staining procedures, and evaluation methods are discussed.

Key words: Angiogenesis, Cerebral ischemia, BrdU, Tomato lectin

1. Introduction

Time is brain for patients suffering a stroke. Unfortunately, very few patients receive effective medicine within 4.5 h after the onset of stroke (1). Recently, scientists have intensified their work on different approaches beyond the hyperacute phase of stroke. Neurorestorative processes, which allow a far greater time window for improving neurological recovery, include angiogenesis, neurogenesis, and synaptic plasticity (2).

Angiogenesis is a physiological process during development; however, during some pathological events such as stroke, endothelial cells become activated and angiogenesis ensues to provide conduits for blood flow (3, 4). Angiogenic vessels provide neurotrophic support to newly generated neurons. Thus, angiogenesis plays a crucial role in the recovery of blood flow in affected brain tissues. It has been suggested that greater microvessel density in the ischemic border correlates with longer survival in stroke patients (5–10). Increasing evidence suggests that angiogenesis plays a key role in neurological recovery after stroke.

Jun Chen et al. (eds.), *Animal Models of Acute Neurological Injuries II*, Springer Protocols Handbooks, DOI 10.1007/978-1-61779-782-8_18,

There are several methods to assess angiogenesis after ischemic stroke, including immunohistochemistry and magnetic resonance imaging (MRI). By both of them, we can observe the situation of angiogenesis. However, MRI needs more conditions and expense. Thus, the current research method for studying angiogenesis after cerebral ischemia mainly is immunohistochemistry. We summarize the morphological assessments of angiogenesis in focal cerebral ischemia as below.

2. Materials and Methods

2.1. Antibodies

1. 5-bromo-2-deoxyuridine (BrdU): BrdU can be incorporated into the newly synthesized DNA of replicating cells, substituting for thymidine during DNA replication. So anti-BrdU can be used to detect the incorporated chemical, thus indicating cells that were actively replicating their DNA after cerebral ischemia.
2. Tomato lectin: Tomato lectin is a very stable glycoprotein containing about 50% arabinose and galactose. It is an effective marker of blood vessels and microglial cells in rodents. Conjugation of the lectin with a fluorophore facilitates fast one-step detection and visualization using intravascular perfusion methods or direct application to tissue sections.

2.2. Staining Procedures

2.2.1. Tissue Preparation

1. Perform transcardial perfusion on the animal using cold phosphate-buffered saline (PBS) and 4% paraformaldehyde. Animals received daily intraperitoneal injections of BrdU (50 mg/kg; Sigma, St. Louis, MO).
2. Remove brains and postfix overnight in 4% paraformaldehyde.
3. Immerse brains in 15% sucrose in PBS for 24 h or until the brains sink.
4. Immerse brains in 30% sucrose in PBS until the brains sink.
5. Cut the brains coronally using a cryostat for frozen sections and store the sections at –20°C until used; or embed brains in paraffin blocks and then cut them for paraffin sections.

2.2.2. Pretreatments of Sections

Frozen sections do not require pretreatment. Paraffin sections need to be deparaffinized as follows.

1. Immerse sections in xylene for 5 min ×2.
2. Immerse sections in 100% ethanol for 3 min ×2.
3. Hydrate with 95% ethanol for 2 min.
4. Rehydrate with 70% ethanol for 2 min.
5. Rinse in distilled water.
6. Optional: Perform antigen retrieval.

2.2.3. Staining

1. Wash sections in 0.01 M PBS (pH 7.4) with 1% Triton X-100 (3× for 5 min).
2. Incubate sections in HCl (1 N) for 60 min at 37°C to break open the DNA structure of the labeled cells.
3. Wash sections for 10 min in 0.1 M boric acid (pH 8.5).
4. Wash sections in 0.01 M PBS (pH 7.4) with 1% Triton X-100 (3× for 5 min) at r.t.
5. Incubate in 0.01 M PBS (pH 7.4) + 0.2% Triton X-100 + 10% normal goat serum (1 h) prior to incubating overnight at 4°C with anti-BrdU or a combination of anti-BrdU and other antibodies.
6. Following the incubation overnight, wash in 0.01 M PBS (pH 7.4) with 0.3% Triton X-100 (3× for 5 min).
7. Samples can then be treated with a variety of secondary antibodies to visualize the anti-BrdU-labeled cells, including fluorescent-conjugated antibodies (Fig. 1).

2.3. Evaluation of Angiogenesis

Vascular density is evaluated by two- or three-dimensional analysis in the peri-infarct regions of ischemic brain.

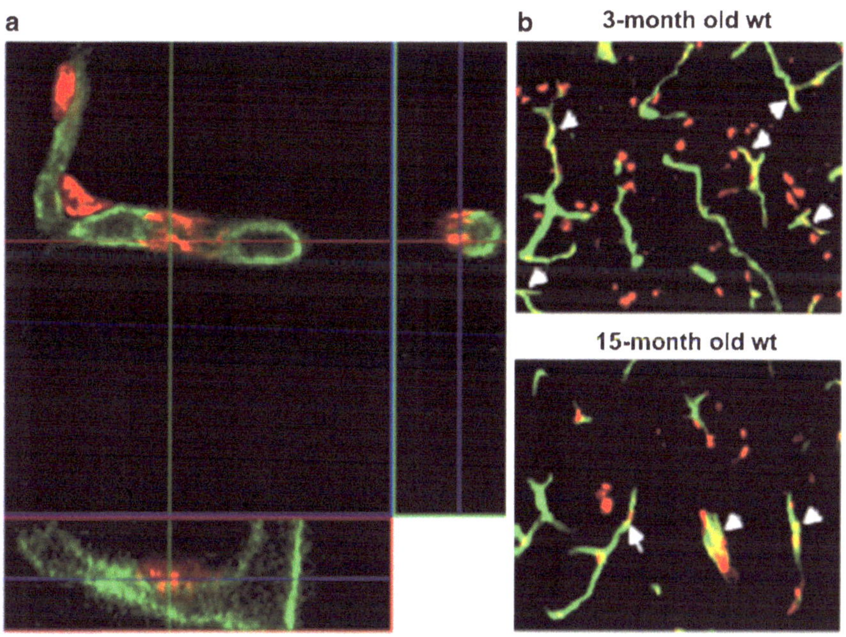

Fig. 1. Representative photos of post-stroke angiogenesis in mouse brain. The mouse was subjected to 90 min of focal cerebral ischemia induced by left middle cerebral artery occlusion. (**a**) Three-dimensional confocal image of double-labeled immunofluorescence for BrdU (*red*) and tomato lectin (*green*). (**b**) Confocal projection images of double-label immunofluorescent staining for BrdU (*red*) and tomato lectin (*green*) in peri-infarct regions 7 days after ischemia. The double-labeled microvessels display *yellow* (*arrows*).

2.3.1. Quantify the Number of Newly Generated Vessels

Staining with BrdU and PECAM-1, an endothelial cell marker, is done on brain sections. Accurate visualization of the newly generated microvascular network is achieved. The number of vessels double labeled with BrdU and PECAM-1, indicating blood vessels that contain newly born vascular, is quantified in the peri-infarct areas (including the ischemic boundary) and the corresponding contralateral areas.

2.3.2. Quantify the Number of Functional Vessels That Show BrdU Labeling

BrdU is administered as described above. Functional vessels are identified by transcardial administration (5 min before euthanasia) of FITC-conjugated tomato lectin (100 μg/μL), which labels endothelial cells only in perfused vessels. Brains are removed and coronal sections are cut using a microtome. BrdU staining is performed on brain sections. Accurate visualization of the newly generated functional microvascular network is achieved by 3D reconstruction. The number of lectin-perfused vessels that show BrdU staining is quantified in the peri-infarct areas and the corresponding contralateral areas.

2.3.3. Three-Dimensional Analysis of Vascular Density

Coronal sections (60 μm thick) are cut on a vibratome and mounted on slides. For each mouse brain, six coronal sections of 0.5-mm intervals are obtained and analyzed with a laser-scanning confocal microscopic (Olympus FluoView 1000) imaging system. Six regions of interest (ROIs) in the ipsilateral and contralateral hemispheres are selected. The ipsilateral ROIs are in the peri-infarct zone, and the corresponding regions in the contralateral hemisphere are used as controls. The ROIs are scanned in 512 × 512-pixel (233 × 233 μm) format in the *x–y* direction, and 1-μm-step-size optical sections along the *z* axis are acquired under a ×40 objective. Three-dimensional reconstruction is performed using an image analysis software package (3D Doctor 3.5, Able software, USA). The vascular surface area and the total vascular length per volume of tissue are calculated using the software, and the number of vascular branch points is counted in the three-dimensional images by an investigator who is blinded to the experiment design.

3. Summary

In summary, the current research methods for studying angiogenesis after cerebral ischemia are morphological staining techniques. In addition, MRI provides a powerful tool to noninvasively assess angiogenesis after stroke. The excellent soft tissue contrast, high spatial and temporal resolution, and versatility render MRI particularly suitable to monitor the dynamic processes involved in vascular remodeling after stroke (11–15). The future development of functional methods, such as positron emission tomography, will be helpful (16).

References

1. Rothwell P (2009) Is intravenous recombinant plasminogen activator effective up to 4.5 h after onset of ischemic stroke? Nat Clin Pract Cardiovasc Med 6:164–165
2. Chen J, Chopp M (2006) Neurorestorative treatment of stroke: cell and pharmacological approaches. NeuroRx 3:466–473
3. Shi H, Li P, Liang W, Chen J, Gao Y (2010) Mechanisms of microRNA-mediated regulation of angiogenesis. Front Biosci 6:1304–1319
4. Pober JS, Sessa WC (2007) Evolving functions of endothelial cells in inflammation. Nat Rev Immunol 7:803–815
5. Rafii S, Lyden D (2003) Therapeutic stem and progenitor cell transplantation for organ vascularization and regeneration. Nat Med 9: 702–712
6. Leventhal C, Rafii S, Rafii D, Shahar A, Goldman SA (1999) Endothelial trophic support of neuronal production and recruitment from the adult mammalian subependyma. Mol Cell Neurosci 13:450–464
7. Thored P, Wood J, Arvidsson A, Cammenga J, Kokaia Z, Lindvall O (2007) Long-term neuroblast migration along blood vessels in an area with transient angiogenesis and increased vascularization after stroke. Stroke 38: 3032–3039
8. Teng H, Zhang ZG, Wang L, Zhang RL, Zhang L, Morris D, Gregg SR, Wu Z, Jiang A, Lu M, Zlokovic BV, Chopp M (2008) Coupling of angiogenesis and neurogenesis in cultured endothelial cells and neural progenitor cells after stroke. J Cereb Blood Flow Metab 28: 764–771
9. Krupinski J, Kaluza J, Kumar P, Kumar S, Wang JM (1994) Role of angiogenesis in patients with cerebral ischemic stroke. Stroke 25: 1794–1798
10. Wei L, Erinjeri JP, Rovainen CM, Woolsey TA (2001) Collateral growth and angiogenesis around cortical stroke. Stroke 32:2179–2184
11. Bardutzky J, Shen Q, Henninger N et al (2007) Characterizing tissue fate after transient cerebral ischemia of varying duration using quantitative diffusion and perfusion imaging. Stroke 38:1336–1344
12. Farr TD, Wegener S (2010) Use of magnetic resonance imaging to predict outcome after stroke: a review of experimental and clinical evidence. J Cereb Blood Flow Metab 30: 703–717
13. Dijkhuizen RM, Nicolay K (2003) Magnetic resonance imaging in experimental models of brain disorders. J Cereb Blood Flow Metab 23:1383–1402
14. Baird AE, Warach S (1998) Magnetic resonance imaging of acute stroke. J Cereb Blood Flow Metab 18:583–609
15. Seevinck PR, Deddens LH, Dijkhuizen RM (2010) Magnetic resonance imaging of brain angiogenesis after stroke. Angiogenesis 13:101–111
16. Cai W, Guzman R, Hsu AR et al (2009) Positron emission tomography imaging of poststroke angiogenesis. Stroke 40:270–277

Part III

Neonatal Hypoxia-Ischemia

Chapter 19

Morphological Assessments of Neonatal Hypoxia–Ischemia: In Situ Cell Degeneration

Changlian Zhu and Klas Blomgren

Abstract

Cell death in the immature brain can be studied in many ways using morphological and biochemical markers. Essential requirements for cell death and degeneration assessment techniques in the brain include sufficient sensitivity and the ability to differentiate between apoptotic and necrotic cell death, and different forms of DNA damage, and to distinguish between different stages in the cell death process. Detection of cell death by morphology still remains the gold standard. Morphological analyses can also be combined with immunohistochemistry or in situ labeling using different types of specific antibodies or markers to identify cell types undergoing degeneration. This chapter introduces widely used methods to detect different types of cell death in the immature brain after hypoxic-ischemic brain injury by using morphological assessment combined with tissue staining. The procedures, including tissue preparation, staining, and evaluation, are described.

Key words: TUNEL, Apoptosis, Necrosis, Autophagy, Cell death, Caspase, AIF

1. Introduction

Cell death is usually classified as apoptotic or necrotic based on biochemical and morphological criteria (1, 2). Necrosis refers to a passive and acute process characterized by irreversible changes in the cytoplasm (condensation and intense eosinophilia, mitochondrial swelling, loss of structure and fragmentation), cytoplasmic membrane (loss of selective permeability), and the nucleus (karyolysis). These morphologies of necrosis are followed by membrane lysis and release of intracellular contents, inducing an inflammatory response (3). Necrosis occurs only under pathological conditions. Apoptosis refers to a selective programmed type of cell death characterized by cell and nuclear shrinkage, chromatin condensation, and nuclear fragmentation to form multiple small membrane-bound bodies that contain intact organelles and/or dense clumps

Jun Chen et al. (eds.), *Animal Models of Acute Neurological Injuries II*, Springer Protocols Handbooks,
DOI 10.1007/978-1-61779-782-8_19,

of condensed chromatin, the so-called apoptotic bodies (4, 5). Apoptotic cells are rapidly removed through phagocytosis without eliciting an inflammatory reaction. This cell death occurs both under physiological and pathological conditions (6, 7).

The immature brain has been suggested to retain part of the developmental cell death "program," and this program is activated following hypoxia–ischemia (HI), more so than in the adult brain (8). Studies using a combination of light and electron microscopy suggest, however, that cell death after HI is different from developmentally regulated cell death in most cases (9, 10). Nevertheless, hypoxic-ischemic cell death shares important morphological and biochemical features with apoptotic cell death.

A human study demonstrated mixed apoptotic and necrotic forms of cell death after perinatal HI. The form of cell death may depend also on the severity of ischemic injury (5, 11). Necrosis predominates after severe HI insults, developing during the insult (primary damage) or during early reperfusion, whereas apoptosis occurs in areas with milder ischemic injury, often appearing days after the initial insult (12). A hybrid morphology of cell death with features of both apoptosis and necrosis, often called pathological apoptosis, was observed in all brain regions at all time points in the neonatal HI model, even in the ischemic core (9, 13). Electron microscopic examination confirmed that a range of morphologies consistent with apoptosis, necrosis, and intermediate hybrid forms was common in the damage area (14–16).

Another form of cell death during ischemia is autophagocytic cell death. The term autophagy refers to a type of cell death in which organelles and other contents of the cytoplasm are actively degraded by lysosomal enzymes well before nuclear changes become visible. The most characteristic feature is the appearance of large autophagic vacuoles of lysosomal origin in the cytoplasm. The change in the intracellular localization of microtubule-associated protein 1 light chain 3 (LC3-II) protein provides the first molecular marker-based method for the detection of autophagic activity (17).

2. Materials and Instruments

2.1. Primary Antibodies

1. Active caspase-3 (BD, Phamingen, Franklin Lakes NJ, USA).
2. Microtubule-associated protein-2 (MAP2) (Sigma, Saint Louis, MO, USA).
3. Apoptosis-inducing factor (sc-9416, Santa Cruz Biotechnology, Santa Cruz, CA, USA).
4. Microtubule-associated protein 1 light chain 3 (LC3) (Cell Signaling, Danvers, MA, USA).

2.2. Secondary Antibodies

1. Biotinylated secondary antibodies and ABC Elite kit (Vector, Burlingame, CA, USA).
2. Fluorescent-conjugated secondary antibodies (Jackson Immunoresearch, West Grove, PA or Molecular Probes, Eugene, OR, USA).

2.3. Other Markers

1. TUNEL kits (Roche-Boehringer Mannheim, Mannheim, Germany).
2. Hairpin oligonucleotide probe (HPP).
3. Fluoro-Jade (Chemicon, Billerica, MA, USA).

2.4. Photography and Image Processing Tools

1. Microscope (electron, light, and epifluorescent).
2. Micro Image software (Olympus, Japan), Stereoinvestigator (MicroBrightField, Germany).

3. Procedures

3.1. Tissue Preparation for Light and Fluorescence Microscopy

1. Pups should be deeply anesthetized with 50 mg/mL pentobarbital.
2. Transcardial perfusion-fixation with cold 4% paraformaldehyde in 0.1 M PBS.
3. Rapidly remove the brains and immersion-fix them in 4% paraformaldehyde at 4°C for 24 h.
4. Dehydrate with graded ethanol and xylene, and embed in paraffin.
5. Cut brains into 5-μm coronal sections.

3.2. Fluoro-Jade Staining

1. Melt the paraffin sections onto the slides in an oven at 60°C for 30 min.
2. Deparaffinize in xylene and rehydrate in graded ethanol (70, 95, and 100%).
3. Rinse with distilled water 3× for 5 min.
4. Block with 0.06% potassium permanganate for 10 min on a shaker at room temp.
5. Rinse with distilled water for 2 min.
6. Incubate with 0.0004% fresh Fluoro-Jade solution for 20 min.
7. Rinse with distilled water 3× for 1 min.
8. Dehydrate with graded ethanol, and then xylene, each for 5 min.
9. Coverslip with mounting medium.

3.3. TUNEL Staining

1. Melt the sections onto the slides in an oven at 60°C for 30 min.
2. Deparaffinize in xylene and rehydrate in graded ethanol as above.
3. Rinse with PBS 3× for 5 min.
4. Heat in boiling 0.01 M citrate buffer, pH 6.0, for 10 min.
5. Rinse with PBS 3× for 5 min.
6. Incubate with 0.1 M Tris–HCl, pH 7.5, containing 3% BSA for 30 min.
7. Rinse with PBS, then add 50 μL TUNEL reaction mixture on each section (450 μL label solution and 50 μL enzyme), and incubate in a humidified box at 37°C for 60 min.
8. Rinse with PBS and coverslip with mounting medium; check with fluorescent microscopy.
9. For 3,3'-diaminobenzidine (DAB) staining, incubate with 3% H_2O_2 in PBS for 10 min.
10. Rinse with PBS three times, and incubate with peroxidase (POD)-conjugated anti-fluorescein 1:5 (100 μL POD and 400 μL buffer) with blocking solution (0.1 M Tris–HCl, pH 7.5, containing 3% BSA) in a humidified box at 37°C for 30 min.
11. Rinse with PBS three times, and incubate with ABC solution for 60 min.
12. Rinse with PBS three times, and incubate with 0.05% DAB for up to 5 min.
13. Stop reaction with 0.1 M NaAc, pH 6.0, for 5 min.
14. Dehydrate with graded ethanol 70, 95, and 100%, and then xylene, each for 5 min.
15. Coverslip with mounting medium.

3.4. HPP Staining

1. Melt the sections onto the slides in an oven at 60°C for 30 min.
2. Deparaffinize in xylene and rehydrate in graded ethanol as above.
3. Rinse with PBS 3× for 5 min.
4. Heat in 0.01 M citrate buffer, pH 6.0, for 90 min at 65°C.
5. Rinse with PBS 3× for 5 min.
6. Incubate with 25 μg/mL proteinase K in PBS at room temperature for 12 min.
7. Rinse with PBS and then add 50 μL HPP solution (50 mmol/L Tris–HCl, pH 7.8, 10 mmol/L, $MgCl_2$, 10 mmol/L dithiothreitol, 1 mmol/L ATP, 15% PEG 6000, HPP 35 μg/mL,

DNA T4 ligase 250 U/mL) in a humidified box at room temperature for 2 h.

8. Rinse with PBS and incubate with 3% H_2O_2 in PBS for 10 min.
9. Rinse with PBS three times, and incubate with ABC solution for 60 min.
10. Rinse with PBS three times, and incubate with 0.05% DAB for up to 5 min.
11. Stop the reaction with 0.1 M NaAc, pH 6.0, for 5 min.
12. Dehydrate with graded ethanol (70, 95, and 100%), and then xylene, each for 5 min.
13. Coverslip with mounting medium.

3.5. MAP2, AIF, Active Caspase-3, and LC3 Staining

1. Melt the sections onto the slides in an oven at 60°C for 30 min.
2. Deparaffinize sections in xylene and rehydrate in graded ethanol as above.
3. Rinse with PBS 3× for 5 min.
4. Heat in boiling 0.01 M citrate buffer, pH 6.0, for 10 min.
5. Rinse with PBS 3× for 5 min.
6. Incubate (block) with 5% normal serum for 30 min.
7. Incubate sections with primary antibody at an appropriate dilution for 60 min at room temperature or overnight at 4°C.
8. Rinse with PBS, and incubate sections with either biotinylated or fluorescent probe-conjugated secondary antibody at an appropriate dilution in PBS. Protect the sections from light when a fluorescent probe-conjugated secondary antibody is used.
9. For fluorescent staining, coverslip with mounting medium after washing.
10. For DAB staining, incubate sections with ABC for 60 min.
11. Rinse with PBS three times, and incubate with 0.05% DAB for up to 5 min.
12. Stop reaction with 0.1 M NaAc, pH 6.0, for 5 min.
13. Dehydrate with graded ethanol 70, 95, and 100%, and then xylene, each for 5 min.
14. Coverslip with mounting medium.

3.6. Transmission Electron Microscopy

1. Cut brain sections (100 μm) with a vibratome, and postfix with 4% glutaraldehyde in 0.1 mM cacodylate buffer (pH 7.4) for 1 h and then with 1% osmium tetroxide in 0.1 M cacodylate buffer for 2 h.

2. After rinsing with distilled water, treat brain sections with 1% aqueous uranyl acetate overnight, dehydrate in an ascending series of ethanol to 100%, followed by dry acetone, and then embed in Durcopan ACM.
3. Cut ultrathin sections (0.1 μm), stain with 3% lead citrate, and subsequently examine using a transmission electron microscope.

4. Evaluation

Brain injury can be evaluated quantitatively by measuring the MAP2-positive or -negative areas of each hemisphere or semi-quantitatively by using a neuropathological score (18). The MAP2-positive and -negative areas in each section (Fig. 1) are measured using Micro Image (Olympus, Japan) or Stereoinvestigator (MicroBrightfield, Germany). The infarct volume is calculated from the MAP2-negative areas according to the Cavalieri's principle using the following formula: $V = \Sigma A \times P \times T$, where V = total volume, ΣA = sum of area measurement, P = the inverse of the sampling fraction, and T = the section thickness. The total hemispheric tissue loss was calculated as the MAP2-positive volume in the contralateral hemisphere minus the MAP2-positive volume in the ipsilateral hemisphere. The neuropathological scores for the cortex, hippocampus, striatum, and thalamus are assessed as described previously (18). Briefly, serial sections from each brain are stained for MAP2 and scored by an observer blinded to the treatment of the animals. The cortical injury is graded from 0 to 4, where 0 indicates no observable injury and 4 confluent infarction encompassing most of the cerebral cortex. The damage in the hippocampus, striatum, and thalamus is assessed both with respect to hypotrophy (0–3) and observable cell injury/infarction (0–3) resulting in a neuropathological score for each brain region (0–6).

Morphological criteria remain the gold standard for detection of necrotic, apoptotic, or hybrid-type cell death, but for practical reasons biochemical surrogate markers are more common (19). The Fluoro-Jade stain is a fluorochrome derived from fluorescein, and is used to label degenerating neurons. The stained cells can be visualized under a fluorescent microscope and quantified within a region or using stereological procedures to estimate the total number of Fluoro-Jade-positive cells within the defined region (20) (Fig. 2). TUNEL (Fig. 3) and HPP (Fig. 4) in situ labeling detect double-stranded DNA damage (7). TUNEL is still a widely used method to detect cell death, both necrosis and apoptosis (not just

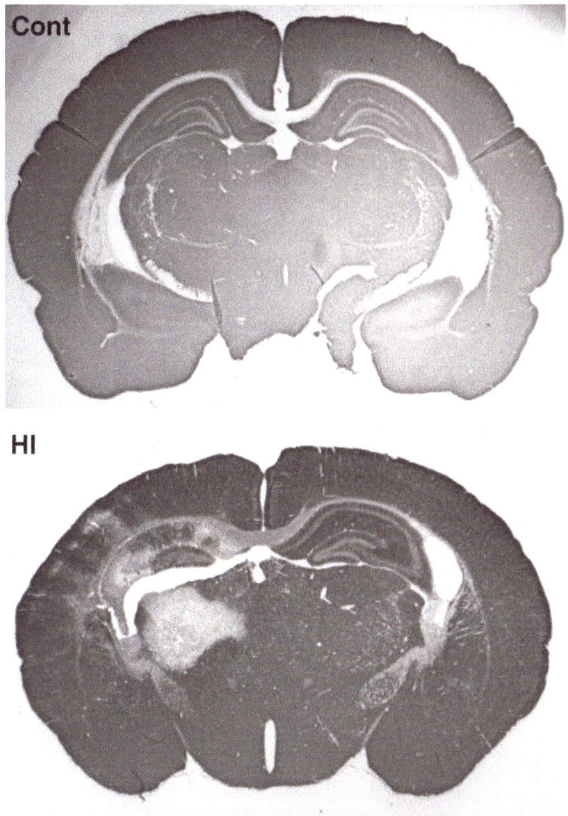

Fig. 1. MAP2 staining in a normal control brain (*upper*) and after HI (*lower*). MAP2-negative areas are located in the ipsilateral hemisphere (*left* in the picture), including the cortex, hippocampus, thalamus, and striatum (magnification ×10).

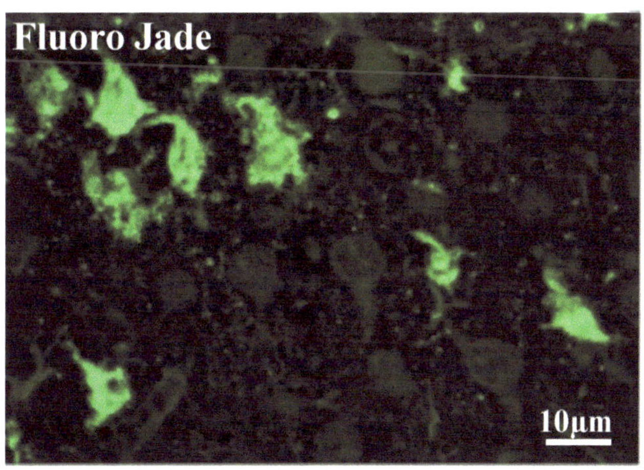

Fig. 2. Fluoro-Jade staining showing injured neurons in the cortex of the ipsilateral hemisphere 24 h after HI (scale bar = 10 μm).

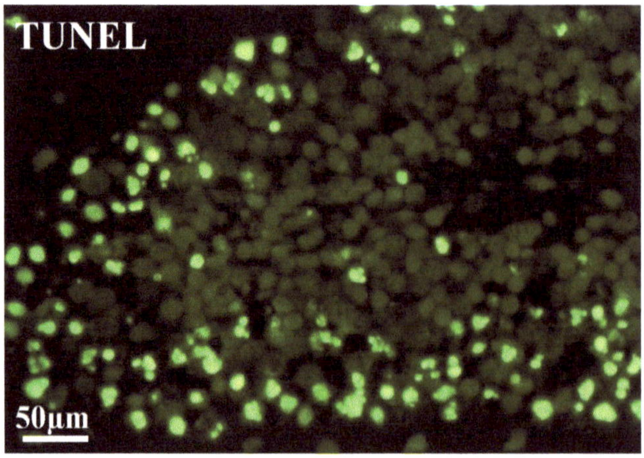

Fig. 3. TUNEL labeling showing DNA damage in granule neurons in the dentate gyrus of the hippocampus of the ipsilateral hemisphere 24 h after HI (scale bar = 50 μm).

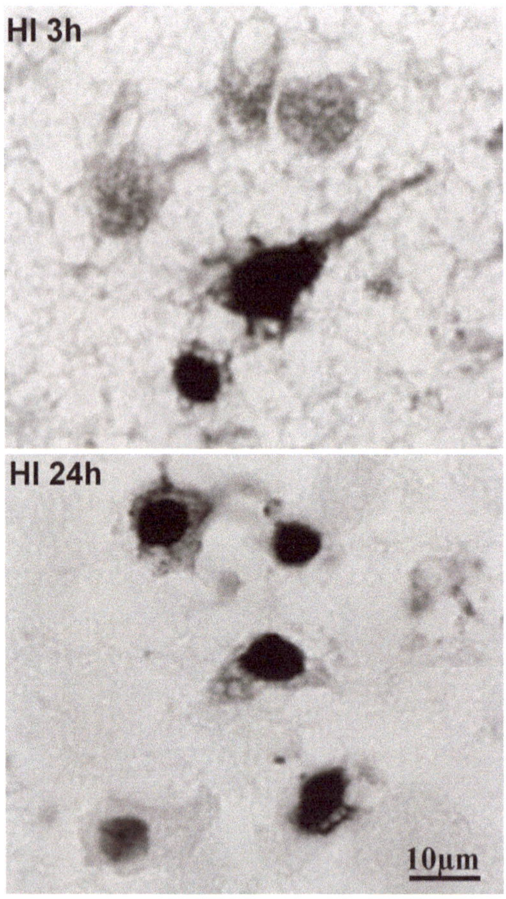

Fig. 4. HPP labeling showing apoptosis-related, double-strand DNA damage in neurons in the cortex 3 h (*upper*) and 24 h (*lower*) after HI (scale bar = 10 μm).

apoptosis as was believed initially). A biotinylated HPP can be used to detect double-stranded DNA breaks with a higher sensitivity and specificity than TUNEL (correlated more closely with caspase-3 activation) (7). An antibody against single-stranded DNA was claimed to be specific to apoptotic cell death, but was not useful in the immature brain (7). Cells immunopositive for active caspase-3 (Fig. 5) and nuclear AIF (Fig. 6) can be counted in different brain regions, either at selected levels only or comprehensively using stereological principles. AIF is located in the mitochondria in normal cells and is translocated to the nucleus in injured cells (21). When using stereology, the counting area can be traced and the positive cells can be expressed as total number or density in a given region. EM identification of autophagosomes is still the gold standard to detect autophagy (Fig. 7) (22). LC3 staining, as indicated by punctate staining in the cytoplasm (Fig. 8), has been accepted as a surrogate marker of autophagy.

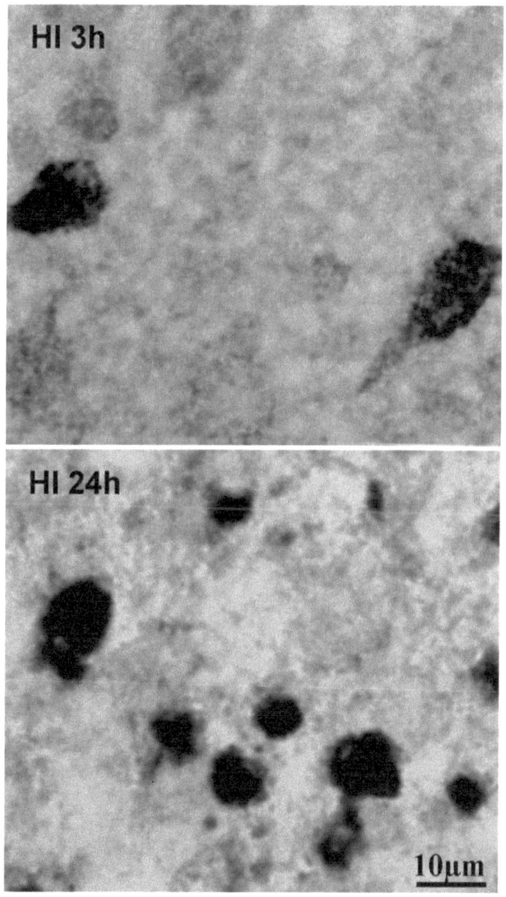

Fig. 5. Active caspase-3 staining showing immunopositive neurons 3 h (*upper*) and 24 h (*lower*) after HI in the cortex (scale bar = 10 μm).

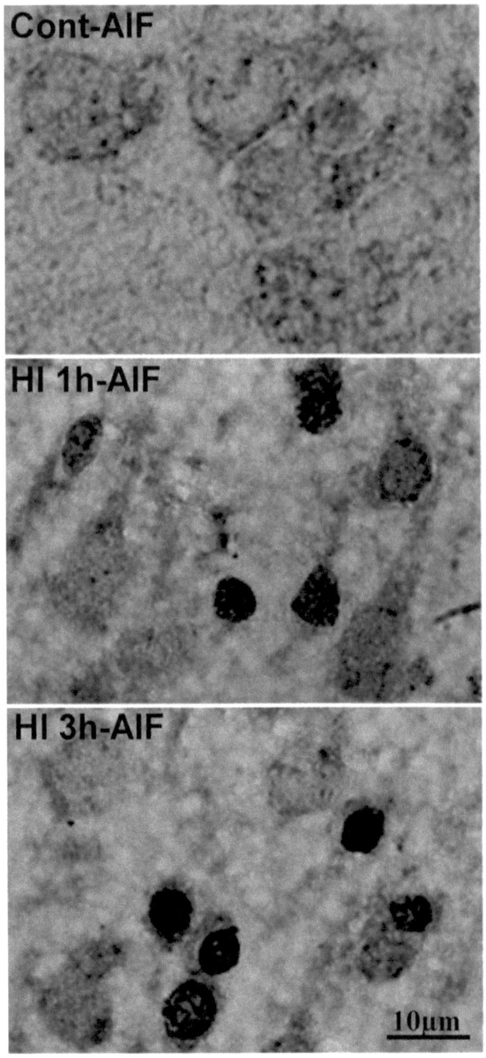

Fig. 6. AIF staining in the normal control (CONT, *upper*), 1 h (*middle*) and 3 h (*lower*) after HI in the cortex (scale bar = 10 μm).

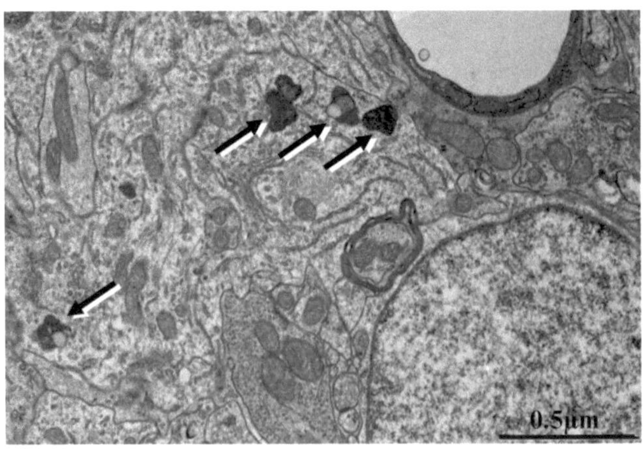

Fig. 7. A transmission electron microscopy image showing dense autophagosomes (*arrows*) in the mouse brain (scale bar = 0.5 μm).

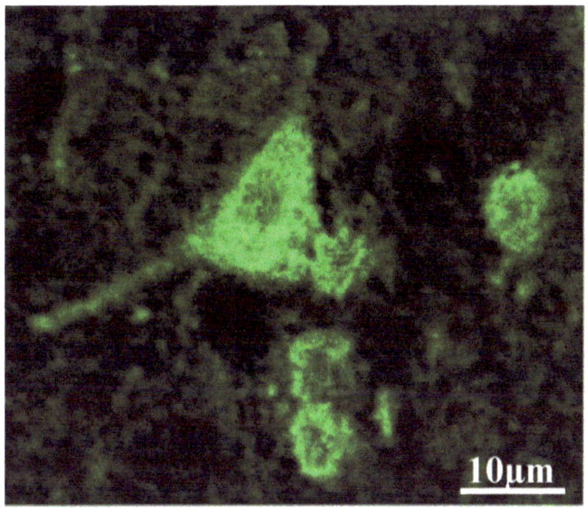

Fig. 8. LC3 staining showing punctate staining typical of autophagy (scale bar = 10 μm).

5. Summary

Morphological criteria still remain the gold standard for detection of different types of cell death, but these do not easily lend themselves to quantification, particularly not for detection of cell death and degeneration in the immature brain after HI. Paraffin-embedded sections are preferred mainly because of superior morphology and because the free-floating approach leads to separation of injured and noninjured tissue in the wells. The Fluoro-Jade stain is a very simple and reliable staining method to detect neuronal degeneration (23). DNA damage is easily detected using TUNEL (general cell death) or HPP (apoptosis). Active caspase-3 detects caspase-dependent apoptosis. Nuclear translocation of AIF indicates caspase-independent apoptosis (24). LC3 staining showing punctate staining in the cytoplasm is now an accepted hallmark of autophagy (25).

References

1. Martin LJ et al (1998) Neurodegeneration in excitotoxicity, global cerebral ischemia, and target deprivation: a perspective on the contributions of apoptosis and necrosis. Brain Res Bull 46(4):281–309
2. MacManus JP, Linnik MD (1997) Gene expression induced by cerebral ischemia: an apoptotic perspective. J Cereb Blood Flow Metab 17(8):815–832
3. Majno G, Joris I (1995) Apoptosis, oncosis, and necrosis. An overview of cell death. Am J Pathol 146(1):3–15
4. Wyllie AH, Beattie GJ, Hargreaves AD (1981) Chromatin changes in apoptosis. Histochem J 13(4):681–692
5. Kerr JF, Wyllie AH, Currie AR (1972) Apoptosis: a basic biological phenomenon with wide-ranging implications in tissue kinetics. Br J Cancer 26(4):239–257
6. Oppenheim RW et al (2008) Developing postmitotic mammalian neurons in vivo lacking Apaf-1 undergo programmed cell death by a caspase-independent, nonapoptotic pathway involving autophagy. J Neurosci 28(6):1490–1497

7. Zhu C et al (2000) Correlation between caspase-3 activation and three different markers of DNA damage in neonatal cerebral hypoxia-ischemia. J Neurochem 75(2):819–829
8. Sidhu RS, Tuor UI, Del Bigio MR (1997) Nuclear condensation and fragmentation following cerebral hypoxia-ischemia occurs more frequently in immature than older rats. Neurosci Lett 223(2):129–132
9. Portera-Cailliau C, Price DL, Martin LJ (1997) Excitotoxic neuronal death in the immature brain is an apoptosis-necrosis morphological continuum. J Comp Neurol 378(1):70–87
10. Ishimaru MJ et al (1999) Distinguishing excitotoxic from apoptotic neurodegeneration in the developing rat brain. J Comp Neurol 408(4):461–476
11. Bonfoco E et al (1995) Apoptosis and necrosis: two distinct events induced, respectively, by mild and intense insults with N-methyl-D-aspartate or nitric oxide/superoxide in cortical cell cultures. Proc Natl Acad Sci U S A 92(16):7162–7166
12. Stroemer RP, Rothwell NJ (1998) Exacerbation of ischemic brain damage by localized striatal injection of interleukin-1beta in the rat. J Cereb Blood Flow Metab 18(8):833–839
13. Nakajima W et al (2000) Apoptosis has a prolonged role in the neurodegeneration after hypoxic ischemia in the newborn rat. J Neurosci 20(21):7994–8004
14. Johnston MV, Nakajima W, Hagberg H (2002) Mechanisms of hypoxic neurodegeneration in the developing brain. Neuroscientist 8(3):212–220
15. Puka-Sundvall M et al (2000) Impairment of mitochondrial respiration after cerebral hypoxia-ischemia in immature rats: relationship to activation of caspase-3 and neuronal injury. Brain Res Dev Brain Res 125(1–2):43–50
16. Puka-Sundvall M et al (2000) Subcellular distribution of calcium and ultrastructural changes after cerebral hypoxia-ischemia in immature rats. Brain Res Dev Brain Res 125(1–2):31–41
17. Gozuacik D, Kimchi A (2004) Autophagy as a cell death and tumor suppressor mechanism. Oncogene 23(16):2891–2906
18. Zhu C et al (2005) The influence of age on apoptotic and other mechanisms of cell death after cerebral hypoxia-ischemia. Cell Death Differ 12(2):162–176
19. Blomgren K, Leist M, Groc L (2007) Pathological apoptosis in the developing brain. Apoptosis 12(5):993–1010
20. Slikker W Jr et al (2007) Ketamine-induced neuronal cell death in the perinatal rhesus monkey. Toxicol Sci 98(1):145–158
21. Zhu C et al (2003) Involvement of apoptosis-inducing factor in neuronal death after hypoxia-ischemia in the neonatal rat brain. J Neurochem 86(2):306–317
22. Galluzzi L, Blomgren K, Kroemer G (2009) Mitochondrial membrane permeabilization in neuronal injury. Nat Rev Neurosci 10(7):481–494
23. Schmued LC, Albertson C, Slikker W Jr (1997) Fluoro-Jade: a novel fluorochrome for the sensitive and reliable histochemical localization of neuronal degeneration. Brain Res 751(1):37–46
24. Zhu C et al (2006) Different apoptotic mechanisms are activated in male and female brains after neonatal hypoxia-ischaemia. J Neurochem 96(4):1016–1027
25. Puyal J et al (2009) Postischemic treatment of neonatal cerebral ischemia should target autophagy. Ann Neurol 66(3):378–389

Chapter 20

Morphological Assessments of Neonatal Hypoxia–Ischemia: White Matter and Blood-Brain Barrier Injury

Dianer Yang and Chia-Yi Kuan

Abstract

Damages to the blood–brain barrier (BBB) and white matter (WM)/oligodendrocytes (OLs) are typical pathological findings in infants or animal models of neonatal hypoxia–ischemia (NHI). These injuries in turn produce severe neurological consequences, such as germinal matrix hemorrhage, periventricular leukomalacia, and cerebral palsy. Hence, the assessment of BBB and WM/OL damage, and its attenuation by experimental therapies, is an important component in the evaluation of NHI-induced brain injury. Immature brains have unique physiology and a different maturation status in the BBB and WM from the adult brain. Accordingly, there is a need of designing detection methodology tailored for the immature brain. This chapter describes several simple but reliable methods as well as some advanced imaging methods to assess the BBB and WM/OL damage following NHI. Specifically, we discuss sodium fluorescein-based assay of BBB permeability, zymography to evaluate matrix metalloproteinases, histological and immunocytochemical analysis of WM/OL injury, and magnetic resonance imaging-based quantification of gray-matter and WM injury.

Key words: Blood–brain barrier, White matter, Matrix metalloproteinases, Diffusion tensor imaging

1. Introduction

The Rice-Vannucci model of rodent neonatal cerebral hypoxia–ischemia (HI) uses a procedure, originally devised by S. Levine in 1960, consisting of ligation of unilateral carotid artery followed by a period of systemic hypoxia (1). In this model, the mean systemic blood pressure decreases 25–30% during hypoxia. The cerebral blood flow decreases to 40–60% of the control value in the hemisphere ipsilateral to carotid ligation, but recovers to baseline immediately upon return to normoxic conditions. As such, the Rice-Vannucci model does not produce total blockade of focal cerebral vessels. Rather, it inflicts a period of hypoxia and

Jun Chen et al. (eds.), *Animal Models of Acute Neurological Injuries II*, Springer Protocols Handbooks,
DOI 10.1007/978-1-61779-782-8_20, © Springer Science+Business Media, LLC 2012

hypoperfusion in immature brains, similar to perinatal asphyxia in human infants. The Rice-Vannucci model is typically performed in 7-day-old rat pups, which are in a stage of brain development similar to near-term gestation in human. The patterns of resultant brain damage are also similar to those observed clinically, such as PVL and cystic encephalomalacia. Moreover, the Rice-Vannucci model uses a simple surgical procedure (common carotid artery ligation) and objective parameters—the age of animals, length of hypoxia, and temperature of the hypoxic chamber are all controllable—to produce persistent brain injury and a low mortality rate (<5%) in challenged animals. These are all important merits for an experiment paradigm of human diseases. A strong testimony for its clinical relevance lies in the fact that hypothermia was first demonstrated neuroprotective in the Rice-Vannucci model, 12 years before its successful application in clinical trial (2).

An important new development of the Rice-Vannuci model is the intraperitoneal (IP) administration of bacterial endotoxin lipopolysaccharide (LPS) to rat pups at 4–6 h before HI to enhance the extent of brain damage (3). It is well known that intrauterine infection and the fetal inflammatory response either directly lead to white-matter (WM) injury or increase the brain vulnerability to secondary perinatal insults, such as HI. The introduction of LPS-sensitized Rice-Vannucci model offers a simple model to study the underlying mechanisms of premature brain injury associated with intrauterine infection and search for more effective therapies.

In Volume I of this book, we have introduced a simple method to quantify HI-induced brain tissue loss based on the Nissl stain (4). Chapter 23 by Iwai and Chen in this volume provides an excellent review of the methods for neurobehavioral assessment after neonatal HI (NHI) (5). Thus, we focus on more specialized methods to assess the damage to blood–brain barrier (BBB) and WM/oligodendrocytes (OLs) after NHI challenges. We have used most of these methods in our recent study of the toxicity of tissue-type plasminogen activator (tPA) in NHI (6, 7).

2. Assessment of BBB Permeability

The most popular method to assess BBB permeability in animal models of adult stroke is by intravenous administration of the fluorescent Evans Blue (EB) dye. The EB dye binds to albumin in the blood stream and leaks to brain parenchyma through damaged vascular wall. However, in our experience, NHI rarely produces massive leakage of EB-bound albumin in newborn rodent brains. In contrast, the use of small fluorescent molecules, such as sodium fluorescein (NaF), is more sensitive (Fig. 1a, b). These differential

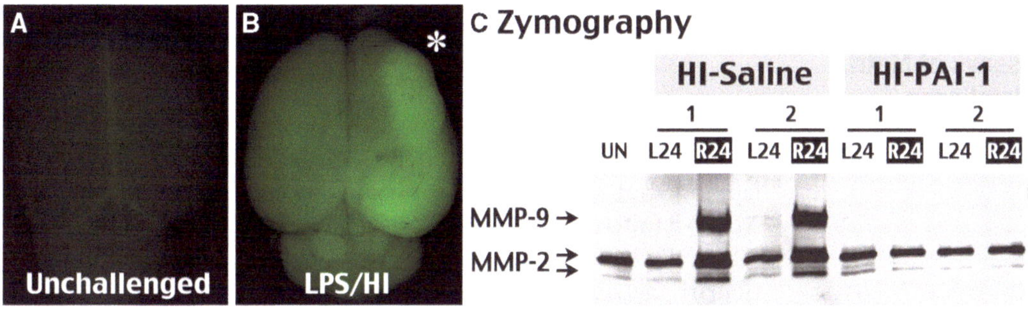

Fig. 1. Assays for the blood–brain barrier permeability and injury following neonatal cerebral hypoxia–ischemia. (**a**, **b**) Images of the brains from unchallenged or LPS/HI-injured mice that received intraperitoneal injection of sodium fluorescein (NaF) at 24 h recovery. *Asterisk*: The unilaterally HI-challenged side of the brain. (**c**) Zymography shows unilateral induction of MMP-9 and MMP-2 at 24 h after HI challenge and cerebroventricular injection of saline (HI saline). In contrast, cerebroventricular injection of plasminogen activator inhibitor 1 (PAI-1) blocked the induction of MMPs (**c** is modified from Yang et al. (7)).

behaviors may relate to the fact that large molecules, such as EB-bound albumin, are mainly transported across the BBB by pinocytotic vesicles in endothelial cells, whereas small molecules, such as NaF, enter the brain in a paracellular route involving passage through interendothelial clefts (8).

2.1. Procedures

1. Prepare 5% NaF (Sigma-Aldrich, F6377) solution in phosphate-buffered saline (PBS).
2. Intraperitoneal injection of 5% NaF at the dose of 10 μL/g of body weight.
3. Wait for 2 h, and then perform transcardial perfusion using PBS to wash out excess NaF in the blood stream.
4. Remove the brain, and inspect it under fluorescence stereomicroscope. The typical excitation-emission wavelength and filter for fluorescein-based secondary antibody or green fluorescence protein (GFP) will work. The excitation wavelength shall be in the region 420–480 nm. The yellow/green fluorescence at about 530–540 nm can be recorded by digital camera.
5. To quantify the relative BBB permeability to contralateral hemisphere, cut the cerebral cortex into halves. Homogenize it with 500 μL trichloroacetic acid (80%) to precipitate the proteins, followed by centrifugation for 10 min at 13,600 × *g*.
6. Add 150 μL supernatant and 50 μL 5 M NaOH in triplicates into 96-well microplate and measure the absorption at 490 nm using a spectrophotometer.

3. Biochemical Assays of Matrix Metalloproteinases

The matrix metalloproteinases (MMPs) constitute a multigene family of zinc- and calcium-dependent endopeptidases with overlapping but preferential substrate specificities. Among them, MMP-2 (also called Gelatinase A or 72 kDa gelatinase) and MMP-9 (also called Gelatinase B or 92 kDa gelatinase) are considered particularly detrimental to cerebral blood vessels. We and others have shown that the timing of MMP-9/2 induction following the Rice-Vannucci model of NHI is at 18–24 h post injury (6, 7). We have also shown that therapeutic administration of neuroprotective agents, such as plasminogen activator inhibitor-1 (PAI-1), mitigates the induction of MMP-9/2 (Fig. 1c).

There are several methods to measure the amount and activity of MMPs in tissues, with substrate gel electrophoresis (zymography) methods being the most commonly employed. In this method, a protein mixture is separated by substrate-containing sodium dodecyl sulfate-polyacrylamide gel electrophoresis (SDS-PAGE) under nonreducing conditions. Following SDS removal and incubation in a buffer that allows enzymatic activity to take place, the gel can be stained for protein, thus revealing proteolytic activity as clear bands against uniformly stained substrate in a dark background. Besides being a highly sensitive method, zymography also provides molecular mass and proteinase class information. Moreover, the denaturation–renaturation process reveals both latent and active forms of MMPs that differ slightly in molecular weight.

3.1. Procedures

1. Prepare brain protein samples using RIPA buffer (50 mM Tris, pH 8.0, 150 mM NaCl, 0.1% SDS, 0.5% sodium deoxycholate, 1% Triton X-100 or NP-40, and cocktail of protease inhibitors), and normalized to equal protein concentration.
2. To enrich MMP-9/2 in assayed samples, aliquots of protein extracts were incubated with gelatin sepharose beads (Amersham 17-0956-01) in RIPA buffer at 4°C overnight.
3. Wash beads twice with RIPA buffer. Mix protein samples with equal volume of the sample buffer (0.5 M Tris–HCl, pH 6.8, 100% glycerol, 0.05% bromophenol blue, and 10% SDS). Heat at 45°C for 5 min and run with 10% SDS-PAGE containing 0.15% porcine skin gelatin (Sigma G1393 liquid or G2500 powder) at low voltage for up to 5 h. Purified MMP-2 (M-1552, Sigma) and MMP-9 (M-7942, Sigma) can be used for positive controls.
4. After electrophoresis, rinse the gel twice in 2.5% Triton X-100 for 30 min each at room temperature, followed by an incubation in reaction buffer (50 mM Tris, pH 7.5, 200 mM NaCl, and 5 mM $CaCl_2$) at 37°C overnight.

5. Stain the gel with Coomassie blue solution for 10–30 min, followed by destain for 2–20 h to reveal the lytic zones of protease activity.
6. Photograph the gel, which can be presented as inverted images to assist the visualization of signal (now it is dark band under a blank background). Quantify the area of signal using the NIH Image J software.

4. Immunocytochemical Analysis of WM and Oligodendrocytes

Damage to WM and OLs is an integrate component in the pathology of perinatal HI brain injury. The old terminology "periventricular leukomalacia (PVL)" was based on the use of cranial sonography to detect severe cystic degeneration in WM. While the incidence of cystic PVL steadily declines in developed countries, the recent use of diagnostic magnetic resonance imaging (MRI) reveals a high frequency of diffuse and subtle lesions in WM (9). Hence, there is a trend of using the term "white-matter injury of prematurity (WMIP)" to encompass all forms of WM/OL lesions in perinatal brain injury. These types of subtle WM injury are contributing factors of cerebral palsy and other neurological consequences of NHI that have no clear GM lesions. Moreover, oligodendrocyte progenitors and nascent oligodendrocytes in neonatal brains are particularly vulnerable to oxidative and nitrative stress due to deficiency of the antioxidant system (10). Together, these considerations stress the importance of assessing WM/OL damage—or protection following experimental therapy—in preclinical models of NHI.

The standard and easiest method to evaluate WM/OL damage following NHI is histo- and immunocytochemistry. There are several excellent reviews with a comprehensive list of antibody markers to the oligodendrocyte lineage cell (11). Table 1 is a selection of a few key reagents covering different stages of maturation along the OL lineage. Of note, anti-NG2 antibody labels early OL progenitors, as well as pericytes, in the newborn brain. Comparison of the percentage of O4+ vs. O1+ cells assesses the rate of maturation in OL lineage (12). Anti-GST-π and CC-1 antibodies are also useful tools for the counting of mature oligodendrocytes in the adult brain. Finally, anti-myelin basic protein (MBP) antibody is a very useful tool to reveal the destruction of oligodendrocyte processes in WM. For example, the Nissl stain showed typical columnar lesions in the cerebral cortex and swelling of the corpus callosum (CC) after HI (Fig. 2a, b). Anti-MBP immunostaining showed reduction of radiating fibers from CC to cerebral cortex, as well as general reduction and fragmentation of OL processes within CC (Fig. 2c, d).

Table 1
Commonly used antibodies to oligodendrocyte lineage cells

Antibody	Antigen	Maturation stage	Source	Host/Ab type
Anti-NG2	NG2 chondroitin sulfate proteoglycan	Early OL progenitors	Millipore	Rabbit/Polyclonal
O4	OL sulfatide	Late OL progenitors	Millipore	Mouse/Monoclonal IgM
O1	OL galactolipid	Immature OLs	Millipore	Mouse/Monoclonal IgM
Anti-GST-π	Glutathione-*S*-transferase isoenzyme pi	Mature OLs	MBL	Rabbit/Polyclonal
CC-1	N-terminal fragment of adenomatous polyposis coli (APC)	Mature OLs	Abcam	Mouse/Monoclonal IgG
Anti-MBP	Myelin basic protein	Mature OLs	Millipore	Rabbit/Polyclonal

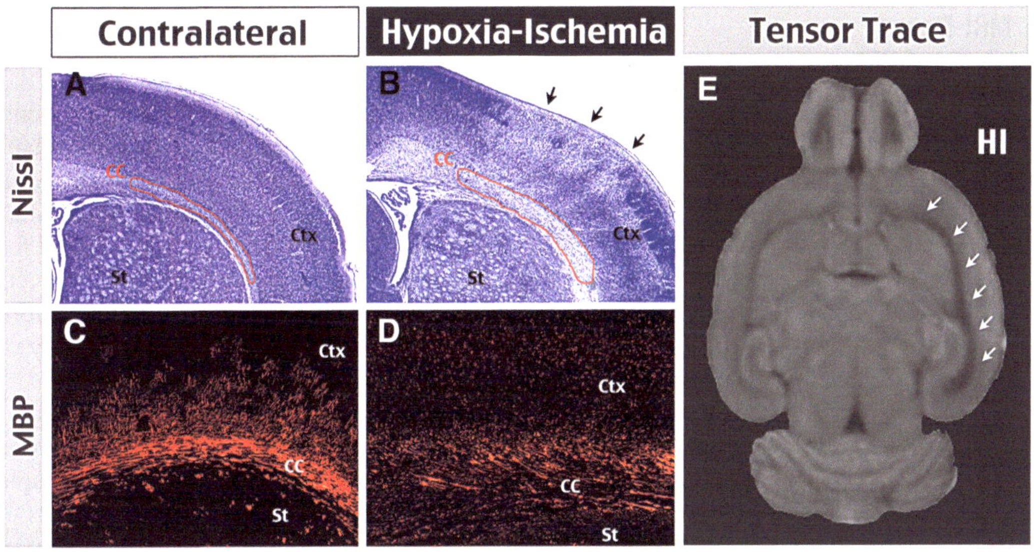

Fig. 2. Assessment of white-matter (WM) and oligodendrocyte injury following neonatal cerebral hypoxia–ischemia. (**a**, **b**) Nissl stain showed typical columnar damage in the cerebral cortex (*arrows* in **b**) and swelling of the corpus callosum (CC) at 24 h after HI. (**c**, **d**) Anti-myelin basic protein (MBP) staining showed fragmentation of oligodendrocyte processes in corpus callosum on the HI-challenged hemisphere. (**e**) The swelling of corpus callosum following HI was clearly indicated by the tensor trace image in magnetic resonance imaging (MRI) (**a**–**d** modified from Yang et al. (7)).

While it is easy to detect OL damage with anti-MBP immunostaining, it is challenging to quantify the extent of WM/OL injury by this immunocytochemistry method. Yet, "Seeing is believing, but measuring is understanding." Without quantification, it is difficult, if not impossible, to compare the efficacy of two therapies or appreciate the synergic effects of various treatments. Towards this end, there are at least two approaches in the literature to quantify WM injury according to anti-MBP staining. The first method uses a semiquantitative scale (13). In this approach, lesion size was scored in three similar sections per animal by a blinded observer using a 1–5 scale: 0, no abnormality; 1, any loss of processes radiating from the capsule into the cortex; 2, some loss of processes with any capsule involvement; 3, loss of processes and fragmenting or thinning of capsule; 4, ragged-appearing capsule and thinning; 5, discontinuity in capsule because of massive thinning, full thickness loss in areas. The second approach measures the mean thickness of the subcortical MBP staining under the forelimb area of the cortex, as determined by the NIH Image J software in a microscopic image (14). Though not ideal, either method appears to work satisfactorily.

5. MRI-Based Analysis of GM and WM Injury Following NHI

MRI is becoming widely used in clinical management of NHI brain injury. For example, the use of MRI revealed a previously unappreciated high incidence of WM injury in premature neonates (9). The use of small-animal MRI systems also contributes to preclinical studies of NHI. For example, T2-weighted imaging is useful for revealing brain edema, and diffusion-weighted imaging detects cytotoxic edema in a quantifiable manner (Fig. 3a–c). In addition, diffusion tensor imaging (DTI) is a powerful tool in the study of WM protection in preclinical models.

DTI is a relatively new MRI technique to visualize axonal organization and measure directional water diffusion in the nervous system. In DTI, measurements of water diffusion along multiple axes are fitted into a 3D model to determine three eigenvalues in the principal (λ_1) and two shorter perpendicular axes (λ_2 and λ_3) of the "diffusion ellipsoid." The three eigenvalues are in turn used to calculate several DTI indexes, including fractional anisotropy (FA), axial/longitudinal diffusivity ($\lambda_{\parallel}$), and radial/transverse diffusivity ($\lambda_{\perp}$) (15). FA, the most frequently used DTI parameter, is a measurement of the relative difference of the three eigenvalues. FA is scaled from 0 (isotropic) to 1 (anisotropic), and large reduction of FA often equals severe structural breakdown of axons (16).

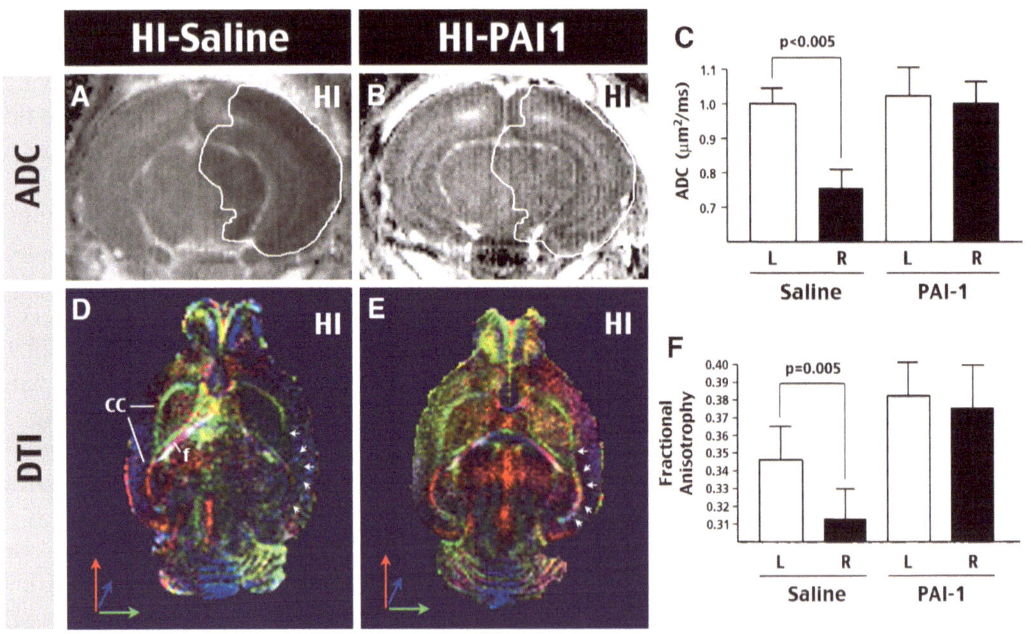

Fig. 3. Quantification of gray-matter (GM) and white-matter (WM) injury by MR imaging. (**a–c**) Apparent diffusion coefficient (ADC) image visualizes and quantifies GM damage after HI, and protection by the therapeutic administration of PAI-1. (**d–f**) Diffusion tensor imaging (DTI) visualizes and quantifies WM injury after HI (modified from Yang et al. (7)).

For example, we have shown that therapeutic administration of PAI-1 prevents the reduction of FA in CC at 24 h post HI (Fig. 3d–f) (7). Another major advantage offered by in vivo DTI is the possibility of longitudinal study of the development–maturation of WM in the same animals, which cannot be achieved by traditional histological methods.

Besides FA, there are several very useful DTI parameters that provide additional information of WM/OL damage. Axial diffusivity, equivalent to λ_1, represents water motion along the length of axons. Axial diffusivity can be reduced by structural breakdown or focal enlargement–constriction of the neurites (16). Radial diffusivity, the average of λ_2 and λ_3, indicates perpendicular water diffusion across the axon, which is hindered by barriers, such as the myelin sheaths (16). Even tensor trace, the sum of three eigenvalues (λ_1, λ_2, and λ_3), can easily reveal the swelling of corpus callosum post HI (Fig. 2e). Together, these promising data suggest that DTI will be widely used in preclinical studies to reveal a fuller picture of WM and OL damage.

6. Limitations and Future Directions

A major success in clinical management of NHI in the past decade is the introduction of hypothermia treatment. However, hypothermia does not solve all clinical problems. The original clinical trials indicated that hypothermia only help infants suffering from moderate HIE, but is ineffective in severely affected HIE (17, 18). Moreover, hypothermia is currently not applied to preterm infants, who have a higher incidence of PVL-like WM/OL injury. Thus, it remains an important task in neonatal care to devise more effective brain protection therapies in NHI. Preclinical studies of NHI play a critical role in accomplishing this goal.

In preclinical studies, there is already richness in various "static" methods to assess brain damage at any given time point after injury. However, there are very few "dynamic" methods that offer longitudinal assessment of the progression of brain injury or recovery following specific medication. In this regard, the recently introduced manganese-enhanced MRI offers in vivo detection of neuroarchitectural details. Combined with 3D reconstruction and volumetric measurement, this method can assess longitudinal progression of brain injury or recovery (19). Another very useful innovation in MRI technology is sodium imaging (20). As neuronal injury progresses, the energy failure attracts more sodium ions from blood into the brain interstitial space. It has been shown that sodium MRI in stroke has a longer duration of linear signal progression than diffusion-weighted imaging, which can be used as a timer of injury onset. Through MRI-histopathology correlation studies that are only feasible in animal models, future studies may establish a threshold of "salvageable" tissue based on regional tissue sodium concentration.

References

1. Rice JE III, Vannucci RC, Brierley JB (1981) The influence of immaturity on hypoxic-ischemic brain damage in the rat. Ann Neurol 9:131–141
2. Yager J, Towfighi J, Vannucci RC (1993) Influence of mild hypothermia on hypoxic-ischemic brain damage in the immature rat. Pediatr Res 34:525–529
3. Eklind S, Mallard C, Arvidsson P, Hagberg H (2005) Lipopolysaccharide induces both a primary and a secondary phase of sensitization in the developing rat brain. Pediatr Res 58:112–116
4. Adhami F, Kuan CY (2009) Cerebral hypoxia-ischemia model in adult mice. In: Chen J, Xu ZC, Xu XM, Zhang JH (eds) Animal models of acute neurological injuries. Humana Press, New York
5. Iwai M, Chen J (2012). Neurobehavioral assessments of NHI. In: Chen J, Xu ZC, Xu XM, Zhang JH (eds) Animal models of acute neurological injuries II: injury and mechanistic assessments. Humana Press, New York
6. Adhami F, Yu D, Yin W, Schloemer A, Burns KA, Liao G, Degen JL, Chen J, Kuan CY (2008) Deleterious effects of plasminogen activators in neonatal cerebral hypoxia-ischemia. Am J Pathol 172:1704–1716
7. Yang D, Nemkul N, Shereen A, Jone A, Dunn RS, Lawrence DA, Lindquist D, Kuan CY (2009) Therapeutic administration of plasminogen activator inhibitor-1 prevents hypoxic-ischemic brain injury in newborns. J Neurosci 29:8669–8674
8. Lenzser G, Kis B, Snipes JA, Gaspar T, Sandor P, Komjati K, Szabo C, Busija DW (2007) Contribution of poly(ADP-ribose) polymerase to postischemic blood-brain barrier damage in rats. J Cereb Blood Flow Metab 27: 1318–1326
9. Khwaja O, Volpe JJ (2008) Pathogenesis of cerebral white matter injury of prematurity. Arch Dis Child Fetal Neonatal Ed 93:F153–F161
10. Haynes RL, Baud O, Li J, Kinney HC, Volpe JJ, Folkerth DR (2005) Oxidative and nitrative injury in periventricular leukomalacia: a review. Brain Pathol 15:225–233
11. Ness JK, Valentino M, McIver SR, Goldberg MP (2005) Identification of oligodendrocytes in experimental disease models. Glia 50: 321–328
12. Segovia KN, McClure M, Moravec M, Luo NL, Wan Y, Gong X, Riddle A, Craig A, Struve J, Sherman LS, Back SA (2008) Arrested oligodendrocyte lineage maturation in chronic perinatal white matter injury. Ann Neurol 63:520–530
13. Gerstner B, Lee J, DeSilva TM, Jensen FE, Volpe JJ, Rosenberg PA (2009) 17beta-estradiol protects against hypoxic/ischemic white matter damage in the neonatal rat brain. J Neurosci Res 87:2078–2086
14. Fan LW, Lin S, Pang Y, Lei M, Zhang F, Rhodes PG, Cai Z (2005) Hypoxia-ischemia induced neurological dysfunction and brain injury in the neonatal rat. Behav Brain Res 165:80–90
15. Mori S, Zhang J (2006) Principles of diffusion tensor imaging and its applications to basic neuroscience research. Neuron 51:527–539
16. Beaulieu C (2002) The basis of anisotropic water diffusion in the nervous system—a technical review. NMR Biomed 15:435–455
17. Gluckman PD, Wyatt JS, Azzopardi D, Ballard R, Edwards AD, Ferriero DM, Polin RA, Robertson CM, Thoresen M, Whitelaw A, Gunn AJ (2005) Selective head cooling with mild systemic hypothermia after neonatal encephalopathy: multicentre randomised trial. Lancet 365:663–670
18. Shankaran S, Laptook AR, Ehrenkranz RA, Tyson JE, McDonald SA, Donovan EF, Fanaroff AA, Poole WK, Wright LL, Higgins RD, Finer NN, Carlo WA, Duara S, Oh W, Cotten CM, Stevenson DK, Stoll BJ, Lemons JA, Guillet R, Jobe AH (2005) Whole-body hypothermia for neonates with hypoxic-ischemic encephalopathy. N Engl J Med 353: 1574–1584
19. Aoki I, Wu YJ, Silva AC, Lynch RM, Koretsky AP (2004) In vivo detection of neuroarchitecture in the rodent brain using manganese-enhanced MRI. Neuroimage 22:1046–1059
20. Hussain MS, Stobbe RW, Bhagat YA, Emery D, Butcher KS, Manawadu D, Rizvi N, Maheshwari P, Scozzafava J, Shuaib A, Beaulieu C (2009) Sodium imaging intensity increases with time after human ischemic stroke. Ann Neurol 66:55–62

Chapter 21

Biochemical and Molecular Biological Assessments of Neonatal Hypoxia–Ischemia: Cell Signaling

Changlian Zhu and Klas Blomgren

Abstract

Neonatal hypoxia–ischemia (HI) induces a series of intracellular signaling events, including destructive and protective mechanisms. The destructive events include neuronal membrane depolarization, excitotoxicity, free radical injury, overactivation of calcium-dependent enzymes like calpains, proapoptotic protein release from mitochondria, and caspase activation. The protective events include upregulation of trophic factors and stimulation of neural stem cell proliferation. The injury after HI depends on the balance between destructive and protective events. Characterizing the mechanisms of injury and repair is important to be able to devise protective treatments or effective rehabilitation strategies. These mechanisms can be detected by analyzing RNA, protein level, or enzyme activity by using PCR, microarrays, ELISA, western blotting, or fluorometric assays of enzyme substrate turnover. This chapter illustrates how HI-related events can be studied using these methods.

Key words: Caspase, Western blotting, Mitochondria, Apoptosis-inducing factor

1. Introduction

During hypoxia–ischemia (HI), the lack of oxygen disrupts neuronal oxidative phosphorylation and the resulting energy depletion triggers a sequence of neurochemical and biological events. Excitotoxicity begins with the HI-induced accumulation of extracellular glutamate and overstimulation of neuronal glutamate receptors (1). This stimulates a massive influx of calcium, sodium, and water, resulting in free radical production, impairment of mitochondrial function, cell swelling, and, ultimately, neuronal cell death if the HI period is extended (2–4). A second wave of neuronal cell damage occurs during the reperfusion phase. Although restoration of blood and oxygen supply to the brain is necessary to limit ischemic neuronal damage, the renewed availability of oxygen during the reperfusion phase gives rise to the activation of multiple

Jun Chen et al. (eds.), *Animal Models of Acute Neurological Injuries II*, Springer Protocols Handbooks,
DOI 10.1007/978-1-61779-782-8_21, © Springer Science+Business Media, LLC 2012

biochemical pathways, such as the postischemic release of oxygen free radicals (5), synthesis of NO (6, 7), inflammatory reactions, and release of excitatory amino acids, particularly glutamate (1, 8, 9). Secondary damage occurs from hours to days after the insult and accounts for a significant proportion of the final cell loss even after severe insults (10). The immature brain is more resistant to hypoxia than the adult brain, presumably because the adult brain has a much more elaborate dendritic tree that needs to be supplied with energy for maintenance of ionic gradients. Many of the effectors of the apoptotic machinery are expressed at higher levels in the immature brain and, consequently, neurons in the immature brain display apoptotic features, such as caspase activation, after HI to a much greater extent than neurons in the adult brain (11, 12).

Oxidative stress is an early feature after cerebral ischemia. The immature brain is more vulnerable to oxidative stress, and mitochondria are prime targets of oxidative stress. Activation of apoptotic mechanisms requiring the release of proapoptotic proteins from mitochondria is much more prominent in the immature brain (13). Apoptotic cell death can be divided into three distinct phases: initiation, commitment, and execution phases. In the initiation phase, the cell receives a trigger signal, e.g., loss of a survival factor, to commit suicide. During the commitment phase, multiple signaling pathways are induced leading to activation of caspases. The execution phase involves multiple events leading to DNA cleavage and chromatin condensation (14, 15). One of the hallmarks of apoptosis is the degradation and concomitant compaction of chromatin. There are at least two pathways leading to chromatin condensation and degradation during apoptosis. One is through activation of caspases as well as downstream effectors, such as caspase-activated DNase (caspase-dependent pathway) (16). Another is via apoptosis-inducing factor (AIF), a mitochondrial intermembrane flavoprotein that has been found to translocate from mitochondria to nuclei in a caspase-independent fashion (caspase-independent pathway) (17, 18).

Caspases (cysteine-dependent aspartate-specific proteinases) belong to a superfamily of cysteine proteases that regulate many aspects of cell survival and death, and they are named in order of identification and publication (19, 20). Caspases are activated by regulated proteolysis of proenzymes. Caspases share a unique requirement of an aspartate residue in their substrate proteins, limiting the number of target proteins (21). Caspases cleave nuclear proteins, cytoskeletal proteins (e.g., actin and fodrin), and cytosolic proteins (16, 18, 22).

The AIF pathway is mitochondria dependent but caspase independent (23). The AIF precursor is synthesized in the cytosol and imported into mitochondria. The mature AIF protein is normally confined to the mitochondrial intermembrane space (24, 25).

Under a variety of different apoptosis-inducing conditions, AIF translocates through the outer mitochondrial membrane to the cytosol, earlier than the release of cytochrome c (18, 24, 26), to the nucleus, where it binds with DNA and causes chromatin condensation as well as large-scale (approximately 50 kb) DNA fragmentation (22, 24).

Calpains, a family of calcium-dependent proteases with a neutral pH optimum, represents a class of cytosolic cysteine proteases which are activated by elevated intracellular calcium concentrations. A number of subcellular targets have been identified as substrates for calpain cleavage. In the brain, potential calpain substrates include a number of cytoskeletal proteins, such as fodrin. It is well established that when cellular calpain is overactivated its substrate protein fodrin is degraded into two proteolytic fragments (150 and 145 kDa). The cleavage products are relatively stable and can be detected long after the injury, and this distinct pattern of fragmentation of fodrin has been used as an indicator of calpain activation in intact cells (27). A specific anti-FBDP150 antibody recognizes only the FBDP150 in injured cells undergoing necrosis (12, 28–30).

2. Materials and Equipment

2.1. Primary Antibodies

1. Caspase-3 (sc-7148, Santa Cruz Biotechnology, Santa Cruz, CA, USA).
2. Fodrin (FG 6090, Affiniti Research Products, Mamhead, UK).
3. AIF (sc-9416, Santa Cruz Biotechnology, Santa Cruz, CA, USA).
4. Microtubule-associated protein 1 light chain 3 (LC3) (Cell Signaling, Danvers, MA, USA).
5. Actin (A2066, Sigma, Stockholm, Sweden).
6. COX I (Molecular probes, Eugene, OR, USA).
7. Cytochrome c (clone 7H8.2C12, Pharmingen, San Diego, CA, USA).
8. Catalase (Lab Frontier, Seoul, Korea).
9. SOD2 (Lab Frontier, Seoul, Korea).
10. Phospho-GSK-3β (Ser9) (#9323, Cell Signaling, Danvers, MA, USA).
11. Phospho-ERK (Thr202/Tyr204) (#9101, Cell Signaling, Danvers, MA, USA).

2.2. Secondary Antibodies

Peroxidase-labeled secondary antibodies (Vector, Burlingame, CA, USA).

2.3. Substrate for Immunoblotting

Super Signal West Dura substrate (Pierce, Rockford, IL, USA).

2.4. Substrates for Caspases Activity Assays

1. Ac-YVAD-AMC (for caspase-1, from Alexis Corp., San Diego, CA, USA).
2. Ac-DEVD-AMC (for caspase-3, Peptide Institute, Osaka, Japan).
3. Ac-VDVAD-AFC (for caspase-2, Enzyme System Products, Livermore, CA, USA).
4. Ac-IETD-AFC (for caspase-8, Enzyme System Products, Livermore, CA, USA).
5. Ac-LEHD (for caspase-9, Enzyme System Products, Livermore, CA, USA).

2.5. Assay and Image Equipment

1. LAS 3000 cooled CCD camera (Fujifilm, Tokyo, Japan).
2. Spectramax Gemini microplate fluorometer (Molecular Devices, Sunnyvale, CA, USA).
3. Spectramax Plus microplate reader (Molecular Devices, Sunnyvale, CA, USA).
   - UV-permeable microtiter plate for caspase activity assay (Microfluor, Dynatech, Haverhill, MA, USA).
   - Glass homogenizers (Merck, Eurolab, Göteborg, Sweden).
   - Sonicator (Branson, Danbury, CT, USA).
   - Centrifuge (Hettich, Tuttlingen, Germany).
   - Electrophoresis system (Invitrogen, Madison, WI, USA).
   - High temperature dry block heater (Grant, Ohio, USA).
   - PCR Thermal Cycler (Labrepco, Horsham, PA, USA).

3. Procedures

3.1. Sample Preparation

1. Decapitate animals at selected time points.
2. Dissect out the brains rapidly on a bed of ice. Different brain regions, e.g., parietal cortex, hippocampus, and diencephalon (including the thalamus, hypothalamus and striatum), can be dissected out from each hemisphere.
3. Add ice-cold isolation buffer (15 mM Tris–HCl, pH 7.6, 320 mM sucrose, 1 mM DTT, 1 mM $MgCl_2$, 3 mM EDTA-K_2, 0.5% protease inhibitor cocktail).
4. Homogenize either by sonication or gently by hand (to preserve mitochondrial integrity) in a 2-mL glass/glass homogenizer (Merck Eurolab) using, sequentially, two different pestles

with a total clearance of 0.12 and 0.05 mm, respectively (ten strokes each).

5. Centrifuge homogenates at 800 × *g* at 4°C for 10 min. Wash the pellets in homogenizing buffer and recentrifuge, producing a crude nuclear pellet (P1).
6. Centrifuge the supernatant further at 9,200 × *g* for 15 min at 4°C, producing a mitochondrial and synaptosomal fraction in the pellet and a crude cytosolic fraction in the supernatant (S2). Wash the pellet in homogenizing buffer and recentrifuge, producing a mitochondrial and synaptosomal fraction (P2).
7. Centrifuge the crude cytosolic fraction (S2) at 20,000 × *g* for 40 min at 4°C, yielding a pure cytosolic fraction in the supernatant (S3). The pellet thus produced (P3) is a membrane fraction consisting of lighter membranes, such as cytosolic membranes and endoplasmic reticulum.

3.2. Protein Determination

1. Determine the protein concentration according to Whitaker and Granum (31), adapted for microplates, using a Spectramax Plus plate reader (Molecular Devices, Sunnyvale, CA). This method is cheap and simple, since it does not require any reagents. Other popular methods include the classical Lowry method, the Bradford method, and the more recent bicinchoninic acid assay.
2. Dilute the samples (10 μL) with 590 μL PBS and add 250 μL of the mixture to parallel wells.
3. The difference in absorbance between 235 and 280 nm is used to calculate the protein concentration using the equation: Protein concentration (mg/mL) = $(A_{235} - A_{280})/2.51$.

3.3. RNA Extraction

1. Remove brains rapidly and freeze in liquid nitrogen.
2. Extract total RNA from the whole brain or selected brain regions using RNeasy mini kit (Qiagen Inc.).
3. Quantify RNA by spectrophotometry at 260 nm, and determine the purity by a 260/280 ratio.
4. The quality of RNA can be further checked by running samples on 1.1% agarose and 2.2 M formaldehyde gels to ascertain that no degradation has taken place.
5. Store samples at −80°C until analysis.

3.4. Caspase Activity Assay

1. Homogenates or cytosolic samples (25 μL) are mixed with 75 μL of extraction buffer (50 mM Tris–HCl (pH 7.3), 100 mM NaCl, 5 mM EDTA, 1 mM EGTA, 3 mM NaN_3, 1 mM PMSF, 1% protease inhibitor cocktail, and 0.2% CHAPS) on a microtiter plate (Microfluor, Dynatech, VA, USA).

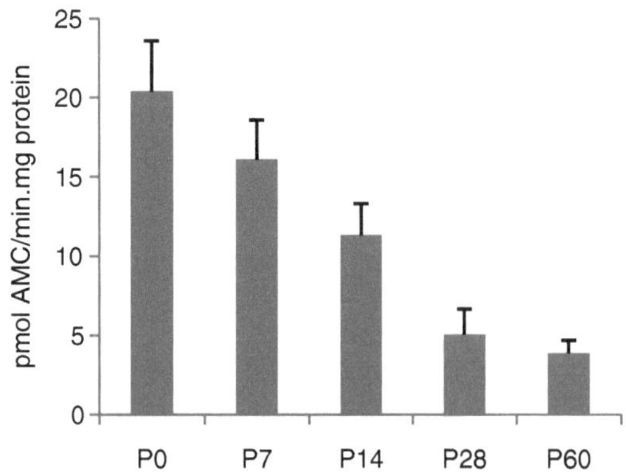

Fig. 1. Caspase-3 activity (DEVD cleavage) in the cytosolic fraction was measured using fluorometric assays in the normal nonischemic rat brain. The activity decreased with brain development, indicating that the baseline caspase-3 activity decreased as brain maturation occurred.

2. After incubation for 15 min at room temperature, 100 μL of assay buffer (50 μM substrate, without inhibitors or CHAPS, but with 4 mM DTT) is added.
3. Cleavage of Ac-DEVD-AMC (for caspase-3/7-like activity, from Peptide Institute, Osaka, Japan) or Ac-YVAD-AMC (for caspase 1, from Alexis Corp., San Diego, CA) is measured at 37°C using a Spectramax Gemini microplate fluorometer (Molecular Devices, Sunnyvale, CA) with an excitation wavelength of 380 nm and an emission wavelength of 460 nm.
4. The degradation is followed at 2-min intervals for 2 h, and the V_{max} is calculated from the entire linear part of the curve.
5. The activity is expressed as pmol AMC released per mg protein and minute (Fig. 1). Cleavage of Ac-VDVAD-AFC (for caspase-2), Ac-IETD-AFC (for caspase-8), and Ac-LEHD-AFC (for caspase-9) (Enzyme Systems Products, Livermore, CA) is measured as for the AMC conjugates, but with an excitation wavelength of 400 nm and an emission wavelength of 505 nm, and expressed as pmol AFC released per mg protein and minute (16).

3.5. Immunoblotting

1. Samples of 65 μL are mixed with 25 μL NuPAGE LDS 4× sample buffer and 10 μL reducing agent and heated (70°C) for 10 min.
2. Individual samples are run on 4–12% NuPAGE Bis-Tris gels (Novex, San Diego, CA) and transferred to reinforced nitrocellulose membranes.
3. Block with 30 mM Tris–HCl (pH 7.5), 100 mM NaCl, and 0.1% Tween 20 (TBS-T) containing 5% fat-free milk powder for 1 h at room temperature.

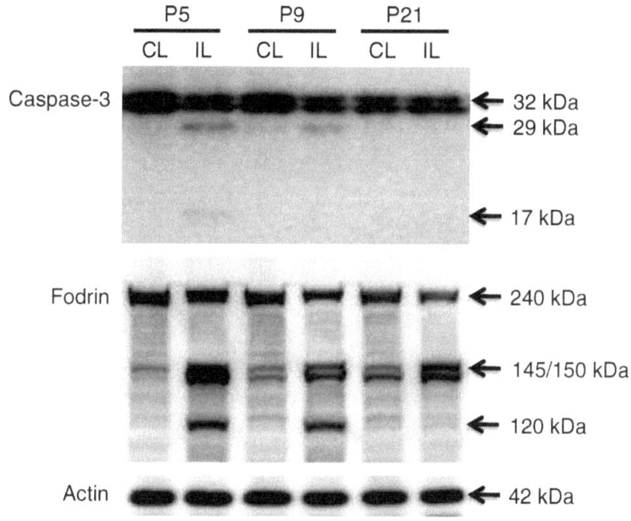

Fig. 2. Immunoblotting of caspase-3 (*top panel*) showed cleavage of the 32 kDa proform, yielding 29- and 17-kDa bands after HI in the ipsilateral hemisphere (IL) in the young animals (P5, P9), but not P21 rats. Note that the amount of caspase-3 proform decreases as brain maturation occurs, in accordance with the decreased baseline activity seen in Fig. 1. The cytoskeletal 240 kDa protein fodrin was cleaved and yielded 145/150 kDa (reflecting calpain activation) and 120 kDa (reflecting caspase-3 activation) fragments in the ipsilateral after HI (*middle panel*). The *bottom panel* shows actin levels, as a loading control.

4. After shaking, wash with TBS-T for 5 min ×3, and then incubate the membranes with primary antibody at an appropriate dilution for 1 h at room temperature or overnight at 4°C.
5. After shaking, wash for 5 min ×3, and incubate the membrane with an appropriate peroxidase-labeled secondary antibody at an appropriate dilution for 30 min at room temperature (Vector, Burlingame, CA).
6. After washing with TBS-T three times, immunoreactive species are visualized using the Super Signal Western Dura substrate (Pierce, Rockford, IL) and an LAS 3000 cooled CCD camera (Fujifilm, Tokyo, Japan). Alternatively, photographic film (X-ray film) can be used instead of a cooled CCD camera (Fig. 2).
7. Immunoreactive bands are quantified using the Image Gauge software (Fujifilm, Tokyo, Japan).

3.6. ELISA

1. The homogenates or appropriate cellular fractions are collected and used for the ELISA.
2. The samples are assayed in duplicate.
3. Create a standard curve and perform the assay according to the manufacturer's instructions.
4. Correct the results for protein concentration.

4. Evaluation of Subcellular Fractions

All investigators should assess the purity of cellular fractions isolated by centrifugation in their hands by western blotting. Useful markers include lamin B for the nuclear fraction, cytochrome oxidase for the mitochondrial fraction, and lactate dehydrogenase for the cytosolic fraction. In our hands, we had tremendous difficulties obtaining a nuclear fraction free from mitochondrial contamination when preparing samples from immature, but not adult, brains. To detect protein release from mitochondria to the cytosol, for example cytochrome c, the mitochondrial and cytosolic fractions can be assayed in parallel to detect relative changes in the two.

References

1. Hagberg H et al (1987) Extracellular overflow of glutamate, aspartate, GABA and taurine in the cortex and basal ganglia of fetal lambs during hypoxia-ischemia. Neurosci Lett 78(3): 311–317
2. Choi DW (1992) Excitotoxic cell death. J Neurobiol 23(9):1261–1276
3. Choi DW (1995) Calcium: still center-stage in hypoxic-ischemic neuronal death. Trends Neurosci 18(2):58–60
4. Zipfel GJ et al (2000) Neuronal apoptosis after CNS injury: the roles of glutamate and calcium. J Neurotrauma 17(10):857–869
5. McCord JM (1985) Oxygen-derived free radicals in postischemic tissue injury. N Engl J Med 312(3):159–163
6. Zhu C et al (2004) Nitrosylation precedes caspase-3 activation and translocation of apoptosis-inducing factor in neonatal rat cerebral hypoxia-ischaemia. J Neurochem 90(2): 462–471
7. Ferriero DM et al (1995) Selective destruction of nitric oxide synthase neurons with quisqualate reduces damage after hypoxia-ischemia in the neonatal rat. Pediatr Res 38(6):912–918
8. Volpe JJ (2001) Perinatal brain injury: from pathogenesis to neuroprotection. Ment Retard Dev Disabil Res Rev 7(1):56–64
9. Bona E et al (1999) Chemokine and inflammatory cell response to hypoxia-ischemia in immature rats. Pediatr Res 45(4 Pt 1):500–509
10. Du C et al (1996) Very delayed infarction after mild focal cerebral ischemia: a role for apoptosis? J Cereb Blood Flow Metab 16(2):195–201
11. Hu BR et al (2000) Involvement of caspase-3 in cell death after hypoxia-ischemia declines during brain maturation. J Cereb Blood Flow Metab 20(9):1294–1300
12. Zhu C et al (2005) The influence of age on apoptotic and other mechanisms of cell death after cerebral hypoxia-ischemia. Cell Death Differ 12(2):162–176
13. Blomgren K, Hagberg H (2006) Free radicals, mitochondria, and hypoxia-ischemia in the developing brain. Free Radic Biol Med 40(3):388–397
14. Hengartner MO (2000) The biochemistry of apoptosis. Nature 407(6805):770–776
15. Yuan J, Yankner BA (2000) Apoptosis in the nervous system. Nature 407(6805):802–809
16. Wang X et al (2001) Caspase-3 activation after neonatal rat cerebral hypoxia-ischemia. Biol Neonate 79(3–4):172–179
17. Susin SA et al (2000) Two distinct pathways leading to nuclear apoptosis. J Exp Med 192(4):571–580
18. Zhu C et al (2003) Involvement of apoptosis-inducing factor in neuronal death after hypoxia-ischemia in the neonatal rat brain. J Neurochem 86(2):306–317
19. Nicholson DW, Thornberry NA (1997) Caspases: killer proteases. Trends Biochem Sci 22(8):299–306
20. Kumar S (1995) ICE-like proteases in apoptosis. Trends Biochem Sci 20(5):198–202
21. Stennicke HR et al (1999) Caspase-9 can be activated without proteolytic processing. J Biol Chem 274(13):8359–8362
22. Zhu C et al (2007) Cyclophilin A participates in the nuclear translocation of apoptosis-inducing factor in neurons after cerebral hypoxia-ischemia. J Exp Med 204(8): 1741–1748

23. Cande C et al (2002) Apoptosis-inducing factor (AIF): key to the conserved caspase-independent pathways of cell death? J Cell Sci 115(Pt 24):4727–4734
24. Susin SA et al (1999) Molecular characterization of mitochondrial apoptosis-inducing factor. Nature 397(6718):441–446
25. Cande C et al (2002) Apoptosis-inducing factor (AIF): a novel caspase-independent death effector released from mitochondria. Biochimie 84(2–3):215–222
26. Daugas E et al (2000) Mitochondrio-nuclear translocation of AIF in apoptosis and necrosis. FASEB J 14(5):729–739
27. Bahr BA et al (1995) Induction of calpain-mediated spectrin fragments by pathogenic treatments in long-term hippocampal slices. J Pharmacol Exp Ther 273(2):902–908
28. Wang KK (2000) Calpain and caspase: can you tell the difference? Trends Neurosci 23(1):20–26
29. Blomgren K et al (2001) Synergistic activation of caspase-3 by m-calpain after neonatal hypoxia-ischemia: a mechanism of "pathological apoptosis?". J Biol Chem 276(13):10191–10198
30. Dutta S et al (2002) Selective release of calpain produced alphaII-spectrin (alpha-fodrin) breakdown products by acute neuronal cell death. Biol Chem 383(5):785–791
31. Whitaker JR, Granum PE (1980) An absolute method for protein determination based on difference in absorbance at 235 and 280 nm. Anal Biochem 109(1):156–159

Chapter 22

Biochemical and Molecular Biological Assessments of Neonatal Hypoxia–Ischemia: Inflammation

Xiaoming Hu and Jun Chen

Abstract

Neonatal hypoxic-ischemic (H-I) brain injury is a leading cause of perinatal mortality. Inflammation contributes substantially to the pathogenesis of perinatal H-I brain injury. The inflammatory response in neonatal H-I model could be assessed by measuring the expression of inflammatory mediators in the blood or organs of interest using quantitative polymerase chain reaction or enzyme-linked immunosorbent assay or by analyzing the number and morphology of inflammatory cells using immunohistochemistry staining, Western blot, and flow cytometry. Unique features of inflammatory response in neonatal H-I brain injury compared to the adult counterpart, which affect the results of inflammation assessments, are discussed in this chapter.

Key words: Inflammation, Neonatal hypoxia–ischemia

1. Introduction

Neonatal hypoxic-ischemic (H-I) brain injury is a leading cause of perinatal mortality. The neonatal H-I injury to the brain induces inflammatory responses similar to those in the adult ischemic injury, which is characterized by activation of microglia, macrophages, and leukocyte infiltration, release of proinflammatory cytokines, and increased expression of endothelial adhesion molecules (1). Inflammation contributes substantially to the pathogenesis of perinatal H-I brain injury due to the exceeding susceptibility of immature brain to the consequences of inflammatory responses compared to its adult counterpart (2). Actually, the extent of inflammatory response has been demonstrated to correlate with the severity of brain injury and neurological deficits in infants suffered from H-I (3).

Despite all the similarities, there are still a few differences in the cerebral inflammation between the neonates H-I model and

Jun Chen et al. (eds.), *Animal Models of Acute Neurological Injuries II*, Springer Protocols Handbooks,
DOI 10.1007/978-1-61779-782-8_22, © Springer Science+Business Media, LLC 2012

the adult focal ischemia model, which will affect the results of inflammation assessments.

2. Regions of Injury

Whereas focal ischemia by middle cerebral artery occlusion induces brain injury mainly in striatum and cortex, neonatal H-I results in more extensive brain injury, as well as cerebral inflammation, in cortex, striatum, and hippocampus.

3. Leukocyte Infiltration

The cellular component and temporal course of leukocyte infiltration are different between adult and neonate brain. For example, in the adult, neutrophils are the earliest leukocyte infiltrated in the brain after ischemic injury and contribute to the opening of the blood–brain barrier (BBB). Neutrophils, however, do not infiltrate into the compromised neonate brain until 42 h after H-I and are present in the parenchyma only briefly (4). Instead, macrophages are more rapidly accumulated in the neonate brain after ischemia compared to the adult brain (5). The cerebral infiltration of T lymphocyte after neonatal H-I is less prominent compared to adult (3).

Similar to the inflammation assessment in adult focal ischemic models, the inflammatory response in neonatal H-I models could be assessed by measuring the expression of inflammatory mediators in the blood or organs of interest using quantitative polymerase chain reaction (PCR) or enzyme-linked immunosorbent assay (ELISA) or by analyzing the number and morphology of inflammatory cells using immunohistochemistry staining, Western blot, and flow cytometry. The technique details about PCR, ELISA, immunohistochemistry staining, Western blot, and flow cytometry are discussed in related chapters.

References

1. Benjelloun N, Renolleau S, Represa A, Ben-Ari Y, Charriaut-Marlangue C (1999) Inflammatory responses in the cerebral cortex after ischemia in the P7 neonatal rat. Stroke 30:1916–1923; discussion 1914–1923
2. Calvert JW, Zhang JH (2005) Pathophysiology of an hypoxic-ischemic insult during the perinatal period. Neurol Res 27:246–260
3. Bona E, Andersson AL, Blomgren K, Gilland E, Puka-Sundvall M, Gustafson K, Hagberg H (1999) Chemokine and inflammatory cell response to hypoxia-ischemia in immature rats. Pediatr Res 45:500–509
4. Hudome S, Palmer C, Roberts RL, Mauger D, Housman C, Towfighi J (1997) The role of neutrophils in the production of hypoxic-ischemic brain injury in the neonatal rat. Pediatr Res 41:607–616
5. Vexler ZS, Yenari MA (2009) Does inflammation after stroke affect the developing brain differently than adult brain? Dev Neurosci 31:378–393

Chapter 23

Neurobehavioral Assessments of Neonatal Hypoxia–Ischemia

Masanori Iwai and Jun Chen

Abstract

Neurobehavioral assessment plays an important role in both adult and neonatal models of cerebral hypoxia–ischemia (NHI), since the final goal of such basic research is to identify new information leading towards novel approaches in human medical treatment. In animal NHI model, behavioral deficit is typically manifested as potential delay in the emergence of neurological reflexes and the time course of reflex performances after the development of neurological reflexes which represent hallmarks of nervous system development (Anim Behav 23:896–920, 1975; J Comp Neurol 175:207–231, 1977; Brain Res 28:85–95, 1971; Brain Res 33:303–314, 1971). Early motor coordination tests are performed on the NHI models because these tests are closely correlated with long-term functional deficits measured by water-maze test (Behav Brain Res 145:209–219, 2003). This chapter introduces a brief history of neurobehavioral assessment of NHI, followed by procedures of representative neurological reflexes (righting reflex, negative geotaxis, and gait) and an early motor coordination test (grid walking and foot-fault test), as well as the required materials and tools to perform such assessments.

Key words: Neonatal hypoxia–ischemia, Rat, Neurological reflex, Early motor coordination test, Maturation

1. Introduction

Animal neonatal hypoxia–ischemia (NHI) models mimic human neonatal hypoxic-ischemic encephalopathy. The most common animal NHI model is the unilateral common carotid artery ligation under anesthesia followed by exposure to hypoxia (8% O_2) after 1–2 h of recovery period in postnatal day-7 rats (6). The concomitant functional neurological deficit of NHI can be determined by performing neurobehavioral assessment through examining neurological signs and reflexes as well as carrying out early motor coordination tests. Assessment of the neurological signs and reflexes typically begins 1 day after the hypoxia (postnatal day 8) and is performed daily at a fixed designated time for 6 or 13 days (postnatal

Jun Chen et al. (eds.), *Animal Models of Acute Neurological Injuries II*, Springer Protocols Handbooks,
DOI 10.1007/978-1-61779-782-8_23, © Springer Science+Business Media, LLC 2012

day 14 or 21). On the other hand, grid walking and foot-fault test, a common early motor coordination test, begins 7 days after hypoxia (postnatal day 14) and is carried out weekly for 3 weeks (postnatal day 35). For these tests, a group of naive (nonischemic) animals must be included to serve as negative control. Reflex-ontogeny and behavioral development of the laboratory animal was first reported by Fox (7). Also, Altman and Sudarshan showed postnatal development of locomotion in neonatal rat in 1975 (1, 2). Later, these methods mentioned above have been used for assessing neurobehavioral development under nutritional deprivation (3, 4), neonatal hypoxic-ischemic encephalopathy (5) X-ray radiation (8), and exposure to alcohol (9) or drugs, such as thyroxin (10), tartrazine (11), and diazepam (12).

In the field of NHI research, Alimi et al. was the first group to perform neurobehavioral test, showing that brain-derived neurotrophic factor (BDNF) was protective against spatial learning/memory dysfunction (13). In today's NHI study, neurobehavioral or neurofunctional test is widely used for evaluation of the efficacy of neuroprotective agents, such as erythropoietin (14, 15), insulin-like growth factor-1 (16), hydrogen saline (17), matrix metalloproteinases inhibitor (18), and cyclooxygenase-2 inhibitor (19). Neurobehavioral test is also used in the neuronal regeneration study (20) and oligodendrocyte regenerative study (21) by erythropoietin after NHI.

On the other hand, it is very important to know how extensive hypoxic-ischemic insult delays the development of neurological reflexes and the time course of reflex performance in neonatal rodents. In these aspects, Lubics et al. have performed detailed investigations into the development of neurological reflexes and motor coordination in rats subjected to NHI (22).

In principle, NHI would retard the development of somatic growth and several neurological reflexes (ear twitch, grasping, gait, and negative geotaxis) as well as the performance in righting, negative geotaxis, and gait reflexes. The retardation of righting reflex lasts at least 2 weeks after NHI. However, negative geotaxis and gait reflexes return to control level within 2 and 1 week after NHI, respectively. On the other hand, hypoxic-ischemic insult affects motor coordination for a longer duration as reflected by rope suspension, rotarod, grid walking and foot-fault, inclined board, vertical screen, walking initiation, elevated body swing, and traveling a square bridge tests, but the subjects should return to normal by 5 weeks with the exception of foot-fault test results (22).

In this chapter, the procedure of three neurological reflexes tests (righting, negative geotaxis, and gait) and an early motor coordination test (grid walking and foot-fault test) are described. In addition, the materials and tools required for these tests are also listed.

2. Righting Reflex

Righting reflex is usually carried out daily between 1 and 7 days post hypoxic-ischemic brain insult.

2.1. Materials and Tools

1. Digital videotape or video disc recorder (preferentially with frame-by-frame slow-motion playback option).
2. Digital stopwatch accurate to one one-hundredth of a second.
3. An instrument for holding a recorder in place, enabling the recording of the movement above the rat.

2.2. Procedure

1. Place a recorder on the instrument above the board, where the rats carry out righting reflex.
2. Start recording.
3. Place the rat on supine position on the board and release your grip on the rat.
4. Wait until the rat turns over to prone position and places all four paws in contact with the surface of the board; repeat step 3 five times.
5. After all the tests are recorded, the duration the rat takes to turn over to prone position and place all four paws in contact with the surface is measured and averaged for each rat.

3. Gait Reflex

Gait reflex is usually carried out daily between 1 and 7 days post hypoxic-ischemic brain insult.

3.1. Materials and Tools

1. Digital videotape or video disc recorder (preferentially with frame-by-frame slow-motion playback option).
2. Digital stopwatch accurate to one one-hundredth of a second.
3. An instrument for holding a recorder in place, enabling the recording of the movement form above the rat.
4. A white paper circle of 13 cm in diameter.

3.2. Procedure

1. Place a recorder on the instrument above the board, where gait reflex is carried out on the rats.
2. Start recording.

3. Place the rat on prone position in the center of a 13-cm-diameter white paper circle on the board and release your grip on the rat.
4. Starting from the first day, record the number of times and the duration in seconds for the rat to move off the circle three to five times daily until the seventh day after hypoxic-ischemic brain insult. If the rat does not move off the circle with both forelimbs within 30 s, the test result is considered to be negative.
5. After all the tests are recorded, measure and average the number of times and duration the rat takes to move off the circle with both forelimbs for each rat.

4. Negative Geotaxis

Negative geotaxis reflex is usually carried out daily between 1 and 7 days after hypoxic-ischemic brain insult.

4.1. Materials and Tools

1. Digital videotape or video disc recorder (preferentially with frame-by-frame slow-motion playback option).
2. Digital stopwatch accurate to one one-hundredth of a second.
3. An instrument for holding a recorder which can record the movement horizontally.
4. An inclined board (45°) of 30 cm covered with No. 400 sandpaper.

4.2. Procedure

1. Place a recorder on the instrument above the horizontal level of the board, where negative geotaxis reflex is performed on the rats.
2. Start recording.
3. Position the rat head down prone with its hind limbs in the middle of the incline.
4. Record the movement of rat from day 1 post hypoxic-ischemic brain insult. Henceforth, record the time in seconds for the rat to climb up the board with its forelimbs reaching the upper end three to five times daily for 7 days. If the rat does not turn around and climb up within 30 s, the test result is considered to be negative.
5. After all tests are recorded, measure and average the duration in seconds for each rat to turn over to the prone position and place all paws in contact with the surface.

5. Grid Walking and Foot-Fault Test

Grid walking and foot-fault test is usually performed weekly between 1 and 4 weeks after hypoxic-ischemic brain insult.

5.1. Materials and Tools

1. Digital videotape or video disc recorder (preferentially with frame-by-frame slow-motion playback option).
2. Digital stopwatch accurate to one one-hundredth of a second.
3. A stainless steel grid floor (40 × 40 cm with a mesh size of 4 cm^2 or 1 $in.^2$) elevated 30–50 cm above the floor.

5.2. Procedure

1. Use a video recorder to pursue and record the movement of rat.
2. Start recording.
3. Place the rat in prone position on the grid.
4. Video record the movement of the rat for 1 min.
5. After all the tests are recorded, the total number of steps of each rat is counted respectively. The number of foot-fault errors, when the rat misplaced a forelimb or hind limb resulting in it falling through the grid, is also recorded.

References

1. Altman J, Sudarshan K (1975) Postnatal development of locomotion in the laboratory rat. Anim Behav 23(4):896–920
2. Donatelle JM (1977) Growth of the corticospinal tract and the development of placing reactions in the postnatal rat. J Comp Neurol 175(2).207–231
3. Smart JL, Dobbing J (1971) Vulnerability of developing brain II. Effects of early nutritional deprivation on reflex ontogeny and development of behaviour in the rat. Brain Res 28(1): 85–95
4. Smart JL, Dobbing J (1971) Vulnerability of developing brain VI. Relative effects of foetal and early postnatal undernutrition on reflex ontogeny and development of behaviour in the rat. Brain Res 33(2):303–314
5. Ten VS, Bradley-Moore M, Gingrich JA, Stark RI, Pinsky DJ (2003) Brain injury and neurofunctional deficit in neonatal mice with hypoxic-ischemic encephalopathy. Behav Brain Res 145(1–2):209–219
6. Rice JE III, Vannucci RC, Brierley JB (1981) The influence of immaturity on hypoxic-ischemic brain damage in the rat. Ann Neurol 9(2): 131–141
7. Fox WM (1965) Reflex-ontogeny and behavioral development of the mouse. Anim Behav 13(2):234–241
8. Jensh RP, Brent RL (1986) Effects of 0.6-Gy prenatal X irradiation on postnatal neurophysiologic development in the Wistar rat. Proc Soc Exp Biol Med 181(4):611–619
9. Cogan DC, Cohen LE, Sparkman G (1983) Effects of gestational alcohol on the development of neonatal reflexes in the rat. Neurobehav Toxicol Teratol 5(5):517–522
10. Chen C, Fuller JL (1975) Neonatal thyroxine administration, behavioral maturation, and brain growth in mice of different brain weight. Dev Psychobiol 8(4):355–361
11. Sobotka TJ, Brodie RE, Spaid SL (1977) Tartrazine and the developing nervous system of rats. J Toxicol Environ Health 2(5): 1211–1220
12. Ryan CL, Pappas BA (1986) Intrauterine diazepam exposure: effects on physical and neurobehavioral development in the rat. Neurobehav Toxicol Teratol 8(3):279–286
13. Almli CR, Levy TJ, Han BH, Shah AR, Gidday JM, Holtzman DM (2000) BDNF protects against spatial memory deficits following

neonatal hypoxia-ischemia. Exp Neurol 166(1): 99–114

14. Kumral A, Uysal N, Tugyan K, Sonmez A, Yilmaz O, Gokmen N, Kiray M, Genc S, Duman N, Koroglu TF, Ozkan H, Genc K (2004) Erythropoietin improves long-term spatial memory deficits and brain injury following neonatal hypoxia-ischemia in rats. Behav Brain Res 153(1):77–86
15. Demers EJ, McPherson RJ, Juul SE (2005) Erythropoietin protects dopaminergic neurons and improves neurobehavioral outcomes in juvenile rats after neonatal hypoxia-ischemia. Pediatr Res 58:297–301
16. Lin S, Fan LW, Pang Y, Rhodes PG, Mitchell HJ, Cai Z (2005) IGF-1 protects oligodendrocyte progenitor cells and improves neurological functions following cerebral hypoxia-ischemia in the neonatal rat. Brain Res 1063(1):15–26
17. Cai J, Kang Z, Liu K, Liu W, Li R, Zhang JH, Luo X, Sun X (2009) Neuroprotective effects of hydrogen saline in neonatal hypoxia-ischemia rat model. Brain Res 1256:129–137
18. Chen W, Hartman R, Ayer R, Marcantonio S, Kamper J, Tang J, Zhang JH (2009) Matrix metalloproteinases inhibition provides neuroprotection against hypoxia-ischemia in the developing brain. J Neurochem 111(3): 726–736
19. Fathali N, Ostrowski RP, Lekic T, Jadhav V, Tong W, Tang J, Zhang JH (2010) Cyclooxygenase-2 inhibition provides lasting protection against neonatal hypoxic-ischemic brain injury. Crit Care Med 38(2):572–578
20. Iwai M, Cao G, Yin W, Stetler RA, Liu J, Chen J (2007) Erythropoietin promotes neuronal replacement through revascularization and neurogenesis after neonatal hypoxia/ischemia in rats. Stroke 38(10):2795–2803
21. Iwai M, Stetler RA, Xing J, Hu X, Gao Y, Zhang W, Chen J, Cao G (2010) Enhanced oligodendrogenesis and recovery of neurological function by erythropoietin after neonatal hypoxic/ischemic brain injury. Stroke 41(5):1032–1037
22. Lubics A, Reglodi D, Tamás A, Kiss P, Szalai M, Szalontay L, Lengvári I (2005) Neurological reflexes and early motor behavior in rats subjected to neonatal hypoxic-ischemic injury. Behav Brain Res 157(1):157–165

Chapter 24

Assessment of Neurogenesis and White Matter Regeneration

Masanori Iwai, Hiroko Tajiri, Shiro Matsumoto, Hiroshi Mitsubuchi, and Fumio Endo

Abstract

Persistent neurogenesis occurs in discrete regions of the brain through neonatal to adult period, including the subgranular zone (SGZ) of the hippocampal dentate gyrus (DG) (J Comp Neurol 124:319–335, 1969) and the subventricular zone of the lateral ventricle (J Comp Neurol 137:433–458, 1969). After Eriksson et al. showed that the human hippocampus retains an ability to generate neurons throughout life (Nat Med 4:1313–1317, 1998), it is now expected that neurogenesis has a potential to compensate for and recover neural functions that were destroyed by central nervous system disorders. In fact, since Liu et al. (J Neurosci 18:7768–7778, 1998) first showed that ischemia stimulates neurogenesis in the SGZ of the DG using gerbil forebrain transient ischemia model, the ability of neurogenesis and neuronal repair in the ischemic brain has been reported not only in adult ischemia models (Nat Med 8:963–970, 2002; Ann Neurol 52:802–813, 2002; Cell 110:429–441, 2002; J Clin Invest 114:330–338, 2004), but also in neonatal hypoxia–ischemia (NHI) models (Neurobiol Dis 16:585–595, 2004; Brain Res 1038:41–49, 2005; Neurol Res 28:461–468, 2006). Neurogenesis can be divided into three stages: (1) neural stem/progenitor cells proliferation, (2) neural stem/progenitor cells migration, and (3) differentiation to mature neurons, astrocytes, and oligodendrocytes (OLs) (Science 287:1433–1438, 2000). In ischemic brain, the period of increasing neural stem/progenitor cells' proliferation and migration is very limited, so it is very important to choose the adequate period to investigate the stimulation of neurogenesis under postischemic condition (Neurol Res 28:461–468, 2006; Brain Res 902:288–293, 2001; J Cereb Blood Flow Metab 22:411–419, 2002; J Cereb Blood Flow Metab 23:331–341, 2003). OLs' proliferation (oligodendrogenesis) is quite different from neuronal and glial cell proliferation because early oligodendrocyte progenitor cells (OPCs), which can proliferate under normal and pathogenic conditions, exist not only in white matter, but also in gray matter (J Neurosci Res 69:826–836, 2002). In the developing brain, OLs can be divided into four stages during its maturation: (1) early OPCs, (2) late OPCs, (3) immature OLs, and (4) mature OLs (Trends Cell Biol 3:191–197, 1993; J Neurosci 22:455–463, 2002). A lot of late OPCs in the developing brain, including neonatal brain, are very vulnerable to hypoxia or ischemia (J Neurosci 22:455–463, 2002). Therefore, the investigation of how early OPCs increase and reproduce myelin sheath after lack of late OPCs post NHI (Stroke 41:1032–1037, 2010) becomes significant.

This chapter introduces a brief history of assessment of neurogenesis and white matter regeneration in models of NHI, as well as the required materials and tools for performing the assessment using methods, such as immunohistochemistry and immunofluorescence.

Key words: Neonatal hypoxia–ischemia, Rat, Neurogenesis, Oligodendrogenesis, White matter regeneration

Jun Chen et al. (eds.), *Animal Models of Acute Neurological Injuries II*, Springer Protocols Handbooks,
DOI 10.1007/978-1-61779-782-8_24, © Springer Science+Business Media, LLC 2012

1. Introduction

After neonatal hypoxia–ischemia (NHI), neurogenesis and white matter regeneration start in the first week during ischemic lesion and last for 1 month. To assess neurogenesis and white matter regeneration, histological methods (immunohistochemistry and immunofluorescence) are commonly used.

The first step of neurogenesis, or neural stem cell proliferation, is impossible to be detected directly due to the lack of specific marker for neural stem cell (1–3). Hence, the thymidine analog bromodeoxyuridine (BrdU), which is incorporated into the DNA of dividing cells during S phase of the mitotic cycle, is used as a labeling marker for newly generated cells (cell proliferation) (12). In the subventricular zone (SVZ) of neonatal brain, the BrdU-labeled cells decrease continuously after birth (11). In the ischemic injured neonatal brain, the number of BrdU-labeled cells in the SVZ increases compared to the same time point control between 1 and 14 days after NHI, and reaches its peak in day 7 (Figs. 1 and 2). Therefore, to evaluate cell proliferation, one or several times of BrdU intraperitoneal injection are normally carried out 1 week after NHI, and animals are usually sacrificed 1 day after the last injection. After serially cutting the removed brain in 10–20-μm-thick coronal sections with a cryostat or sliding microtome, BrdU-labeled cells are revealed following the incubation with anti-BrdU antibody.

Neural progenitor cells' migration, which is the second step of neurogenesis, is commonly analyzed using anti-doublecortin (DCX) (14, 15) or anti-highly polysialylated neural cell adhesion molecule (PSA-NCAM) antibodies (12, 13) 1 week after NHI. Double immunofluorescence technique of BrdU and DCX or PSA-NCAM is widely used to confirm whether newly generated cells are migrating neural progenitor cells.

The final step of neurogenesis, or neuronal differentiation, is studied with double immunofluorescence method using anti-BrdU, anti-neuronal nuclei (NeuN), anti-microtubule-associated protein 2 (MAP2), anti-calbindin D28K, or anti-DARPP-32 antibodies approximately 1 month after NHI.

After Liu et al. (4) proposed that ischemic brain injury stimulates neurogenesis in the subgranular zone (SGZ) of the hippocampus in adult gerbils, neuronal repair in the ischemic lesion has been reported using adult stroke model (5, 6). The administration of EGF and FGF-2 (7) or stimulation of angiogenesis with endothelial progenitor cells (8) also promotes neuronal repair in adult ischemic brain. In the NHI brain injury, on the other hand, Plane et al. (9) reported that neurogenesis was stimulated, but neuronal repair was not detected using mouse NHI model. However, the cell proliferation in the SVZ is stimulated and the number of migrating neural progenitor cells is increased after NHI (10, 11). In the ischemic

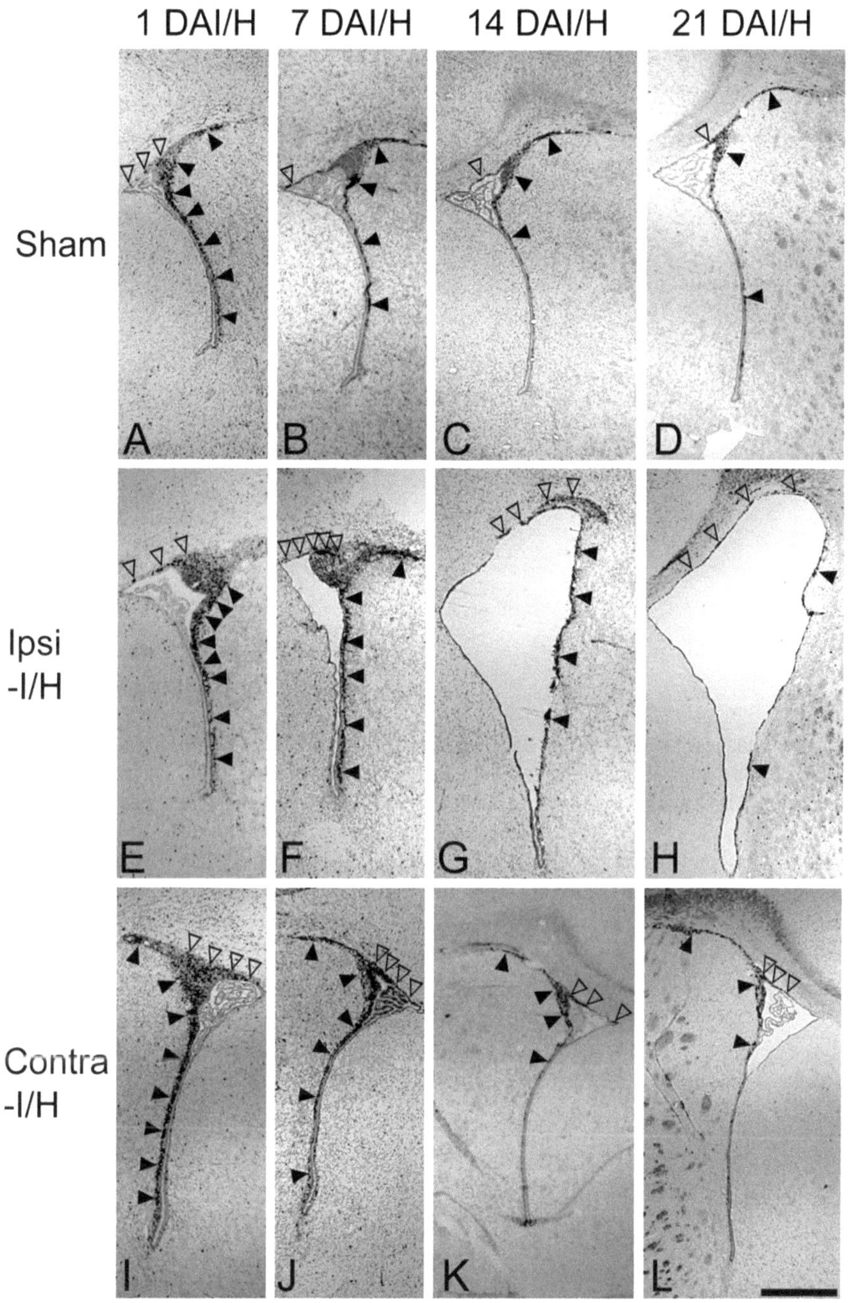

Fig. 1. Immunohistochemistry for bromodeoxyuridine (BrdU) of SVZ of sham control (**a**–**d**), ipsilateral side of hypoxia/ischemia (**e**–**h**), and contralateral side of hypoxia/ischemia (**i**–**l**) at 1 day (**a**, **e**, **i**), 7 days (**b**, **f**, **j**), 14 days (**c**, **g**, **k**), and 21 days after hypoxia/ischemia (**d**, **h**, **l**).

lesion, migrated neural progenitor cells are impossible to survive and differentiate into mature neuron (9, 20). Although the reason is unclear, it is suggested that suppressed angiogenesis fails to provide a suitable environment for the survival of neural progenitor cells (21). According to Iwai et al., erythropoietin (EPO) can stimulate

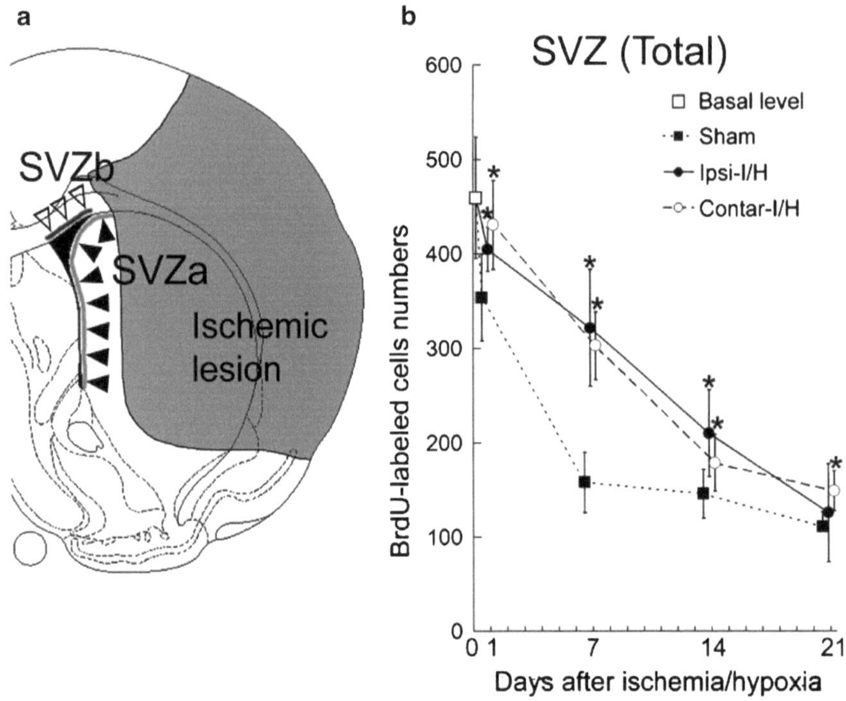

Fig. 2. Chronological changes of the numbers of BrdU-labeled cells in the SVZ after NHI. A, Indication of the SVZ; the lateral wall of the LV (SVZa) and the upper wall of the LV (SVZb). B. Quantification of BrdU-labeld cell number of the SVZ (SVZa + SVZb) of sham control (filled aquares), ipsilateral side (Ipsi-I/H, filled circles) and contralateral side (Contra-I/H, open circles) after NHI.

angiogenesis and promote neuronal replacement in the ischemic lesion after NHI (21). Thus, the neonatal brain has the potential to replace neuronal tissues for cell survival and neural functions' recovery in an appropriate environment. However, further analysis is still needed to support this idea.

The restoration of postischemic brain involves the process of neurogenesis and oligodendrocytes (OLs)' regeneration, which occurs in both white matter and gray matter. Four stages of OLs' maturation are as described above (16, 17). To detect each stage of OLs, specific markers are used as follows:

Early OPCs: Anti-NG2 (positive)

Late OPCs: Anti-O4 (positive), anti-O1 (negative)

Immature OLs: Anti-O4 (positive), anti-O1 (positive)

Mature OLs: Anti-myelin basic protein (MBP) (positive), anti-adenomatous polyposis coli (APC) (positive)

According to Back et al., late oligodendrocyte progenitor cells (OPCs) are vulnerable to hypoxia–ischemia, and their absence causes the premature infant's major pathogenesis of white matter injury (known as periventricular leukomalacia) (18). The importance of both protecting the late OPCs against hypoxia–ischemia and regenerating OLs has been recognized.

The OLs' regeneration studied by Zaidi et al. has showed that many newly generated mature OLs are in the injured corpus callosum and striatum of rats 4 weeks post NHI (22). Iwai et al. (19) recently showed that EPO enhances oligodendrogenesis and promotes white matter regeneration after NHI.

The experiments, sections fixed with paraformaldehyde, and staining are performed using immunohistochemistry or immunofluorescence. Before moving onto the next section, the general materials and tools are described as follows.

1. To make sections

   Heparinized saline

   4% paraformaldehyde phosphate-buffered saline (PBS) (4% PFAPBS)

   10, 20, 30% sucrose PBS

   Cryostat

   Slide glasses

   Sliding microtome

   Storage buffer

2. Mounted section (10–20-mm-thick section)

   Glass jar with basket for slide glasses

   Humidity chamber

   PBS

   5% bovine serum albumin (BSA)

   0.1% BSA–PBS

   Dako PEN or PAP PEN

3. Free floating (40–60-mm-thick section)

   12-well plate

   Spuit (syringe)

   PBS

   PBS with Triton X-100

   5% BSA

   0.1% BSA–PBS

4. Immunohistochemistry

   Methanol with 0.3% H_2O_2

   Avidin biotin complex (ABC) kit

   0.02% Diaminobenzidine

   0.02% H_2O_2

5. Immunofluorescence

   Mounting medium

2. Neural Stem/ Progenitor Cells' Proliferation

The thymidine analog BrdU is commonly used as the marker for proliferated cells as a means to analyze neural stem/progenitor cells' proliferation. BrdU is incorporated into the DNA of dividing cells during S phase of the mitotic cycle. Due to the lack of specific marker for neural stem cell, BrdU is usually used to detect proliferating cell at the SGZ of the hippocampal dentate gyrus (DG) and the SVZ of the LV.

Be aware that BrdU-labeled cells are not always newly generated neural stem/progenitor cells.

2.1. Material and Tools

To inject BrdU

BrdU

Saline

Syringe (1 mL)

Needle (27 or 30G)

To perform immunohistochemistry with mounted section

Mouse anti-BrdU antibody

Biotinylated secondary antibody for mouse IgG

1 N hydrochloric acid (HCl)

If free floating method is used, spread shrunken sections with 0.1 N boric acid (pH 8.5) after HCl treatment.

2.2. Procedure

After inducing NHI, a single dose of BrdU (dissolved in saline with concentration of 10 mg/mL, usually 50 mg/kg/dose) is injected intraperitoneally 24 h before each sampling. The animals were sacrificed in deep anesthesia, and were perfused transcardinally with heparinized saline followed by 4% PFAPBS (pH 7.4). The brains were removed, postfixed overnight, and rapidly frozen (for serial cut of 10-μm-thick coronal sections on a cryostat) or stored in 30% sucrose solution (for serial cut of 20–60-μm-thick coronal sections on a sliding microtome) after cryoprotection (10, 20, and 30% sucrose PBS solution).

Suggested immunohistochemical procedure is described below (we recommend using mounted 10-μm-thick coronal sections due to their ease of counting).

1. Rinse sections with PBS for 5 min using glass jar ×3.
2. Incubate sections in 1 N HCl for 1 h at 37°C (followed with 0.1 N boric acid (pH 8.5] for 10 min at room temperature if free floating method is used).
3. Rinse with PBS for 5 min ×3.
4. Blocking with 5% BSA in PBS for 1 h.

5. Incubate sections in anti-BrdU antibody in 0.1% BSA/PBST at 4°C overnight (mouse anti-BrdU antibody is recommended).
6. Rinse with PBS for 5 min ×3.
7. Incubate sections in biotinylated secondary antibody in PBST for 1 h at room temperature.
8. Rinse sections with PBS for 5 min ×3.
9. Incubate sections in horseradish peroxidase–streptavidin–biotin complex solution (ABC kit) for 30 min.
10. Rinse section with PBS for 5 min ×3.
11. Incubate sections for 2–5 min in peroxidase reaction solution (0.02% diaminobenzidine + 0.02% H_2O_2). Stop the reaction with water.
12. Mount cover glass on slides. Observe slides under a light microscope.

3. Neural Progenitor Cells' Migration

To study neural progenitor cells' migration, anti-DCX or anti-highly PSA-NCAM antibodies are usually used. We recommend using anti-DCX antibody for double staining with BrdU because PSA-NCAM antigen is vulnerable to HCl, which is required to reveal the antigen of BrdU. According to previous studies (9, 10), 1 week after NHI is usually the best time for neural precursor migration analysis. Because free floating method is typically used to obtain fine image under confocal microscope, 40–60-μm-thick coronal sections are suggested to carry out double immunofluorescence study.

This chapter briefly describes double staining technique using free floating method for BrdU and DCX to detect newly generated neural precursor migration. Extra care should be taken while handling the sections because ischemic lesion is vulnerable to washing. Selected images are shown in Fig. 3.

3.1. Material and Tools

In addition to the materials and tools mentioned in introduction, four antibodies are needed as described below.

Mouse anti-BrdU antibody

Anti-mouse IgG conjugated with FITC

Goat anti-DCX antibody

Anti-goat IgG conjugated with Cy3

3.2. Procedure

1. Rinse sections with PBS for 5 min using 12-well plate ×3.
2. Incubate sections in 1 N HCl for 1 h at 37°C (sections shrink immediately in HCl).

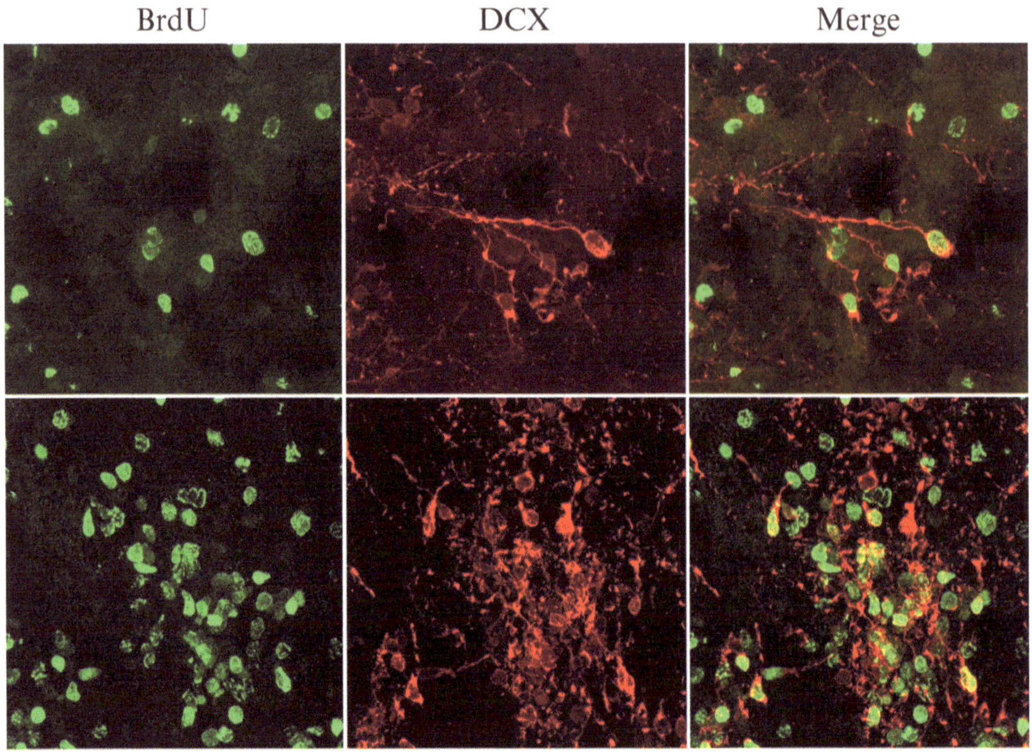

Fig. 3. Representative images of double immunofluorescence for BrdU and DCX in cerebral cortex (*upper row*) and striatum (*lower row*).

3. Incubate sections in 0.1 N boric acid (pH 8.5) for 10 min at room temperature (sections spread slowly).
4. Rinse with PBST for 5 min ×3.
5. Block with 5% BSA/PBST for 1 h.
6. Incubate sections in mouse anti-BrdU and goat anti-DCX antibodies in 0.1% BSA/PBST at 4°C overnight.
7. Rinse with PBST for 5 min ×3.
8. Incubate sections in anti-mouse IgG conjugated with FITC and anti-goat IgG conjugated with Cy3 antibodies in PBST for 1 h at room temperature.
9. Rinse with PBST for 5 min ×3.
10. Mount sections on slides and cover with cover glass.
11. Observe under confocal microscope.

4. Neuronal Differentiation

To analyze neuronal differentiation, anti-NeuN, anti-MAP2, anti-calbindin D28K, or anti-DARPP-32 antibodies are used to detect

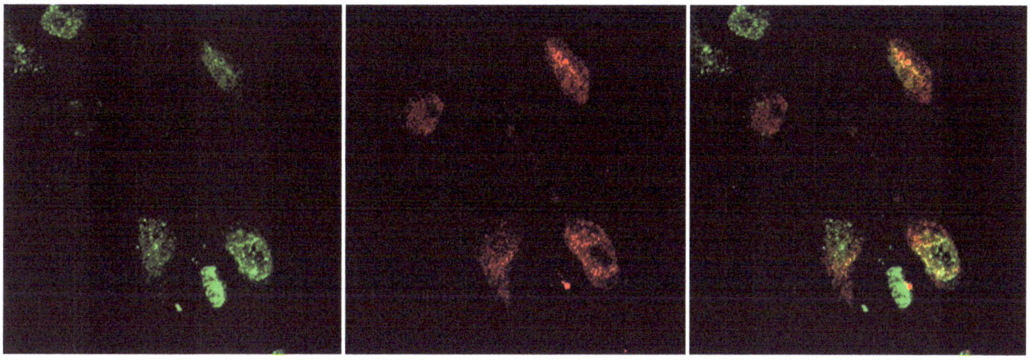

Fig. 4. Representative images of double immunofluorescence for BrdU and NeuN.

mature neurons. Anti-glial fibrillary acidic protein (GFAP) and anti-familial polyposis coli (AFP) are used to detect astrocytes and mature OLs, respectively. We recommend using anti-NeuN antibody for double staining with BrdU because of NeuN's ease of staining and HCl resistance. The time to investigate neuronal differentiation is usually 2–4 weeks after NHI, according to previous reports (Ikeda, Iwai). To obtain fine image under confocal microscope with free floating method, 40–60-μm-thick coronal sections are typically recommended for double immunofluorescence technique.

This chapter briefly describes the double staining method using free floating for BrdU and NeuN to detect newly generated mature neuron. Extra care should be taken while handling sections because ischemic lesion is vulnerable to washing.

Selected images are shown in Fig. 4.

4.1. Materials and Tools

In addition to the materials and tools mentioned in introduction, four antibodies are needed as described below.

Mouse anti-BrdU antibody

Fluorescent streptavidin

Mouse anti-NeuN antibody

Anti-mouse IgG conjugated with Cy3

Also, the M.O.M kit (Vector laboratories) is needed to avoid cross-reaction of BrdU and NeuN double immunofluorescence.

4.2. Procedure

1. Rinse sections with PBS for 5 min using 12-well plate ×3.
2. Incubate sections in 1 N HCl for 1 h at 37°C (sections shrink immediately in HCl).
3. Incubate sections in 0.1 N boric acid (pH 8.5) for 10 min at room temperature (sections spread slowly).
4. Rinse with PBST for 5 min ×3.
5. Block with 5% BSA/PBST for 1 h.

6. Incubate sections in mouse anti-BrdU antibody + 0.1% BSA/PBST at 4°C overnight.
7. Rinse with PBST for 5 min ×3.
8. Incubate sections in anti-mouse IgG conjugated Cy3 antibody in PBST for 1 h at room temperature.
9. Rinse with PBST for 5 min ×3.
10. Incubate sections for 1 h in working solution of M.O.M.™ mouse Ig blocking reagent, prepared according to manufacturer's instruction.
11. Rinse with PBST for 2 min ×2.
12. Incubate sections for 5 min in working solution of M.O.M.™ diluents, prepared according to manufacturer's instruction.
13. Tip excess of M.O.M.™ diluent off sections. Dilute anti-BrdU antibody in M.O.M.™ diluent to the appropriate concentration. Incubate sections in diluted anti-NeuN antibody for 30 min.
14. Rinse with PBST for 2 min ×2.
15. Apply working solution of M.O.M.™ Biotinylated Anti-Mouse IgG Reagent, prepared according to manufacturer's instruction. Incubate sections for 10 min.
16. Rinse with PBST for 2 min ×2.
17. Apply the appropriate amount of streptavidin–Cy3 in PBST.
18. Rinse with PBST for 5 min ×3.
19. Mount sections on slides and cover with cover glass.
20. Observe under confocal microscope.

5. Regeneration of Oligodendrocyte

To study OLs' regeneration, their maturation stages are matched with antibodies to detect developing OLs:

1. Early OPCs: Antigen; NG2+, O4−.
2. Late OPCs: Antigen; O4+, O1−.
3. Immature OLs: Antigen; O4+, O1+.
4. Mature OLs: Antigen; MBP+, APC+.

By double staining BrdU and NG2, it is verified that early OPCs' generation reaches its peak within both white and gray matter 1 week after NHI. Although the duration for early OPCs to become mature OLs remains unclear, it is proposed that newly generated mature OLs (BrdU + APC + cells) are observed in both white and gray matter 2 weeks after NHI. To detect mature OLs, anti-APC antibody is usually used because it stains only cell body (neither axon nor dendrite).

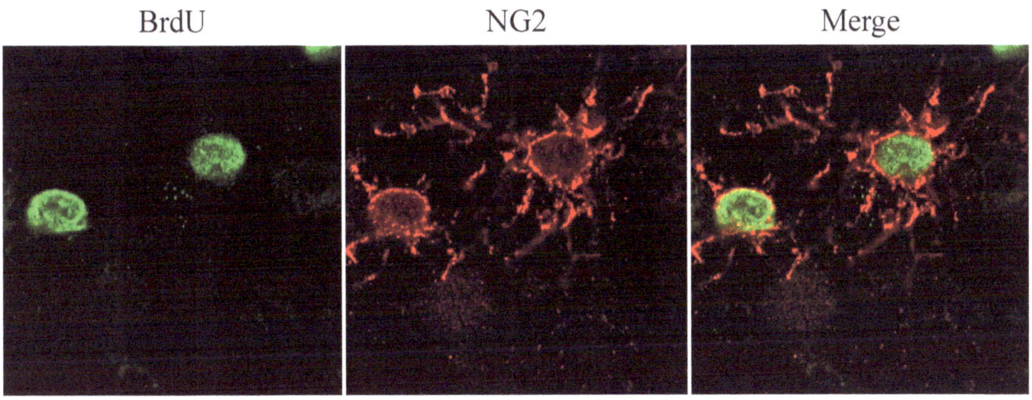

Fig. 5. Representative images of double immunofluorescence for BrdU and NG2.

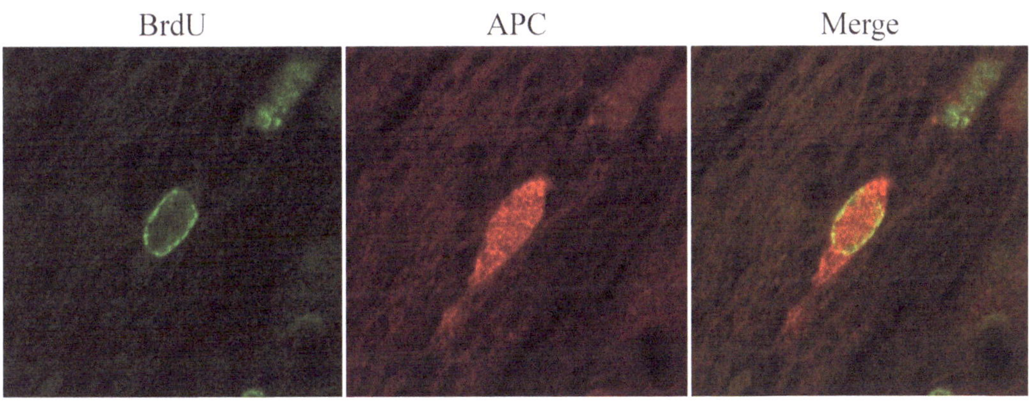

Fig. 6. Representative images of double immunofluorescence for BrdU and APC.

In order to obtain the best image results under confocal microscope, it is recommended to use 20-μm-thick coronal sections for BrdU/NG2 or BrdU/APC double immunofluorescence.

Selected images are shown in Figs. 5 (BrdU and NG2) and 6 (BrdU and APC).

5.1. Materials and tools

For double immunofluorescence with BrdU and NG2, in addition to the materials and tools mentioned in introduction, four antibodies are needed as described below.

Mouse anti-BrdU antibody

Anti-mouse IgG conjugated with FITC

Rabbit anti-NG2 antibody

Anti-rabbit IgG conjugated with Cy3

For double immunofluorescence with BrdU and APC, in addition to the materials and tools mentioned in introduction, four antibodies are needed as described below.

Mouse anti-BrdU antibody

Fluorescent streptavidine

Mouse anti-APC antibody

Anti-mouse IgG conjugated with Cy3

Also, the M.O.M kit (Vector laboratories) is needed to avoid cross-reaction of APC and BrdU double immunofluorescence.

5.2. Procedure

5.2.1. Double Staining for BrdU and NG2

1. Rinse sections with PBS for 5 min ×3.
2. Incubate sections in 1 N HCl for 1 h at 37°C (followed with 0.1 N boric acid [pH 8.5] for 10 min at room temperature, if free floating method is used).
3. Rinse with PBS for 5 min ×3.
4. Block with 5% BSA in PBS for 1 h.
5. Incubate sections in mouse anti-BrdU and rabbit anti-NG2 antibodies in 0.1% BSA/PBST at 4°C overnight.
6. Rinse with PBS for 5 min ×3.
7. Incubate sections in anti-mouse IgG conjugated with FITC and anti-rabbit IgG conjugated with Cy3 antibodies in PBST for 1 h at room temperature.
8. Rinse with PBS for 5 min ×3.
9. Mount sections on slide and cover with cover glass.
10. Observe under confocal microscope.

5.2.2. Double Staining for BrdU and APC

1. Rinse sections with PBS for 5 min using 12-well plate ×3.
2. Incubate sections in 1 N HCl for 1 h at 37°C (sections shrink immediately in HCl).
3. Incubate sections in 0.1 N boric acid (pH 8.5) for 10 min at room temperature (sections spread slowly).
4. Rinse with PBST for 5 min ×3.
5. Block with 5% BSA/PBST for 1 h.
6. Incubate sections in mouse anti-BrdU antibody in 0.1% BSA/PBST at 4°C overnight.
7. Rinse with PBST for 5 min ×3.
8. Incubate sections in anti-mouse IgG conjugated Cy3 antibody in PBST for 1 h at room temperature.
9. Rinse with PBST for 5 min ×3.
10. Incubate sections for 1 h in working solution of M.O.M.™ mouse Ig blocking reagent, prepared according to manufacturer's instruction.
11. Rinse with PBST for 2 min ×2.
12. Incubate sections for 5 min in working solution of M.O.M.™ diluents, prepared according to manufacturer's instruction.

13. Tip excess of M.O.M.™ diluent off sections. Dilute anti-BrdU antibody in M.O.M.™ diluent to the appropriate concentration. Incubate sections in diluted anti-APC antibody for 30 min.
14. Rinse with PBST for 2 min ×2.
15. Apply working solution of M.O.M.™ Biotinylated Anti-Mouse IgG Reagent, prepared according to manufacturer's instruction. Incubate sections for 10 min.
16. Rinse with PBST for 2 min ×2.
17. Apply appropriate amount of streptavidin–Cy3 in PBST.
18. Rinse with PBST for 5 min ×3.
19. Mount sections on slide and cover with cover glass.
20. Observe under confocal microscope.

6. Remyelination

Anti-neurofilament 200 (NF200) and MBP are usually used to detect axon or dendrite and myelin sheath, respectively. The pretreatment of sections with citrate buffer (pH 6.0) helps to retrieve antigen of axon or dendrite. However, care should be taken while pretreating the sections. The longer the sections are immersed in citrate buffer, the less the MBP antigen will be retrieved. One to four weeks after NHI is usually the optimal time for remyelination study, according to previous reports (19). In order to achieve the best image results under confocal microscope, it is recommended to use 20-μm-thick coronal sections for double immunofluorescence of NF200 and MBP.

Selected images are shown in Fig. 7.

6.1. Materials and Tools

In addition to the materials and tools mentioned in introduction, other materials are needed as described below.

Rabitt anti-neurofilamment 200 (NF200) antibody.

Anit-rabbit IgG conjugated with Cy3.

Mouse anti-MBP antibody.

Anti-mouse IgG conjugated with FITC.

Citrate buffer (pH 6.0).

6.2. Procedure

1. Rinse sections with PBS for 5 min ×3.
2. Incubate sections in citrate buffer (pH 6.0) for 20 min at 95–100°C.
3. Cool down in citrate buffer (pH 6.0) for 20 min.
4. Rinse with PBS for 5 min ×3.

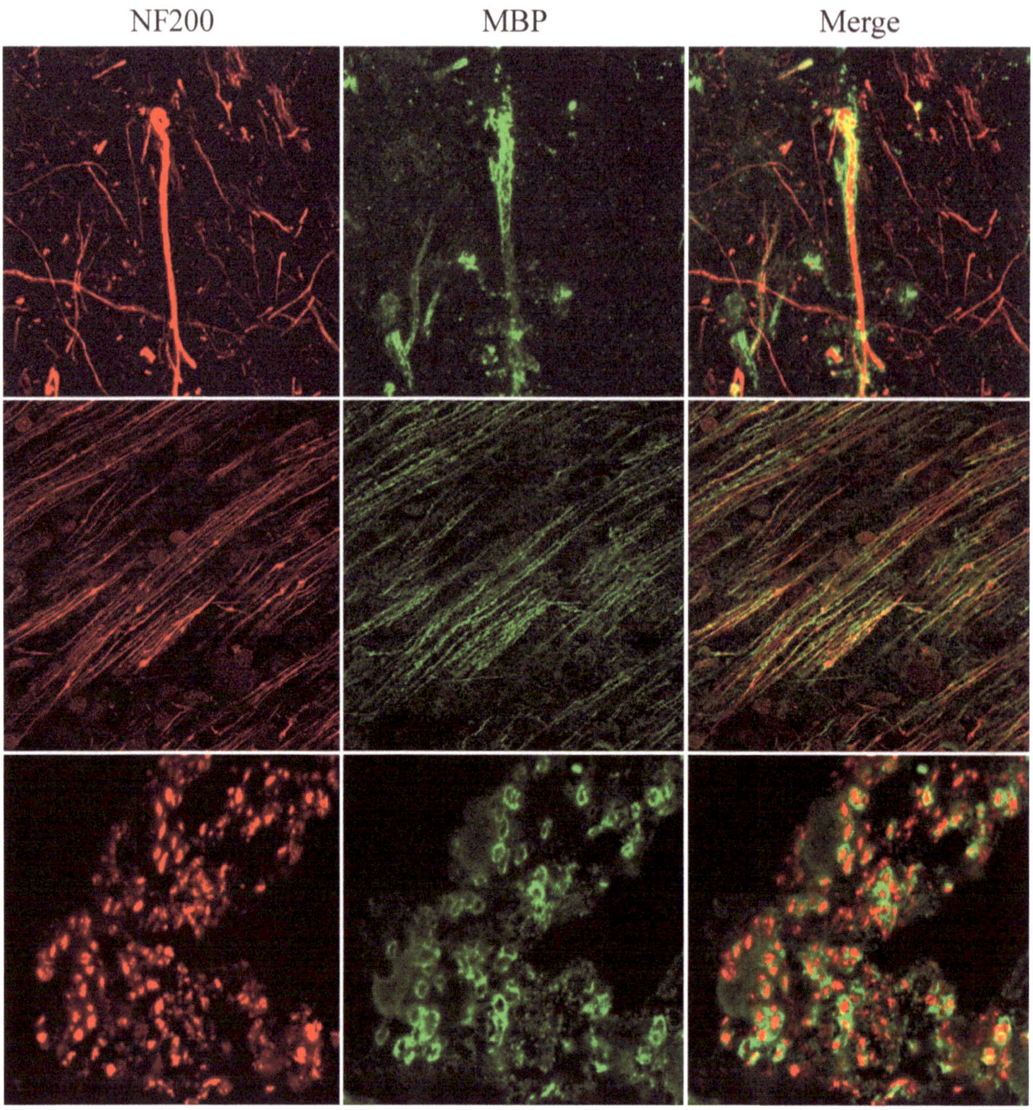

Fig. 7. Representative images of double immunofluorescence for NF200 and MBP in cerebral cortex (*upper row*), corpus callosum (*middle row*), and striatum (*lower row*).

5. Incubate sections for 30 min in 1 N HCl at 37°C.
6. Incubate sections in 0.1 N boric acid (pH 8.5) for 10 min at room temperature.
7. Rinse with PBS for 5 min ×3.
8. Block with 5% BSA/PBST for 1 h.
9. Incubate sections in rabbit anti-NF 200 and mouse anti-MBP antibodies in 0.1% BSA/PBST at 4°C overnight.
10. Rinse with PBS for 5 min ×3.

11. Incubate sections in anti-mouse IgG conjugated with FITC and anti-rabbit IgG conjugated with Cy3 antibodies in PBST for 1 h at room temperature.
12. Rinse with PBS for 5 min ×3.
13. Mount sections on slide glass and cover with cover glasses.
14. Observe under confocal microscope. Obtain TIFF images of the corpus callosum, cerebral cortex, and striatum.
15. Calculate the mean intensity value of MBP and NF200 staining in the corpus callosum, cerebral cortex, and striatum.
16. For densitometry, use a computerized video camera-based image-analysis system (with NIH Image software).
17. Digitize unaltered TIFF images. Pick out the component of red color for NF200 and component of green color for MBP.
18. Segment (using a consistent arbitrary threshold 50%) and binarize (black vs. white) red-component TIFF images. Count the total black pixels per red images.
19. Count the total black pixels per green-component TIFF images the same way as in step 18.
20. Myelination is expressed as a relative ratio of the number of pixels obtained from green-component TIFF image (MBP) divided by the number of pixels obtained from red-component TIFF image (NF200).

    *Photomontages are prepared from images processed using Photoshop 5.0 (Adobe, Mountain View, CA, USA).

References

1. Altman J, Das GD (1969) Autoradiographic and histological evidence of postnatal hippocampal neurogenesis in rats. J Comp Neurol 124:319–335
2. Altman J (1969) Autoradiographic and histological studies of postnatal neurogenesis. J Comp Neurol 137:433–458
3. Eriksson PS, Perfilieva E, Bjork-Eriksson T, Alborn A, Nordborg C, Peterson DA, Gage FH (1998) Neurogenesis in the adult human hippocampus. Nat Med 4:1313–1317
4. Liu J, Solway K, Messing RO, Sharp FR (1998) Increased neurogenesis in the dentate gyrus after transient global ischemia in gerbils. J Neurosci 18:7768–7778
5. Arvidsson A, Collin T, Kirik D, Kokaia Z, Lindvall O (2002) Neuronal replacement from endogenous precursors in the adult brain after stroke. Nat Med 8:963–970
6. Parent JM, Vexler ZS, Gong C, Derugin N, Ferriero DM (2002) Rat forebrain neurogenesis and striatal neuron replacement after focal stroke. Ann Neurol 52:802–813
7. Nakatomi H, Kuriu T, Okabe S, Yamamoto S, Hatano O, Kawahara N et al (2002) Regeneration of hippocampal pyramidal neurons after ischemic brain injury by recruitment of endogenous neural progenitors. Cell 110:429–441
8. Taguchi A, Soma T, Tanaka H, Kanda T, Nishimura H, Yoshikawa H, Tsukamoto Y, Iso H, Fujimori Y, Stern DM, Naritomi H, Matsuyama T (2004) Administration of CD34+ cells after stroke enhances neurogenesis via angiogenesis in a mouse model. J Clin Invest 114:330–338
9. Plane JM, Liu R, Wang TW, Silverstein FS, Parent JM (2004) Neonatal hypoxic–ischemic injury increases forebrain subventricular zone neurogenesis in the mouse. Neurobiol Dis 16:585–595
10. Hayashi T, Iwai M, Ikeda T, Jin G, Deguchi K, Nagotani S, Zhang H, Sehara Y, Nagano I,

Shoji M, Ikenoue T, Abe K (2005) Neural precursor cells division and migration in neonatal rat brain after ischemic/hypoxic injury. Brain Res 1038:41–49

11. Iwai M, Ikeda T, Hayashi T, Sato K, Nagata T, Nagano I, Shoji M, Ikenoue T, Abe K (2006) Temporal profile of neural stem cell proliferation in the subventricular zone after ischemia/hypoxia in the neonatal rat brain. Neurol Res 28:461–468
12. Gage FH (2000) Mammalian neural stem cells. Science 287:1433–1438
13. Iwai M, Hayashi T, Zhang WR, Sato K, Manabe Y, Abe K (2001) Induction of highly polysialylated neural cell adhesion molecule (PSA-NCAM) in postischemic gerbil hippocampus mainly dissociated with neural stem cell proliferation. Brain Res 902:288–293
14. Iwai M, Sato K, Omori N, Nagano I, Manabe Y, Shoji M, Abe K (2002) Three steps of neural stem cells development in gerbil dentate gyrus after transient ischemia. J Cereb Blood Flow Metab 22:411–419
15. Iwai M, Sato K, Kamada H, Omori N, Nagano I, Shoji M, Abe K (2003) Temporal profile of stem cell division, migration, and differentiation from subventricular zone to olfactory bulb after transient forebrain ischemia in gerbils. J Cereb Blood Flow Metab 23:331–341
16. Watanabe M, Toyama Y, Nishiyama A (2002) Differentiation of proliferated NG-2 positive glial progenitor cells in a remyelinating lesion. J Neurosci Res 69:826–836
17. Pfeiffer SE, Warrington AE, Bansal R (1993) The oligodendrocyte and its many cellular processes. Trends Cell Biol 3:191–197
18. Back SA, Han BH, Luo NL, Chricton CA, Xanthoudakis S, Tam J, Arvin KL, Holtzman DM (2002) Selective vulnerability of late oligodendrocyte progenitors to hypoxia-ischemia. J Neurosci 22:455–463
19. Iwai M, Stetlar RA, Xing J, Hu X, Gao Y, Zhang W, Chen J, Cao G (2010) Enhanced oligodendrogenesis and recovery of neurological function by erythropoietin after neonatal hypoxic/ischemic brain injury. Stroke 41:1032–1037
20. Ikeda T, Iwai M, Hayashi T, Nagano I, Shogi M, Ikenoue T, Abe K (2005) Limited differentiation to neurons and astroglia from neural stem cells in the cortex and striatum after ischemia/hypoxia in the neonatal rat brain. Am J Obstet Gynecol 193:849–856
21. Iwai M, Cao G, Yin W, Stetler RA, Liu J, Chen J (2007) Erythropoietin promotes neuronal replacement through revascularization and neurogenesis after neonatal hypoxia/ischemia in rats. Stroke 38:2795–2803
22. Zaidi AU, Bessert DA, Ong JE, Xu H, Barks JD, Silverstein FS, Skoff RP (2004) New oligodendrocytes are generated after neonatal hypoxic-ischemic brain injury in rodents. Glia 46:380–390

Part IV

Traumatic Brain Injury

Chapter 25

Assessments for Traumatic Brain Injury: An Introduction

Kathryn E. Saatman

Abstract

Traumatic brain injury (TBI) is a complex disease, altering numerous intracellular cascades which affect neuronal, vascular, and glial elements of the brain. Structural damage is coupled with cellular- and systems-level functional impairment. The array of assessments used in state-of-the-art investigations of TBI in experimental models is necessarily diverse, encompassing morphological, physiological, biochemical and molecular, neurobehavioral, and neuroimaging tools. Selecting an appropriate assessment approach and utilizing it in a manner which optimizes insights into the pathophysiology of TBI can be a daunting task. This introduction discusses how selection of assessments might be influenced by the model and severity of TBI to be studied and the species, sex, and age of the experimental subjects. The importance of matching outcome measurements to the pathophysiological events being studied and the selection of appropriate time points for assessment are then highlighted. Finally, considerations surrounding the variability, validity, and sensitivity of assessment tools are discussed.

Key words: Animal models, Head injury, Outcome measures, Sensitivity, Species, Validity, Variability

1. Introduction

Traumatic brain injury (TBI) is a heterogeneous disease affecting all walks of life, men and women, the young and the elderly, and those previously healthy or with preexisting health problems. The type and intensity of the traumatic insult that induces any given head injury can vary widely, from mild impact during sporting events to severe acceleration/deceleration during high-speed automobile collisions. Not surprisingly, then, the presentation and pathology of head injury span a spectrum that poses a challenge for clinical and basic researchers interested in determining causative factors of brain damage and identifying therapeutic targets for

Jun Chen et al. (eds.), *Animal Models of Acute Neurological Injuries II*, Springer Protocols Handbooks,
DOI 10.1007/978-1-61779-782-8_25,

intervention. It is increasingly appreciated that the pathobiology of subdural hematoma, diffuse brain swelling, or focal contusion, for example, is distinctly different, and may require unique intervention strategies (1). The development of a number of different in vivo models of TBI has facilitated a greater understanding of the cellular mechanisms of neuronal, glial, and vascular damage after different types and severities of traumatic insult, the effect of secondary insults in amplifying TBI-induced pathology, and the ability of these cellular events to be slowed or stopped through therapeutic manipulations.

One challenge in designing experiments in TBI is selecting the appropriate assessments with which to measure outcome. Clinically, head-injured patients are monitored using radiological, physiological, neurological, and neuropsychological tools and assessments. Samples of blood and cerebrospinal fluid are often screened for levels of critical ions, metabolites, and proteins. In many cases, animal studies can utilize parallel assessments, such as intracerebral microdialysis to evaluate the temporal course of changes in interstitial proteins, amino acids, or ions (2, 3); magnetic resonance imaging to evaluate the location, size, and progression of mass lesions (4–6); or blood pressure and blood gas measurements to monitor systemic physiological parameters. Other parallels are more difficult to draw. Widely used outcome measures, such as the Glasgow Outcome Scale and the Disability Rating Scale, do not have clear surrogates in animal studies. Animal models, on the other hand, allow more invasive sampling and manipulation, providing a window into the intricacies of cellular and molecular events that is not readily available in the clinical setting.

2. Traumatic Brain Injury in Animals

Selection of outcome measures is often guided and customized based on the experimental TBI model and severity of injury to be employed and the species, age, and sex of the experimental subjects. Each of these parameters should be considered as to how it may impact on the feasibility and relevance of any given assessment measure. Existing data in the literature should be useful in determining how best to apply or interpret a selected assessment.

One of the first considerations is the species. While rats and mice are by far the most commonly used species for experimental TBI, laboratories are also studying pigs (7, 8), sheep (9, 10), nonhuman primates (11, 12), guinea pigs (13, 14), and rabbits (15). Early characterization of some of the most commonly used models of TBI was done in the ferret (16) and cat (17, 18). Important considerations related to species include (1) species size, (2) brain size, (3) proportion of gray matter to white matter, (4) cost,

(5) availability of reagents, and (6) collective base of knowledge. Larger brain size is typically associated with a greater fidelity to human brain anatomy and is an asset for imaging studies, providing improved neuroanatomical resolution. Because acceleration/deceleration injury is dependent on brain mass, inertial brain injury is easier to induce in larger animals. Rodents have a much smaller proportion of white matter to gray matter compared to larger mammals, such as pig or sheep, and therefore may be a less attractive species for the study of diffuse axonal injury. Cost is an obvious consideration in selecting a species for study. Rodents are widely used as they are far cheaper to purchase, house, and maintain postoperatively than larger mammals, and are almost exclusively used for studies requiring large cohorts. Small animals also require lower quantities of drugs given on a mg/kg body weight basis, a potentially important cost consideration in preclinical studies utilizing systemic administration. However, physiological responses of larger mammals may more closely resemble humans, providing significant advantage. Availability and specificity of reagents may be a concern for lesser studied species. Many antibodies have been characterized for cross-reactivity in mouse and rat while fewer are tested in guinea pig or sheep, for example. The collective body of work for rat and mouse models is certainly much deeper than for larger mammals, reducing the amount of characterization needed when contemplating applying a post-TBI assessment. For example, with respect to cognitive and motor function assessments, far less is known in larger animals (19) than has been characterized for rat and mouse (20). However, there is a growing consensus that initial findings in preclinical testing in rodent models should be translated to models utilizing larger animals with a gyroencephalic brain (e.g., pig, nonhuman primate) prior to launching clinical trials (21).

The age and sex of the animal to be studied are also important to consider when selecting both a TBI model and associated assessments. Pediatric and neonatal models of TBI are being increasingly utilized, but they present unique challenges as to how to appropriately scale the traumatic insult to a smaller brain with different tissue properties (22). Furthermore, age at injury may affect the "baseline" for many parameters, such as gene or protein expression, metabolic status, receptor function, mitochondrial function, or myelination (23, 24), leading to vastly different responses to TBI (25). The effect of sex on TBI assessments may depend on the model used and the particular assessment. For example, using calpain-mediated spectrin proteolysis as an assessment of TBI-induced neuronal damage, Hall and colleagues showed that males and females differed in the time course of posttraumatic spectrin proteolysis following a diffuse brain injury (26) but not a focal contusion injury (27).

Selection of injury model may be based on the type of injury one is trying to model, ease of use of the model, cumulative data available for that model, or suitability of the model to the specific assessment in which the investigator is interested. Other considerations may include the reproducibility of the model, its ability to produce injury across a range of injury severities, or its applicability across species. The most commonly used models of focal contusive TBI are lateral (or parasagittal) fluid percussion and controlled cortical impact. Both are used in rats and larger animals, and controlled cortical impact is also widely used in mice. Closed skull weight drop injury can also produce focal contusion when employed at more severe levels that result in skull fracture (28). Common models of diffuse brain injury include midline (or central) fluid percussion, impact acceleration, and weight drop closed head injury. The use of midline fluid percussion and impact acceleration injury has largely been limited to the rat. Weight drop closed head injury is a widely used TBI model due to its ease of use and applicability for mice. In rodents, these focal and diffuse TBI models have been extensively characterized using physiological, histological/morphological, biochemical and molecular, neuroimaging, and neurobehavioral assessments. The nonimpact rotational acceleration model of diffuse axonal injury, applied to nonhuman primates and pigs, is less commonly used but is highly clinically relevant as the only model to date to produce prolonged posttraumatic coma (hours) (8, 12).

3. Pathophysiology of Traumatic Brain Injury

The pathophysiology of TBI is highly complex, with multiple interdependent acute events triggering a broad set of downstream consequences. Initial events, such as altered membrane permeability (Chap. 28), excessive neurotransmitter release, and ion dysregulation (29), contribute to mitochondrial dysfunction (30), oxidative damage (Chap. 32), and alterations in cerebral metabolism (31). These acute changes in cellular homeostasis trigger additional intracellular events, such as activation of proteases, kinases, and phosphatases, perturbing cell signaling cascades and facilitating further damage to cytoskeletal and membrane elements. Changes in protein, mRNA, and microRNA expression (Chap. 31) contribute to acute damage and lay the groundwork for more lasting modifications in cell structure and function, associated with either cell death or plasticity. Cell death, disrupted neuronal circuitry, and altered cell signaling ultimately manifest in neurobehavioral dysfunction which may encompass changes in motor, sensorimotor, vestibular, learning, memory, or anxiety-related behaviors (Chaps. 33 and 34).

Neuronal damage may include axonal injury and Wallerian degeneration, dendritic beading or retraction, loss of synapses, or cell death by necrosis, apoptosis, or other means (32, 33). Posttraumatic adaptations, such as axonal or dendritic sprouting, synaptic plasticity, and alterations in neurogenesis (Chap. 29), may promote recovery of function or exacerbate neuronal circuit dysfunction, possibilities that can be explored through the use of electrophysiological assessments (Chap. 30). Concomitant with neuronal damage, TBI produces alterations in cerebral blood flow and vascular damage (Chap. 27), including breakdown of the blood–brain barrier and hemorrhage. Vasogenic and cytotoxic edema (34) contributes to increased intracranial pressure and, hence, decreased cerebral perfusion pressure and brain tissue oxygenation, compromising survival of injured neurons with increased metabolic demands. Neuroinflammation (35) and reactive gliosis (Vol. 1, Chap. 4) are potentially deleterious responses after TBI that paradoxically may also have beneficial effects in the traumatically injured brain.

4. Selection of Outcome Assessment Measures

4.1. Aspect of Pathophysiological Response to Be Assessed

The choice of outcome assessment depends heavily on the scientific question posed. That is, what aspect of the pathophysiology of TBI is to be evaluated in the experiment? At a tissue level, brain damage may be explored through neuroimaging, which can provide regional and temporal information regarding events, such as cerebral swelling or atrophy, edema, hematoma, contusion, axonal injury, and tissue metabolites (4). At a cellular level, examination of structural damage is most often coupled with morphological assessments of neurovascular damage and gliosis (Chap. 26). Probing cellular mechanisms of damage or mechanisms of neuroprotective effects typically requires the use of targeted biochemical and molecular tools for assessing changes in gene and protein expression or posttranslational modifications that affect signaling pathways (Chap. 31). Physiological assessments, on the other hand, yield information regarding the integrative response of the system. For example, measurements of blood flow reflect vascular structure and function intertwined with neural regulation. Similarly, neurobehavioral assessments provide information about the central nervous system as a complex network coordinating sensory input and motor outputs, and are influenced by numerous environmental and intrinsic factors (Chaps. 33 and 34). As mentioned above, some assessments have strong correlates to clinical research and practice and are ideal for translational studies while others may be far more mechanistic and better suited to focused basic research questions.

4.2. Time Point(s) of Assessment

In selecting assessment tools, it is important to understand the time course of the pathophysiological event(s) that the outcome is intended to measure. For example, blood–brain barrier breakdown, increases in neuronal membrane permeability, and oxidative damage occur very rapidly, within minutes to hours after focal TBI. Secondary axotomy, reactive astrocytosis, inflammation, and tissue atrophy are events that are more delayed in the onset and can be persistent for weeks to months after TBI. Therefore, one must carefully select the time point(s) at which to apply a given assessment tool to optimize sensitivity. A notable caveat is that the time course for a given outcome parameter may differ substantially for different models of TBI and species employed.

Consideration should be given as to whether the outcome measure provides a "snapshot" in time or a picture of the cumulative process. For example, detection of cell death by Fluorojade staining or active caspase-3 immunolabeling yields important information on the location and amount of cell death occurring within a fixed time frame, and cannot be used to assess cumulative cell death at the time of euthanasia. Therefore, evaluation of the neuroprotective effects of an experimental treatment regimen at a single time point using Fluorojade or active caspase-3 is not sufficient, as it leaves open the possibility that the treatment regimen merely shifted the time course of cell death. In this case, quantification of cell death would need to be performed over multiple time points or, alternatively, assessed through counts of surviving neurons at a single late time point after the majority of cell death had taken place, as predetermined for the given TBI model. Stereological cell counts provide a powerful means for quantifying cell survival and, on the flip side, cumulative cell death (Vol. 1, Chap. 3). Certain assessments, such as neuroimaging, physiological monitoring, and many behavioral tests, have the distinct advantage of allowing repeated sampling within a given subject, increasing power and sensitivity to detect temporal patterns of damage or recovery.

4.3. Reproducibility, Variability, and Validity

Before implementing assessments for TBI, the reliability or reproducibility of the measurement technique should be taken into account. It is important to perform pilot studies or consult the literature to obtain an estimate of variability, which will encompass not only the experimental variability of the assessment tool, but also the biological variability of the measured parameter for the age/sex/species of interest and the variability in induction of TBI for a given model. In this manner, the group size ("*n*") can be customized through power analysis. Potential sources of variability in the assessment should be identified and minimized. While assessments across experimental groups should ideally be performed at the same time, this is often not logistically possible. For example, large group sizes may necessitate measuring outcomes in subsets of animals across different days, especially with time-consuming electrophysiological or behavioral assessments or neuroimaging sessions on equipment with limited availability. Therefore, sources of variability across experimental sessions, such as changes in baseline signals of equipment, changes in environmental cues/conditions, or differences in reagents/solutions, should be minimized. Variability and bias can also be introduced by the evaluator, particularly for assessments with subjective components (e.g., scoring for some behavioral tests, selection of sampling sites in tissue, selection of image analysis parameters, identification of objects for manual counting). In such cases, intra-rater and inter-rater reliability should be determined and evaluations should be conducted in a blinded fashion. Technique-specific suggestions to minimize variability are provided in individual chapters.

The issue of validity for an assessment is centered on the question: Are you actually measuring what you think you are measuring? With biochemical and molecular tools, it is critical that the specificity of reagents or protocols is established through characterization and the use of proper controls. Certainly, an improperly characterized antibody may provide erroneous or misleading results when used for immunohistochemistry or immunoblotting. Typical negative controls using approaches, such as omission or preadsorption of antibody, use of missense probes or empty vectors, are important in ruling out false positives. Positive controls are also important in many assessments to guard against false negatives. Assessment-specific controls should be incorporated along with experimental controls, such as surgical controls and treatment (vehicle) controls. The validity of an assessment may also rest with the interpretation. For example, terminal deoxynucleotidyl transferase-mediated biotinylated deoxyuridine triphosphate nick-end labeling (TUNEL) staining for damaged DNA was characterized in a developmental context as specific for programmed cell death (apoptosis) (36). Despite clear evidence that TUNEL stains both necrotic and apoptotic cells in experimental TBI (37), TUNEL staining is still often misinterpreted as a specific marker of posttraumatic apoptosis. It is imperative that assessment approaches are evaluated and applied in concert with critical analysis, particularly when measures are adapted from non-trauma models.

4.4. Assessment Sensitivity

Sensitivity is an important determination in selecting an assessment measure for use. This is especially critical in experiments where effect sizes may be small, such as mild TBI, or where incremental responses need to be differentiated, such as dose–response curves or comparisons of single and combination therapies. Biochemical and molecular assessments performed in homogenates may have inadequate sensitivity to measure changes in low-abundance genes or proteins. However, in some cases, it is advantageous to select an assessment with a low baseline signal (i.e., in the uninjured animal) and an expected increase in signal or response with injury.

This provides a high signal-to-noise ratio, facilitating detection of small injury-induced changes. For example, a small number of dying cells can be easily detected using an assessment, such as active caspase-3 or Fluorojade staining, which produces minimal signal in an uninjured brain while a small number of dead cells is difficult to detect by looking for an absence of Nissl-stained cells, which are highly abundant in the normal CNS. Assessments that utilize the appearance of a signal with injury are also highly useful in models of diffuse brain injury, where small subsets of damaged cells or axons may be interspersed with many normal appearing cells. Certain assessments may be limited in their application due to "floor" or "ceiling" effects. For example, rodents exhibit a remarkable degree of spontaneous recovery of motor function after TBI; therefore, many motor function assessments do not have adequate sensitivity to detect delayed or chronic improvements in motor function in response to an experimental manipulation.

4.5. Implementation and Analysis

An obvious practical consideration in the selection of assessments is the ease of use and cost of implementing the assessment. The sensitivity and reliability of the measure may need to be weighed against the time intensiveness of data collection or analysis, the need for specialized training, and the need for specialized or costly equipment or reagents. Some tissue assessments require the dedicated use of a single animal, such as measurement of edema by the wet weight/dry weight technique, while other assessments can be combined to maximize data gathering from a given animal. For histological, biochemical, and molecular assessments, attention should be paid to the specifics of tissue preparation to determine which assessments may be compatible in terms of tissue processing.

Finally, each assessment should be carefully evaluated to determine the type of data it produces. Is the assessment qualitative, semiquantitative, or quantitative, and is this appropriate for your needs? If the assessment is qualitative, how can outcomes be standardized and effectively summarized? If the data is quantitative, is it nonparametric or parametric and what statistical tools are most appropriate for analyses? An appreciation of the type of data generated through the use of a given assessment aids in experimental design and accurate interpretation of results.

5. Summary

The growing armamentarium of assessments for TBI presents challenges for selecting and implementing appropriate outcome measures. Choice of assessment should be based in part on the applicability to the TBI model and severity being utilized (or vice versa), as not all assessments are equally suitable for focal or diffuse

TBI, or mild or severe TBI. The age, sex, and species employed for study may also influence which tools are optimal or how the data obtained with the assessment is interpreted. Knowledge of the nature and time course of the pathophysiologic event to be studied helps guide the investigator in selecting rational time points for assessing outcomes. To improve assessment utility, sources of variability should be minimized, validity of the assessment tool verified, and sensitivity maximized. Practical considerations of time and cost need to be balanced with the sensitivity and power of the measurement tools. Finally, the methodology for data collection and analyses specific to the assessment should be considered in advance of implementation. The following chapters provide detailed guidance on the implementation of protocols for assessing morphological, physiological, biochemical and molecular, neurobehavioral, and neuroimaging outcomes for TBI.

References

1. Saatman KE, Duhaime AC, Bullock R, Maas AI, Valadka A, Manley GT (2008) Classification of traumatic brain injury for targeted therapies. J Neurotrauma 25:719–738
2. Hillered L, Vespa PM, Hovda DA (2005) Translational neurochemical research in acute human brain injury: the current status and potential future for cerebral microdialysis. J Neurotrauma 22:3–41
3. Bentzer P, Davidsson H, Grande PO (2000) Microdialysis-based long-term measurements of energy-related metabolites in the rat brain following a fluid percussion trauma. J Neurotrauma 17:441–447
4. Le TH, Gean AD (2009) Neuroimaging of traumatic brain injury. Mt Sinai J Med 76:145–162
5. Kawamata T, Katayama Y (2007) Cerebral contusion: a role model for lesion progression. Prog Brain Res 161:235–241
6. Onyszchuk G, Al-Hafez B, He YY, Bilgen M, Berman NE, Brooks WM (2007) A mouse model of sensorimotor controlled cortical impact: characterization using longitudinal magnetic resonance imaging, behavioral assessments and histology. J Neurosci Methods 160: 187–196
7. Armstead WM (2000) Age-dependent cerebral hemodynamic effects of traumatic brain injury in newborn and juvenile pigs. Microcirculation 7:225–235
8. Smith DH, Nonaka M, Miller R et al (2000) Immediate coma following inertial brain injury dependent on axonal damage in the brainstem. J Neurosurg 93:315–322
9. Vink R, Bahtia KD, Reilly PL (2008) The relationship between intracranial pressure and brain oxygenation following traumatic brain injury in sheep. Acta Neurochir Suppl 102: 189–192
10. Van den Heuvel C, Blumbergs PC, Finnie JW et al (1999) Upregulation of amyloid precursor protein messenger RNA in response to traumatic brain injury: an ovine head impact model. Exp Neurol 159:441–450
11. King C, Robinson T, Dixon CE, Rao GR, Larnard D, Nemoto CE (2010) Brain temperature profiles during epidural cooling with the ChillerPad in a monkey model of traumatic brain injury. J Neurotrauma 27(10): 1895–1903
12. Gennarelli TA, Thibault LE, Adams JH, Graham DI, Thompson CJ, Marcincin RP (1982) Diffuse axonal injury and traumatic coma in the primate. Ann Neurol 12:564–574
13. Gennarelli TA, Thibault LE, Tipperman R et al (1989) Axonal injury in the optic nerve: a model simulating diffuse axonal injury in the brain. J Neurosurg 71:244–253
14. Inci S, Ozcan OE, Kilinc K (1998) Time-level relationship for lipid peroxidation and the protective effect of alpha-tocopherol in experimental mild and severe brain injury. Neurosurgery 43:330–335; discussion 335–336
15. Hartl R, Medary M, Ruge M, Arfors KE, Ghajar J (1997) Blood-brain barrier breakdown occurs early after traumatic brain injury and is not related to white blood cell adherence. Acta Neurochir Suppl 70:240–242

16. Lighthall JW (1988) Controlled cortical impact: a new experimental brain injury model. J Neurotrauma 5(1):1–15
17. Sullivan HG, Martinez J, Becker DP, Miller JD, Griffith R, Wist AO (1976) Fluid-percussion model of mechanical brain injury in the cat. J Neurosurg 45:521–534
18. Hayes RL, Stalhammar D, Povlishock JT et al (1987) A new model of concussive brain injury in the cat produced by extradural fluid volume loading: II. Physiological and neuropathological observations. Brain Inj 1:93–112
19. Friess SH, Ichord RN, Owens K et al (2007) Neurobehavioral functional deficits following closed head injury in the neonatal pig. Exp Neurol 204:234–243
20. Fujimoto ST, Longhi L, Saatman KE, Conte V, Stocchetti N, McIntosh TK (2004) Motor and cognitive function evaluation following experimental traumatic brain injury. Neurosci Biobehav Rev 28:365–378
21. Margulies S, Hicks R (2009) Combination therapies for traumatic brain injury: prospective considerations. J Neurotrauma 26:925–939
22. Duhaime AC (2006) Large animal models of traumatic injury to the immature brain. Dev Neurosci 28:380–387
23. Gilmer LK, Ansari MA, Roberts KN, Scheff SW (2010) Age-related mitochondrial changes after traumatic brain injury. J Neurotrauma 27:939–950
24. Giza CC, Kolb B, Harris NG, Asarnow RF, Prins ML (2009) Hitting a moving target: basic mechanisms of recovery from acquired developmental brain injury. Dev Neurorehabil 12:255–268
25. Onyszchuk G, He YY, Berman NE, Brooks WM (2008) Detrimental effects of aging on outcome from traumatic brain injury: a behavioral, magnetic resonance imaging, and histological study in mice. J Neurotrauma 25:153–171
26. Kupina NC, Detloff MR, Bobrowski WF, Snyder BJ, Hall ED (2003) Cytoskeletal protein degradation and neurodegeneration evolves differently in males and females following experimental head injury. Exp Neurol 180:55–73
27. Hall ED, Gibson TR, Pavel KM (2005) Lack of a gender difference in post-traumatic neurodegeneration in the mouse controlled cortical impact injury model. J Neurotrauma 22: 669–679
28. Chen Y, Constantini S, Trembovler V, Weinstock M, Shohami E (1996) An experimental model of closed head injury in mice: pathophysiology, histopathology and cognitive deficits. J Neurotrauma 13:557–568
29. Katayama Y, Maeda T, Koshinaga M, Kawamata T, Tsubokawa T (1995) Role of excitatory amino acid-mediated ionic fluxes in traumatic brain injury. Brain Pathol 5:427–435
30. Soustiel JF, Larisch S (2010) Mitochondrial damage: a target for new therapeutic horizons. Neurotherapeutics 7:13–21
31. Hovda DA, Lee SM, Smith ML et al (1995) The neurochemical and metabolic cascade following brain injury: moving from animal models to man. J Neurotrauma 12:903–906
32. Stoica BA, Faden AI (2010) Cell death mechanisms and modulation in traumatic brain injury. Neurotherapeutics 7:3–12
33. Raghupathi R (2004) Cell death mechanisms following traumatic brain injury. Brain Pathol 14:215–222
34. Donkin JJ, Vink R (2010) Mechanisms of cerebral edema in traumatic brain injury: therapeutic developments. Curr Opin Neurol 23: 293–299
35. Ziebell JM, Morganti-Kossmann MC (2010) Involvement of pro- and anti-inflammatory cytokines and chemokines in the pathophysiology of traumatic brain injury. Neurotherapeutics 7:22–30
36. Gavrieli Y, Sherman Y, Ben-Sasson SA (1992) Identification of programmed cell death in situ via specific labeling of nuclear DNA fragmentation. J Cell Biol 119:493–501
37. Rink AD, Fung KM, Trojanowski JQ, Lee VMY, Neugebauer E, McIntosh TK (1995) Evidence of apoptotic cell death after experimental traumatic brain injury in the rat. Am J Pathol 147:1575–1583

Chapter 26

Morphological Assessments of Traumatic Brain Injury

Daniel van Bregt, Theresa Currier Thomas, Rachel K. Rowe, and Jonathan Lifshitz

Abstract

This chapter reviews documented methods for tracking morphological damage across the neurovascular unit following experimental models of focal or diffuse traumatic brain injury. Sections within this chapter focus on the chronology of morphological change in specific components of the neurovascular unit (neurons, astrocytes, vasculature, and microglia). Morphological disruption of circuit connectivity is briefly addressed in the last section. Efforts have been made to reference relevant external review articles and appropriate internal textbook references for in-depth methodological detail.

Key words: Cellular morphology, Neurovascular unit, Neuron, Astrocyte, Vasculature, Microglia, Circuit connectivity

1. Morphological Assessment of the Neurovascular Unit After TBI

Histopathological outcomes following focal and diffuse brain injury are characteristic of the particular models employed, broadly categorized into focal and diffuse injury models. *In vivo* focal injuries (e.g., controlled cortical impact) primarily lead to the development of a cortical contusion that evolves into a prominent cavity by 1 week post-injury. In addition, the core and penumbra of the impact site demonstrate markers of neuroinflammation (early to chronic phase, hours to weeks), reactive gliosis (early to chronic phase, days to weeks), primary and secondary axonal damage (early to chronic phase, days to weeks), and compromised blood–brain barrier (early phase, minutes to hours). The extent of cellular loss and neuropathology depends on the biomechanical parameters of the injury (e.g., velocity and depth of the injury).

Jun Chen et al. (eds.), *Animal Models of Acute Neurological Injuries II*, Springer Protocols Handbooks,
DOI 10.1007/978-1-61779-782-8_26,

Diffuse brain injuries (e.g., impact acceleration) produce diffuse and multifocal damage that is uncomplicated by contusion. Primary histopathological characteristics include cellular damage, neuroinflammation, reactive gliosis, primary and secondary axonal damage, and compromised blood–brain barrier throughout the brain, occurring at similar time points to focal brain injury (1). Again, the extent and duration of cellular damage and dysfunction depend on the magnitude of the biomechanical parameters used to induce the injury. Since biomechanical forces are applied to the intact brain or skull, it is essential to assess changes in morphology over time, including the size, shape, structure, and viability of neurons and supporting cells. This chapter serves as an overview of neuropathological markers for experimental models of traumatic brain injury (TBI), with references to relevant chapters throughout this publication.

1.1. Neurons

Neurons are the key component to neuronal network function within the nervous system. As such, neurons have received the majority of the attention in histopathological evaluations after all models of acute neurological injuries. The neuron can be separated into the soma, axon, and synapse when evaluating injury-induced damage. Observations of the soma provide indications of cell number, pathological changes in morphology, and structural viability. Design-based stereology (Stereology—Part I:Chapter 3, Volume 1) can be conducted to assess injury-induced alterations in neuronal number, regional volume, and individual neuronal nuclear volume over time (2), depending on the model neuronal death occurs from 1 to 4 weeks post-injury. Moreover, DNA damage markers combined with Nissl or hematoxylin and eosin staining (Histology—Part IX: Chap. 103) can identify apoptosis and necrosis based on the irregularity of the nucleus, often detectable within 1 h post-injury (3, 4). Silver staining identifies degenerating neurons and their processes from the first day, persisting up to months post-injury (5), providing a time course of retrograde degeneration and secondary denervation. As an alternative to silver stain, the fluoro-jade C stain indicates degenerating neurons, likely beyond repair (6). Immunohistochemistry can visualize and quantify injury-induced changes in protein expression within neuronal networks (Immunohistochemistry—Part I:Chapter 1, Volume 1). To focus on the neuronal processes, immunohistochemistry for amyloid precursor protein (APP) identifies axonal injury by the accumulation of anterograde transported protein in sheared axons (7), with a similar time course to silver stain or fluoro-jade stain. Similarly, a loss of neurofilament or microtubule-associated protein indicates axonal and dendritic pathology, respectively. Synaptophysin and growth-associated protein-43 (GAP-43) levels, in addition to others, can demonstrate injury-related neuroplastic response, typically in a delayed time course (8, 9). Furthermore, these same markers could be interpreted to indicate synaptic terminal number (Synaptic Plasticity—Part II:Chapter 8, Volume 1).

1.2. Astrocytes

Astrocytes are key components of the neurovascular unit. Their primary role is to regulate ion and neurotransmitter concentrations in the extracellular environment. In addition, glial end feet form the blood–brain barrier. In the presence of a focal contusion, astrocytes form the boundary of the eventual cavity by establishing a glial scar which can be visualized through expression of vimentin or glial fibrillary acidic protein (GFAP) at 2 weeks, becoming extensive by 4 weeks (Reactive Astrogliosis—Part I:Chapter 4, Volume 1). Primarily, GFAP immunohistochemistry is used to evaluate the astrocytic response to experimental TBI (10, 11). Another marker of astrocytes includes the S100 proteins, which have been shown to colocalize with GFAP-expressing astrocytes (10).

1.3. Vasculature

Historically, each component of the neurovascular unit has been studied in isolation. Interactions among neurons, glia, and vascular components are essential to maintain homeostasis. For example, nutrient delivery is mediated by neuronal activity through vasoconstriction and/or dilation while astrocytes maintain the integrity of the blood–brain barrier to exclude toxins from the parenchyma. Immediately following TBI, mechanical forces can disrupt the vasculature (Microvascular Damage and Dysfunction—Part IV:Chapter 27, Volume 2), opening the blood–brain barrier and allowing cytotoxic blood components to come in contact with brain tissue for up to 6 h post-injury (12). Intraparenchymal hemorrhage can be visualized in unstained brain tissue following injury by the presence of blood with either the light or fluorescence microscopy. Evidence of serum leakage can be immunohistochemically observed by staining for host IgG or albumin in the parenchyma (12) (Histological and Immunochemical Methods—Part I:Chapter 1, Volume 1). To evaluate the time course of blood–brain barrier disruption and resealing, intravenous injection of molecular tracers, such as Evans blue-albumin (68 kDa), microperoxidase (20 kDa; 2 nm diameter) or horseradish peroxidase (40 kDa; 6 nm diameter), can be evaluated for extravasation at various times post-injury and following injection (13–16) (Microvascular Damage and Dysfunction—Part IV:Chapter 27, Volume 2). Functional assessment of the pial vasculature reactivity has been performed using the cranial window technique coupled to microperfusion, where measurement of pial vessel diameters indicates vasoreactivity (17, 18).

1.4. Microglia

Microglia are the resident macrophages of the central nervous system, becoming activated in response to signaling cascades initiated by brain injury (19). Upon activation, these CNS macrophages release proinflammatory cytokines which lead to the recruitment of peripheral white blood cells through areas of blood–brain barrier compromise. Microglia morphology changes with injury, as evidenced by an increase in size of the soma, thickening of proximal processes, and decrease in the ramification of the distal branches (20). Microglial activation is observed within the first 24 h post-injury, concomitant with increases in circulating levels of cytokines (20). In focal brain injuries, the activation of microglia is localized to the area of injury, in contrast to diffuse injuries in which microglial activation is evident throughout the brain. Activated microglia can be measured using immunohistochemistry to stain for ED-1, a cellular marker that is specific to activated microglia. CD-11b (OX-42) is an antigen that identifies activated macrophages (21). Alternatively, a recent marker of inflammation is the translocator protein 18 kDa (TSPO), which is normally expressed at low levels in the brain, but increases at sites of brain injury and inflammation (22). TSPO is present in microglia and following an injury, microglia proliferate and migrate to the injury site, thereby increasing TSPO levels (22).

2. Morphological Circuit Assessment

After TBI, damage to the neurovascular unit manifests in circuit disruption that can contribute to the development of cognitive, motor, and sensory dysfunction. The onset and duration of specific injury-induced deficits depend on the experimental model of brain injury. Anatomical integrity of the neuron, axon, and synapse within defined circuits can be assessed using tract-tracing techniques. Numerous retrograde and anterograde tracers (Retrograde Axonal Tract Tracing and Anterograde Axonal Tract Tracing—Part V: Chapters 38, 39, Volume 2, respectively) are available to evaluate post-injury connectivity (e.g., biotinylated dextran amines, horseradish peroxidase, Fluoro-Gold, and *Phaseolus vulgaris* leucoagglutinin [PHA-L]) (23). Biotinylated dextran amines are retrograde and anterograde transported, providing good morphological detail of neurons as has been demonstrated across species (23). Several tracers are amenable to electron microscopy for detailed synaptic morphology studies (Electron Microscopy—Part IX: Chap. 105). Detailed tract tracing, in comparison to uninjured control, can indicate maladaptive plasticity associated with acute pathology and chronic repair. The complexity of brain connectivity necessitates discrete anatomical localization of tracers, rather than diffuse exploration across the brain.

3. Summary

Assessing the morphological changes occurring after TBI involves a combination of immunohistochemical, anatomical, and stereological methods in an attempt to develop a comprehensive neurovascular analysis of the ensuing histopathology. The morphological markers highlighted in this introduction are all expressed at differing magnitudes and durations depending on the model and severity of injury. The following chapters provide more detailed explanations of how the injuries are assessed with regard to vasculature, membrane permeability, and neurogenesis. Other chapters define how physiological, biochemical, molecular, neurobehavioral, and neuroimaging assessments combine to explain the encompassing effects of injury-induced changes in morphology within the neurovascular unit.

References

1. Huh JW, Widing AG et al (2008) Midline brain injury in the immature rat induces sustained cognitive deficits, bihemispheric axonal injury and neurodegeneration. Exp Neurol 213(1): 84–92
2. Lifshitz J, Kelley BJ et al (2007) Perisomatic thalamic axotomy after diffuse traumatic brain injury is associated with atrophy rather than cell death. J Neuropathol Exp Neurol 66(3): 218–229
3. Hong J, Cui J et al (2002) Study on cognition disorder and morphologic change of neurons in hippocampus area following traumatic brain injury in rats. Chin J Traumatol 5(1):36–39
4. Adelson PD, Jenkins LW et al (2001) Histopathologic response of the immature rat to diffuse traumatic brain injury. J Neurotrauma 18(10):967–976
5. Tenkova TI, Goldberg MP (2007) A modified silver technique (de Olmos stain) for assessment of neuronal and axonal degeneration. Methods Mol Biol 399:31–39
6. Schmued LC, Albertson C et al (1997) Fluoro-Jade: a novel fluorochrome for the sensitive and reliable histochemical localization of neuronal degeneration. Brain Res 751(1):37–46
7. Itoh T, Satou T et al (2009) Expression of amyloid precursor protein after rat traumatic brain injury. Neurol Res 31(1):103–109
8. Gobbel GT, Bonfield C et al (2007) Diffuse alterations in synaptic protein expression following focal traumatic brain injury in the immature rat. Childs Nerv Syst 23(10):1171–1179
9. Marklund N, Bareyre FM et al (2007) Cognitive outcome following brain injury and treatment with an inhibitor of Nogo-A in association with an attenuated downregulation of hippocampal growth-associated protein-43 expression. J Neurosurg 107(4):844–853
10. Hill SJ, Barbarese E et al (1996) Regional heterogeneity in the response of astrocytes following traumatic brain injury in the adult rat. J Neuropathol Exp Neurol 55(12):1221–1229
11. Takamiya Y, Kohsaka S et al (1988) Immunohistochemical studies on the proliferation of reactive astrocytes and the expression of cytoskeletal proteins following brain injury in rats. Brain Res 466(2):201–210
12. Hicks RR, Baldwin SA et al (1997) Serum extravasation and cytoskeletal alterations following traumatic brain injury in rats. Comparison of lateral fluid percussion and cortical impact models. Mol Chem Neuropathol 32(1–3):1–16
13. Taya K, Marmarou CR et al (2010) Effect of secondary insults upon aquaporin-4 water channels following experimental cortical contusion in rats. J Neurotrauma 27(1):229–239
14. Sharma HS, Johanson CE (2007) Intracerebroventricularly administered neurotrophins attenuate blood cerebrospinal fluid barrier breakdown and brain pathology following whole-body hyperthermia: an experimental study in the rat using biochemical and morphological approaches. Ann N Y Acad Sci 1122: 112–129

15. Drummond JC, Patel PM et al (1998) The effect of the reduction of colloid oncotic pressure, with and without reduction of osmolality, on post-traumatic cerebral edema. Anesthesiology 88(4):993–1002
16. McIntosh TK, Noble L et al (1987) Traumatic brain injury in the rat: characterization of a midline fluid-percussion model. Cent Nerv Syst Trauma 4(2):119–134
17. Povlishock JT, Kontos HA (1992) The role of oxygen radicals in the pathobiology of traumatic brain injury. Hum Cell 5(4): 345–353
18. Levasseur JE, Wei EP et al (1975) Detailed description of a cranial window technique for acute and chronic experiments. Stroke 6(3): 308–317
19. Perry VH, Nicoll JA et al (2010) Microglia in neurodegenerative disease. Nat Rev Neurol 6(4):193–201
20. Raivich G, Jones LL et al (1999) Molecular signals for glial activation: pro- and anti-inflammatory cytokines in the injured brain. Acta Neurochir Suppl 73:21–30
21. Bellander BM, Bendel O et al (2004) Activation of microglial cells and complement following traumatic injury in rat entorhinal-hippocampal slice cultures. J Neurotrauma 21(5):605–615
22. Chen MK, Guilarte TR (2008) Translocator protein 18 kDa (TSPO): molecular sensor of brain injury and repair. Pharmacol Ther 118(1):1–17
23. Reiner A, Veenman CL et al (2000) Pathway tracing using biotinylated dextran amines. J Neurosci Methods 103(1):23–37

Chapter 27

Assessment of Cerebral Vascular Dysfunction After Traumatic Brain Injury

Douglas S. DeWitt and Donald S. Prough

Abstract

Today's cerebral vascular physiologists have a variety of techniques available to measure cerebral blood flow (CBF) and assess cerebral vascular reactivity. CBF measurement techniques can be grouped based on anatomical considerations (e.g., local, regional, global), number of measurements permitted (e.g., single, multiple, continuous), and/or the types of data provided (e.g., images, absolute vs. relative CBF values). In addition to the methods that measure CBF, others are designed to measure the responses of individual or groups of cerebral arteries to changes in intravascular pressure or extracellular oxygen or carbon dioxide levels (e.g., cranial windows, isolated arterial segments). However, since each of these myriad techniques has advantages and limitations, investigators must choose the most appropriate based on when and where they want to measure cerebral vascular function. This chapter provides a brief discussion of the physiological principles on which the majority of CBF measurement methods are based as well as descriptions of the advantages and disadvantages of the methods most often used in the current measurements of cerebral vascular function and dysfunction after traumatic brain injury.

Key words: Cerebral blood flow, Cerebral vascular reactivity, Cerebral blood flow imaging, Traumatic brain injury, Cerebral blood flow measurement methods

1. Introduction

Cerebral hypoperfusion and ischemia, among the numerous factors that contribute to mortality and morbidity after traumatic brain injury (TBI), are caused by complex interactions among reduced blood pressure, impaired cerebral vasodilatory responsiveness, cerebral swelling, and elevated intracranial pressure (ICP). Therapies that restored normal compensatory vasodilatory properties of the cerebral vasculature would improve cerebral blood flow (CBF) and help to protect the brain from secondary injury due to systemic hypotension or hypoxemia. However, the development of

Jun Chen et al. (eds.), *Animal Models of Acute Neurological Injuries II*, Springer Protocols Handbooks,
DOI 10.1007/978-1-61779-782-8_27, © Springer Science+Business Media, LLC 2012

"vasoprotective" therapies requires an understanding of the mechanisms of traumatic cerebral vascular injury, an understanding that can only be acquired through the use of techniques that accurately assess the effects of TBI on CBF and cerebral vascular reactivity.

The capability of the cerebral circulation to change vascular resistance in response to alterations in local and systemic physiological parameters has been recognized since 1890, when Roy and Sherrington (1) observed that "…the brain possesses an intrinsic mechanism by which its vascular supply can be varied locally in correspondence with local variations of functional activity." Although Roy and Sherrington (1) theorized that the cerebral circulation was capable of reacting to changes in systemic physiology in order to minimize changes in the brain's external environment, they lacked the methodologies to test their landmark hypothesis. Today's cerebral vascular physiologists have a variety of techniques available to measure CBF and assess cerebral vascular reactivity. However, since each of these myriad techniques has advantages and limitations, investigators must choose the most appropriate based on when and where they want to measure cerebral vascular function. Since a thorough discussion of all of the techniques that have been used to measure CBF is beyond the scope of this chapter, the methods described here are those that currently are the most widely used to evaluate cerebral perfusion and cerebral vascular reactivity after TBI.

2. Types of CBF Measurement Methods

Although CBF measurements can only be made in vivo, cerebral vascular reactivity can be assessed either in vivo or in vitro. Measurements of cerebral vascular reactivity (e.g., cranial windows) are described in a separate section. The majority of CBF measurement methods can be categorized based on the frequency of measurements: continuous, repeated, or single (Table 1). For example, laser Doppler (2) and transcranial Doppler flowmetry (3) provide continuous recordings of cerebral perfusion, which changes in parallel with CBF as long as hematocrit remains constant. CBF can be measured repeatedly, though not continuously, using radioactive (4) or nonradioactive microspheres (5), radioactive (6) or nonradioactive xenon (7), single-photon emission computed tomography (SPECT) (8), or positron emission tomography (PET) (9). Finally, autoradiographic methods using tracers, such as iodo-(^{14}C)antipyrine (10), ^{125}I- or ^{14}C-isopropyliodoamphetamine (11, 12), or ^{125}I-iodoamphetamine (11), provide single, high-resolution measurements of local CBF.

Table 1
Cerebral blood flow (CBF) methods by number of flow measurements

Frequency	Method	1° Use	References
Continuous	Laser Doppler flowmetry	Experimental	(2)
	Transcranial Doppler flowmetry	Clinical	(3)
	Thermal diffusion flowmetry	Experimental	(57)
Repeated	Hydrogen clearance	Experimental	(53)
	Radioactive microspheres	Experimental	(4)
	Nonradioactive microspheres	Experimental	(5)
	^{133}Xenon clearance	Both	(6)
	Nonradioactive (stable) xenon CT	Clinical	(7)
	Contrast perfusion CT	Clinical	(29, 56)
	Single-photon emission tomography	Clinical	(8)
	Positron emission tomography	Both	(9)
Single	Iodo-(^{14}C)antipyrine AR	Experimental	(10)
	^{125}I-isopropyliodoamphetamine AR	Experimental	(11)
	^{14}C-isopropyliodoamphetamine AR	Experimental	(12)
	^{125}I-iodoamphetamine AR	Experimental	(11)

1° use—experimental and clinical refer to the setting in which the method is usually employed. "Both" means that the method is used in experimental animals and humans
AR autoradiography, *CT* computerized tomography

3. Theoretical Issues

With the exception of the laser/transcranial Doppler flowmetry methods, nearly all CBF measurement techniques are based on the work of Bell (13), Kety, and Schmidt (14, 15). The Fick principle stated that the total oxygen absorbed by an organ divided by the uptake of oxygen (the arteriovenous difference) is equal to the blood flow through the organ. Restated as a dilution principle, the tissue concentration per unit time of a tracer is the product of tissue blood flow and the amount of tracer entering the tissue (i.e., the arterial blood concentration) minus the amount of tracer leaving the tissue (i.e., the venous blood concentration) (13, 16). Kety and Schmidt reasoned that the amount of an inhaled, highly diffusible, inert gas (nitrous oxide) taken up by the brain per unit time is equal to the amount of that gas brought to the brain by the arterial blood minus the amount carried away in the cerebral venous blood (14, 15, 17). In their measurements of total CBF, Kety and Schmidt made the assumptions that CBF was in a steady state during the period of study and was not affected by the tracer, that the venous blood sample was not contaminated by extracerebral blood, that the period of measurement was long enough to allow equilibration of gas in the brain with the cerebral venous blood, that no significant arterial-venous shunts were present, and that the value of the partition coefficient of the inert gas was representative of the entire brain (14, 15). Their initial studies, using N_2O as the tracer, proved these assumptions to be valid in most circumstances and the Kety–Schmidt N_2O method yielded accurate measurements of total CBF. The substitution of radiolabeled gases (e.g., ^{85}Kr, ^{133}Xe) permitted the measurement of regional CBF but required intracarotid injection of the tracer (18, 19). The Kety–Schmidt method was subsequently modified to use inhaled ^{133}Xe as the tracer, but the effects of arterial recirculation of the ^{133}Xe complicated the measurements (20). Inaccuracies associated with recirculation and extracranial contamination were largely, though not entirely, corrected by Obrist et al. (21) using a multicompartmental analysis of the Kety–Schmidt equation. These refinements and the commercial availability of ^{133}Xe CBF measurement systems with multiple, well-collimated detector arrays led to the widespread use of the method for measurements of CBF under physiological conditions as well as in numerous disease states (22).

Currently, there are numerous techniques for measuring CBF using diffusible tracers in variations of the Kety–Schmidt equations. Examples include nonradioactive (stable) Xe with computerized tomography (7), ^{99}technetium SPECT (8), ^{15}O-labeled carbon monoxide PET (9), and iodo-(^{14}C)antipyrine autoradiography (10). Simplified versions of the Kety–Schmidt equations are used to calculate CBF using diffusible tracers that are trapped in a single passage through the cerebral circulation. Examples include ^{125}I-isopropyliodoamphetamine, ^{125}I-iodoamphetamine (11), and ^{14}C-isopropyliodoamphetamine (12) autoradiography. Considering these and many other CBF measurement methods that rely on the contributions of Kety and Schmidt, their studies remain landmarks.

4. Cerebral Blood Flow Imaging

The first CBF images, that is, two-dimensional representations of CBF values in specific anatomical regions, were autoradiographs (23). Since CBF autoradiography was described by Landau et al. (23), the method has been modified to use radiolabeled, freely diffusible indicators, such as ^{14}C-iodoantipyrine (10) and ^{14}C- or ^{125}I-iodoamphetamine (11, 12). CBF autoradiography currently is the most widely used method for measurements of absolute CBF with high spatial resolution in experimental animals. Despite the value of the technique for CBF measurements in experimental animals, the need to remove and section the brain for exposure to X-ray precludes the clinical use of conventional autoradiographic methods. This limitation has led to the development of "autoradiographic" methods using tracers (e.g., ^{18}F, ^{13}N, ^{11}C, and ^{15}O) that can be detected using PET (24, 25). Although the resolution of the PET "autoradiographs" is not equal to standard ^{14}C autoradiography in experimental animals, the short half-life of PET isotopes permits repeated measurements of CBF and glucose utilization (25) in experimental animals and humans (9).

A variety of CBF measurement techniques are appropriate for human studies. Those using inert diffusible tracers include stable Xe-computed tomography (CT) (7), technetium-99m-hexamethyl-propylamine-oxime (^{99}TC-HMPAO) SPECT (8), ^{15}O PET ((26), p. 53; (27), p. 7), and arterial spin labeling MRI (28). CBF imaging methods employing intravascular (nondiffusible) tracers include contrast enhanced CT ((29), p. 54) and MRI ((30), p. 55). Although the theories and detailed methodologies of MRI perfusion imaging in patients are beyond the scope of this chapter, they are available in some excellent reviews ((29), p. 54; (31), p. 56; (32), p. 57).

5. Cerebral Vascular Reactivity

Although dilation and constriction in pial arteries in response to changes in arterial blood pressure were reported by Fog in 1939, the concept of CBF autoregulation was first described by Lassen (33), based on the studies conducted in humans using the Kety–Schmidt nitrous oxide technique. Lassen defined autoregulation as the ability of the cerebral circulation to vary cerebral vascular resistance in order to maintain a constant level of CBF despite changes in cerebral perfusion pressure (CPP—mean arterial pressure - ICP or central venous pressure) (33). In addition to responding to changes in CPP, cerebral arteries are capable of significant dilation or constriction in response to changes in $PaCO_2$ (34), PaO_2 (34), extracellular pH (35), brain metabolic activity (36), and hematocrit (37). More thorough discussions of normal cerebral vascular reactivity are available elsewhere (38, 39).

Although measurements of CBF in vivo have resulted in valuable information about the properties and mechanisms of cerebral vascular reactivity in health and disease or injury, an equal or perhaps greater body of information has been acquired by direct measurements of cerebral arterial diameter changes in vivo using cranial window preparations or in vitro using isolated, pressurized, or perfused cerebral arterial preparations.

The advantage of cranial window preparations is the ability to directly observe cerebral arterial and arteriolar responses to physiological stimuli or injury in vivo and to measure responses in cerebral vessels of different sizes. This is a potentially important advantage since arteries of different diameters may respond differently to either physiological or pathological stimuli. For example, large cerebral arteries respond differently than smaller arterioles to changes in arterial blood pressure (40) and CO_2 levels (41). Similarly, the effects of TBI on cerebral arteries are related to the size of the vessel. Posttraumatic cerebral vasoconstriction is greater in arterioles than in small pial arteries (42, 43). An additional substantive advantage of cranial window preparations is that the local effects of agents that may also have systemic effects can be perfused through the window in concentrations that are physiologically relevant locally while being too low to exhibit systemic effects (42). Combined studies using cranial window preparations and regional CBF measurements (radioactive microspheres) after fluid percussion TBI demonstrated that the responses of pial arteries/arterioles and regional CBF to TBI were similar (44).

6. Measurements of Cerebral Vascular Function In Vitro

One of the difficulties associated with studying the effects of TBI on the cerebral circulation is separating the direct effects of TBI on the cerebral vasculature from the indirect vascular effects of substances produced by the injured brain. This potential confound can be circumvented, in part, by studying the vasodilator/vasoconstrictor responses of isolated, pressurized, and/or perfused cerebral arteries ex vivo. Admittedly, cerebral arteries would be exposed to substances produced by traumatic injury to the underlying brain prior to harvest, but the magnitude of this effect can be minimized by harvesting cerebral arteries immediately after TBI, thereby limiting the time available for vasoactive substances to be released locally and the duration of exposure of the arteries in vivo. Studies of isolated cerebral arteries, in vitro, also provide the opportunity to test mechanisms of traumatic vascular injury using specific receptor agonists and antagonists or putative agents of vascular injury that cannot be examined as effectively or selectively in vivo.

The effects of TBI on cerebral vascular reactivity have been studied in vitro in cerebral arterial segments harvested from rats subjected to TBI. Middle or posterior cerebral arterial rings harvested from rats 5 or 30 min after paramedian fluid percussion injury (FPI) responded normally to acetylcholine, serotonin, and potassium (45) but exhibited reduced vasodilatory responses to reduced intravascular pressure (46). (47), p. 38) reported that middle cerebral arterial segments taken from rats subjected to controlled cortical impact (10) showed impaired vasoconstrictor responses to increases in intravascular pressure at 24 h but not at 2 or 120 h after injury. Vasodilator responses to reduced intravascular pressure or to arterial hypercapnia, in vitro, were not altered by CCI (48).

These and other (47), p. 38, 42; (49), p. 41; (50) in vitro studies demonstrate the value of isolated arterial preparations for studies of the effects of TBI on cerebral vascular function.

7. Advantages/ Disadvantages of Cerebral Blood Flow and Cerebral Vascular Reactivity Measurement Techniques

Each of the techniques used to measure CBF has strengths and limitations (Table 2) and investigators choose methods based on the relative importance of these properties in each experimental setting. For example, autoradiographic methods are well suited for discrete measurements of CBF at a single time point in small brain regions, such as the neurohypophysis (12) in rodents and other small experimental animals. However, radioisotope autoradiography requires harvest and sectioning of the brain. In addition, although it is possible to perform concurrent autoradiographic CBF measurements (51), most autoradiographic CBF methods permit only one measurement per animal. Laser Doppler flowmetry is suitable for continuous measurements of cerebral perfusion (2, 52), but LDF values are arbitrary perfusion units rather than absolute CBF. Hydrogen clearance methods provide repeated (though not continuous) measurements of absolute CBF (53) but only in small regions adjacent to the electrode. Therefore, the choice of techniques for CBF measurements in experimental animals is based on the size and numbers of regions of interest, as well as the type (relative vs. absolute) and number of CBF measurements required.

There are several methods suitable for CBF measurements in humans and they too have strengths and weaknesses. Nonradioactive (stable) ^{131}Xenon is a radio-opaque diffusible indicator that, in combination with CT scanning and the appropriate data acquisition and processing software, permits repeated CBF measurements in multiple brain regions with good spatial resolution (7). Stable Xe CT CBF values correlate closely with values acquired using other CBF measurement methods (54, 55). The limitations of the method include the need for long scanning times (4–5 min) that increase the likelihood of motion artifacts in confused or semiconscious patients and inaccuracies induced by impaired pulmonary function in patients with pulmonary trauma or other lung diseases (31). Contrast-enhanced perfusion CT offers many of the advantages of the stable Xe CT method in terms of accuracy and spatial resolution with shorter scanning times (<1 min) (29, 56). However, perfusion CT assumes that the tracer remains intravascular, an assumption that may be invalidated by blood–brain barrier compromise, not uncommon after TBI (31). As noted above, several other imaging techniques provide qualitative or quantitative measurements of CBF. The strengths and weaknesses of CBF imaging methods are described in detail elsewhere (31, 56, 32).

Table 2
Advantages and disadvantages of CBF techniques

Method	Specific examples	Advantage(s)	Disadvantage(s)	References
Flowmetry	Laser Doppler	Continuous measurements in real time	LDF and TCD do not measure absolute CBF	(2, 44)
	Transcranial Doppler	Repeated measurements	LDF, thermal diffusion, and H_2 clearance limited to small regions	(3)
	Thermal diffusion	Repeated CBF measurements		(57)
	H_2 clearance	Repeated CBF measurements		(53)
Autorad.	Iodo-(^{14}C) antipyrine	Precise local and/or regional CBF	Limited number (1–2) of CBF measurements; requires brain harvest	(10)
	Iodoamphetamine			(11, 12)
Microsph.	Radioactive	Repeated measurements of absolute regional CBF	Number of measurements limited to available radionuclide or fluorescent labels or colors	(4)
	Fluorescent			(5)
	Colored			(53)
Imaging	Stable Xe CT	Images and absolute regional CBF values	Long (5 min) scan times; radiation exposure	(7)
	Contrast perfusion CT	Images and absolute regional CBF values	BBB rupture affects results; radiation exposure	(29, 56)
	Single-photon emission computed tomography (SPECT)	CBF images	Limited spatial resolution; nonquantitative	(8)
	Positron emission tomography (PET)	Images and absolute regional CBF values	Limited spatial resolution; requires specialized equipment (cyclotron); radiation exposure	(9)
	Arterial spin labeling MRI (3)	Images and absolute regional CBF values	Limited spatial resolution	(28)
	Flow-sensitive alternating inversion-recovery MRI (FAIR)	Images and absolute regional CBF values	Limited spatial resolution	(59)

(continued)

Table 2 (continued)

Method	Specific examples	Advantage(s)	Disadvantage(s)	References
Inert gas	Nitrous oxide (N_2O)	Repeated measures of absolute global CBF	N_2O only measures global CBF	(14, 15)
	^{133}Xe ^{85}Kr	Repeated measures of absolute global or regional CBF	Limited spatial resolution; radiation exposure	(21) (18)

Autorad autoradiography, *BBB* blood–brain barrier, *CBF* cerebral blood flow, *CT* computerized tomography, *LDF* laser Doppler flowmetry, *Microsph.* microspheres, *TCD* transcranial Doppler

8. Summary

Unfortunately, there is no ideal method for measurements of CBF. The highest-resolution method, autoradiography, for single measurements in small brain regions in experimental animals is unsuitable for repeated measurements of regional CBF; conversely, methods that permit temporal resolution provide less anatomic resolution. Fortunately, there are a wide variety of different methods that provide accurate reproducible measurements of CBF and cerebral vascular reactivity, making it likely that one or more available methods will be satisfactory for physicians and basic scientists investigating the effects of traumatic or ischemic brain injury on the cerebral vasculature.

References

1. Roy CS, Sherrington MB (1890) On the regulation of the blood-supply of the brain. J Physiol 11:85–108
2. Lauritzen M, Fabricius M (1995) Real time laser-Doppler perfusion imaging of cortical spreading depression in rat neocortex. Neuroreport 6:1271–1273
3. Aaslid R, Huber P, Nornes H (1984) Evaluation of cerebrovascular spasm with transcranial Doppler ultrasound. J Neurosurg 60:37–41
4. Rudolph AM, Heymann MA (1967) The circulation of the fetus in utero. Methods for studying distribution of blood flow, cardiac output and organ blood flow. Circ Res 21:163–184
5. Powers KM et al (1999) Cerebral blood flow determinations using fluorescent microspheres: variations on the sedimentation method validated. J Neurosci Methods 87(2):159–165
6. Overgaard J, Tweed WA (1974) Cerebral circulation after head injury. Part 1: CBF and its regulation after closed head injury with emphasis on clinical correlations. J Neurosurg 41: 531–541
7. Latchaw RE et al (1986) Xenon/CT cerebral blood flow determination following cranial trauma. Acta Radiol 369:370–373
8. Roper SN et al (1991) An analysis of cerebral blood flow in acute closed-head injury using technetium-99m-HMPAO SPECT and computed tomography. J Nucl Med 32:1684–1687
9. Diringer MN et al (2002) Regional cerebrovascular and metabolic effects of hyperventilation after severe traumatic brain injury. J Neurosurg 96:103–108
10. Ringel M, Bryan RM, Vannucci RC (1991) Regional cerebral blood flow during hypoxia-ischemia in

the immature rat: comparison of iodoantipyrine and iodoamphetamine as radioactive tracers. Brain Res Dev Brain Res 59:231–235

11. Rapin JR et al (1984) Iodoamphetamine as a new tracer for local cerebral blood flow in the rat: comparison with isoproylidoamphetamine. J Cereb Blood Flow Metab 4:270–274
12. Bryan RM, Myers CL, Page RB (1988) Regional neurohypophysial and hypothalamic blood flow in rats during hypercapnia. Am J Physiol 255:R295–R302
13. Bell BA (1984) A history of the study of the cerebral circulation and the measurement of cerebral blood flow. Neurosurgery 14(2): 238–246
14. Kety SS, Schmidt CF (1945) The nitrous oxide method for the quantitative determination of cerebral blood flow in man: theory, procedure, and normal values. J Clin Invest 27:476–483
15. Kety SS, Schmidt CF (1948) The determination of cerebral blood flow in man by the use of nitrous oxide in low concentrations. Am J Physiol 143:53–66
16. Kety SS (1960) Blood-tissue exchange methods: theory of blood-tissue exchange and its application to measurements of blood flow. Methods Med Res 8:223–227
17. Traystman RJ (2004) The paper that completely altered our thinking about cerebral blood flow measurement. J Appl Physiol 97: 1601–1602
18. Lassen NA, Ingvar DH (1961) The blood flow of the cerebral cortex determined by radioactive krypton85. Experientia (Basel) 17:42–43
19. Mallett BI, Veall N (1965) Measurement of regional cerebral clearance rates in man using xenon-133 inhalation and extracranial recording. Clin Sci 29:179–191
20. Veall N, Mallett BI (1966) Regional cerebral blood flow determination by 133Xe inhalation and external recording: the effect of arterial recirculation. Clin Sci 30(3):353–369
21. Obrist WD et al (1967) Determination of regional cerebral blood flow by inhalation of 133-xenon. Circ Res 20(1):124–135
22. Obrist WD, Wilkinon WE (1990) Regional cerebral blood flow measurement in humans by xenon-133 clearance. Cerebrovasc Brain Metab Rev 2(4):283–327
23. Landau WH et al (1955) The local circulation of the living brain: values in the unanaesthetised and anaesthetised cat. Trans Am Neurol Assoc 80:125–129
24. Levin CS, Hoffman EJ (1999) Calculation of positron range and its effect on the fundamental limit of positron emission tomography system spatial resolution. Phys Med Biol 44:781–799
25. Schmidt KC, Smith CB (2005) Resolution, sensitivity and precision with autoradiography and small animal positron emission tomography: implications for functional brain imaging in animal research. Nucl Med Biol 32(7): 719–725
26. Derdeyn CP (2005) Positron emission tomography imaging of cerebral ischemia. Neuroimaging Clinics of North America. 15:341–350
27. Diringer MN, Videen TO, Yundt K, Zazulia AR, Aiyagari V, Dacey RG, Grubb RL and Powers WJ (2002) Regional cerebrovascular and metabolic effects of hyperventilation after severe traumatic brain injury. Journal of Neurosurgery. 96:103–108
28. Barbier EL et al (2001) Perfusion imaging using dynamic arterial spin labeling (DASL). Magn Reson Med 45(6):1021–1029
29. Miles KA (2004) Brain perfusion: computed tomography applications. Neuroradiology 48(suppl 2): S194–S200
30. Perkio J, Soinne L, Østergaard L, Helenius J, Kangasmaki A, Martinkauppi S, Salonen O, Savolainen S, Kaste M, Tatlisumak T and Aronen HJ (2005) Abnormal intravoxel cerebral blood flow heterogeneity in human ischemic stroke determined by dynamic susceptibility contrast magnetic resonance imaging. Stroke. 36:44–49
31. Coles JP (2006) Imaging of cerebral blood flow and metabolism. Curr Opin Anaesthesiol 19:473–480
32. Latchaw RE et al (2003) Guidelines and recommendations for perfusion imaging in cerebral ischemia: a scientific statement for healthcare professionals by the writing group on perfusion imaging, from the Council on Cardiovascular Radiology of the American Heart Association. Stroke 34: 1084–1104
33. Lassen NA (1959) Cerebral blood flow and oxygen consumption in man. Physiol Rev 39: 183–238
34. Wolff HG, Lennox WG (1930) Cerebral circulation. XII. The effect on pial vessels of variations in the oxygen and carbon dioxide content of the blood. In: Payne JP, Hill DW (eds) Oxygen measurements in blood and tissues. Churchill, London, pp 205–214
35. Kontos HA, Raper AJ, Patterson JL (1977) Analysis of vasoactivity of local pH, PCO_2, and bicarbonate on pial vessels. Stroke 8: 358–360
36. Kuschinsky W (1982) Coupling between functional activity, metabolism and blood flow in the brain: state of the art. Microcirculation 2:357–378

37. Hudak ML et al (1986) Effect of hematocrit on cerebral blood flow. Am J Physiol 251: H63–H70
38. Busija DW (1993) Cerebral autoregulation. In: Phillis JW (ed) The regulation of the cerebral circulation. CRC Press, Boca Raton, pp 46–61
39. Kontos HA (1981) Regulation of the cerebral circulation. Annu Rev Physiol 43: 397–407
40. Kontos HA et al (1978) Responses of cerebral arteries and arterioles to acute hypotension and hypertension. Am J Physiol 234: H371–H383
41. Wei EP, Kontos HA, Patterson JL (1980) Dependence of pial arteriolar response to hypercapnia on vessel size. Am J Physiol 238:H697–H703
42. Wei EP et al (1980) Functional, morphological, and metabolic abnormalities of the cerebral microcirculation after concussive brain injury in cats. Circ Res 46:37–47
43. Armstead WM, Kurth CD (1994) Different cerebral hemodynamic responses following fluid percussion brain injury in the newborn and juvenile pig. J Neurotrauma 11(5): 487–497
44. DeWitt DS et al (1986) Effects of fluid-percussion brain injury on regional cerebral blood flow and pial arteriolar diameter. J Neurosurg 64:787–794
45. Bukoski RD et al (1997) Traumatic brain injury does not alter cerebral artery contractility. Am J Physiol 272:H1406–H1411
46. Mathew BP et al (1999) Traumatic brain injury reduces myogenic responses in pressurized rodent middle cerebral arteries. J Neurotrauma 16:1177 1186
47. Golding EM, Contant CF, Robertson CS and Bryan RM (1998) Temporal effect of severe controlled cortical impact injury in the rat on the myogenic response of the middle cerebral artery. Journal of Neurotrauma. 15:973–984
48. Golding EM et al (1999) Cerebrovascular reactivity to CO_2 and hypotension after mild cortical impact injury. Am J Physiol 277: H1457–H1466
49. Golding EM (2002) Sequelae following traumatic brain injury. The cerebrovascular perspective. Brain Research Reviews. 38:377–388
50. DeWitt DS et al (2001) Peroxynitrite reduces vasodilatory responses to reduced intravascular pressure, calcitonin gene-related peptide and cromakalim in isolated middle cerebral arteries. J Cereb Blood Flow Metab 21:253–261
51. Lear JL et al (1981) Use of 123I and 14C in a double radionuclide autoradiographic technique for simultaneous measurement of LCBF and LCMRgl. Theory and method. Stroke 12(5):589–597
52. Muir JK, Boerschel M, Ellis EF (1992) Continuous monitoring of posttraumatic cerebral blood flow using laser-Doppler flowmetry. J Neurotrauma 9:355–362
53. Aukland K (1965) Hydrogen polarography in measurement of local blood flow; theoretical and empirical basis. Acta Neurol Scand Suppl 14:42–45
54. DeWitt DS et al (1989) Stable xenon versus radiolabeled microsphere cerebral blood flow measurements in baboons. Stroke 20:1716–1723
55. Johnson DW et al (1991) Stable xenon CT cerebral blood flow imaging: rationale for and role in clinical decision making. AJNR Am J Neuroradiol 12(2):201–213
56. Wintermark M et al (2001) Quantitative assessment of regional cerebral blood flows by perfusion CT studies at low injection rates: a critical review of the underlying theoretical models. Eur Radiol 11(7):1220–1230
57. Gibbs FH (1933) A thermoelectric blood flow recorder in the form of a needle. Proc Soc Exp Biol Med 31:141–146
58. Bauer R et al (1996) Colored microsphere technique as a new method for quantitative-multiple estimation of regional hepatic and portal blood flow. Exp Toxicol Pathol 48(5): 415–420
59. Kim SG (1995) Quantification of relative cerebral blood-flow change by flow-sensitive alternating inversion-recovery (FAIR) technique — application to functional mapping. Magn Reson Med 34(3):293–301

Chapter 28

Assessment of Membrane Permeability After Traumatic Brain Injury

Michelle C. LaPlaca and Marcus Chris Lessing

Abstract

Traumatic brain and spinal cord injury manifests following structural compromise of the tissue, including neurons and their axons, glial cells, blood vessels and extracellular components, and subsequent secondary injury cascades. The degree of primary injury depends on the cellular orientation within the tissue, local tissue mechanical properties, strain experienced at the cell level, and rate at which the insult is delivered. The plasma membrane may be directly stretched to a point of failure in either a permanent (lethal) or transient (initially sublethal) manner. The acute evidence of membrane damage is a nonspecific increase in membrane permeability immediately following a traumatic insult. Increased membrane permeability has been observed in several in vitro and in vivo models of traumatic neural injury; therefore, understanding the events that occur in the acute and subacute time periods may provide a preclinical basis for protective and/or reparative therapeutic targets. The implementation of a strategy to repair damaged membranes in the injured nervous system, however, requires continued study of the temporal and spatial extents of membrane damage, biomolecules that are acutely damaged, role of ongoing membrane damage due to free radical attacks and other degradative processes, and potential for endogenous repair. Experimental procedures to detect the presence and mechanisms of plasma membrane damage and repair, therefore, should be established in order to scrutinize the link between primary insult parameters and membrane damage and the balance between secondary membrane damage and endogenous repair attempts. The use of various markers to label cells with compromised membranes in TBI models is described, as well as limitations and considerations for future studies.

Key words: Mechanoporation, Plasmalemma, Axolemma, Plasma membrane, Traumatic brain injury, Spinal cord injury, Permeability marker, Dextran

1. Introduction

1.1. The Neuronal Plasma Membrane: Physiology and Damage

The cellular membrane separates the extracellular and intracellular spaces and is designed for controlled separation and transport of gases, ions, nutrients, and waste products. Small penetrating molecules (such as urea, MW ~60 Da) can diffuse through the

Jun Chen et al. (eds.), *Animal Models of Acute Neurological Injuries II*, Springer Protocols Handbooks, DOI 10.1007/978-1-61779-782-8_28, © Springer Science+Business Media, LLC 2012

membrane given a concentration gradient, but ions and proteins require channels and transporters. The plasma membrane comprises various molecules that not only provide a barrier between the cytosol and the extracellular fluid, but also play critical roles in ion transport, receptor-mediated signaling, growth, and secretion. Phospholipids and sphingolipids serve as structural elements as well as important signaling molecules in the mediation of cell growth, migration, and survival. A delicate balance exists between membrane damage and repair, allowing the cell to respond appropriately to environmental cues (1). For example, an increase in phospholipid production was noted in astrocytes following mechanical injury, suggesting that neural cells may respond to membrane damage through endogenous mechanisms aimed at repair (2). Loss of structural integrity of the neuronal somatic or axonal membrane, however, will initially hinder the ability to maintain ion homeostasis and generate action potentials, leading to energy deficits, secondary signaling alterations, enzymatic degradation, and many other deleterious events that act in a positive feedback manner. Understanding the mechanisms of acute and secondary membrane damage and uncovering the associated endogenous repair response are necessary to control damage and focus membrane repair on the appropriate cellular target(s).

It is generally agreed upon that the primary mechanism of acute membrane damage following traumatic injury to the brain and spinal cord is tissue strain that causes the plasma membrane to be stretched beyond its structural limit, altering membrane permeability. The traumatic insult itself may induce pores in the phospholipid bilayer, alter channel permeability, or produce gaps between the phospholipid bilayer and membrane proteins/glycoproteins/cholesterol while ongoing damage is likely due to biochemical imbalance (e.g., phospholipase activity and lipid peroxidation (3, 4)) (Fig. 1). The degree of permeability changes is a function of the nature of the damage (size of tears, pores, membrane components affected) and the ability of the cell to repair. Thermodynamic principles may dictate reorientation of the phospholipids and return of the membrane to its normal tension and fluidity while active repair of damaged membranes may involve membrane patching and/or synthesis of new membrane components (5). Mechanoporation is defined as the process by which a traumatic insult produces sublethal pores or defects in the membrane. This is distinct from primary axotomy or neuron–neuron disconnection, in the context of size of defects, as poration implies that the general anatomy of the cell and its processes remains intact, but that nanometer-scale membrane damage has occurred. Left unchecked, the membrane is capable of repair, but recovery is likely linked to restoration of ion imbalances, adequate cell energy, and an ability to counteract the effects of continued membrane attack by free radicals, phospholipases, and other sources of biomolecular

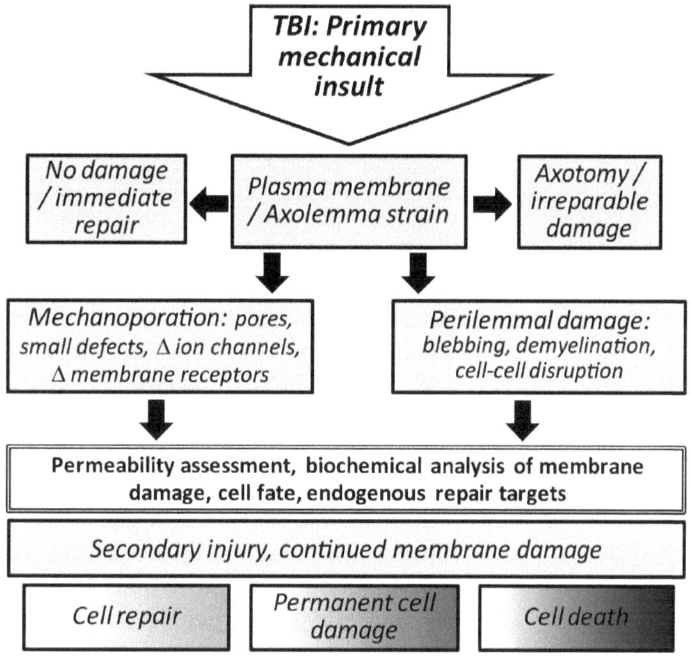

Fig. 1. Schematic of the primary and secondary phases of traumatic membrane damage, including the various general methods that may be used to assess mechanoporation and overt membrane damage. The response to injury may also include repair (*upper left*) or catastrophic failure (*upper right*). A better understanding of the type of damage may help to mitigate secondary injury.

degradation. The ability to assess mechanoporation using membrane permeability as a functional cell measure has gained attention in the last decade and continues to be an area of study, as plasma membrane integrity and function are key to a neuron's function.

1.2. Historical Perspective of Membrane Damage Following Trauma

Physical damage to the spinal cord was recorded on papyrus scrolls by the Egyptians in 1700 BC in what is considered the first medical neuroscience text. These pioneering observations also suggest that the damage was irreparable, a general belief until Ramon y Cajal described sprouting in central nervous system (CNS) cells in response to transection (Degeneration and Regeneration of the Nervous System, 1928). Toward the end of the twentieth century, the repair capacity in severed axons was further developed in the seminal work of David and Aguayo (6), which focused the field of regeneration on reversing the inhibitory microenvironment of damaged neurons. These observations all bring attention to the damaged neurolemma, whether caused by overt disruption, stretching, or some membrane adaptation and highlight that the damaged neuronal membranes are capable of being restored, even in instances of complete transection or crush.

In the 1990s, further studies examined more subtle forms of membrane damage and established assessment methods that have been used in numerous experimental models of traumatic brain and spinal cord damage. In nonhuman primates, ultrastructural axolemma disruptions were observed in the brain following severe acceleration injury (7), a model that is characterized by coma, diffuse axonal injury, and persistent neuronal dysfunction. Axonal injury has also been observed in severely injured human brains; however, the time course of axolemma disruption is difficult to discern. It is widely accepted that primary axotomy at the time of a traumatic brain insult is rare, and that the majority of axonal damage is the result of cumulative ongoing secondary injury cascades (8). In order to assess axonal permeability, Pettus et al. (9) introduced horseradish peroxidase (HRP, MW ~40 kDa) to the cerebrospinal fluid (CSF) prior to fluid percussion injury in the cat to observe whether this normally impermeant molecule entered the axolemma. Mild injury did not result in HRP uptake, but at more severe levels intra-axonal HRP colocalized with neurofilament packing and subsequent cytoskeletal disassembly, supporting the theory that primary axonal damage initiated secondary events following TBI. The observation that exogenous extracellular molecules could be used to assess membrane damage following neural trauma in animals was significant and this general methodology has been used in numerous experimental preparations, both in vivo and in vitro.

1.3. Experimental Evidence of Membrane Permeability Alterations

Damage to the neuronal membrane is associated with poor cellular outcome in numerous experimental models of TBI and SCI (9–24). Specifically, plasma membrane disruptions have been correlated with cell death (22), structural damage (10), and impaired axonal transport (16). While the type of injury model and method of assessment of membrane damage differ from study to study, some general conclusions can be made: (1) changes in acute (<10 min post injury) membrane integrity are associated with high strain rate physical insults; (2) membrane damage observed hours following injury in vivo is likely an indication of cumulative damage from primary and secondary insults to the cell membrane; (3) membrane permeability that follows in vitro trauma appears to resolve much more rapidly than permeability changes seen in vivo, pointing to potential limitations of simplified cellular preparations; (4) assessment of time points varies greatly between models, and therefore the time course of membrane damage is still not well understood; and (5) permeability markers (and methods) used vary among studies and may lead to ambiguity in determining membrane defect size and resealing dynamics.

2. General Considerations in Measuring Membrane Permeability

Plasma membrane damage can be assessed with various methods: (1) adding a normally impermeant marker or dye to the extracellular space before injury, rinsing away the excess dye, and visualizing or counting the cells that are positive; (2) examining membranes using electron microscopy at specific time points following injury; and/or (3) measuring changes in membrane components, including breakdown products, with molecular analysis methods, such as liquid chromatography and electrospray tandem mass spectrometry (LC ESI MS/MS) (25). Membrane permeability is a dynamic cell state, and therefore discrete analyses such as these should be interpreted as such. Nonetheless, these methods use standard lab procedures and provide a means to quantify cells with membrane defects at specified time points. We focus on the first of these three methods.

2.1. Use of Permeability Markers

2.1.1. Assumptions Made in Using Markers to Assess Membrane Permeability

Permeability markers must not normally cross the plasma membrane and preferably do not become sequestered by intracellular organelles once inside the cells. A marker should not interfere with normal cell function, but may load a cell in a reversible (e.g., dextran and Lucifer yellow) or irreversible (e.g., propridium iodide and ethidium bromide) manner. In addition, a good marker should be a substance that is not metabolized and remains stable for cell/tissue fixation and observation. It is assumed that the marker, or dye, of choice enters the cells through nonspecific membrane openings or pores, and not through ion channels, via membrane carriers, or from neighboring cells. Depending on the size and other properties of the marker, there is a possibility of abnormal flux through membrane channels, and, as such, this should be considered in data interpretation. Blocking of channels with specific pharmacological blockers or toxins is an approach to discern the route of membrane flux, but it must be kept in mind that altered ion channels may not have the same affinity for blockers as normal ion channels. The marker must also be stable and easily detected for data analysis, with minimal or no degradation or fading. One must also be aware of the properties of the particular marker (e.g., size, configuration, charge, etc.) in interpreting the data, as under- or overestimation of membrane damage can result if there is artifact from the dye (e.g., if the molecular shape of the marker is different than the shape of the defect, then there may be an underestimation of permeability).

2.1.2. General Approach for Assessing Membrane Permeability In Vitro

Cell culture studies can take advantage of direct access to the extracellular fluid to easily add and remove markers that are normally impermeant (e.g., calcein, Lucifer yellow, and dextrans). Following the insult, excess markers are removed, leaving behind those

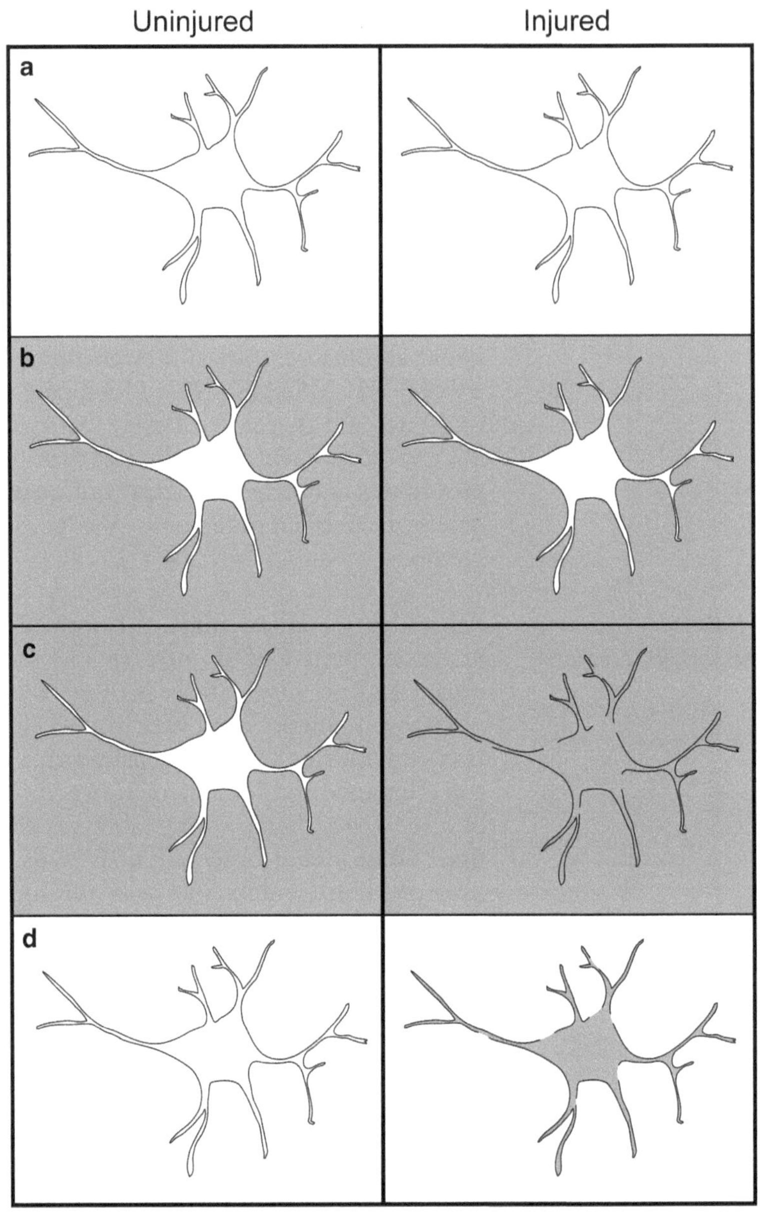

Fig. 2. The general methodology for cell exclusion assays in the assessment of membrane permeability. In the *top row* (**a**), cells are initially unlabelled. In (**b**), permeability marker solution (depicted in *gray*) is added extracellularly to the culture (or animal). Following injury (**c**), damage (*right side*) allows the dye to enter the cell, whereas the cell on the left is undamaged and no dye enters. (**d**) The extracellular fluid is rinsed away, leaving the dye in the damaged cell, but not tagging the undamaged cell.

trapped inside the cells (Fig. 2). Cells are then either observed live (or fixed) with microscopy or flow cytometry. The assumption is that the markers are permanently loaded inside the cell, either through intracellular binding or membrane resealing. There are some limitations to this assumption, since dye may leak out of the cell if the membranes do not reseal or if the membrane stability is such that opening and closing of pores occur. Cell culture studies have several advantages over animal studies, including the ability to view cells during and immediately after the insult, as well as better control over the volume and concentration of dye solution. Diffusion within the extracellular space, however, is drastically different than in brain or spinal cord tissue, where the volume of the extracellular space is much smaller than that of the (collective) intracellular space and diffusion in vivo has a more complex 3D pattern than in most cultures. A potential advantage to in vitro preparations is that the dye solution is well mixed, whereas uniform distribution in tissue is more difficult to achieve and verify due to tissue tortuosity (degree of intercellular path turns).

The steps are straightforward and can be varied depending on experimental needs. (1) Culture neural cells according to desired protocol, making sure that cells are well adhered and have formed a typical culture network. (2) Prepare permeability dye buffer. Use a warmed, osmotically balanced solution with buffer of choice (a good buffer, such as HEPES, is especially important if cells will be outside the incubator for any length of time, in order to avoid large pH swings) to rinse media from the cells. (3) Add an equal volume of buffer with permeability marker (be sure to include separate cultures with the same buffer minus the permeability dye as controls). (4) Injure cell culture following a 10-min incubation period. Permeability marker can also be added at set time points following the insult to assess membrane resealing. (5) Replace permeability buffer with fresh buffer or media and return to incubator (except cells to be fixed or counted immediately). (6) Observe cells and quantify.

2.1.3. Assessing Membrane Permeability In Vivo

The approach for using membrane permeability following injury in vivo is essentially the same as described for the cellular experiments (above), but several unique systemic issues (such as volume and pressure) must be considered when using intact animals. The composition, volume, route of administration, and rate of delivery all potentially affect dye distribution and availability, and should be chosen carefully to avoid methodological artifact. Permeability buffer/solution is added to the extracellular space, best reached via the CSF intrathecally or intracerebroventricularly rather than directly to the interstitial fluid space. Intraperitoneal infusion is another route that has been reported (22) and results in equilibrium across the extracellular fluid compartments, given that the dye freely diffuses or is transported unhindered. Rapid volume changes in the CSF volume may increase the intracranial pressure (ICP), thus compounding the injury, and therefore the rate of infusion (regardless of the compartment) must not exceed the rate of CSF outflow. The osmolarity of the dye solution is also important and should result in an isotonic equilibrium; tonicity depends

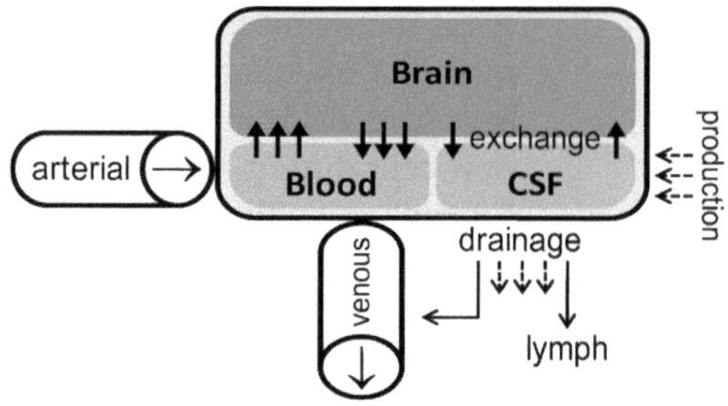

Fig. 3. The balance of intracranial space. The brain occupies roughly 80% of the volume while the blood and CSF take up approximately 10% each. Blood volume is regulated by the resistance of the arterioles, and the compliance of the venous returns, as well as other systemically controlled parameters. CSF is produced in the epithelium of the ventricles and drains into the blood and lymph. Both of these fluid compartments have microexchange systems within the brain and work together to maintain a constant volume and pressure during normal function and during small shifts in homeostatic balance.

on the amount added and buffer composition. To emphasize, the addition of permeability marker must be carefully considered in light of the closed CSF compartment and injury-associated edema, as one does not want to artificially elevate ICP.

According to the Monro-Kellie doctrine, the cranium maintains a balance of constant brain volume (80%), CSF volume (10%), and blood volume (10%). An increase in one compartment will initially cause adjustment in either CSF or blood volume in order to avoid pressure increases (26) (Fig. 3). The nominal weight of a rat brain (350–400 g rat) is 2 g and the volume of CSF is approximately 250 μl. A rule of thumb suggested for lumbar injection into the subarachnoid space that the injection rate should be <10 μl/min and the volume into the CSF space should be not more than 10% (~25 μl) of the total volume (27), but these guidelines are conservative for other infusion routes. We routinely add 15 μl to the cistern magna at a rate of 2–3 μl/min with adequate dye distribution (24), as also reported by Choo et al. (21). Infusion of 75 μl of dextran solution into the lateral ventricle (0.5 mm posterior, 1.3 mm lateral to bregma, at a rate of 2 μl/min) did not result in significant ICP elevation (18) (see Table 1 for a summary of doses, rates, and other administration variables).

Normally, CSF is produced at approximately 2 μl/min and the CSF outflow (to the venous sinuses, lymph, blood–brain barrier, choroid plexus, and tissue) is matched (28). When fluid is added to the CSF compartment, compliant blood vessels will permit an increase in venous outflow and CSF production will decrease and resistance to CSF drainage will decrease. Therefore, slow, steady matched infusion should not be problematic, as homeostatic mass balance should maintain production and drainage rates to accommodate an increase in volume to the CSF. A delayed ICP rise (2 h, 400% of baseline), however, has been noted in rats following an

Table 1
In vivo studies that have used permeability markers to assess mechanoporation following brain and spinal cord injury

Marker	Concentration	Volume	Rate	Administration route	Administration time	Model	Reference
HRP40k	166.7 mg/ml	300 µl	NR	Cisterna magna	1 h pre-injury	Cat, FPI	(9)
HRP40k/fluorescein dextran40k/Texas Red dextran10k	50 mg/ml in saline	~100 µl	333 µl/kg in 20–30 min	ICV (0.5 mm posterior, 1.4 mm lateral to bregma, 4.0 mm deep)	2 h pre-injury	Rat, FPI	(15)
EB394, HRP40k	90 µM in saline 30% *w/v*	110 µl/150 µl	30 min	ICV	2 h pre-injury	Rat, weight drop	(17)
Alexa Fluor 488 dextran10k/Texas Red dextran10k	26.67 mg/ml	75 µl	2 µl/min	ICV (0.5 mm posterior, 1.3 mm lateral to bregma)	2 h pre-injury, 2, 6 h post injury	Rat, weight drop	(18)
PI368/YOYO-1	1 mg/kg in 0.9% saline (i.p.), 50 µg/ml, ICV/10 µg/ml	100 µl, 2 µl/4 µl	NR	ICV (0.1 mm posterior, 1 mm lateral to bregma, 2.5 mm deep), intraperitoneal	1 h pre-injury, 5, 23, 47 h post injury, 20 min pre-injury	Mouse CCI	(22)
Fluorescein dextran10k	25 mg/ml in H_2O	15 µl	NR	Cisterna magna	1.5 h pre-injury	Rat, SCI	(21)
LY457/Fluorescein dextran3k/Fluorescein dextran10k	25 mg/ml in H_2O	15 µl	1.5 µl/min	Cisterna magna	3 h pre-injury	Rat, SCI	(24)

PI propidium iodide, *EB* ethidium bromide, *LY* Lucifer yellow, *HRP* horseradish peroxidase, *ICV* intracerebroventricular, *CCI* controlled cortical impact, *FPI* fluid percussion injury, *SCI* spinal cord injury, *NR* not reported

80-μl infusion of artificial CSF into the lateral ventricle over 10 min, followed by 0.5 μl/min for 24 h (29), suggesting the potential importance of biomolecules absent in the artificial CSF that may be necessary for volume equilibrium over time. In another approach, Pettus et al. (9) removed 300 μl from the CSF of a cat, mixed dye (HRP) in that volume, and injected the mixture back into the same animal. While this methodology may be difficult in the rodent, the principle behind it is rational and rodent-appropriate adaptations should be considered. In summary, differences in parameters of permeability dye administration point to the need for careful consideration of volumes and rates in both the design of studies and in the interpretation, especially since TBI itself can affect ICP, cerebral perfusion pressure, and venous drainage (30).

2.2. Permeability Marker Selection

2.2.1. General Considerations in Choosing a Permeability Marker

The selection of permeability marker is important when comparing size of the membrane defects, as well as when performing temporal studies and comparing to other studies. Factors that affect marker uptake include size, shape, and charge of the molecule (Table 2). We found a size distribution of membrane defects during mechanoporation in stretched neurons (12), with larger (10 kDa dextran) molecules crossing the membrane in smaller numbers than smaller (down to 380 Da) molecules, suggesting a range of pore size and/or resealing of larger pores, which are likely to be more thermodynamically unstable versus smaller pores. Similarly, in an electroporated squid giant axon, sulforhodamine 101 (650 Da) entered the axon, but Texas Red dextran (40 kDa) did not (31). Since there may be a distribution of defect sizes, a variety of marker molecular sizes is recommended while characterizing the spatial profile and degree of damage.

The shape of the molecule may alter permeability assessment. The molecular size does not necessarily scale with molecular radius, as the Stoke's radius approximation assumes that the molecular shape is spherical. Dextrans, for example, are polysaccharide chains that are shaped more like flexible rods rather than spheres; therefore, the effective diffusion radius of a dextran is about sevenfold that of an equal molecular weight molecule of HRP (32). The charge may affect dye uptake as well. For example, negatively charged dyes did not enter the ends of cut crayfish giant axons, whereas neutral dyes of the same molecular size and shape did cross the axolemmal barrier that formed (33). A neutrally charged particle is desirable as marker movement at the cell surface may be influenced by local charge differences and electrostatic interactions. In addition, the smaller the membrane defects, the more electrostatic effects will dominate the solute diffusion across the membrane. Despite this potential factor (i.e., charge of the permeability marker molecule), TBI studies of membrane permeability have not

Table 2
Different molecules that have been used, or have potential use, for assessing membrane permeability in traumatic brain and spinal cord injury

Marker	Abbreviation	Molecular weight (Da)
Carboxyfluorescein	CBF	380
Ethidium bromide	EB	394
Lucifer yellow	LY	457
Alexa 488 hydrazide	A488	570
Calcein	Calcein	623
Propidium iodide	PI	668
YOYO-1	YOYO-1	1,323
Dextran	D3	3,000
Dextran	D10	10,000
Dextran	D40	40,000
Horseradish peroxidase	HRP	40,000
Dextran	D70	70,000
Lactate dehydrogenase	LDH	140,000
Dextran	D150	150,000
Dextran	D464	464,000

Common abbreviations and nominal molecular weight are given. Of note, nucleic acid-binding molecules (EB, PI) irreversibly bind, whereas others require a fixative (LY, dextran). In addition, equivalent molecular weights do not necessarily translate to equal molecular radii (e.g., Dextran 40 kDa and HRP 40 kDa). Dextrans form chains, whereas other molecules are better approximated as spheres

typically considered charge, nor has a side-by-side comparison of different charges for a given molecule been performed.

Other considerations when choosing a membrane permeability marker are concentration, solvent, and binding properties. Hydrophilic properties are desired to minimize binding to or passing through the membrane under normal conditions. Permeability markers like propidium iodide and ethidium bromide bind irreversibly to nucleic acids, eliminated artifact caused by leakage. Alternatively, aldehyde fixable dyes, such as Lucifer yellow, are convenient, since 0.1% glutaraldehyde is easily added to a parafomaldehyde or formalin fixative.

3. Working Protocols for Membrane Permeability Assessment

3.1. Materials

3.1.1. Animals

Rats are preferred over mice due to the size, but mice have successfully been used to measure membrane permeability (22). Procedures here are for adult male rats (275–325 g; Sprague Dawley, Charles River Laboratories Inc., Wilmington, MA). Rats are anesthetized, positioned on a stereotactic frame, and permeability marker delivered into the CSF via intrathecal injection. Methods described are for CCI, but other injury models can be used in place of this method.

3.1.2. Instruments and Materials

1. Anesthetic (injectable or inhalant) (ketamine/xylazine, Nembutal, and isoflurane).
2. Hair clippers.
3. Cyclohexamide antiseptic or a similar solution.
4. Rodent stereotaxic frame.
5. Sterile PE-10 catheter tubing attached to 10-μl Hamilton syringe with 26-G needle.
6. Syringe pump.
7. Sterile gauze and cotton-tipped applicators.
8. Sterile surgical tools.
   (a) Scalpel.
   (b) Scraper.
   (c) Hemostats.
   (d) Forceps.
   (e) Sutures.
   (f) Surgical clips.
   (g) 6-mm (outer diameter) trephine.
   (h) Hand-operated or powered drill.
9. Permeability marker solution.
10. Pneumatic controlled cortical impact device (e.g., CCI, Pittsburgh Precision Instruments) or other injury device, according to experimental needs.

3.2. Surgical Procedures

3.2.1. Injection of Permeability Marker (Fig. 4)

1. Anesthetize the rat with isoflurane-inhalant anesthetic (induction at 3–5% and maintenance at 2–3%), ketamine/xylazine/acetylpromazine (50 mg/kg/10 mg/kg/1.67 mg/kg, i.p. or s.c.), or Nembutal (50 mg/kg, i.p.).
2. Shave the hair on the dorsal neck using hair clippers to expose the skin used for all procedures (dye injection and injury). Hair should be clipped from just above the eyes, through the space between the ears, and caudal to the base of the skull.

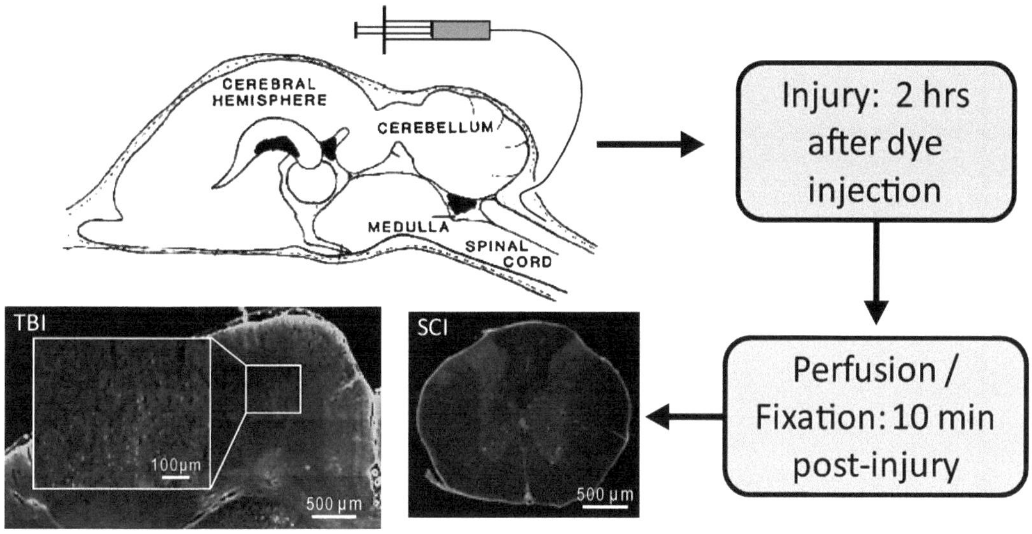

Fig. 4. Schematic of the preinjury injection method for permeability dye into the cisterna magna. Following circulation throughout the CSF, the selected injury procedure proceeds, followed by fixation and histological assessment. Insets are for TBI (CCI) and SCI (contusion).

3. Mount the animal in the stereotaxic frame to stabilize the skull or vertebral column (with proper spinal adaptor). (Note: We have injected dye into the cisterna magna with no stereotaxic support and find that neck flexion is preferable to expose the dura. For intracerebroventricular injection, stereotaxic support and 3D coordinate control are necessary).
4. Apply protective eye ointment to prevent drying out of the eyes during the surgical procedure.
5. Clean the exposed skin and surrounding areas with a chloroheximide or betadine solution and allow to dry.
6. Prepare the injection by loading the syringe with (premade) permeability marker solution and priming the catheter or syringe.
7. Access the dura. There are two basic methods for injection into the cistern magna.
   (a) Direct access: The musculature at the base of the skull is cut parasagitally (i.e., in the rostral-caudal direction) to avoid damaging the midline muscle/tendon and blunt dissected to reveal the atlantooccipital membrane. This can be done by making an incision (approx. 1.5 cm in length) through the skin and the muscle at the base of the skull. A sterile cotton-tipped applicator can be used to retract the muscles and control bleeding. If the experimental is terminal, a medial–lateral cut across the dorsal neck muscles and tendons can be performed for more

rapid access. (Caution: This cut will severely limit the ability for neck extension, however, which will alter the rat's ability to maintain normal posture and is only recommended for terminal experiments.) The dura mater appears smooth and shiny with a gray tint.

(b) Blind access: The cistern magna can be accessed without cutting the skin and the muscle (34). With the neck flexed, a 27-G needle is inserted through the skin along the skull base, through the muscle and other connective tissue, and slowly until resistance at the atlantooccipital membrane is felt. The cisterna magna is reached following puncture, with a 1–2-mm clearance.

(c) Puncture the dura mater using a microscalpel or needle and inject the dye via a polyethylene catheter (or small, 26–30-G needle) inserted into the subdural spinal canal using forceps. Gently slide the catheter into the subdural space (cistern magna). (Note: We have inserted the catheter in the caudal direction for approx. 1–2 cm into the subarachnoid space of the dorsal cord for terminal experiments, but recommend practice if this is to be used in survival animals.) CSF leakage may occur, and therefore the smallest gauge needle/tubing should be used and multiple punctures should be avoided.

(d) Practice for either of these procedures can easily be done with a needle (and no syringe or tubing attached) by puncturing the membrane and observing CSF fill the plastic of the needle bore. The fluid enters quickly (due to the pressure difference) and should contain no blood or debris.

(e) If resistance is encountered, it is likely that the catheter has reached an obstruction and the catheter should be removed and reinserted.

(f) Applying marks to the catheter at 1-mm increments prior to insertion assists with determination of inserted length.

8. Inject permeability marker solution (10–100 μl) in sterile physiological saline (or other buffer) into the CSF (0.5–5 μl/min, preferably 2–2.5 μl/min, see above discussion on ICP and CSF turnover) through the catheter. Chaser (or pulse) labeling at various times post injury with a second permeability marker has been used (15, 18, 22) in order to double label cells. This method is useful to examine resealing, secondary mechanoporation, and/or multiple-sized pores. Care must be taken in interpretation since following the initial physical insult, it is likely that multiple cellular mechanisms (in the form of secondary injury cascades) dictate delayed membrane permeability.

9. Leave the needle or catheter in place for 5–10 min following dye injection and slowly remove.

10. Following the injection, the surgery site is closed with sutures or surgical clips. Remove rat from the stereotaxic frame and return it to its cage. Observe the animal as it recovers from anesthesia and the injection.
11. Wait for 2 h for the dye to travel through the CSF and to diffuse throughout the CNS prior to performing injury procedures.

3.2.2. Controlled Cortical Impact (Adapt Method for Different Injury Method)

1. Anesthetize the rat with isoflurane-inhalant anesthetic (induction at 3–5% and maintenance at 2–3%), ketamine/xylazine/acetylpromazine (50 mg/kg/10 mg/kg/1.67 mg/kg, i.p. or s.c.), or Nembutal (50 mg/kg, i.p.).
2. Mount the animal in the stereotaxic frame with the body supported and the skull flat.
3. Perform a midline incision from above the eyes to between the ears. Use hemostats to retract the skin and expose the top of the skull.
4. Retract the muscle and remove the periosteum from the left side of the skull (for left unilateral injury) using a blunt instrument or scraper.
5. Perform a 6-mm-diameter craniectomy over the left frontoparietal cortex with the center 3.0 mm posterior to the Bregma and 3.0 mm lateral to the midline using a trephine.
   (a) The 6-mm-diameter craniectomy should be performed in two phases: the top of the skull and the side of the skull.
   (b) Care should be taken not to cut the dura beneath the skull or the vasculature within the skull at the sutures.
6. Align the animal on the stereotaxic frame within the CCI device.
   (a) For rats, the CCI device is used with a 5-mm-diameter impactor tip.
   (b) The device is set to impact at 15° from vertical for a unilateral injury.
7. Deliver the controlled cortical impact injury.
   (a) Depending on desired injury severity, the velocity is 2 or 5 m/s and the depth should be set to 0.5, 1.5, or 2.5 mm.
   (b) The CCI device is set to 250 ms injury duration.
8. Observe the animal after the injury and determine the need for resuscitation.
9. Remove hemostats and retractors and suture the scalp closed.
10. Remove the animal from the stereotaxic frame and return it to its cage for recovery in a temperature- and humidity-controlled chamber.
11. Perform postoperative animal care according to institutional approved animal care and use protocols.

3.3. Tissue Collection and Processing

1. At the desired time point, deeply anesthetize the animal with an overdose of pentobarbital and perfuse transcardially with 0.1 M phosphate-buffered saline (pH 7.4), followed by 4% paraformaldehyde plus 0.1% glutaraldehyde in cold 0.1 M phosphate-buffered saline (4°C, pH 7.4).
2. Using rongeurs, scissors, and a spatula, and carefully remove the brain tissue from the skull. (Using a guillotine to detach the head helps with collection of the tissue.) Place the tissue in 4% paraformaldehyde with 0.1% glutaraldehyde in cold 0.1 M phosphate-buffered saline (pH 7.4) and store at 4°C overnight.
3. For cryoprotection, transfer the tissue to 20% sucrose and store at 4°C for 2–3 days until the tissue sinks to the bottom of the storage container.
4. Embed the tissue using cryoembedding compound and use standard techniques to prepare sections using a cryostat (we have found that 20-μm sections work well). Sections should be collected from 2 mm anterior through 2 mm posterior to the impact site in the area of interest (e.g., ipsilateral to impact, contralateral to impact, cortex at impact site, peri-injury cortex, hippocampus subregions CA1, CA2, and CA3, dentate gyrus, striatum, and subthalamic areas).
5. Tissue sections can be counterstained for evaluation of specific cell types or anatomical structures according to the experimental requirement. For example, neurons and astrocytes phenotypic markers can be detected with immunohistochemistry using monoclonal antibodies against NeuN (1:500 dilution; Millipore, Billerica, MA) and GFAP (1:500 dilution, Millipore), respectively. We have also used Fluoro-Jade B staining (Millipore, Billerica, MA) as described previously (35) to label degenerating neurons (see applications below).

3.4. Histological Procedures

3.4.1. Direct Observation

1. Coverslip sections (contain fluorescence tracers) with an anti-fading fluorescent mounting medium.
2. View fluorescence labeling of the tracer under a fluorescent microscope.

3.4.2. Fluoro-Jade Labeling

1. Thaw and dry the sections for 30–60 min.
2. Rinse to remove cryoembedding compound and then heat to 50°C for 30 min.
3. Soak slides in 100% ethanol for 3 min, 70% ethanol for 1 min, and deionized water for 1 min.
4. Add Fluoro-Jade staining solution (0.0001% Fluoro-Jade B in 0.1% acetic acid with 0.00001% DAPI to visualize cell nuclei) and submerge slides in the solution for 1 h in the dark at 4°C.

5. Rinse slides three times in deionized water, dry, followed by a xylene rinse, and coverslip.
6. Notes: Make solution immediately before staining. A background quenching step was eliminated from the procedure to minimize the reduction of permeability marker fluorescence.

4. Quantification of Permeable Cells

4.1. Notes Related to Quantification

The tissue background (when using fluorescent markers) is especially important (see Sect. 6 for more info) and must be uniform across the tissue of interest. Background is acceptable as long as the dye uptake within the cells is clearly delineated. Related to this issue, it is also crucial that sham-operated rats injected with marker do not show significant cellular staining (<5% total permeable cells from an injured brain) and that selective cellular staining occurs in injured animals (i.e., verify cell morphology at high magnification before quantifying cells). In order to ensure that negative and positive signals are true to the state of the cell, it is also critical to confirm widespread diffusion of the dye throughout the CNS. We found that dye diffused evenly between 1 and 3 h post injection. Cells can be quantified by stereology, manual counting, and/or intensity analysis. Colabeling with histological stains or immunohistochemistry is feasible with tissue sections, if interference from the permeability dye is absent. Fixation techniques that incorporate glutaraldehyde may require harsher tissue permeabilization or antigen retrieval used according to individual needs.

4.2. Permeability Marker Visualization

Stained tissue sections can be visualized on upright microscope with fluorescein and/or Texas red filters (or other appropriate filters). We have used a Texas red filter for imaging of dextran and Fluoro-Jade double labeling to eliminate bleed through of Fluoro-Jade emission signal into the red channel. Serial section reconstruction (such as Neurolucida, MicroBrightField, Inc. Williston, VT) can be used to examine the distribution of dextran uptake and identify regions of interest. We have counted cells in the CA1/CA2, CA3, and dentate gyrus subregions of the hippocampus that were dextran positive, Fluoro-Jade positive, or showed evidence of colocalized dextran and Fluoro-Jade staining, as these were areas that consistently showed permeability marker uptake following TBI (see below). Alternatively, standard stereology (e.g., Stereo-Investigator, MicroBrightField, Inc.) can be used for counting and logging positive cells. Furthermore, histogram analysis can be used to examine dye intensity from cell to cell, as this may be an additional measure of degree of permeability and possibly cell injury severity.

5. Applications

5.1. In Vitro Assessment of Injury-Associated Membrane Permeabilty

Injuries to primary cortical neurons in the presence of permeability markers with biaxial stretch and fluid shear stress devices have revealed nonspecific increases in membrane permeability and changes in network activity (12, 36). Various sized fluorescent molecules were added to the media either immediately before injury or 1, 2, 5, or 10 min after injury and removed 1 min later. The degree of membrane permeability depended on the applied strain rate, strain magnitude, and molecular size. Severe stretch (10/s, 0.30) resulted in significant uptake of all tested molecules (ranging between 0.5 and 8.9 nm radii) with up to 60% of cells positively stained. Furthermore, the neurons remained permeable to the smallest molecule (carboxyfluorescein) up to 5 min after severe stretch but were only permeable to larger molecules (≥10 kDa) immediately after stretch. These transiently formed membrane defects may be the initiating mechanism that translates mechanical stretch to cellular dysfunction. Similarly, a loading rate-dependent increase in acute membrane permeability was measured using lactate dehydrogenase (LDH) release following shear stress injury (11). Acute LDH release may be an indication of transient increases in membrane permeability not necessarily associated with cell death pathways. Both biaxial stretch device and fluid shear stress device lead to immediate cytosolic calcium increases following traumatic loading that are dependent on the magnitude and rate of loading, suggesting a nonspecific increase in membrane permeability. In addition, when membrane strain and calcium were measured simultaneously in the same cultures of NTera2 neurons subjected to high-rate hydrodynamic deformation (in the first 100 ms of loading), membrane strains were correlated with increases in $[Ca^{2+}]_i$, suggesting nonspecific membrane poration.

More recently, we investigated plasma membrane disruption following either shear or compressive strain fields on 3D neuronal-astrocytic cocultures. The mechanical insult (0.50 shear or compressive strain at 1, 10, or 30/s strain rate) was applied in the presence of calcein (non-AM conjugated), which enters the cell upon a breach in membrane integrity. Following both bulk shear and compressive loading, permeability markers entered cells in a strain rate-dependent manner, but cells were more permeable following shear strain than compressive strain (Fig. 5). We are currently examining the fate of the cells over time, as well as the mechanics behind the differential response.

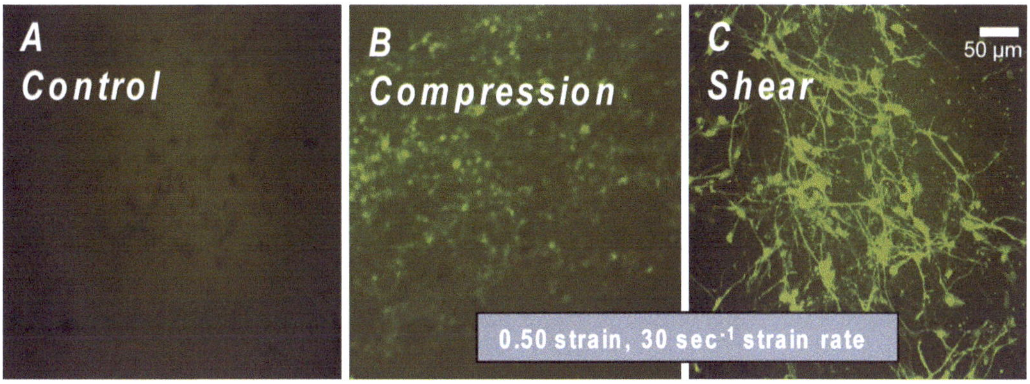

Fig. 5. Permeability assessed in vitro for injury of primary cortical neuronal-astrocytic cocultures (2,500 cells/mm^3 at a 1:1 neuron:astrocyte ratio grown within a 500-μm-thick layer of Matrigel™; final Matrigel™ concentration 7.5 mg/ml). Shown are a control 3D culture (**a**), a 3D culture subjected to compression injury (**b**), and a 3D culture subjected to shear injury (**c**). Note the background staining of permeability marker in (**a**) and the distinct cellular uptake in (**b**), and especially (**c**). Assessment was performed 10 min post injury, following rinsing of excess dye.

5.2. In Vivo Assessment of Injury-Associated Membrane Permeability

We have also assessed membrane permeability following TBI (CCI) in rats at 10 min ($n=12$) and 24 h ($n=12$) post injury. Adult male rats (Sprague Dawley, 300–350 g) were prepared for preinjury injection of impermeant markers and subsequent cortical impact as described above. Rats were anesthetized using isofluorane (induction at 3% and maintenance at 1.5%), mounted in a stereotaxic frame, and the atlantooccipital membrane exposed. PE-10 tubing attached to a 26-G needle was inserted through the atlantooccipital membrane and a rhodamine-conjugated 3 kDa dextran solution was injected into the cistern magna at a rate of 5 μl/min with a syringe pump and the needle and tubing slowly removed (37, 38). Three hours post injection, rats were injured over the left lateral fronto-parietal cortex (CCI, 3 m/s velocity, 2 mm depth). Following sacrifice at the specified time point, brain tissue was fixed, frozen, and sectioned, and membrane permeability was quantified. Permeable cells were located primarily in the ipsilateral hippocampus of injured rats at 10 min post injury; however, by 24 h, there was a significant increase in the number of permeable cells in the contralateral hippocampus of injured rats as well. Fluoro-Jade staining revealed degenerating neurons in the same regions of the hippocampus in which permeable cells were identified; however, only a subset of permeable cells showed signs of degeneration at 24 h (Fig. 6). Conversely, plasma membrane damage was evident in most degenerating cells. Together, these results indicate that the response is complex, both spatially within the brain and over time, suggesting that membrane permeability may partially contribute to cell death but is likely part of a multifaceted dynamic course of events contributing to secondary injury.

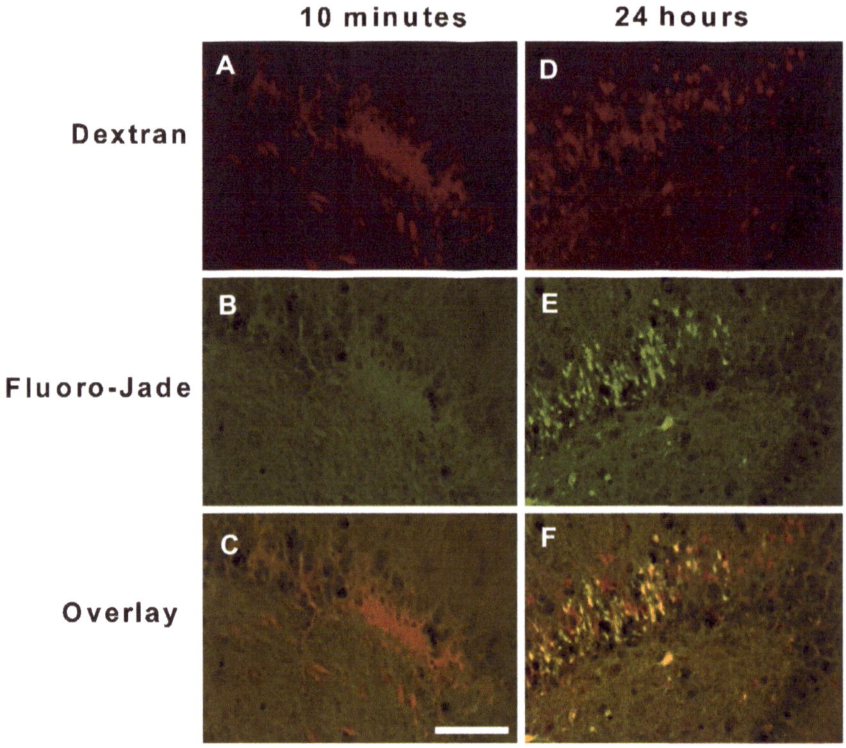

Fig. 6. Permeability marker uptake in the dentate gyrus of the ipsilateral hippocampus 10 min (**a**) and 24 h (**d**) post injury. Fluoro-Jade staining shows degenerating neurons in the same tissue sections at 10 min (**b**) and 24 h (**e**) post injury, respectively. Overlays of these images (**c**, **f**) show that most Fluoro-Jade-stained cells contain permeability marker (FJ w/Dex), whereas only a subset of permeable cells are Fluoro-Jade positive (Dex w/FJ) (scale bar = 100 μm).

6. Troubleshooting Guide

Problem	Potential reason(s)	Solution/suggestion
Permeability marker is not visible in tissue sections	Circulation time is too short (permeability marker requires additional time to circulate through CSF and to penetrate tissue)	Extend circulation time
	Circulation time is too long (permeability marker clears CSF and tissue)	Shorten circulation time
	Adequate amount of permeability marker is not being delivered to CSF	Check catheter placement, permeability marker concentration, and injection rate
	Permeability markers are photosensitive	Protect tissues and slides from light to prevent photobleaching

(continued)

(continued)

Problem	Potential reason(s)	Solution/suggestion
Excess permeability marker in background is observed	Excess permeability marker is not being adequately rinsed out prior to fixation	Perfuse animal with more saline prior to fixative
Excess cellular uptake of permeability marker is observed	Tissue is being disturbed during harvest, causing nonspecific plasma membrane damage	Take care to gently harvest tissue
	Tissue is not adequately fixed	Confirm perfusion setup (blood clears from lungs and liver when perfusing with saline; limbs become rigid when perfused with fixative)
		Use cold, freshly made fixative
	Secondary plasma membrane damage (not a direct result of the mechanical insult but rather secondary injury mechanisms) occurs in the subacute time frame	Collect tissue in the acute time frame post insult
Insertion of the catheter is technically challenging	PE10 tubing must be inserted into the subdural space, which is limited	Consider alternate delivery routes, including intraventricular injection or direct injection into cisterna magna

7. Conclusions and Future Directions

The assessment of membrane permeability has the potential to isolate a critical component of traumatic injury. The plasma membrane is the ultimate barrier to the intracellular machinery, and the integrity of its structural and functional roles is essential to survival. To this end, the mechanisms of both primary and secondary damage to this barrier must be better understood in order to establish relationships among the complex injury cascades and to develop upstream therapeutic targets. The methods presented in this chapter outline the direct assessment of membrane permeability by adding impermeant fluorescent dye solutions to the extracellular space of the CNS. Cell entry is indicative of nonspecific cellular uptake, but it must be kept in mind that injury may cause alterations of channels and other membrane components, permitting access. Injury models include both in vitro and in vivo preparations and care must be taken when transferring methods (and results) from one platform to another. Emphasis was given here to animal studies, since the majority of studies in the field currently take place in vivo. Other methods of membrane function and membrane breakdown are considered complementary and would enhance our understanding of membrane pathophysiology. In addition, studies using other organ systems and other forms of tissue damage can shed some light on mechanisms of membrane damage and repair and should be explored (i.e., lung injury (39)).

Additional studies may include adding different dyes at the same time or comparing different dyes in the same model, varying the size, shape, or charge. Some potential limitations to the permeability marker protocols are the procedural compromise of the dura mater (and brain, in the case of ICV injection) to enter the intrathecal or intracerebroventricular space, as well as the marker interactions with the plasma membrane. Related is the risk of artificially altering ICP, either directly because of the dye solution or indirectly by disturbing CSF outflow regulatory mechanisms. Because of the need to add dye solution to the extracellular space in order to introduce impermeant molecules, less invasive measures, such as imaging or postmortem analysis of membrane damage, may partially address this issue.

Future studies in the area of injury-induced membrane permeability are needed to advance our current understanding of the time course of membrane damage, the related cell processes which exacerbate damage or are downstream from membrane damage, and the ability to reverse or repair the damage. In addition, we found the time course of increased axolemmal permeability and mechanoporation of the cell bodies following SCI to vary, with prolonged permeability marker uptake in axons as compared to cell bodies (24), suggesting either differing resealing mechanisms or cell death patterns. Moreover, the diameter of the axon as well as the myelination state may alter the response at the membrane level and would be interesting to probe in light of other aspects of axonal injury and white matter tract functions. And finally, damage to the nodal regions of axons has been observed to have acute (15 min) and subacute (6–24 h) postinjury phases (40), suggesting that nodal vulnerability may play a role in the axolemmal response to injury. In summary, assessment of membrane permeability is of interest to the neurotrauma community and the results from such studies can help to determine the role that the membrane plays in the complex injury response.

Acknowledgments

The authors thank Dr. Donna Geddes-Klein, Dr. Gustavo Prado, Dr. Crystal Simon, and Dr. D. Kacy Cullen for assistance with experiments highlighted within the methods, as well as trouble-shooting and protocol development.

References

1. Farooqui AA, Horrocks LA, Farooqui T (2007) Interactions between neural membrane glycerophospholipid and sphingolipid mediators: a recipe for neural cell survival or suicide. J Neurosci Res 85:1834–1850
2. Lamb RG, Harper CC, McKinney JS, Rzigalinski BA, Ellis EF (1997) Alterations in phosphatidylcholine metabolism of stretch-injured cultured rat astrocytes. J Neurochem 68:1904–1910
3. Deng Y, Thompson BM, Gao X, Hall ED (2007) Temporal relationship of peroxynitrite-induced oxidative damage, calpain-mediated cytoskeletal degradation and neurodegeneration after traumatic brain injury. Exp Neurol 205:154–165
4. Braughler JM, Hall ED (1992) Involvement of lipid peroxidation in CNS injury. J Neurotrauma 9(Suppl 1):S1–S7
5. McNeil PL, Vogel SS, Miyake K, Terasaki M (2000) Patching plasma membrane disruptions with cytoplasmic membrane. J Cell Sci 113(Pt 11):1891–1902
6. David S, Aguayo AJ (1981) Axonal elongation into peripheral nervous system "bridges" after central nervous system injury in adult rats. Science 214:931–933
7. Maxwell WL, Watt C, Graham DI, Gennarelli TA (1993) Ultrastructural evidence of axonal shearing as a result of lateral acceleration of the head in non-human primates. Acta Neuropathol 86:136–144
8. Adams JH, Graham DI, Gennarelli TA, Maxwell WL (1991) Diffuse axonal injury in non-missile head injury. J Neurol Neurosurg Psychiatry 54:481–483
9. Pettus EH, Christman CW, Giebel ML, Povlishock JT (1994) Traumatically induced altered membrane permeability: its relationship to traumatically induced reactive axonal change. J Neurotrauma 11:507–522
10. Pettus EH, Povlishock JT (1996) Characterization of a distinct set of intra-axonal ultrastructural changes associated with traumatically induced alteration in axolemmal permeability. Brain Res 722:1–11
11. LaPlaca MC, Lee VM, Thibault LE (1997) An in vitro model of traumatic neuronal injury: loading rate-dependent changes in acute cytosolic calcium and lactate dehydrogenase release. J Neurotrauma 14:355–368
12. Geddes DM, Cargill RS II, LaPlaca MC (2003) Mechanical stretch to neurons results in a strain rate and magnitude-dependent increase in plasma membrane permeability. J Neurotrauma 20:1039–1049
13. Geddes DM, LaPlaca MC, Cargill RS II (2003) Susceptibility of hippocampal neurons to mechanically induced injury. Exp Neurol 184:420–427
14. Shi R (2004) The dynamics of axolemmal disruption in guinea pig spinal cord following compression. J Neurocytol 33:203–211
15. Singleton RH, Povlishock JT (2004) Identification and characterization of heterogeneous neuronal injury and death in regions of diffuse brain injury: evidence for multiple independent injury phenotypes. J Neurosci 24: 3543–3553
16. Stone JR, Okonkwo DO, Dialo AO et al (2004) Impaired axonal transport and altered axolemmal permeability occur in distinct populations of damaged axons following traumatic brain injury. Exp Neurol 190:59–69
17. Koob AO, Duerstock BS, Babbs CF, Sun Y, Borgens RB (2005) Intravenous polyethylene glycol inhibits the loss of cerebral cells after brain injury. J Neurotrauma 22:1092–1111
18. Farkas O, Lifshitz J, Povlishock JT (2006) Mechanoporation induced by diffuse traumatic brain injury: an irreversible or reversible response to injury? J Neurosci 26: 3130–3140
19. Geddes-Klein DM, Schiffman KB, Meaney DF (2006) Mechanisms and consequences of neuronal stretch injury in vitro differ with the model of trauma. J Neurotrauma 23:193–204

20. Shi R, Whitebone J (2006) Conduction deficits and membrane disruption of spinal cord axons as a function of magnitude and rate of strain. J Neurophysiol 95:3384–3390
21. Choo AM, Liu J, Lam CK, Dvorak M, Tetzlaff W, Oxland TR (2007) Contusion, dislocation, and distraction: primary hemorrhage and membrane permeability in distinct mechanisms of spinal cord injury. J Neurosurg Spine 6:255–266
22. Whalen MJ, Dalkara T, You Z et al (2007) Acute plasmalemma permeability and protracted clearance of injured cells after controlled cortical impact in mice. J Cereb Blood Flow Metab 28:490–505
23. Kilinc D, Gallo G, Barbee KA (2008) Mechanically-induced membrane poration causes axonal beading and localized cytoskeletal damage. Exp Neurol 212:422–430
24. Simon CM, Sharif S, Tan RP, LaPlaca MC (2009) Spinal cord contusion causes acute plasma membrane damage. J Neurotrauma 26:563–574
25. Ivanova PT, Milne SB, Byrne MO, Xiang Y, Brown HA (2007) Glycerophospholipid identification and quantitation by electrospray ionization mass spectrometry. Methods Enzymol 432:21–57
26. Andrews PJ, Citerio G (2004) Intracranial pressure. Part one: historical overview and basic concepts. Intensive Care Med 30:1730–1733
27. Rieselbach RE, Di Chiro G, Freireich EJ, Rall DP (1962) Subarachnoid distribution of drugs after lumbar injection. N Eng J Med 267:1273–1278
28. Davson H, Hollingsworth G, Segal MB (1970) The mechanism of drainage of the cerebrospinal fluid. Brain 93:665–678
29. Morrow BA, Holt MR, Starcevic VP, Keil LC, Severs WB (1992) Mechanism of delayed intracranial hypertension after cerebroventricular infusions in conscious rats. Brain Res 570: 218–224
30. Bouma GJ, Muizelaar JP (1992) Cerebral blood flow, cerebral blood volume, and cerebrovascular reactivity after severe head injury. J Neurotrauma 9:S333–S348
31. Gallant PE, Galbraith JA (1997) Axonal structure and function after axolemmal leakage in the squid giant axon. J Neurotrauma 14:811–822
32. Venturoli D, Rippe B (2005) Ficoll and dextran vs. globular proteins as probes for testing glomerular permselectivity: effects of molecular size, shape, charge, and deformability. Am J Physiol Renal Physiol 288:F605–F613
33. Eddleman CS, Smyers ME, Lore A, Fishman HM, Bittner GD (1998) Anomalies associated with dye exclusion as a measure of axolemmal repair in invertebrate axons. Neurosci Lett 256: 123–126
34. Nirogi R, Kandikere V, Mudigonda K et al (2009) A simple and rapid method to collect the cerebrospinal fluid of rats and its application for the assessment of drug penetration into the central nervous system. J Neurosci Methods 178:116–119
35. Schmued LC, Hopkins KJ, Fluoro-Jade B (2000) a high affinity fluorescent marker for the localization of neuronal degeneration. Brain Res 874:123–130
36. Prado GR, Ross JD, DeWeerth SP, LaPlaca MC (2005) Mechanical trauma induces immediate changes in neuronal network activity. J Neural Eng 2:148–158
37. Okonkwo DO, Pettus EH, Moroi J, Povlishock JT (1998) Alteration of the neurofilament sidearm and its relation to neurofilament compaction occurring with traumatic axonal injury. Brain Res 784:1–6
38. Howard MJ, David G, Barrett JN (1999) Resealing of transected myelinated mammalian axons in vivo: evidence for involvement of calpain. Neuroscience 93:807–815
39. Vlahakis NE, Schroeder MA, Pagano RE, Hubmayr RD (2002) Role of deformation-induced lipid trafficking in the prevention of plasma membrane stress failure. Am J Respir Crit Care Med 166:1282–1289
40. Gennarelli TA, Tipperman R, Maxwell WL, Graham DI, Adams JH, Irvine A (1993) Traumatic damage to the nodal axolemma: an early, secondary injury. Acta Neurochir Suppl (Wien) 57:49–52

Chapter 29

Assessment of Neurogenesis by BrdU Labeling After Traumatic Brain Injury

Jinhui Chen and Xiang Gao

Abstract

Traumatic brain injury (TBI) is the leading cause of death in children and young adults. It is particularly worth noting that an estimated 7,500 soldiers have suffered TBI caused by improvised explosive device explosions in the Iraq war. TBI represents a significant socioeconomic burden. TBI not only results in immediate CNS tissue disruption (primary injury), but also causes secondary damage among the surviving cells via complex mechanisms triggered by the primary event. Currently, there is no effective treatment for this disorder. Fortunately, recent research has identified neural stem/progenitor cells (NSCs) in the adult mammalian hippocampus of rat, primate, and human brain. Adult-born new neurons are continuously generated from NSCs, develop into mature granular neurons, and integrate into the existing neural network. These adult-born neurons are a potential resource for repairing the damaged hippocampus following TBI. This chapter briefly introduces the adult neurogenesis and typical procedures to assess adult-born new neurons following TBI.

Key words: Neural stem cells, Proliferation, Neurogenesis, Traumatic brain injury, Thymidine analogues

1. Introduction

1.1. Adult Neurogenesis

The discovery of adult neurogenesis and stem cells in the adult brain has changed our view of the mature brain. Although the bulk of neurogenesis in the mammalian brain occurs during embryonic development (13–15), a significant number of newborn neurons were generated and integrated into the existing neural network in the adult brain (16, 17). In the late 1990s, Gage and colleagues (11) and later several other groups revealed that Hippocampus is one of the regions of the adult brain that can support neurogenesis

Jun Chen et al. (eds.), *Animal Models of Acute Neurological Injuries II*, Springer Protocols Handbooks,
DOI 10.1007/978-1-61779-782-8_29,

throughout life (12, 18–20). A large number of newly generated granular neurons were found in the adult rat (21), primate (22–24), and human (25) dentate gyrus (DG) (~9,000/day) (21). Alternatively, the neurons generated from neural precursor cells in the subventricular zone (SVZ) migrate through the rostral migratory stream (RMS) into the olfactory bulb and turn into interneuron (30,000/day) (26–32). These postnatal-born granular neurons are electrically active and are capable of firing action potentials and receiving synaptic inputs (33–35). Adult neurogenesis represents a striking example of structural plasticity in the mature CNS that may be compromised by disease or injury, such as traumatic brain injury (TBI). These adult-born neurons are potential resources for repairing the damaged hippocampus following TBI.

1.2. Tritiated Thymidine

A great deal of knowledge has been known on the adult neurogenesis, although the method of proving of neurogenesis is still majorly limited to the "birth-date" making of cells using nucleoside analogues to integrate into the DNA of dividing cells during the S phase of the cell cycle. The first widely used substance that allowed permanent labeling of cell divisions was tritiated thymidine from 1950s to 1969s (17). Tritiated thymidine was injected systematically to compete with endogenous thymidine in all cells in the S phase of cell division. It is permanently incorporated into the DNA. At a later time point, the cells with tritiated thymidine integration in the tissue sections could be detected autoradiographically with a photo emulsion. This thymidine autoradiography-based technique initiated the description of adult neurogenesis from 1962 to 1993 (17, 36–44). Nonetheless, the tritiated thymidine is rarely used recently. The main disadvantage is that tritiated thymidine autoradiography cannot be easily combined with the immunostaining with cell type-specific markers that would allow a deeper and three-dimensional analysis of the labeled cells. Also, labeled thymidine has a short half-life in vivo.

1.3. Bromodeoxyuridine

The next generation of the "birth-date" making of cells is bromodeoxyuridine (BrdU) (45), a synthetic nucleoside that is an analogue of thymidine. BrdU can be incorporated into the newly synthesized replicating cells (during the S phase of the cell cycle), substituting for thymidine during DNA replication. Antibody specific for BrdU can then be used to detect the incorporated chemical, thus indicating cells that were actively replicating their DNA. BrdU could be detected with antibody. Its fluorescent visualization can be combined with two or more other cell type-specific markers. Thus, it makes it possible to visualize the adult-born neurons with imaging technologies. In addition, the advantages of BrdU labeling also include low cost and simplicity. Two other thymidine analogues with a structure similar to BrdU, iododeoxyuridine (IrdU) and chlorodeoxyuridine (CdU), were used to label newly generated neurons in several recent papers as well (46–50). These analogues can be detected individually by different antibodies that have been reported to recognize IdU and CldU. Nonetheless, antibodies against IdU and CldU are not as sensitive as antibody against BrdU and stained fewer cells when compared to BrdU labeling (51). For most researchers, BrdU has replaced tritiated thymidine as the method of choice. As a matter of fact, in the last 2 decades, in most publications, demonstration of adult neurogenesis is based on the BrdU labeling. Therefore, this chapter briefly outlines the materials and methods for BrdU administration, tissue processing, staining, and photographing brain sections.

1.4. Other Strategies

Several other strategies have been developed, and successfully applied for studying adult neurogenesis. One of the unique features of the retrovirus is that it can insert its own genome into the host DNA during cell division. The engineered retroviral vectors carrying maker protein, such as fluorescent protein, provide alternative and complementary strategies for studying adult rodent (20, 35, 52–56). In human, most of these strategies are not applicable or, like BrdU labeling, limited to cases when patients treated with BrdU as part of their cancer therapy donated tissue samples for research investigations (22). Most studies in human rely, therefore, on the use of markers of the cell cycle, like PCNA and Ki-67, to study adult neurogenesis (57, 58). Retrospective birth dating using [^{14}C] may provide an additional mean to this aim (59).

1.5. Pitfalls

1. Integration during DNA repair and cell death: Because DNA repair involves DNA synthesis and BrdU is not a marker of cell proliferation, but a marker of DNA synthesis. The major concern against the BrdU labeling method has been the suspicion that BrdU would not only label dividing cells, but also pick up the dying cells and cells undergoing DNA repair. DNA repair is a normal process in the life of a cell. Damaged or degenerating terminally differentiated postmitotic neurons in the adult brain can reenter the cell cycle and initiate abortive DNA synthesis without cell division before undergoing cell death (60–65). The appearance of nuclear BrdU labeling in dying cells is typically that of minor nuclear speckling rather than the diffuse labeling of the entire nucleus seen in mitotic cells. However, these in vitro data call attention to the need for caution with BrdU labeling.
2. Specificity of markers: Prospectively labeling newly generated cells, using multiple antibodies against cell type-specific markers and BrdU combined with microscopy imaging, has been proven valuable for characterizing adult neurogenesis. The immunohistological identification of newborn neurons is usually based on the detection of BrdU in a cell that expresses

neuronal markers. This approach sounds straightforward; however, there are two main caveats in this approach. The first one relates to the specificity of the markers used for identifying neural progenitor and stem cells, and immature neurons. Nestin, musashi 1, sox-2, oct3/4, doublecortin, PSA-NCAM, and Tuj-1 are markers of neural precursor cells. They have been also detected in glial cells, as in non-neuronal cells (66–68) and tumor cells (68–71), and are reexpressed in preexisting neurons (72). Particularly, it has been well known that TBI reactivates astrocytes, which express Nestin, musashi 1, sox-2, and oct3/4. Therefore, the phenotypic identification of a cell type based solely on immunohistochemistry may not reveal its exact identity. Morphological analysis and functional study with electrophysiology may permit to define the specified cell types.

3. Challenges of multichannel immunofluorescence: Analysis of double- and triple-strained fluorescent sections poses several challenges. Classic fluorescent microscopy is widely used to detect adult-born neurons with multiple labeling. In addition to limited spatial resolution, there are several occasions conventional fluorescent microscopy has difficulty to distinguish a true adult-born neuron with an existing neuron localized close to another BrdU-positive cell. For example, one common case is the cross talk between the channels. Two superposed cells, one a BrdU-positive cell and the other an existing neuron with positive neuronal marker staining, depending on the direction of the observer may be erroneously determined as one cell with BrdU and neuronal marker double-positive staining and counted as a false-positive adult-born neuron. In the hippocampus following TBI, there are so many BrdU-positive reactive astrocytes that are proximitically located to the newborn neurons in the subgranular zone. Especial carefulness and higher standard of imaging are required to evaluate the neurogenesis in this region. In another case, satellite cells, a type of glial cells, are tightly attached to the soma of neurons, especially in cortical regions. They often engulf one side of the neuronal-like cap. At the certain observing angle, because of the close proximity and low spatial resolution, the BrdU-labeled nucleus of the satellite cell can be mistaken as the nucleus of the neurons and falsely identified as an newly generated neuron in the adult brain. Thus, when identifying the double- or multiple-strained fluorescent cells, confocal microscope has become the gold standard and some of classic fluorescent microscopes can be avoided with this technique. The advantage of confocal microscopy lies in its sophisticated optical construction with its sequential scanning which avoids the cross talk between different channels using conventional fluorescent microscope.

The laser light and aperture to exclude out of focus light produce a sharp image of the optical plane of focus. The laser beam can scan the specimen in *x*-, *y*-, and *z* dimensions. The stack of images allows to reconstruct three-dimensional image of cells with imaging software, which allows to distinguish closely neighboring structure.

4. BrdU is a toxic substance. The integration of halogenated thymine analogues into the DNA alters its stability, increasing the risk of sister-chromatid exchanges, mutations, and DNA double-strand breaks, and lengthens the cycle of cells that incorporate it (73–76). In rodent, when injected into pregnant animals, BrdU affects the litter size and causes several developmental defects, including reduced body weight, reduced in the size of the cerebellum, exencephaly, cleft palate, limb abnormalities, teratogenic malformation, and behavior changes in the progenies (73, 77, 78). When administrated postnatally, BrdU may cause lung changes and alter the development of cerebellum (79–81). Nonetheless, in the adult, injection of BrdU 50 mg/kg (body weight, intraperitoneal, i.p.) as well as dose as high as 300 mg/kg do not show obvious physiological side effects, like weight loss or behavioral changes, as seen in the injection during pregnancy and early postnatal pups (21, 45, 82). Administration of BrdU would have less toxicity for the adult CNS than embryonic or neonatal CNS, as the blood–brain barrier (BBB) is more developed in the adult.

2. Materials

1. 5-Bromo-2-deoxyuridine (Sigma-Aldrich, St. Louis, MO, Cat. No. B5002). BrdU is a suspected mutagen when administered to prenatal or neonatal animals and should be handled with care. Wear proper protective gear and a lab coat, and weigh under a fume hood.
2. Rat anti-BrdU primary monoclonal antibody (clone BU1/75; Accurate Chemical & Scientific, Westbury, NY. Cat. No. OBT0030; optimal dilution ratio 1:200; rat-derived antibodies are relatively uncommon, making it easier to perform double or triple labeling by using antibodies derived from other species).
3. Rabbit anti-Nestin primary polyclonal antibody (Covance, Princeton, NY, Cat. No. PBR-315c; optimal dilution ratio 1:500). Nestin is a marker for neural stem cell (NSC)/neural progenitor cells (NPCs). The antibody against Nestin can be used to detect NPCs. Guinea pig anti-Doublecortin (DCX) primary polyclonal antibody (Millipore, Billerica, MA, Cat. No. AB2253; optimal dilution ratio 1:500) is widely used to

detect the immature newborn neurons and show their dendritic morphology. Guinea pig-derived antibody is very rare, making it easier to perform the multiple labeling by combining with the antibodies derived from other species. Mouse anti-NeuN primary monoclonal antibody (Millipore, Billerica, MA, Cat. No. MAB377; optimal dilution ratio 1:1,000) is a marker for mature neuron. Mouse anti-GFAP primary monoclonal antibody (Sigma, St. Louis, MO, Cat. No. G3893; optimal dilution ratio 1:1,000) is a marker for astrocyte/quiescent NSC.

4. Secondary antibody, Alexa 488 goat anti-rat (Invitrogen, Carlsbad, CA, Cat. No. A11006), Alexa 546 goat anti-rabbit, mouse, and guinea pig (Invitrogen, Cat. No. A11035, A11030 and A11074) are used in dilution ration 1:1,000.
5. Blocking solution: 5% goat serum, Invitrogen, Cat. No. 16210-072; 1% bovine serum albumin, Sigma-Aldrich, Cat. No. A6793; and 0.1% Triton X-100, Sigma-Aldrich, Cat. No. T9284. Dissolve serum and Triton in phosphate-buffered saline (PBS).
6. PBS: 2.625 g/L monobasic sodium phosphate (anhydrous), Sigma-Aldrich, St. Louis, MO, Cat. No. S0751; 11.5 g/L dibasic sodium phosphate (anhydrous), Sigma-Aldrich, Cat. No. S0876; and 9 g/L sodium chloride, Sigma-Aldrich, Cat. No. S9888.
7. 4% Paraformaldehyde in PBS: Make an 8% solution of paraformaldehyde (Electron Microscopy Sciences, Hatfield, PA, Cat. No. 19202) in ddH_2O. Paraformaldehyde does not enter solution easily. To dissolve it more quickly, add 4 mL of 1 M NaOH for each 100 mL of solution, and heat the solution to 50°C while stirring constantly. Add an equal volume of 2 × PBS to make a 4% solution of paraformaldehyde in PBS. Adjust the pH to 7.35 and filter the solution through Fisherbrand Qualitative P8 filter paper, Fisher Scientific, Pittsburg, PA, Cat. No. 09-795C. Paraformaldehyde is stable for several weeks.

3. Procedures

3.1. Prepare BrdU Solution

Warm saline solution (0.9% w/v NaCl in sterile H_2O) to 40–50°C, slowly dissolve BrdU in saline solution by gently vortexing, and adjust pH to 7.35. Filter sterilize through a 0.2-μm filter. It is recommended to use the BrdU injection solution immediately. However, when stored at 4°C, BrdU is a stable solution for months. A 20 mg/mL stock is good for most cases that require intraperitoneal injections. It is recommended to make sure that there is no formation of white BrdU precipitates (crystals) after storage.

3.2. BrdU Administration

BrdU can be administered via intraperitoneal injection, intracerebroventricular (i.c.v.) injection, intravenous (i.v.) injection, or oral ingestion. The most common method of BrdU administration is through i.p. injection. The i.p. injection is technically simple and easily temporally controlled. BrdU is metabolized rapidly through dehalogenation in the plasma. The concentration that reaches the brain is, therefore, only a fraction of the administered dose. It is estimated that BrdU is available for labeling in the adult brain for approximately a 2-h period, after systemic injection. After 2 h, the level of BrdU available for labeling in the brain drops sharply (83, 84). A single BrdU injection labels cells for approximately 5–6 h (12), although this may vary depending on the initial dose given. Thus, labeling cells over an extended period requires repeated injections, a potentially time-consuming and invasive procedure. BrdU doses ranging from 50 mg/kg up to 200 mg/kg are common for rodents.

3.3. Tissue Processing

1. *Anesthesia*. Animals are deeply anesthetized with an overdose of phenobarbital, and then monitored to determine whether they are fully anesthetized and ready for transcardial perfusion. Fully anesthetized animals should display the following properties: respiratory rate (breathing) should be regular and relaxed; withdrawal reflexes should be absent (this can be tested by pinching one of the paws); and there should be no response to external stimuli (e.g., blowing air on the eye).
2. *Transcardial perfusion*. Expose the heart using sharp dissecting tools. Insert the needle connected to the pump into the left ventricle. Make an incision in the right atrium to allow blood to flow out. Wash out blood with 0.9% saline at pH 7.4. When the draining blood becomes clear, perfuse the animal with 4% paraformaldehyde in 0.1 M PBS. The amount of 4% paraformaldehyde in 0.1 M PBS required for perfusion varies with the weight of the animal. Typically, a mouse weighing 30 g requires 30 mL solution.
3. *Dissection*. Using appropriate dissecting tools, remove the head, then remove the muscle and membranous tissue from the top part of the skull, and gently extract the brain from the skull. Cut the trigeminal and optic nerves, and let the brain soak into 4% paraformaldehyde in 0.1 M PBS.
4. *Postfixation*. Immerse the removed brain in the 4% paraformaldehyde in 0.1 M PBS for 24 h. Overfixation may result in a lack of staining due to the unavailability of antigens.
5. *Sectioning*. Use the cryostat to cut tissue into 10–40-μm slices.

3.4. DNA Denaturation

Incubate in HCl (2 N) for 30 min at room temperature (R.T.) to open the DNA structure of the cells, then wash with borate buffer for 10 min, and wash with PBS for 10 min for three times.

3.5. Immunohistochemistry

Day 1:

(a) Block tissues in 5% NGS and 1% BSA with 0.1% Triton in PBS for 2 h at room temperature.

(b) Drain slides for a few seconds (do *not* rinse) and wipe round sections.

(c) Apply primary antibody made up in blocking buffer.

(d) Incubate overnight at 4°C, preferably on an orbital shaker (*gently*!).

Day 2:

(a) Rinse 3× for 10 min with PBS.

(b) Apply secondary antibody made up in blocking buffer for 1 h at room temperature.

(c) Rinse 3× for 10 min with PBS.

(d) Apply DAPI made up in PBS for 2 min at room temperature (optional).

(e) Rinse 3× for 10 min with PBS.

(f) Clear with d2H_2O and mount with antifade reagent.

3.6. Image Analysis

Images of adult-born neurons in the hippocampal dentate gyrus (Fig. 1) were acquired using a Zeiss (Oberkochen, Germany) conventional inverted microscopy system Axiovert 200 M equipped with ApoTome, with which microscope can achieve the single plane image with 0.6-μm optical thickness. For example, to assess the colocalization of BrdU and mature neuron, the sections which double immunostained with antibodies against BrdU and mature neuron marker NeuN were used. First, the slid mounted with stained sections was face down put on the stage, and the DG of hippocampus was found under the blue channel with ×10 object (DAPI staining shows the whole structure of brain). Then, the channel was changed to green to locate the BrdU-positive cells in DG. Afterward, change object to oil immersion ×63 or higher, and find single BrdU-positive cells in focus. Finally, serials *z*-stacked images with three channels, green, red, and blue, which represented the BrdU, NeuN, and DAPI signals, respectively, were taken to cover the whole BrdU-positive cell. Cutview through *x*, *y*, and *z* three dimensions was generated to show the BrdU and NeuN colocalization.

3.7. General Guidelines

Carry out all incubations in a humidified chamber so that the sections do not dry out. At no stage during the protocol should drying of the sections be allowed to occur. Drying out leads to

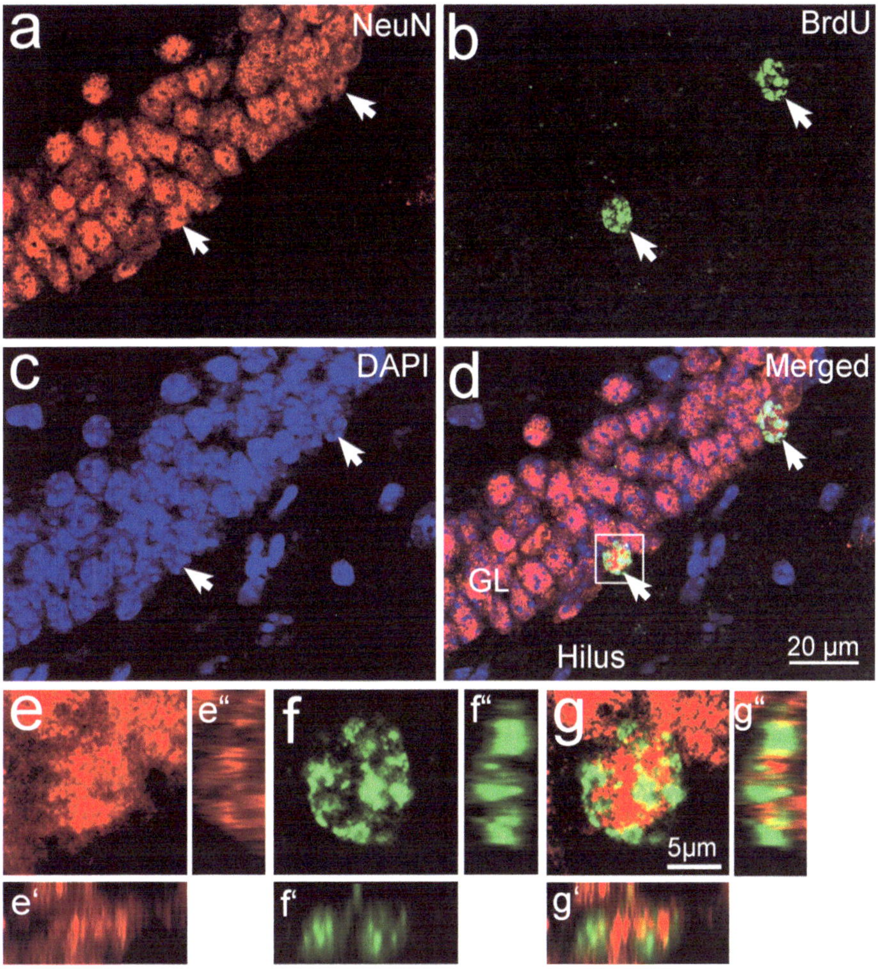

Fig. 1. Adult-born neurons in the hippocampal dentate gyrus. (**a**–**d**) Immunostaining to trace the fates of the BrdU labeled cells in the DG one month after TBI. (**a**) NeuN. (**b**) BrdU. (**c**) DAPI. (**d**) Merge of (**a**) to (**c**). (**e**–**g**). Confocal microscopy to verify the colocalization of NeuN with BrdU in the cell within the white box in panel (**d**). Confocal image to observe the NeuN signal in the cell at the x-axis (**e**), y-axis (**e'**), and z-axis (**e"**). Confocal image to observe the BrdU signal in the cell at the x-axis (**f**), y-axis (**f'**), and z-axis (**f"**). (**g**) Merged image of (**e**) and (**f**). (**g'**) Merged image of (**e'**) and (**f'**). (**g"**) Merged image of (**e"**) and (**f"**).

nonspecific binding and ultimately high background staining. A shallow, plastic, "sandwich type" box with a sealed lid and wet tissue paper in the bottom is adequate, just as long as the slides can lay flat so that the reagents do not drain off.

A better solution is to cut a plastic serological pipette into lengths to fit your incubation chamber. Glue them in pairs to the bottom of the chamber, with the two individual pipette tubes of each pair being placed about 4.0 cm apart. This provides a level and raised surface for the slides to rest on away from the wet tissue paper.

3.8. Troubleshooting

Problem	Possible reasons	Solution
BrdU: Poor positive staining and/or no positive staining with little or no background staining	Denaturation of DNA by HCl was not sufficient	Use longer incubation time. In addition, increase temperature with HCl incubation
BrdU/NeuN: Poor positive staining and/or no positive staining with little or no background staining	BrdU/NeuN primary and secondary antibody concentration was not optimal	Test BrdU/NeuN primary and secondary antibody when it arrives, as concentrations may vary across batches
	Incubation time with primary and/or secondary antibody was too short	Use longer incubation times for primary antibody (up to 48 h) and/or secondary antibodies
	The primary antibody does not recognize the antigen due to incorrect or overfixation	Note: For BrdU primary antibody, overfixation does not effect BrdU immunostaining
BrdU/NeuN: Nonspecific and/or high background staining	Nonspecific binding of primary or secondary antibody	Increase the number and time of washes in between steps
	Incubation time with BrdU/NeuN primary antibody was too long	Reduce BrdU/NeuN primary antibody incubation time (no more than 72 h)

References

1. Stiles J et al (2005) Cognitive development following early brain injury: evidence for neural adaptation. Trends Cogn Sci 9(3):136–143
2. Cicerone KD et al (2005) Evidence-based cognitive rehabilitation: updated review of the literature from 1998 through 2002. Arch Phys Med Rehabil 86(8):1681–1692
3. Prigatano GP (2005) Impaired self-awareness after moderately severe to severe traumatic brain injury. Acta Neurochir Suppl 93:39–42
4. Salmond CH, Sahakian BJ (2005) Cognitive outcome in traumatic brain injury survivors. Curr Opin Crit Care 11(2):111–116
5. McCarthy ML et al (2005) The Pediatric Quality of Life Inventory: an evaluation of its reliability and validity for children with traumatic brain injury. Arch Phys Med Rehabil 86(10):1901–1909
6. Coronado VG et al (2005) The CDC traumatic brain injury surveillance system: characteristics of persons aged 65 years and older hospitalized with a TBI. J Head Trauma Rehabil 20(3):215–228
7. Langlois JA, Rutland-Brown W, Thomas KE (2005) The incidence of traumatic brain injury among children in the United States: differences by race. J Head Trauma Rehabil 20(3): 229–238
8. Minino AM et al (2006) Deaths: injuries, 2002. Natl Vital Stat Rep 54(10):1–124
9. Hall ED et al (2005) Spatial and temporal characteristics of neurodegeneration after controlled cortical impact in mice: more than a focal brain injury. J Neurotrauma 22(2):252–265
10. Fujimoto ST et al (2004) Motor and cognitive function evaluation following experimental

traumatic brain injury. Neurosci Biobehav Rev 28(4):365–378
11. Kuhn HG, Dickinson-Anson H, Gage FH (1996) Neurogenesis in the dentate gyrus of the adult rat: age-related decrease of neuronal progenitor proliferation. J Neurosci 16(6):2027–2033
12. Kempermann G, Gage FH (2000) Neurogenesis in the adult hippocampus. Novartis Found Symp 231:220–235; discussion 235–241, 302–306
13. Rakic P (1972) Mode of cell migration to the superficial layers of fetal monkey neocortex. J Comp Neurol 145(1):61–83
14. Altman J (1969) Autoradiographic and histological studies of postnatal neurogenesis. IV. Cell proliferation and migration in the anterior forebrain, with special reference to persisting neurogenesis in the olfactory bulb. J Comp Neurol 137(4):433–457
15. Walsh C, Cepko CL (1992) Widespread dispersion of neuronal clones across functional regions of the cerebral cortex. Science 255(5043): 434–440
16. Altman J (1962) Are new neurons formed in the brains of adult mammals? Science 135: 1127–1128
17. Altman J, Das GD (1966) Autoradiographic and histological studies of postnatal neurogenesis. I. A longitudinal investigation of the kinetics, migration and transformation of cells incorporating tritiated thymidine in neonate rats, with special reference to postnatal neurogenesis in some brain regions. J Comp Neurol 126(3):337–389
18. Ming GL, Song H (2005) Adult neurogenesis in the mammalian central nervous system. Annu Rev Neurosci 28:223–250
19. Shapiro LA, Ribak CE (2005) Integration of newly born dentate granule cells into adult brains: hypotheses based on normal and epileptic rodents. Brain Res Brain Res Rev 48(1): 43–56
20. Zhao C et al (2006) Distinct morphological stages of dentate granule neuron maturation in the adult mouse hippocampus. J Neurosci 26(1):3–11
21. Cameron HA, McKay RD (2001) Adult neurogenesis produces a large pool of new granule cells in the dentate gyrus. J Comp Neurol 435(4):406–417
22. Eriksson PS et al (1998) Neurogenesis in the adult human hippocampus. Nat Med 4(11): 1313–1317
23. Gould E et al (1999) Neurogenesis in the neocortex of adult primates. Science 286(5439): 548–552
24. Kornack DR, Rakic P (1999) Continuation of neurogenesis in the hippocampus of the adult macaque monkey. Proc Natl Acad Sci U S A 96(10):5768–5773
25. Eriksson PS et al (1998) Neurogenesis in the adult human hippocampus. Nat Med 4(11): 1313–1317
26. Alvarez-Buylla A, Garcia-Verdugo JM, Tramontin AD (2001) A unified hypothesis on the lineage of neural stem cells. Nat Rev Neurosci 2(4):287–293
27. Doetsch F, Garcia-Verdugo JM, Alvarez-Buylla A (1997) Cellular composition and three-dimensional organization of the subventricular germinal zone in the adult mammalian brain. J Neurosci 17(13):5046–5061
28. Alvarez-Buylla A, Garcia-Verdugo JM (2002) Neurogenesis in adult subventricular zone. J Neurosci 22(3):629–634
29. Alvarez-Buylla A, Seri B, Doetsch F (2002) Identification of neural stem cells in the adult vertebrate brain. Brain Res Bull 57(6):751–758
30. Carleton A et al (2003) Becoming a new neuron in the adult olfactory bulb. Nat Neurosci 6(5):507–518
31. Doetsch F et al (1999) Subventricular zone astrocytes are neural stem cells in the adult mammalian brain. Cell 97(6):703–716
32. Lois C, Alvarez-Buylla A (1994) Long-distance neuronal migration in the adult mammalian brain. Science 264(5162):1145–1148
33. Carlen M et al (2002) Functional integration of adult-born neurons. Curr Biol 12(7):606–608
34. Markakis EA, Gage FH (1999) Adult-generated neurons in the dentate gyrus send axonal projections to field CA3 and are surrounded by synaptic vesicles. J Comp Neurol 406(4):449–460
35. van Praag H et al (2002) Functional neurogenesis in the adult hippocampus. Nature 415(6875):1030–1034
36. Anderson MJ, Waxman SG (1985) Neurogenesis in adult vertebrate spinal cord in situ and in vitro: a new model system. Ann N Y Acad Sci 457:213–233
37. Das GD, Altman J (1971) Postnatal neurogenesis in the cerebellum of the cat and tritiated thymidine autoradiography. Brain Res 30(2): 323–330
38. Goffinet AM et al (1986) Neurogenesis in reptilian cortical structures: 3H-thymidine autoradiographic analysis. J Comp Neurol 243(1):106–116
39. McConnell JA, Sechrist JW (1980) Identification of early neurons in the brainstem and

spinal cord: I. An autoradiographic study in the chick. J Comp Neurol 192(4):769–783
40. Polleux F, Dehay C, Kennedy H (1997) The timetable of laminar neurogenesis contributes to the specification of cortical areas in mouse isocortex. J Comp Neurol 385(1):95–116
41. Rakic P, Nowakowski RS (1981) The time of origin of neurons in the hippocampal region of the rhesus monkey. J Comp Neurol 196(1): 99–128
42. Rebiere A, Dainat J, Bisconte JC (1983) Autoradiographic study of neurogenesis in the duck olfactory bulb. Brain Res 282(2): 113–122
43. Richter W, Kranz D (1978) [Autoradiographic studies of the neurogenesis and morphogenesis of the cingulate region in the rat. I. Proliferation patterns in different pre- and postnatal stages]. Z Mikrosk Anat Forsch 92(2):222–240
44. Zimmerman RP, Polley EH, Fortney RL (1988) Cell birthdays and rate of differentiation of ganglion and horizontal cells of the developing cat's retina. J Comp Neurol 274(1):77–90
45. Cameron HA, Woolley CS, Gould E (1993) Adrenal steroid receptor immunoreactivity in cells born in the adult rat dentate gyrus. Brain Res 611(2):342–346
46. Bauer S, Patterson PH (2005) The cell cycle-apoptosis connection revisited in the adult brain. J Cell Biol 171(4):641–650
47. Burns KA, Kuan CY (2005) Low doses of bromo- and iododeoxyuridine produce near-saturation labeling of adult proliferative populations in the dentate gyrus. Eur J Neurosci 21(3):803–807
48. Vega CJ, Peterson DA (2005) Stem cell proliferative history in tissue revealed by temporal halogenated thymidine analog discrimination. Nat Methods 2(3):167–169
49. Thomas RM, Hotsenpiller G, Peterson DA (2007) Acute psychosocial stress reduces cell survival in adult hippocampal neurogenesis without altering proliferation. J Neurosci 27(11):2734–2743
50. Dupret D et al (2007) Spatial learning depends on both the addition and removal of new hippocampal neurons. PLoS Biol 5(8):e214
51. Leuner B, Glasper ER, Gould E (2009) Thymidine analog methods for studies of adult neurogenesis are not equally sensitive. J Comp Neurol 517(2):123–133
52. Yamaguchi M et al (2000) Visualization of neurogenesis in the central nervous system using nestin promoter-GFP transgenic mice. Neuroreport 11(9):1991–1996
53. Tashiro A, Zhao C, Gage FH (2006) Retrovirus-mediated single-cell gene knockout technique in adult newborn neurons in vivo. Nat Protoc 1(6):3049–3055
54. Laplagne DA et al (2006) Functional convergence of neurons generated in the developing and adult hippocampus. PLoS Biol 4(12):e409
55. Tashiro A et al (2006) NMDA-receptor-mediated, cell-specific integration of new neurons in adult dentate gyrus. Nature 442(7105): 929–933
56. van Praag H et al (2005) Exercise enhances learning and hippocampal neurogenesis in aged mice. J Neurosci 25(38):8680–8685
57. Curtis MA et al (2003) Increased cell proliferation and neurogenesis in the adult human Huntington's disease brain. Proc Natl Acad Sci U S A 100(15):9023–9027
58. Sanai N et al (2004) Unique astrocyte ribbon in adult human brain contains neural stem cells but lacks chain migration. Nature 427(6976): 740–744
59. Spalding KL et al (2005) Retrospective birth dating of cells in humans. Cell 122(1):133–143
60. Yang Y, Geldmacher DS, Herrup K (2001) DNA replication precedes neuronal cell death in Alzheimer's disease. J Neurosci 21(8): 2661–2668
61. Herrup K et al (2004) Divide and die: cell cycle events as triggers of nerve cell death. J Neurosci 24(42):9232–9239
62. Herrup K, Yang Y (2007) Cell cycle regulation in the postmitotic neuron: oxymoron or new biology? Nat Rev Neurosci 8(5):368–378
63. Wang L, Wang R, Herrup K (2007) E2F1 works as a cell cycle suppressor in mature neurons. J Neurosci 27(46):12555–12564
64. Yang Y, Herrup K (2005) Loss of neuronal cell cycle control in ataxia-telangiectasia: a unified disease mechanism. J Neurosci 25(10): 2522–2529
65. Yang Y, Herrup K (2007) Cell division in the CNS: protective response or lethal event in post-mitotic neurons? Biochim Biophys Acta 1772(4):457–466
66. Frisen J et al (1995) Rapid, widespread, and longlasting induction of nestin contributes to the generation of glial scar tissue after CNS injury. J Cell Biol 131(2):453–464
67. Komitova M, Eriksson PS (2004) Sox-2 is expressed by neural progenitors and astroglia in the adult rat brain. Neurosci Lett 369(1):24–27
68. Florenes VA et al (1994) Expression of the neuroectodermal intermediate filament nestin in human melanomas. Cancer Res 54(2):354–356

69. Kanemura Y et al (2001) Musashi1, an evolutionarily conserved neural RNA-binding protein, is a versatile marker of human glioma cells in determining their cellular origin, malignancy, and proliferative activity. Differentiation 68(2–3):141–152
70. Ignatova TN et al (2002) Human cortical glial tumors contain neural stem-like cells expressing astroglial and neuronal markers in vitro. Glia 39(3):193–206
71. Katsetos CD, Herman MM, Mork SJ (2003) Class III beta-tubulin in human development and cancer. Cell Motil Cytoskeleton 55(2):77–96
72. Charles P et al (2002) Re-expression of PSA-NCAM by demyelinated axons: an inhibitor of remyelination in multiple sclerosis? Brain 125(Pt 9):1972–1979
73. Bannigan J, Langman J (1979) The cellular effect of 5-bromodeoxyuridine on the mammalian embryo. J Embryol Exp Morphol 50:123–135
74. Saffhill R, Ockey CH (1985) Strand breaks arising from the repair of the 5-bromodeoxyuridine-substituted template and methyl methanesulphonate-induced lesions can explain the formation of sister chromatid exchanges. Chromosoma 92(3):218–224
75. Morris SM (1991) The genetic toxicology of 5-bromodeoxyuridine in mammalian cells. Mutat Res 258(2):161–188
76. Morris SM et al (1992) Effect of bromodeoxyuridine on the proliferation and growth of ethyl methanesulfonate-exposed P3 cells: relationship to the induction of sister-chromatid exchanges. Cell Biol Toxicol 8(1):75–87
77. Sekerkova G, Ilijic E, Mugnaini E (2004) Bromodeoxyuridine administered during neurogenesis of the projection neurons causes cerebellar defects in rat. J Comp Neurol 470(3):221–239
78. Kolb B et al (1999) Embryonic and postnatal injections of bromodeoxyuridine produce age-dependent morphological and behavioral abnormalities. J Neurosci 19(6):2337–2346
79. Yu WH (1976) The effect of 5-bromodeoxyuridine on the postnatal development of the rat cerebellum: a biochemical study. Brain Res 118(2):281–291
80. Yu WH (1979) Restitution of bromodeoxyuridine-induced changes in cerebellar development. Neuropathol Appl Neurobiol 5(1):41–48
81. Nagai A et al (1993) Administration of bromodeoxyuridine in early postnatal rats results in lung changes at maturity. Exp Lung Res 19(2):203–219
82. Cooper-Kuhn CM, Kuhn HG (2002) Is it all DNA repair? Methodological considerations for detecting neurogenesis in the adult brain. Brain Res Dev Brain Res 134(1–2):13–21
83. Packard DS Jr, Menzies RA, Skalko RG (1973) Incorporation of thymidine and its analogue, bromodeoxyuridine, into embryos and maternal tissues of the mouse. Differentiation 1(6):397–404
84. Hayes NL, Nowakowski RS (2000) Exploiting the dynamics of S-phase tracers in developing brain: interkinetic nuclear migration for cells entering versus leaving the S-phase. Dev Neurosci 22(1–2):44–55

Chapter 30

Electrophysiological Approaches in Traumatic Brain Injury

Thomas M. Reeves and Beverly S. Colley

Abstract

This chapter briefly reviews electrophysiological applications in experimental models of traumatic brain injury (TBI). The review was restricted to studies where rodents were used in one of the more common in vivo models of TBI and where electrophysiological recordings were then conducted either in vivo or in brain slices. This summary discussion is intended for researchers new to the topic of electrophysiology in TBI, who may be interested in knowing which specific recording techniques have been adapted in the various models. The chapter concludes with a protocol for recording compound action potentials in the corpus callosum of rats which have previously undergone fluid percussion TBI.

Key words: Electrophysiology, Traumatic brain injury, Axonal dysfunction, Corpus callosum

1. Introduction

In laboratory studies of traumatic brain injury (TBI), the use of electrophysiological outcome measures has lagged behind that of morphological and molecular approaches. This is likely due, in part, to perceived methodological or interpretive challenges, which may be compounded when recordings are conducted in the injured brain. Nevertheless, there is a prevalent recognition that bioelectric assessments address fundamental properties of neural tissue, when implemented alone or as correlative observations. In fact, recent trends do suggest an increasing number of laboratories incorporating electrophysiological recordings as core measurement strategies, with findings providing insights into TBI pathophysiology and generating hypotheses to guide further investigations.

The following brief review is especially directed toward researchers who may be considering applying an electrophysiological approach in the study of TBI but have questions regarding how to pursue

Jun Chen et al. (eds.), *Animal Models of Acute Neurological Injuries II*, Springer Protocols Handbooks,
DOI 10.1007/978-1-61779-782-8_30,

the topic within the practical framework of ongoing TBI research. It is hoped that such individuals will benefit from a summary of how various recording techniques have been employed in some major models of experimental TBI. The discussion is focused on experimental paradigms, where a rodent model of TBI is applied in vivo, and subsequent electrophysiological recordings are made either in vivo or through the use of ex vivo brain slices. This provides a practical limit to the scope of the chapter, and necessarily excludes models of neurotrauma applied to cell culture preparations, whether dissociated or organotypic (for a review of in vitro trauma models, see ref. (1)). The present information is largely organized in tabular form, showing representative studies, where selected bioelectric recording techniques have been used in the more prevalent experimental TBI models. The publications in the tables were selected only as representatives of particular recording and injury techniques, and by no means exhaust the numbers of excellent studies that have been conducted in this field. The review material is followed by a detailed protocol now used in the authors' laboratories, where properties of axonal conduction are assessed in brain slices following the induction of an in vivo fluid percussion TBI.

As a practical matter, it is likely that a new researcher wishing to incorporate electrophysiological methods will find it most expedient to use one of the more well-characterized rodent models, including lateral fluid percussion (LFP), central fluid percussion (CFP), controlled cortical impact (CCI), or weight drop (WD) models. This is not to discourage the use of alternative models, but rather to exploit the wide availability and accessibility of instrumentation and laboratories that currently implement these injury models. The impact and significance of electrophysiological results may be enhanced by their interpretation within an extensive knowledge base already accumulated for these models.

The researcher should consider how the neuropathology induced by a given model may dictate the optimal recording strategy. If the objective is to characterize a pathological process, a preliminary question will be whether this is predicted to occur in a focal or diffuse manner. Addressing this constrains the choice of model and suggests anatomical regions to observe. Similarly, a new research initiative should consider if a pathology would be primarily expressed in a specific subcellular compartment (soma, dendrites, axons, and terminals), as this should also inform the choice of a TBI model, as well as the electrophysiological probe applied to the tissue.

A number of published reviews are available that compare various experimental TBI models with regard to anatomical regions affected, fidelity to pathophysiology observed in the human clinic, and applications in studies of pharmacological interventions (or other neuroprotective strategies), and include assessments of the validity and reliability of the models (e.g., refs. (2–4)). Common themes emerge from these reviews, useful to new recording projects in TBI. First, a particular injury model occupies a position on several dimensions: type of injury (contact vs. acceleration); severity of injury (usually mild, moderate, or severe); and distribution of injury (focal vs. diffuse). An investigator interested in a diffuse pathological process might explore a CFP. Alternatively, if pathology is predicted at focal injury sites, then CCI may be better suited. Secondly, phase of injury (primary vs. secondary) should be considered, which dictates the timing of recordings. It is noteworthy that one of the advantages of electrophysiological approaches in TBI is the possibility to assess the functionality of surviving neurons, which may exhibit secondary changes, and also represent salvageable cell populations amenable to therapeutic interventions. Thirdly, new electrophysiological applications require decisions regarding anatomical regions of interest, and each of the models considered here permits some degree of freedom regarding the precise cranial position where force is applied.

1.1. In Vivo Recording Following TBI

Table 1 lists representative studies, where in vivo electrophysiological measurements were implemented after application of one of the major experimental injury models: LFP, CFP, CCI, and WD. Also listed for each study is the anatomical region recorded and the type of electrophysiological signal. Under "Other Measures and Outcomes" are listed the major correlative measures obtained with the electrophysiology, and range widely from histological assessments to behavioral observations to microdialysis sampling of extracellular constituents. Although not listed, many of the studies also used common histological staining, such as cresyl violet or hematoxylin and eosin, to enable stereological counting of cells to quantify cell death.

Referring again to Table 1, the combination of "region" and "signal" determines the configuration of recording electrodes employed. For example, EEG measurements of neocortical activity typically employ miniature skull screws as electrodes, with attached wires leading to a multiconductor socket cemented to the skull. After recovery from electrode implantation, repeated measurements can then be obtained in awake behaving rats by attaching recording cables to the cranial socket. This "chronic recording" arrangement is well suited for longitudinal projects which monitor, for example, the incidence of posttraumatic seizures. A contrasting set of constraints would apply, where "region" and "signal" were hippocampal CA1 and field excitatory postsynaptic potentials (fEPSPs), respectively. This latter experiment would necessitate a deeper penetration of a stimulating electrode, to activate CA1 afferents, along with the deep placement of a recording electrode to detect the fEPSPs, which are generated by populations of neurons and measured extracellularly. The recording electrodes are

Table 1
In vivo recording

	Electrophysiological measurement				
TBI model	**Region**	**Signal**	**Other measures, outcomes**	**Study**	**Notes**
LFP	Neocortex	Epidural ECoG	GFAP IHC, behavioral seizures	D'Ambrosio et al. (14)	Also assessed neocortical brain slices
LFP	Neocortex, CA3	Epidural ECoG and depth recording	GFAP IHC, behavioral seizures	D'Ambrosio et al. (15)	Rostrolateral injury placement
LFP	DG	fEPSP, pop spike	Silver impregnation, toluidine blue stain, somatostatin IHC	Lowenstein et al. (16)	
LFP	CA1, DG	fEPSP, pop spike	GABA IHC	Reeves et al. (17)	
LFP	Neocortex	ECoG, DC potential	ICP, CBF, K^+ microdialysis	Rogatsky et al. (18)	
LFP	Barrel fields	Sensory-evoked field potential	Open-field water maze	Sanders et al. (19)	
CFP	CA1	fEPSP, pop spike	LTP	Miyazaki et al. (6)	
CFP	CA1	fEPSP, pop spike	NMDA receptor IHC, GABA IHC, LTP	Reeves et al. (8)	Combined CFP and deafferenting lesions
CCI	Neocortex	EEG	Glutamate microdialysis	Sakowitz et al. (20)	
CCI	Neocortex	EEG	Laser Doppler flowmetry	Stover et al. (21)	
CCI	Striatum	Voltammetry	Rotation behavior (6-OHDA induced), WB (DAT, TH, VMAT, DRD2)	Wagner et al. (9)	
WD	Neocortex	EEG	Serum melatonin, electron microscopy	Ucar et al. (22)	Melatonin treatment

All experimental subjects are rats unless otherwise noted

CBF cerebral blood flow, *CCI* controlled cortical impact, *CFP* central fluid percussion, *DAT* dopamine transporter, *DG* dentate gyrus, *DRD2* dopamine D2 receptor, *ECoG* electrocorticogram, *fEPSP* field excitatory postsynaptic potential, *GFAP* glial fibrillary acidic protein, *ICP* intracranial pressure, *IHC* immunohisotchemistry, *LFP* lateral fluid percussion, *Pop spike* population spike, *TH* tyrosine hydroxylase, *VMAT* vesicular monoamine transporter, *WB* Western blot, *WD* weight drop

generally of low impedance (e.g., 2–12 MΩ) fabricated from either electrolyte-filled micropipettes or teflon-insulated tungsten. Such "acute recording" techniques are usually made in anesthetized animals stabilized in a stereotaxic frame.

It is useful to consider the contrasting biopathological features of the TBI models. Within the TBI literature, it is common to see an injury model selected because it is either a "focal" or "diffuse" model of TBI. While it is true that the more common TBI models—fluid percussion, cortical impact, and WD—differ with respect to their tendency to induce focal or diffuse tissue damage, it must be kept in mind that the degree to which these patterns of pathology are expressed is strongly contingent on the magnitude of the applied force. Each of the injury devices in these respective models can be adjusted to impart sufficiently high energy to brain tissue to induce frank contused areas, and, conversely, can be adjusted downward to elicit mild sublethal cellular reactions more detectable with functional measurements than with histopathological end points. For example, it has long been appreciated that a mild CFP injury can induce memory impairments in the absence of cell death (5). It is a central thesis of this chapter that electrophysiological assessments comprise underutilized functional probes which may be especially valuable when applied to mild/moderate injuries or in the context of diffuse brain damage. New researchers should also be alerted to the sometimes confusing terminology regarding "diffuse" brain injury, which often refers specifically to the evidence of diffuse axonal injury (DAI) detected with specific markers, such as amyloid precursor protein. However, as is argued in the protocol section of the present chapter, quantification of electrophysiological responses, evoked in axonal populations, provides a sensitive indicator of postinjury dysfunction which may be detected in widespread regions, in injury conditions where conventional histopathological markers are not strongly positive.

As indicated in Table 1, the LFP model has been used more frequently than CFP for in vivo recording following brain injury. Accounting for this preference may be that several changes, induced in the hemisphere ipsilateral to the injury site, present well-defined histopathological correlates to electrophysiological measures. LFP reliably elicits cell death in the ipsilateral hippocampus, and induces more unilateral cortical change. Recordings made in LFP models have emphasized how cortical and hippocampal electrophysiological changes correlate with the appearance of posttraumatic behavioral seizures, with markers of astrocytic activation, or with immunohistochemical changes that may reflect abnormalities in the normal balance of excitatory and inhibitory neurotransmission. While both LFP and CFP are associated with petechial hemorrhage, observations in our laboratory indicate hemorrhage to be more sparsely distributed in CFP, and, for equivalent magnitudes of applied pressure, cerebral histopathology is less focally expressed throughout the cerebrum in CFP than in LFP. For these reasons, earlier studies in our laboratories (6–8) explored CFP as a model of functional impairments to hippocampal neuroplasticity, reflected in the capacity for long-term potentiation (LTP), that may be related to learning and memory deficits which often persist after TBI. As suggested by Tables 1 and 2, electrophysiological assessments have not been prominent in CCI models of injury, although with the expanding use of this model this is anticipated to change. Because the CCI pattern of pathology does include pronounced cortical contusion and ventricular expansion, investigators planning electrophysiological assessments should carefully consider how electrode placement would be standardized relative to the focal injury site. Notably, a promising ongoing program of functional assessments uses voltametric measurements of monoamine release in the striatum after CCI injury (9), although the specialized instrumentation of this approach might present technical challenges to a researcher newly applying electrophysiology in the field of TBI.

1.2. Brain Slice Recording Following TBI

As suggested by the larger number of studies listed in Table 2, there is a strong trend toward in vitro assessments of electrophysiological dysfunction following TBI. Multiple factors underlie this trend, with the common elements being issues of practicality. The use of brain slices obviates the need for anesthesia during recording sessions, avoids respiration and muscle movement artifacts, facilitates visual placement of electrodes and the use of patch clamping, and integrates more readily with imaging and fluorescent methodologies. While these gains in control are achieved at the cost of increased artificiality, the consensus among researchers is that the compromise is justified. The most obvious caveat to the new electrophysiological researcher in TBI would be difficulties in separating the effects of trauma, associated with decapitation and slice preparation, from the specific neurotrauma effects induced by the experimental injury model. Ideally, functional differences between slices harvested from controls and slices from injured rats would reflect only the experimental injury. However, TBI may render neural tissue more susceptible to trauma incident to slice preparation, and, accordingly, observed electrophysiological deficits may reflect to some degree a serial injury effect. In spite of such challenges, posttraumatic slice recording has achieved a widespread acceptance as a valid functional assay added to the array of methodologies in TBI.

Preceding comments regarding the various TBI models, offered in the context of Table 1 in vivo recordings, apply equally well to the studies listed in Table 2, and are not repeated here. In terms of the region studied and the electrophysiological signal acquired, the applications in LFP models strongly reflect the interest in the hippocampal role in the pathophysiology of TBI. At least two reasons account for this. The hippocampus is preferentially vulnerable to TBI, and hippocampal neurons exhibit a robust

Table 2
Brain slice recording

TBI model	Electrophysiological measurement		Other measures, outcomes	Study	Notes
	Region	Signal			
LFP	DG	Perforated patch	KCC2 (WB), Cl fluorescence	Bonislawski et al. (23)	Mice
LFP	Cerebellum	Cerebellar CAPs	NF200 IHC, NF200 WB, alpha-spectrin WB	Park et al. (24)	
LFP	CA1	fEPSP, pop spike, PV		Reeves et al. (25)	
LFP	CA1	fEPSP	LTP	Sanders et al. (26)	
LFP	DG	Patch clamp, CA1 fEPSP	Timm's stain (mossy fiber)	Santhakumar et al. (27)	
LFP	CA1	PV, fEPSP	LTP	Sick et al. (28)	
LFP	DG	fEPSP	Fluoro-Jade	Tran et al. (29)	Mice
LFP	DG, CA1	fEPSP, patch (IPSCs)	Context fear behavior	Witgen et al. (10)	DHK, BMI treatment, mice
CFP	Corpus callosum	CAP	APP	Ai et al. (30)	Calpain inhibitor
CFP	Corpus callosum	CAP	APP	Baker et al. (31)	
CFP	Corpus callosum	CAP		Colley et al. (32)	Cyclosporin treatment
CFP	CA3	Patch clamp (glia), fEPSPs	GFAP IHC, [K]o (ion-selective electrode)	D'Ambrosio et al. (33)	
CFP	Corpus callosum	CAP		Reeves et al. (13)	
CFP	Corpus callosum	CAP	APP	Reeves et al. (34)	FK506 treatment
CFP	Corpus callosum	CAP	Glutathione levels, myeloperoxidase activity	Schultke et al. (35)	Quercetin treatment

(continued)

Table 2 (continued)

	Electrophysiological measurement				
TBI model	**Region**	**Signal**	**Other measures, outcomes**	**Study**	**Notes**
CCI	CA1	Pop spike	LTP	Albensi et al. (36)	Cyclosporin treatment
CCI	DG	fEPSP, pop spike	Timm's stain (mossy fiber)	Hunt et al. (11)	Mice
CCI	DG	Patch clamp	Focal glutamate photostimulation	Hunt et al. (12)	Mice
FP (applied to cerebellum)	Cerebellum	PV, CAP		Ai et al. (37)	
WD	CA3, DG	fEPSP, pop spike	Fluoro-Jade, Timm stain, optical recording, behavioral seizures	Golarai et al. (38)	Evaluated GABA antagonists

Experimental subjects are rats, unless otherwise noted

APP amyloid precursor protein, *BMI* bicuculline methiodide, *CAP* compound action potential, *CCI* controlled cortical impact, *CFP* central fluid percussion, *DG* dentate gyrus, *DHK* dihydrokainic acid, *fEPSP* field excitatory postsynaptic potential, *FP* fluid percussion, *IHC* immunohistochemistry, *KCC2* K-Cl cotransporter 2, *LFP* lateral fluid percussion, *LTP* long-term potentiation, *NF200* neurofilament 200 kDa subunit, *Pop spike* population spike, *PV* presynaptic volley, *WB* Western blot, *WD* weight drop

capacity for forms of neuroplasticity that may subserve learning and memory. The hippocampus has also been implicated in inhibitory synaptic dysfunction (10). Turning to slice recording following CFP, the frequent listings of corpus callosum recordings reflect the continued use of a functional assessment technique that was first applied to TBI by Dr. Andrew Baker and his colleagues at the University of Toronto in 2002. In this approach, compound action potentials (CAPs) are evoked in corpus callosum axons, and various parameters of the CAP waveforms are used to estimate the effects of injury on axonal conduction, and the efficacy of neuroprotective compounds. One of the advantages of CAP recording is that it is comparatively easy to implement, and a high-quality evoked signal can be obtained in the majority of slices recorded. For this reason, it was selected as the protocol described in the second part of this chapter, and is considered to be an accessible methodology for those interested in axonal injury to incorporate a functional outcome measure. The brain slice applications listed in Table 2, for the CCI models, also reflect the importance placed in the hippocampus not only for the structure's links to memory and cognition, but also for its possible role in the development of posttraumatic seizures. The recent studies of Hunt et al. (11, 12) are noteworthy, especially for the integrated recording and photostimulation techniques implemented in a mouse model of CCI.

2. Protocol for Assessing Postinjury Axonal Dysfunction

A protocol is described below for the recording of CAPs evoked in corpus callosum axons in brain slices from rats having previously undergone a fluid percussion injury or sham injury. Representative publications are listed in Table 2, from work in our laboratory and that of Dr. Andrew Baker. Compared to many alternative electrophysiological recording approaches, the acquisition and analysis of CAPs are relatively uncomplicated and straightforward. Our experience with CAP recording leads us to conclude that it is well suited for applications in TBI for the following reasons. Each evoked CAP is the summation of unitary action potentials generated in large numbers of axons. Accordingly, each CAP is already a "sample" of axons, which presumably reflect a spectrum of injury status from uninjured to mild to more severely affected. In this sense, there is a reduced potential for a selection bias, in comparison to more advanced intracellular or single-unit approaches, where the investigator must be vigilant that cells are not selected because they are more active, larger, easier to patch, or otherwise unequally accessible to the recording probe. Multiple studies have now documented that callosal CAPs are altered by the CFP injury model, as observed in the suppression of CAP amplitude, and injury-induced changes to other CAP parameters as duration and refractoriness. CAP recording provides a window on the functional state of myelinated axons separately from that of the unmyelinated axons, and in repeated experiments we observe dissimilar responses of these fiber populations to injury and to the administration of neuroprotective compounds. We are confident that an increased understanding of how these fiber subtypes differ, in their postinjury alterations, will translate to new strategies to treat axonal injury.

2.1. Materials

2.1.1. Equipment

1. Pipette puller: Stoelting Model APP-1 (Wood Dale, IL, USA).
2. Micromanipulator: WPI Model M3301 (Sarasota, FL).
3. Electrometer: Axoclamp 2B (Molecular Devices Inc., Sunnyvale, CA, USA).
4. Analog-to-digital converter: Digidata 1322A and pClamp v9 software (Molecular Devices Inc.).
5. Stimulator: MCS 4002 (Multichannel Systems, Reutlingen, Germany).

2.1.2. Electrode Fabrication Supplies

1. Glass capillaries: Kwik-FilTM Borosilicate, 100 mm, 1.5 mm OD, w/filament (WPI, Sarasota, FL, USA).
2. Stimulating electrodes: Teflon-insulated tungsten, 0.010″ diam., 12 MΩ, 8°, A-M Systems #576000 (Sequim, WA, USA).

2.2. Procedures

2.2.1. Animals and Central Fluid Percussion

Adult male rats (350–400) are subjected to moderate CFP (2.0 ± 0.1 atm) and allowed a variable survival interval. To date, we have used 3 h up to 15 days. With a 4.0-mm craniotomy centered between bregma and lambda (Fig. 1), callosal recordings may be conducted directly beneath the impact (e.g., splenium or in the middle portion of the callosum) and compared to recordings outside the impact axis (genu).

2.2.2. Electrode Fabrication

Bipolar stimulating electrodes are prepared in advance of recording sessions. For callosal CAP recording, we glue two tungsten electrodes together to ensure an intertip distance of 0.25–0.30 mm, which approximately spans the dorsoventral width of the rat callosum (Fig. 2, also see Note 1). The teflon insulation is abraded only from the extreme tips by gently drawing the electrode across a sheet of fine wet/dry sandpaper (1,000 grit) and checking the resistance after each pass, until a low resistance (<0.2 MΩ) is achieved. Recording micropipettes are pulled from the 1.5-mm capillary tubes, backfilled with artificial cerebrospinal fluid (ACSF), and the tip broken back to achieve an impedance of 2–8 MΩ.

2.2.3. Artificial Cerebrospinal Fluid Preparation

The ACSF contains (in mM) NaCl 124, KCl 5, NaH_2PO_4 1.25, $NaHCO_3$ 26, $MgSO_4$ 1.3, $CaCl_2$ 2, and glucose 10, pH 7.4, saturated with a 95% O_2/5% CO_2 gas mixture. Our laboratory maintains

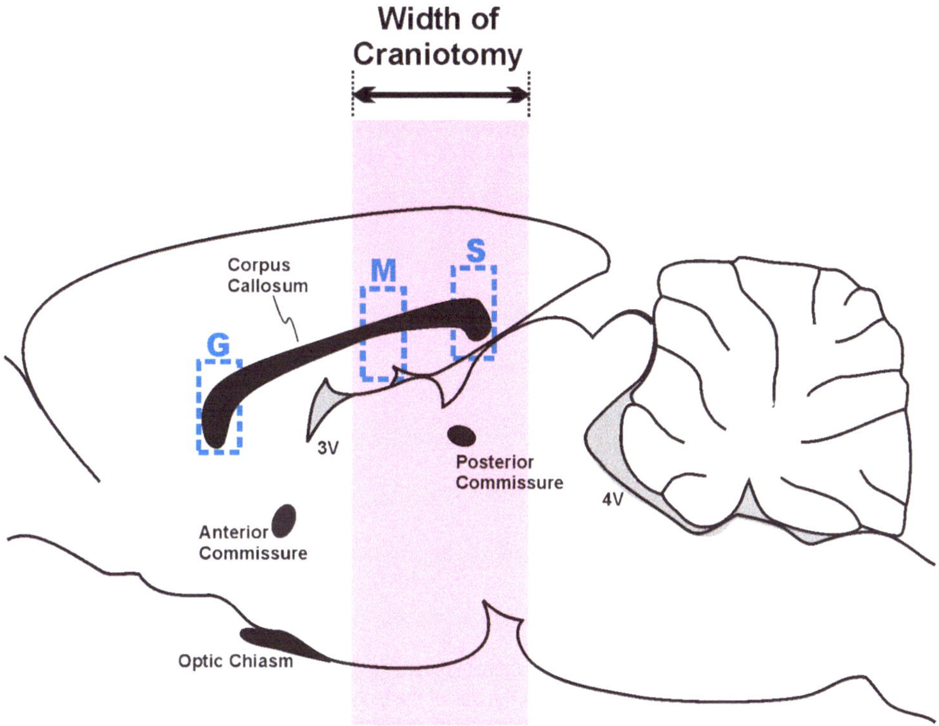

Fig. 1. Midsagittal view showing compound action potential recording sites in the genu (G), midportion (M), and splenium (S) of the corpus callosum. The location of the overlying craniotomy used for the CFP injury is also shown. Axon diameter and the proportions of myelinated and unmyelinated axons differ along the rostrocaudal axis of the corpus callosum. Ongoing studies in our laboratory are comparing how these subregions of the hippocampus may exhibit different injury responses.

stock solutions (up to 2 weeks at 4°C) at the following concentrations (in mM): NaCl 2,000, KCl 500, NaH_2PO_4 125, $NaHCO_3$ 650, $MgSO_4$ 200, $CaCl_2$ 200. Glucose aliquots (1,000 mM) are kept frozen. ACSF is made fresh daily, avoiding calcium precipitation by adding $CaCl_2$ lastly after the solution is thoroughly saturated with the 95/5 gas mix. A total volume of 2 L is usually prepared for a recording session, with 500 ml chilled to 2–4°C for use during slice preparation. All ACSF uses mentioned below are continuously oxygenated using fine-dispersal air stones.

2.2.4. Slice Preparation

A rat is anesthetized in a sealed chamber containing 4% isoflurane in a carrier gas of 70% N_20 and 30% O_2 for 4 min. The rat is then decapitated, the skull removed with rongeurs, and the brain gently removed using a spatula. This step is critical with the objectives being speed and gentleness and, with practice, can be accomplished in approximately 1 min. Brain removal is done under a drip of ice-cold ACSF: at the earliest penetration of the cranium, the cold solution is allowed to flow over the brain, and continue during the entire extraction. The removed brain is placed in a bath of ice-cold ACSF for 1 min. The brain is then blocked, caudally at the junction

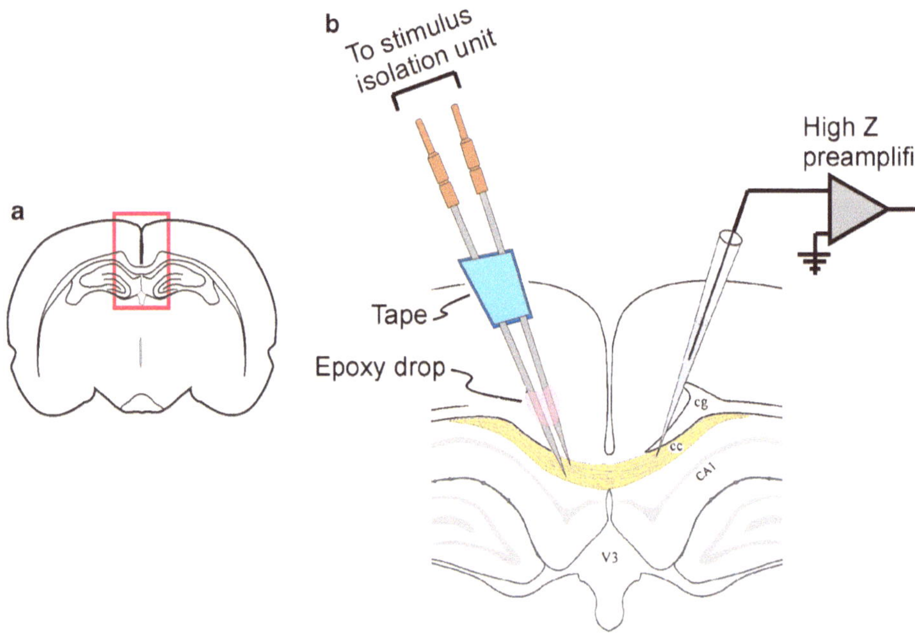

Fig. 2. Placement of electrodes for compound action potential recording. Midline crossing region of the corpus callosum outlined in red (**a**) is enlarged in (**b**). Distance between the tips of the bipolar stimulating electrode is fixed and stabilized with a drop of epoxy glue and tape (see Note 1). Artificial cerebrospinal fluid-filled recording micropipette is placed at a distance of 1.0–1.5 mm from stimulating electrode.

of the cerebrum and cerebellum and rostrally by removing frontal poles anterior to the genu of the callosum. All blocking is performed on an ice-cold platform, with chilled blades and tools. The blocked cerebrum is glued (cyanoacrylate) to a vibratome stage, and 450-μm-thick coronal slices are cut in ice-cold ACSF, in an antero-posterior sequence, saving all sections with midline-crossing callosal tissue present (Note 2). Slices are allowed to equilibrate in room-temperature ACSF for 1 h prior to transfer to a submersion-type recording chamber (Note 3).

2.2.5. Electrode Placement and CAP Acquisition

While we have explored recording in various bath temperatures (see Fig. 3), most of our CAP recordings are conducted at room temperature (22–23°C), with slices perfused with ACSF at a rate of 2–3 ml/min. In comparison to most electrophysiological approaches, technical requirements for acquisition of evoked CAPs are minimal. Micromanipulators to secure and position electrodes can be hand operated. For CAPs, we use the WPI model M3301, secured to a metal table with a magnetic base. Cables from an isolated constant current stimulator are attached to the stimulating electrode, and the recording pipette is inserted into the head stage of the electrometer. Under a dissecting microscope, the bipolar stimulating electrode is positioned over the corpus callosum at approximately 0.5 mm from the midline (Fig. 2). The recording micropipette is positioned in the contralateral hemisphere, at a distance

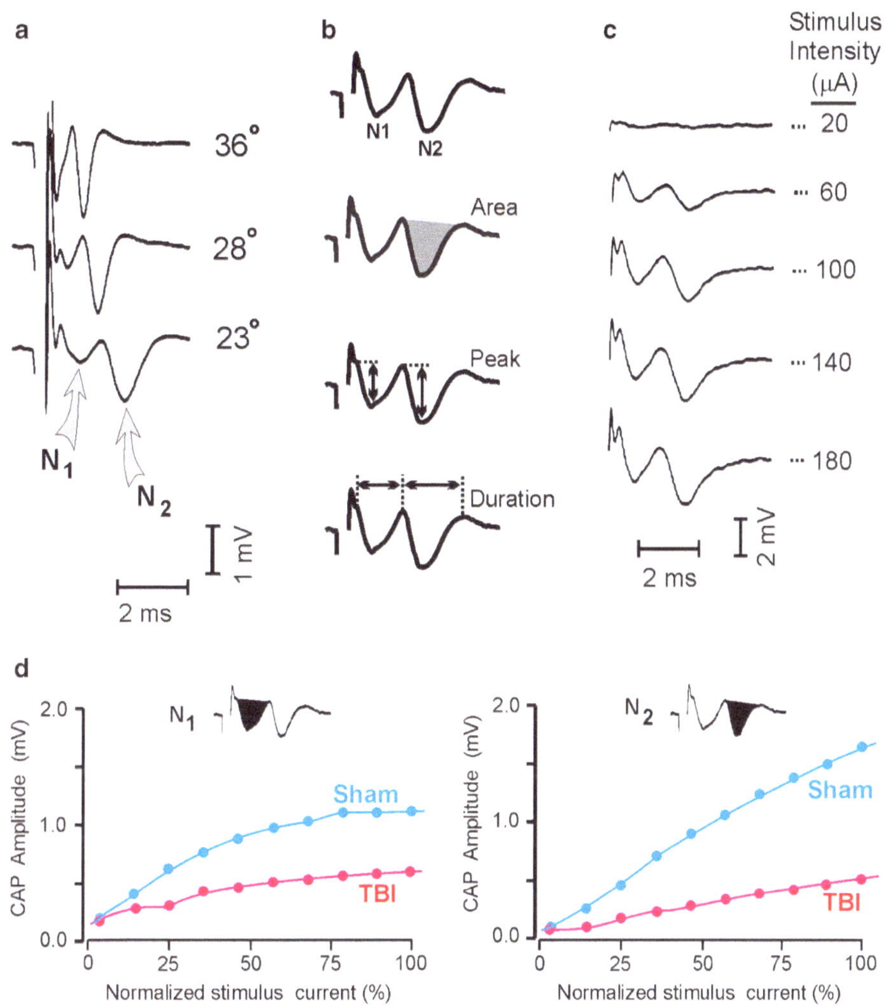

Fig. 3. Evoked callosal compound action potential (CAP) signals. (**a**) Reducing temperature of the perfusing artificial cerebrospinal fluid slows axonal conduction and allows the early N1 signal (generated by myelinated axons) to separate from the stimulus artifact. This facilitates a more precise quantification of N1. (**b**) CAP measurement parameters. Traumatic brain injury (TBI) effects on amplitude parameters (area, peak) are not always the same as injury effects to duration. (**c**) Representative CAPs obtained in input–output (IO) testing using graded stimulus current. (**d**) Example group data (means adapted from ref. (13)) to show the use of IO curves in evaluating the effects of TBI.

of 1.0—1.5 mm from the stimulating electrode. Small petechial hemorrhages are sometimes seen in callosum after CFP, and we avoid placing electrodes in or near such zones. Both stimulating and recording probes are lowered to a depth of approximately 200 μm below the surface of the slice. The preparation is stimulated using monophasic square wave pulses (200 μs duration, with initial amplitude of 250 microamp, applying one stimulus pulse each 10 s). A high-quality CAP is a biphasic negative-going signal with the duration of the total waveform between 6 and 8 ms (see typical examples in Fig. 3). The early negative peak (N1) is generated by fast-conducting fibers, primarily myelinated axons, while the following

N2 peak reflects the activity of slower, predominantly unmyelinated, axons. Fine adjustments may be made in the depths of the stimulating and recording probes to achieve the maximal amplitude of the CAP, but these movements must be minimal (approx. 10 μm per move, and allowing several minutes to elapse between each move) (Note 4).

2.2.6. Digital Processing of CAPs

While a number of manufacturers offer analog-to-digital converters that could readily process CAP signals, our laboratory uses the Digidata 1322A and pClamp v9 software listed above. We amplify (×500), filter (DC—10 kHz), and digitize the CAP waveform at 25 kHz for off-line analysis. CAPs should be examined using a systematic range of stimulation intensities, from low-current pulses near threshold to asymptotic maximum responses. This graded stimulus–response series, referred to as an input–output (IO) function (or "IO curve"), demonstrates the recruitment of individual fiber action potentials into the aggregate CAP waveform. Theoretically, TBI pathophysiology may be revealed at different points in the IO curves, for example if injury differentially alters fibers which are

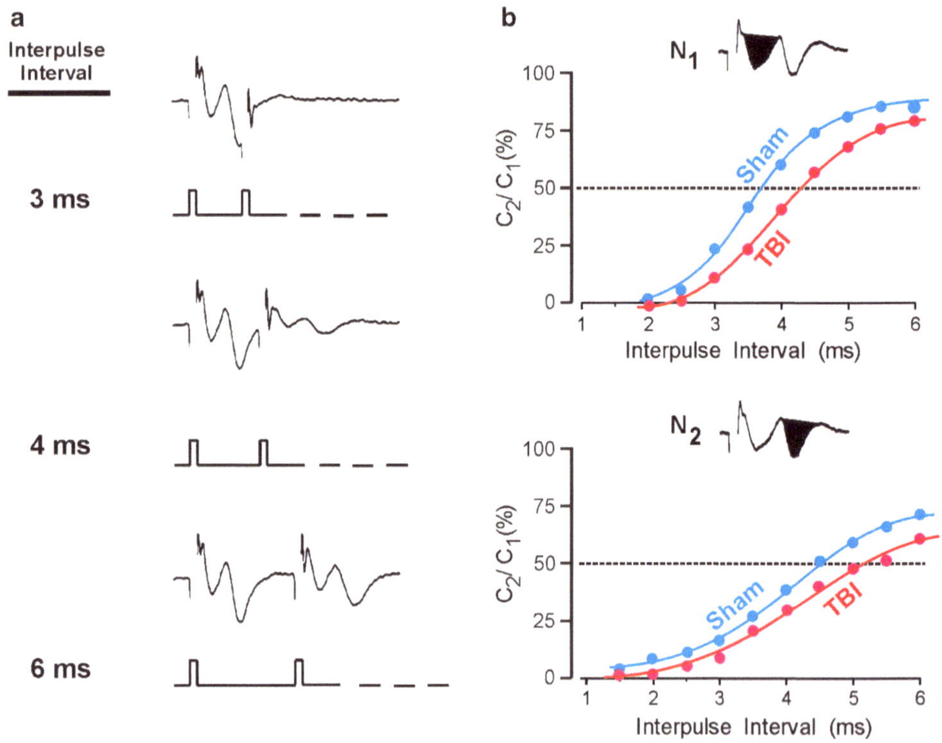

Fig. 4. Compound action potentials (CAPs) evoked with paired stimulus pulses. (**a**) Identical stimulus pulses produce different CAP responses, depending on interpulse interval. As interpulse interval is increased, amplitude of the second CAP also increases, reflecting "refractoriness" of axons. (**b**) Example group data showing that traumatic brain injury (TBI) shifts the curve relating the ratio (CAP 2/CAP 1) to interpulse interval (means adapted from ref. (13)). TBI effects on CAP timing functions, such as refractoriness, are distinct from injury effects on CAP amplitude.

activated at low current levels. We standardize the generation of IO curves by finding the current level evoking a maximum N1 CAP component (100% stimulation) and gathering CAPs at 10, 20, 30, ..., 100% of this level for both N1 and N2 (see examples of IO curves in Fig. 3d). In one respect, this represents a compromise approach because this stimulus range remains below the asymptotic maximum of N2, but does avoid excessive stimulation at supramaximal levels for N1. Injury and drug effects can be assessed statistically by group differences in these IO curves. In addition to accumulating IO curves, we routinely examine responses to pairs of stimulus pulses, with brief (1.5–6 ms) interpulse intervals. Here, it is observed that the amplitude of the second CAP increases as the interpulse interval increases, reflecting the refractoriness of the activated axonal population (Fig. 4). The effects of TBI, and the potential efficacy of neuroprotective treatments, may be evaluated by comparing CAP parameters—amplitude, slope, duration, and area—from different analytic groups: injured versus sham and treated versus untreated. Our laboratory is especially interested in differential susceptibility of the N1 and N2 CAP components to TBI, and to pharmacological treatment, because this may reflect a differential role of myelinated and unmyelinated axons in injury and point to new treatment strategies.

3. Notes

1. Preparation of the bipolar stimulating electrodes entails gluing two teflon-coated wires (diameter approximately 0.010) in such a way that the tips are separated by a specific distance. After experience with several approaches, we now secure one of the wires to a platform with a small amount of modeling clay. The second wire is placed in a micromanipulator and, under a dissecting microscope, is moved into the desired position next to the fixed wire. A drop of 5-min epoxy is placed near the tip. When the glue dries, we place a piece of tape on the shafts of the electrodes (see Fig. 2) to help stabilize the electrode. The stimulating electrodes can be reused for multiple slices by checking the impedance routinely and retouching the tips to the fine sandpaper if needed.
2. When securing the blocked brain to the vibratome cutting platform, we install a "backstop" made of a slab of agar gel (2%) approximately 5 mm thick. Thus, when the blade has passed through the brain tissue, it enters the agar gel without damaging the blade and helps stabilize the brain tissue. When cutting at 450 μm, we usually obtain five or six slices with midline-crossing callosal axons in adult rats. Of course, fewer slices of this thickness would be obtained from mice. The slices

span the rostrocaudal extent of the corpus callosum, and our lab is making systematic electrophysiological and ultrastructural comparisons of these callosal subregions.

3. We strongly recommend that slices are not transferred using a brush. It is more gentle to the slices to aspirate them into a large blunt "eye dropper" made from Pasteur pipettes. We use the larger disposable pipettes (outer diameter approx. 9 mm) (Fisher catalog # 13-678-206). The tip is broken back, and fire polished. This allows a rubber bulb (Fisher #03-448-22) to be placed on the narrow end of the pipette and to use the wide end for slice transfer.
4. We recommend only minimal repositioning of the electrode. Extensive experience has shown that there is a slow "spatial rate of change" of the CAP signal, meaning that the CAP waveform, measured with electrode tips near the middle of the slice, is generally the same throughout most of the slice thickness. Therefore, there is usually little to be gained by repeated adjustments of electrode depth, and excessive movement can disturb the tissue or even cut axons.

Acknowledgments

Supported by NIH research grants NS-057758 (TMR), and by a Virginia Commonwealth Neurotrauma Initiative Award 07-302F (TMR).

References

1. Cohen AS, Pfister BJ, Schwarzbach E, Grady MS, Goforth PB, Satin LS (2007) Injury-induced alterations in CNS electrophysiology. Prog Brain Res 161:143–169
2. Gennarelli TA (1994) Animate models of human head injury. J Neurotrauma 11:357–368
3. Povlishock JT, Hayes RL, Michel ME, McIntosh TK (1994) Workshop on animal models of traumatic brain injury. J Neurotrauma 11: 723–732
4. Morales DM, Marklund N, Lebold D, Thompson HJ, Pitakanen A, Maxwell WL, Longhi L, Laurer H, Maegele M, Neugebauer E, Graham DI, Stocchetti N, McIntosh TK (2005) Experimental models of traumatic brain injury: do we really need to build a better mouse trap? Neuroscience 136:971–989
5. Lyeth BG, Jenkins LW, Hamm RJ, Dixon CE, Phillips LL, Clifton GL, Young HF, Hayes RL (1990) Prolonged memory impairment in the absence of hippocampal cell death following traumatic brain injury in the rat. Brain Res 526:249–258
6. Miyazaki S, Katayama Y, Lyeth BG, Jenkins LW, DeWitt DS, Goldberg SJ, Newton PG, Hayes RL (1992) Enduring suppression of hippocampal long-term potentiation following traumatic brain injury in rat. Brain Res 585: 335–339
7. Reeves TM, Lyeth BG, Povlishock JT (1995) Long-term potentiation deficits and excitability changes following traumatic brain injury. Exp Brain Res 106:248–256
8. Reeves TM, Zhu J, Povlishock JT, Phillips LL (1997) The effect of combined fluid percussion and entorhinal cortical lesions on long-term potentiation. Neuroscience 77:431–444
9. Wagner AK, Sokoloski JE, Ren D, Chen X, Khan AS, Zafonte RD, Michael AC, Dixon CE (2005) Controlled cortical impact injury affects

dopaminergic transmission in the rat striatum. J Neurochem 95:457–465
10. Witgen BM, Lifshitz J, Smith ML, Schwarzbach E, Liang S-L, Grady MS, Cohen AS (2005) Regional hippocampal alteration associated with cognitive deficit following experimental brain injury: a systems, network and cellular evaluation. Neuroscience 133:1–15
11. Hunt RF, Scheff SW, Smith BN (2009) Posttraumatic epilepsy after controlled cortical impact injury in mice. Exp Neurol 215:243–252
12. Hunt RF, Scheff SW, Smith BN (2010) Regionally localized recurrent excitation in the dentate gyrus of a cortical contusion model of posttraumatic epilepsy. J Neurophysiol 103: 1490–1500
13. Reeves TM, Phillips LL, Povlishock JT (2005) Myelinated and unmyelinated axons of the corpus callosum differ in vulnerability and functional recovery following traumatic brain injury. Exp Neurol 196:126–139
14. D'Ambrosio R, Fairbanks JP, Fender JS, Born DE, Doyle D, Miller JW (2004) Posttraumatic epilepsy following fluid percussion injury in the rat. Brain 127:304–314
15. D'Ambrosio R, Fender JS, Fairbanks JP, Simon EA, Born DE, Doyle DL, Miller JW (2005) Progression from frontal–parietal to mesial–temporal epilepsy after fluid percussion injury in the rat. Brain 128:174–188
16. Lowenstein DH, Thomas MJ, Smith DH, McIntosh TK (1992) Selective vulnerability of dentate hilar neurons following traumatic brain injury: a potential mechanistic link between head trauma and disorders of the hippocampus. J Neurosci 12:4846–4853
17. Reeves TM, Lyeth BG, Hamm RJ, Phillips LL, Povlishock JT (1997) The effects of traumatic brain injury on inhibition in the hippocampus and dentate gyrus. Brain Res 757:119–132
18. Rogatsky GG, Sonn J, Kamenir Y, Zarchin N, Mayevsky A (2003) Relationship between intracranial pressure and cortical spreading depression following fluid percussion brain injury in rats. J Neurotrauma 20:1315–1325
19. Sanders MJ, Dietrich WD, Green EJ (2001) Behavioral, electrophysiological, and histopathological consequences of mild fluid-percussion injury in the rat. Brain Res 904:141–144
20. Sakowitz OW, Unterberg AW, Stover JF (2002) Neuronal activity determined by quantitative EEG and cortical microdialysis is increased following controlled cortical impact injury in rats. Acta Neurochir Suppl 81:221–223
21. Stover JF, Sakowitz OW, Unterberg AW (2004) Neuronal activity and cortical perfusion determined by quantitative EEG analysis and laser Doppler flowmetry are uncoupled in brain injured rats. Acta Neurochir Suppl 89:81–85
22. Ucar T, Ozkaya G, Demir N, Gurer I, Akyuz M, Onal MZ (2005) The effects of environmental light–dark changes on experimental mild traumatic brain injury. Acta Neurol Scand 112: 163–172
23. Bonislawski DP, Schwarzbach EP, Cohen AS (2007) Brain injury impairs dentate gyrus inhibitory efficacy. Neurobiol Dis 25:163–169
24. Park E, Liu E, Shek M, Park A, Baker AJ (2007) Heavy neurofilament accumulation and alpha-spectrin degradation accompany cerebellar white matter functional deficits following forebrain fluid percussion injury. Exp Neurol 204:49–57
25. Reeves TM, Kao CQ, Phillips LL, Bullock MR, Povlishock JT (2000) Presynaptic excitability changes following traumatic brain injury in the rat. J Neurosci Res 60:370–379
26. Sanders MJ, Sick TJ, Perez-Pinzon MA, Dietrich WD, Green EJ (2000) Chronic failure in the maintenance of long-term potentiation following fluid percussion in the rat. Brain Res 861:69–76
27. Santhakumar V, Ratzliff ADH, Jeng J, Toth Z, Soltesz I (2001) Long-term hyperexcitability in the hippocampus after experimental head trauma. Ann Neurol 50:708–717
28. Sick TJ, Pqrez-Pinzn MA, Feng ZZ (1998) Impaired expression of long-term potentiation in hippocampal slices 4 and 48 h following mild fluid-percussion brain injury in vivo. Brain Res 785:287–292
29. Tran LD, Lifshitz J, Witgen BM, Schwarzbach E, Cohen AS, Grady MS (2006) Response of the contralateral hippocampus to lateral fluid percussion brain injury. J Neurotrauma 23: 1330–1342
30. Ai J, Liu E, Wang J, Chen Y, Yu J, Baker AJ (2007) Calpain inhibitor MDL-28170 reduces the functional and structural deterioration of corpus callosum following fluid percussion injury. J Neurotrauma 24:960–978
31. Baker AJ, Phan N, Moulton RJ, Fehlings MG, Yucel Y, Zhao M, Liu E, Tian GF (2002) Attenuation of the electrophysiological function of the corpus callosum after fluid percussion injury in the rat. J Neurotrauma 19:587–599
32. Colley BS, Phillips LL, Reeves TM (2010) The effects of cyclosporin-A on axonal conduction deficits following traumatic brain injury in adult rats. Exp Neurol 224:241–251
33. D'Ambrosio R, Maris DO, Grady MS, Winn HR, Janigro D (1999) Impaired K^+ homeostasis and altered electrophysiological properties of post-traumatic hippocampal glia. J Neurosci 19:8152–8162

34. Reeves TM, Phillips LL, Lee NN, Povlishock JT (2007) Preferential neuroprotective effect of tacrolimus (FK506) on unmyelinated axons following traumatic brain injury. Brain Res 1154:225–236
35. Schultke E, Kamencic H, Zhao M, Tian G-F, Baker AJ, Griebel RW, Juurlink BHJ (2005) Neuroprotection following fluid percussion brain trauma: a pilot study using quercetin. J Neurotrauma 22:1475–1484
36. Albensi BC, Sullivan PG, Thompson MB, Scheff SW, Mattson MP (2000) Cyclosporin ameliorates traumatic brain-injury-induced alterations of hippocampal synaptic plasticity. Exp Neurol 162:385–389
37. Ai J, Baker AJ (2002) Presynaptic hyperexcitability at cerebellar synapses in traumatic injury rat. Neurosci Lett 332:155–158
38. Golarai G, Greenwood AC, Feeney DM, Connor JA (2001) Physiological and structural evidence for hippocampal involvement in persistent seizure susceptibility after traumatic brain injury. J Neurosci 21: 8523–8537

Chapter 31

Biochemical and Molecular Biological Assessments of Traumatic Brain Injury

Coleen M. Atkins and Anthony Andrew Oliva Jr.

Abstract

The mechanical forces imparted on brain tissue during trauma trigger changes in proteins, membranes, DNA, and mRNA expression patterns of many different cell types, such as neurons, astrocytes, microglia, and oligodendrocytes. Biochemical and molecular biological assessments have thus become powerful, important tools for yielding significant insights into the pathomechanisms of traumatic brain injury (TBI). Using Western blotting and enzyme-linked immunosorbent assays (ELISAs), assessment of changes in proteins in large areas of tissue can identify inflammatory signaling, dysfunction in synaptic signaling, and endogenous reparative strategies. Using quantitative RT-PCR, changes in DNA and mRNA expression patterns can be identified, providing additional understanding into how areas of the brain are responding after TBI. In situ hybridization and immunohistochemistry can further refine our understanding by identifying the cell types and brain regions affected. Through such biochemical and molecular biological assessments, we can understand the mechanisms that create pathology in the brain, and potentially develop therapeutic strategies to target these molecules and improve outcome after TBI. In this chapter, we outline the steps required to analyze brain-injured tissue at the protein level by ELISA, Western blotting, and immunohistochemistry.

Key words: ELISA, Fluid-percussion brain injury, Immunohistochemistry, In situ hybridization, RT-PCR, Western blot

1. Introduction

Traumatic brain injury (TBI) elicits many injury cascades, ranging from inflammation, apoptosis, and synaptic dysfunction to progressive atrophy. All of these pathological changes in the brain are mediated by changes in the underlying proteins, lipids, DNA, and mRNA. Biochemical and molecular biological techniques are important sets of tools in the repertoire of a TBI researcher to understand the pathomechanisms of TBI and are the key to

Jun Chen et al. (eds.), *Animal Models of Acute Neurological Injuries II*, Springer Protocols Handbooks,
DOI 10.1007/978-1-61779-782-8_31, © Springer Science+Business Media, LLC 2012

developing targeted therapies. In this chapter, we focus on protein assessment techniques, since nearly all cellular functions are mediated by proteins, and a wide range of therapeutic strategies can target protein expression and functioning.

Protein changes can be studied at three major levels: enzyme-linked immunosorbent assays (ELISAs), Western blotting, and immunohistochemistry. Each of these levels has advantages and disadvantages. ELISAs provide a rapid screening method for detecting changes in one specific protein in an area of brain-injured tissue. An ELISA is very rapid to perform, yielding results within less than a day and providing a high degree of sensitivity for detecting the predicted protein changes. However, ELISAs require an antibody that is highly specific for the protein of interest, and this is often not available. Western blotting does not require the high antibody specificity that an ELISA requires, broadening the number of proteins that can be analyzed. In addition, posttranslational modifications, such as phosphorylation, acetylation, ubiquitination, amidation, glycosylation, and proteolysis, are among the many changes that occur in proteins during TBI and there are a number of commercialized tools to detect these changes using Western blotting. However, like an ELISA, the Western blot detects changes in a large area of tissue, leaving unanswered questions, such as in which cell types and in which subregions of the brain did these protein changes occur. Immunohistochemistry does provide answers to which cell-type and subregions of the brain these protein changes occur. However, immunohistochemistry has the same limitation as an ELISA in that the antibody must be highly specific for the protein of interest. Furthermore, immunohistochemistry is not easily quantifiable, unlike an ELISA and Western blot. Thus, the techniques of ELISA, Western blot, and immunohistochemistry are best used in combination, and can yield important, critical insights into the biochemical mechanisms of TBI.

With the advent of commercialization of many biochemistry techniques, new resources have opened for researchers. Commercialized antibodies, ELISA kits, and precast polyacrylamide gels are just a few of the resources now available. This has made it easier for a novice in these techniques to perform them with reliability and speed. In addition, this has provided greater consistency in results across laboratories. Another area of commercialization that is increasingly providing new insights into the TBI field is the outsourcing of techniques. For example, protein arrays, which detect changes in total protein levels and some posttranslational modifications, are now easily outsourced to companies and the TBI researcher simply has to prepare the brain-injured tissue and in return, receives a vast amount of information about the protein changes occurring after TBI. Protein biochemistry techniques are in continual evolution and new protein techniques already in use in other fields have yet to even be explored in the TBI field.

In this chapter, we focus on three major techniques that have already been successfully used to understand the pathomechanisms of TBI at the protein level: ELISAs, Western blots, and immuno-histochemistry.

2. Materials

2.1. Animals

Adult male Sprague-Dawley rats (Charles River Laboratories, Wilmington, MA) weighing 280–320 g are typically used in our experiments. Age and duration of anesthesia are important factors in the expression levels of many proteins (1, 2). To ensure that the biochemical changes observed are due to the trauma and not a confounding variable, age-matched sham animals that receive the same duration of anesthesia as the TBI animals need to be included for all biochemical assessments.

2.2. Anesthesia

Rats are induced with isoflurane (3%) and a mixture of N_2O (70%) and O_2 (30%) for 5 min in an induction chamber.

2.3. Dissection Equipment (Fig. 1)

Forceps, Moloney, 4.5″ (Roboz RS-8254).

Forceps, 2 × 3 teeth, length 5″ (Roboz RS-8210).

Glass beaker, 20 ml (VWR, 89000-194).

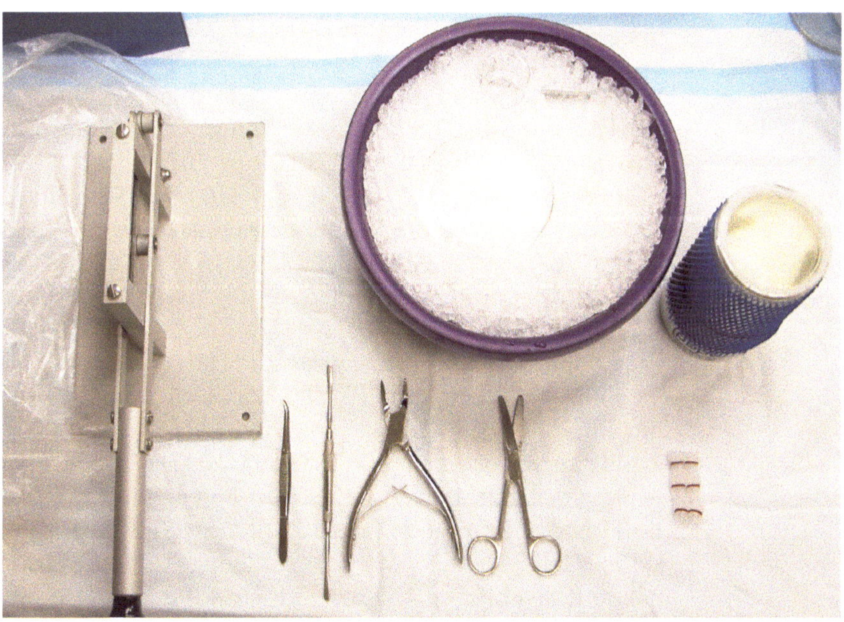

Fig. 1. Setup for removing the brain and dissecting out the parietal cortex, hippocampus, and brain. The dissection tools required are a forceps, spatula, rongeurs, and scissors. Other equipment needed are a guillotine, ice bucket, petri dish lid with a filter paper wetted with saline, a 20 ml beaker filled with chilled saline, low temperature freezer tubes, and a container of liquid nitrogen.

Glass petri dish, top half, 10 cm (VWR 89000-204).

Guillotine (VWR 50000-450).

Hemostatic forceps, MIXTER, 5½″ (Roboz RS-7291).

Isotonic sodium chloride solution, 0.9% (*w/v*) (Ricca Chemical Company 7210-5).

Low temperature freezer vials, 2.0 ml, round bottom (VWR 16001-114).

Liquid nitrogen.

Razor blade, no. 9, 0.22 mm thick (VWR 55411-050).

Rongeurs, Blumenthal, 4.5″ (Roboz RS-8340).

Scissors (Roboz RS-6849).

Spatula (Roboz RS-8820).

Whatman circle filter paper, 70 mm, grade 1 (Whatman 1001-070).

2.3.1. ELISA Materials

Coomassie Plus protein assay kit (Pierce 23236).

Wheaton tissue grinder, 2 ml capacity (Wheaton 358003).

ELISA kit (available from many manufacturers: Assay Designs, R&D Systems, Pierce).

Horizontal orbital shaker (VWR 5000).

Microplate reader (Perkin Elmer 1420 Multilabel Victor^{3}counter).

Overhead stirrer (Wheaton 903475).

Tissue grinder, 2 ml capacity (Wheaton 358003).

2.3.2. Western Blotting Materials

Anti-secondary antibody, conjugated to horseradish peroxidase (HRP, anti-rabbit IgG, Cell Signaling Technology 7074, anti-mouse IgG, Cell Signaling Technology 7076).

Blocking solution: 10 mM Tris–HCl pH 7.5 (Sigma-Aldrich T3253), 150 mM sodium chloride (Sigma-Aldrich S7653), 0.05% Tween 20 (Sigma-Aldrich P1379), 5% *w/v* powdered milk (Carnation), and 0.01% thimerosal (Sigma-Aldrich T5125). Store at 4°C and use within 30 days.

Coomassie Plus protein assay kit (Pierce 23236).

Enhanced chemiluminescence kit (Cell Signaling Technology 7003).

Heating block, dry (VWR 12621-104).

High current power supply (Bio-Rad 164-5052).

Homogenization buffer: 15 mM Tris–HCl pH 7.6 (Sigma-Aldrich T3253), 0.25 M sucrose (Sigma-Aldrich 84097), 1 mM magnesium chloride (Sigma-Aldrich M9272), 1 mM ethylene glycol-bis(2-aminoethylether)-*N,N,N′,N′*-tetraacetic acid (Sigma-Aldrich E0396), 1 mM dithiothreitol (DTT, Sigma-Aldrich 43817), 1.25 μg/ml pepstatin (Sigma-Aldrich 77170), 10 μg/ml leupeptin (Sigma-Aldrich 62070), 25 μg/ml aprotinin (Sigma-Aldrich 10820), 1 mM benzamidine hydrochloride (Sigma-Aldrich 434760),

0.1 mM sodium orthovanadate (Sigma-Aldrich S6508), 50 mM sodium fluoride (Sigma-Aldrich S6521), 2 mM sodium pyrophosphate decahydrate (Sigma-Aldrich 221368), 0.5 mM phenylmethanesulfonyl fluoride (PMSF, Sigma-Aldrich P7626). Add all protease and phosphatase inhibitors to solution when on ice. Add PMSF last and immediately use while keeping homogenization buffer on ice. PMSF is only effective for up to 30 min at 4°C.

Horizontal orbital shaker (VWR 5000).

Immobilon-P PVDF, 0.45 μm (Millipore IPVH00010).

Film processor (Kodak).

Film, 8″ × 10″ blue X-ray film, sharp contrast (Phenix Research Products F-BX810).

Filter paper, extra thick (Bio-Rad 1703965).

Microplate, 96 well.

Mini-PROTEAN Tetra cell gel electrophoresis system (Bio-Rad 165-8000).

Mini-PROTEAN Trans-blot module (Bio-Rad 165-8062).

Mini-PROTEAN TGX precast gels (Bio-Rad).

Overhead stirrer (Wheaton 903475).

PBS with 0.05% Tween 20, prepare 1 L: 8 g sodium chloride (Sigma-Aldrich S7653), 0.2 g potassium chloride (Sigma-Aldrich P9333), 1.44 g sodium phosphate dibasic anhydrous (Sigma-Aldrich S9763), 0.24 g potassium phosphate dibasic (Sigma-Aldrich P3786), 0.05% Tween 20 (Sigma-Aldrich P1379), adjust pH to 7.4.

PerfectWestern containers (Genhunter B101).

Precision Plus protein dual color standards (Bio-Rad 161-0374).

Running buffer: 25 mM Tris–HCl (Sigma-Aldrich T3253), 200 mM glycine (Sigma-Aldrich G7126), 0.1% *w/v* sodium dodecyl sulfate (SDS, Sigma-Aldrich 436143), prepare 2 L.

Sample buffer 4×: 0.3 M Tris–HCl pH 6.8 (Sigma-Aldrich T3253), 8% SDS (Sigma-Aldrich 436143), 40% glycerol (Sigma-Aldrich G5516), 0.04 mg/ml bromophenol blue (Sigma-Aldrich 114391), and 200 mM DTT (Sigma-Aldrich 43817). Warm in microwave to dissolve SDS. Aliquot and store at −20°C. Warm in heating block at 95°C prior to using.

Tissue grinder, 2 ml capacity (Wheaton 358003).

Transfer buffer: 25 mM Tris–HCl (Sigma-Aldrich T3253), 200 mM glycine (Sigma-Aldrich G7126), 20% methanol (Sigma-Aldrich 322415), prepare 2 L. Cool to 4°C prior to use.

Vortex Mixer (VWR 58816-121).

2.3.3. Immunohistochemistry Materials

Antigen retrieval solution: 10 mM citric acid trisodium salt pH 9.0 (Sigma-Aldrich C3674) Store at room temperature (RT).

Blocking solution, 5% normal goat serum, 0.4% Triton X-100 in PBS: 2.5 ml normal goat serum (Invitrogen, PCN5000), 2 ml 10% Triton X-100 (Sigma-Aldrich T8532, diluted 1:10 in H_2O), 5 ml 10× PBS (Invitrogen, 70011-044), and 40.5 ml

H_2O. Note: the type of normal serum chosen must match with the species that the secondary antibody is made in.

Brain sectioning matrix, coronal, 1 mm (Ted Pella 15007).

Cryo-preservation solution, for 1 L: 1.57 g sodium phosphate monobasic monohydrate (Sigma-Aldrich S9638), 5.46 g sodium phosphate dibasic anhydrous (Sigma-Aldrich S9763), 400 ml H_20, 300 ml glycerol (Sigma-Aldrich G5516), 300 ml ethylene glycol (Sigma-Aldrich 102466). Store at –20°C.

Fluorescently labeled secondary antibodies. We generally use Alexa Fluor-labeled secondary antibodies (Invitrogen) at 1:1,000–1:2,000 dilution, but a number of other vendors (e.g., Jackson ImmunoResearch and Cell Signaling) also carry labeled secondary antibodies. It is important that if you are running multiple primary antibodies (e.g., anti-NeuN and anti-ERK raised in mouse and rabbit, respectively), the secondary antibodies should be from the same species (e.g., Alexa Fluor-488 goat anti-mouse and Alexa Fluor-594 goat anti-rabbit) and highly cross-reacted to prevent cross-species reactivity.

Isotonic sodium chloride solution, 0.9% *w/v* (Ricca Chemical Company 7210-5).

Low temperature freezer vials, 2.0 ml, round bottom (VWR 16001-114).

Paraformaldehyde, 4%, pH 7.4, in 0.1 MPB, 350 ml per animal needed. Note: measure and dissolve paraformaldehyde in fume hood. Warm up 400 ml of H_2O to 65°C. Do not exceed 70°C. Add 40 g of paraformaldehyde (JT Baker S898-07). Stir and add drop wise 4 M NaOH (Sigma-Aldrich 35274) until solution clears. Add 300 ml of 0.2 M PB pH 7.4. Mix and filter. Adjust pH to 7.4. Store at 4°C protected from light and use with 7 days.

PBS.

PBS with 0.4% Triton X-100 (Sigma-Aldrich T8532), prepare 1 L.

Tissue adhesive, Loctite® 404 (Ted Pella 10035).

Vibratome (Leica).

Vibratome dissection blades, stainless steel, PTFE-treated (Ted Pella 121-4).

3. Methods

3.1. ELISA

1. Anesthetize the rat.
2. Decapitate the rat with a guillotine.
3. Remove skin over skull using scissors (Fig. 2a).
4. Place rongeurs in eye sockets, snap together to break skull bone (Fig. 2b).

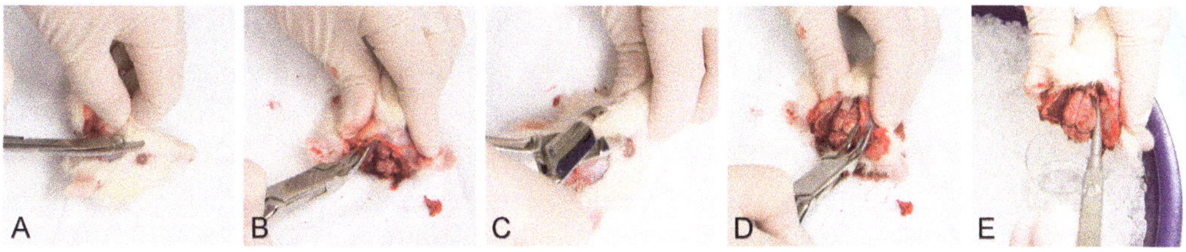

Fig. 2. Removal of the brain from the skull. (**a**) Remove skin overlying skull. (**b**) Snap skull bone between eye sockets with rongeurs. (**c**) Remove skull bone at lateral sides of cerebellum. (**d**) Lift skull bone with rongeurs away from brain. (**e**) Scoop brain out of skull with spatula.

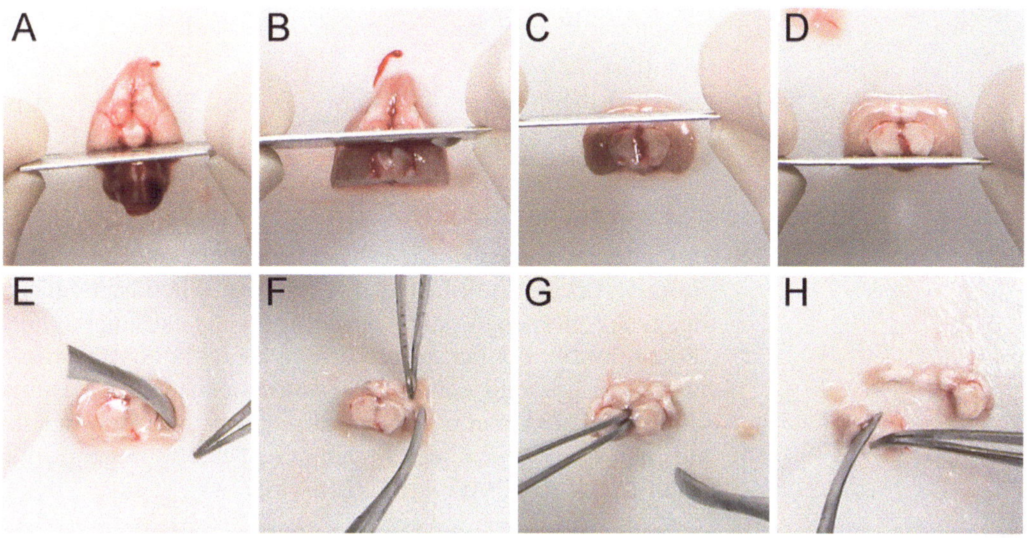

Fig. 3. Dissection of the parietal cortex, hippocampus, and thalamus. (**a**) Dissect caudal brain at circle of Willis and remove. (**b**) Dissect rostral brain at circle of Willis and remove. (**c**) Dissect frontal cortex and remove. (**d**) Dissect tips of temporal cortex and remove. (**e**) Peel ipsilateral and contralateral parietal cortex from hippocampus and midbrain. (**f**) Dissect ipsilateral and contralateral hippocampus. (**g**) With forceps, separate ipsilateral and contralateral thalamus. (**h**) Dissect thalamus. Place tissue in low temperature freezer vials, cap, and drop into liquid nitrogen.

5. With rongeurs, remove skull bone at both sides of cerebellum (Fig. 2c).
6. Slowly and carefully use rongeurs to remove skull bone (Fig. 2d).
7. Once brain is completely revealed, use spatula to scoop brain out of skull (Fig. 2e).
8. Place in isotonic saline in 20 ml beaker at 4°C.
9. Place brain on petri dish with Whatman filter paper wetted with isotonic saline with ventral side facing up. With ice-cold razor blade, dissect away the caudal brain at the circle of Willis (Fig. 3a).

10. Dissect away the rostral brain at circle of Willis (Fig. 3b).
11. Rotate brain so that the caudal side is facing up, and the dorsal side is facing away. Dissect away the frontal cortex placing razor blade just at the dorsal hippocampus (Fig. 3c).
12. Dissect away the tips of the temporal cortex placing razor blade just at the ventral tips of the hippocampus (Fig. 3d).
13. Peel parietal cortex away from hippocampus using spatula (Fig. 3e). Using forceps, place in low temperature freezer vial sitting on ice.
14. Peel hippocampus away from remaining midbrain (Fig. 3f). Place in low temperature freezer vial sitting on ice.
15. Separate ipsilateral and contralateral thalamus with forceps (Fig. 3g).
16. Dissect thalamus from brain stem (Fig. 3h). Place in low temperature freezer vial sitting on ice.
17. Place low temperature freezer vials in liquid nitrogen. Store at –80°C until ready for ELISA.
18. Homogenize tissue using overhead stirrer and tissue grinder in buffer according the ELISA kit directions. When homogenizing, avoid removing pestle from solution and causing frothing of the solution. For best results, use tissue homogenates within 30 min of homogenization. Keep homogenates on ice until added to ELISA plate. We typically weigh the tissue rapidly, and then homogenize in ten volumes of the homogenization buffer that is specified by the ELISA kit.
19. Set aside 10 μl of homogenate to perform a Coomassie plus protein assay. If necessary, dilute homogenate to stay within standard curve.
20. Follow ELISA kit instructions to complete protocol.

3.2. Western Blot

1. Remove brain from animals and dissect tissue as described above. Homogenize tissue using overhead stirrer and 2 ml tissue grinder. For parietal cortex and thalamus, homogenize in 2 ml volume. For hippocampus, homogenize in 750 μl volume.
2. Set aside 10 μl of homogenate to perform Coomassie plus protein assay. Dilute homogenate 1:10 to remain within standard curve of protein assay.
3. Add 4× sample buffer to remaining homogenate for 1× final concentration. Vortex, then boil 7–9 min at 95°C using dry heating block. Samples can be stored at –20°C for up to 1 year. If using frozen samples, rewarm in heating block at 95°C for 2 min, centrifuge (20,800 g, 2 min, RT), and then briefly vortex. Vortexing is critical to ensure that the sample is homogenous prior to loading.

4. Using results from Coomassie plus protein assay, load 50 μg protein/lane.
5. Electrophorese samples using precast polyacrylamide gels.
6. Transfer proteins from gels to Immobilon PVDF using transfer apparatus.
7. Incubate Immobilon PVDF membrane in blocking solution for 1 h at RT.
8. Dilute primary antibody in blocking solution. Apply to membrane for 2 h at RT or overnight at 4°C using Perfect Western containers. Ensure that antibody solution completely covers membrane and is continually shaking during application (need approx. 10 ml/membrane).
9. Wash 4 × 15 min each in PBS with Tween 20 while shaking at RT.
10. Dilute secondary antibody in blocking solution. Apply to membrane for 2 h at RT or overnight at 4°C using Western blot containers.
11. Wash 4 × 15 min each in PBS with Tween 20 while shaking at RT.
12. Develop using enhanced chemiluminescence kit by following manufacturer's directions (Cell Signaling, Amersham, and Pierce).

3.3. Immunohistochemistry (Table 1)

1. Prepare perfusion lines by filling with isotonic saline at 4°C.
2. Anesthetize the rat.
3. Make a V-shaped cut below sternum, grab the skin with the forceps, and cut along the whole sternum. Grab lower part of sternum and make two cuts along the ribs.
4. Secure heart with forceps using the teeth of the forceps.
5. Cut into right atrium to allow blood to flow freely out of atrium. This step is critical because it prevents the blood vessels from swelling.
6. Cut tip of left ventricle and introduce the perfusion cannula into the aorta.
7. Fix cannula into aorta with hemostatic forceps. Do not push cannula too far into aorta, stop within 2–3 mm of heart. Otherwise, the cannula can occlude artery branches that perfuse the brain.
8. Perfuse with isotonic saline at 4°C for 2 min (approx. 40 ml/min).
9. Perfuse with 4% paraformaldehyde at 4°C for 30 min (10 ml/min).
10. Remove brain as described above. Store in 4% paraformaldehyde at 4°C for 4–24 h.

Table 1
Representative protocol form for immunohistochemistry of free-floating brain sections

FREE-FLOATING SECTIONS IMMUNOHISTOCHEMISTRY Date______________________

1. Washed sections free-floating 3 x 5 min at **25** /_____°C with **PBS** /__________ (Use ~500 µL per 3 slices)__________

2. Incubated **20** / **60** /_______min at **RT** /________°C w/ **Blocking Soln** ______________________

3. Incubated with 1° Ab(s) diluted in **Blocking Soln** ______________ for ___________**hr** at **4** / **25** /______°C.________

1° Ab #1: Anti-____

Dilution:________. Species: **rb** / **ms** / **ch** _____**mAb** / **pAb**. Cat #______________ Lot #_________

Vendor: **Cell Sig** / **Cell Tech** / **Chemicon** / **BD Pharm** / **Mol Probes** /______________________

and / **or** 1° Ab #2: Anti-____

Dilution:________. Species: **rb** / **ms** / **ch**______**mAb** / **pAb**. Cat #______________ Lot #_________

Vendor: **Cell Sig** / **Cell Tech** / **Chemicon** / **BD Pharm** / **Mol Probes** /______________________

and / **or** 1° Ab #3: Anti-____

Dilution:________. Species: **rb** / **ms** / **ch** _____**mAb** / **pAb**. Cat #______________ Lot #_________

Vendor: **Cell Sig** / **Cell Tech** / **Chemicon** / **BD Pharm** / **Mol Probes** /______________________

4. Rinsed 3 x 5 min at **25** /_____°C with **Blocking Soln** ______________________

5. Incubated with following diluted in **Blocking Soln** _______________ for ______**min** / **hr** at **4** / **25** /______°C._______

Lectin:____ Dilution: ______________

Cat #_____________ Lot #________________ Label: **Biotin** /______________________

Vendor: **Cell Sig** / **Cell Tech** / **Chemicon** / **BD Pharm** / **Mol Probes** /______________________

and / **or** 2° Ab #2: Dilution:________. Species: **goat** / **donkey** /________ anti- **ms** / **rb** / **ch** /_______________

Label: **Alexa 350** / **488** / **555** / **568** / **594** / **647** / **700** / **biotin** / ______________________

and / **or** 2° Ab #3: Dilution:________. Species: **goat** / **donkey** /________ anti- **ms** / **rb** / **ch** /_______________

Label: **Alexa 350** / **488** / **555** / **568** / **594** / **647** / **700**______________________

and / **or** 2° Ab #3: Dilution:________. Species: **goat** / **donkey** /________ anti- **ms** / **rb** / **ch** /_______________

Label: **Alexa 350** / **488** / **555** / **568** / **594** / **647** / **700**______________________

and / **or** **Phalloidin**-Alexa **488** / **546**_______________. **Hoechst 33342 dye**. ______________________

SKIP TO STEP #8 IF NOT PERFORMING A TERTIARY REACTION STEP; ELSE PROCEED TO STEP #6.

6. Rinsed 3 x 5 min at **25** /_____°C with **Blocking Soln** ______________________

7. Incubated with following diluted in **Blocking Soln** _____________ for ______**min** / **hr** at **4** / **25** /______°C._________

Steptavidin: Dilution:______________. Label: **Alexa 350** / **488** / **555** / **568** / **594** / **647** / **700** /______________

and / **or** 2° Ab #2: Dilution:________. Species: **goat** / **donkey** /________ anti- **ms** / **rb** / **ch** /_______________

Label: **Alexa 350** / **488** / **555** / **568** / **594** / **647** / **700** /______________________

and / **or** 2° Ab #3: Dilution:________. Species: **goat** / **donkey** /________ anti- **ms** / **rb** / **ch** /_______________

Label: **Alexa 350** / **488** / **555** / **568** / **594** / **647** / **700** /______________________

and / **or** **Phalloidin**-Alexa **488** / **546**_______________. **Hoechst 33342 dye**. ______________________

8. Rinsed 3 x 5 min at **25** /_____°C with **PBS** /______________________

9. Mounted with **ProLong Gold Antifade** / **Elvanol** /______________________

Notes:____

____

Blocking Solution: 5% normal goat serum in 1X PBS (2.5 mL normal goat serum, 5 mL 10X PBS, 42.5 mL H_2O)

11. Transfer to PBS and 30% sucrose for 3–4 days until brain sinks to the bottom of the tube.
12. Block brain around injury site. For fluid-percussion brain injury, blocking at bregma levels −1 and −8 mm ensures that the entire injury area will be sectioned.
13. Secure brain to vibratome sectioning block with tissue adhesive.
14. Section brain into 50 μm sections in PBS. Use paintbrush to remove sections.
15. Place sections into cryo-preservation solution in low temperature freezer vials and store at −20°C until ready for immunohistochemistry.
16. Wash free-floating sections 3×5 min (minimum time) at RT with PBS to remove cryo-preservation solution. Use a 6-well plate with 2 ml of PBS per well.
17. Perform antigen retrieval by incubating sections with sodium citrate solution while in water bath heated to 80°C for 30 min.
18. Rinse 3×20 min each at RT with PBS with Triton X-100 while shaking using orbital shaker.
19. Incubate 20 min at RT with blocking solution. Note: the normal serum used to block should be chosen based on the species used for the secondary antibody. For example, if using anti-mouse made in goat, use normal goat serum for the blocking solution.
20. Incubate with primary antibody diluted in blocking solution for >12 h at 4°C while shaking on an orbital shaker. Use at least 150 μl per slice. Slices can be placed in a 24-well plate or within a low temperature freezer vial. Note: some antibodies are sensitive to Triton X-100. In such cases, it may be necessary to perform an extended step 19 (e.g., >1 h), followed by PBS rinses without Triton X-100, and incubation with blocking solution containing a lower concentration, or even no, Triton X-100 throughout the remainder of the procedure.
21. Rinse 3×20 min each at RT with PBS.
22. Incubate with secondary antibody (Alexa Fluor-conjugated goat anti-rabbit IgG (H+L), highly cross-adsorbed, Invitrogen or Alexa Fluor-conjugated goat anti-mouse IgG (H+L), highly cross-adsorbed, Invitrogen) diluted in blocking solution for >12 h at 4°C. Optional: if needed, add Hoeschst, DAPI, labeled-lectin, or labeled-phallodin in this step. If secondary antibody needs to be amplified, use biotin-conjugated secondary antibody in this step (Biotinylated anti-rabbit IgG (H+L), made in goat, Vector Labs BA-1000 or biotinylated anti-mouse IgG (H+L), made in goat, Vector Labs BA-9200).

23. Rinse 3 × 20 min each at RT with PBS. If not doing a secondary amplification step using biotin/streptavidin, proceed to step 26.
24. If secondary antibody requires amplification, use streptavidin-labeled conjugated antibody (Invitrogen, Alexa conjugated) diluted in blocking solution for >12 h at 4°C. If needed, add Hoeschst, DAPI, labeled-lectin, or labeled-phallodin in this step.
25. Rinse 3 × 20 min each at RT with PBS.
26. Mount with ProLong Gold Antifade (Invitrogen, P36930).
27. Store microscope slides flat and in dark. Image within 7 days of mounting.

4. Notes

Using the techniques of ELISA, Western blotting, and immunohistochemistry can provide detailed knowledge regarding the level of changes in proteins, alterations in posttranslational modifications, and the cell types and brain regions that exhibit these changes. However, these techniques are typically not interchangeable, and thus, require separate sets of animals for each technique.

In the analysis of the results, it is important to keep in mind that the pathology of TBI is not homogenous and different mechanisms are at play in different regions of the brain. For example, when performing Western blotting, the parietal cortex is the main contused area, whereas the hippocampus and thalamus exhibit more diffuse and delayed injury (3, 4). Inflammatory molecules and cellular death markers will be found much more prominently in the contused cortex as compared to the thalamus (Fig. 4).

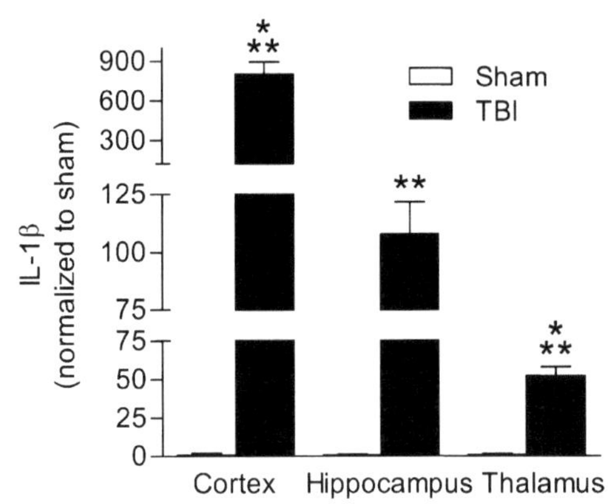

Fig. 4. Differential changes in the inflammatory molecule interleukin-1β in the ipsilateral parietal cortex, hippocampus, and thalamus 3 h after fluid-percussion brain injury. Adapted from ref. (12). Mean ± SEM, $n = 3$–8, $^{**}p < 0.01$, $^{***}p < 0.001$.

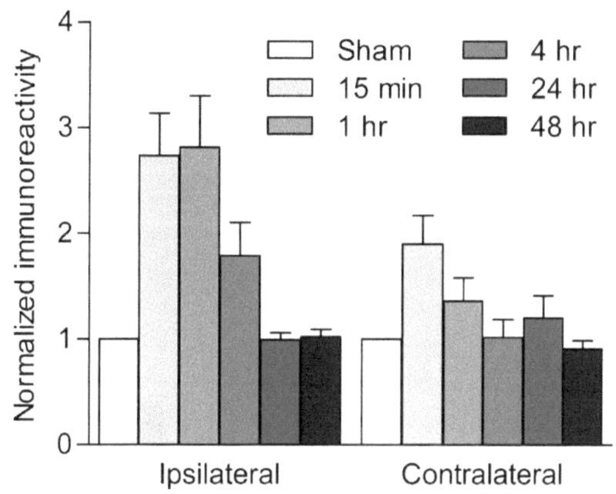

Fig. 5. Extracellular signal-regulated protein kinase 2 (ERK2) phosphorylation increased in both the ipsilateral and contralateral hippocampus after fluid-percussion brain injury, although the magnitude and duration of ERK2 phosphorylation was attenuated in the contralateral hippocampus. Mean ± SEM, $n = 4–7$.

Thus, the dilution of the homogenate for the ELISA or the primary antibody dilution for the Western blot will be significantly different depending upon the area of tissue that is being assessed.

Despite all best efforts, there is variability not only between the level and location of the brain trauma, but also between animals. Thus, proper controls within all experiments are important for the analysis. An age-matched sham animal remains the best control for a TBI animal. The contralateral area of the brain also responds to the trauma, albeit to much greater levels, due to spreading cortical depression, loss of connections from the ipsilateral side, and action potential bursting due to the mechanical trauma (5, 6). Thus, using contralateral tissue as a control can be very misleading, in particular at earlier time points after brain trauma (Fig. 5). Additionally, anesthesia itself can cause memory dysfunction and changes in a number of synaptic proteins (2, 7–9); thus, using naïve, nonoperated animals as control can also yield spurious results. The time under anesthesia as well as the depth of anesthesia needs to remain constant between all animal groups.

Sampling frequency is another important consideration when using biochemical techniques. Each time point assessed is another animal group, and it is tempting to reduce sampling frequency. However, in the minutes to hours after TBI, the changes in proteins are often very rapid and transient and an infrequent sampling frequency can easily miss these changes (10–12). A time course that has worked well for us is 15 min, 1, 4, 24 h, and 3 days (Fig. 6). However, the ideal sampling frequency needs to be tailored for the individual molecule that will be assessed.

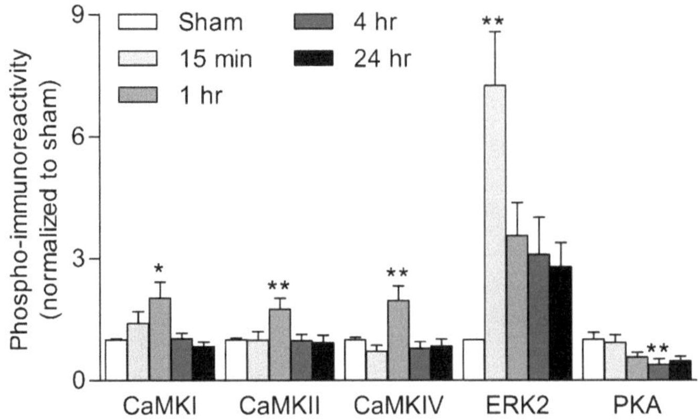

Fig. 6. Rapid and transient alterations in cellular signaling pathways after TBI. CaMKI, CaMKII, CaMKIV, and ERK2 phosphorylation increased, whereas PKA phosphorylation was reduced in the hippocampus after fluid-percussion brain injury. Adapted from refs. (10–12). Mean ± SEM, $n = 5–9$/group, $^{*}p < 0.05$, $^{**}p < 0.01$.

4.1. ELISA

An ELISA is the best method to measure fold levels of changes in specific proteins. The linearity of the standard curve can be several orders of magnitude. However, to avoid extrapolation errors, it is important to perform a pilot experiment with several dilutions of a representative sample to ensure that the measurements stay within the linear curve. Since ELISAs are highly sensitive, performing at least two replicates of each sample and standard will provide a measure of consistency.

4.2. Western Blot

The selection of the polyacrylamide gel percentage needs to be based on the molecular weight of the protein of interest. Gradient gels are now commercially available and offer the ability to resolve a very broad range of molecular weight proteins. If subcellular localization is important to assay, the Western blot homogenization can easily be modified into a fractionation protocol (11–13). Resolving fractions down to presynaptic and postsynaptic components can reveal important insights about the nature of the changes in the proteins (14).

If detecting the protein of interest is difficult, due to its low abundance or the lack of a high affinity antibody, fractionation of the samples can help purify the protein and make it easier to assay. Other practical tips to improving the Western blot signal include optimizing the dilution of the primary and secondary antibodies, changing the blocking solution so that it masks fewer antigens (e.g., substituting 3% bovine serum albumin for the 5% powdered milk), or using an enhanced chemiluminescence kit that has a higher degree of sensitivity (Amersham Biosciences ECL plus RPN2133). A number of new kits are now available, allowing detection of only 60 fg of a protein (Pierce SuperSignal West Femto Substrate 34095).

4.3. Immunohistochemistry

Perhaps the most important consideration of any immunohistochemistry result is the antibody specificity. The antibody must be able to recognize the one native protein among all others. Pre-absorbing the primary antibody against the antigen used to make the antibody prior to application of the primary antibody provides a measure of assurance that the observed signal is specific, but does not rule out that the signal observed is due to another protein with a similar epitope. Some antibodies also react nonspecifically to the contused area, thus, including a TBI section which receives parallel processing but without the primary antibody also confers a level of control to the analysis. One can further improve the specificity of the signal observed by using commercially available highly cross-adsorbed antibodies. Finally, if available, the primary antibodies can be pre-adsorbed to tissue from knockout animals for the protein of interest, which should eliminate nonspecific binding and will indicate if other proteins have an equivalent epitope by examining sections from the null animals.

References

1. Lister JP, Barnes CA (2009) Neurobiological changes in the hippocampus during normative aging. Arch Neurol 66:829–833
2. Zhan X, Fahlman CS, Bickler PE (2006) Isoflurane neuroprotection in rat hippocampal slices decreases with aging: changes in intracellular Ca^{2+} regulation and *N*-methyl-D-aspartate receptor-mediated Ca^{2+} influx. Anesthesiology 104:995–1003
3. Bramlett HM, Dietrich WD (2002) Quantitative structural changes in white and gray matter 1 year following traumatic brain injury in rats. Acta Neuropathol (Berl) 103:607–614
4. Bramlett HM, Kraydich S, Green EJ, Dietrich WD (1997) Temporal and regional patterns of axonal damage following traumatic brain injury: a β-amyloid precursor protein immunocytochemical study in rats. J Neuropathol Exp Neurol 56:1132–1141
5. Chen X, Li Y, Kline AE, Dixon CE, Zafonte RD, Wagner AK (2005) Gender and environmental effects on regional brain-derived neurotrophic factor expression after experimental traumatic brain injury. Neuroscience 135:11–17
6. Enomoto T, Osugi T, Satoh H, McIntosh TK, Nabeshima T (2005) Pre-injury magnesium treatment prevents traumatic brain injury-induced hippocampal ERK activation, neuronal loss, and cognitive dysfunction in the radial-arm maze test. J Neurotrauma 22:783–792
7. Culley DJ, Baxter MG, Yukhananov R, Crosby G (2004) Long-term impairment of acquisition of a spatial memory task following isoflurane-nitrous oxide anesthesia in rats. Anesthesiology 100:309–314
8. Culley DJ, Yukhananov RY, Xie Z, Gali RR, Tanzi RE, Crosby G (2006) Altered hippocampal gene expression 2 days after general anesthesia in rats. Eur J Pharmacol 549:71–78
9. Dong Y, Zhang G, Zhang B, Moir RD, Xia W, Marcantonio ER, Culley DJ, Crosby G, Tanzi RE, Xie Z (2009) The common inhalational anesthetic sevoflurane induces apoptosis and increases beta-amyloid protein levels. Arch Neurol 66:620–631
10. Atkins CM, Oliva AA Jr, Alonso OF, Chen S, Bramlett HM, Hu BR, Dietrich WD (2007) Hypothermia treatment potentiates ERK1/2 activation after traumatic brain injury. Eur J Neurosci 26:810–819
11. Atkins CM, Chen S, Alonso OF, Dietrich WD, Hu BR (2006) Activation of calcium/calmodulin-dependent protein kinases after traumatic brain injury. J Cereb Blood Flow Metab 26:1507–1518
12. Atkins CM, Oliva AA Jr, Alonso OF, Pearse DD, Bramlett HM, Dietrich WD (2007) Modulation of the cAMP signaling pathway after traumatic brain injury. Exp Neurol 208:145–158
13. Chen S, Atkins CM, Liu CL, Alonso OF, Dietrich WD, Hu BR (2006) Alterations in mammalian target of rapamycin signaling pathways after traumatic brain injury. J Cereb Blood Flow Metab 27:939–949
14. Phillips GR, Huang JK, Wang Y, Tanaka H, Shapiro L, Zhang W, Shan WS, Arndt K, Frank M, Gordon RE, Gawinowicz MA, Zhao Y, Colman DR (2001) The presynaptic particle web: ultrastructure, composition, dissolution, and reconstitution. Neuron 32:63–77

Chapter 32

Assessments of Oxidative Damage and Lipid Peroxidation After Traumatic Brain Injury and Spinal Cord Injury

Jeffrey M. Bosken, Juan A. Wang, and Edward D. Hall

Abstract

Free radical-induced oxidative damage to proteins and lipids is one of the most convincingly validated secondary injury mechanisms following traumatic brain injury or spinal cord injury. In particular, central nervous system (CNS) tissue is exquisitely sensitive to the process of lipid peroxidation (LP) due to the high content of peroxidizable polyunsaturated acids and a high content of iron which is a catalyst for the propagation of LP reactions in cell membranes. This chapter is focused on a presentation of the principal contemporary methods for monitoring LP in injured CNS tissue, including immunoblotting measurement of aldehydic breakdown products such as 4-hydroxynonenal (4-HNE) and acrolein as well as the gas chromatographic–mass spectrophotometric (GC–MS) analysis of the arachidonic acid-derived F_2 isoprostanes and the docosahexaenoic acid (DHA)-derived F_4 isoprostanes (neuroprostanes). Protein oxidative damage produced by direct free radical damage to basic amino acids (lysine, arginine, histidine, and cysteine) as well as the covalent binding of 4-HNE and acrolein to proteins can be collectively measured using the carbonyl immunoblot assay. Additionally, nitrative damage to protein tyrosine residues produced by peroxynitrite-generated nitrogen dioxide radicals can be assessed by immunoblotting of 3-nitrotyrosine (3-NT) levels in CNS tissue samples. Lastly, immunohistochemical procedures for measuring 4-HNE and 3-NT are detailed.

Key words: Lipid peroxidation, 4-Hydroxynonenal, 3-Nitrotyrosine, Acrolein, Protein carbonyl, Isoprostane, Neuroprostane

1. Introduction

Oxidative damage is defined as *the biomolecular damage caused by the attack of reactive species upon the constituents of living organisms* (1). The “reactive species” are more commonly called reactive oxygen species (ROS) which include compounds, such as hydrogen peroxide (H_2O_2) or peroxynitrite (various forms include peroxynitrite anion, $ONOO^-$; peroxynitrous acid, ONOOH; nitrosoperoxocarbonate,

Jun Chen et al. (eds.), *Animal Models of Acute Neurological Injuries II*, Springer Protocols Handbooks,
DOI 10.1007/978-1-61779-782-8_32, © Springer Science+Business Media, LLC 2012

$ONOOCO_2^-$), that can give rise to highly reactive oxygen free radicals, such as hydroxyl radical (·OH), nitrogen dioxide radical (·NO_2), and carbonate radical (·CO_3). These radicals can then induce oxidative damage by attacking lipids, proteins, carbohydrates, and nucleic acids (DNA and RNA). However, in the case of acute central nervous system (CNS) injury, the primary targets include polyunsaturated fatty acids (e.g., arachidonic, docosahexaenoic or DHA, linoleic, and linolenic) and certain amino acids in proteins (e.g., lysine, histidine, arginine, cysteine, and tyrosine).

The process of free radical-induced lipid damage is called lipid peroxidation (LP) which first of all causes a disruption in cell and organellar membrane function by damaging the membrane phospholipid architecture and hence the structure and functionality of phospholipid-dependent membrane proteins. Secondly, the process of LP can lead to the formation of aldehydic peroxidation by-products, such as malondialdehyde (MDA), 4-hydroxynonenal (4-HNE), and 2-propenal (acrolein), which can covalently bind to proteins and are most commonly detected by immunoblotting or immunohistochemistry. The latter two protein-bound aldehydes, 4-HNE and acrolein, are themselves highly neurotoxic and are thus more pathophysiologically important to monitor than MDA which is relatively nontoxic. Moreover, MDA is not only produced by free radical-induced LP, but also during enzymatically induced LP (i.e., the arachidonic acid cascade), whereas 4-HNE and acrolein are selectively formed during free radical-induced LP. Thus, this chapter only presents methods for the immunoblotting or immunohistochemical measurement of 4-HNE or acrolein. Although MDA has long been a favorite choice focus for quantitative measurement of LP most often using the thiobarbituric acid (TBA) spectrophotometric method, the current authors do not recommend its use due to the more recent development of more sensitive immunochemical methods that provide more selective indices of the tissue levels of free radical-induced LP by-products. In addition to the formation of LP-derived aldehydic products, LP can also be assessed by instrumental measurement of free radical-induced arachidonic acid peroxidation products, such as the F_2 isoprostanes or the docosahexaenoic acid (DHA)-derived F_4 isoprostanes (aka. neuroprostanes), the methods for which are included in this chapter.

Direct attack of free radicals on proteins leads to the formation of protein carbonyls, tyrosine nitration, and cysteine oxidation. The methods that are detailed in this chapter include the immunoblotting measurement of protein carbonyl content and the immunoblot or immunohistochemical measurement of 3-nitrotyrosine (3-NT).

1.1. Oxidative Damage in CNS Injury

Free radical-induced oxidative damage has been shown to play a major role in the acute pathophysiology of traumatic brain injury

(TBI) and spinal cord injury (SCI). In the case of SCI, the initial demonstrations of free radical formation and LP in the injured spinal cord were provided by Demopoulos et al. (2, 3) and Hall and Braughler (4, 5). In regards to TBI, the first work that documented a role of oxygen radical was provided by Kontos et al. (6, 7) who demonstrated an almost immediate postinjury increase in brain microvascular superoxide radical production associated with compromise of autoregulatory function in fluid percussion TBI models. They also demonstrated that pharmacological scavengers of superoxide radical ($O_2^{\cdot-}$) decrease the posttraumatic superoxide levels and protect against the loss of autoregulatory competency. Using the salicylate trapping method, a rapid post-TBI increase in brain hydroxyl radical ($^{\cdot}OH$) levels has also been documented in a mouse diffuse and rat focal TBI models (8, 9) and others (9–11). As with the work of Kontos, the cerebral microvasculature appears to be the initial source of posttraumatic radical production (12). More recent work shows that a major source of posttraumatic brain radical production is increased ROS leakage from injured brain or spinal cord mitochondria (13–17). Spinal cord mitochondria display a higher baseline production of oxygen radicals compared to brain mitochondria (18). In addition, recently published experiments show that spinal cord mitochondria are more sensitive to functional impairment by oxidative damage (19). For a detailed discussion of the role of free radicals and oxidative damage in the pathophysiology of acute CNS injury, the reader is referred to selected reviews on the topic (9, 20–23).

1.2. Chemistry of Lipid Peroxidation and Formation of Measurable By-Products

The most studied mechanism of oxidative damage in models of TBI and SCI concerns ROS-induced LP. CNS tissue is exquisitely sensitive to LP due to the high content of peroxidizable polyunsaturated fatty acids as well as a high content of iron which is a catalyst of LP reactions (9) as explained below.

"Initiation" of lipid peroxidation occurs when a highly reactive radical (e.g., $^{\cdot}OH$) attacks and removes an allylic hydrogen from a polyunsaturated fatty acid (LH) resulting in a radical chain reaction: $LH + R^{\cdot} \rightarrow L^{\cdot} + RH$.

In the process, the initiating radical is quenched by receipt of an electron (hydrogen) from the polyunsaturated fatty acid. This, however, converts the latter into a lipid or alkyl radical. This sets the stage for a series of "propagation" reactions which begin when the alkyl radical reacts with molecular oxygen creating a lipid peroxyl radical ($LOO^{\cdot}$): $L^{\cdot} + O_2 \rightarrow LOO^{\cdot}$.

The peroxyl radical then reacts with a neighboring LH within the membrane and steals its electron forming a lipid hydroperoxide (LOOH) and a second alkyl radical ($L^{\cdot}$): $LOO^{\cdot} + LH \rightarrow LOOH + L^{\cdot}$.

Once LP begins, iron may participate in driving the process as LOOH formed through initiation are decomposed by reactions with either ferrous iron (Fe^{++}) or ferric iron (Fe^{+++}). In the case of Fe^{++}, the reaction results in the formation of a lipid alkoxyl radical ($LO^{\cdot}$): $LOOH + Fe^{++} \rightarrow LO^{\cdot} + OH^{-} + Fe^{+++}$.

If, however, the reaction involves Fe^{+++}, the LOOH is converted back into a $LOO^{\cdot}$: $LOOH + Fe^{+++} \rightarrow LOO^{\cdot} + Fe^{++}$.

Both of the reactions of LOOH with iron have acidic pH optima. Either alkoxyl ($LO^{\cdot}$) or peroxyl ($LOO^{\cdot}$) radicals arising from LOOH decomposition by iron can initiate the so-called LOOH-dependent LP resulting in "chain branching" reactions: $LOO^{\cdot} + LH \rightarrow LOOH + L^{\cdot}$ or $LO^{\cdot} + LH \rightarrow LOH + L^{\cdot}$.

Ultimately, the LP process leads to "fragmentation" or "scission" reactions in which the peroxidized polyunsaturated fatty acid (e.g., arachidonic acid) breaks down to give rise to the neurotoxic aldehydes *acrolein* or *4-HNE*. Figure 1 shows the hydroxyl radical ($^{\cdot}OH$)-induced initiation, propagation, and fragmentation reactions for arachidonic acid leading to the formation of 4-HNE (Fig. 1) or acrolein (Fig. 2). Acrolein and 4-HNE produce their neurotoxicity by binding to basic amino acids, such as lysine or histidine as well as sulfhydryl-containing cysteine residues in cellular proteins, the chemistry of which is illustrated in Fig. 3. The levels of 4-HNE and acrolein are best measured by dot, slot, or Western immunoblotting or immunohistochemistry using the methods detailed below. Alternatively, arachidonic acid peroxidation can lead to the generation of F_2 isoprostanes (Fig. 4) which are best measured by gas chromatography with detection and quantitation by mass spectrometry (GC–MS), although less precise ELISA kits are also commercially available. Furthermore, peroxidation of DHA can generate a variety of F_4 isoprostanes (aka. neuroprostanes; Fig. 5) that are also analyzed by GC–MS.

1.3. Chemistry of Protein Oxidative Damage

Free radicals can cause various forms of oxidative protein damage. A major mechanism involves carbonylation by reaction of various free radicals with susceptible amino acids. Figure 6 shows the chemistry of hydroxyl radical-induced histidine oxidation which results in carbonylation of the two position of histidine's imidazole ring. The protein carbonyls thus formed are measured through immunoblotting after derivatization of the carbonyl groups with dinitrophenylhydrazine (DNPH). Although the measurement of protein carbonyls by the so-called DNPH assay has long been used, a commercially available protein carbonyl Oxyblot® assay is widely used by present-day investigators. It should be noted that the carbonyl assay also picks up protein carbonyls that are present due to covalent binding of LP-derived 4-HNE and acrolein to proteins (see Fig. 3), in addition to those resulting from direct free radical-induced amino acid oxidation. Thus, as a result, the carbonyl assay

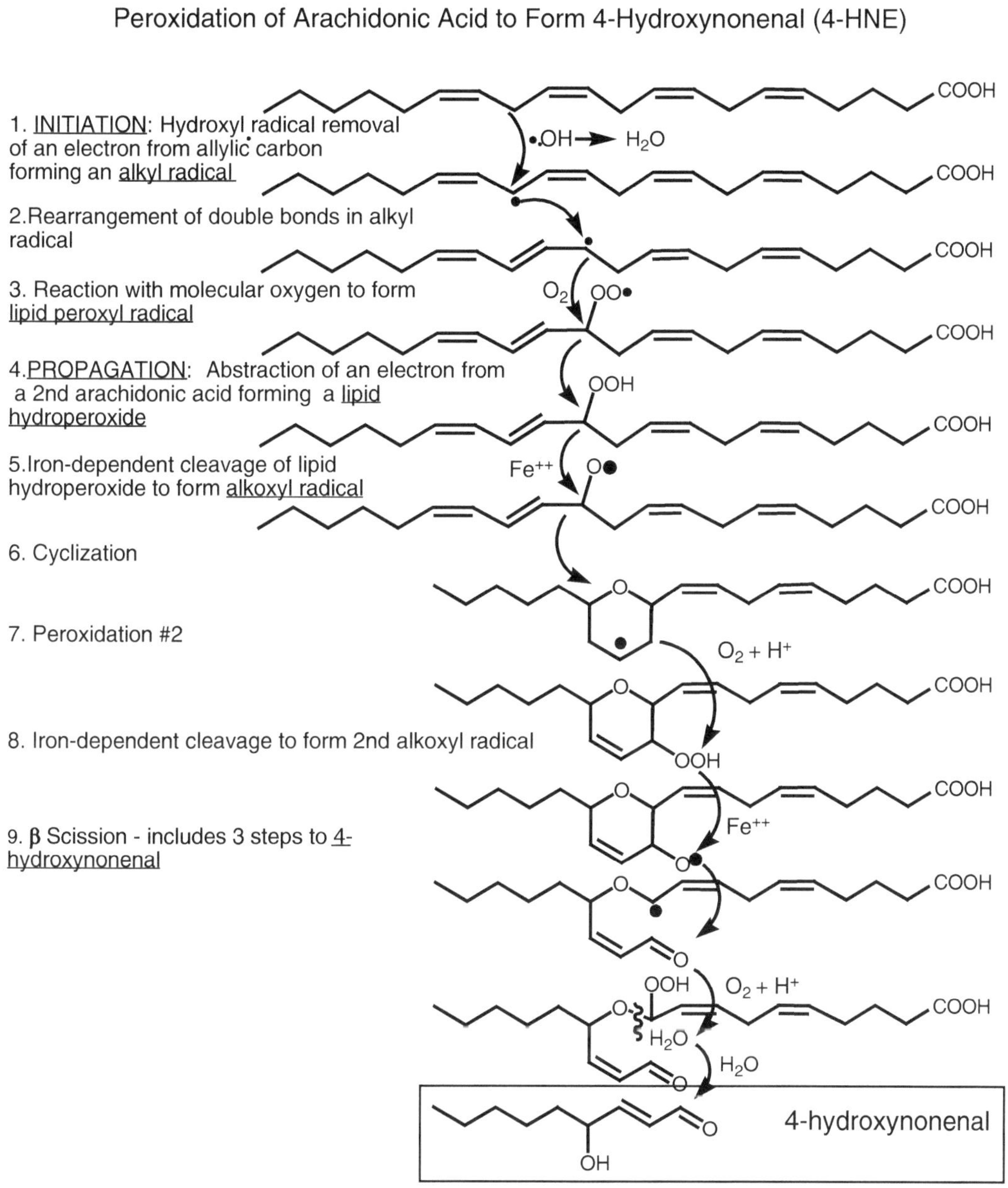

Fig. 1. Chemistry involved in the initiation, propagation, and termination reactions of arachidonic acid with the resulting formation of 4-HNE. Reprinted with permission from Hall et al. (2010).

is as much an indirect index of LP as it is of direct protein oxidation. Secondly, $^{\cdot}NO_2$ can nitrate the third position of tyrosine residues in proteins; 3-NT is a specific footprint of PN-induced cellular damage (see Fig. 7) which is best measured by immunoblotting or immunohistochemistry as detailed below.

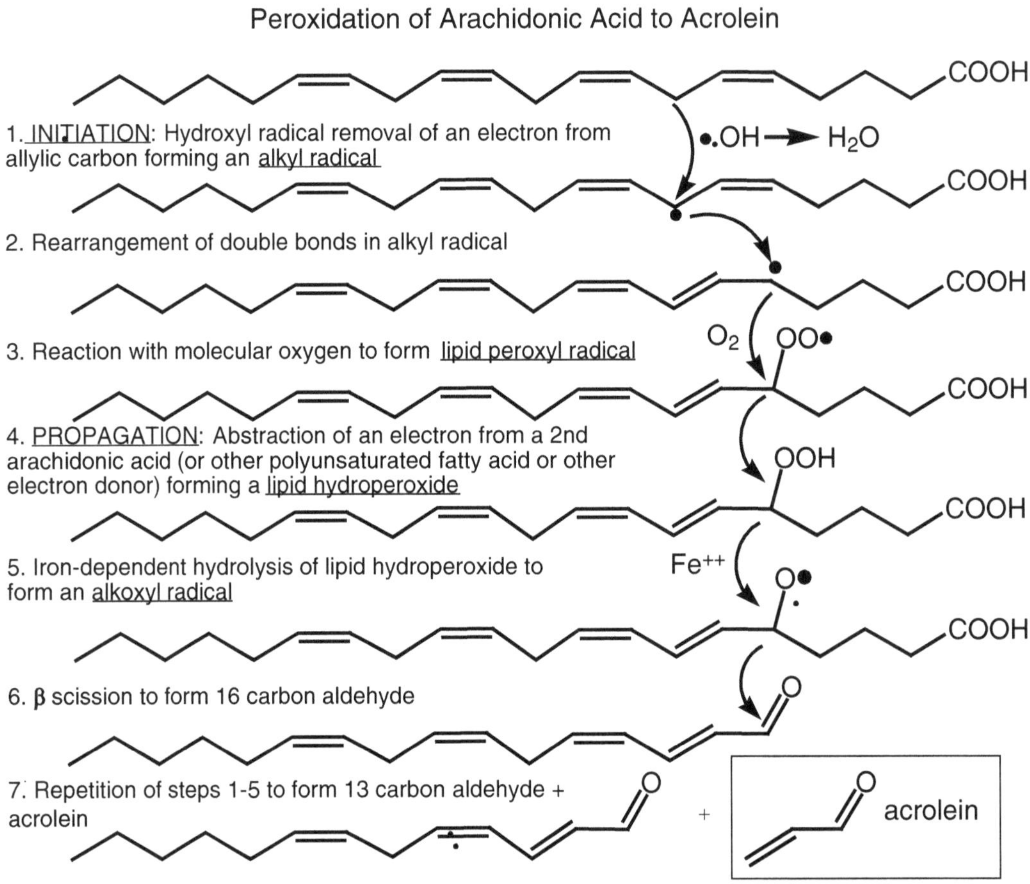

Fig. 2. Chemistry involved in the initiation, propagation, and termination reactions of arachidonic acid with the resulting formation of 2-propenal (acrolein).

2. Immunoblotting and Immunohistochemical Measurements of Lipid Peroxidation and Protein Oxidation

2.1. Western Immunoblotting of 4-Hydroxynonenal, Acrolein, 3-Nitrotyrosine, or Protein Carbonyl

2.1.1. Introduction

4-HNE is an amphiphilic product of LP which has a molecular weight of 156. The 4-HNE is produced due to peroxidation of omega-6 polyunsaturated fatty acids, including arachidonic (Fig. 1) and linoleic acids. It is known to concentrate in biomembranes. At higher than normal concentrations, 4-HNE has affinity for amino groups, thereby forming adducts to DNA, protein, or even phospholipids. 4-HNE is found in most tissues at a low concentration, 1 μM or less; this basal level is likely protein bound. The prominent reactions between 4-HNE and biomolecules are the Michael addition reaction between 4-HNE and biomolecular amino or thiol groups and the Schiff base reaction between amino groups and the carbonyl group of 4-HNE (Fig. 3). The reaction between phosphatidylethanolamine and 4-HNE involves both reactions. Increased levels of protein-bound 4-HNE have been detected by

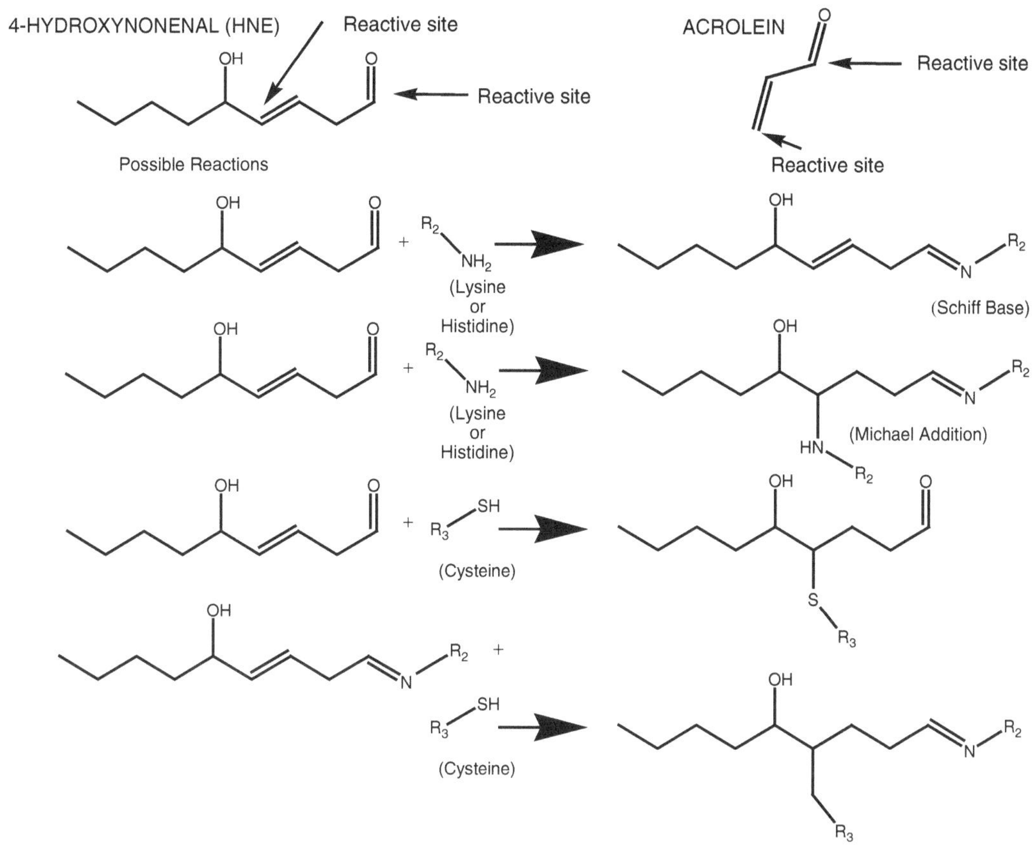

Fig. 3. Chemical reactions of the lipid peroxidation-derived aldehydes 4-HNE or acrolein with proteins by formation of either Schiff bases or Michael adducts. Reprinted with permission from Vaishnav et al. (25).

immunoblotting and immunohistochemistry in TBI (24) and SCI (25), supporting its utility as a biomarker for these conditions.

Acrolein. As noted in the overall Introduction to this chapter, another of the aldehydic LP products is 2-propenal which is more commonly known as acrolein. Recent work by Shi and colleagues (26, 27) has been focused on the role of acrolein in the pathophysiology of acute SCI. Moreover, acrolein and 4-HNE have recently been shown to be acutely toxic to spinal cord and brain mitochondria with acrolein being an order of magnitude more potent in that regard (19). Thus, there is increased reason to assay acrolein levels in models of CNS injury in addition to 4-HNE, discussed above.

3-NT is a product of tyrosine nitration mediated by reactive nitrogen species, such as peroxynitrite-derived nitrogen dioxide ($^{\cdot}NO_2$). It is detected in a number of pathological conditions and is considered a marker of $^{\cdot}$NO-dependent oxidative stress as occurs in TBI and SCI.

Protein carbonyl groups are introduced into proteins as a consequence of oxidative modification of certain residues as a hallmark

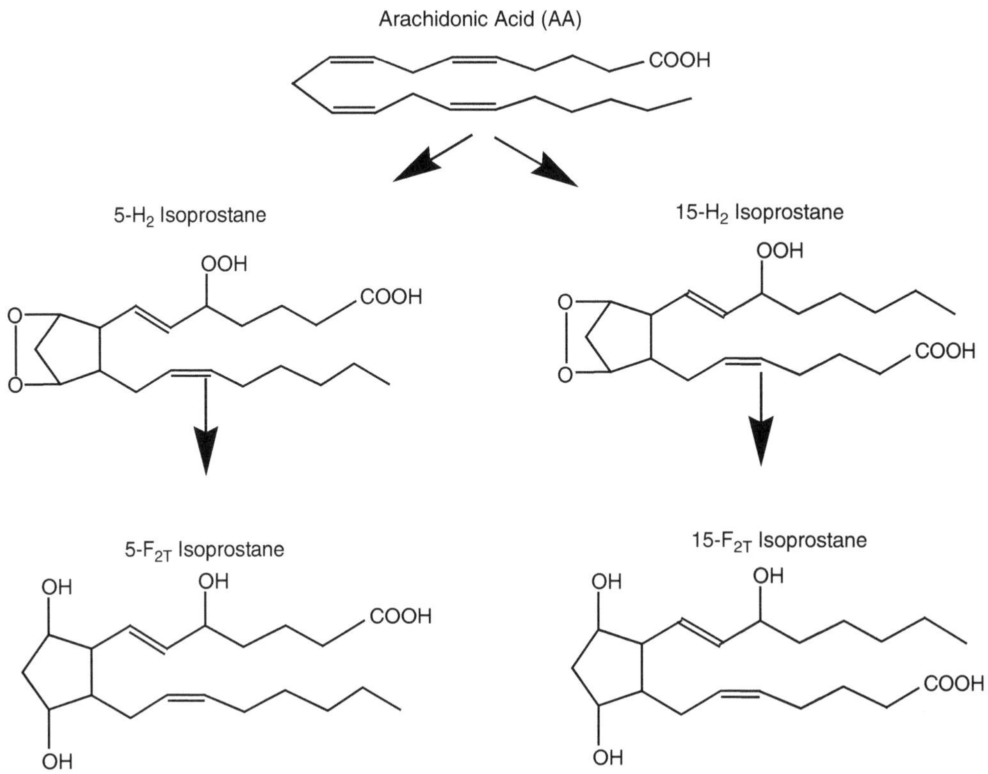

Fig. 4. Lipid peroxidation-derived formation of isoprostanes from arachidonic acid.

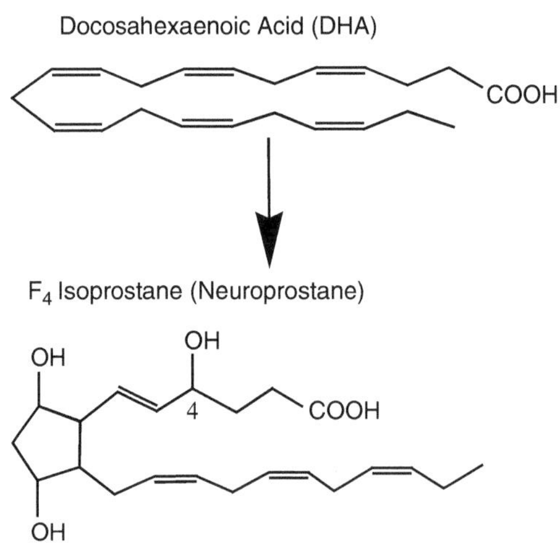

Fig. 5. Lipid peroxidation-derived formation of neuroprostanes from docosahexaenoic acid.

of the oxidation status of protein. The carbonyl groups are derivatized to 2,4-dinitrophenylhydrazone (DNP-hydrazone) by reaction with 2,4-dinitrophenylhydrazine (DNPH). The DNP-derivatized protein samples can be separated by polyacrylamide gel electrophoresis followed by Western immunoblotting or directly slot blotted onto a membrane to look at total carbonyl content.

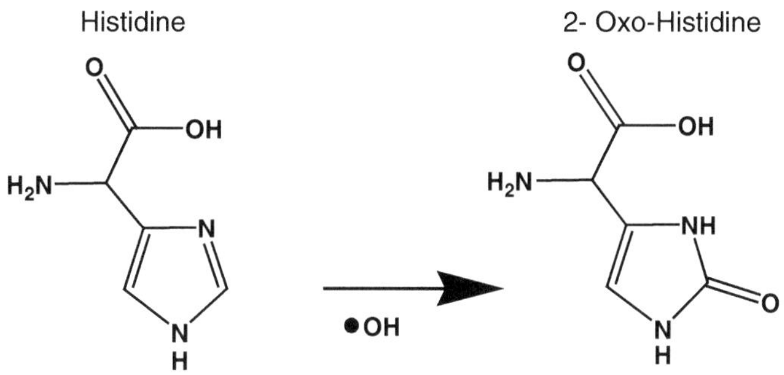

Fig. 6. Chemistry of oxidative modification of histidine to 2-oxohistidine: Carbonyl can be reacted with dinitrophenylhydrazine (DNPH) to produce a product that can be detected by either spectrophotometry or immunoblotting.

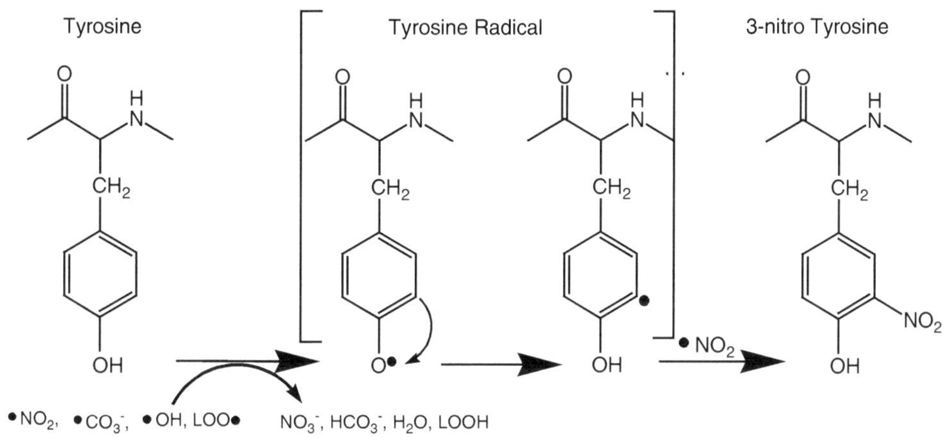

Fig. 7. Chemistry of tyrosine nitration: The phenolic hydroxyl moiety if oxidized by nitrogen dioxide, carbonate, hydroxyl or lipid peroxyl radical creating a tyrosyl radical. Next, the radical is relocated to the third position (meta) of the phenolic ring, where it is then able to react with a nitrogen dioxide radical creating a 3-nitrotyrosine.

2.1.2. Solutions and Reagents

- Triton lysis buffer for brain or spinal cord sample collection.

	Stock [ ]	Amount needed for 1 L	Final [ ]
Tris–HCl	1 M	20 ml	20 mM
NaCl	1 M	150 ml	150 mM
Triton X-100	20%	50 ml	1%
EGTA		1.902 g	5 mM
EDTA	500 mM	20 ml	10 mM
Glycerol		100 ml	10%
HEPES		5.528 g	20 mM

Note: Raise the final volume to 1,000 ml with deionized H_2O (pH of 7.4) and keep at 4°C (shelf life: 6 months)

  - EDTA—Sigma cat# E-7889.
  - EGTA—Sigma cat# 3889.
  - HEPES—Sigma cat# H0527.
- Protease inhibitor (PI) buffer:
  - Add one PI tablet (Roche Applied Science cat #. 1836 153) to 10 ml of Triton lysis buffer.
- Bio-Rad DC protein assay kit reagents.
- Reagent A cat.# 500-0113.
- Reagent B cat.#. 500-0114.
- Reagent S cat.# 500-0115.
- Bio-Rad Criterion™ XT precast gel, 26-well comb, 15 μl capacity, 1.0 mm thickness (12% Bis-Tris gel Cat# 345-0119).
- Bio-Rad 4× loading buffer Cat# 161-0791).
- Bio-Rad 20× reducing Agent Cat# 161-0792).
- Bio-Rad Precision-Plus Dual Color Standard Cat# 161-0374.
- Bio-Rad Criterion Blotting Sandwiches (0.2 μm Nitrocellulose) Cat# 162-0233.
- Bio-Rad Criterion™ extra thick blotting paper Cat# 1703967.
- XT MEX Running Buffer (Bio-Rad Cat# 161-0789), 20×.
- Transfer buffer recipe:

500 ml	Invitrogen NuPage® Transfer Buffer (20×) cat# NP0006
500 ml	Methanol
4,000 ml	Deionized water
5,000 ml	Total volume

- TBS:
  - Tris USP (MW = 121.14): 4.84 g.
  - NaCl: 58.48 g.
- In ddH_2O, pH to 7.5 to a final volume = 2 L.
- BLOTTO (5% milk + TBS: blocking buffer):
- Dissolve 5 g of dry nonfat milk per 100 ml TBS.
- TBST: Sigma Tris-buffered saline with Tween 20, pH 8.0 Cat# T-9039.
- 1° Antibodies:
  - 4-HNE: Monoclonal antibody (JaICA, Cat# MHN-020P).
  - Acrolein: Rabbit polyclonal anti-acrolein (Novus Biologicals Inc., Cat# NB200-556).

  - 3-NT: Nitrotyrosine Mouse IgG2b IgG monoclonal Antibody (GenWay Biotech, Inc. Cat#: 20-321-175259).
  - Protein carbonyl: Rabbit α-DNP antibody (Chemicon international, Cat: 90451, Chemicon International Oxyblot® Protein Oxidation Detection Kit S7150 or Sigma Rabbit α-DNP antibody: D9656).
- 2° Antibodies:
  - For 4-HNE and 3-NT: Rockland Immunochemicals IRDye800CW Conjugated Goat Anti-Mouse IgG (H&L) (Cat# 610-131-121).
  - For acrolein and protein carbonyl: Rockland Immunochemicals; IRDye800CW Conjugated GOAT Anti-Rabbit IgG (H&L) (Cat# 611-131-122).
- Invitrogen PowerEase 500 power supply.
- Sonic Dismembrator, Model 100, Fisher Scientific.
- Trans-Blot SD Semi-dry transfer cell (Bio-Rad).
- Li-Cor Odyssey Infrared Imager or equivalent.

2.1.3. Working Protocol

Collect Samples

(a) In a 1.5- or 2.0-ml centrifuge tube, add 500 μl of chilled PI buffer for cortex and 400 μl for hippocampus. Keep all sample tubes on ice.

(b) Add tissue and sonicate tube for 10 s on ice at a setting of 2 to completely disperse the tissues.

(c) Place tubes containing sonicated tissues on ice for 1 h, vortexing the tubes every 15 min to provide time for the lysis buffer and detergent to work. These samples can be stored for up to 3 months at –80°C until assayed.

(d) Centrifuge samples for 30 min at 14,000×*g* or maximum at 4°C.

(e) Keep the supernatant and store at –80°C.

Perform Protein Analysis

Determine the protein concentration of the samples, e.g., using a protein assay kit, such as the Bio-Rad DC protein assay kit.

Sample Preparation for Western Blot

1. Prepare samples:

   Sample: 14 μl/20 μg.

   4× Loading buffer: 5 μl.

   20× Reduce agent: 1 μl.

   Total volume: 20 μl.

   Load only 15 μl/15 μg/well without boiling the samples.

   Also use loading control to reduce variation among the blots (our lab uses pooled TBI samples as loading control). Mix samples thoroughly.

2. Prepare gels for electrophoresis:
   (a) Remove the comb and protective tape.
   (b) Fill the upper buffer chamber completely and the lower buffer chamber to the fill line.
   (c) Do not allow the middle space between gels to fill with buffer above the small partition. Doing so will cause short circuit of the apparatus and your gels will run very slowly.
   (d) If at all possible, do NOT use the outside lanes for important samples or load controls since these lanes often experience fading effects and do not quantify as accurately as inner lanes.
   (e) Run the gel under the following conditions:
      - 160 V constant.
      - 200 mA.
      - 25 Watts.
      - 2 h or run off the front dye.
      - The gel cassette should sit in ice.
3. Transfer to the membrane:
   (a) Remove the gel very carefully from the cassette and cut off the gel comb; cut the corner as a gel direction indicator. Submerge the gel into a tray of transfer buffer.
   (b) Equilibrate each gel for 10 min with gentle agitation on the rotator at RT. In a separate tray, also equilibrate enough membranes as needed for transfer.
   (c) After equilibrating the gels and membranes, wet two pieces of extra-thick blotting paper in transfer buffer and assemble the gel sandwich so that the membrane is beneath the gel and the two together are between the pieces of blotting paper.
   (d) Transfer the protein to the nitrocellulose membrane for 30 min at 15 V constant in Trans-Blot SD Semi-dry transfer cell (Bio-Rad). Be sure to thoroughly wet the blot to avoid the membrane to dry during the transfer.
4. Immunoblotting procedure:
   (a) When removing the membranes from the gel sandwich, it is easiest to turn the sandwich upside down and peel the membrane up from the gel bottom first. In case some of the gel sticks to the membrane, it will usually be at the top of the gel where the samples were loaded. The top part of the gel and nitrocellulose membrane can simply be trimmed off.

(b) Block for 1 h at RT in gel trays with 5% milk in TBS (blocking buffer).

(c) Apply 1° antibody:

- 4-HNE: Monoclonal antibody (JaICA, Cat# MHN-020P)—reconstitute antibody to concentration of 1 μg/10 μl with ddH_2O. Then, dilute with TBS-T 1:100 for working solution.
- Acrolein: Rabbit polyclonal anti-acrolein (Novus Biologicals Inc., Cat# NB200-556). Dilute in TBST 1:2,000 for working solution.
- 3-NT: Nitrotyrosine Mouse IgG2b IgG monoclonal Antibody (GenWay Biotech, Inc. Cat#: 20-321-175259). Dilute in TBS-T 1:2,000 for working solution.
- Protein carbonyl: Rabbit α-DNP antibody (Chemicon international, Cat: 90451, Chemicon International Oxyblot® Protein Oxidation Detection Kit S7150 or Sigma Rabbit α-DNP antibody: D9656). Dilute in TBS-T 1:2,000 for working solution.

(d) Wash the blot for 3 × 5 min in wash buffer (TBS-T).

(e) Apply 2° antibody: 1:5,000 with 5% milk in TBS-T for 1 h at RT. Be sure to keep blot box in dark.

- For 4-HNE and 3-NT: Rockland Immunochemicals IRDye800CW Conjugated Goat Anti-Mouse IgG (H&L) (Cat# 610-131-121).
- For acrolein and protein carbonyl: Rockland Immunochemicals; IRDye800CW Conjugated GOAT Anti-Rabbit IgG (H&L) (Cat# 611-131-122).

(f) Wash for 3 × 10 min in washing buffer (TBS-T).

(g) Scan the wet blot or dry blot using Li-Cor Odyssey Infrared Imager. It can be kept for 6 months.

Figure 8 displays Western blots for 4-HNE and 3-NT from hippocampal tissue taken from a TBI experiment in mice. It should be noted that the 3-NT and 4-HNE in the injured mouse brain samples parallel one another consistent with the likelihood that the tyrosine nitration at least partially involves LP-derived lipid peroxyl radicals as illustrated in the previous Fig. 7.

2.2. Slot Immunoblotting of 4-Hydroxynonenal, Acrolein, 3-Nitrotyrosine, or Protein Carbonyl

Slot blot is a tool used to detect the total content of the oxidatively modified 4-HNE, acrolein, 3-NT, or carbonylated proteins in a protein sample. This section outlines the procedure this laboratory uses with special reference to pitfalls and helpful hints.

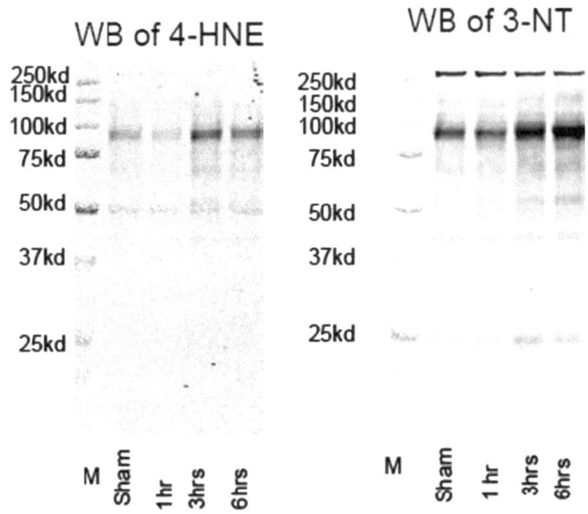

Fig. 8. Examples of 4-HNE and 3-NT Western blot from uninjured and controlled cortical impact (CCI) TBI mouse brains. Two things are notable: (1) for both 4-HNE and 3-NT, several proteins of various molecular weights show increase in posttraumatic oxidative damage reflecting the rather indiscriminate (shot gun) nature of free radical-induced oxidative damage and (2) there is considerable overlap in regards to the proteins that show 4-HNE and 3-NT immunostaining indicating the relationship of the two forms of oxidative damage and perhaps their each having a common initiator (e.g., peroxynitrite). For quantitative analysis of the blots, one can analyze individual protein bands at particular molecular weights, a portion of the blot within a certain range of molecular weights or the entire lane.

2.2.1. Solutions and Reagents

Same as in Sect. 2.1.2, except as follows:

1. Denaturing solution: 12% SDS.
2. Derivatizing solution: 0.1% DNP in 2 N HCl solution (store in light-protected bottle at 4 C).
3. Neutralization solution: 2 M Tris (pH 7.4).
4. 1° Antibodies—same as for Western blot protocol:
   (a) 4-HNE: Monoclonal antibody (JaICA, Cat# MHN-020P).
   (b) Acrolein: Rabbit polyclonal anti-acrolein (Novus Biologicals Inc., Cat# NB200-556).
   (c) 3-NT: Nitrotyrosine Mouse IgG2b IgG monoclonal Antibody (GenWay Biotech, Inc. Cat#: 20-321-175259).
   (d) Protein carbonyl: Rabbit α-DNP antibody (Chemicon international, Cat: 90451, Chemicon International Oxyblot® Protein Oxidation Detection Kit S7150 or Sigma Rabbit α-DNP antibody: D9656).
5. 2° Antibodies—same as for Western blot protocol:
   (a) For 4-HNE and 3-NT: Rockland Immunochemicals IRDye800CW Conjugated Goat Anti-Mouse IgG (H&L) (Cat# 610-131-121).

(b) For acrolein and protein carbonyl: Rockland Immunochemicals; IRDye800CW Conjugated GOAT Anti-Rabbit IgG (H&L) (Cat# 611-131-122).

6. Bio-Rad SF Filter Paper (Cat: 1620161).
7. Bio-Dot SF Microfiltration Apparatus (Cat: 170-6542).

2.2.2. Working Protocol

Same as in Sect. 2.1.3.1.

Collect Samples

Same as in Sect. 2.1.3.2.

Perform Protein Analysis

Sample Preparation for Slot Blot

1. Prepare samples for slot blot of 3-NT and 4-HNE.
   (a) Prepare the slot blot apparatus by thoroughly washing it out with soap and dH_2O. Wipe down with 95% ethanol and then dry.
   (b) Equilibrate the nitrocellulose membrane for 10 min in TBS.
   (c) Prepare the 2 μg protein sample per well with total volume of 30 μl TBS. Be sure to thoroughly mix each sample without creating bubbles.
   (d) Wet three pieces of slot-blot filter paper in TBS.
   (e) Assemble the apparatus as instructed in the slot-blot manual from Bio-Rad. Hand tightening of the screws is sufficient as using pliers can damage both the apparatus and the membrane.
   (f) Load the samples into the apparatus and triplicate the wells; allow them to filter through the membrane by gravity (NO VACUUM). Keep an eye on the samples and do not let membrane dry out completely until all samples have filtered through the membrane.
   (g) Add another 200 μl of TBS to each well, and let it filter through.
   (h) Disassemble the apparatus and proceed exactly as a normal immunoblot. Note: For slot blot of carbonyl:
      - To tube, add 2 μg/5 μl sample.
      - Add 5 μl 12% SDS.
      - Add 10 μl 0.1% DNP solution (stored in a light-protected bottle at 4°C).
      - Incubate samples for 10 min at RT.
      - Neutralize samples using 10 μl 2 M Tris, pH 7.4.
      - After this preparation, you can follow the preparation for slot blot of 3-NT and 4-HNE given immediately above.

2. Immunoblotting procedure:

   (a) Block for 1 h at RT in a tray with 5% milk in TBS (blocking buffer).

   (b) Apply 1° antibody:

   - 4-HNE: Monoclonal antibody (JaICA, Cat# MHN-020P)—reconstitute antibody to concentration of 1 μg/10 μl with ddH_2O. Then, dilute with TBS-T 1:100 for working solution.
   - Acrolein: Rabbit polyclonal anti-acrolein (Novus Biologicals Inc., Cat# NB200-556). Dilute in TBST 1:2,000 for working solution.
   - 3-NT: Nitrotyrosine Mouse IgG2b IgG monoclonal Antibody (GenWay Biotech, Inc., Cat#: 20-321-175259). Dilute in TBS-T 1:2,000 for working solution.
   - Protein carbonyl: Rabbit α-DNP antibody (Chemicon International, Cat: 90451, Chemicon International Oxyblot® Protein Oxidation Detection Kit S7150 or Sigma Rabbit α-DNP antibody: D9656). Dilute in TBS-T 1:2,000 for working solution.

   (c) Wash the blot for 3 × 5 min in wash buffer (TBS-T).

   (d) Apply 2° antibody: 1:5,000 with 5% milk in TBS-T for 1 h at RT. Be sure to keep blot box in dark.

   - For 4-HNE and 3-NT: Rockland Immunochemicals IRDye800CW Conjugated Goat Anti-Mouse IgG (H&L) (Cat# 610-131-121).
   - For acrolein and protein carbonyl: Rockland Immunochemicals; IRDye800CW Conjugated GOAT Anti-Rabbit IgG (H&L) (Cat# 611-131-122).

   (e) Wash for 3 × 10 min in washing buffer (TBS-T).

   (f) Scan the wet blot or dried blot using Li-Cor Odyssey Infrared Imager. It can be kept for at least 6 months.

Figure 9 shows examples of slot blots of 4-HNE, 3-NT, and protein carbonyl content as a function of time after contusion SCI in rat thoracic spinal cord.

2.3. 4-Hydroxynonenal and 3-Nitrotyosine Immunohistochemistry

2.3.1. Protocol for 4-HNE and 3-NT (2 Days)

Equipment and Reagents

1. Pipette aid and 50-ml pipettes.
2. 12-Well plates.
3. Tissue wells.
4. Sharpies.
5. Paint brush.
6. Beakers and stir bars.

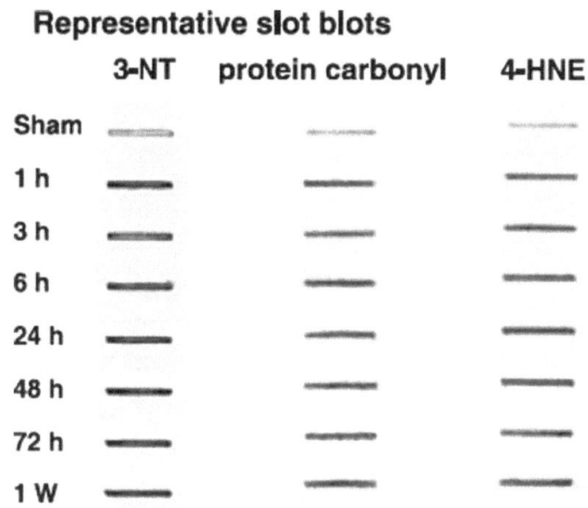

Fig. 9. Examples of protein carbonyl, 4-HNE, and 3-NT slot immunoblots showing a time course of posttraumatic oxidative damage in rat thoracic spinal cord following contusion spinal cord injury. Blots are reproduced from ref. 25 with permission.

7. Tape.
8. 1 M stock to make 0.2 M PBS.
9. 0.1 M MOPS, pH 8.0.
10. $NaBH_4$.
11. H_2O_2.
12. Dry milk.
13. Goat serum.
14. Triton X-100.
15. Primary antibodies.
16. Secondary antibody kit.
17. Vector Blue Kit.
18. Tris–HCl.
19. Tap water.
20. Aluminum foil.

Day One

1. Set up dishes and lids. Fill plates with 0.2 M PBS. Transfer tissue into new plates.
2. Rinse sections for 3 × 5 min in 0.2 M PBS.
3. For 4-HNE only:
   - Reduce tissue for 10 min in 0.1 M $NaBH_4$ in 0.1 M MOPS, pH = 8.0.
   - To make 1 L 0.1 M MOPS, add one packet MOPS powder to 1 L dH_2O.

	$NaBH_4$	0.1 M MOPS
100 ml	0.38 g	100 ml
200 ml	0.76 g	200 ml
400 ml	1.52 g	400 ml
800 ml	3.04 g	800 ml
1 L	3.8 g	1 L

- Rinse sections for 3 × 5 min in 0.2 M PBS.

4. Incubate sections in 0.3% H_2O_2 in 0.2 M PBS for 30 min.

	30% H_2O_2	0.2 M PBS
100 ml	1 ml	99 ml
200 ml	2 ml	198 ml
400 ml	4 ml	396 ml
800 ml	8 ml	792 ml
1 L	10 ml	980 ml

5. Make blocking buffer and store extra at 4°C.

	1% Dry milk	Triton X-100	Goat serum	0.2 M PBS
100 ml	1 g	250 μl	5 ml	100 ml
200 ml	2 g	500 μl	10 ml	200 ml
400 ml	4 g	1 ml	20 ml	380 ml
800 ml	8 g	2 ml	40 ml	760 ml
1 L	10 g	2.5 ml	50 ml	950 ml
1.2 L	12 g	3 ml	60 ml	1,150 ml
1.4 L	14 g	3.5 ml	70 ml	1,330 ml

6. Rinse sections for 3 × 5 min in 0.2 M PBS.
7. Block sections for 2 h in blocking buffer:

   0.2 M PBS, 5% goat serum, 0.25% Triton X-100, and 1% dry milk.
8. Incubate sections for 24 h in primary antibody solution (blocking buffer + 1° Ab).

   - Rabbit Anti-4-HNE 1:5,500 (Calbiochem in −80°C).
   - Rabbit Anti-3-NT 1: 1,200 (Upstate in −20°C).

Day Two

1. Rinse sections for 6 × 5 min with 0.2 M PBS.

2. Incubate sections with Secondary Antibody Solution (Blocking Buffer + Biotinylated Goat anti-rabbit 2° Antibody) for 2 h at room temperature 1 drop/20 ml (Vector ABC-AP kit, cat# AK-5001).

	2° Ab	Blocking buffer
100 ml	5 drops	100 ml
200 ml	10 drops	200 ml
400 ml	20 drops	400 ml
800 ml	40 drops	800 ml
1 L	50 drops	1 L

3. Prepare ABC reagent.
4. Rinse for 3 × 10 min with 0.2 M PBS.
5. ABC reagent made 30 min before use; incubate section for 1 h 1 drop/5 ml.

	A	B	0.2 M PBS
100 ml	20 drops	20 drops	100 ml
200 ml	40 drops	40 drops	200 ml
400 ml	80 drops	80 drops	400 ml
800 ml	160 drops	160 drops	800 ml
1 L	200 drops	200 drops	1 L

6. Prepare 0.1 M Tris–HCl using 1 M stock; fill the plates with it.
7. Rinse sections for 3 × 5 min with 0.1 M Tris–HCl, pH = 8.2.
8. Vector blue method: incubate sections for 10–20 min, covered with foil (light sensitive) two drops/5 ml (cat# SK-5300).

	Reagent 1	Reagent 2	Reagent 3	dH_2O
100 ml	40 drops	40 drops	40 drops	100 ml
200 ml	80 drops	80 drops	80 drops	200 ml
400 ml	160 drops	160 drops	160 drops	400 ml
800 ml	320 drops	320 drops	320 drops	800 ml
1 L	400 drops	400 drops	400 drops	1 L

9. Fill the plates with tap water.
10. Rinse for 3–4 × 5 min in tap water.
11. Mount sections on slides, let dry, counterstain very lightly with nuclear fast red (optional) for 1–10 min, dehydrate, and cover lid.

2.3.2. Nuclear Fast Red Counter Staining Protocol

Equipment and Materials Needed

Slide holders and handles.

Glass pipette and bulb.

Permount.

Coverslips.

Spatula.

Dental bib.

Tray.

100 and 95% EtOH.

dH_2O.

Fast red stain.

Citrisolv.

Timer.

Working Protocol

1. Refill the staining wells with the appropriate solutions. Always use fresh EtOH and dH_2O. The Nuclear Fast Red Stain and the Citrisolv can be reused.
2. Place the slides into the slide holders and then allow them to sit in the following:

95% EtOH	5 min
70% EtOH	5 min
50% EtOH	5 min
dH_2O	5 min
Nuclear fast red	2 min
dH_2O	8 dips
50% EtOH	8 dips
70% EtOH	8 dips
95% EtOH	8 dips
100% EtOH	5 min
Citrisolv I	5 min
Citrisolv II	5 min

3. Allow the slides to remain in Citrisolv until you are ready to coverslip. "The Citrisolv and the Permount should only be used in the hood."
4. To coverslip, run a thin line of permount along the bottom edge of the slide using the glass pipette.
5. Use the spatula to gently lower the coverslip onto the slide making sure that there are no air bubbles. If bubbles are present, press gently with the spatula to remove them.

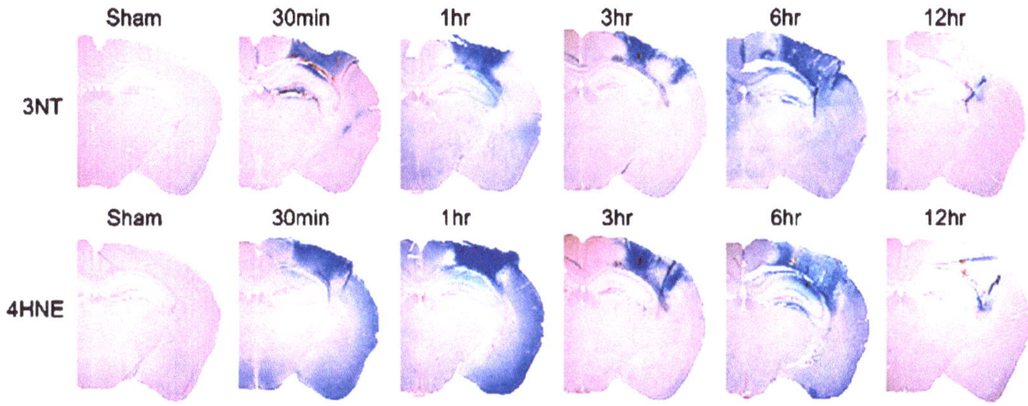

Fig. 10. Examples of 4-HNE and 3-NT immunochemistry from mice subjected to controlled cortical impact (CCI) TBI. Reproduced with permission from ref. 24.

6. Place the slides on a tray lined with a dental bib to dry overnight.

Figure 10 shows examples of 4-HNE and 3-NT immunostaining as a function of time after controlled cortical impact TBI in mice.

2.4. Instrument-Based Measurement of F_{2t}-Isoprostanes and/or F_4-Neuroprostanes

2.4.1. Introduction

F_2-Isoprostanes and F_4-neuroprostanes are products of in situ free radical-induced oxidation of esterified unsaturated fatty acids. In addition to representing markers of LP, some of the isomeric products of this oxidation have pathophysiological activities (28). Isoprostanes are the product of the free radical-induced oxidation of arachidonic acid (20:4, Ω-6). F_4-neuroprostanes are the products of the free radical-induced oxidation of DHA (22:6, Ω-3) (29). Both isoprostanes and neuroprostanes are biomarkers of nervous system injury.

In this assay, isoprostanes and/or neuroprostanes are released from cerebral tissue by use of Folch extraction followed by base hydrolysis. The hydrolysates next undergo solid-phase extraction to remove interfering substances. The extracts are next reacted to form pentafluoro-benzyl ester derivatives. These derivatives are further purified by the use of thin-layer chromatography (TLC). The esters of interest are then scraped from the plates and extracted with ethyl acetate. Finally, the evaporated esters are further derivatized to form trimethyl-silyl ethers by the addition of bis-silyltrifluoroacetamide. An aliquot of the final sample is injected onto the gas chromatographic system to separate its substituents which then can be analyzed by the use of negative ion chemical ionization to form diagnostic fragments that are detected by the mass spectrometer and reported. The areas of departure between the analysis of isoprostanes and neuroprostanes are designated with an (*) and the neuroprostane instruction will follow.

2.4.2. Materials and Solutions

1. Solid phase extraction column, C_{18}—#WAT020805, Waters Corporation, Milford, CT.
2. Solid phase extraction column, Silica—#WAT020810, Waters Corporation.
3. TLC plates, channeled, Silica Gel HL, 250 μM thickness, #43711, Analtech.
4. Folch solution: Chloroform:methanol 2:1, *v/v* containing 0.005% BHT. If preparing 1 L, use 667 ml of chloroform with 333 ml of methanol—dissolve 50 mg of BHT in the solution. Store at −20°C in a dark bottle.
5. 0.9% Saline: Commercially prepared 0.9% saline is available. If preparing 1 L, dissolve 9 g of sodium chloride in 1 L of deionized water.
6. Methanol containing 0.005% BHT: It is suggested to prepare 100 ml—add 5 mg of BHT to methanol and bring up to 100-ml volume.
7. 15% KOH: Carefully dissolve 15 g of KOH in methanol (caution—hot!) and bring up to 100-ml final volume.
8. pH 3 solution: Make up 1 L of 0.001 M hydrochloric acid, and adjust to pH 3.
9. 2 mM BHT: Dissolve 44 mg of BHT (MW = 220.4) in methanol, and make up to final volume of 100 ml.
10. Ethyl acetate:heptane 1:1: *v/v*; mix equal volumes of ethyl acetate and heptane. Prepare fresh on the day of assay.
11. Sodium sulfate: Preferably powdered.
12. Ethyl acetate:methanol, 1:1: *v/v*; mix equal volumes of ethyl acetate and methanol. Prepare fresh on the day of assay.
13. Ethyl acetate:heptane 3:1: *v/v*; mix three volumes of ethyl acetate with one volume of heptane. Prepare fresh on the day of assay.
14. Pentafluorobenzyl bromide (PFBB) 10%: Add one volume of PFBB to nine volumes of acetonitrile, and mix thoroughly. Prepare fresh before use. *Rarely needed in volumes greater than 1 ml.
15. Diisopropylethanolamine (DIPE) 10%: Add one volume of DIPE to nine volumes of acetonitrile, and mix thoroughly. Prepare fresh before use. *Rarely needed in volumes greater than 1 ml.
16. Thin-layer chromatographic wash solution: Ethyl acetate: ethyl alcohol 9:1: For conventional TLC developing tank

(~11″×10″×2.5″) add 10 ml of ethyl alcohol to 90 ml of ethyl acetate, mix in a 250 ml beaker before decanting into tank. *Be sure to cover tank to reduce evaporation. Prepare fresh before washing plates.

17. TLC developing solution: chloroform:ethyl alcohol 93:7: For conventional TLC developing tank (~11″×10″×2.5″), add 7 ml of ethyl alcohol to 93 ml of chloroform, and mix in 250-ml beaker before decanting into tank. *Be sure to cover the tank to reduce evaporation. Prepare fresh before developing plates.
18. 10% Phosphomolybdic acid solution: This solution is commercially available, i.e., Sigma #P4869. Store at 4°C, shield from light with aluminum foil as it is light sensitive.
19. *N,O*-bis[Trimethylsilyl]trifluoroacetamide (BSTFA): Available in different sized vials, i.e., Thermo-Fisher #TS-38830. Store at 4°C.

2.4.3. Working Protocol

Tissue Extraction and Hydrolysis

1. Weigh either freshly collected tissue or tissue which has been stored at −80°C and place in 10 ml of ice-cold Folch solution. Record the weight of sample.
2. While sample tube is immersed in ice, disrupt tissue using homogenizer for at least 30 s (such as Fisher Scientific Power Gen Model 125 Homogenizer, Fisher Scientific, #14-261-02 and Flat-bottom stator, Fisher Scientific, #14-261-11).
3. Introduce nitrogen into the sample via gentle flow from nitrogen source, quickly cap the sample, and allow it to sit at room temperature for 30–60 min.
4. Add 4 ml of 0.9% NaCl to each sample, recap sample, and vortex vigorously.
5. Centrifuge samples at room temperature at 1,800×*g* for 10 min.
6. After centrifugation, carefully remove and discard the upper layer.
7. Carefully aliquot the lower organic layer of each sample to 10-ml centrifuge tubes; do not transfer any of the solid intermediate layer.
8. Evaporate the contents of the 10-ml centrifuge tubes under nitrogen at 37°C.
9. Perform the hydrolysis step by adding 1.0 ml of methanol containing BHT (0.005%) and 1.0 ml of KOH (15%), mixing thoroughly and heating at 37°C for 30 min.
10. After cooling the samples to room temperature, adjust each sample to pH 3 by use of 1 M HCl.

11. Dilute each sample with pH 3 solution to a volume appropriate to dilute the methanol to less than 5% of the total volume (i.e., for 1 ml as above, dilute the sample to at least 20 ml).
12. Add 1 ng of internal standard 15-F_{2t}-isoprostane-d_4 to each sample.

Solid-Phase Extraction

1. Use Waters C_{18} columns; prepare columns with 5 ml of methanol and allow the columns to clear, but do not let C_{18} columns dry out at any time.
2. Add 7 ml of pH 3 solution to each column.
3. Add samples to columns; flow rate through columns should be approximately 1 ml/min.
4. Wash columns with 10 ml of pH 3 solution.
5. Add 10 ml of heptane to columns.
6. Collect the elutions from the columns and then collect eluent after passing 10 ml of ethyl acetate:heptane 1:1 through them.
7. Pour each eluent sample into a beaker containing sodium sulfate; this serves to scavenge any remaining water from the sample eluent. Carefully transfer the liquid from the beaker into a different container so that the sample is ready for the next extraction.
8. Replace the C_{18} columns with Waters Silica columns.
9. Prepare the silica columns with 5 ml of ethyl acetate.
10. Apply the samples from the C_{18} columns to the columns at a flow rate of ~1 ml/min.
11. Add 5 ml of ethyl acetate to the columns*. (Neuroprostane: Add 5 ml of mixture of ethyl acetate:heptane 3:1, *v*/*v*.)
12. Elute the analytes of interest with 5 ml of ethyl acetate:methanol, 1:1 (*v*/*v*).
13. Evaporate the final eluted samples under nitrogen at 37°C.

First Derivatization, Pentafluorobenzyl Ester Formation

1. Make up the two solutions necessary: the 10% PFBB and the 10% DIPE.
2. Add 50 μl of the PFBB and 25 μl of DIPE to each sample, and vortex thoroughly.
3. React for 20 min at 37°C.
4. After the reaction time is complete, evaporate the reaction mixture under nitrogen at 37°C.

Thin-Layer Chromatographic Purification

1. The TLC plates used in this lab provide four lanes per plate. Determine how many plates are required for samples and add one for the standard plate. Visually inspect the plates for any

defects, such as poor surface consistency; do not use any lanes which have inconsistencies.

2. Prewash the plates to be used with a solvent system of ethyl acetate:ethyl alcohol, 90:10 (*v/v*). Allow the solvent to travel to the top of each plate before removing it from the chamber.
3. Allow the plates to air dry for a few minutes and then put the plates into an oven of at least 200°C for 15 min to complete drying. After drying, place the plates in a desiccator to reduce moisture.
4. Reconstitute the evaporated samples from Sect. 3.1.3.3, step 4, with 50 μl of a mixture of methanol:chloroform 3:2, *v/v*.
5. Remove the plates from the desiccator. Mark each plate with a pencil line 13 cm above the application area at the bottom of the plate. Label the top of each lane, in pencil, with its sample number. Reserve one plate to run the standard solution.
6. Add the developing mobile-phase solution ($CHCl_3$:EtOH, 93:7, *v/v*) to the developing chamber along with a piece of filter paper large enough to occupy the back wall of the camber, cover, and seal.
7. Carefully apply each sample to its lane using the top half of the application area. Allow the area to dry.
8. Apply 2–5 μg of Prostaglandin $F_2\alpha$, methyl ester to the application area of at least two of the lanes on the standard solution plate.
9. Verify that the samples are dry on the plates and put them into the chamber leaning forward. Allow the plates to develop until the solvent reaches the 13-cm line.
10. Remove the plates from the chamber and allow them to air dry.
11. Set the sample plates aside and then spray the standard solution plate with the 10% phosphomolybdic acid solution.
12. Allow the sprayed plate to sit and dry for a minute and then gently place the TLC standard plate on a heating plate. Do not preheat the heating plate! After a few minutes of heating, the distinctive blue bands of the standard should appear. Remove the TLC standard plate from the heat.
13. Make a mark at the corresponding point on the sample plates equal to the development point of the standard band.
14. Remove the silica from the sample lane from a point 1 cm below to 1 cm above the standard band point. Collect the silica on a piece of glassine weighing paper*. (Neuroprostane: Remove the silica from the sample lane from a point 1 cm below to 4 cm above the standard band point).

15. Transfer the silica from each sample into a corresponding 1.5-ml microcentrifuge tube. Add 1 ml of ethyl acetate to each sample tube, cap, and vortex thoroughly.
16. Centrifuge the samples at 13,000 × *g* at 4°C. Transfer each sample supernatant to 10-ml centrifuge tubes. It is advisable to perform two ethyl acetate extractions to optimize recovery. Note: The samples can be stored in ethyl acetate at −20°C overnight if necessary.

Final Derivatization

1. Evaporate the ethyl acetate under nitrogen at 37°C.
2. Add 20 μl BSTFA to each sample, cap, vortex thoroughly, and react at 60°C for 20 min.
3. Allow the samples to cool and then transfer to appropriate vials for GC/MS analysis.

Gas Chromatographic/ Mass Spectrometric Analysis

The only requirement regarding the gas chromatographic column to be used for this analysis is that the stationary phase should reflect a polarity close to 5% diphenyl/95% dimethyl polysiloxane. This laboratory uses a column with an initial length of 30 m. Two microliters of sample are injected onto the column in splitless mode which is maintained for 2 min. Injector temperature is 250°C. The column is set to a temperature of 170°C and held for 2 min and then the temperature is increased at a rate of 18°C/min to 300°C and held there for 7 min. The temperature of the interface is held at 300°C and the ion source temperature is set to 250°C.

Analysis is carried out in selected ion monitoring (SIM) mode. The quantitation fragment ion for isoprostanes is *m/z* 569.3, and for the deuterated internal standard the quantitation fragment ion is *m/z* 573.4. A typical chromatogram is shown in Fig. 11a. The deuterated internal standard quantitation ion should be detected seconds ahead of the analyte diagnostic ion. Quantitation is performed by comparing the abundance of the known quantity of the internal standard to the unknown*. The quantitation fragment ion for neuroprostanes is *m/z* 593.4. The neuroprostanes elute over 1 min; an example of integrated area is provided in Fig. 11b.

2.5. Enzyme-Linked Immunoassay of Isoprostanes

It is strongly recommended that whenever possible the best technique to employ for measurement of the isoprostanes in regards to sensitivity and selectivity is GC–MS. However, there are less labor-intensive commercial EIA kits (e.g., Cayman Chemical Co.) that have been used in CNS injury studies that probably provide reasonable estimates of the levels of LP-derived F_2 isoprostanes in biofluids and tissues (30–32).

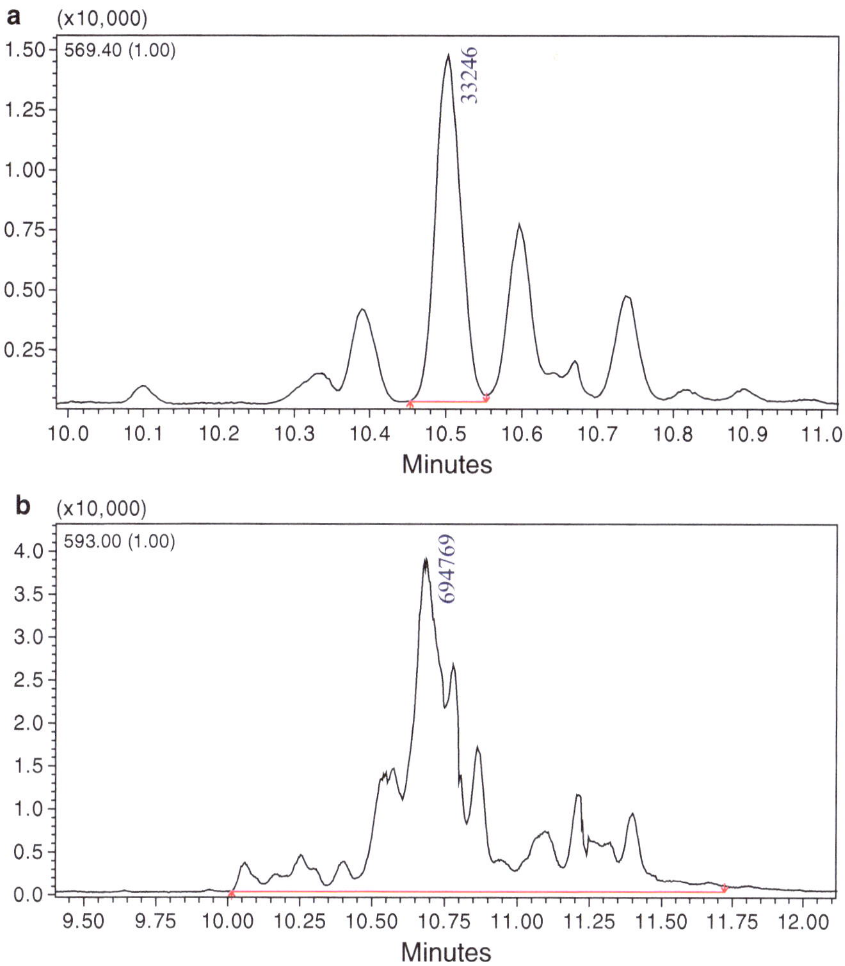

Fig. 11. Examples of GC chromatograms of 5 and 15 F_{2t}-isoprostanes and F_4-neuroprostanes in mouse cerebral tissue. (**a**) Representative ion chromatogram displaying integrated peak representing 5 and 15 F_{2t}-isoprostanes (*m/z* 569) in murine cerebral cortex. The deuterated internal standard has a similar retention time, yet has an *m/z* value of 573. (**b**) Representative ion chromatogram displaying integrated peak representing F_4-neuroprostanes (*m/z* 593) in murine cerebral cortex. The deuterated internal standard has a retention time of 9.6 min. The *horizontal red lines* show the portion of the chromatograms that were integrated.

References

1. Halliwell B, Gutteridge J (2008) Free radicals in biology and medicine, 3rd edn. Oxford University Press, Oxford
2. Demopoulos HB, Flamm ES, Pietronigro DD, Seligman ML (1980) The free radical pathology and the microcirculation in the major central nervous system disorders. Acta Physiol Scand Suppl 492:91–119
3. Demopoulos HB, Flamm ES, Seligman ML, Pietronigro DD, Tomasula J, DeCrescito V (1982) Further studies on free-radical pathology in the major central nervous system disorders: effect of very high doses of methylprednisolone on the functional outcome, morphology, and chemistry of experimental spinal cord impact injury. Can J Physiol Pharmacol 60:1415–1424
4. Hall ED, Braughler JM (1982) Glucocorticoid mechanisms in acute spinal cord injury: a review and therapeutic rationale. Surg Neurol 18: 320–327
5. Hall ED, Braughler JM (1982) Effects of intravenous methylprednisolone on spinal cord lipid peroxidation and Na++ K+)-ATPase activity.

Dose–response analysis during 1st hour after contusion injury in the cat. J Neurosurg 57:247–253
6. Kontos HA, Povlishock JT (1986) Oxygen radicals in brain injury. Cent Nerv Syst Trauma 3:257–263
7. Kontos HA, Wei EP (1986) Superoxide production in experimental brain injury. J Neurosurg 64:803–807
8. Hall ED, McCall JM, Means ED (1994) Therapeutic potential of the lazaroids (21-aminosteroids) in acute central nervous system trauma, ischemia and subarachnoid hemorrhage. Adv Pharmacol 28:221–268
9. Hall ED, Braughler JM (1993) Free radicals in CNS injury. Res Publ Assoc Res Nerv Ment Dis 71:81–105
10. Smith SL, Andrus PK, Zhang JR, Hall ED (1994) Direct measurement of hydroxyl radicals, lipid peroxidation, and blood–brain barrier disruption following unilateral cortical impact head injury in the rat. J Neurotrauma 11:393–404
11. Globus MY, Alonso O, Dietrich WD, Busto R, Ginsberg MD (1995) Glutamate release and free radical production following brain injury: effects of posttraumatic hypothermia. J Neurochem 65:1704–1711
12. Hall ED, Andrus PK, Yonkers PA (1993) Brain hydroxyl radical generation in acute experimental head injury. J Neurochem 60:588–594
13. Azbill RD, Mu X, Bruce-Keller AJ, Mattson MP, Springer JE (1997) Impaired mitochondrial function, oxidative stress and altered antioxidant enzyme activities following traumatic spinal cord injury. Brain Res 765:283–290
14. Sullivan PG, Thompson MB, Scheff SW (1999) Cyclosporin A attenuates acute mitochondrial dysfunction following traumatic brain injury. Exp Neurol 160:226–234
15. Sullivan PG, Bruce-Keller AJ, Rabchevsky AG et al (1999) Exacerbation of damage and altered NF-kappaB activation in mice lacking tumor necrosis factor receptors after traumatic brain injury. J Neurosci 19:6248–6256
16. Matsushita M, Xiong G (1997) Projections from the cervical enlargement to the cerebellar nuclei in the rat, studied by anterograde axonal tracing. J Comp Neurol 377:251–261
17. Sullivan PG, Krishnamurthy S, Patel SP, Pandya JD, Rabchevsky AG (2007) Temporal characterization of mitochondrial bioenergetics after spinal cord injury. J Neurotrauma 24:991–999
18. Sullivan PG, Rabchevsky AG, Keller JN et al (2004) Intrinsic differences in brain and spinal cord mitochondria: Implication for therapeutic interventions. J Comp Neurol 474:524–534
19. Vaishnav RA, Singh IN, Miller DM, Hall ED (2010) Lipid peroxidation-derived reactive aldehydes directly and differentially impair spinal cord and brain mitochondrial function. J Neurotrauma 27:1311–1320
20. Braughler JM, Hall ED (1989) Central nervous system trauma and stroke. I. Biochemical considerations for oxygen radical formation and lipid peroxidation. Free Radic Biol Med 6:289–301
21. Hall ED, Braughler JM (1989) Central nervous system trauma and stroke. II. Physiological and pharmacological evidence for involvement of oxygen radicals and lipid peroxidation. Free Radic Biol Med 6:303–313
22. Kontos HA (1989) Oxygen radicals in CNS damage. Chem Biol Interact 72:229–255
23. Povlishock JT, Kontos HA (1992) The role of oxygen radicals in the pathobiology of traumatic brain injury. Hum Cell 5:345–353
24. Deng Y, Thompson BM, Gao X, Hall ED (2007) Temporal relationship of peroxynitrite-induced oxidative damage, calpain-mediated cytoskeletal degradation and neurodegeneration after traumatic brain injury. Exp Neurol 205:154–165
25. Xiong Y, Rabchevsky AG, Hall ED (2007) Role of peroxynitrite in secondary oxidative damage after spinal cord injury. J Neurochem 100:639–649
26. Hamann K, Shi R (2009) Acrolein scavenging: a potential novel mechanism of attenuating oxidative stress following spinal cord injury. J Neurochem 111:1348–1356
27. Luo J, Uchida K, Shi R (2005) Accumulation of acrolein-protein adducts after traumatic spinal cord injury. Neurochem Res 30:291–295
28. Roberts LJ 2nd, Morrow JD (2002) Products of the isoprostane pathway: unique bioactive compounds and markers of lipid peroxidation. Cell Mol Life Sci 59:808–820
29. Roberts LJ 2nd, Montine TJ, Markesbery WR et al (1998) Formation of isoprostane-like compounds (neuroprostanes) in vivo from docosahexaenoic acid. J Biol Chem 273: 13605–13612
30. Bayir H, Marion DW, Puccio AM et al (2004) Marked gender effect on lipid peroxidation after severe traumatic brain injury in adult patients. J Neurotrauma 21:1–8
31. Bayir H, Adelson PD, Wisniewski SR et al (2009) Therapeutic hypothermia preserves

antioxidant defenses after severe traumatic brain injury in infants and children. Crit Care Med 37:689–695

32. Kochanek PM, Berger RP, Bayir H, Wagner AK, Jenkins LW, Clark RS (2008) Biomarkers of primary and evolving damage in traumatic and ischemic brain injury: diagnosis, prognosis, probing mechanisms, and therapeutic decision making. Curr Opin Crit Care 14:135–141

Chapter 33

Neurobehavioral Assessments of Traumatic Brain Injury

James W. Bales, Kristin Macfarlane, and C. Edward Dixon

Abstract

Assessment of neurologic deficit following traumatic brain injury (TBI) is integral to effective evaluation of injury severity, course, and recovery. Paradigms that involve training regimens can be difficult to employ and may not be feasible for short survival time points or simple monitoring of injury level. Acute reflex suppression testing and chronic composite neurobehavioral scoring rely upon innate animal responses that allow for quick, reproducible, and quantifiable measurements of injury severity and recovery. The utilization of these examinations grants researchers the ability to monitor the effect of injury upon behavior and assign injury severity without the necessity of ongoing histological characterization. Daily examinations of simple neurological deficits are also useful in monitoring animal status postinjury.

Key words: Traumatic brain injury, Neuroscore, Reflex suppression, Recovery

1. Introduction

1.1. Historical Perspective

Examination of neurologic function following an experimental traumatic brain injury (TBI) has been integral to TBI research since suppressed reflex testing was first described by Denny-Brown and Russell (1). Early paradigms limited themselves to the assessment of unconsciousness as a marker for neurologic function following injury (2–4). Most criteria for establishing the level or severity of injury was based predominantly upon a calculation of the forces involved or a histological analysis of traumatic damage (5, 6). While important to evaluating the neuronal, axonal, and vascular damage that occurred due to a TBI, the necessity of dissection made these assessments difficult to translate to clinical outcomes of surviving patients. In the early 1980s, more detailed cardiovascular monitoring (heart rate, respiratory rate, blood pressure, perfusion pressure) coupled with tests of neurologic deficit (consciousness, corneal reflex) were used in monkeys to enhance the clinical applicability of injury models (7, 8). In 1974, Teasdale

Jun Chen et al. (eds.), *Animal Models of Acute Neurological Injuries II*, Springer Protocols Handbooks,
DOI 10.1007/978-1-61779-782-8_33, © Springer Science+Business Media, LLC 2012

and Jennett (9) introduced the Glasgow Coma Scale (GCS) to provide a reproducible, consistent scoring system for TBI. Based upon the GCS a battery, of acute neurological tests was developed to examine the immediate severity and scope of experimental TBI in felines (10) and rats (11). To evaluate gross neurological motor function, over a period of recovery McIntosh et al. (12) developed a composite neuroscore to evaluate the chronic neurobehavioral state of animals following a TBI. The composite neuroscore utilizes a series of tests designed to evaluate motor and somatosensory components of rodent behavior.

Both the acute examination of reflex suppression (11) and the chronic evaluation of motor neurologic function (12) are utilized to evaluate and track the status of experimental TBI. This allows researchers to determine injury severity, recovery, and outcome without relying purely upon pathology and histology. Furthermore, reflex suppression and the composite neuroscore are more direct measures of neurologic damage than physiologic monitoring that involves cardiodynamic processes.

2. General Considerations of Using Acute and Chronic Neurobehavioral Monitoring Following a Traumatic Brain Injury

2.1. Acute Evaluation of Traumatic Injury

Acute neurological evaluation of TBI in rodents relies upon a battery of tests developed to be analogous to the motor components of the GCS. These tests rely upon the quantification of the suppression of animals' responsiveness to external stimuli and the relationship between duration of behavioral suppression and the magnitude of injury. Acute reflex suppression is based upon the classical conditioning paradigm of an unconditioned stimulus. An unconditioned stimulus is defined as any stimulus naturally capable of eliciting a specific response. The acute suppression of an unconditioned stimulus is therefore indicative of an alteration in the ability of the stimulus to elicit a response (i.e., unconsciousness).

Variations of acute reflex suppression have been evaluated in cats (10, 13), monkeys (3), and rodents (11). The battery discussed in this chapter is utilized in rodents as described by Dixon et al. (11), and allows quantification of various features of suppression of the animal's responsiveness to external stimuli and the relationships between durations of behavioral suppression and magnitudes of injury.

2.2. Chronic Neurobehavioral Monitoring of Traumatic Injury

Daily neurologic assessments provide researchers with the opportunity to evaluate the health status of their subjects and monitor recovery. The composite neuroscore is a combination of motor and somatosensory examinations designed to evaluate rodent recovery from a traumatic injury. Similar to reflex suppression, the neuroscore utilizes innate reflexes and rodent behavior so there is no need to pretrain the animals and evaluation can begin as early as 1 h postinjury. The techniques discussed in this chapter are based upon the composite neuroscore described by McIntosh et al. (12) and expanded in Ling et al. (14). Many variations of the composite neuroscore are described in the literature (15, 16). In general, the composite neuroscore relies upon a series of examinations ranked from 0 (nonfunctional) to 4 (normal) that are incorporated into the final score. This allows for a quantification of deficit that is consistently measured over time and across all levels of injury.

3. Working Protocol for Neurobehavioral Assessment Following an Experimental Traumatic Brain Injury

The following protocols are written for rats and mice and have been evaluated in these two species. Lateral and midline fluid percussion (17), controlled cortical impact (11), and weight drop (18) models have all utilized reflex suppression to monitor injury severity.

3.1. Acute TBI-Induced Reflex Suppression

3.1.1. Animals and Injury

3.1.2. Materials

1. Cotton swab.
2. Forceps (without teeth).

3.1.3. Working Protocol

For each reflex record, the duration of time post the cessation of gas anesthesia that it takes for the animal to recover the reflex. This will indicate the length of reflex suppression relative to the level of injury. Simple nonpostural reflexes are assessed continuously until recovery and other reflexes should be assessed every minute until recovery (Fig. 1).

Simple nonpostural somatosensory functions

1. The pinna reflex is assessed by touching the auditory meatus of the animal's ear to elicit a vigorous headshake.
2. The corneal reflex is assessed by lightly touching the cornea with a cotton swab to elicit an eyeblink.

Simple postural somatomotor functions

1. Assessment of the paw and tail flexion, consists of gradual application of pressure on the hind paw or tail with forceps and

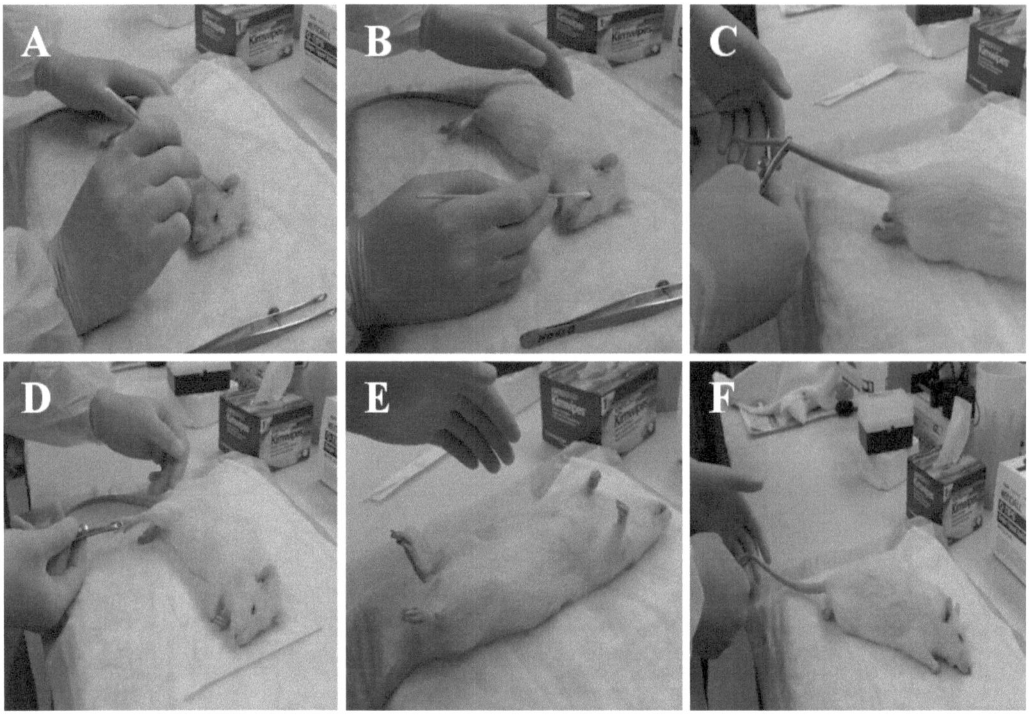

Fig. 1. Demonstration of acute animal reflex suppression examination on an isoflurane-anesthetized animal. (**a**) Pinna reflex: Elicit a headshake by placing the wooden end of a cotton swab into the animals' auditory meatus. (**b**) Corneal reflex: Gently touch the animals' cornea with a cotton swab to elicit an eyeblink. (**c**) Tail pinch: Begin by gently squeezing the tip of the animals' tail with forceps and progressively applying more pressure to elicit tail withdrawal. *Avoid tissue damage.* (**d**) Hind paw pinch: Begin by gently squeezing the hind paw of the animal with forceps and progressively applying more pressure to elicit limb withdrawal. *Avoid tissue damage.* (**e**) Righting response: Place the animal on its back and note if it is able to return to standing position three consecutive times. (**f**) Escape response: With the animal on its feet pinch its tail with forceps and observe if the animal attempts to move away from the noxious stimuli.

the presence or absence of paw or tail withdrawal is noted. *Do not cause tissue damage.*

(a) For hind paw withdrawal, note the duration of suppression in both the ipsilateral and contralateral (relative to injury) withdrawal reflex.

2. Head support is assessed by noting the duration of suppression of an animal's ability to support the weight of its head. While in a supine position, the head is lifted and released. The ability of animal to even briefly support its head is noted.

Complex postural somatosensory function

1. Righting response is measured as the duration of suppression of the animal's ability to right itself three times consecutively after being place on its back.
2. Escape response is assessed by briefly pinching the tail of the animal to elicit locomotive activity away from the noxious stimulus.

3.2. Composite Neurobehavioral Score

3.2.1. Animals and Injury

The composite neuroscore was designed to examine rats (12), but has been modified for mice (19, 20). The composite neuroscore has been employed in lateral fluid percussion (15) and controlled cortical impact (21).

3.2.2. Materials

1. A hinged wooden plane that can be adjusted in 5° increments covered with a rubber mat with 2 mm horizontal ribs.
2. Enclosed area that allows for observation of the animals normal gait.
3. Novel open-field enclosure.

3.2.3. Working Protocol

The original neuroscore composite is a battery of tests scored from 4 (uninjured) to 0 (nonfunctional) by a blinded observer. It is composed of five independent tests for a total possible score of 20. Depending upon if your injury is midline or lateral, the components of this test may need to be amended (see Sect. 4.4).

1. Suspend the animal by its tail and determine the forelimb extension towards floor.
   4. Normal: Rat extends both forelimbs fully and equally towards floor.
   3. Slightly impaired: There is slight forelimb flexion.
   2. Moderately impaired: There is moderate forelimb flexion.
   1. Severely impaired: There is severe forelimb flexion.
   0. Nonfunctional: Forelimb remains tucked close to body.
2. During free walking, gently pull the rat to the side by each forelimb and determine the decrease in resistance to lateral pulsion.
   4. Normal: Rat should resist equally when pulled to each side.
   3. Slightly impaired
   2. Moderately impaired
   1. Severely impaired
   0. Nonfunctional: Rat does not resist when pulled
3. Place the rat on the ground and observe the rats ability to walk straight (i.e., does not circle).
   4. Normal: Rat can walk straight even with partial circling.
   3. Slightly impaired: Rat always walks towards the paretic side.
   2. Moderately impaired: Rat shows partial circle, always towards the paretic side.
   1. Severely impaired: Rat circles to the paretic side when attempting to walk.
   0. Nonfunctional: Rat is unable to walk, spins in place.

4. Place the rat horizontally on an inclined plane for 5 s. Adjust by 5° increments between placements.
   4. Normal: Ability to stand for 5 s on 45–50° plane.
   3. Slightly impaired: Ability to stand for 5 s on 40–45° plane.
   2. Moderately impaired: Ability to stand for 5 s on 35–40° plane.
   1. Severely impaired: Ability to stand for 5 s on 30–35° plane.
   0. Nonfunctional: Unable to stand on a >30° plane for 5 s.
5. Place the rat into a novel open-field environment and observe exploratory behavior.
   4. Normal: Rat explores the area actively walking and sniffing.
   3. Slightly impaired: Rat explores the area and sniffs, but not to the extent of a normal rat.
   2. Moderately impaired: Rat either does not sniff or does not actively walk and explore.
   1. Severely impaired: Rat neither sniffs nor explores.
   0. Nonfunctional: Rat does not move.

4. Limitations, Potential Problems and Solutions

4.1. Advantages of Acute Reflex Suppression Testing

Acute reflex suppression testing post-TBI relies upon innate reflexes inherent to the animal. These tests allow for the assessment of multiple components of the animals neurologic function. Animals do not need to be trained in order to evaluate post-TBI reflexes. By monitoring these simple neurologic functions postinjury, the researcher is able to effectively evaluate intact neural circuitry and determine, grossly, the level of injury acutely and the extent of damage. This allows the researcher to effectively monitor the consistency of injury within groups of animals and assign severity based upon an output parameter rather than purely upon the initial forces applied.

4.2. Limitations of Acute Reflex Suppression Testing

When considering the evaluation of acute reflex testing the researcher must consider certain caveats. Many of the responses involved in the examination may be partially spinally mediated. Thus, suppression of the reflex may not represent a pure analysis of brain damage. This caveat is partially overcome by utilizing multiple reflex tests to assure that the brain injury is properly assessed.

Reflex testing is also not an assay of higher order cognitive or behavioral function. This is important to consider given that the presence or absence of a reflex may not accurately predict cognitive dysfunction within the animal given the severity of injury. To overcome this limitation, reflex suppression should be coupled with examinations of higher order cognitive and behavioral function.

There are also methodological limitations. Acute reflex testing can only be performed in protocols, where short-acting anesthetics are used because the duration of reflex suppression rarely exceeds 10–20 min. Since an investigator must be able to rapidly suture the skin, discontinue anesthesia, and place the animals on a flat surface for testing within minutes, acute reflex suppression testing is not recommended in cannulated animals.

4.3. Advantages of Composite Neurobehavioral Score

The composite neuroscore offers chronic evaluation of traumatic injury by utilizing a series of tests that, like acute reflex suppression, are dependent upon innate animal behavior and thus do not require a training period. This allows for a fairly simple and consistent measurement of neurologic deficit over time and across injury parameters. Furthermore, modifications of the general neuroscore paradigm allow for the inclusion or exclusion of varying tests to better meet the requirements of the researchers (15, 22, 23).

4.4. Limitations of Composite Neurobehavioral Score

The original composite neuroscore (12) relied predominantly upon motor recovery. More recent variations have included some forms of somatosensory recovery; however, the majority of utilized examinations analyze the level of motor ability in the animal. This limits the composite neuroscore in much the same way as acute reflex suppression. Many of the aspects of the neuroscore may be mediated by spinal injury and are unable to assay higher order brain function. These limitations can be addressed through the use of multiple components of the reported neuroscore allowing for a more complete examination of rodent recovery. Furthermore, the addition of higher order cognitive tasks should also be utilized when evaluating recovery and function following an experimental TBI. Finally, the composite neuroscore as presented assumes laterality to the injury; a midline injury may not produce circling behavior or noticeable deficits in one paw vs. another. To address this issue, the neuroscore can be modified with the addition of other tests (i.e., beam balance) or the modification of the presented tests (i.e., measure each paw independently using the forelimb reach task).

References

1. Denny-Brown D, Russell WR (1940) Experimental cerebral concussion. J Physiol 99:153
2. Ommaya AK, Hirsch AE, Flamm ES, Mahone RH (1966) Cerebral concussion in the monkey: an experimental model. Science 153:211–212
3. Ommaya AK, Gennarelli TA (1974) Cerebral concussion and traumatic unconsciousness. Correlation of experimental and clinical observations of blunt head injuries. Brain 97:633–654
4. Yarnell P, Ommaya AK (1969) Experimental cerebral concussion in the rhesus monkey. Bull N Y Acad Med 45:39–45
5. Brown WJ, Yoshida N, Canty T, Verity MA (1972) Experimental concussion. Ultrastructural and biochemical correlates. Am J Pathol 67:41–68

6. Saunders ML, Miller JD, Stablein D, Allen G (1979) The effects of graded experimental trauma on cerebral blood flow and responsiveness to CO_2. J Neurosurg 51:18–26
7. Brown FD, Johns LM, Mullan S (1980) Dimethyl sulfoxide in experimental brain injury, with comparison to mannitol. J Neurosurg 53: 58–62
8. Gennarelli TA, Adams JH, Graham DI (1981) Acceleration induced head injury in the monkey. I. The model, its mechanical and physiological correlates. Acta Neuropathol Suppl 7:23–25
9. Teasdale G, Jennett B (1974) Assessment of coma and impaired consciousness. A practical scale. Lancet 2:81–84
10. Hayes RL, Pechura CM, Katayama Y, Povlishock JT, Giebel ML, Becker DP (1984) Activation of pontine cholinergic sites implicated in unconsciousness following cerebral concussion in the cat. Science 223:301–303
11. Dixon CE, Lyeth BG, Povlishock JT, Findling RL, Hamm RJ, Marmarou A, Young HF, Hayes RL (1987) A fluid percussion model of experimental brain injury in the rat. J Neurosurg 67:110–119
12. McIntosh TK, Vink R, Soares H, Hayes R, Simon R (1989) Effects of the N-methyl-D-aspartate receptor blocker MK-801 on neurologic function after experimental brain injury. J Neurotrauma 6:247–259
13. Hayes RL, Stalhammar D, Povlishock JT, Allen AM, Galinat BJ, Becker DP, Stonnington HH (1987) A new model of concussive brain injury in the cat produced by extradural fluid volume loading: II. Physiological and neuropathological observations. Brain Inj 1:93–112
14. Ling GS, Lee EY, Kalehua AN (2004) Traumatic brain injury in the rat using the fluid-percussion model. Curr Protoc Neurosci Chapter 9:Unit 9 2
15. Riess P, Bareyre FM, Saatman KE, Cheney JA, Lifshitz J, Raghupathi R, Grady MS, Neugebauer E, McIntosh TK (2001) Effects of chronic, post-injury cyclosporin A administration on motor and sensorimotor function following severe, experimental traumatic brain injury. Restor Neurol Neurosci 18:1–8
16. Smith DH, Okiyama K, Thomas MJ, Claussen B, McIntosh TK (1991) Evaluation of memory dysfunction following experimental brain injury using the Morris water maze. J Neurotrauma 8:259–269
17. Hamm RJ (2001) Neurobehavioral assessment of outcome following traumatic brain injury in rats: an evaluation of selected measures. J Neurotrauma 18:1207–1216
18. Shohami E, Novikov M, Mechoulam R (1993) A nonpsychotropic cannabinoid, HU-211, has cerebroprotective effects after closed head injury in the rat. J Neurotrauma 10:109–119
19. Murai H, Pierce JE, Raghupathi R, Smith DH, Saatman KE, Trojanowski JQ, Lee VM, Loring JF, Eckman C, Younkin S, McIntosh TK (1998) Twofold overexpression of human beta-amyloid precursor proteins in transgenic mice does not affect the neuromotor, cognitive, or neurodegenerative sequelae following experimental brain injury. J Comp Neurol 392:428–438
20. Raghupathi R, Fernandez SC, Murai H, Trusko SP, Scott RW, Nishioka WK, McIntosh TK (1998) BCL-2 overexpression attenuates cortical cell loss after traumatic brain injury in transgenic mice. J Cereb Blood Flow Metab 18:1259–1269
21. Scherbel U, Raghupathi R, Nakamura M, Saatman KE, Trojanowski JQ, Neugebauer E, Marino MW, McIntosh TK (1999) Differential acute and chronic responses of tumor necrosis factor-deficient mice to experimental brain injury. Proc Natl Acad Sci U S A 96:8721–8726
22. Hoover RC, Motta M, Davis J, Saatman KE, Fujimoto ST, Thompson HJ, Stover JF, Dichter MA, Twyman R, White HS, McIntosh TK (2004) Differential effects of the anticonvulsant topiramate on neurobehavioral and histological outcomes following traumatic brain injury in rats. J Neurotrauma 21:501–512
23. Longhi L, Perego C, Ortolano F, Zanier ER, Bianchi P, Stocchetti N, McIntosh TK, De Simoni MG (2009) C1-inhibitor attenuates neurobehavioral deficits and reduces contusion volume after controlled cortical impact brain injury in mice. Crit Care Med 37:659–665

Chapter 34

Vestibular Assessments Following Traumatic Brain Injury

James W. Bales, Kristin Macfarlane, and C. Edward Dixon

Abstract

Motor function is a common deficit in patients who have experienced a traumatic brain injury (TBI). These deficits include difficulties with balance, increased dizziness, reductions in strength, and fine motor difficulties. Clinically relevant rodent models of TBI have utilized complex vestibulomotor and strength tasks to evaluate similar deficits in experimental TBI. The utilization of these tasks allows researchers to determine the causes of motor deficits in TBI and evaluate therapeutic manipulations in animal models.

Key words: Traumatic brain injury, Beam balance, Beam walking, Rotarod, Wire grip

1. Introduction

1.1. Historical Perspective

Traumatic brain injury (TBI) causes persistent dizziness and disturbances in motor function and balance (1). This study also noted disturbances in fine motor skills in even mild levels of TBI. The examination of complex motor tasks is an important part of the evaluation of experimental injury. The combination of beam-balance, to examine dizziness and balance, and beam-walking, to examine components of fine motor coordination, has been used extensively in rodent models of TBI (2–4).

The utilization of a beam-walking task in experimental brain injury was first described by Feeney et al. (5) while the beam balance task was described by Dixon et al. (2). Both beam tasks (beam balance and beam walking) were used to characterize the fluid percussion (2) and cortical impact (3) models of injury. In addition to the beam tasks, the rotarod task was modified by Hamm et al. (6) to evaluate deficits following injury. The rotarod task was first developed by Dunham and Miya (7) to evaluate the neurological effects of drugs and is the equivalent of a forced walking task often

Jun Chen et al. (eds.), *Animal Models of Acute Neurological Injuries II*, Springer Protocols Handbooks,
DOI 10.1007/978-1-61779-782-8_34,

employed in patient recovery. Vestibulomotor testing in rodent models of TBI shows excellent sensitivity and reliability in predicting severity and monitoring outcomes (8). The rotarod and beam tests are also valuable in the evaluation of therapeutic interventions.

In addition to the beam-balance, beam-walking, and rotarod tasks, other measurements of motor coordination and strength have been developed to better assess multiple aspects of motor dysfunction. The grid walk task (9) is a useful method for assessing motor coordination dysfunction and can be modified for both rats and mice. Hall (10) described a method of measuring motor strength in addition to fine motor control utilizing a wire grip strength task for mice. The grip strength test has also been modified for use in rats (11).

The combination of vestibular and motor examinations following experimental TBI in rodents is particularly useful given that the parietal cortex, the location of the primary motor and sensory cortices, is often the cortex of greatest impact. The examination of complex vestibulomotor testing following experimental brain injury allows for a clinically relevant paradigm of recovery, one which is both reliable and sensitive to therapeutic manipulation.

2. General Considerations of Using Vestibular Motor Testing Posttraumatic Brain Injury

2.1. Beam-Balance, Beam-Walking, and Rotarod Tasks

The beam-balance task is used to assess motor and vestibular functioning by observing the animal's ability to balance on a stationary beam. The beam-walking task is a more refined locomotor evaluation that employs the context of a learned avoidance task. It is important to note that both of these tasks have been employed as a component of the composite neuroscore (12) testing paradigm discussed in the previous chapter. Each of these tasks also requires a pretraining and evaluation period prior to injury. In doing so, the experimenter is able to determine the baseline level of function for each animal.

The rotarod task, modified to examine chronic neurological deficits following TBI (6), is a sensitive and efficient index for determining motor impairment produced by varying magnitudes of brain injury. The TBI-modified rotarod employs a series of rods fashioned into a cylinder rather than a contiguous cylindrical surface. The reason for this modification is that grip strength is known to be impaired following TBI (13) and the added requirement of having the rats grip the rods may increase sensitivity of the device to injury.

2.2. Wire Grip Task

The wire grip task (10) assesses fine motor strength and control and has been used to monitor therapeutic intervention and recovery (14, 15). Similarly, to the rotarod and beam tasks, the wire grip task requires a degree of training and setup. It has been used most extensively in mouse models, given their small size and relative ease in training on wire grip compared to rats. The description provided in this chapter is of the mouse task.

3. Working Protocol for Vestibulomotor Testing: Beam Balance, Beam Walking, Rotarod, Wire Grip

The beam balance task has been used in both rats (3) and mice (16). Beam balance has been evaluated in fluid percussion (8), controlled cortical impact (3), and weight drop (17) models of TBI.

3.1. Beam Balance

3.1.1. Animals and Injury

3.1.2. Materials

1. Stopwatch.
2. Sheet to record times.
3. Appropriate apparatus (described below for rats and mice) (Fig. 1).
   (a) Material used for construction of the apparatus described in this protocol: Sealed wood.
4. Cushioned pad to prevent injury from falling off beam.

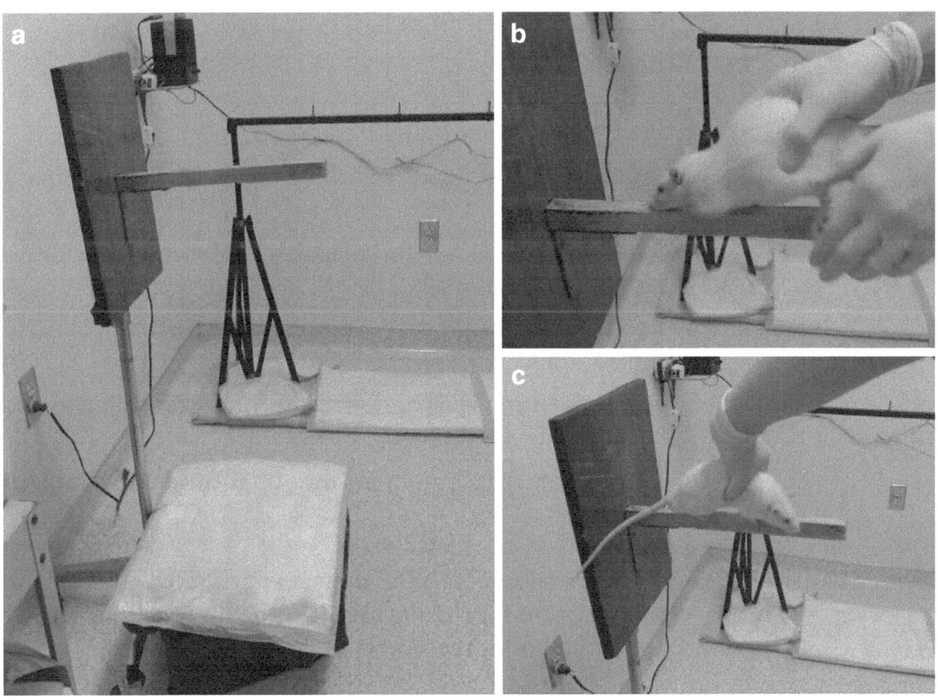

Fig. 1. Rat beam balance apparatus. (**a**) Rat beam balance setup. Animals placed onto the beam must remain there for 60 s during preassessment. During testing record the amount of time, an animal is able to remain on the beam. The pad below protects the animal from injury should it fall. *Do not catch the animal if it should fall.* (**b**) First trial, place the animal nose towards the backboard. (**c**) Subsequent trials place the animal nose away from the backboard.

Rat Beam Balance Apparatus

Beam dimensions
Height: 2.5 cm.
Length from backboard: 35 cm.
Width: 2 cm.
Height from ground to bottom of beam: 92 cm.

Backboard dimensions
Height: 40 cm.
Length: 26 cm.
Width: 2.5 cm.

Mouse Beam Balance Apparatus

Beam dimensions
Diameter: 0.5 cm.
Length from backboard: 21 cm.
Height from ground to bottom of beam: 70 cm.

Backboard dimensions
Height: 23 cm.
Length: 20 cm.
Width: 1 cm.

3.1.3. Working Protocol

Preassessing (Day of Surgery)

Rat must sit on beam for 60 s for three consecutive trials. If rat falls off, start over with trial one. Repeat until completion.

1. Place rat lengthwise on beam facing backboard.
2. After 60 s have elapsed, raise rat to eye level and place rat on beam facing away from backboard.
3. Repeat step 2.

Testing

1. Place rat on beam facing backboard.
2. Record time that the rat remained on the beam with a maximum time of 60 s.
3. Raise rat to eye level and place rat on beam facing away from backboard.
4. Record time that the rat remained on the beam with a maximum time of 60 s.
5. Raise rat to eye level and place rat on beam facing away from backboard.
6. Record time that the rat remained on the beam with a maximum time of 60 s.

Beam Balance Notes

- If the rat falls, record the time at which it fell and place the rat on the beam for the next trial. If the rat remains on the beam for the full 60 s, record 60.

3.2. Beam Walking

3.2.1. Animals and Injury

The beam walking task has been used in both rats (2) and mice (18). Beam walking has been evaluated in fluid percussion (8), controlled cortical impact (3), and weight drop (19) models of TBI.

3.2.2. Materials

1. Stopwatch.
2. Sheet to record times.
3. Appropriate apparatus (described below, same for rats and mice) (Fig. 2).
   (a) Material used for construction of the apparatus described in this protocol: Sealed wood.
4. White noise generator.

Beam Walking Apparatus

Beam dimensions
Height: 2 cm.
Length: 122.5 cm.
Width: 2.5 cm.
Height from ground to bottom of beam: 88 cm.
Distance from white noise to beam: 8 cm.

Goal/retreat box dimensions
Height: 18 cm.
Length: 27 cm.
Width: 21 cm.

Obstacle dimensions
Composition: Nut at base around bolt.
Diameter of nut: 1 cm.
Diameter of bolt: 0.5 cm.
Height: 3 cm.

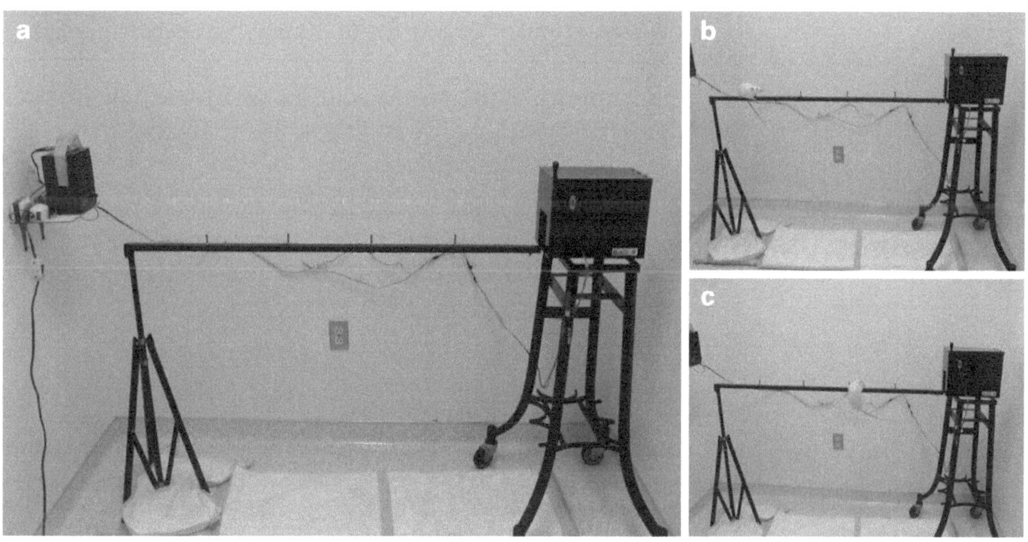

Fig. 2. Rat beam walk apparatus. (**a**) Rat beam walk setup. At the far right is the escape box. On the far left is the noise system. At four points along the beam are placed bolts on alternating sides. The inside of the box should be associated with quiet and the adverse noise stimulus should only be engaged when the rat is removed from the box. Padding below the beam walk protects the animal should it fall. *Do not catch the animal if it falls.* (**b**) Demonstration of where the rat is placed at the beginning of the beam walk task. If the rat falls from this point, it receives a score of 0 and a latency of 60F (60 s and fall). (**c**) Demonstration of a rat falling from a point prior to the escape box. At this point, the rat would receive a score of 3 and a latency of 60F (60 s and fall). If the animal retreats to the escape box, it receives a score of 5 and the recorded latency that it took to arrive there.

Placement of obstacle: At the edge of the beam so that the bolt encompasses 1 cm of the beam width. Alternate sides of bolt placement. Each bolt should be placed approximately 24 cm apart for a total of four bolts along the beam.

3.2.3. Working Protocol

Training (Day Before Surgery)

For training completion, rat must traverse beam with a latency under or equal to 5 s for three consecutive trials. If rat fails to complete the course within time limit or falls off of beam, start over with trial one. Repeat until completion.

1. Place rat in dark box for 60 s.
2. Turn on noise stimulus and remove the rat from the goal box. Place rat on beam close to white noise generator.
3. After entry into the goal box, turn off noise and allow rat to remain in box for 30 s after entry.
4. If rat successfully entered the box on the previous trial, place rat at a starting point farther from the box.
5. Repeat steps 3 and 4 until the rat is at the starting point farthest from the box.
6. Place rat at starting point. Rat must traverse the beam and enter box with latency under or equal to 5 s.
7. Allow rat to remain in box for 30 s.
8. Repeat steps 6–8 until rat traverses beam with a latency under or equal to 5 s for three consecutive trials.
9. Allow rat to remain in box for 60 s after last successful trial.

Training Notes

- Let the rat learn at its own pace. If the rat successfully makes it to the box from where it was placed, it is ready to advance to a starting spot further from the box.
- In order to aid learning of task, let the rat fall off of the beam. If the rat falls, place the rat at the place where it fell and walk the rat into the box. If the rat has trouble learning, walk the rat across the beam for a few trials.
- The rat must complete the task without help to successfully train.

Preassessing (Day of Surgery)

Rat must traverse beam with a latency under or equal to 5 s for three consecutive trials. If rat fails to complete the course within time limit or falls off of beam, start over with trial one. Repeat until completion.

1. Place rat in dark box for 60 s.
2. Place rat at starting point. Rat must traverse the beam and enter box with latency under or equal to 5 s.
3. Record latency and score if the rat has entered the box within time limit.

4. Allow rat to remain in box for 30 s.
5. Repeat steps 2–4 until rat traverses beam with a latency under or equal to 5 s for three consecutive trials.
6. Allow rat to remain in box for 60 s after last successful trial.

Testing

1. Place rat in dark box for 60 s.
2. Place rat at starting point.
3. Record latency and score.
4. Allow rat to remain in box for 30 s.
5. Repeat steps 2–4 twice for a total of three trials.
6. Allow rat to remain in box for 60 s.

Beam Walking Notes

- White noise should be turned on when rat is not in dark box between the first and last time the rat is in the black box. Turn on white noise when taking the rat out of the box and turn off white noise when the rat enters the box.
- Rat should remain in dark box for 60 s only on the first and last period in the box. Rat should remain in dark box for 30 s in between trials.

Scoring for Motor Testing

- If at any point the rat falls off of the beam, the rat receives a latency of 60(F). Place the rat in the box for 30 s and move on to the next trial.
- If the rat remains on the beam for a maximum of 60 s, the rat receives a latency of 60. Place the rat in the box for 30 s and move on to the next trial.

Score Descriptions

0—If the rat falls or remains on the beam for 60 s before traversing the first obstacle.

1—If the rat traverses only the first obstacle, but falls or remains on the beam for 60 s.

2—If the rat traverses only the first and second obstacles, but falls or remains on the beam for 60 s.

3—If the rat traverses the first, second, and third obstacles, but falls or remains on the beam for 60 s.

4—If the rat traverses all the four obstacles and does not enter the box, but falls or remains on the beam for 60 s.

5—The rat enters the box in 60 s or under. The time at which the rat enters the box is recorded as the latency.

Beam-Walking Scoring Variations

A scoring scale may also be used to assess hind limb function:

7—The rat completed the task with two foot slips or fewer

6—The rat completed the task and used its hind/affected limb for at least 50% of steps

5—The rat completed the task and used its hind/affected limb for less than 50% of steps

4—The rat competed the task and used one hind/affected limb on the surface of the beam

3—The rat completed the task and dragged its hind/affected limb

2—The rat was unable to complete the task, but remained on beam and placed hind/affected limb on the surface of beam

1—The rat was unable to complete the task and unable to use hind/affected limb on the surface of the beam

3.3. Rotarod

3.3.1. Animals and Injury

The rotarod task has been evaluated in both rats (6) and mice (20). The protocol presented in this chapter was modified for experimental TBI in the rat by Hamm et al. (6). Rotarod has been used in fluid percussion (8) and controlled cortical impact (20) models of experimental TBI.

3.3.2. Materials

1. Stopwatch.
2. Sheet to record times.
3. Appropriate apparatus (described below for rats).

Rat Rotarod Apparatus: TBI-Modified Version

Consists of a metal frame with a motorized (Gearmotor, Bodine Electric Company), rotating assembly of rods. Two large plexiglass disks are positioned on each side of the frame.

Walking area
Cylinder of 18 rungs (1-mm stainless steel rods).

Control box
Bodine Electric Company.
Rotational speed range from 0 to 30 rpm.
Must be able to change direction or rotation and have a brake.

Rat Rotarod Apparatus: Commercially Available

The commercial rotarod apparati (IITC Rotarod assembly (21)) are similar in composition to the TBI-modified version; however, they do not utilize an assembly of rods, but rather a single solid axel typically covered by a rubberized material. This limits the motor strength component of the task.

3.3.3. Working Protocol

Protocol is the same for the TBI version and the commercial version.

1. Place the animal on the device facing the wall (forward direction).
2. Allow the animal to remain stationary for 10 s at 0 rpm.
3. Start the device rotating in the opposite direction than the animal is facing.

4. Slowly increase the speed to 3 rpm for 10 s.
5. Steadily increase the speed by 3 rpm every 10 s until the maximum speed of 30 rpm is reached.
6. The animal remains on the device at 30 rpm for 20 s, at which time the 2 min test period elapses.
7. Animals are tested twice per day and the mean duration (in seconds) on the device is recorded.
8. If the animal falls from the device or grips, the device spins around twice consecutively without walking the trial ends and the time is recorded.

3.4. Wire Grip

3.4.1. Animals and Injury

The wire grip task has been evaluated in both rats (11) and mice (10). It is most commonly employed in mice and the protocol presented in this chapter is for the mouse version. Wire grip has been used in fluid percussion (10) and controlled cortical impact (15) models of experimental TBI.

3.4.2. Materials

1. Stopwatch.
2. Sheet to record times.
3. Appropriate apparatus (described below for mice).

Mouse Wire Grip Apparatus

Height from ground to wire: 58 cm.
Height from padding to wire: 38 cm.
Length of wire: 38 cm.
Diameter of vertical point: 2.5 cm.

3.4.3. Working Protocol

Training (Day Prior to Surgery)

1. Place the mouse onto the center of the wire so that it hangs by its front feet.
   (a) Let it hang for 30 s.
   (b) If the mouse falls or does not move, guide the mouse to either of the end points. Orient the mouse so that it travels the wire using all four paws and its tail.
   (c) Guide the mouse to the top of the endpoint and let it sit for 15 s.
   (d) Alternate between the two endpoints each time the mouse is guided (the mouse does not need to alternate between points to complete trials).
2. Repeat until the mouse is able to perform the task in 30 s or less for three consecutive trials.
   (a) If the mouse fails the task, start over from first trial.

Testing

1. Follow wire grip training procedure.
2. If the mouse completes the task, record each individual latency.

3. For each successful completion, record the appropriate score (0–5).
   (a) If the mouse falls, place the mouse at the point from which it fell and guide to the end point.
   (b) Record 30F (30 s and fall) for the time and the appropriate score.
4. Allow the mouse to remain at the endpoint for 15 s.
5. Repeat until a total of three trials are completed.

Score Descriptions

The score denotes how far the mouse traveled in its attempt to traverse the wire and mount the vertical point.

0—If the mouse falls from the wire within the 30 s period.

1—If the mouse holds on in some way for the full 30 s period, or if the mouse mounts the vertical point but only uses two paws.

2—If the mouse holds on with four paws for greater than or equal to 5 s.

3—If the mouse holds on with four paws and places tail on wire for greater than or equal to 5 s.

4—If the mouse holds on with four paws, places tail on wire, and travels along wire for greater than or equal to 5 s.

5—If the mouse holds on with four paws, places tail on wire, and travels to one of the vertical points within the 30 s time period.

4. Limitations, Potential Problems and Solutions

4.1. Advantages of Vestibulomotor and Grip Strength Testing

Deficits in balance and gross and fine motor skills are common following TBI. Utilizing the beam tasks and rotarod examinations, it is possible to effectively model these deficits in an experimental animal model. These tasks reliability and sensitivity to injury have been examined in rodent models of TBI (8). This allows researchers to evaluate the level of deficit in complex vestibulomotor function associated with the level of injury in their TBI models of choice. Furthermore, the benefit of any therapeutic manipulations on clinically relevant tasks can be examined utilizing vestibulomotor testing. The wire grip and TBI rotarod modification employ an evaluation of grip strength in addition to the motor coordination required to complete the task. This is an important addition to motor testing as fine motor control and strength deficits are also common in clinical TBI.

Though more experimentally difficult to employ than simple reflex testing, these tasks are still relatively simple and do not require complex setup compared to tasks designed to assess higher order behavior.

4.2. Limitations of Vestibulomotor and Grip Strength Testing

Vestibulomotor tasks rely upon an intact motor system to evaluate behavior, therefore any injury that may involve the spinal cord may cause deficits unrelated to brain damage. These tasks are also examinations of complex motor function and thus it may be difficult to assess the direct cause of failure on any given task. This can be partially ameliorated by employing more than one vestibulomotor task in combination with simple motor reflex testing and an evaluation of grip strength.

One particular concern is how to differentiate vestibulomotor function from motivation. An animal that does not walk from a start position may lack the motivation to do so instead of suffering from a deficit in motor function. Alternative or additional motivational modifications can be employed in order to address this such as the inclusion of a bright light in addition to the loud noise or gentle prodding of the animal to elicit movement on a beam walking task (5). There is also the concern of quality of movement. For example, the experimenter may have two rats who receive the same score and latency on the beam tasks, however one of the rats slips or shows less coordinated movement. To address this issue, the researcher may want to employ the grid walking task (9) or a variation of the beam walk (11). Both the grid walk and beam walk modification employ the counting of foot faults while the animal walks to examine ipsilateral vs. contralateral (with reference to injury) and forelimb vs. hindlimb function.

Vestibulomotor function and grip strength testing are excellent ways to experimentally examine motor dysfunction similar to that experienced by TBI patients, however, much like reflex testing it cannot assay higher order cognitive functions.

References

1. Rutherford WH (1977) Sequelae of concussion caused by minor head injuries. Lancet 1:1–4
2. Dixon CE, Lyeth BG, Povlishock JT, Findling RL, Hamm RJ, Marmarou A, Young HF, Hayes RL (1987) A fluid percussion model of experimental brain injury in the rat. J Neurosurg 67:110–119
3. Dixon CE, Clifton GL, Lighthall JW, Yaghmai AA, Hayes RL (1991) A controlled cortical impact model of traumatic brain injury in the rat. J Neurosci Methods 39:253–262
4. Kline AE, Bolinger BD, Kochanek PM, Carlos TM, Yan HQ, Jenkins LW, Marion DW, Dixon CE (2002) Acute systemic administration of interleukin-10 suppresses the beneficial effects of moderate hypothermia following traumatic brain injury in rats. Brain Res 937:22–31
5. Feeney DM, Gonzalez A, Law WA (1982) Amphetamine, haloperidol, and experience interact to affect rate of recovery after motor cortex injury. Science 217:855–857
6. Hamm RJ, Pike BR, O'Dell DM, Lyeth BG, Jenkins LW (1994) The rotarod test: an evaluation of its effectiveness in assessing motor deficits following traumatic brain injury. J Neurotrauma 11:187–196
7. Dunham NW, Miya TS (1957) A note on a simple apparatus for detecting neurological deficit in rats and mice. J Am Pharm Assoc Am Pharm Assoc (Baltim) 46:208–209
8. Hamm RJ (2001) Neurobehavioral assessment of outcome following traumatic brain injury in rats: an evaluation of selected measures. J Neurotrauma 18:1207–1216

9. Hernandez TD, Schallert T (1988) Seizures and recovery from experimental brain damage. Exp Neurol 102:318–324
10. Hall ED (1985) High-dose glucocorticoid treatment improves neurological recovery in head-injured mice. J Neurosurg 62:882–887
11. Goldstein LB (1995) Right vs. left sensorimotor cortex suction-ablation in the rat: no difference in beam-walking recovery. Brain Res 674: 167–170
12. McIntosh TK, Banbury M, Smith D, Thomas M (1990) The novel 21-aminosteroid U-74006F attenuates cerebral oedema and improves survival after brain injury in the rat. Acta Neurochir Suppl (Wien) 51:329–330
13. Hall ED, Yonkers PA, McCall JM, Braughler JM (1988) Effects of the 21-aminosteroid U74006F on experimental head injury in mice. J Neurosurg 68:456–461
14. Varma MR, Dixon CE, Jackson EK, Peters GW, Melick JA, Griffith RP, Vagni VA, Clark RS, Jenkins LW, Kochanek PM (2002) Administration of adenosine receptor agonists or antagonists after controlled cortical impact in mice: effects on function and histopathology. Brain Res 951:191–201
15. Whalen MJ, Clark RS, Dixon CE, Robichaud P, Marion DW, Vagni V, Graham SH, Virag L, Hasko G, Stachlewitz R, Szabo C, Kochanek PM (1999) Reduction of cognitive and motor deficits after traumatic brain injury in mice deficient in poly(ADP-ribose) polymerase. J Cereb Blood Flow Metab 19:835–842
16. Goldstein FC, Levin HS, Goldman WP, Clark AN, Altonen TK (2001) Cognitive and neurobehavioral functioning after mild versus moderate traumatic brain injury in older adults. J Int Neuropsychol Soc 7:373–383
17. Adelson PD, Dixon CE, Robichaud P, Kochanek PM (1997) Motor and cognitive functional deficits following diffuse traumatic brain injury in the immature rat. J Neurotrauma 14:99–108
18. Shelton SB, Pettigrew DB, Hermann AD, Zhou W, Sullivan PM, Crutcher KA, Strauss KI (2008) A simple, efficient tool for assessment of mice after unilateral cortex injury. J Neurosci Methods 168:431–442
19. Mikawa S, Kinouchi H, Kamii H, Gobbel GT, Chen SF, Carlson E, Epstein CJ, Chan PH (1996) Attenuation of acute and chronic damage following traumatic brain injury in copper, zinc-superoxide dismutase transgenic mice. J Neurosurg 85:885–891
20. Fox GB, Fan L, LeVasseur RA, Faden AI (1998) Effect of traumatic brain injury on mouse spatial and nonspatial learning in the Barnes circular maze. J Neurotrauma 15:1037–1046
21. Jones BJ, Roberts DJ (1968) The quantitative measurement of motor in coordination in naive mice using an accelerating rotarod. J Pharm Pharmacol 20:302–304

Part V

Spinal Cord Injury

Chapter 35

Introduction on Assessments for Spinal Cord Injury

Theo Hagg

Abstract

This chapter introduces the chapters of Part VIII, which describe a variety of assessment methods to determine the outcomes after spinal cord injury (SCI) in animal models. The SCI community should be encouraged by the view that SCI has certain advantages for translation of therapies in humans compared to other neurological conditions. The strengths and shortcomings of a number of assessment categories are discussed. These include characterization of the lesion size, spared tissue, inflammation, scarring, proliferation of oligodendrocyte precursors and vascular changes at the injury epicenter, as well as axonal tracing to identify neuron populations. Combined with behavioral analyses, these measures can be used to identify the role of degeneration, regeneration, and plasticity in the functional outcomes. Comments are made about the need of functional testing appropriate for the spinal level and type of injury, and several behavioral motor tests are discussed. The power and need for acute invasive and real-time noninvasive electrophysiology and physiological measurements are also highlighted. Methods for important, clinically relevant, noninvasive imaging are also discussed, as well as the fact that few groups are in a position to use them regularly. Finally, the sparse attention that functions other than locomotor function have received is noted.

Key words: Electrophysiology, Epicenter, Function, Histology, Imaging, Spinal cord injury

Research and clinical translation of treatments for spinal cord injury (SCI) appear to have a disadvantage compared to traumatic brain injury (TBI), stroke, and neurodegenerative disorders because of the relatively low number of patients. This underlies the low level of interest of most pharmaceutical companies to develop treatments for SCI. However, as Naomi Kleitman taught me, SCI has several unique "advantages" and might be one of the most approachable neurological disorders in terms of interventions. As an animal model, SCI has the same "advantages." First, compared to the other disorders, SCI is most often very focal, affecting a segment, with stroke often involving large volumes of brain tissue and TBI often being very diffuse in nature. Second, the exact time of injury is essentially always known for acute SCI. This is often not the case for stroke. Thus, SCI has a clear window of opportunity for intervention. Third, the development of secondary damage most often follows a

Jun Chen et al. (eds.), *Animal Models of Acute Neurological Injuries II*, Springer Protocols Handbooks,
DOI 10.1007/978-1-61779-782-8_35, © Springer Science+Business Media, LLC 2012

temporally identifiable course, whereas most neurodegenerative disorders have a very variable progressive nature. Fourth, the functional deficits are well described and are the result of injury of well-known axonal pathways and loss of distinct populations of neurons. This is less clear in TBI and many of the slow degenerative disorders. Besides this bit of encouragement, the SCI community has a sound reason to continue its research into therapies: SCI is a devastating condition affecting and disabling people for decades of their lives, costing them and society dearly.

One important caveat in our assessments is that we do not know whether our animal, including primate, studies are predictive for humans until treatments are validated in human phase III trials, including safety and efficacy. The fact that we collectively assume that animal models are relevant to humans should also be kept in mind when the term "more clinically relevant" is used.

The outcome of therapies is heavily dependent on understanding the pathophysiology of the condition and the efficacy of the experimental treatments. Thus, every laboratory has a set of outcome measures it uses, often times ones it itself developed. Assessing outcomes following SCI in animal models is often driven by what is possible and easy, and clearly subserving scientific inquiry, including mechanistic studies. However, these assessments are not necessarily good mimics of what is feasible or reasonable in humans. Thus, in developing assessments in animal models, we should increasingly consider potential human applications, including their widespread applicability in medical centers. For discussions on the validity and need for further improvement of current functional assessment tools for humans with SCI, see Anderson et al. (1) and Alexander et al. (2).

The cornerstone for interpretation of all outcome measures is the characterization of the injury epicenter. There are major differences between, on the one hand, contusion or compression models and, on the other hand, laceration (transection) models with regards to their usefulness for studying mechanisms and their relevance to human SCI. Contusion/compression injuries are most common in humans, most often affecting the cervical level, but lacerations are also commonly seen in humans. Injuries can also consist of combinations of contusion, compression, and laceration. Thus, all of these models might be relevant to human disorders. Generally speaking, contusion and compression models are most useful for neuroprotective studies, whereas transection models are advised for pure axonal regeneration studies. The focal nature of the injury makes characterization of the epicenter pathology a bit easier. In Chap. 37, Jakeman describes uncomplicated unbiased stereological methods to estimate the volume of lesions and spared tissue, as well as their cellular components. Importantly, she addresses the issue of tissue shrinkage. Walker and Xu provide insight into a variety of additional morphological assessments, including myelination, reactive gliosis, axonal damage, protection, and regeneration (Chap. 36). Depending on the level of injury, white matter damage, with or without gray matter damage, is important. For example, midthoracic lesions are often assessed by measuring the spared white matter. For descending and ascending systems, this measure is very important, as it often correlates well with the extent of spared locomotor function. Thus, if a treatment improves function without obviously affecting the epicenter, there is a good likelihood that it increased plasticity, possibly below the injury site. Determining the neuronal cell types involved in affecting functional deficits and recovery is also important. Thus, axonal tracing is used to assess degeneration and regeneration. This is nicely addressed by Deng et al. and Wang et al. (Chaps. 38 and 39) who describe retrograde and anterograde axonal tracing, respectively. They each have their specific application and when combined can provide insight into the occurrence of protection, plasticity, and regeneration. For example, when combined, these methods can define whether fiber growth is due to sprouting or regeneration. Retrograde tracing has the advantage of providing data on the cell morphology and phenotype of the neurons involved. Anterograde tracing is most useful for documenting axons. Other repair-related or detrimental processes occur following SCI, including inflammation, neurogenesis, and microvascular degeneration and angiogenesis. Zhang and colleagues provide a detailed description of assessment tools used to characterize the inflammatory response, including various methods to quantify cytokines, leukocytes, resident microglia, and macrophages (Chap. 45). Progenitors, including oligodendrocyte precursors, proliferate and differentiate in response to the injury to contribute to functional recovery. Sellers and Horner (Chap. 40) describe methods to specifically label and study these cells. Hagg and colleagues address a wide variety of assessment tools for the microvascular aspects of SCI in Chap. 41. It is not efficient or perhaps even wise to study everything in and around the injury site. However, it should be understood that every cell in that environment is important and contributes to the final outcome.

Obviously, people with SCI want recovery of function without detrimental side effects. In rodent models, behavioral testing is used to confirm the efficacy of treatments. Quite often, there is a disconnect between the morphology at the epicenter, the fiber systems studied, and the functional recovery reported. Thus, the morphological methods described, including the axonal tracing, are of great importance. One way to confirm the involvement of a particular fiber pathway or nucleus is to lesion or re-lesion that system. Such experiments are rarely performed. Behavioral analyses have historically focused on the hind limbs following a thoracic injury. Recently, increasingly more studies involve the cervical level, assessing forelimb function. The cervical injury might be most relevant to humans, although few have ventured to completely mimic the injuries with a ventral injury approach. One should understand that the sensory system is very important for the functioning of the locomotor system and that not all descending pathways are equally important. For example, the corticospinal tract has little influence on hindlimb movement. Equally important is to realize that there can be redundancy between systems, where one takes over for the loss of another in the process of plasticity. A wide variety of behavioral tests are used in rodent studies, ranging from standard, more general, tests such as the BBB and BMS to much more refined ones such as hand movement and gait analysis. Chapter 46 by Shumsky and Houle nicely points out the importance of defining the nature and level of the injury which should direct the choice of appropriate neurobehavioral tests and can provide insight into the mechanisms of functional loss and recovery. Burke and Magnuson (Chap. 47) describe the utility of the global BBB test for overground locomotor function, and its ability to predict whether an animal can perform more difficult and more specific tests. Combined, the BBB plus other tests provide a broad view of the animal's functional status. As they point out, it is important to have consistency within a laboratory but also between research groups within a center and elsewhere. Brown et al. describe a unique Louisville Swim Scale in Chap. 51, which is a sensitive and robust assessment tool for hindlimb function. One advantage of this test is the ability of testing rats which can otherwise not support their weight. The addition of 2D kinematics adds an objective and efficient way to quantify the locomotor function. Forelimb function is much more precise and skillful than hindlimb function, being affected by cervical lesions and mimicking the predominant levels of human injuries. Onifer describes several tests for rats and mice, including general usage, and skilled movements in Chap. 50. This is complemented by Chap. 48 by Whishaw and Sacrey, who describe the use of high-speed video plus kinematics to document hand shaping in a variety of tests, including skilled reaching. They also describe hand shaping in other behaviors, such as walking, each of which provides additional insight into motor function and influence of sensory control. Lastly, a plethora of measurements can be made by using automated gait analysis on a translucent treadmill using Treadscan software, as described in Chap. 49 by Beare et al. One issue that remains to be resolved is how many of these tests are needed in each study and how big of an effect is necessary to convincingly show an improvement after a treatment. In other words, how robust an effect do we need before moving therapies to the clinic?

Electrophysiological methods constitute powerful assessment tools to understand the function of specific pathways affected by injury and repair. Electrophysiology can also determine, in combination with tracing and myelination, whether the improvements or deficits are due to such specific pathways or due to plasticity of other compensating systems. Unfortunately, not many groups have the ability to perform or have invested in setting up electrophysiology.

Zhang et al. describe state-of-the-art methods for electrophysiological testing using noninvasive magnetic and electrical stimulation methods for motor and sensory systems, respectively (Chap. 43). Importantly, this can be done in awake animals. In Chap. 44, Chen and Wolpaw describe a unique and powerful method for determining the role of the CST in plastic functional changes and adaptation by measuring the operantly conditioned H-reflex in rats. Importantly, this method can also be part of a rehabilitation strategy, including in humans. Shields and colleagues (Chap. 42) describe intraspinal electrophysiology and EMG as addition methods to assess spinal function. They also describe several methods to determine spinal blood flow and oxygenation. This adds an important technique, particularly if combined with systemic blood pressure measurements and core body temperature as they are critical to the outcome in human SCI.

One of the most directly clinically translatable assessment tools is imaging. Repetitive noninvasive imaging is important for characterization of the pathology and monitoring of therapeutic efficacy. In humans, this will also enhance success in diagnosis and classification of the injury severity and localization. Despite the promise of noninvasive imaging, especially when guided by ultimate human applications, very few groups have access to small animal MRI and microCT equipment. Chapter 52 by Bilgen describes the application of rodent MRI and correctly points out that although it is challenging rodent MRI is feasible. Of particular interest is the ability to image myelinated and demyelinated axons at submicrometer resolution using CARS microscopy as shown by Cheng in Vol. 1, Chap. 18.

The focus of most SCI research and assessment has been on locomotor and, less so, on sensory systems and their function. This is in part due to historical reasons and in part due to the accessibility of these systems for experimental manipulation. However, people with SCI rank several other functions more important than locomotor function (3). One of them is pain, which is present in over half of people with SCI and is often debilitating. Pain is an area of investigation in SCI models (4), but "Pain models have relied on measurements of the shifts in behavioral hypersensitivity to tactile and thermal stimuli, tests that are not used quantitatively in human patients" (5). Moreover, many of the pain tests, including vocalization, do not necessarily measure perceived pain (involving the brain), but measure reflexes (6). Assessment of sexual dysfunction, one of the more important issues for people with SCI, is applied by a few laboratories (7, 8). Another important aspect of human SCI is bowel and bladder control, and assessment tools are also available for rodents (8, 9). For people with high tetraplegic injuries, respiratory and pulmonary function is also of great importance. Assessment following SCI in rodents is performed by a few groups (10). Autonomic dysreflexia, a common and potentially life-threatening condition for humans with SCI, is also studied in rodents by a few groups (11, 12).

References

1. Anderson K et al (2008) Functional recovery measures for spinal cord injury: an evidence-based review for clinical practice and research. J Spinal Cord Med 31:133–144
2. Alexander MS et al (2009) Outcome measures in spinal cord injury: recent assessments and recommendations for future directions. Spinal Cord 47:582–591
3. Anderson KD (2004) Targeting recovery: priorities of the spinal cord-injured population. J Neurotrauma 21:1371–1383
4. Hall BJ, Lally JE, Vukmanic EV, Armstrong JE, Fell JD, Gupta DS, Hubscher CH (2010) spinal cord injuries containing asymmetrical damage in the ventrolateral funiculus is associated with a higher incidence of at-level allodynia. J Pain 11(9):864–875
5. Eaton M (2003) Common animal models for spasticity and pain. J Rehabil Res Dev 40(4 Suppl 1):41–54
6. Woolf CJ (1984) Long term alterations in the excitability of the flexion reflex produced by peripheral tissue injury in the chronic decerebrate rat. Pain 18:325–343
7. Hubscher CH, Johnson RD (2000) Effects of acute and chronic midthoracic spinal cord injury on neural circuits for male sexual function. II. Descending pathways. J Neurophy-siol 83:2508–2518
8. Nout YS, Schmidt MH, Tovar CA, Culp E, Beattie MS, Bresnahan JC (2005) Telemetric monitoring of corpus spongiosum penis pressure in conscious rats for assessment of micturition and sexual function following spinal cord contusion injury. J Neurotrauma 22:429–441
9. Gondim FA et al (2001) Neural mechanisms involved in the delay of gastric emptying and gastrointestinal transit of liquid after thoracic spinal cord transection in awake rats. Auton Neurosci 87:52–58
10. Doperalski NJ, Sandhu MS, Bavis RW, Reier PJ, Fuller DD (2008) Ventilation and phrenic output following high cervical spinal hemisection in male vs. female rats. Respir Physiol Neurobiol 162:160–167
11. Fleming JC, Bao F, Chen Y, Hamilton EF, Relton JK, Weaver LC (2008) Alpha4beta1 integrin blockade after spinal cord injury decreases damage and improves neurological function. Exp Neurol 214:147–159
12. Rabchevsky AG, Patel SP, Duale H, Lyttle TS, O'Dell CR, Kitzman PH (2011) Gabapentin for spasticity and autonomic dysreflexia after severe spinal cord injury. Spinal Cord 49(1):99–105

Chapter 36

Morphological Assessments Following Spinal Cord Injury

Chandler L. Walker and Xiao-Ming Xu

Abstract

Spinal cord injury (SCI), by name alone, infers damage to structural elements of the spinal cord. The spinal white matter, composed of ascending (sensory) and descending (motor and autonomic) axonal tracts, and gray matter, which encompasses columns of neural cell bodies and glial cells, can become disrupted by a myriad of physical insults resulting in dysregulation of functions related to these anatomical components. As the extent and types of physical damage to the spinal cord reflect the ensuing level of functional disruption, characterizing alterations in the physical structure of spinal cord elements, such as lesion volume, astrogliosis, axonal dieback, cell death, and regrowth, following injury, is important for better understating when and by what means to most effectively approach the treatment for neuroprotection and repair. The following section provides a brief introduction to some of the morphological assessments available and commonly used in SCI research. Succeeding chapters elaborate and provide further details on these and other important measures in the field.

Key words: Spinal cord injury, Morphological assessments, Lesion, Reactive gliosis, Axonal tracing, Axon regeneration, Neuroprotection

1. Introduction

Following spinal cord injury (SCI), a variety of biochemical and morphological changes occur within the spinal cord resulting in various degrees of functional deficits. Disruption of long ascending and descending axonal pathways within the spinal cord leads to dysfunction of sensory, motor, and autonomic activities. The morphological outcomes following SCI are influenced by many factors, including the type, location, and severity of injury and the age and health status of the individual. Combined, these factors influence the survival and functional prognosis in affected individuals. In particular, the extent of the primary injury and the successive biochemical and immunological responses leading to enhancement of the SCI lesion highly correlate with the degree of

Jun Chen et al. (eds.), *Animal Models of Acute Neurological Injuries II*, Springer Protocols Handbooks,
DOI 10.1007/978-1-61779-782-8_36, © Springer Science+Business Media, LLC 2012

the resulting functional impairment. For such reasons, the assessment of morphological alterations and damage resulting from injury to the spinal cord has long been of prime importance in SCI research. Many possible tools and methods exist for examining such physical characteristics of the injured spinal cord, and the type of method selected is often decided from the type of injury being addressed.

Several potential primary mechanical insults to the cord are possible, including penetrating, transecting, or contusive injuries, with the latter being the most common (1). Experimentally, researchers also use chemical agents, such as kainic acid, to induce a chemotoxic lesion in order to replicate certain biochemical events, like excitotoxicity, involved in secondary damage following SCI. However, this is not a common form of diagnosed primary SCI, and this method is essentially used for investigative purposes. Secondary events following primary injury often lead to extensive spread of inflammatory, excitotoxic, and oxidative tissue degradation, affecting initially spared tissue, activating local glial cells and greatly enhancing the lesion. As a result of the combined primary and secondary damage, a lesion is formed, bordered by a cellular and chemical barrier formation known as the glial scar. The lesion differs somewhat between injury types, both the initial and chronic lesion (Fig. 1), but the general inhibitory nature of the scarred injury site is universal across the various forms of SCI. Such factors are primary contributors to the lack of axonal regeneration and functional recovery following SCI. The resulting sensorimotor disruption ranges from mild to more severe outcomes, such as paralysis or death, depending on the location and severity of the

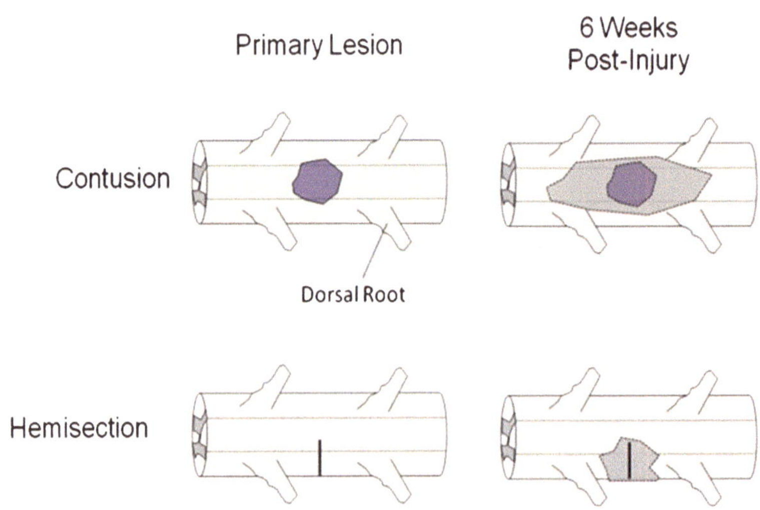

Fig. 1. Examples of lesion variations between mechanically induced SCI types. Contusive lesions are initially and chronically more extensive in size, resulting in cavity formation. Following transection and hemisection, the lesion is more localized adjacent to the primary injury site.

injury. Structural, biochemical, and functional alterations in spinal tissue integrity following SCI are not independent of one another, with each coinciding, causing, or resulting from the others.

Many biochemical and immunohistochemical methods can be used for many or all types of SCIs, and certain forms of imaging, such as magnetic resonance imaging (MRI) and positron emission tomography (PET), can be highly effective for a wide variety of injury types in vivo. Light and electron microscopies are often utilized for the analysis of fixed tissue, rather than in the assessment of living specimens. Some assessments, or certain aspects of a given assessment, may be more suited for particular types of SCI. Due to the importance of structural integrity in the spinal cord to functional integrity and outcome, several morphological parameters relating to the primary mechanical or secondary biochemical and inflammatory damage resulting from SCI are considered.

The amount of tissue spared, the size of the lesion, as well as the response of neural and glial cells, and the extent of axonal damage and regeneration following injury, are all parameters commonly assessed in SCI research. In addition to neural and glial damage, the microvasculature at and around the injury site is severely disrupted following SCI. The result is a combined hemorrhagic and ischemic environment, exposing lesion-adjacent tissue with blood-borne immune agents and inflammatory mediators while blocking blood flow to surrounding areas and contributing to secondary tissue damage. Therefore, assessment of vascular integrity, function, and regeneration can also provide key insights into the mechanisms of tissue damage and ways to enhance protection and repair of damaged tissue following SCI. In accordance with the variety of morphological pathology analyzed, many laboratory techniques are employed to perform such analyses. In the following sections, both general and specific morphological assessments and the experimental methods commonly used for such evaluation are described.

2. Assessment of Lesion Volume and Tissue Sparing

2.1. Myelin and Neural Tissue Damage

As mentioned, following SCI, a lesion is formed of varying size, orientation, and shape dependent on the type and magnitude of the mechanical injury inflicted. Transecting injuries tend to sever far more axonal tracts than contusive or compressive SCI, which usually spare many axons. However, as shown in Fig. 1, the secondary damage following the primary contusive insult results in expansive cavitation that can extend more than one segment of the spinal cord, causing extensive scarring within the surrounding tissue and presenting a large barrier for regenerating axons. After complete transection injury, all spinal tissue is separated, leaving a gap between a rostral and caudal stump of the cord. Partial transection will leave some tracts or tissue intact, though this injury type will produce a gap lesion where the cord was severed.

Despite these differences, some form of lesion occurs, axons and myelin are disrupted, and a dense glial scar at the margins of the lesion will form following SCI. As such, the borders of the lesion can be defined using many histological techniques that rely on chemical staining for identification of particular morphological aspects of the tissue. By identifying the boundaries, and thus the extent of the lesion, an accurate assessment of lesion size and volume can be obtained.

Procedures for histological assessment of spinal cord lesion include staining serial cross- or longitudinal sections of injured cord tissue for identifying myelination or demyelination, and neuronal somas and nuclei. Commonly used methods of examining demyelination in histological preparations of SCI tissue involve chemical staining methods specific for myelin, which in turn allows one to compare normal myelinated white matter to damaged demyelinated areas, which typically demarcate the extent of the lesion damage. In turn, quantification of lesion parameters, such as volume or area, can be determined. Classic methods, which are still used in modern neurological research, are the Weil and Luxol Fast Blue (LFB) myelin stains (2, 3). Demyelination can be determined by identifying areas of white matter devoid of stain, indicating myelin loss. With myelin staining techniques, rostral, caudal, and lateral boundaries of the lesion can be defined when only solid nondisrupted LFB stain is visible through cross- or longitudinal tissue sections.

Another commonly used chemical stain for determining and assessing the SCI lesion is the Cresyl Violet (CV) stain, also known as the Kluver–Barrera staining method when counterstained with LFB (4). CV stain is specific for Nissl substance, which is simply rough endoplasmic reticulum in neural cells, and the specificity of this stain for neurons makes it very useful for assessing gray matter damage. CV, as well as the basophilic nuclear stain, hematoxylin, is often used in conjunction with eosin counterstain of the cellular cytoplasm. When combined with LFB, CV staining provides histological details of gray and white matter integrity and damage.

Concerning the above staining techniques, as with most chemical stains used in histological examination of lesion size and volume, their usefulness is usually a product of their absence in the tissue section being observed. In other words, chemical stains mostly label intact and undamaged tissue and disturbed and disrupted tissue usually lacks the stain allowing for identification of such areas in sections. Other staining methods, however, allow for labeling certain cell types and therefore, visualization and characterization of structures present in histological specimens. Such labeling methods are often utilized to assess activity, response, and distribution of cell types or anatomical structures post SCI. Part VIII, Chaps. 37–41 provide greater details concerning various morphological assessments following SCI.

2.2. Reactive Gliosis

In normal CNS tissue, glial cells play important supportive roles, participating in maintaining the blood–brain barrier (BBB), keeping the balance of extracellular excitatory glutamate and its derivatives in the case of astrocytes, and serving in CNS immune response in the case of microglial cells, which typically reside more within gray than white matter of the spinal cord (5). However, after SCI, local glial cells respond to the injury by migrating towards the lesion, undergoing hypertrophy, and producing chemicals, such as chondroitin sulfate proteoglycans (CSPGs) produced by activated astrocytes. The assessment of reactive gliosis is a useful method of cell-based lesion identification and measurement. As previously mentioned, the end result of this glial response to injury is the glial scar, which serves to prevent the movement of chemical and cellular material into and out of the vulnerable site of injury, but also prohibits axonal regeneration into and through the injury site. Activated astrocytes, a main constituent of the glial scar, thus play both beneficial and detrimental roles following SCI (6, 7).

As the lesion is consistently bordered by this reactive glial cell network and their biochemical products in the glial scar, immunological labeling of markers for injury-activated glial cells and secretions, such as astrocytes ($GFAP^{+}$ vimentin), microglia ($OX\text{-}42^{+}$), and CSPGs ($CS\text{-}56^{+}$), can also be used in defining the boundaries of the SCI lesion. Once the boundaries of the lesion are determined, the cross-sectional or longitudinal area and volume can be measured providing an accurate assessment of lesion size, depending on the goal of the particular study. See Chap. 40 for more detailed description of glial responses after SCI.

2.3. Calculation of Lesion Volume and Spared Tissue

With the lesion margins identified, evaluation of lesion volume provides a well-utilized measurement for quantifying the extent of primary and secondary damage following SCI. Behavioral outcome following SCI can be influenced by the extent of damage encompassed by the lesion (8, 9). The volume of the lesion is primarily used in assessment of contusive SCI, as such an injury leads to extensive cystic cavity formation within the damaged spinal cord. Hemi- or transectional forms of SCI cleanly sever the cord into a rostral and caudal component, and therefore lesion volume may not be appropriate for use in these models. Other methods characterizing the extent of demyelination or gray matter damage are often used in quantification of the extent of such forms of SCI. Obtaining an accurate lesion volume calculation depends on many variables, though several sections through the lesion are needed to obtain the best quantification of this injury parameter.

Conversely, the same methods used to define lesion volume can allow for assessment of the amount of spared tissue following SCI. Analysis can be performed using both light and electron microscopy depending on the level of detail and the morphological characteristics specified for assessment in the given study. Lyn Jakeman provides a thorough reference on the many aspects of determining the extent of spared and damaged tissue in Chap. 37 of this section.

3. Assessment of Axonal Damage, Protection, and Regeneration

Though important, simply assessing what tissue is damaged and spared is insufficient for the investigation of effective treatments for SCI. To regain function that is lost following injury, it is essential that neuronal and axonal damage caused by SCI is repaired allowing for reestablishment of adequate communication between the CNS and the autonomic and lower motor neurons responsible for driving muscular action. Therefore, assessing the levels of axonal regeneration and degeneration within the spinal cord that takes place following SCI is a very important area of SCI research. In order to perform such experiments, methods have been established that take advantage of the retrograde and anterograde transport mechanisms of conveying materials to and from the CNS neural soma. The reasoning for the use of such methods stems from our current understanding of the pathological processes damaged spinal axons undergo following SCI.

Once an axon is severed or damaged, a process known as Wallerian degeneration occurs in which the segment of the injured axon distal to the site of injury swells, in the case of CNS axons, and undergoes anterograde degradation, leaving the proximal portion of the neuron's axon shortened, but intact (10, 11). However, as noted by Ramon y Cajal, this state is only temporary as the remaining free end of the axon typically undergoes cytoskeletal and biochemical reorganization, resulting in "dieback" or retraction of the injured axon towards the cell body. Cajal termed this proximal stump of injured spinal cord axons the retraction ball or bud (12). Experimentally transected axons have been shown to retract hundreds of microns in vivo from the injury site within 1 h after injury (13). The distance injured axons may retract is variable; however, it has been demonstrated that following contusive SCI axonal retraction is significantly greater than that observed following transection or hemisection injuries (14).

Methods by which to assess axonal degradation following SCI are many, but commonly used methods include immunolabeling tissue for amyloid precursor protein (APP), which is an indicator of disturbed axonal transport and accumulates at the damaged end of the axon (15), or using silver impregnation methods, such as the Fink-Heimer method (16) which allows for the identification of degenerated axons and their terminal endings. Quantifying degradation or axon dieback using such labeling is often dependent on the type of injury and the point of reference chosen for the assessment. Regeneration of the injured axons towards, into, and beyond the spinal cord lesion is inherently prohibited in the mammalian CNS, as many factors at and around the site of injury, such as myelin-associated Nogo, MAG, and Omgp (17–19), are highly inhibitory for the progress of this regrowth. However, in the past several years, a variety of treatments, ranging from cellular and tissue transplantation (20, 21) to direct application of pharmacological and chemical agents to the injured cord (22), have provided promising results concerning the regeneration of damaged axons.

To study axonal regeneration, it is crucial to be able to track single-axon pathways in the spinal cord both spatially and temporally. This analysis can be successfully achieved by supplying the neurons of interest in the CNS with specific substances that will be taken into the cell either through the neuron cell body for transport to the axon terminus, known as anterograde transport, or via the terminus and relayed back to the cell body, or retrograde transport. By performing postlabeling histological analysis, it can be determined whether axons regenerated through or around the site of injury following SCI.

Anterograde axonal tracing is especially useful for the study of axonal regeneration, as it is only possible to detect the label within and distal to the site of injury if the axon regenerated distances in accordance with the location of such detection. Biotinylated dextran amine (BDA) tracers are common agents used in anterograde axonal transport analysis, though many methods can be potentially utilized. Retrograde labeling is less reliable for regenerative assessment, especially concerning contusive SCI, as it is difficult to establish whether the labeled axons are regenerated or spared axons. As such, retrograde labels, such as FluoroGold, are commonly used in the quantification of spared axons in combination with other chemical methods, such as myelin staining. Chapters 38 and 39 address the types and qualities of retrograde and anterograde labeling techniques that can be employed for the assessment of axonal regeneration and protection following SCI.

4. Assessment of Microvasculature Following SCI

In normal healthy spinal tissue, as with all tissues, blood must regularly deliver vital nutrients and oxygen via capillaries for proper cellular function. Within the spinal cord and the brain, the vascular interaction at the terminal capillary networks with the nervous tissue is referred to as the neurovascular unit (NVU), and a variety of cellular interactions play roles in forming and maintaining this highly regulated system of exchange within the immune-privileged CNS (23). Following SCI, this unit is highly disrupted leading to a variety of biochemical, immunological, and other physiological dysregulations at and around the site of injury (24).

Soon after the initial insult, hemorrhage is observed locally at the lesion, as blood leaks from the damaged microvasculature into the surrounding spinal tissue. As a result, numerous erythro- and leukocytes, as well as inflammatory cytokines and other blood-borne substances, enter the lesion area contributing to the secondary damage described previously. Astrocytes, participants of the BBB and the blood–spinal cord barrier (BSCB), are also disrupted and reactive gliosis leading to the glial scar occurs in part to reestablish a blockade of foreign and harmful substances from entering the delicate injured spinal tissue (Fig. 2). In light of the importance of the microvasculature to neural tissue health and functioning, investigating the damage and potential for protection and repair of this essential transport system following SCI is crucial.

Hagg et al. (Chap. 78) provide extensive details on the types and variety of markers used for histological assessment of the neurovascular system following SCI, either as general assessment or comparative assessment following pharmacological or other therapeutic treatments. As mentioned, astrocyte (GFAP^{+}) location and activation status can be determined, as well as for endothelial

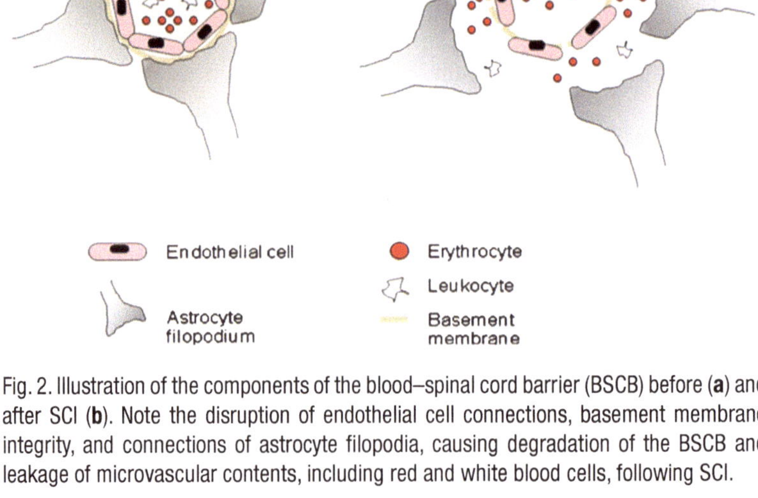

Fig. 2. Illustration of the components of the blood–spinal cord barrier (BSCB) before (**a**) and after SCI (**b**). Note the disruption of endothelial cell connections, basement membrane integrity, and connections of astrocyte filopodia, causing degradation of the BSCB and leakage of microvascular contents, including red and white blood cells, following SCI.

(ex. RECA-1+) cells and pericytes (NG2+), using specific antibodies or markers dependent on the type and purpose of the assessment. In order to find appropriate ways to protect or repair the vasculature that is damaged by the primary or secondary injury process following SCI, quantification of the microvasculature is necessary to provide a basis for comparison of the efficacy of a neuroprotective or reparative treatment agent. Also, status of vascular function is very important in making such determinations.

It is critical to determine whether protected or regenerated vasculature is functional following SCI and after treatment in order to appropriately judge the overall status of the NVU during such assessments. As such, imaging modalities, such as PET imaging, can provide an in vivo method of localizing and viewing vascular function within the CNS, giving an overall picture of blood flow in a given region of interest (ROI). However, limitations exist as to the level of detail that can be determined from imaging techniques; therefore, careful selection of the type of imaging used for certain types of assessments is needed. Chapter 41 highlights the advantages and disadvantages of methods to quantify and verify the status of the microvasculature after SCI, as well as insights into other aspects and techniques of assessing the NVU following SCI.

5. In Vitro and In Vivo Imaging and Assessment

As mentioned several times, appropriate imaging of the morphological characteristics of either processed histological specimens or of the spinal cord and related structures in vivo following SCI is a necessity for accurate assessment of such parameters. Selection of the imaging method is dependent on the type of analysis desired by the investigator. Most histological assessments, either with chemical or immunological labeling methods, are carried out using some form of light or fluorescent microscopy. For most low-magnification assessments, i.e., calculation of lesion volume and spared tissue, chemical or cellular distribution, standard light microscopy utilizing the appropriate full-spectrum or fluorescent illumination is sufficient. For certain types of assessments, such as determining cellular colocalization of certain antigenic markers via fluorescent labeling, higher magnification is often used to provide more details for such cell-specific analyses.

Confocal microscopy, which is typically used in visualizing fluorescent labeling in histological sections, can be used for live cell and real-time tissue imaging, as well as providing high-magnification and -resolution two- and three-dimensional images throughout several planes of a given specimen. This type of microscopy has become very common for the quality and details it can provide of specimens of interest. To obtain fine details, such as synaptic interactions, density, and myelin sheath thickness, one must use electron microscopy, as the resolution capabilities of light and confocal microscopes are not sufficient for such assessments.

Stereology, an unbiased method of random ROI selection of a histological section, allows for objective quantification of parameters including cell or axon abundance in a given specimen. This process takes into account the three-dimensional characteristics of the specimen under analysis. This method of quantification is critical for reducing investigator and sample bias in the assessment of physical characteristics of tissue samples obtained following SCI. Design-based stereology, in which a particular stereological method is selected for use in given assessments, does not rely on assumption of size, number or other specific characteristics of a specimen (25). However, there are certain precautions that must be taken to further reduce bias and confounding data, such as ensuring that samples are collected and processed under the same conditions (26). Nonetheless, design-based stereology can be very useful for accurate quantification of selected morphological parameters. Chapter 3 in Volume I, Part 1, describes this important tool in neuroscience research in greater depth.

In vivo imaging offers many benefits, such as allowing for longitudinal long-term analyses of given morphological and physiological characteristics in a single animal following SCI. Such imaging methods also allow for reduced animal usage and time requirements for processing and analyzing histological specimens while potentially enhancing the quality of given assessments. Additionally, in vivo imaging is the most appropriate method of obtaining diagnostic and morphological data for assessing severity, progress, or deterioration of spinal tissue in humans within a clinical setting. Recent studies utilizing MRI (27, 28), PET imaging (29), and coherent anti-Stokes Raman spectroscopy (CARS) (30) have demonstrated the usefulness of such techniques in the study of tissue morphology following experimental injury to the spinal cord. There are merits and disadvantages to all methods of assessing morphological characteristics following SCI, but the wide variety of potential techniques available allows for more complete analyses and interpretations, which enhance the quality of science produced and disseminated to the scientific and lay communities alike.

6. Summary

SCI causes much damage to spinal and vascular tissues, leading to a myriad of motor, sensory, and autonomic dysfunctions in affected individuals. Developing an understanding of the anatomical and histological outcome following injury is critical not only to correlate the amount of morphological damage or repair to functional deficits or recovery witnessed, but also to provide a wider knowledge base of the intricacies and unknowns of the molecular and physiological mechanisms by which the complex network of tissues and structures in the spinal cord interact. The frailty and importance of the CNS in the context of the mobility, social interaction, and overall quality of life more than justify the need for useful, replicable, and widely accepted methods of characterizing types and the extent of anatomical damage observed following damage to this system. In addition, the application of such assessments for determination of the efficacy of potential therapies following injury only enhances this need. In the sections that follow, a full and extensive reference is provided for the battery of morphological assessments available and utilized in SCI research.

References

1. Jackson AB, Dijkers M, Devivo MJ, Poczatek RB (2004) A demographic profile of new traumatic spinal cord injuries: change and stability over 30 years. Arch Phys Med Rehabil 85: 1740–1748
2. Weil A (1945) Textbook of neuropathology. Grune and Stratton, New York, pp 328–329
3. Salthouse TN (1964) Luxol fast blue "G" as a myelin stain. Stain Technol 39:123
4. Kluver H, Barrera E (1953) A method for the combined staining of cells and fibers in the nervous system. J Neuropathol Exp Neurol 12:400–403
5. Yong VW, Rivest S (2009) Taking advantage of the systemic immune system to cure brain diseases. Neuron 64(1):55–60
6. Sofroniew MV (2009) Molecular dissection of astrogliosis and glial scar formation. Trends Neurosci 32(12):638–647
7. Sofroniew MV, Vinters HV (2010) Astrocytes: biology and pathology. Acta Neuropathol 119(1):7–35
8. Basso DM, Beattie MS, Bresnahan JC (1995) A sensitive and reliable locomotor rating scale for open field testing in rats. J Neurotrauma 12(1):1–21
9. Hoschouer EL, Basso DM, Jakeman LB (2010) Aberrant sensory responses are dependent on lesion severity after spinal cord contusion injury in mice. Pain 148(2):328–342
10. Waller A (1850) Experiments on the section of the glossopharyngeal and hypoglossal nerves of the frog, and observations of the alterations produced thereby in the structure of their primitive fibres. Philos Trans R Soc Lond 140:423–429
11. Coleman M (2005) Axon degeneration mechanisms: commonality amid diversity. Nat Rev Neurosci 6(11):889–898
12. Cajal Ramon y (1991) Degeneration and regeneration of the white matter. In: DeFelipe J, Jones EG (eds) Degeneration and regeneration of the nervous system. Oxford University Press, New York, p 489
13. Kerschensteiner M, Schwab ME, Lichtman JW, Misgeld T (2005) In vivo imaging of axonal degeneration and regeneration in the injured spinal cord. Nat Med 11(5):572–577
14. Iseda T, Okuda T, Kane-Goldsmith N, Mathew M, Ahmed S, Chang YW, Young W, Grumet M (2008) Single, high-dose intraspinal injection of chondroitinase reduces glycosaminoglycans in injured spinal cord and promotes corticospinal axonal regrowth after hemisection but not contusion. J Neurotrauma 25(4): 334–349
15. Li GL, Farooque M, Holtz A, Olsson Y (1995) Changes of beta-amyloid precursor protein after compression trauma to the spinal cord: an experimental study in the rat using immunohistochemistry. J Neurotrauma 12(3): 269–277
16. Fink RP, Heimer L (1967) Two methods for selective silver impregnation of degenerating axons and their synaptic endings in the central nervous system. Brain Res 4:369–374
17. Xie F, Zheng B (2008) White matter inhibitors in CNS axon regeneration failure. Exp Neurol 209(2):302–312
18. Yang LJ, Schnaar RL (2008) Axon regeneration inhibitors. Neurol Res 30(10): 1047–1052

19. Cao Z, Gao Y, Deng K, Williams G, Doherty P, Walsh FS (2010) Receptors for myelin inhibitors: structures and therapeutic opportunities. Mol Cell Neurosci 43(1):1–14
20. Xu XM, Onifer S (2009) Transplantation mediated strategies to promote axonal regeneration following spinal cord injury. Respir Physiol Neurobiol 169(2):171–182
21. Tom VJ, Sandrow-Feinberg HR, Miller K, Santi L, Connors T, Lemay MA, Houlé JD (2009) Combining peripheral nerve grafts and chondroitinase promotes functional axonal regeneration in the chronically injured spinal cord. J Neurosci 29(47):14881–14890
22. Kwon BK, Okon EB, Plunet W, Baptiste DC, Fouad K, Hillyer J, Weaver L, Fehlings M, Tetzlaff W (2011) A systematic review of directly applied biologic therapies for acute spinal cord injury. J Neurotrauma 28(8):1589–1610
23. Choi YK, Kim KW (2008) Blood-neural barrier: its diversity and coordinated cell-to-cell communication. BMB Rep 41(5):345–352
24. Mautes AE, Weinzierl MR, Donovan F, Noble LJ (2000) Vascular events after spinal cord injury: contribution to secondary pathogenesis. Phys Ther 80(7):673–687
25. West MJ (2002) Design-based stereological methods for counting neurons. Prog Brain Res 135:43–51
26. Schmitz C, Hof PR (2005) Design-based stereology in neuroscience. Neuroscience 130(4):813–831
27. Sydekum E, Baltes C, Ghosh A, Mueggler T, Schwab ME, Rudin M (2009) Functional reorganization in rat somatosensory cortex assessed by fMRI: elastic image registration based on structural landmarks in fMRI images and application to spinal cord injured rats. Neuroimage 44(4):1345–1354
28. Kozlowski P, Raj D, Liu J, Lam C, Yung AC, Tetzlaff W (2008) Characterizing white matter damage in rat spinal cord with quantitative MRI and histology. J Neurotrauma 25(6):653–676
29. Tewarie RD, Yu J, Seidel J, Rahiem ST, Hurtado A, Tsui BM, Grotenhuis JA, Pomper MG, Oudega M (2009) Positron emission tomography for serial imaging of the contused adult rat spinal cord. Mol Imaging 9(2):108–116
30. Ouyang H, Sun W, Fu Y, Li J, Cheng JX, Nauman E, Shi R (2010) Compression induces acute demyelination and potassium channel exposure in spinal cord. J Neurotrauma 27:1–12

Chapter 37

Assessment of Lesion and Tissue Sparing Volumes Following Spinal Cord Injury

Lyn B. Jakeman

Abstract

Spinal cord trauma causes acute hemorrhage and ischemia, which in turn initiate a cascade of secondary events resulting in cell death and loss of neural tissue at the site of impact. In contusion or compression injuries, the ensuing tissue damage can extend several segments rostral and caudal to the initial insult. Careful quantitative evaluation of the lesion site and measurement of damaged and spared tissue volumes are essential if one is to fully understand the substrates that underlie functional loss and recovery. For example, neuroprotective strategies can improve the sparing of gray and white matter regions that support recovery. Alternatively, reparative strategies may reduce cavitation, alter cellular invasion or tissue contracture, or enhance axonal growth or regeneration. Interpretation of each of these possibilities depends on accurate estimation of the tissue reference volumes. This chapter provides a brief review of the historical context for measuring tissue volumes following spinal cord injury (SCI) in rodent models. Then, procedures are given for determining unbiased estimates of the volume of SCI lesions and spared tissue as well as cellular component volume fractions. The described methods utilize digital images and simple stereological tools that can be applied without sophisticated three-dimensional reconstruction or dedicated stereology software. Finally, the effects of tissue shrinkage on other outcome measures that might be relevant in SCI research studies are discussed.

Key words: Cavalieri method, Secondary injury, Contusion, Cavitation, Gray matter, White matter, Sparing, Tissue shrinkage, Volume component analysis

1. Introduction

1.1. Spinal Cord Injury Pathology and Functional Recovery

Chronic functional loss after spinal cord injury (SCI) is due to loss of neurons and axons found within the gray matter of the spinal cord and loss of conduction of ascending and descending fibers that span the level of injury. The vascular and axonal damage caused by shearing and compressive force initiates a cascade of secondary injury events, including blood–brain barrier disruption, excitotoxicity, oxidative tissue damage, inflammation, protease activation,

Jun Chen et al. (eds.), *Animal Models of Acute Neurological Injuries II*, Springer Protocols Handbooks, DOI 10.1007/978-1-61779-782-8_37,

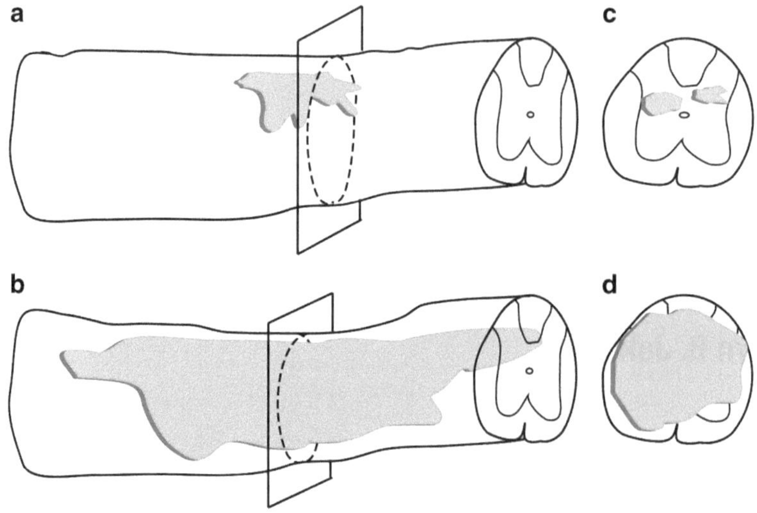

Fig. 1. Progression of SCI pathology over time. (**a**) Drawing of the site of initial impact as typically seen in the first 24 h following a spinal contusion lesion. The initial site of damage (*gray-shaded area*) occupies a small proportion of the cord volume, illustrated in the drawing of a section taken in the transverse plane (**c**). Gray matter regions and white matter regions are minimally affected by the initial hemorrhage and shearing. (**b**) Secondary injury events progressing over the first days to weeks after injury result in the expansion of the area of damage to encompass most of the centrally located gray matter and portions of the peripheral white matter, as illustrated in a drawing of a transverse section in (**d**). Damage can extend for several spinal segments rostral and caudal to the initial site of impact.

immune cell invasion, and activation of cell death cascades (reviewed in refs. 1–4). Together, these secondary mechanisms cause the gradual expansion of tissue damage such that an initially localized contusion or compression can create a lesion spanning several segments above and below the original site of impact (Fig. 1 (5–7)). In the clinical setting, the size and distribution of the resulting spinal cord lesion is highly variable due to differences in the spinal level of the injury, the type of insult (vascular damage, laceration, contusion, compression, distention, or closed cord injury), and the wide range of biomechanical forces and physiological complications associated with the insult (8–10).

Under controlled experimental settings in animal models, the severity of the induced SCI is predictive of the final extent of functional recovery. Contusion or compression injuries are graded by controlling either the impact momentum (force and/or speed of impact) (11) or the displacement distance (12, 13), or the speed and/or maximal extent of compression (14–17). Chemical lesions can be controlled by standardizing either the volume or concentration of neurotoxic agent (18), and partial transection injuries can be graded by controlling the distribution and amount of spinal tissue severed by a knife or microscissors (19). The functional effects of injury can then be measured in all of these models by using appropriate behavioral or electrophysiological tests. Most functional measures reveal an improvement over time that reaches a plateau within a few weeks after the injury (e.g., 17, 19–22). Thus, the continued pathology of secondary injury is superimposed upon a functional curve of improved recovery that reflects the ongoing resolution of ionic and physiological blockade and restoration and plasticity of residual circuitry.

Both the anatomical extent and the distribution of tissue damage reflect and predict the extent and nature of functional recovery in SCI models (5, 16, 23–25). For instance, mild SCI that produces damage limited to the gray matter of the midthoracic spinal cord is associated with near-normal locomotor recovery, while damage at the same spinal level that extends to encompass the ascending and descending white matter will disrupt hindlimb function and preclude the recovery of voluntary locomotion and normal sensation. Injuries that occur at spinal segments in the cervical or lumbar enlargements result in loss of the primary motor and sensory neurons that innervate the forelimbs or hind limbs (25–27), while lesions extending into rostral cervical segments compromise respiratory capacity (28). Thus, it is important to document both the *extent* and *distribution* of tissue damage in order to interpret functional and cellular results in SCI research because interventions or genetic manipulations that alter the secondary injury or repair mechanisms in different ways can result in improvement or exacerbation of functional recovery.

1.2. Historical Perspective for Measuring Spared Tissue Volume

1.2.1. Cavity Volume and Spared White Matter Area

The earliest modeling studies of compression and contusion injuries noted a positive correlation between the size of the resultant lesion and functional outcomes. Initial studies focused on either the area or volume of the cavity produced at chronic time points in these models (29, 30). The focus on spared white matter (SWM) became paramount after Andrew Blight demonstrated elegantly in 1983 that intact, healthy, myelinated axons located primarily in the peripheral rim of the cat spinal cord survive for several months to more than a year following a severe spinal contusion injury. Blight estimated axon numbers using unbiased sampling methods to show that relatively few such axons are required to enable effective locomotion (31). Several labs have since confirmed these observations and extended this work to show a close positive correlation between the measured cross-sectional area of normal myelin-stained or "spared" white matter obtained from transverse sections at the injury site and the extent of functional recovery following midthoracic contusion SCI (5, 22). The greatest damage is located in or near the site of impact and is commonly referred to as the lesion "epicenter." In our lab, we refer to this area measurement as SWM area. SWM can be expressed in units of area (mm^2) or as a proportion or percentage of the total cross-sectional area at the lesion epicenter (%SWM = SWM/total cross-sectional area × 100%). The amount of SWM at the site of greatest injury is a simple and robust outcome measure that reflects the continuity of uninjured ascending and descending myelinated axons.

1.2.2. Longitudinal Mapping of SCI Site

In 1990, Sydney Finkelstein asserted that the longitudinally asymmetric contusion or compression cavity was better described by illustrating a series of cross-sectional profiles than by a single cross section area measured only at the lesion epicenter (32). He examined serial stained transverse sections through the extent of an SCI and identified the epicenter section for each spinal cord. He then used the estimate of distances between sections based on microtome cutting thickness to select five equally spaced sections spanning the entire lesion. These sections were traced to create line drawings of regions of spared tissue gray and white matter for each specimen. The "cross-sectional area of the spared white and gray matter" was measured in each of these images, and these values were then corrected to account for the shape of the thoracic spinal cord and summed to provide a composite estimate of residual gray and white matter tissue volume. The two major contributions of this paper were the following: first, although cross-sectional area closely correlates with hindlimb functional recovery, the volumetric measure was better correlated to the biomechanical injury severity of the model than a single cross-sectional area. Second, the use of multiple section drawings to illustrate the amount of spared white and gray matter provided a more thorough view of the variations in lesion and spared tissue distribution between specimens. This approach to illustration and measurement of spared tissue along the rostrocaudal lesion extent is now widely used in SCI research labs (e.g., 33–36). The procedures described below incorporate documentation of the spared tissue areas or "lesion mapping" as part of the volume estimate approach.

1.2.3. Methods for Volume Estimation

Geometric Formulas and 3D Reconstruction

Early estimates of lesion and spared tissue volumes incorporated geometric estimates of the shape of a contusion injury site as a biconical lesion, with the volume estimated by using the formula for the volume of a frustrum of a cone (22) and the normal thoracic spinal cord volume estimated similarly (32). However, the actual lesion shape is much more variable and complex than a simple biconical model would reflect. Three-dimensional (3D) reconstruction can be used to obtain a full and accurate representation of the extent and distribution of an irregular lesion. For good resolution using this approach, a total of 20–30 sections are required, and they must be precisely digitally aligned and carefully outlined. Then, a representation of the full volume of the spinal cord is calculated and filled in by computerized extension of thousands of 2D pixels into 3D voxels. Because it is so labor intensive and adds primarily descriptive information to the volume estimate, 3D reconstruction is usually used to illustrate a pattern of lesion topography from one or two representative specimens in model development (6, 12) or to illustrate treatment comparisons (37), but has not been found to be practical for routine estimation of lesion and spare tissue volumes across large groups of specimens.

Cavalieri Method

Responding to a need for reliable, but easily obtained, measures of component and spared tissue volumes, recent studies have taken advantage of the use of unbiased sampling approaches and stereology tools. By employing a point-counting area estimate procedure termed the Cavalieri method, accurate volume estimates of a highly irregular object can be obtained by preparing lesion maps of 10–12 equally spaced tissue sections per specimen (38, 39). The volumes of intrinsic components of the lesion and spared tissues can be determined quickly and easily from the lesion maps. With the availability of these simple quantitative outcome measures, it is no longer sufficient for interpretation of functional or cellular response data to simply present a photomicrograph of a single stained tissue section from selected treated and untreated specimens to document the effects of experimental manipulations. The Cavalieri method can be used to report spared gray matter and white matter volumes as well as volumes and the volume fraction of glial cells, inflammatory cells, connective tissues, and other cellular elements at the lesion site. In addition, measuring the volumes of lesion and spared tissue regions is essential in order to obtain reference volumes needed for comparing cell numbers and other quantitative histological measures across treatment groups (40).

2. Measuring the Volume of SCI Regions

2.1. Importance of Using an Unbiased Approach

Using modern immunohistochemical and histological methods, a seemingly infinite number of features can be identified and colocalized with specific markers at the cellular or subcellular level. While it is tempting to collect and stain hundreds of tissue sections and then select a few seemingly comparable or "representative" sections for comparisons, many of the markers and features of interest are subject to substantial variations both within and between individual samples. Notably, the segmental organization of the spinal cord affects the patterns of gray and white matter loss. In addition, the distribution of primary afferent and motoneuron projections confers additional variations from section to section. To obtain meaningful quantitative outcomes, it is important that sampling procedures ensure that all of the variation within a population has an equal chance of being measured (40).

The process of obtaining unbiased 3D measures, such as region volumes, from principally 2D sections is referred to as stereology. Volume estimates can be obtained by applying systematic point counting to equally spaced tissue sections. The theoretical basis for this point-counting method has been described elsewhere (38, 40, 41). The approach is based on the principle that if an exhaustive series of thin parallel sections is taken through an object, then the volume is estimated by multiplying the area of the object in each section by the distance between sections (Fig. 2a, b). Mathematical models and empirical studies show that it is not necessary to obtain a precise area measure for each section through the object. Instead, one can estimate the area of the defined region in each section by counting the number of times the region of interest intersects a site on a fixed-point grid. The area per section is then calculated as a product of the area/point (a/p) times the number of points for each section (P). The volume estimate is now equal to the sum of the number of points multiplied by a/p times the thickness of the sections: $V = T \times a/p \times \sum P$ (Fig. 2c). There are no other conditions, except that the position of the first section must be random in the object. This is met by complete sampling of the spinal segments from tissue blocks that are imperfectly aligned, and then using position on the slide to select the starting section. Thus, it is not necessary to perform time-consuming 3D reconstruction, which does not add additional quantitative data.

2.2. Requirements for Obtaining Unbiased Volume Estimates

The following concepts should be incorporated in the design of the histological tissue preparation.

1. A series of sections must be obtained that spans the entire volume of interest. Prior knowledge or estimate of the lesion volume is very helpful for determining the size of the block required to exhaustively sample the lesion. Good sources for initial estimates for midthoracic contusion injuries can be found in Ma et al. (42) (mouse contusion injuries) and Scheff et al. (11), 2003 (rat contusion injuries). Alternatively, estimates of lesion length can be determined by performing a pilot experiment in which a few specimens are cut in the longitudinal plane.
2. Sections must be collected so that the distance between sampled sections is known. The easiest way to do this is to collect equally spaced sections in series such that every *n*th section of a fixed thickness is selected for volume estimates. Adjacent series can then be used for qualitative or quantitative documentation of other cellular staining patterns. These distances are only estimates; they do not account for compression and expansion that certainly occur with cutting.
3. The stain used to define the volume of interest must be of sufficient contrast to easily define the regions for measurements. For example, we use eriochrome cyanine (EC) to stain

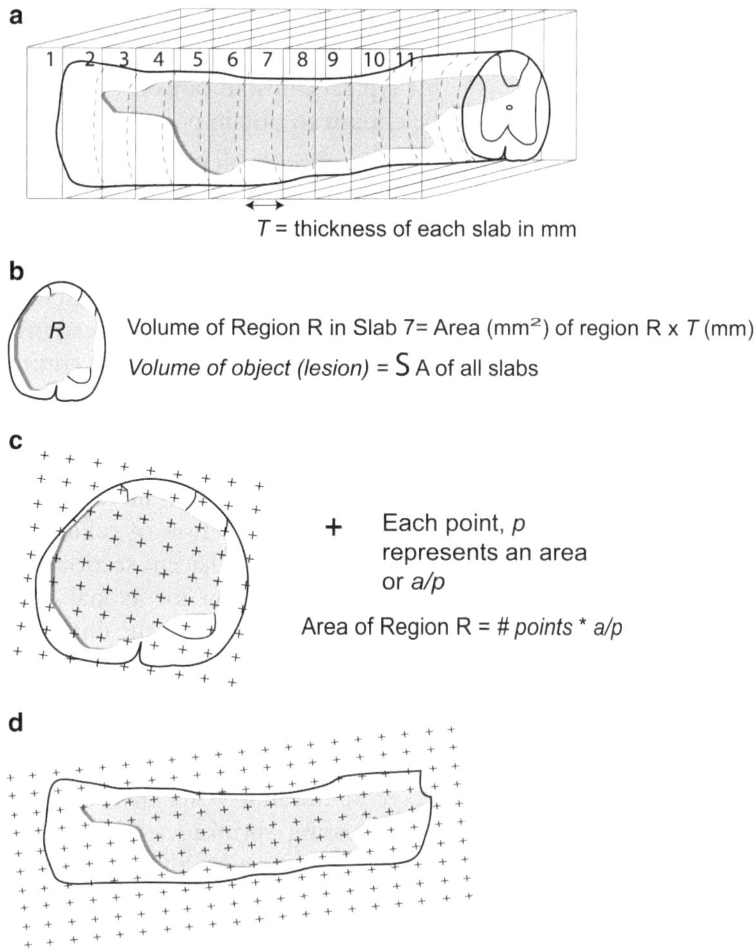

Fig. 2. The Cavalieri method is used to estimate the volume of an irregular object. (**a**) An object of any shape can be divided into an infinite number of parallel slabs of section thickness, *T*. (**b**) The volume of a defined region in each slab can be estimated by multiplying the area of the object (*A*) by the distance between sections, or slab thickness (*T*). (**c**) The area of an object in a planar section does not have to be measured with high resolution, but can be estimated reliably by multiplying the number of evenly spaced points (*P*) that are superimposed on that object, if each point represents a fixed area or *a/p*. (**d**) This method should yield the same volume estimate regardless of whether the sections are cut in a transverse (D) or longitudinal (D) plane.

phospholipids in white matter and then often use cresyl violet as a counterstain to identify cell nuclei and neuronal cell bodies.

4. The starting section and orientation of the counting grid must be random. To ensure that differences in topography do not induce systemic bias to the volume estimates, each region of the structure of interest must have an equal likelihood of being included.

5. Theoretical considerations have shown that for an object with a complex 3D shape a maximum of 200 points on 10–15 equally

spaced sections are sufficient to obtain a volume estimate with a coefficient of error of 5–10% while fewer points are sufficient for typical ellipsoid objects (40). The size of the grid lines can be chosen to obtain points in this general range.

6. The theoretical calculation of lesion volumes predicts that the orientation of the sections is not important for estimating volume. Application of the Cavalieri approach should give the same volume whether applied to transverse or longitudinal sections of a cylindrical structure, such as the spinal cord (Fig. 2d). In practice, compression and expansion of tissue sections prepared on microtome or cryostat cutting devices induce variations that are specific for different cutting orientations. Therefore, it is advised to cut all specimens for volume estimates using the same plane of section. In most of our studies, the spinal cord lesion site is cut in the transverse plane to allow analysis of spinal laminar distribution of cellular elements (neurons, axons, and glial cells) in adjacent section series. However, the favored plane of orientation for axonal regeneration studies may be longitudinal to facilitate the illustration of axons in the rostrocaudal axis.

2.3. Corrections for Tissue Shrinkage in SCI

In time after an SCI, the loss of tissue and removal of debris followed by increased cellular contraction lead to profound tissue shrinkage at the injury site that is biological and independent of tissue processing shrinkage. Direct measures of spared tissue or lesion area can prove highly variable as they incorporate different degrees of shrinkage and cavity collapse. Measures that use the total cross-sectional area measured from the injured specimen (%SWM) reduce this variability (22), but may result in an underestimate of the tissue loss or a high estimation of tissue sparing when comparing between labs or across ages or time points (34). To express spared tissue as a proportion of the corresponding region and level of the normal spinal cord without incurring these errors, one can use tissue sections from the most rostral and caudal regions of the spinal cord block to establish a correction factor that is applied to the measured areas. Methods and tables for correction factors that account for the changing diameter of the thoracic spinal cord are described in detail elsewhere (39, 43, 44).

3. Methods for Preparation of Tissue Sections for Volume Analysis

There are a wide range of methods used to obtain quality series of equally spaced sequential tissue sections that span the lesion site of an SCI tissue block. An overview of SCI morphological approaches can be found in chapter 36, Vol 2. Featured steps of tissue processing include tissue fixation, dissection, postfixation, blocking, and sectioning. Our procedures are described only briefly in Sects. 3.1 and 3.2 below. Detailed methods for a good contrast myelin and Nissl stain series and the methods for measuring volumes are described in greater detail in Sect. 3.3.

3.1. Tissue Collection and Postfixation

1. Where cross-linking is not contraindicated and immunohistochemistry is employed as one of the histological outcome measures, the subjects are anesthetized deeply with a high dose of anesthetic and perfused transcardially with 50–100 ml of 0.9% saline or 0.1 M phosphate-buffered saline followed by 100–300 ml of an aldehyde fixative, such as buffered 4% paraformaldehyde in 0.1 M phosphate-buffered saline.
2. For cryosectioning or vibratome sections, the spinal cord tissues are dissected from the vertebral column, taking care to mark the site of vertebral or spinal root landmarks on the tissue prior to removal (34). The tissue blocks are then postfixed in a specimen bottle containing fresh perfusion fixative for 1 or 2 h, rinsed overnight in 0.2 M phosphate buffer, and finally immersed in 20–30% sucrose as a cryoprotectant until they equilibrate. Aldehyde fixation continues over time, so it is important to include controls and experimental specimens together for all steps.

Note that in some situations, cryosections are not ideal. For example, if precise vertebral localization is required to co-register histology with imaging data, the spinal cord can be dissected with the vertebrae intact and the specimens embedded in paraffin or resin and cut on a microtome (13). Likewise, if the cells of primary interest are found within the loose tissue present in the acute lesion epicenter, then paraffin- or resin-embedded sections may be desired so that the cells are not lost with subsequent washing steps.

3.2. Blocking and Sequential Sectioning Through the Lesion Site

Volume measures can be obtained using any method for sectioning that allows the preparation of a series of equally spaced sections with a known distance between each stained section. In our SCI research labs, we have adopted a procedure that allows routine staining and stereological measures that span the length of an SCI site using cryosections.

1. Cut equivalent-length tissue blocks from three to five rat or mouse spinal cords and embed them within the same specimen block (Fig. 3a). The tissues from different treatment groups can be embedded together and then cut and mounted on the same slides. Thus, the researcher is able to ensure identical processing for all treatment groups.
2. Embed and rapidly freeze the tissue blocks in optimal cutting temperature (OCT) freezing compound (CRYO-OCT Tissue-Tek; Andwin Scientific #4583), which is used to fill plastic tissue

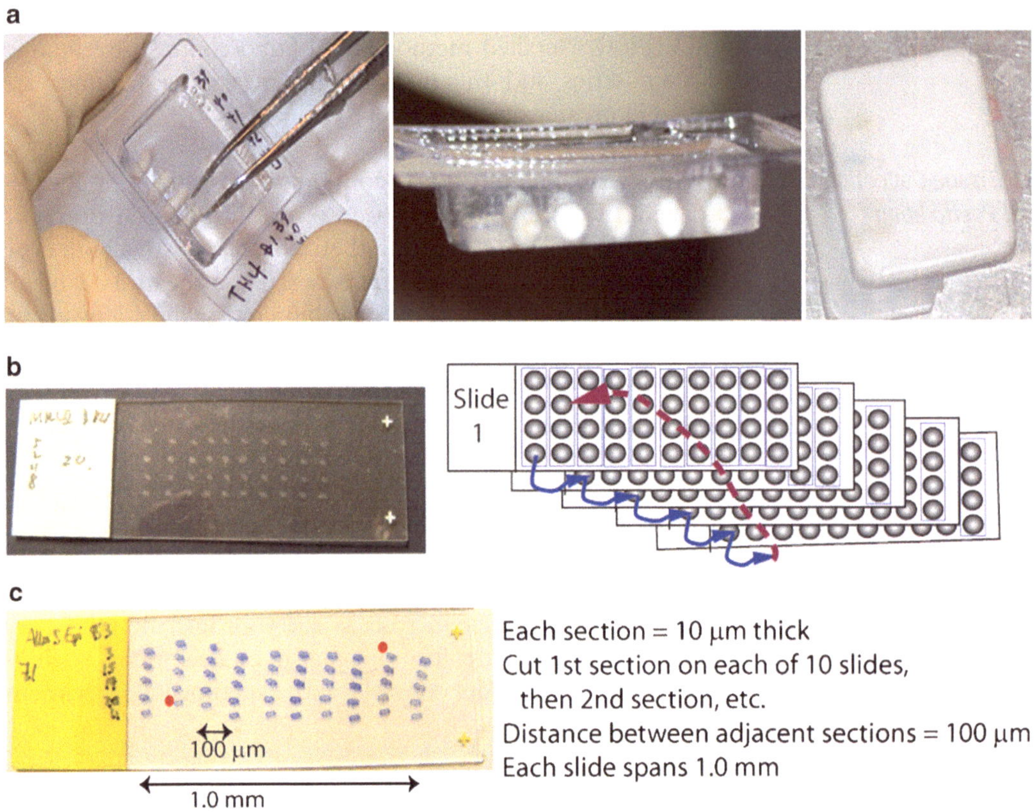

Fig. 3. Methods for obtaining evenly spaced tissue section series. (**a**) Align tissue blocks from three to five rat or mouse spinal cords and embed them within the same specimen block. Adapted with permission from unpublished protocols of the Spinal Cord Injury Training Manual, NINDS, Facilities of Research Excellence in Spinal Cord Injury, © The Ohio State University, 2004–2010. (**b**) Cut ten sets of slides with ten sections from the block on each slide to create adjacent series of sections, each spaced 100 μm apart. Each slide contains sections spanning 1 mm of spinal cord tissue. The *photo* illustrates an unstained slide with ten columns of sections from a tissue block with four specimens. The *Drawing* shows the sequence of section collection. (**c**) Photo of a slide containing 10-μm-thick sections stained with EC/CV. Each adjacent section is 100 μm apart and the slide spans a total of 1.0 mm. The *red dots* illustrate markings on the slide to indicate the start of the lesion site in two of the specimens.

Cryomolds (Tissue-Tek; Andwin Scientific #4557). It is essential to freeze the tissue blocks rapidly to prevent the formation of water crystals, which lead to the development of large "swiss cheese" holes in the sections. Rapid freezing can be done by immersing the cryomolds in isopentane cooled to −30 to −40°C or in a closed chamber surrounded by powdered dry ice.

3. Cut section blocks with a cryostat at 10–20-μm thickness onto electrostatically charged slides (e.g., Superfrost or Color frost+Plus®; Fisher Scientific #12550-xx). Collect five or ten series of sections in order such that each section is obtained at 100 μm from the adjacent section and a series of ten adjacent sections on a slide represent regions spaced 1.0 mm apart (Fig. 3b, c). Some research groups discard intervening sections (34), but

we find that the intervening sections can be used to obtain a wealth of data regarding other features of the lesion site (45).

3.3. Staining with Iron Eriochrome Cyanine and Cresyl Violet

3.3.1. Materials

1. Slide warmer (e.g., EM Sciences #71319 or 713-20; found in most histology or cytology labs).
2. Acetone.
3. Coplin jars or staining dishes (e.g., Wheaton 900200; 20 slides with rack and cover).
4. Eriochrome (solochrome) cyanine solution.

Eriochrome (solochrome) cyanine RS 0.2 g (Sigma E2502; C.I. 43820)
Sulfuric acid (H_2SO_4) concentrated 0.5 ml, *HIGHLY CORROSIVE
Distilled water, 96.0 ml
10% Iron alum, 4.0 ml (see below)
Add H_2SO_4 to EC in a glass flask
Add distilled water and stir until dye goes into solution
Add 10% iron alum
Mix well. Filter into screw-top bottle. Protect from light. *Filter this solution before use
This solution may be used for about 6 months, depending on the no. of slides processed

5. 5% Iron alum. Prepare from 10% stock solution.

   10% Iron alum stock (ferric ammonium sulfate) differentiating solution.

Ferric ammonium sulfate	10.0 g ($NH_4Fe(SO_4)2 \times 12H_2O$; Sigma F1668)
Distilled water	100.0 ml

   Stir until dissolved. Protect solution from light. Good if stored for years, but use only once.

6. Borax-ferricyanide differentiator.

Borax	1.0 g (Sigma, B3545)
Potassium ferricyanide	1.25 g (Sigma, P8131)
Distilled water	100.0 ml

   Stir until dissolved. Protect solution from light. *Filter this solution before use.

7. 1% Cresyl violet solution.

Cresyl violet acetate	1.0 g (Sigma C1761)
Distilled water	100.0 ml

8. Distilled water.
9. 80, 95, and 100% Ethanol.
10. 95% Ethanol with 0.1% glacial acetic acid (0.6 ml/60 ml 95% ethanol).
11. Clearing agent (e.g., Xylene, Histoclear, SafeClear, Citrisolv).
12. Coverslips (24×50 mm).

3.3.2. Staining Procedures for Eriochrome Cyanine/Cresyl Violet

Note: All staining step times (** below) are determined empirically, must be optimized for each new set of tissue sections, and each time new solutions are prepared.

Eriochrome Cyanine Stain

1. Remove slides from freezer and dry for 1–2 h on a slide warmer at 37°C.
2. Place in fresh 100% acetone for 5–10 min at room temperature.
3. Let the acetone dry on the slide (place in fume hood for at least 30 min).
4. Stain in freshly filtered EC solution for ~30** min at room temperature.
5. Wash in gently running tap water for 4–5 min.
6. Differentiate in 5% iron alum for 5–15** min at room temperature (usually, ~15 min for rat spinal cord, 10 min for mouse). The gray matter will be gray or light blue and white matter will be dark blue after this step. Do not reuse this solution.
7. Rinse briefly in gently running tap water.
8. Complete the differentiation in the borax-ferricyanide solution at room temperature for 3–10 min** (usually, 10 min for rat spinal cord, 5 min for mouse). The gray matter should be cream colored or clear.
9. Rinse briefly in gently running tap water for 1–2 min and dry slides well.
10. At this point, the EC stain should be uniform in SWM regions, while the injury site and substantia gelatinosa should be unstained or very pale.

Cresyl Violet (1%) Counterstain (Optional)

11. Place slides in 80% alcohol for 3 min.
12. Stain in 1% Cresyl Violet solution for 7–10 min** with gentle agitation.
13. Wash in running tap water for a few seconds and dry well.
14. Place slides in 95% alcohol, including 0.1% acetic acid for 5 min**.
15. Place slides in 95% alcohol for 2×2 min (neuronal cytoplasm light violet, nuclei dark blue).
16. Place slides in 100% alcohol for 2×5 min to complete dehydration.

17. Place slides in Histo-clear for 3 × 5 min each.
18. Coverslip using a permanent mounting medium. Sections will appear as shown in Fig. 3d.

4. Methods for Determining Volumes of Selected Regions

4.1. Mapping 10–15 Equally Spaced Sections

4.1.1. Materials

1. Slides with appropriate stain at high contrast.
2. Microscope with 2.5 or 5× objective. The section should fill most of the field of view. Mouse and rat spinal cord sections are too small to obtain sufficient resolution for regional volume analysis using regular flatbed scanners.
3. Digital or CCD camera and software for image capture. A low-cost CCD camera is sufficient for capturing images of EC/CV or other brightfield stains, such as diaminobenzidine and chromagens. We use a Sony 970 CCD camera. Cameras with higher sensitivity may be required for low-power fluorescence images. The camera is mounted onto a microscope and the image can be captured using an imaging board in a desktop personal computer. Options for image capture programs range from simple free capture software, such as Image J (available for all common operating system platforms at http://rsbweb.nih.gov/ij/), to expensive imaging programs, such as MCID (Imaging Research, http://www.mcid.co.uk/software/mcid_core/), or other specialized image capture software available from microscope vendors. Images of individual sections are collected and stored as .TIF, .RAW, or .JPG format files.
4. Presentation/image display and printing software: If volume estimates are done manually, then each of the images used for analysis should be printed onto white paper for archival purposes. We typically import 6–12 images into a single page using a slide presentation software, such as Microsoft Powerpoint. All images should be imported using the same enlargement and a scale bar should be included on at least one image per page.
5. Computer printer: If standard presentation software is used, any color printer will work.
6. Fine point colored markers: Felt-tip permanent markers (Sharpie Brand) work well. These are used to identify the sections for mapping.

4.1.2. Procedures

1. Determine the range of sections that encompass the regions targeted for measurement. Mark the lesion start and end on the slide using a fine-tipped permanent marker (Fig. 3d). To measure spared gray and white matter regions from moderate to severe injuries, the range of sections should extend at least 10–25% beyond the most rostral and caudal sections with damage.

2. Note that it is NOT necessary to determine the lesion "epicenter" section in order to calculate lesion volumes using the Cavalieri method. However, identifying this section would be required for the recommended related measurement of the epicenter cross-sectional areas of the lesion, SWM, and spared gray matter.
3. Determine the interval for section imaging. Recall that 10–15 sections through the region of interest are sufficient to obtain reliable volume estimates. Determine the appropriate section interval that provides this number of sections spanning the full volume of tissue block of interest (Table 1). For example, a typical mouse moderate contusion lesion may range from 4 to 6 mm in length. If each section in the series is 100 μm apart, then imaging every third section will provide approximately 12 section images spanning the lesion volume.
4. Mark the slide next to the sections that are chosen for imaging. Ideally, all chosen sections are flat and evenly stained. Sometimes, this is impossible. In those situations, an adjacent section can be chosen, but this does introduce unwanted error into the volume estimate.
5. Digitize microscope images of the selected sections and save each to an image file. Collect all images at the same magnification. *Remember to include a scale bar for at least one image per specimen*. This is enlarged or reduced with the image and is used to calculate the conversion of grid point distances to microns. Depending on your methods for image collection, you will probably wish to subtract background illumination.
6. Using a presentation software or contact sheet-type software, assemble all of the collected images for each specimen into one or more pages and print a full set of images onto paper using a color printer. We import images into Microsoft Powerpoint with 20–25 images (one specimen) per page (Fig. 4a). Note: While volume estimates can feasibly be done directly on a computer monitor without printing the images on paper, we find the time taken to create these printed image maps to be very useful for laboratory notebook archives to document the lesion for each specimen.

4.2. Region Maps and Regional Point Counting

4.2.1. Materials

1. Printed maps of 15–25 equally spaced sections per specimen.
2. Superfine-tipped colored markers or pens, one color for each region of interest.
3. Starting grid overlay.
4. Enlarging/reducing copier.
5. Clear transparency sheets.
6. Graph paper and pen/notebook or spreadsheet for recording counts.

Table 1
Examples of section intervals for lesion maps

Lesion severity and species	Maximum lesion length (mm)	Distance between adjacent sections (μm)	Section interval	*T* (thickness) (μm)	No. of sections sampled per specimen
Moderate contusion (mouse)	2.0–3.0	100	Every other section	200	10–15
Severe contusion (mouse)	6.0–8.0	100	Every fifth section	500	12–16
Severe contusion (rat)	15–18	200	Every sixth section	1,200	12–15

These values represent empirical estimates of the lengths of typical contusion injury lesions and the section interval and slab thickness (*T*) that could be chosen to obtain 10–20 sections for Cavalieri analysis of region volumes

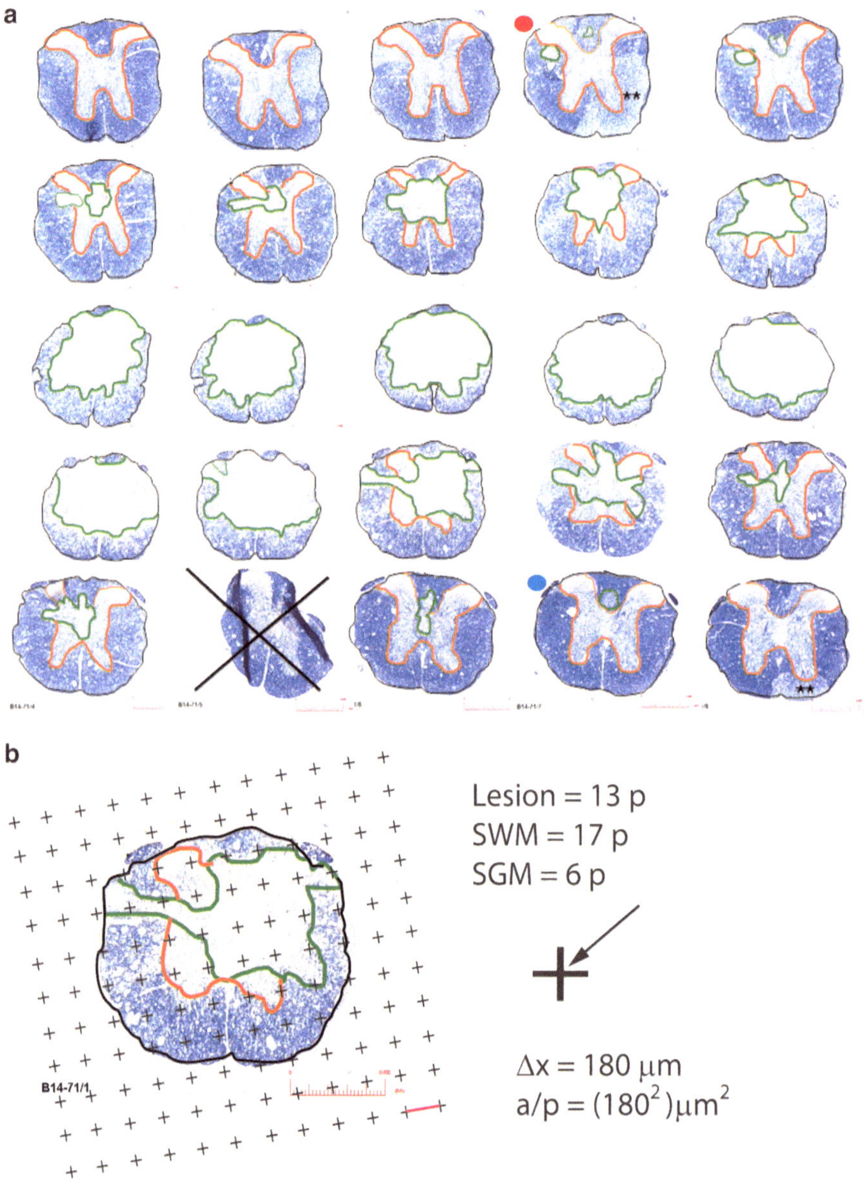

Fig. 4. (**a**) Illustration of a lesion map prepared using presentation software. A total of 25 sections were observed and images captured using digital image software and a Sony CCD camera with a 5× microscope objective. Scale bars are included with the images as shown in the bottom row. One of the sections is folded and is not used for point counting (*black X*). On each of the other 24 images, colored pens are used to outline the regions of interest, including the total cross-sectional area (*black*), spared gray matter (*orange*), and the damaged or lesion area (*red*). The first section containing clear tissue damage is marked with a red dot, and the last section with a *blue dot*. Some imperfections in staining are found (**). These are recognized by studying adjacent sections, and are ignored. (**b**) Point counting method used to determine region areas from one section. A grid of the appropriate point spacing is superimposed over one section to illustrate approximately 30–40 points within the tissue section. The points are counted if the open point in the top-right quadrant of the + is inside of the designated region of interest. Scale = 200 μm. By including a scale bar with the image, the actual distance between points (ΔY) can be determined and a/p calculated easily.

4.2.2. Procedures

1. Using superfine-tipped markers or colored pens, outline the regions of interest for volume measurements directly on the paper printouts. Each region should be outlined on every section (Fig. 4a).
   (a) Total spinal cord cross section (TCSA)—Outline the peripheral rim of the spinal cord, excluding dorsal roots and meninges.
   (b) Normal white matter (SWM)—Tissue found in the peripheral rim of a spinal cord transverse section; staining with EC resembles that of uninjured spinal cord in staining pattern and intensity (pallor); regions lack vacuoles, amorphous tissue debris, or clusters of cells or tissues not normally found in white matter; dark staining Schwann cell-myelinated regions are not included (9).
   (c) Normal gray matter (SGM)—Tissues in the central regions of the spinal cord that contain neuronal cell bodies as seen with cresyl violet counterstaining. Note that with the exception of the most superficial laminae of the dorsal horn these regions also contain myelinated fibers; so the staining pattern includes both violet and blue.
   (d) Lesion (L)—All tissue and cavities that lack normal neuropil architecture. These include cyst or cavities, regions filled with multinuclear macrophages and connective tissue, infiltrating peripheral cells, and connective tissue elements. In rat SCI, cystic cavities are predominant, while mouse SCI evolves with a dense cellular and connective tissue scar in the lesion center.
   (e) Other cellular regions may also be identified as desired.
2. Determine a grid size for counting. A low coefficient of error can be obtained with a total of ~200 points for the total region of interest (40, 41). As a starting point, consider that the ends of the biconical shape are much smaller than the center. Assuming a biconical approximation with ten sections spanning the injury site, 200 points can be obtained with approximately 30–40 points in each of the three centermost sections. Therefore, as a rule of thumb, one can enlarge or reduce the grid so that there are approximately 6×6 or ~36 points that fall within the lesion site of the centermost section. This grid size works for both lesion and spared tissue regions (Fig. 4b).
3. Make the grid onto a transparent overlay sheet. Enlargements or reductions of a standard template can be made using a copier machine or computer program to prepare a grid of evenly spaced sections of any size. Examples of grids can be found in ref. 40 or Fig. 5.
4. Place the transparent grid in a random orientation over the region of interest for each section, and count the number of

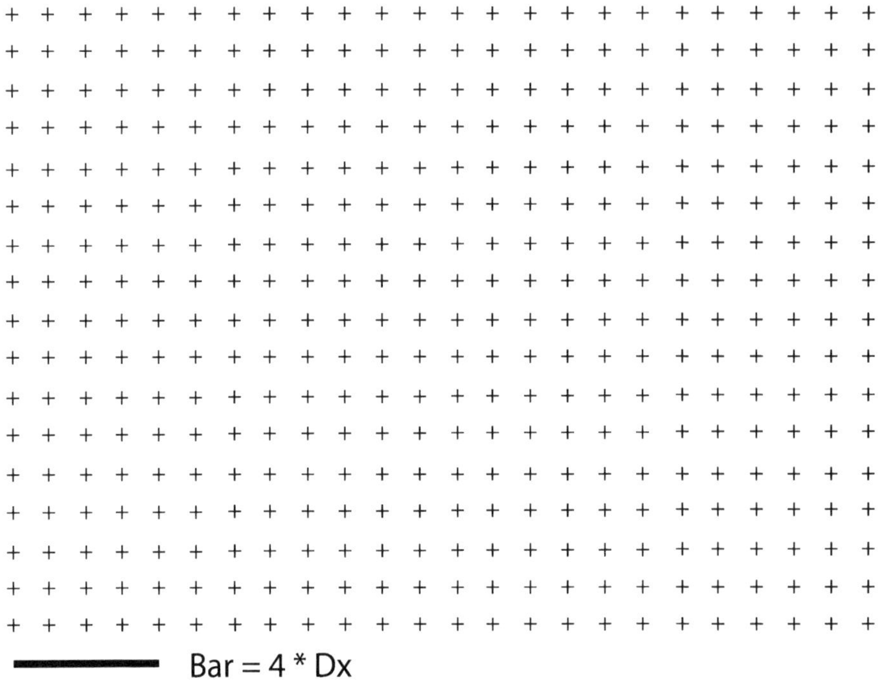

Fig. 5. Grid template to use for Cavalieri measurements. This grid can be copied and enlarged or reduced before printing onto a clear sheet of transparency paper (designed based on ref. 40).

points that lie within the outlined structures. Count those points only if the same site is within the region borders. We use the point that is located at the precise right-top quadrant of the plus symbol (Fig. 4b) because it is easy to see over many different backgrounds. Enter the counts into a notebook or spreadsheet table (see Table 2).

5. Multiple regions can be counted using the same printed images. To ensure that sufficient points are included, different sized grids can be used for the different regions if necessary.

4.3. Calculating Volume Estimates from Point-Count Records

1. Determine the area per point (a/p) for the grid that was used. This is the area of tissue that is represented by a point on the grid that is overlaid on each enlarged image. The value of a/p is calculated by first determining the magnification of the images and distance between points. The scale bar printed with the image sets provides the actual distance in the x or y axis in microns. This can be used to determine the distance between points used in the grid.
   Example: See Fig. 4b.
   (a) Measure the length of the scale bar that is printed at the same magnification as the section images.
   (b) Convert scale bar measurement to microns.
   (c) Calculate linear magnification of the scale bar.

Table 2
Sample volume calculations for mouse spinal cord contusion

Section	Total spinal cord		Lesion		SWM		SGM	
	Points # P	Area (mm^2) $P \times a/p$	Points # P	Area (mm^2) $P \times a/p$	Points # P	Area (mm^2) $P \times a/p$	Points # P	Area (mm^2) $P \times a/p$
1	38	1.231	0	0.000	30	0.972	8	0.259
2	36	1.166	1	0.032	33	1.069	9	0.292
3	38	1.231	2	0.065	30	0.972	6	0.194
4	34	1.102	18	0.583	16	0.518	0	0.000
5	32	1.037	21	0.680	8	0.259	0	0.000
6	33	1.069	19	0.616	4	0.130	0	0.000
7	30	0.972	23	0.745	7	0.227	0	0.000
8	29	0.940	27	0.875	6	0.194	1	0.032
9	31	1.004	15	0.486	12	0.389	3	0.097
10	35	1.134	10	0.324	18	0.583	5	0.162
11	32	1.037	6	0.194	20	0.648	7	0.227
12	39	1.264	2	0.065	32	1.037	7	0.227
13	41	1.328	2	0.065	30	0.972	10	0.324
ΣP	448		146		246		56	
$a/p \times \Sigma P$ (mm^2)	14.52		4.73		7.97		1.81	

(continued)

Table 2 (continued)

Section	Total spinal cord		Lesion		SWM		SGM	
	Points $\#P$	Area (mm^2) $P \times a/p$	Points $\#P$	Area (mm^2) $P \times a/p$	Points $\#P$	Area (mm^2) $P \times a/p$	Points $\#P$	Area (mm^2) $P \times a/p$
T (mm)		0.2		0.2		0.2		0.2
Volume (mm^3)		2.90		0.95		1.59		0.36

This data table illustrates the counts and calculations that would be typical for a mouse tissue specimen with a moderate contusion injury. The values are hypothetical and do not represent real data from other figures or actual studies. Block and lesion lengths are estimated by multiplying the number of sections measured by the distance (thickness, T) between sections, in this case 0.2 mm. The area of each region is estimated by multiplying the # of points (P) contacted by the area/point (a/p)

Volume (V) = $a/p \times \Sigma P \times T$

Tissue block length = 13 sections × 0.2 mm/section = 2.6 mm

Lesion length = 12 sections × 0.2 mm/section = 2.4 mm

ΔY = 180 μm a/p = 32,400 μm^2 = 0.0324 mm^2

(d) Measure area per point on the grid and divide by the enlargement magnification.

For example: Distance between points in x direction = 5 mm divided by 170 X = 0.029 mm.

Distance between points in y direction = 5 mm divided by 170 X = 0.029 mm.

$a/p = x\ mm \times y\ mm = 8.65 \times 10^{-4} mm^2\ (\times 10^6 \mu m^2/mm^2) = 865\ \mu m^2$.

2. Sum the number of points for all sections in the defined regions for each specimen (Table 2).
3. Calculate the volume of the region using the relationship:

$$V = T \times a/p \times \sum P,$$

where T = the distance between the top of each section used for counting in micrometer,

a/p = the area per point in μm^2,

and P = the total number of points within the region.

Following completion of these steps, the investigator will have a full lesion map with regional outlines and an accurate volume measure for each region examined. With experience, this full dataset can be obtained in roughly 20–30 min per specimen.

5. Additional Considerations and Further Directions

5.1. Component Volumes and Volume Density (V_v) Measures

The methods highlighted in this chapter are used to define the volume of lesion and spared tissue following SCI, especially as seen with contusion or compression injuries. However, these identical point-counting methods can be used to measure the volume of any component within the lesion site. For example, the volume of axons that occupy a transplant or lesion site can be determined by counting the number of axons that cross the grid points as described for lesion volumes (Fig. 6). Detailed procedures for this application can be found in ref. 40. The volume fraction of component *Y* within a reference space, such as the lesion, *L*, is the proportion of the volume of the lesion that is taken up by the component. For example, the volume fraction of axons in a lesion is designated as:

V_v *(Axons, lesion) = Volume of the Axons* $= \sum P_{Axons}$.
Volume of the lesion (reference space) $\sum P_{Lesion}$.

The volume fraction is, therefore, a proportion measure, ranging from 0 to 1. Note that both the numerator and denominator can be estimated simply by counting the total number of points where the object (or reference space) crosses the grid. These two measures are done using the same images so that the only parameter that must be determined for each is *P*. The magnification and area

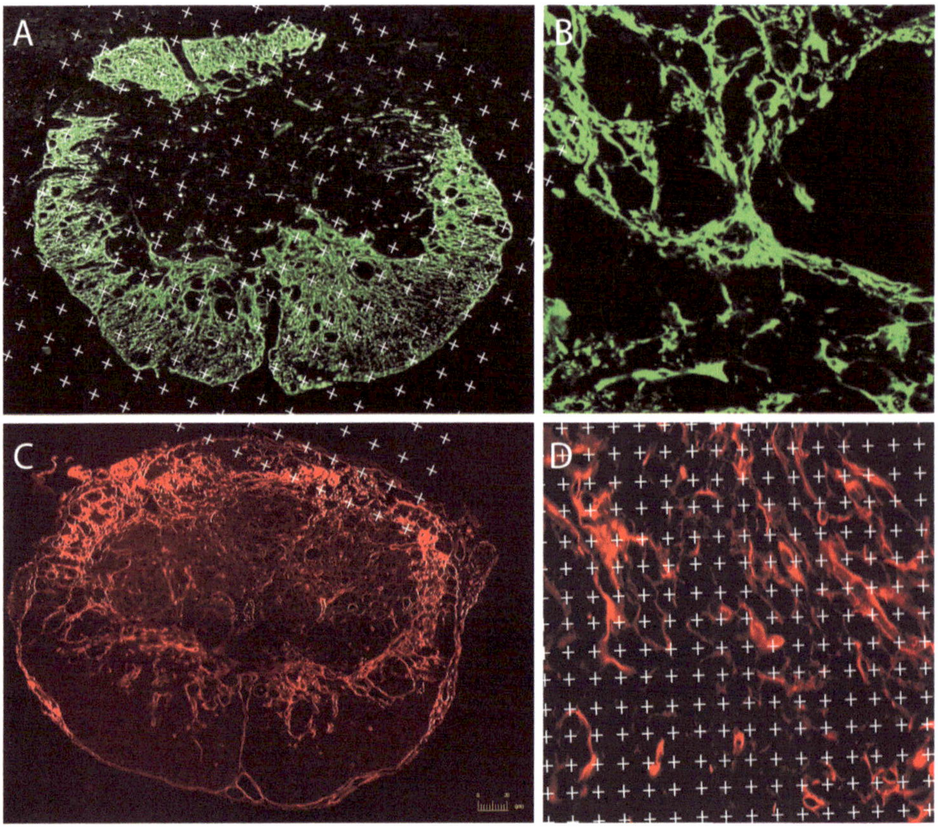

Fig. 6. Examples of lesion components that can be measured and described using the volume fraction. (**a**, **b**) Low-power and high-power images of astrocytes, labeled with an antibody to glial fibrillary acidic protein. Astrocytes within the spinal cord can be estimated as V_v (*GFAP, spinal cord*) by counting the points that coincide with green staining as a fraction of all the points that lie over the section. (**b**) V_v (laminin, lesion site) can be determined in the same manner, but counting the points that overlie red laminin staining/all points within the lesion or tissue section.

per point are not needed. An example of the V_v measure used to compare the proportion of the spinal cord lesion occupied by axons between two strains of mice is shown in Ma et al. (46). The advantage of V_v for reporting axon densities is that the tortuosity of axons does not confound the results, as might be the case with simply counting axon profiles. Other components that might be reported in this manner include V_v (*astrocyte processes, spinal cord*), V_v (*laminin, lesion site*), or V_v (*terminal fields, spared gray matter*).

5.2. Experimental Effects on Tissue Shrinkage and Resulting Volume Estimates

As described briefly above, SCI results in substantial shrinkage of tissue at the injury site as the lesion evolves. It is important to evaluate how experimental effects on the amount of tissue shrinkage might influence the measurements and interpretations of changes in lesion or spared gray and white matter volumes. For example, large cellular transplants may fill the center of the injury site and even infiltrate the host spared white and gray matter regions. One might find that a transplant has a greater volume compared to a cyst while the amount of gray and white matter is unchanged.

Alternatively, the transplant itself may not survive, but might reduce the extent of secondary injury and protect the surrounding gray or white matter. Similar differences could result when these approaches are applied to knife-cut lesions, in which close approximation of two sides of the cut would result in a smaller lesion volume due to either greater contracture of the wound or more tissue sparing. In both cases, the measurement of region volumes should be compared with uninjured controls to provide a clear interpretation of the differences between the lesions.

5.3. Other Stereological Outcome Measures That Depend on Knowing the Reference Volume

Perhaps the most compelling reason to evaluate lesion and spared tissue volumes is to establish a reference for many other histological and morphological outcomes that may be directly targeted by experimental manipulations. This is particularly critical in relation to tissue shrinkage at the lesion site. Using sampling methods to determine numbers of objects (e.g., cells and nuclei) requires knowing the reference volume of the region where the objects are contained. If different manipulations result in changes in reference volume, then counting cell numbers without correcting for the change in volume will give spurious results. Other outcome measures that depend on knowing the reference volume include estimations of the length of fibers or capillaries or the density of axons, axon terminals, or synapses. In a recent study from our laboratory, we had direct experience with the errors that can be incurred by counting cells without measuring changes in the reference volume. We were examining the distribution and counts of astrocytes that arise from cells born in the first week after a mouse contusion SCI. We counted cells that had incorporated a proliferation marker (bromodeoxyuridine or BrdU), and found that the number of new cells per unit area had increased or stayed the same over time in spared gray and white matter regions. However, when we corrected these cell counts for the shrinking volume of the spared tissue regions, we learned that the total number of labeled cells in these regions was actually decreasing over time. Thus, our interpretation of cell proliferation and survival was changed with accurate estimation of total cell numbers (47).

5.4. Summary

This chapter has described the history and procedures that can be used to obtain unbiased measures of the volumes of an SCI lesion site and spared gray and white matter regions. These approaches are simple and reliable and should be completed as a basic documentation of the possible anatomical basis of recovery in most, if not all, SCI studies. In addition to documenting differences or changes in spared tissue volumes to support neuroprotection mechanisms of action, these same approaches can be used to document the volume of glial infiltration (45), inflammation (48), or other important cellular components that could be altered in the experimental or therapeutic manipulation of SCI.

Acknowledgments

Funding for development of these protocols in our laboratory has been provided by NINDS and Spinal Research (International Spinal Research Trust) and the OSU Center for Brain and Spinal Cord Repair. The author would like to thank a number of people for discussions and technical assistance in developing these methods, including Dana McTigue, Ping Wei, Ying Chen, and Feng Qin Yin. Methods and three pictures in Fig. 3a were adapted with permission from unpublished protocols of the Spinal Cord Injury Training Manual, NINDS, Facilities of Research Excellence in Spinal Cord Injury, © The Ohio State University, 2004–2010.

References

1. Young W (1993) Secondary injury mechanisms in acute spinal cord injury. [Review; 33 refs.]. J Emerg Med 11(Suppl 1):13–22
2. Donnelly DJ, Popovich PG (2008) Inflammation and its role in neuroprotection, axonal regeneration and functional recovery after spinal cord injury. Exp Neurol 209(2): 378–388
3. Kwon BK, Tetzlaff W, Grauer JN, Beiner J, Vaccaro AR (2004) Pathophysiology and pharmacologic treatment of acute spinal cord injury. Spine J 4(4):451–464
4. Tator CH, Fehlings MG (1991) Review of the secondary injury theory of acute spinal cord trauma with emphasis on vascular mechanisms. J Neurosurg 75:15–26
5. Noble LJ, Wrathall JR (1985) Spinal cord contusion in the rat: morphometric analyses of alterations in the spinal cord. Exp Neurol 88:135–149
6. Bresnahan JC, Beattie MS, Stokes BT, Conway KM (1991) Three-dimensional computer-assisted analysis of graded contusion lesions in the spinal cord of the rat. J Neurotrauma 8(2):91–101
7. Balentine JD (1978) Pathology of experimental spinal cord trauma. I. The necrotic lesion as a function of vascular injury. Lab Invest 39: 236–253
8. Bunge RP, Puckett WR, Becerra JL, Marcillo A, Quencer RM (1993) Observations on the pathology of human spinal cord injury: a review and classification of 22 new cases with details from a case of chronic cord compression with extensive focal demyelination. In: Seil FJ (ed) Advances in neurology. Raven Press, New York, pp 75–89
9. Kakulas BA (1984) Pathology of spinal injuries. CNS Trauma 1:117–129
10. Kakulas BA (1999) A review of the neuropathology of human spinal cord injury with emphasis on special features. J Spinal Cord Med 22(2):119–124
11. Scheff SW, Rabchevsky AG, Fugaccia I, Main JA, Lumpp JE Jr (2003) Experimental modeling of spinal cord injury: characterization of a force-defined injury device. J Neurotrauma 20(2):179–193
12. Jakeman LB, Guan Z, Wei P et al (2000) Traumatic spinal cord injury produced by controlled contusion in mouse. J Neurotrauma 17(4):299–319
13. Kim JH, Loy DN, Liang HF, Trinkaus K, Schmidt RE, Song SK (2007) Noninvasive diffusion tensor imaging of evolving white matter pathology in a mouse model of acute spinal cord injury. Magn Reson Med 58(2):253–260
14. Kearney PA, Ridella SA, Viano DC, Anderson TE (1988) Interaction of contact velocity and cord compression in determining the severity of spinal cord injury. J Neurotrauma 5:187–208
15. Rivlin AS, Tator CH (1978) Effect of duration of acute spinal cord compression in a new acute cord injury model in the rat. Surg Neurol 10:38–43
16. Fehlings MG, Tator CH (1995) The relationships among the severity of spinal cord injury, residual neurological function, axon counts, and counts of retrogradely labeled neurons after experimental spinal cord injury. Exp Neurol 132(2):220–228
17. Gruner JA, Yee AK, Blight AR (1996) Histological and functional evaluation of experimental spinal cord injury: evidence of a

stepwise response to graded compression. Brain Res 729(1):90–101

18. Yezierski RP, Liu S, Ruenes GL, Kajander KJ, Brewer KL (1998) Excitotoxic spinal cord injury: behavioral and morphological characteristics of a central pain model. Pain 75(1):141–155
19. Hill RL, Zhang YP, Burke DA et al (2009) Anatomical and functional outcomes following a precise, graded, dorsal laceration spinal cord injury in C57BL/6 mice. J Neurotrauma 26(1): 1–15
20. Gale K, Kerasidis H, Wrathall JR (1985) Spinal cord contusion in the rat: behavioral analysis of functional neurologic impairment. Exp Neurol 88:123–134
21. Basso DM, Beattie MS, Bresnahan JC (1995) A sensitive and reliable locomotor rating scale for open field testing in rats. J Neurotrauma 12(1):1–21
22. Bresnahan JC, Beattie MS (1987) Todd FD3, Noyes DH. A behavioral and anatomical analysis of spinal cord injury produced by a feedback-controlled impaction device. Exp Neurol 95:548–570
23. Metz GA, Curt A, van de Meent MH, Klusman I, Schwab ME, Dietz V (2000) Validation of the weight-drop contusion model in rats: a comparative study of human spinal cord injury. J Neurotrauma 17(1):1–17
24. Basso DM, Beattie MS, Bresnahan JC (1996) Graded histological and locomotor outcomes after spinal cord contusion using the NYU weight-drop device versus transection. Exp Neurol 139(2):244–256
25. Magnuson DSK, Trinder TC, Zhang YP, Burke D, Morassutti DJ, Shields CB (1999) Comparing deficits following excitotoxic and contusion injuries in the thoracic and lumbar spinal cord of the adult rat. Exp Neurol 156:191–204
26. Schrimsher GW, Reier PJ (1992) Forelimb motor performance following cervical spinal cord contusion injury in the rat. Exp Neurol 117:287–298
27. Pearse DD, Lo TP Jr, Cho KS et al (2005) Histopathological and behavioral characterization of a novel cervical spinal cord displacement contusion injury in the rat. J Neurotrauma 22(6):680–702
28. Fuller DD, Sandhu MS, Doperalski NJ et al (2009) Graded unilateral cervical spinal cord injury and respiratory motor recovery. Respir Physiol Neurobiol 165(2–3):245–253
29. Dohrmann GJ, Panjabi MM (1976) "Standardized" spinal cord trauma: biomechanical parameters and lesion volume. Surg Neurol 6:263–267
30. Means ED, Anderson DK, Waters TR, Kalaf L (1981) Effect of methylprednisolone in compression trauma to the feline spinal cord. J Neurosurg 55:200–208
31. Blight AR (1983) Cellular morphology of chronic spinal cord injury in the cat: analysis of myelinated axons by line-sampling. Neuroscience 10:521–543
32. Finkelstein SD, Gillespie JA, Markowitz RS, Johnson DD, Black P (1990) Experimental spinal cord injury: qualitative and quantitative histopathologic evaluation. J Neurotrauma 7:29–40
33. Wrathall JR, Choiniere D, Teng YD (1994) Dose-dependent reduction of tissue loss and functional impairment after spinal cord trauma with the AMPA/kainate antagonist NBQX. J Neurosci 14(11:Pt 1):6598–6607
34. Rabchevsky AG, Fugaccia I, Turner AF, Blades DA, Mattson MP, Scheff SW (2000) Basic fibroblast growth factor (bFGF) enhances functional recovery following severe spinal cord injury to the rat. Exp Neurol 164(2):280–291
35. Ankeny DP, McTigue DM, Jakeman LB (2004) Bone marrow transplants provide tissue protection and directional guidance for axons after contusive spinal cord injury in rats. Exp Neurol 190(1):17–31
36. McTigue DM, Tripathi R, Wei P, Lash AT (2007) The PPAR gamma agonist pioglitazone improves anatomical and locomotor recovery after rodent spinal cord injury. Exp Neurol 205(2):396–406
37. Jones TB, Basso DM, Sodhi A et al (2002) Pathological CNS autoimmune disease triggered by traumatic spinal cord injury: implications for autoimmune vaccine therapy. J Neurosci 22(7):2690–2700
38. Michel RP, Cruz-Orive LM (1988) Application of the Cavalieri principle and vertical sections method to lung: estimation of volume and pleural surface area. J Microsc 150(Pt 2):117–136
39. Rabchevsky AG, Fugaccia I, Sullivan PG, Scheff SW (2001) Cyclosporin A treatment following spinal cord injury to the rat: behavioral effects and stereological assessment of tissue sparing. J Neurotrauma 18(5):513–522
40. Howard CV, Reed MA (1998) Unbiased stereology: three-dimensional measurements in microscopy. Bios Scientific Publishers Limited, New York
41. Gundersen HJ, Bendtsen TF, Korbo L et al (1988) Some new, simple and efficient stereological methods and their use in pathological research and diagnosis. [Review]. APMIS 96(5):379–394
42. Ma M, Basso DM, Walters P, Stokes BT, Jakeman LB (2001) Behavioral and histological outcome following graded contusion injury in C57Bl/6 mice. Exp Neurol 169(2): 239–254

43. Olby NJ, Blakemore WF (1996) A new method of quantifying the extent of tissue loss following spinal cord injury in the rat. Exp Neurol 138(1):82–92
44. Rabchevsky AG, Fugaccia I, Sullivan PG, Blades DA, Scheff SW (2002) Efficacy of methylprednisolone therapy for the injured rat spinal cord. J Neurosci Res 68(1):7–18
45. White RE, Yin FQ, Jakeman LB (2008) TGF-alpha increases astrocyte invasion and promotes axonal growth into the lesion following spinal cord injury in mice. Exp Neurol 214(1):10–24
46. Ma M, Wei P, Wei T, Ransohoff RM, Jakeman LB (2004) Enhanced axonal growth into a spinal cord contusion injury site in a strain of mouse (w129X1/SvJ) with a diminished inflammatory response. J Comp Neurol 474(4):469–486
47. White RE, McTigue DM, Jakeman LB (2010) Regional heterogeneity in astrocyte responses following contusive spinal cord injury in mice. J Comp Neurol 518(8):1370–1390
48. Beck KD, Nguyen HX, Galvan MD, Salazar DL, Woodruff TM, Anderson AJ (2010) Quantitative analysis of cellular inflammation after traumatic spinal cord injury: evidence for a multiphasic inflammatory response in the acute to chronic environment. Brain 133(Pt 2):433–447

Chapter 38

Retrograde Axonal Tract Tracing

Lingxiao Deng, Xiaofei Wang, Chandler L. Walker, Yiwen Ruan, and Xiao-Ming Xu

Abstract

Deficits in sensory and motor function after spinal cord injury are attributable primarily to the interruption of long sensory and motor axonal tracts in the spinal cord. Different spinal cord tracts display different vulnerability to the injury, capability to regenerate, and contribution to functional recovery. Due to the complexity of the central nervous system (CNS) tracts, it is worthwhile to exploit how to exquisitely label them within the spinal cord and examine their connections to the brain or periphery. Retrograde axonal tract tracing technologies are powerful tools for identifying axonal connections in the normal or injured spinal cord. With appropriate injury models and retrograde tracing techniques to transport tracers from axon terminal to the soma, neuronal origin of targeted axons can be determined. After tracing, tissue is collected following appropriate survival time. To detect tracers inside the neuronal soma, nucleus, and corresponding dendrites, various labeling techniques are used which include direct fluorescence, immunohistochemistry, immunofluorescence, and autoradiography. When combined with anterograde tracing techniques, retrograde tracers may be used to establish anatomical reorganizations of neural networks. When combined with immunohistochemistry, retrograde tracing may be used for neurochemical characterization of specific neuronal pathways or specific cell types.

Key words: Retrograde tracing, Horseradish peroxidase, Fluoro-Gold, Fast Blue, Dextran amine, Biocytin, Neurobiotin

1. Introduction

1.1. Historical Perspective

Over the last 3 decades, the advancement of neuronal tracing techniques has produced a range of different methodologies, all of which are based on the common principle of axonal transport or *axoplasmic flow* (1, 2). Axons can transport materials either from the termini of their processes to the cell body, known as retrograde transport, or vice versa, termed anterograde transport. In some cases, such transport can work in both directions simultaneously. These mechanisms can involve either materials produced or contained within the cell (organelles or macromolecules) or taken

Jun Chen et al. (eds.), *Animal Models of Acute Neurological Injuries II*, Springer Protocols Handbooks,
DOI 10.1007/978-1-61779-782-8_38, © Springer Science+Business Media, LLC 2012

up from the cell's surroundings. In neuroanatomical research, after providing some labeling materials which can be taken into the termini of neurons' axons, retrograde axonal transport allows identification of the cells of origin of afferent nerve fibers to a particular target zone. Such labeling and identification of neurons and their processes are referred to as *tracing*.

The first attempts at retrograde tracing were fiber degeneration studies carried out in combination with tissue impregnation methods (3). A major breakthrough was, undoubtedly, the discovery that axons possess the ability to take up and actively translocate the plant enzyme, horseradish peroxidase (HRP) via retrograde transport to the cell soma. Upon uptake, this enzyme's activity may then be used to stain the cell body and its major dendrites (4). Subsequently, many new substances and molecules have been identified, developed, or used in retrograde tracing studies. Of particular importance are fluorescent dyes of various colors, which can be combined in the same experiment for multiple retrograde labeling (5).

Early retrograde labeling experimentation with fluorescent transport substances utilized Evans Blue (EB) Dye combined with cattle albumin. The efficacy and usefulness of these methods prompted a "neuroanatomical revolution," which enabled the various advantageous properties of using fluorescent dyes to help researchers tackle a wide range of unanswered questions. The number of dyes available and the fields of application have increased rapidly. The nuclear marker diamidinophenylindol (DAPI), Fast Blue (FB), Nuclear Yellow (NY), and Lucifer Yellow (LY) are typical of such tracers developed in the late 1970s. These have now been largely superseded by the use of Fluoro-Gold (FG), cholera toxin subunit B (CTB), fluorescent-tagged beads, and, most recently, Mini Ruby (MR) and fluorescent-tagged dextran amines. Such developments have been motivated partly by whether or not the axon of a parent neuron can branch to more than one target, leading to the current array of different substances that are available for double (and multiple)-retrograde labeling studies, including modification of various HRP techniques (6).

As described for anterograde tracing, recently, viruses have been utilized for retrograde tracing studies. As such, the benefits of this approach, such as the ability to traverse multisynaptic pathways and replicate for signal amplification (7) and limitations, including increased death of infected cells, are similar between the two methods of tract tracing. However, recent developments in genetic approaches to viral-mediated tracing have improved upon these and other disadvantages of viral-mediated tract tracing. An advanced novel method was recently introduced involving single-cell electroporation of engineered rabies virus designed to genetically and selectively target synapses of particular neuron groups and, via retrograde tracing, allowing for the analysis of interactions and constructs within specific single-neural networks in the brain (8), thus reducing or eliminating many of such limitations of viral tract tracing. As a result of such advances, modern neuroanatomical and physiological research relies on and benefits heavily from the various abilities of the neuron to transport labeling substances of interest throughout its own cellular confines, as well as within and through its network with neighboring neurons.

1.2. Retrograde Axonal Tract Tracing for Spinal Cord Injury Research

Deficits in sensory and motor functions after spinal cord injury (SCI) are caused primarily by the interruption of the long sensory and motor tracts within the spinal cord. Tremendous efforts have been made to promote either protection of spared axons or regeneration of damaged axons or both. However, different spinal cord tracts display differences in vulnerability to the injury, capability to regenerate, as well as contribution to functional recovery. Therefore, assessments of both strategies require exclusive labeling of different and specific axons or displaying distinguishable neuronal sources. Retrograde tracing is particularly suitable in such type of research since it allows investigators to trace the trajectory of axons back to their neuronal soma, and is often useful for studying neuronal protection effects in animal models of SCI. However, effective selection, detection, and analysis of tracers following application in SCI research are critical for even the most rudimentary of axonal tract tracing studies.

2. General Considerations of Using Retrograde Tracing: Tracer Application and Cellular Incorporation

Three different methods are used to introduce tracer material into the target tissue: pressure injection, iontophoretic injection, and mechanical insertion of dye crystals. See Chap. 39 describing anterograde tracing, as the methods for applying tracers are similar in both approaches, though the chosen site of injection and desired outcome are different. Also, the method by which cells take up tracers, be it passive or active uptake or through intracellular injection, is comparable irrespective of the direction of transport once taken into the cell. After the tracer has entered the neuron, two methods of transport are used by anterograde and retrograde tracers: active transport in vesicles and lateral diffusion within the cellular membranes. The direction the tracer is transported, either toward the axon terminal from the cell body or from the terminal to the cell body, is dependent on molecular properties, such as size and weight of the tracer itself.

3. Detection of the Tracing Signal: Fluorescent Tracers and Nonfluorescent Tracers

As with anterograde tracers, fluorescent tracers can be directly visualized under a fluorescent microscope. The advantages that fluorescent anterograde tracers afford, such as in vivo or fixed tissue observation of neurons and their processes, as well as allowing one to utilize multiple fluorescent tracers on the same tissue, are also provided by fluorescent retrograde tracers. The disadvantages are same as well. Details, such as fine neuronal dendrites, can be difficult to visualize, fluorescent signals fade after long or repeated exposure to light, called photobleaching, as well as there is difficulty discerning background from actual signal due to autofluorescence of some tissues. A useful mounting medium to reduce photobleaching is 0.1–0.25 M *n*-propyl gallate in glycerol (9). Commercially available mounting media like Fluoromount (Diagnostic BioSystems, In., Pleasanton, CA) can be used in place of glycerol-based media for some purposes. Permanent staining can be achieved by a fluorescent tracer, such as FG which can be detected with specific antibodies incubated with diaminobenzidine (DAB). This allows for visualization of the tracer in a nonfluorescent state under brightfield microscopy.

Immunohistochemistry involves using such a permanent staining technique without the use of a fluorescent tracer, but rather primary antibodies against a tracer and reacted through DAB and peroxidase–anti-peroxidase (PAP) and avidin-biotin-peroxidase complex (ABC) methods. Dextran amines can be conjugated with for detection via the high-affinity binding of avidin or streptavidin that has been labeled with a visible substrate (10). In general, both fluorescent and nonfluorescent methods of tracing can provide benefits and present challenges or obstacles, depending on the experiment and the selected tracer.

4. Selection of Retrograde Tracers

Choosing an appropriate retrograde tracer for SCI research depends on several factors, including animal species and age, physical properties of the tracer, and experimental design, much like for the selection of anterograde tracer. A variety of fluorescent and nonfluorescent probes, biotin derivatives, protein conjugates (including lectins), dextras, and viruses have been used as retrograde tracers. Dissolution and preparation of the tracers for use are highly important, and each tracer has its own properties that must be considered prior to application for an experimental study. Also, the necessary tracer concentration depends upon the sensitivity of the technique selected for use. Such issues are discussed in the following sections describing different retrograde tracers, their preparation, and use in neuroanatomical investigation.

4.1. Horseradish Peroxidase

The plant enzyme, HRP, was the first retrograde neuroanatomical tracer to be used. It may be pressure injected using 20–50% solution in 2% dimethylsulfoxide (DMSO), iontophoresed, or applied as lyophilized enzyme. Typically, the tracer is driven out of micropipettes that are used for simultaneous electrophysiological recordings by gas pressure or negative currents. In some cases, the use of large amounts of concentrated tracer compromises precision. The enzyme is taken up by nerve terminals and incorporated into vesicles, in which HRP is transported toward the cell body of the neuron. The enzymatic properties of HRP allow for histological detection of the tracer. Reaction of HRP with hydrogen peroxide and either DAB or tetramethylbenzidine (TMB) forms a brown precipitate that is stable and electron dense, making HRP-labeled neurons amenable to ultrastructural analysis. As a major drawback, however, the morphological staining produced using HRP is somewhat incomplete, usually being limited to the cell soma and primary dendrites. Also, this method of detection does not permit vital staining, limiting its use for analysis of nonliving tissue (4, 11, 12).

4.2. Biocytin and Neurobiotin

Biocytin and biotinamide (neurobiotin) were introduced as intracellular markers, applied by electrophoresis through glass micropipettes. For use, both compounds can be diluted to 2–5%, injected by extracellular iontophoresis with high currents and large bore pipettes, and are applicable for retrograde tracing. Intracellular transport of these substances is very effective, allowing for labeling of the finest axonal arborizations. Compared to other tracers, biocytin and neurobiotin are relatively small (molecular weight of 372 and 286 kDa, respectively) and, as their major advantage, have a high affinity for avidin, which provides the chemical basis for their detection. After tracing in vivo, brains are usually perfused, sectioned, and processed with an ABC kit, and then subjected to the enzymatic reaction with DAB. As with HRP, this procedure enables the performance of ultrastructural studies and can be combined easily with immunohistochemistry, but it precludes visualization of the tracer in culture or living tissue. However, a variety of fluorescent dyes coupled to avidin or streptavidin are now commercially available, thus allowing a wider range and application of fluorescent visualization of the biotin derivatives for double-labeling experiments.

4.3. Fluorescent Dyes

4.3.1. Fast Blue

FB is applied by pressure injection as a 3% aqueous solution. It is taken up by axon terminals or injured axons, allowing for specification of cells labeled via retrograde transport. Retrograde tracing leads to uniform labeling of the cell soma without much staining of dendrites. The blue fluorescence requires excitation in the ultraviolet range (360 nm) and is subject to high phototoxicity at longer illumination times. Longer survival of the subjected animals increases fluorescence; however, dye leakage occurs over time, resulting in staining of surrounding glial cells and a decrease in the number of labeled neurons (13–16).

4.3.2. Fluoro-Gold

When dissolved to a concentration of 1–4% in buffer, this yellow-golden-color dye can be applied via either iontophoretic or pressure injections. After incorporation by injured axonal profiles and terminals, the dye accumulates in vesicles, is retrogradely transported, and often gives a granular appearance to the labeled cell soma. Hydroxystilbamidine, which has been identified as the active component of FG, has different optical characteristics depending upon pH. Under physiological conditions, optimal excitation occurs at 325 nm, when the emitted light has a maximum wavelength of 440 nm. Antibodies raised against FG are also available, allowing the dye to be converted into a permanent substrate and making it possible to embed sections containing FG-labeled tissue in plastic without reduction of tracer identification. Despite its excellent tracing properties, FG is not suitable for long-term labeling, where survival times of several months are required (11, 17, 18).

4.3.3. Dextran Amines

The water-soluble dextran conjugates can be diluted to 2–4%, applied by any of the injection techniques, or as crystals, and air dried onto needles. Compared to most other fluorescent dyes like FG, carbocyanines, and RITC, dextrans have high molecular weights (10 kDa). They are incorporated into injured neurons at the injection site and are rapidly transported. Uptake can be enhanced by coinjection of permeabilizing agents, like 5% Triton X-100. The dextrans are considered among the best neuroanatomical tracers available, as they provide highly detailed labeling of neuronal processes. In addition, these tracers can be detected with a variety of methods, for example through conjugation to fluorescent dyes or biotinylation, and are suitable for electron microscopy (EM) without prior alteration. Comparatively, however, the persistence of labeling attained using dextran amines is not as suitable as dextran conjugates, like FG, for short-term experiments (19, 20).

4.3.4. Cholera Toxin B Subunit

The cholera toxin (a bacterial toxin produced by *Vibrio cholerae*) is an oligomeric complex made up of six protein subunits: a single copy of the A subunit (CTA) and five copies of the B subunit (CTB). A disulfide bond connects the A and B subunits. CTA is responsible for the toxic effect, whereas the nontoxic CTB allows the protein to bind to cellular surfaces and be transported in axons and cell bodies (21).

CTB is mainly used as retrograde tracer (22, 23), though it can also be used anterogradely, opening up the possibility of simultaneous identification of afferents as well as efferents of a group of cells, such as dorsal root ganglion (DRG) neurons (24). CTB may be delivered by iontophoresis or pressure injection methods.

CTB requires a survival time of 1–10 days, with adequate labeling depending on the length of pathway of interest and the age of the animal. Nonetheless, CTB is a very sensitive tracer which can be detected either by a primary antibody or by the use of CTB conjugates, such as CTB-HRP and CTB-AF488 or CTB-AF594 (25), which can be visualized after histochemical detection or by fluorescence microscopy, respectively.

5. Working Protocols for Retrograde Tract Tracing

5.1. Materials

5.1.1. Animals

The choice of animal (species, age, sex) depends on several factors, including the type of tracer(s) to be used (species and age may differ for density and distribution of viral or tracer receptors, and for the amount of uptake and transport of the dye), the pathway(s) intended for tracing (species and age may be anatomically different), the feasibility of surgery (it is easier to perform tracing in larger animals), and the overall design of experiment (injury model, treatment, etc.). The animals are allowed to acclimatize to the holding facility for at least 1 week before tracing or injury.

5.1.2. Instruments and Materials (Pressure Injection as an Example)

1. Surgical microscope.
2. Rat, mouse, or neonatal rat adaptor.
3. Patch clamp glass (PG10150-4, World Precision Instruments Inc., Sarasota, FL).
4. Micropipette puller (P-1000, Sutter Instrument, Novato, CA).
5. Microbeveler (SYS-48000, World Precision Instruments Inc., Sarasota, FL).
6. High-pressure nitrogen gas.
7. Pneumatic PicoPump with vacuum (PV-830, World Precision Instruments Inc., Sarasota, FL).
8. Digital Lab Standard™ Stereotaxic with LED Display (Stoeling, Wood Dale, IL).
9. Cryostat (CM 1950, Leica Microsystems Inc., Bannockburn, IL).
10. Tracers.

5.2. Surgical Procedures

5.2.1. General Surgical Procedures

1. Anesthetize an animal with an intraperitoneal injection of a mixture of ketamine (87.7 mg/kg; Ben Venue Labs, OH) and xylazine (12.3 mg/kg; Butler, OH).
2. Shave the desired skin area for exposure using an electric clipper.
3. Stabilize the vertebral column using a stereotaxic device with a proper adaptor.

4. Use an iodine swab to sanitize the shaved skin area.
5. Use a scalpel to incise the skin; dissect and retract the connective tissue and/or muscle to expose the vertebral laminae.
6. Stop bleeding with sterile gauze squares or cotton-tipped applicators when needed.
7. Perform a laminectomy in the desired area; care should be taken not to damage the dura mater.
8. Use a 30-gauge (G) needle to poke a small hole through the dura mater.
9. Place the tracer-filled micropipette into the pipette holder carefully. Because the pipette tip is fragile, it should not be touched or manipulated during procedures.
10. Lower the pipette holder so that the pipette tip penetrates (through the poked hole in the dura) into a desired injection site of the brain or spinal cord.
11. Inject the tracer into the desired site at a steady rate and keep the pipette in place for 2–3 min. Repeat steps 10 and 11 for multiple injections.
12. Close the wound in layers.
13. Place the animal in a temperature- and humidity-controlled chamber for recovery.
14. Perform postoperative animal care according to institutional approved animal care and use protocols. For animals receiving SCI, manual bladder emptying should be performed three times daily until reflex bladder emptying is established.

5.2.2. Tracer Injection into the Spinal Cord (FG Injection as an Example)

1. Prepare the tracer: 1% FG in ddH_2O.
2. Draw the tracer into a sharpened glass micropipette attached to a stereotaxic device.
3. Anesthetize the rat.
4. Expose the spinal cord. For thoracic to lumbar spinal exposure, perform a midline incision, part the paraspinous muscles to define the prominent T10 spinous process which is vertical to the horizontal plane or the spinous process of T13 by tracing the lowest rib to the vertebrae.
5. With this region of vertebrae exposed, perform a laminectomy to expose the dorsal aspects of the thoracic- or lumber-level segments of the spinal cord.
6. Poke a hole through the dura, stereotaxically insert the glass micropipette (tip inner diameter 10–15 μm) into appropriate sites of the spinal cord, and deliver the FG using a Midguard Current Device (Midgard Electronics, Inc., Newton, MA) for 20 min (5 μA positive current, 7 s on/7 s off).
7. Suture the skin and muscle in layers.

5.2.3. Tracer Injection into the Sciatic Nerve (CTB as an Example)

1. Prepare the CTB solution (1%): Dilute CTB powder (Sigma, St. Louis, MO; C9903, 2 mg) into 200 μL distilled water.
2. Incise the skin at the level of the hip and femur of the rat, separate the gluteus maximus muscle at the caudal border with scissors, carefully remove the connective tissue surrounding the nerve, and expose the sciatic nerve. The nerve can be dissected without damaging muscle fibers.
3. Draw desired 1% CTB solution into a 25-μL Hamilton syringe. Insert syringe needle tip obliquely into the sciatic nerve underneath the epineurium. Inject 4–5 μL of CTB into the sciatic nerve, hold the syringe in place for 1–2 min, and then remove the needle. After the CTB injection, crush the nerve injection site with a 1-mm-wide forceps for better transportation of the CTB along the nerve.
4. Close the skin with clips.

5.3. Survival Time

The survival time after tracer introduction varies and is dependent on the species and age of the animal, type and length of the pathway, type of the tracer, and purpose and design of the experiment.

5.4. Perfusion and Fixation

After an appropriate survival period:

1. Deeply anesthetize with pentobarbital (60 mg/kg, ip).
2. Intracardially perfuse the animal with 0.9% saline followed by a solution of 4% paraformaldehyde (PFA) in 0.1 M PB (pH 7.4) or by modified Zamboni's fixative (26).
3. After perfusion, remove the brains and spinal cords and postfix tissue in a fresh fixative of 4% PFA at 4°C for 2 h (for light microscopy) or postfix 1-mm^3-size tissue in a fixative solution containing 2.5% glutaraldehyde and 5% sucrose in 0.1 M cacodylate buffer (pH 7.4, for electron microscopy) at 4°C for over 24 h (for electron microscopy).

5.5. Tissue Processing

5.5.1. Tissue Preparation for Light Microscopy

1. Transfer the specimen to a solution containing 30% sucrose in 0.1 M phosphate buffer (PB, pH 7.4) for cryoprotection for 4–5 days at 4°C.
2. Block the specimen into segments, and cut sections into desired thickness (10–40 μm) and orientation.

5.5.2. Tissue Preparation for Electron Microscopy

1. Fix tissue in 2.5% glutaraldehyde in 0.1 M sodium cacodylate buffer at 4°C, for a minimum of 4 h. Tissue should be cut into approximately 1-mm cubes for fixing.
2. Wash tissue in 0.1 M sodium cacodylate buffer—1 h ×2 (or overnight at 4°C).
3. Postfix tissue in osmium tetroxide in 0.2 M sodium cacodylate buffer—1 h (mix equal quantities of 2% aqueous osmium

tetroxide (OsO_4) and 0.4 M sodium cacodylate buffer and use immediately).

4. Rinse tissue in 0.2 M sodium cacodylate buffer—5 min ×2.
5. Dehydrate tissue in 70% ethanol—20 min ×2.
6. Dehydrate tissue in 90% ethanol—10 min ×2.
7. Dehydrate tissue in 100% ethanol—20 min ×2.
8. Place tissue into propylene oxide (PO)—40 min ×2.
9. Place tissue into propylene oxide/epoxy resin mixture (1:1)—1 h.
10. Place tissue into propylene oxide/epoxy resin mixture (1:3)—overnight with caps removed from vials (allows any remaining propylene oxide to evaporate).
11. Embed tissue in labeled capsules with freshly prepared *Spurr's epoxy resin*:
    (a) Vinylcyclohexene dioxide (ERL 4206)—20 g.
    (b) Diglycidyl ether of polypropylene glycol (DER 4360)—10 g.
    (c) Nonenyl succinic anhydride (NSA)—52 g.
    (d) DMAE-dimethylaminoethanol (S-1)—0.8 g.
12. Polymerize embedded tissue at 60°C for 48 h.

5.6. Detailed Histological Procedures

5.6.1. Direct Observation

1. Coverslip sections (containing fluorescence tracers) with an antifading mounting medium, such as Fluoromount (Diagnostic BioSystems, Inc., Pleasanton, CA).
2. View fluorescence labeling of the tracer under a fluorescent microscope.

5.6.2. Immunohistochemistry with a Primary Antibody and Avidin–Biotin–Peroxidase Kit

1. Air dry sections for 30–60 min. Circle sections with a PAP pen (Invitrogen, Carlsbad, CA).
2. Transfer slides into 0.01 M PBS buffer for 5 min ×2.
3. Block and pre-permeate the sections with 1.5% normal donkey serum (NDS) in 0.01 M PBS containing 0.3% Triton X-100 (PBS-T) for 30 min on an orbital shaker at room temperature (RT).
4. Incubate sections with a rabbit anti-FG primary antibody (1:1,000, in 1.5% NRS and 0.3% Triton X-100 in 0.01 M PBS) for 48 h at 4°C.
5. Rinse sections with 0.01 M PBS for 10 min ×3.
6. Incubate sections with biotinylated donkey anti-rabbit secondary antibody (1:200, 1.5% NDS in 0.01 M PBS) for 1 h at RT.
7. Rinse sections with PBS for 10 s ×2; wash sections with 0.01 M PBS for 10 min ×3.

8. Incubate sections for 60 min with VECTASTAIN® EliteABC Reagent (allow this reagent to stand for about 30 min before use):
   (a) 10 μL.
   (b) 10 μL B.
   (c) 2 mL PBS-T.
9. Wash sections for 15 min ×3 in 0.01 M PBS.
10. Incubate sections in peroxidase substrate solution until desired stain intensity develops (10–15 min) VECTASTAIN® EliteABC Reagent:
   (a) 1 mL ddH_2O.
   (b) 10 μL buffer.
   (c) 20 μL DAB.
   (d) 10 μL H_2O_2.
   (e) 10 μL nickel (optional).
11. Wash sections in 0.01 M PBS buffer for 5 min ×3.
12. Rinse sections in ddH_2O for 10 s ×2.
13. Dry and mount sections with DPX.

5.6.3. Immunofluorescence Labeling with a Primary and Fluorescence-Conjugated Secondary Antibodies

1. Air dry the sections for 30–60 min. Circle the sections with a PAP pen.
2. Transfer slides into 0.01 M PBS buffer for 5 min ×2.
3. Block and pre-permeate the sections with 1.5% NDS and 0.01 M PBS containing 0.3% Triton X-100 (PBS-T) for 30 min on an orbital shaker at RT.
4. Incubate sections with a primary antibody (rabbit anti-FG as an example, 1:1,000, 1.5% NDS in PBS-T) for 1.5 h at RT
5. Rinse sections with 0.01 M PBS for 10 min three times.
6. Incubate sections with a fluorescence-conjugated secondary antibody (FITC donkey anti-rabbit antibody as an example, 1:100, 1.5% NDS in 0.01 M PBS) for 1 h at RT.
7. Rinse sections with PBS for 10 min three times.
8. Rinse sections in ddH_2O ×2.
9. Dry and mount sections with a fluorescent mounting medium.

5.6.4. HRP Staining for Electron Microscopy

1. Anesthetize adult rats with ketamine (80 mg/kg).
2. Make an injection of 20% HRP (1 μL/site) into the spinal cord.
3. Two days after injection, transcardially perfuse the rats with 0.9% saline solution followed with 1% glutaraldehyde and 1% PFA in

0.1 M phosphate buffer (pH 7.4) at 4°C for 1 h, and then with 0.1 M phosphate buffer (pH 7.4) at 4°C 500 μL for 30 min.

4. Remove the spinal cord from the vertebral canal and store it in 0.1 M phosphate buffer at 4°C overnight.
5. Transversely cut the spinal cord blocks at 80-μm thickness with a vibratome. Collect the sections into 0.1 M phosphate buffer.
6. Incubate sections in 0.5% cobalt chloride ($CoCl_2$, Sigma) in 0.1 M PBS buffer for 10 min. Wash the sections in 0.1 M PBS for 5 min ×3 before and after incubation.
7. Incubate the sections in 0.05% DAB + 0.01% H_2O_2 (10 μL H_2O_2 in 30 mL 0.1 M PBS) for 5 min, and wash them in 0.1 M PBS for 5 min ×3.
8. Select sections with labeled cells for EM, and wash them in 0.15 M PB for 5 min ×3.
9. Fix sections in 1% OsO_4 in 0.15 M PB for 1 h, and wash them in 0.15 M PB for 5 min ×3.
10. Dehydration:
    (a) ddH_2O for 5 min ×3
    (b) Ethyl alcohols (30, 50, and 70% for 5 min each, 95% for 5 min ×2, 100% for 10 min ×3)
11. Infiltration:
    (a) Propylene oxide for 15 min ×3
    (b) A mixture of 1:1 solution of propylene oxide:embedding medium for 4 h at room temperature without a cap
    (c) Epon 812 resin overnight at room temperature
12. Embedding:
    (a) Transfer sections from resin to slides (coated with liquid release solution) and cover the sections with a coverslip (coated with liquid release solution). Apply pressure to the coverslip with balance weights.
    (b) Polymerize in an oven at 60°C for 2 days.
13. Remove the coverslips from the slides.
14. Cut pieces of tissue of interest to an area about 1 mm^2 from the embedded section.
15. Mount the small tissue on top of a plastic mode with Crazy Glue.
16. Trim the blocks and make semithin sections (1–2 μm) cutting first with an ultramicrotome.
17. Stain these semithin sections with Toluidine Blue to define the area of the tissue.
18. Cut the ultrathin sections at 80 nm and collect them in 200-mesh grids.

19. Dry the grids in air, and perform the electron staining of ultrathin sections with 2% uranyl acetate (EMS) for 30 min followed by lead citrate (protocol from Reynold's Lead Citrate Stain) for 15 min. Wash the grids in ddH_2O three times before and after lead citrate staining.
20. Observe ultrathin sections under a transmission electron microscope (Philips CM200).

6. Quantification of Retrogradely Labeled Neurons

The aim of using neuronal tracing is not only to disclose neuronal origins of projecting fibers to a target area, but also to provide quantitative data, especially in neuronal protection studies. Because the primary neural structure labeled by a retrograde tracer is the cellular soma, the quantification of labeled neurons is much easier than for anterograde-labeled axons. However, repeated counting should be carefully avoided by choosing the appropriate thickness of section according to cellular soma size. Unbiased stereological and optical fractionation counting methods may be employed to reduce this risk. Other details of the labeled neuron can be visualized and quantified, such as the extent of dendritic arborization or atrophy and synaptic interaction between labeled neuron dendrites and upstream neuronal axons depending on the tracer selected for use.

7. Application of Retrograde Tract Tracing in Spinal Cord Injury Research

As stated, modern retrograde axonal tracing strategies can be combined with a variety of techniques and can, therefore, be widely applied in the SCI research field. In particular, in vivo labeling with fluorescent tracers allows not only the identification and definition of neuronal sources of target axons and areas according to their anatomical projection, but also makes labeled neurons readily accessible for a wide range of additional approaches. Prelabeling of a specific group of neurons after an injury and/or treatment, multiple-labeling strategies using different fluorescent dyes in vivo to visualize specific neuronal origins which bi-project or tri-project their axons to different areas, immunohistochemical techniques to characterize specific neurotransmitters, cell types, and molecules, and electron microscopy for ultrastructural characterization of specific neurons and their synaptic interactions are such approaches that can be employed. The following details some examples of common applications of retrograde tracing in the study of SCI.

7.1. Single Labeling

7.1.1. Neuronal Regeneration Study

Several days or weeks after a neuron suffers axotomy injury, the regenerated axon across a spinal lesion engraftment could take up the tracer loaded in host tissue caudal to the graft and retrogradely label the neuronal soma (30). Neurons labeled in this fashion by tracers, such as FG, are visualized in a fluorescence field by ABC staining or at the EM level (Fig. 1a–d).

7.1.2. Neuronal Protection Study

The retrograde tracing method with FG described briefly: 6 weeks after spinal cord contusion injury, 4% FG was injected into the lumbar enlargement at 0.6 and 1 mm from the midline and a depth of 0.7 and 1.5 mm from the dorsal cord surface on both sides (four injections/side, 0.5 μL/injection) using a glass micropipette attached to a pneumatic picopump, approximately 12 mm distal to the lesion epicenter. Seven weeks after contusion injury, rats were killed and tissues fixed by transcardial perfusion of 4% PFA as described above. Transverse sections (40 μm) from the C6 and T5 segments were chosen as representative cervical and thoracic segments because axons from these regions form propriospinal projections that descend to the spinal cord caudal to the injury (27). In contusion injury, few neurons rostral to injury site were labeled by FG; after neuronal protection, more neurons rostral to injury site were labeled by FG (Fig. 1f, g) (29).

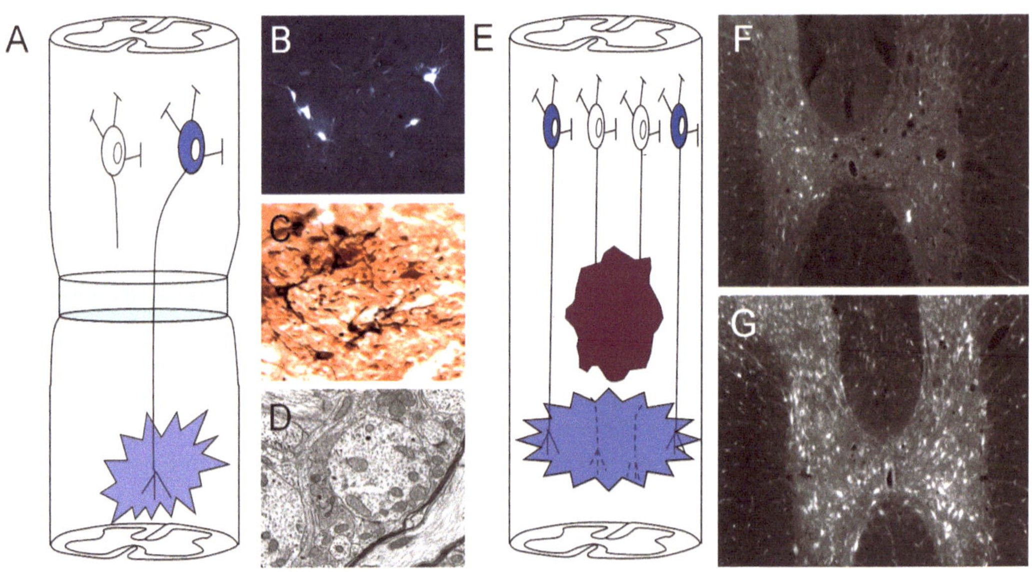

Fig. 1. Simplified presentation of different applications of single-labeling retrograde tracer on CNS injury models: (**a**) Cartoon depicting an axotomized neuron, which, if its axon regenerated across the graft, could take up the tracer loaded in host tissue caudal to the graft and retrogradely label the neuronal soma. (**b**) Neurons labeled by FG were visualized in fluorescence field. (**c**) Neurons labeled by FG visualized by ABC staining. (**d**) At the EM level, axons and dendrites were labeled by HRP. (**e**) Cartoon indicating that after contusion spinal cord injury (SCI) only spared axons take up the retrograde tracer to label the neuronal soma. (**f**) In contusion injury, few neurons rostral to the injury site were labeled by FG. (**g**) After neuronal protection, more neurons rostral to injury site were labeled by FG.

7.2. Double Labeling

7.2.1. Neuronal Regeneration Study

Double-retrograde labeling can be applied to transaction and hemisection injuries and graft models of SCI: rubrospinal cord neuron was prelabeled by one kind of retrograde tracer, for example Fast Blue (FB), and a hemisection was performed; several weeks after injury, regenerated axons crossing or circumventing the graft can pick up another tracer with a different color, loaded into host tissue caudal to the graft, and are doubly labeled. Neurons which failed to regenerate axons will be still single labeled or unlabeled if not in the vicinity of the initial tracer injection (Fig. 2) (28).

7.2.2. Divergent or Convergent Projection Study

Different tracers can be loaded in different levels of the spinal cord. Neurons which projected to different levels can be retrogradely labeled by different tracers. When different tracers are loaded on each side of the unilateral spinal cord, neurons that project unilaterally are labeled by a single tracer, whereas those project bilaterally are labeled by two different tracers (Fig. 3).

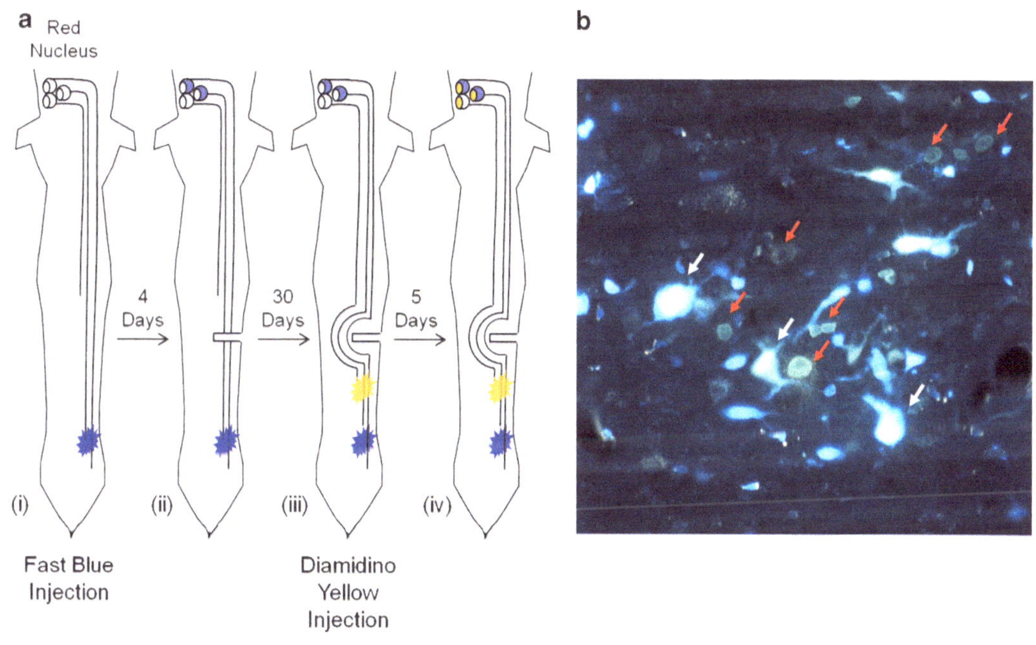

Fig. 2. Double-retrograde labeling applied in a hemisection injury for study of regeneration plasticity following SCI: (**a**) (i) Fast Blue (FB) is injected in the caudal thoracic spinal cord to be taken up by nearby rubrospinal axons; (ii) 4 days later, the FB tracer reached and labeled the cytoplasm of only the cell body of these axons in the contralateral red nucleus, and a unilateral hemisection was made many segments rostral to the injection site; (iii) 30 days later, diamidino yellow (DY), a retrograde nuclear marker, was injected between the FB injection site and the lesion; (iv) approximately 5 days later, new axons that sprouted or extended around the lesion had taken up the DY, as indicated by the labeled yellow nuclei. The two previously FB-labeled nuclei are now double labeled due to such growth, while the single DY-labeled neuronal axon had extended only since the FB injection and hemisection. (**b**) Propriospinal cord neurons retrogradely double labeled by DY and FG. Propriospinal cord neurons were first prelabeled by DY. In a regeneration study after SCI, neurons with axons regenerated across the lesion were doubly labeled by FG and DY after a caudal FG injection (*white arrows*) while neurons that failed to regenerate axons remained single labeled with DY (*red arrows*).

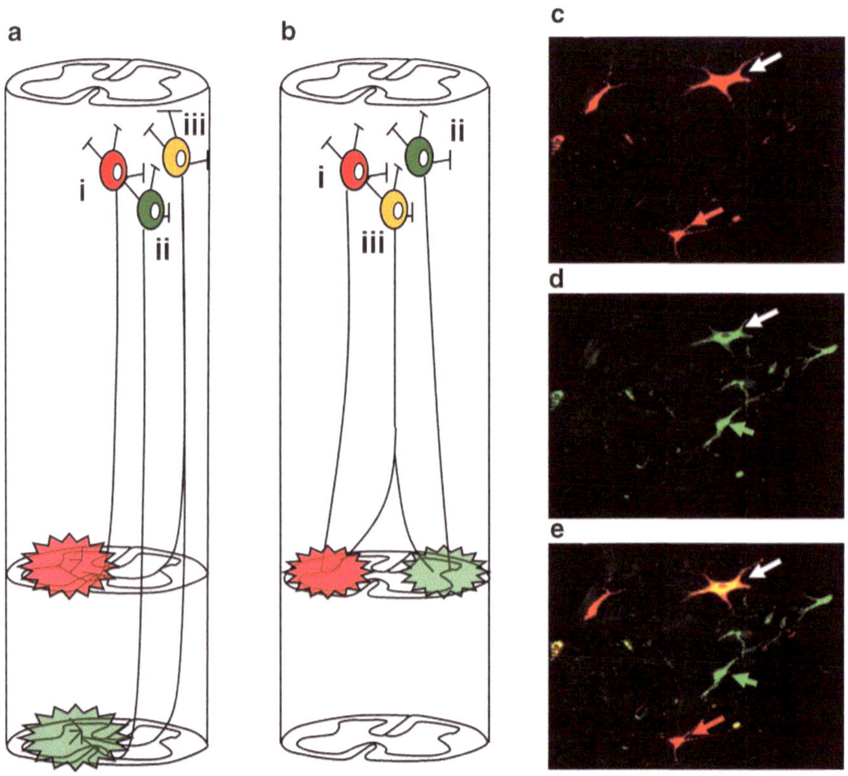

Fig. 3. Multiple tracer labeling to study divergent or convergent projections. (**a**) Different tracers were loaded into different spinal levels, and neurons which projected to different levels (**ai**, **aiii**) were retrogradely single labeled by the corresponding separate tracer, while the neuron projecting axons to each of the two levels (**aiii**) was double labeled. (**b**) Different tracers were loaded unilaterally into the spinal cord, with unilaterally projecting neurons labeled by a single tracer (**bi**, **bii**, **e**, *green* and *red arrow* indicated), while bilaterally projecting neurons were double labeled by different tracers (**biii**, **e**, *white arrow* indicated). (**c**), (**d**) Labeled cells were visualized by single fluorescence color. *Red arrow* in (**c**) indicates a cell labeled by only red tracer. *Green arrow* in (**d**) indicates a cell labeled by only green tracer. *White arrow* in (**c**) and (**d**) indicates a cell co-labeled by two different tracers.

8. Limitations, Potential Problems, and Solutions

1. Different retrograde tracers display different labeling efficiency due to their different transportation mechanisms. For those that are lipophobic and do not easily diffuse through membranes, simple air pressure injection causes low labeling rate. Instead, iontophoretic injection or combination with Triton solution may be a good choice.
2. Different retrograde tracers display different degradation rates in vivo. For example, FG can remain in vivo for over 1 month, while HRP cannot be detected after several days.
3. Most retrograde tracers cannot disclose the details of dendrites of a targeted cell. If fluorescent dyes are used for axonal tracing, it is better to transform the fluorescence to a permanent staining

product (such as through ABC staining) as photobleaching can make density quantification inaccurate. Also, it is usually difficult to obtain an adequate image resolution with a video camera (the sensitivity is usually low for single processes) or camera lucida.

4. Potential problems, possible causes, and solutions in the application of retrograde tracers, as described by Conte et al. (31), are similar to that for anterograde tracers, and are therefore listed in the chapter describing anterograde tracing.

References

1. Dahlstrom A (1971) Axoplasmic transport (with particular respect to adrenergic neurons). Philos Trans R Soc Lond B Biol Sci 261:325–358
2. Jeffrey PL, Austin L (1973) Axoplasmic transport. Prog Neurobiol 2:207–255
3. Nauta WJ, Gygax PA (1954) Silver impregnation of degenerating axons in the central nervous system: a modified technique. Stain Technol 29:91–93
4. Kristensson K, Olsson Y (1971) Retrograde axonal transport of protein. Brain Res 29: 363–365
5. Akintunde A, Buxton DF (1992) Quadruple labeling of brain-stem neurons: a multiple retrograde fluorescent tracer study of axonal collateralization. J Neurosci Methods 45:15–22
6. LaVail JH, LaVail MM (1972) Retrograde axonal transport in the central nervous system. Science 176:1416–1417
7. Kuypers HG, Ugolini G (1990) Viruses as transneuronal tracers. Trends Neurosci 13:71–75
8. Marshel JH, Mori T, Nielsen KJ, Callaway EM (2010) Targeting single neuronal networks for gene expression and cell labeling in vivo. Neuron 67(4):562–574
9. Giloh H, Sedat JW (1982) Fluorescence microscopy: reduced photobleaching of rhodamine and fluorescein protein conjugates by n-propyl gallate. Science 217:1252–1255
10. Deller T, Naumann T, Frotscher M (2000) Retrograde and anterograde tracing combined with transmitter identification and electron microscopy. J Neurosci Methods 103:117–126
11. Mason A, Larkman A, Eldridge JL (1988) A method for intracellular injection of horseradish peroxidase by pressure. J Neurosci Methods 22:181–187
12. Xu XM, Martin GF (1989) Developmental plasticity of the rubrospinal tract in the North American opossum. J Comp Neurol 279:368–381
13. Imura K, Rockland KS (2007) Giant neurons in the macaque pulvinar: a distinct relay subpopulation. Front Neuroanat 1(2):1–8
14. Xu XM, Martin GF (1989) The response of rubrospinal neurons to axotomy in the adult opossum, *Didelphis virginiana*. Exp Neurol 108: 46–54
15. Xu XM, Martin GF (1992) The response of rubrospinal neurons to axotomy at different stages of development in the North American opossum. J Neurotrauma 9:93–105
16. Xu XM, Chen A, Guenard V, Kleitman N, Bunge MB (1997) Bridging Schwann cell transplants promote axonal regeneration from both the rostral and caudal stumps of transected adult rat spinal cord. J Neurocytol 26:1–16
17. Iannotti C, Li H, Yan P, Lu X, Wirthlin L, Xu XM (2003) Glial cell line-derived neurotrophic factor-enriched bridging transplants promote propriospinal axonal regeneration and enhance myelination after spinal cord injury. Exp Neurol 183(2):379–393
18. Schmued LC, Fallon JH (1986) Fluoro-Gold: a new fluorescent retrograde axonal tracer with numerous unique properties. Brain Res 377: 147–154
19. Wessendorf MW (1991) Fluoro-Gold: composition, and mechanism of uptake. Brain Res 553:135–148
20. Reiner A, Veenman CL, Medina L, Jiao Y, Del Mar N, Honig MG (2000) Pathway tracing using biotinylated dextran amines. J Neurosci Methods 103(1):23–37
21. Brandt HM, Apkarian AV (1992) Biotin-dextran: a sensitive anterograde tracer for neuroanatomic studies in rat and monkey. J Neurosci Methods 45:35–40
22. Rajakumar N, Elisevich K, Flumerfelt BA (1993) Biotinylated dextran: a versatile anterograde and retrograde neuronal tracer. Brain Res 607:47–53
23. Angelucci A, Clasca F, Sur M (1996) Anterograde axonal tracing with the subunit B of cholera toxin: a highly sensitive immunohistochemical protocol for revealing fine axonal morphology in adult and neonatal brains. J Neurosci Methods 65:101–112

24. Luppi PH, Sakai K, Salvert D, Fort P, Jouvet M (1987) Peptidergic hypothalamic afferents to the cat nucleus raphe pallidus as revealed by a double immunostaining technique using unconjugated cholera toxin as a retrograde tracer. Brain Res 402:339–345
25. Luppi PH, Fort P, Jouvet M (1990) Iontophoretic application of unconjugated cholera toxin B subunit (CTb) combined with immunohistochemistry of neurochemical substances: a method for transmitter identification of retrogradely labeled neurons. Brain Res 534: 209–224
26. Ericson H, Blomqvist A (1988) Tracing of neuronal connections with cholera toxin subunit B: light and electron microscopic immunohistochemistry using monoclonal antibodies. J Neurosci Methods 24:225–235
27. Conte WL, Kamishina H, Reep RL (2009) Multiple neuroanatomical tract-tracing using fluorescent Alexa Fluor conjugates of cholera toxin subunit B in rats. Nat Protoc 4:1157–1166
28. Holets VR, Hokfelt T, Ude J, Eckert M, Penzlin H, Verhofstad AAJ, Visser TJ (1987) A comparative study of the immunohistochemical localization of a presumptive proctolin-like peptide, thyrotropin-releasing hormone and 5-hydroxytryptamine in the rat central nervous system. Brain Res 408:141–153
29. Liu NK, Zhang YP, Han S, Pei J, Xu LY, Lu PH, Shields CB, Xu XM (2007) Annexin A1 reduces inflammatory reaction and tissue damage through inhibition of phospholipase A2 activation in adult rats following spinal cord injury. J Neuropathol Exp Neurol 66:932–943
30. Xu XM, Martin GF (1991) Evidence for new growth and regeneration of cut axons in developmental plasticity of the rubrospinal tract in the North American opossum. J Comp Neurol 313:103–112
31. Conte WL, Kamishina H, Reep RL (2009) Multiple neuroanatomical tract-tracing using fluorescent Alexa Fluor conjugates of cholera toxin subunit B in rats. Nat Protoc 4:1157–1166

Chapter 39

Anterograde Axonal Tract Tracing

Xiaofei Wang, Lingxiao Deng, and Xiao-Ming Xu

Abstract

Deficits in sensory and motor function after spinal cord injury (SCI) are attributable primarily to the interruption of long sensory and motor axonal tracts in the spinal cord. How to exquisitely label these tracts within the spinal cord and their connections to the brain or periphery continues to be an important concern for SCI research. Axonal tract tracing technologies are powerful tools for identifying axonal connections in the normal or injured spinal cord. With appropriate injury models and tracing techniques, status of axons such as sparing, die-back, sprouting, regeneration, or synaptogenesis can be readily identified. The tracing techniques can also be used to determine axonal projections from neurons that are grafted into the lesion site. For anterograde tracing, tracers are transported by attaching to different vectors with different speed from the soma to axon terminal. After tracing, tissue is collected following appropriate survival time. To detect tracers along axons or at their terminals, various labeling techniques are used which include direct fluorescence, immunohistochemistry, immunofluorescence, and autoradiography. When combined with retrograde tracing techniques, anterograde tracers may be used to establish anatomical reorganizations of neural networks. When combined with immunohistochemistry, anterograde tracing may be used for neurochemical characterization of specific neuronal pathways or specific cell types.

Key words: Anterograde tracing, Biotinylated dextran amine, Fluoro-Ruby, *Phaseolus vulgaris* leucoagglutinin

1. Introduction

1.1. Historical Perspective

The most early anterograde tracing techniques were based on Wallerian degeneration, characterized by the degenerative change of an axon in the distal segment after traumatic separation from the soma. The degeneration methods introduced by Glees (1) and Nauta (2) in the 1940s allowed to the study of long projections from a nucleus to other regions anterogradely within the central nervous system (CNS). In the 1970s, Cowan et al. developed an autoradiographic technique to study axonal connection by anterograde transport of radio-labeled amino acids (3) which was a further

Jun Chen et al. (eds.), *Animal Models of Acute Neurological Injuries II*, Springer Protocols Handbooks,
DOI 10.1007/978-1-61779-782-8_39, © Springer Science+Business Media, LLC 2012

technical improvement. However, both neuronal degeneration and autoradiographic techniques suffered from poor resolution and lack of sensitivity (4). In the 1980s, fluorescent dyes became available for anterograde tracing. Among them were rhodamine-isothiocyanate (RITC) and 1,1′-dioctadecyl-3,3,3′,3′-tetramethyl-indocarbo-cyanine perchlorate (DiI) (5). The list has since lengthened with the introduction of new classes of anterograde tracers, including plant lectins, notably *Phaseolus vulgaris* leucoagglutinin (PHA-L) (6). After that, dextran amines have become the most powerful tools for anterograde tracing (7) because they are relatively simple to use, and a great variety of detection methods have been developed for them. Dextrans can now be purchased in biotinylated form as biotinylated dextran amine (BDA) (8–10) or conjugated to various fluorescent dyes (11), thus allowing immediate visibility, or the production of permanent preparations after immunocytochemical processing.

In recent years, viruses have been developed for tract tracing studies. Compared to conventional tracers, viruses have the ability to traverse multisynaptic pathways and replicate to amplify signals at each step in the process (12). Depending on the species and strain of the virus, viruses can travel preferentially in the anterograde or the retrograde direction. Limitations include a lack of synaptic specificity of spread, the speed of spread depends on both strength and numbers of connections, and induction of cell death. More recently, genetic modifications have allowed for many improvements. These include reduced pathogenicity, addition of marker genes, control of synaptic spread, pseudotyping for infection of selected cells, and addition of ancillary genetic elements for combining circuit tracing with manipulation of activity or functional assays.

1.2. Axonal Transport and Axonal Tracing

Tracing is managed through a phenomenon called axonal transport or *axoplasmic flow* (13, 14). Many membranous organelles (e.g., mitochondria) and vesicles as well as of macromolecules, such as actin and myosin of the neuron can be transported by axoplasmic flow and it can proceed in both directions.

Based on axonal transport, a long series of tracers has been developed as anterograde tracing (labeling begins at the dendrites and cell bodies and the tracer is transported out to the axons and their terminals) or retrograde tracing (labeling begins at the axon termini and the tracer is transported back through the axon to the cell bodies) according to the preferential direction of their transport in the axon. In some cases, tracers work in both ways simultaneously whereas in other cases the tracers only work one way or the other.

1.3. Anterograde Axonal Tract Tracing for Spinal Cord Injury Research

Deficits in sensory and motor functions after spinal cord injury (SCI) are caused primarily by the interruption of the long sensory and motor tracts within the spinal cord. Tremendous efforts have been made to promote either protection of spared axons or regeneration of damaged axons or both. Assessments of both strategies require labeling of these axons. Anterograde tracing is particularly suitable in such type of research since it allows investigators to follow the trajectory of axons as they spare, die-back, regenerate, sprout, or form synapses on target neurons. Anterograde tracing may also be used for studying axonal projections from transplanted neurons in animal models of SCI.

2. General Considerations of Using Anterograde Tracing

2.1. Tracer Application

Three different methods are used to introduce tracer material into the target tissue: pressure injection, iontophoretic injection, and mechanical insertion of dye crystals.

2.1.1. Pressure Injection

The dissolved tracer is usually delivered through a glass micropipette of the type produced in a standard micropipette puller and sharpened on a microbeveler. Injection of the tracer solution can be achieved either with gas pressure (by control of pneumatic pump), or by means of a hydraulic system, which enables the injection volume to be controlled more accurately (15).

2.1.2. Iontophoretic Injection

Tracer molecules usually carry an electrical charge and can therefore be ejected from the micropipette by an applied electrical current. Iontophoretic injection can be used for both intracellular and extracellular applications. An advantage of using iontophoretic injection is that the application site can be precisely located prior to iontophoresis by conventional electrophysiological techniques, using the same micropipette that will deliver the dye (15).

2.1.3. Insertion of Dye Crystals

The insertion of crystal dyes is feasible when the target is readily accessible. The placement of dye crystals can be carried out under microscopic control and the labeling efficiency is very high. An example for this method is the focal application of carbocyanine dyes, such as 1,1′-dioctadecyl-3,3,3′,3′-tetramethylindocarbocyanine perchlorate (DiI) and 4-4-didecylamino styryl-*N*-methyl-pyridinium iodide (DiAsp) (15).

2.2. Incorporation Mechanisms

For tracer incorporation into a neuron, one of the following three mechanisms may be responsible.

2.2.1. Active Uptake

Active uptake is the mediated process of moving tracer across the lipid bilayer membrane of neuron using energy. Tracers can enter axons or dendrites by active uptake via nerve terminals, or through severed axons. In general both pathways work, but dextran amines are taken up more efficiently by injured axons than by terminals. Active uptake is one of the advantages of the lectins. These tracers can bind to the sugar moieties of membrane-bound glycoproteins and glycolipids (6).

2.2.2. Passive Incorporation

Tracer can passively enter into neurons with no energy because of the diffusion toward local concentration gradient. They must therefore be present at higher concentrations at the application site (16).

2.2.3. Intracellular Injection

If single cells are accessible, tracers such as BDA or lucifer yellow (LY) can be introduced directly into the cytoplasm of one cell, yielding the complete description of dendritic arborizations and the axonal projection of individual neurons. Since this injection is more applicable to brain or spinal cord slices in vitro, long-distance tracing is limited by the disruption of axons when the brain or spinal cord is cut up into slices (15). Serial sections with neurolucida 3D reconstruction technology, however, may be used to mimic the cellular morphology.

2.3. Transport and Sequestration within Neurons

Tracers are distributed within the cells in two ways: active transport in vesicles and lateral diffusion within the cellular membranes.

2.3.1. Active Transport in Vesicles

By means of pH-trapping or the ingestion of labeled membrane fragments, the tracer accumulates in vesicles, and then they are actively carried out along the cytoskeleton of axons and dendrites. Whether transport occurs in the anterograde or retrograde direction depends upon the uptake mechanism and tracer types. Active vesicular transport seems to be the only efficient transfer mechanism for longer distances tracing in vivo. Fast axonal transport can reach velocities in the speed of 2 cm/day. The anterograde transport of radioactively tagged proteins is somewhat slower, with velocities varying between 2 and 400 mm/day (15).

2.3.2. Lateral Diffusion Within the Cellular Membranes

Some carbocyanine dyes like DiI possess long lipophilic tails that can be inserted into cell membrane. These tracers can spread through lateral diffusion within the plane of the membrane, resulting in a much smoother labeling of the complete extension of the cellular processes. Passive diffusion proceeds at speeds of a few millimeters per day and therefore the method is confined to tracing distances of a few centimeters. Passive diffusion mechanism also allows the tracers to be exploited in fixed tissue. Tracing in fixed tissues may be performed under refrigeration, at room temperature (RT), or at higher temperatures in an incubator (e.g., 37°C). Tracing after fixation with paraformaldehyde (PFA) can be performed faster when the tissue is kept in an incubator (15).

2.4. Detection of the Tracing Signal

2.4.1. Fluorescent Tracers

Fluorescent tracers can be directly visualized under a fluorescent microscope. The major advantage is that the labeled neurons and their processes can be observed in living or fixed tissue in vivo or in vitro. It is feasible to perform additional experiments on the same tissue. There are a large number of fluorescent tracers available with different absorption spectra permitting simultaneous labeling with multiple tracers. Relative disadvantages include that a strong fluorescence signal may obscure the details of smaller processes, frequent illumination of short-wavelength fluorescence (e.g., fast blue, FB) leads to photo-toxicity of otherwise nontoxic substances, some fixatives have auto-fluorescent properties and tissue fixation cannot be optimized, some fluorescent tracers diffuse out of the cells and are therefore not suitable for long-term studies, and photo-bleaching or fading of the signal after light exposure during examination and storage. Since photo-bleaching is believed to be caused by oxygen and oxygen-induced free radicals, antioxidants can reduce fading. One successful formula, which can be used as a mounting medium, is 0.1–0.25 M *n*-propyl gallate in glycerol (17). For fluorescence labeled latex beads, commercially available mounting media like Fluoromount (Diagnostic BioSystems, Inc., Pleasanton, CA) must be used rather than glycerol.

To achieve a permanent staining, the fluorescence tracer like LY can be detected with specific antibodies that can be visualized after incubation with diaminobenzidine (DAB). This reaction requires great caution since the DAB solution is toxic, carcinogenic, and sticks to contaminated surfaces.

2.4.2. Nonfluorescent Tracers

Most nonfluorescent tracers can be observed with specific antibodies that can be visualized with routine immunohistochemical (IHC) or immunofluorescence staining. There are many techniques that can be employed in combination with anterograde tracers, including signal amplification with secondary antibodies, indirect methods, such as the peroxidase–anti-peroxidase (PAP) and avidin–biotin–peroxidase (ABC) methods, and detection with peroxidase or alkaline phosphatase reaction. Progress in IHC staining has therefore also improved the sensitivity of neuronal tracing. Dextran amines, for instance, are conjugated with biotin so that they can be detected via the high-affinity binding of avidin or streptavidin that has been labeled with a substrate of some kind (e.g., HRP or a fluorescent tag) (18). The combination of tracing studies with ultrastructural investigations involves the production of an electron-dense DAB precipitate after photoconversion of fluorescence or gold-labeled secondary antibodies (19).

3. Selection of Anterograde Tracers

Choosing an appropriate anterograde tracer for SCI research depends on several factors, including animal species (species may differ for density and distribution of viral or tracer receptors and for the amount of uptake and transport of the dye), age (some tracers do not work well in neonatal, adult or aged animals), pathway (some tracers are pathway-sensitive), tracing direction, and overall design of the experiment (multiple pathway tracing or combination with other methods).

Anterograde axonal tracing has been performed using a wide variety of fluorescent and nonfluorescent probes, including lipophilic tracers such as DiI, biotin derivatives, polar fluorescent dyes, protein conjugates (including lectins), dextran conjugates, and various viruses. The solvents required for preparing the tracer injection vary depending upon the tracer. Some tracers are water-soluble, for instance FB, diamidino yellow (DY) or dextran amines, while others, like the carbocyanine dyes, require organic solvents. Suspensions of insoluble dyes must be sonicated and filtered prior to placing them into the micropipette. The concentration used also varies considerably between dyes, depending upon the sensitivity of the technique. 1% FG, 10% BDA, 10% FR, and 1% PHA-L, are routine concentration used in SCI study. The following are a list of tracers used most commonly in the studies of experimental SCIs.

3.1. Biotinylated Dextran Amines

Dextrans are hydrophilic polysaccharides characterized by their moderate to high molecular weight, good water solubility, and low toxicity. They are widely used and can function efficiently as anterograde or retrograde tracers, depending on the study method, and tissue type used. Dextrans are biologically inert and have low immunogenicity. There are about 100 fluorescent and biotinylated dextran conjugates in several molecular weight ranges. Fluorescent and biotinylated dextrans (BDA) are routinely employed to trace axonal tracts.

BDA is a highly sensitive tool for anterograde axonal tract tracing studies (20). BDA can be reliably delivered into the nervous system by iontophoretic or pressure injection and visualized with an avidin-biotinylated HRP (ABC) procedure, followed by a DAB or immunofluorescence reaction. High molecular weight BDA (10,000 MW) yields sensitive and exquisitely detailed anterograde labeling of axons and terminals, while low molecular weight BDA (3,000 MW) yields sensitive and detailed retrograde labeling of neuronal cell bodies (21). Owing to its flexibility as to fixation, BDA can be visualized at the light microscopic (LM) or electron microscopic (EM) level and it can also be combined with various other pathway tracing or IHC methods.

3.2. Fluorescent Dyes

3.2.1. Fluoro-Ruby

Fluoro-Ruby (FR), a red-orange fluorescent, aldehyde-fixable 10,000 MW dextran labeled with both tetramethylrhodamine and lysine, is a quick, sensitive, reliable, and technically simple fluorescent anterograde tracer (22). The compound was first synthesized in 1985 (23). FR labeling is resistant to photo-bleaching and does not seem to leak out from labeled processes. An additional advantage is the tracer's compatibility with other histochemical techniques, including HRP histochemistry (24). If required, FR can also be photo-converted (25) or labeled with a primary antibody.

3.2.2. Lipophilic Carbocyanine Dyes

Lipophilic carbocyanine dyes are a series of tracers, of which DiI, DiA, and DiO (3,3′-dioctadecyloxacarbocyanine perchlorate) are the most frequently used. They label both live and fixed cells. Besides being used in tracing connections in situ, carbocyanine dyes are also useful in the study of connections in slice cultures, in dissociated cell culture (usually DiI and DiO) and in fixed cultures (DiI and DiA). Moreover, they are used in vivo for fate mapping and axonal tracing (26). When used in living slice cultures, DiI is very efficient in putting in evidence the development of axons observed under time-lapse videomicroscopy. In vivo the carbocyanine dyes diffuse rapidly, 6 mm/day for DiI (27), and the dyes are not toxic for the cell. Carbocyanine dyes can label cells anterogradely, retrogradely, and also they can be used to label dissociated cells following a brief incubation in the dye before transplantation or seeding. A critical problem is the fading of the dye labeling in sections, which may be only in part prevented with anti-fading mounting mediums. Carbocyanine dyes may be used in combination with other tracers, such as fluorescent microspheres, fluorescent dextrans, LY, or with other techniques, such as in situ hybridization or immunohistochemistry (15).

3.3. Plant Lectins and Bacterial Toxins

3.3.1. Phaseolus Vulgaris Leucoagglutinin

Phaseolus vulgaris leucoagglutinin is a plant lectin that is selectively taken up by neurons at injection sites and, when immunohistochemically demonstrated, can label individual neurons completely, including their dendrites, axons, and terminal boutons. An estimate speed for PHA-L anterograde transport is about 4–6 mm/day and the survival time is quite long, up to 7 weeks. The quality of labeling does not degrade with time. PHA-L is taken up by fibers passing through the injection site with lower efficiency and, because it produces a Golgi-like staining of even very fine axons over long distances, it is possible to light microscopically reconstruct individual neurons and their entire axon terminal arbors. When prepared for electron microscopy, the PHA-L-labeled terminals are densely and completely stained, allowing synaptic relationships to be defined. All these properties make PHA-L advantageous for studying the termination of select groups of neurons and the patterns of projection in their targets (28). But iontophoretic injection method is required for PHA-L delivery.

3.3.2. Cholera Toxin B Subunit

The cholera toxin (CT) (a bacterial toxin produced by *Vibrio cholerae*) is an oligomeric complex made up of six protein subunits: a single copy of the A subunit (CTA), and five copies of the B subunit (CTB). The two parts are connected by a disulfide bond. CTA is responsible for the toxic effect, whereas the nontoxic CTB allows the protein to bind to cellular surfaces and transport in axons and cell bodies (29).

CTB is mainly used as retrograde tracer (30, 31), but CTB can also be anterogradely transported with an extensive filling of axons and axon terminals and thereby opens up the possibility of identifying simultaneously the afferents as well as the efferents of the group of cells, for instance dorsal root ganglion (DRG) neurons (32). CTB may be delivered by iontophoresis or pressure injection. CTB requires a survival time of 1–10 days, depending on the length of pathway and the age of the animal. CTB is a very sensitive tracer which can be detected either by a primary antibody, or by the use of CTB conjugates, such as CTB-HRP and CTB-AF488 or CTB-AF594 (33), which can be visualized after histochemical detection or by fluorescence microscopy, respectively.

4. Working Protocols for Anterograde Tract Tracing

4.1. Materials

4.1.1. Animals

The choice of animal (species, age, and sex) depends on several factors, including the type of tracer(s) to be used (species and age may differ for density and distribution of viral or tracer receptors, and for the amount of uptake and transport of the dye), the pathway(s) intend to trace (species and age may be anatomically differ), the feasibility of surgery (it is easier to perform tracing in big size animals), and the overall design of experiment (injury model, treatment, etc.). The animals are allowed to acclimatize to the holding facility for at least 1 week before tracing or injury.

4.1.2. Instruments and Materials (Pressure Injection as an Example, Fig. 1)

1. Surgical microscope.
2. Rat, mouse, or neonatal rat adaptor.
3. Patch clamp glass (PG10150-4, World Precision Instruments Inc., Sarasota, FL).
4. Micropipette puller (P-1000, Sutter Instrument., Novato, CA).
5. Microbeveler (SYS-48000, World Precision Instruments Inc.).
6. High pressure nitrogen gas.
7. Pneumatic PicoPump with vacuum (PV-830, World Precision Instruments Inc.).
8. Digital Lab Standard™ Stereotaxic with LED Display (Stoeling, Wood Dale, IL).
9. Cryostat (CM 1950, Leica Microsystems Inc., Bannockburn, IL).
10. Tracers.

4.2. Surgical Procedures

4.2.1. General Surgical Procedures

1. Anesthetize an animal with an intraperitoneal injection of a mixture of ketamine (87.7 mg/kg; Ben Venue Labs, OH) and xylazine (12.3 mg/kg; Butler, OH).
2. Shave the desired skin area for exposure using an electric clipper.

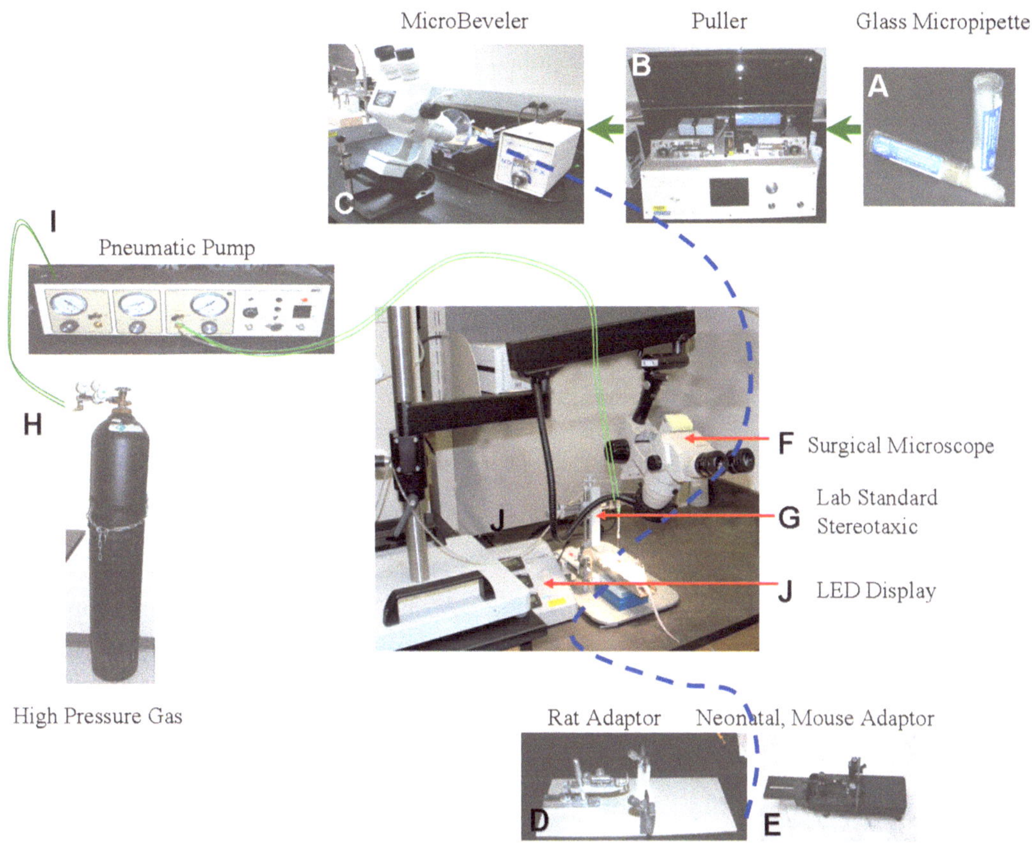

Fig. 1. Overall setup of the pressure injection system. Before surgery, a glass micropipette (**a**) is pulled in a standard puller (**b**) and is sharpened on a microbeveler (**c**). After anesthesia, the skull of an animal is stabilized with an appropriate adaptor (rat in **d**, mouse and neonatal rat in **e**) and ear bars on a stereotaxic device (**g**). Surgeries are performed under a surgical microscope (**f**). A sharpened micropipette containing a tracer is attached to the arm of a micromanipulator (**g**) and is connected to a pneumatic pump (**i**) which is further connected to a high pressure gas tank (**h**). With the guide of LED digital display (**j**), injection sites can be positioned and tracer can be injected accurately.

3. Stabilize the skull or vertebral column using a stereotaxic device with a proper adaptor.
4. Use an iodine swab to sanitize the shaved skin area.
5. Use a scalpel to incise the skin; dissect and retract the connective tissue and/or muscle to expose the underneath skull or vertebral laminae.
6. Stop bleeding with gauze squares or cotton-tipped applicators when needed.
7. Perform a craniotomy or laminectomy in desired area; care should be taken not to damage the dura matter.
8. Using a 30 gauge (G) needle to poke a small hole through the dura matter.
9. Place the tracer-filled micropipette into the pipette holder carefully. Because the pipette tip is fragile, it should not be touched at all times.

10. Lower the pipette holder so that the pipette tip penetrates (through the poked hole in dura) into a desired injection site of the brain or spinal cord.
11. Inject the tracer into desired site at a steady speed and keep the pipette in place for 2–3 min. Repeat steps 10 and 11 for multiple injections.
12. Close the wound in layers.
13. Place the animal in a temperature- and humidity-controlled chamber for recovery.
14. Perform postoperative animal care according to institutional approved animal care and use protocols. For animals receiving SCI, manual bladder emptying should be performed three times daily until reflex bladder emptying is established.

4.2.2. Tracer Injection into the Motor Cortex of a Rat (BDA/FR as an Example, Fig. 2)

1. Prepare 10% BDA: dilute 25 mg BDA with 250 μl Tris buffer (PH 8.0).
2. Draw 3.0 μl 10% BDA into a sharpened glass micropipette attached to a stereotaxic device.
3. Anesthetize the rat.

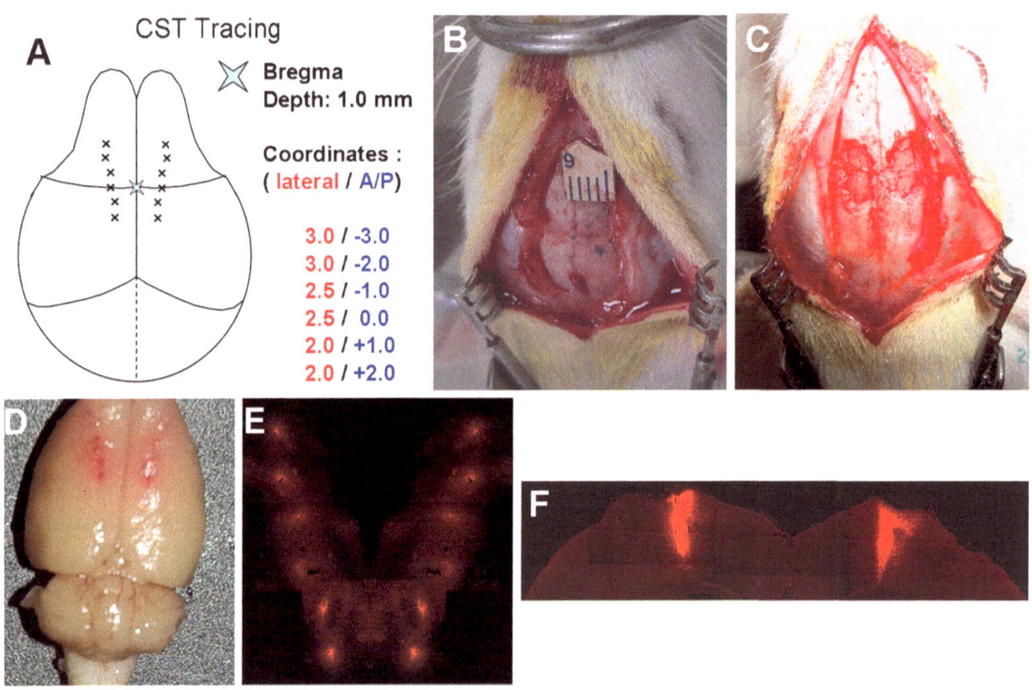

Fig. 2. Anterograde tracing of the corticospinal tract with BDA or FR. (**a**) Schematic drawing of the injection sites on the brain of an adult rat. (**b**) The skin was incised to expose the skull. A segment of a ruler with scales in mm was shown. (**c**) A craniotomy was performed to expose the motor cortex. Note that the dura remained intact. FR or BDA was slowed injected into the motor cortex at a depth of 1.0 mm according to the coordination provided in (**a**). (**d**) After perfusion fixation the brain was removed at 4 weeks postinjection and the sites of FR injection were seen. (**e**, **f**) Immunofluorescent staining of BDA showed tracer injection sites within the motor cortex in both horizontal (**e**) and transverse sections (**f**).

4. Fix one ear bar to the stereotaxic frame and gradually slide the ear into the ear bar. Often, the eardrum will break and a slight snap will be heard.
5. Holding the head, put the free ear bar into the other ear, and tight the ear bar to the frame. Confirm placement in the frame by moving the head of the rat laterally. Make sure the head of the animal is straight and not tilted on its side.
6. Tight the mouth and nose holders in front of the eyes of the rat. Location and position of the mouth piece may vary depending on the stereotaxic atlas used.
7. Incise the skin, retract the connective tissue, and perform craniotomy to expose the desired area.
8. Inject 0.5 μl of 10% BDA into each site of the motor cortex at a depth of 1.0 mm according to the coordination shown in Fig. 2a (six injections/side). Maintain the tip of the glass pipette at site for 2 min before withdrawal.
9. Suture the skin and connective tissue in layers.

4.2.3. Tracer Injection into the Red Nucleus (FR as an Example, Fig. 3)

1. Draw 1.0 μl tracer solution into a sharpened glass micropipette attached to a stereotaxic device.
2. Anesthetize the rat, incise the skin, open the skull over the injected region, pock a hole(s) in the dura, and inject 0.5 μl of the tracer into bilateral red nuclei at a depth of 6.1–6.2 mm according to the coordination shown in Fig. 3a. Maintain the glass pipette tip at site for 5 min before withdrawal.
3. Suture the skin and connective tissue in layers.

4.2.4. Tracer Injection into the Spinal Cord (PHA-L as an Example)

1. Prepare the tracer: 2.5% PHA-L in 15 mM Tris buffer, pH 8.0.
2. Draw the tracer into a sharpened glass micropipette attached to a stereotaxic device.
3. Anesthetize the rat.
4. Expose the spinal cord. For cervical spinal exposure, the head of the rat is angled down and, through a midline incision, the paraspinous muscles are parted to define the prominent T2 spinous process which is clamped and axially distracted. This allows a laminectomy to be performed to expose the dorsal aspects of the cervical segment of the spinal cord. For thoracic or lumbar spinal exposure, the spinous process of T13 was identified by tracing the lowest rib to the vertebrae. Through a midline incision, the paraspinous muscles were detached from T12 to L2 and a laminectomy was performed to allow access to the spinal cord at the lumber level.
5. Poke a hole through the dura, insert the glass micropipette (tip inner diameter 10–15 μm) into appropriate sites of the spinal cord stereotaxically, and deliver the PHA-L using a Midguard

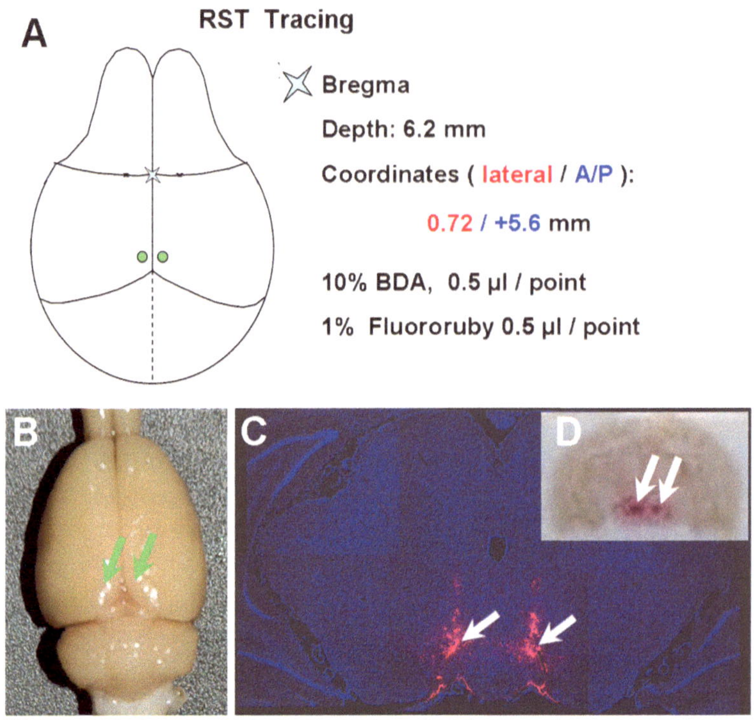

Fig. 3. Anterograde tracing of the rubrospinal tract (RST) with Fluoro-Ruby (FR). (**a**) Schematic drawing shows bilateral injection sites of the red nucleus (*green dots*) according to the coordinations provided. (**b**) *Green arrows* indicate the dorsal surface of the brain where two injections into the red nuclei were made. (**c**, **d**) In both thin (**c**) and thick (**d**) coronal sections of the midbrain, FR injections into bilateral red nuclei were clearly seen (*arrows*) at 4 weeks postinjection.

Current Device (Midgard Electronics, Inc., Newton, MA) for 20 min (5 μA positive current, 7 s on/7 s off).

6. Suture the skin and muscle in layers.

4.2.5. Tracer Injection into the Sciatic Nerve (CTB as an Example, Fig. 4)

1. Prepare the CTB solution (1%). Dilute CTB powder (Sigma-Aldrich, St. Louis, MO; C9903, 2 mg) into 200 μl distilled water.
2. Incise the skin at the level of the hip and femur of the rat (Fig. 4a), separate the gluteus maximus muscle at the caudal border with scissors (Fig. 4b), and carefully remove the connective tissue surrounding the nerve and expose the sciatic nerve. The nerve can be dissected without damaging muscle fibers (Fig. 4c).
3. Prepare 1% CTB solution in a 25 μl Hemilton syringe. Insert syringe needle tip obliquely into the sciatic nerve underneath the epineurium (Fig. 4d). Inject 4–5 μl of CTB into the sciatic nerve, hold the syringe for 1–2 min (Fig. 4e), and then remove the needle. After CTB injection, crush the nerve injection site with a 1 mm wide forceps for better transportation of the CTB along the nerve (Fig. 4f).
4. Close the skin with clips (Fig. 4g).

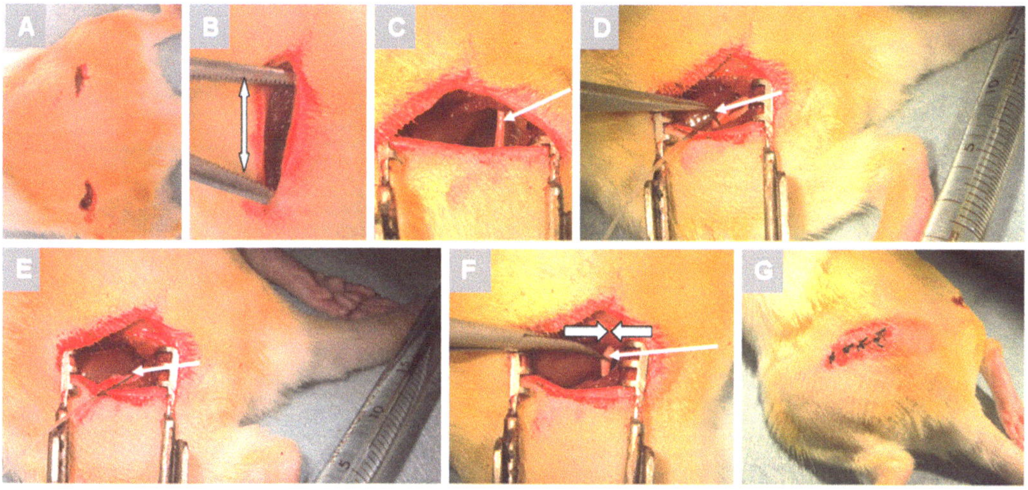

Fig. 4. Procedures for CTB injection into the sciatic nerve. (**a**) Incise the skin. (**b**) Separate the muscle. (**c**) Expose the sciatic nerve. (**d**) Insert the needle into the sciatic nerve. (**e**) Inject the tracer. (**f**) Crush the nerve at the injection site. (**g**) Close the skin.

4.3. Survival Time

The survival time after tracer introduction varies and is dependent on the species and age of the animal, the type and length of the pathway, the type of the tracer, and the purpose and design of the experiment.

4.4. Perfusion and Fixation

After an appropriate survival period, animals are deeply anesthetized with pentobarbital (60 mg/kg, i.p.) and are perfused intracardially with 0.9% saline followed by a solution of 4% PFA in 0.1 M PB (pH 7.4) or by modified Zamboni's fixative (34). After perfusion, the brains or spinal cords are removed and postfixed in a fresh fixative of 4% PFA (for light microscopy) or 1 mm^3 size tissue postfixed in a fixative solution containing 2.5% glutaraldehyde and 5% sucrose in 0.1 M cacodylate buffer (pH 7.4, for electron microscopy) at 4°C for over 24 h (for electron microscopy).

4.5. Tissue Processing

4.5.1. Tissue Preparation for Light Microscopy

1. Transfer the specimen to a solution containing 30% sucrose in 0.1 M phosphate buffer (PB, pH 7.4) for cryoprotection for 4–5 days at 4°C.
2. Block the specimen into segments, cut sections into desired thickness (10–40 μm) and orientation (Fig. 5).

4.5.2. Tissue Preparation for Electron Microscopy

1. Fix tissue in 2.5% glutaraldehyde in 0.1 M sodium cacodylate buffer at 4°C, for a minimum of 24 h. Tissue should be cut into approximately 1 mm cubes for fixing.
2. Wash tissue in 0.1 M sodium cacodylate buffer—1 h × 2 (or overnight at 4°C).
3. Postfix tissue in osmium tetroxide in 0.2 M sodium cacodylate buffer—1 h [mix equal quantities of 2% aqueous osmium tetroxide (OsO4) and 0.4 M sodium cacodylate buffer and use immediately].

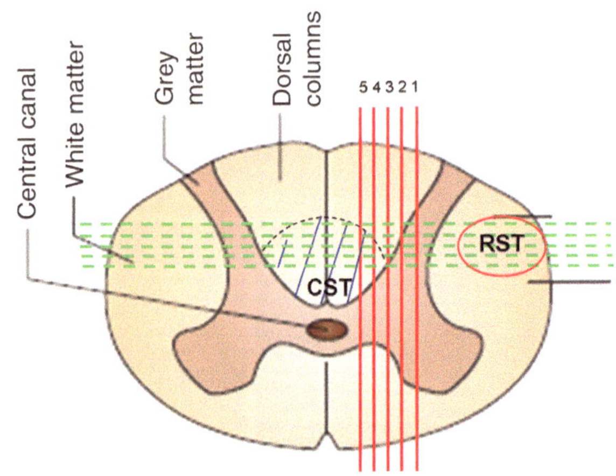

Fig. 5. A specimen, such as the spinal cord, can be cut transversely (shown as a cross-section), para-sagittally (*red lines*), or horizontally (*green lines*) according to the experimental design. Sections can be mounted directly onto glass slides or introduced into buffer solutions for free-floating reactions. *RST* rubrospinal tract, *CST* corticospinal tract.

4. Rinse tissue in 0.2 M sodium cacodylate buffer—5 min × 2.
5. Dehydrate tissue in 70% ethanol—20 min × 2.
6. Dehydrate tissue in 90% ethanol—10 min × 2.
7. Dehydrate tissue in 100% ethanol—20 min × 2.
8. Place tissue into propylene oxide (PO)—40 min × 2.
9. Place tissue into PO/epoxy resin mixture (1:1)—1 h.
10. Place tissue into PO/epoxy resin mixture (1:3)—overnight with caps removed from vials (allows any remaining PO to evaporate).
11. Embed tissue in labeled capsules with freshly prepared *Spurr's epoxy resin.*

    Vinylcyclohexene dioxide (ERL 4206) 20 g.

    Diglycidyl ether of polypropylene glycol (DER 4360) 10 g.

    Nonenyl succinic anhydride (NSA) 52 g.

    DMAE—dimethylaminoethanol (S-1) 0.8 g.
12. Polymerize embed tissue at 60°C for 48 h.

4.6. Detailed Histological Procedures

4.6.1. Direct Observation

1. Coverslip sections (contain fluorescence tracers) with an anti-fading mounting medium, such as Fluoromount (Diagnostic BioSystems, Inc.).
2. View fluorescence labeling of the tracer under a fluorescent microscope.

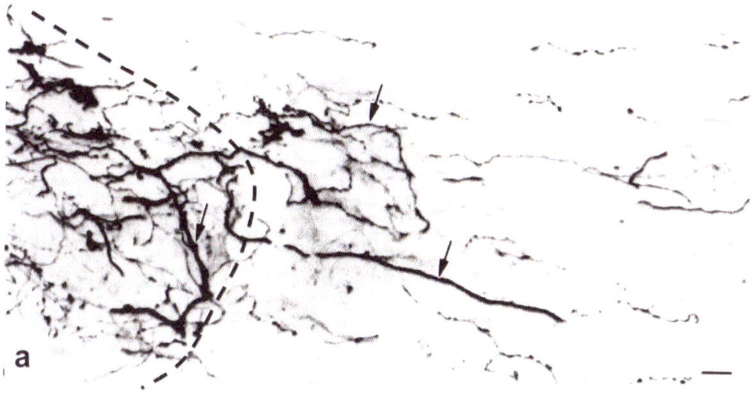

Fig. 6. Anterograde tracing of regenerating axons, labeled with PHA-L (*arrows*), were found to extend through and beyond the graft-host interface (*dashed line*). Scale bar, 10 μm.

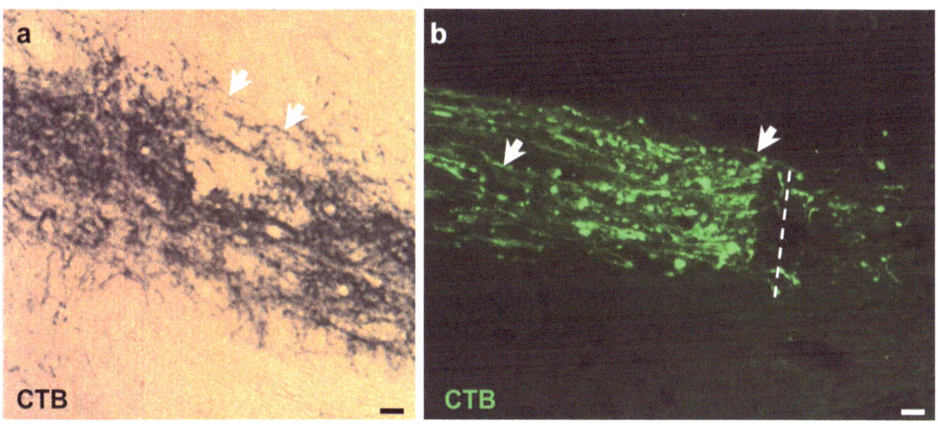

Fig. 7. CTB anterogradely labeled ascending sensory axons (*arrows*) were found in the dorsal column of a normal adult rat spinal cord using immunohistochemistry (**a**) or an adult rat that received a dorsal hemisection at C4 using immunofluorescence staining (**b**). *Dashed line* in (**b**) indicates the lesion site. Scale bar, 50 μm (**a–b**).

4.6.2. Immunohistochemistry with a Primary Antibody and Avidin–Biotin–Peroxidase Kit (Figs. 6 and 7a)

1. Air dry sections for 30–60 min. Circle sections with a PAP pen (Invitrogen, Carlsbad, CA).
2. Transfer slides into 0.01 M PBS buffer for 5 min × 2.
3. Block and prepermeate the sections with 1.5% normal donkey serum (NDS) in 0.01 M PBS containing 0.3% Triton X-100 (PBS-T) for 30 min on orbital shaker at RT.
4. Incubate sections with a rabbit anti-PHA-L primary antibody (1:1,000, in 1.5% NRS and 0.3% Triton X-100 in 0.01 M PBS) or goat anti-CTB (1:10,000, in 1.5% NGS and 0.3% Triton X-100 in 0.01 M PBS) for 48 h at 4°C.
5. Rinse sections with 0.01 M PBS for 10 s × 2; wash sections with 0.01 M PBS for 10 min × 3.

6. Incubate sections with biotinylated donkey anti-rabbit second antibody (1:200, 1.5% NDS in 0.01 M PBS) or donkey anti-goat (1:200, 1.5% NGS in 0.01 M PBS) for 1 h at RT.
7. Rinse sections with PBS for 10 s × 2; wash sections with 0.01 M PBS for 10 min x 3.
8. Incubate sections for 60 min with VECTASTAIN® EliteABC Reagent (allow this reagent to stand for about 30 min before use).

   10 μl A.

   10 μl B.

   2 ml PBS-T.
9. Wash sections for 15 min × 3 in 0.01 M PBS.
10. Incubate sections in peroxidase substrate solution until desired stain intensity develops (10–15 min) VECTASTAIN® EliteABC Reagent.

    1.0 ml dd H_2O.

    10 μl Buffer.

    20 μl DAB.

    10 μl H_2O_2.

    10 μl Nikel (optional).
11. Wash sections in 0.01 M PBS buffer for 5 min × 3.
12. Rinse sections in ddH_2O for 10 s × 2.
13. Dry and mount sections with DPX.

4.6.3. Immunofluorescence Labeling with a Primary Antibody (Fig. 7b)

1. Air dry the sections for 30–60 min. Circle the sections with a PAP pen.
2. Transfer slides into 0.01 M PBS buffer for 5 min × 2.
3. Block and prepermeate the sections with 1.5% NDS and 0.01 M PBS containing 0.3% Triton X-100 (PBS-T) for 30 min on orbital shaker at RT.
4. Incubate sections with a primary antibody (goat anti-CTB as example, 1:10,000, 1.5% NDS in PBS-T) for 48 h at 4°C.
5. Rinse sections with 0.01 M PBS for 10 s two times, wash sections with PBS for 10 min three times.
6. Incubate sections with a fluorescence-conjugated second antibody (FITC donkey anti-goat antibody as an example, 1:200, 1.5% NDS in 0.01 M PBS) for 1 h at RT.
7. Rinse sections with PBS for 10 s two times, wash sections with PBS for 10 min three times.
8. Rinse sections in ddH_2O × 2.
9. Dry and mount sections with a fluorescent mounting medium.

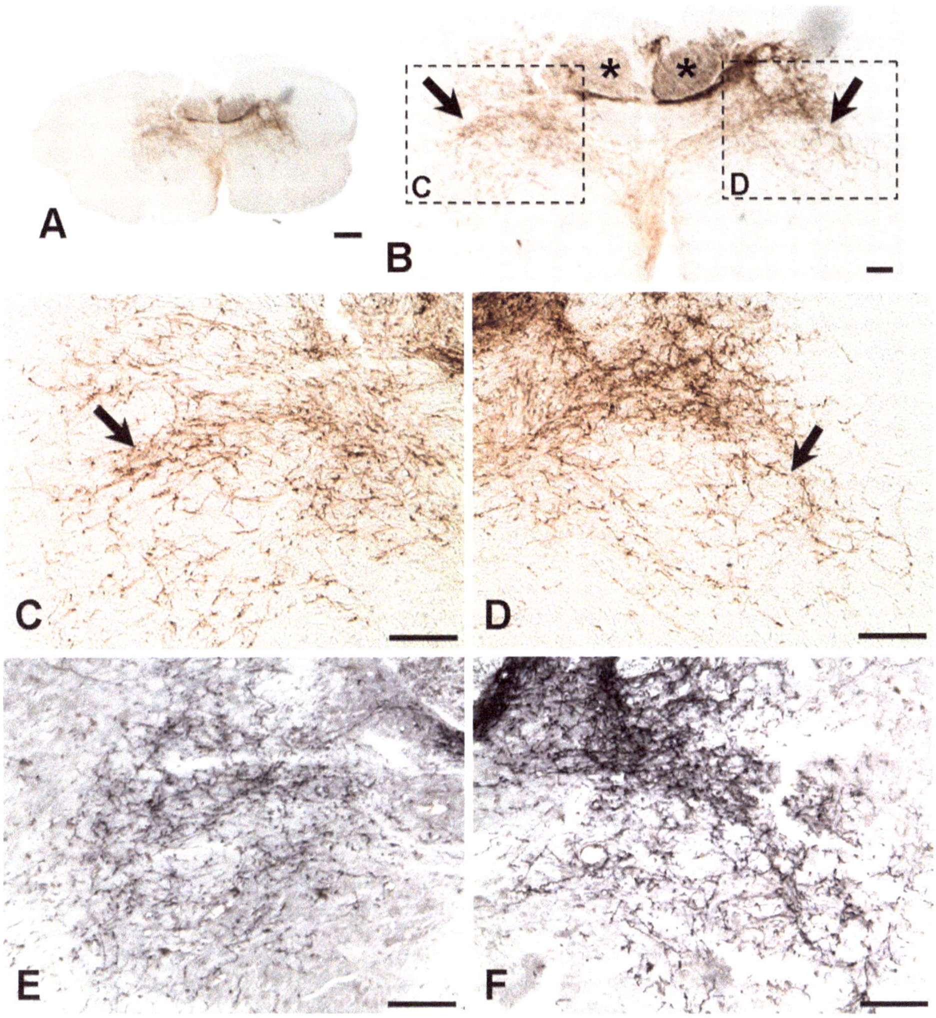

Fig. 8. Representative images of immunohistochemical staining of BDA anterogradely labeled CST axons in the spinal cord using an avidin–biotin–peroxidase (ABC) kit. (**a**) A cross section of the spinal cord at the cord-medullary junction. (**b**) High magnification of (**a**) shows bilateral dorsal CSTs (*asterisk*) and their arborization in the gray matter (*arrows*). (**c**, **d**) High magnification of boxed areas in (**b**) shows BDA staining without Nickel enhancement. (**e**, **f**) Adjacent sections of (**c**) and (**d**) show BDA staining with Nickel enhancement. Reaction was performed in free-floating sections. Scale bars, 250 μm (**a**) and 100 μm (**b**–**f**).

4.6.4. BDA Staining with an Avidin–Biotin–Peroxidase Kit (Fig. 8)

The following steps are carried out on free-floating sections (40 μm thick) under constant agitation at RT unless otherwise specified.

1. Rinse sections in 0.01 M phosphate-buffered saline (pH 7.4 PBS) for 5 min two times.
2. Rinse sections in 0.3% H_2O_2 in 0.01 M PBS for 30 min and then wash in 0.01 M PBS for 5 min × 3.
3. Prepermeate sections in PBS-T (0.01 PBS pH 7.4; 0.3% *v/v* Triton X-100) on orbital shaker for 1 h at RT.

4. Incubate sections with VECTASTAIN® EliteABC Reagent A and B (allow this reagent to stand for about 30 min before use) for 60 min.

   10 μl A.

   10 μl B.

   2 ml PBST.

5. Wash slides in 0.01 M PBS buffer for 15 min × 3.
6. Wash slides in Tris buffer for 10 min × 1.

   Tris–HCl 4.44 g.

   Tris base 2.65 g.

   To 1.0 L H_2O.

7. Incubate sections in peroxidase substrate solution until desired stain intensity develops (10–15 min) VECTASTAIN® EliteABC Reagent.

   1.0 ml dd H_2O.

   10 μl Buffer.

   20 μl DAB.

   10 μl H_2O_2.

   10 μl Nickel (optional).

8. Wash sections in 0.01 M PBS buffer for 5 min × 3.
9. Rinse sections in ddH_2O for 10 s × 2.
10. Dry and mount the sections with DPX.

4.6.5. BDA Immunofluorescence Staining (Figs. 9–13)

The following steps describe immunofluorescence staining of sections (20 μm thick) on slides under constant agitation at RT unless otherwise specified:

1. Air dry the sections for 30–60 min. Circle the sections with a PAP pen.
2. Transfer slides into 0.01 M PBS for 5 min × 2.
3. Wash sections in PBS-T (0.01 M PBS pH 7.4; 0.3% *v*/*v* Triton X-100) for 30 min on an orbital shaker at RT.
4. Incubate sections for 30 min with VECTASTAIN® EliteABC Reagent (allow this reagent to stand for about 30 min before use).

   20 μl A.

   20 μl B.

   1 ml 0.01 M PBS.

   Then, dilute 1:5 in 0.01 M PBS.

5. Wash sections with 0.01 M PBS 5 min × 3.

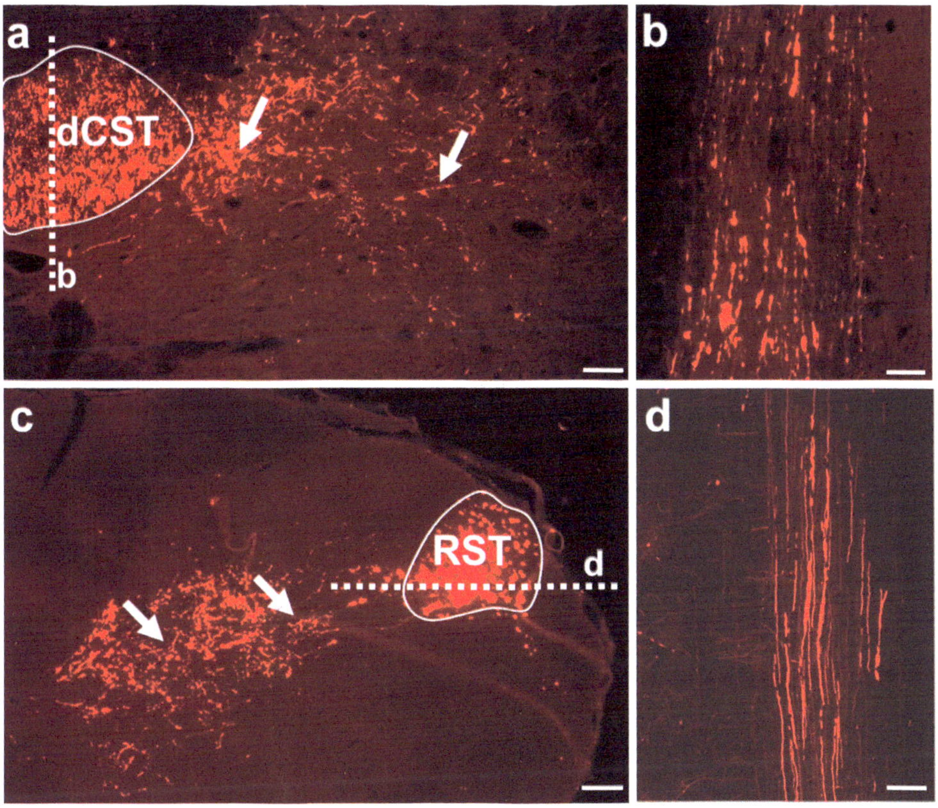

Fig. 9. BDA anterogradely labeled axons in the CST and RST of the spinal cord (**a**, **c**) In cross-sections of the spinal cord at C2, immunofluorescent labeling of BDA was visualized in the dorsal corticospinal tract (**a**, dCST) or rubrospinal tract (**c**, RST) with robust axonal arborization into the gray matter (*arrows*). (**b**, **d**) Longitudinal sections of adjacent segments (*dashed lines* shown in **a**, **c**) show labeled CST (**b**) and RST (**d**) axons in a rostrocaudal direction. Scale bars, 100 μm (**a**–**d**).

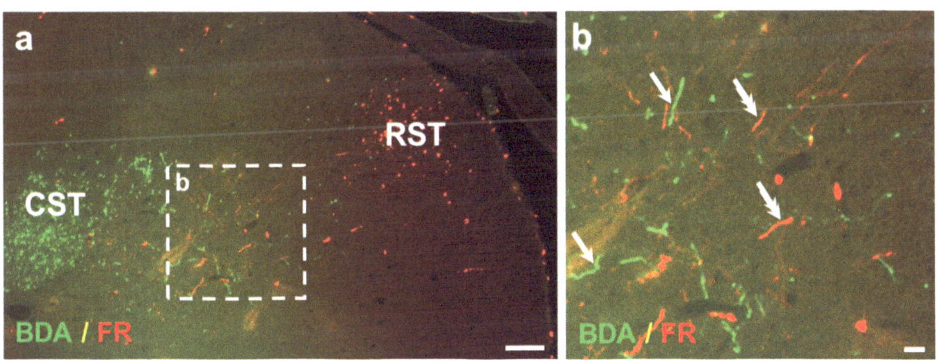

Fig. 10. Projections of anterogradely labeled CST (BDA-labeled; *green*) and RST axons (FR-labeled, *red*) into the gray matter of the C4 spinal cord at 4 weeks after a C5 spinal cord dorsal hemisection. (**a**) Labeled CST and RST were located, respectively, in the ventral portion of the dorsal funiculus (CST) and the dorsal portion of the lateral funiculus (RST). (**b**) High magnification of the *boxed area* in (**a**) clearly shown the overlapping distribution of CST (*arrows*) and RST (*double arrows*) axons in the intermediate region of the spinal cord. Scale bar, 100 μm (**a**) and 50 μm (**b**).

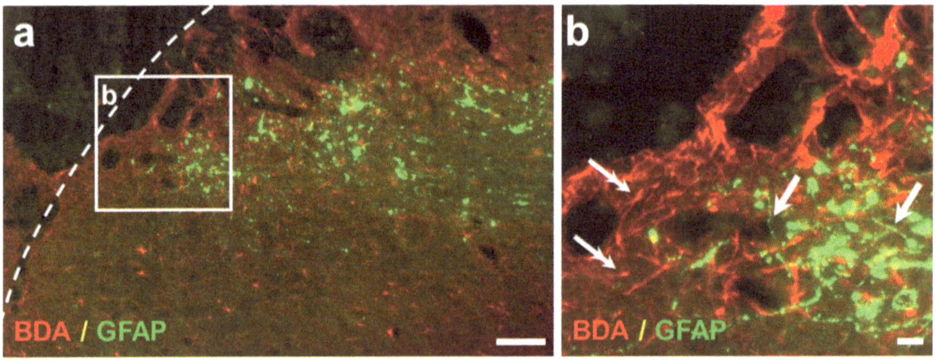

Fig. 11. (**a**) A representative image shows immunofluorescence staining of BDA-positive CST axons (*green*) and GFAP-positive astrocytes (*red*) in an adult rat following a C5 dorsal hemisection. A *dashed line* outlines the lesion border. (**b**) High magnification of *boxed area* in (**a**) shows relationship between BDA-labeled CST axons (*arrows*) and GFPA-labeled reactive astrocytes (*double arrows*) at the lesion border. Scale bar, 100 μm (**a**); 50 μm (**b**).

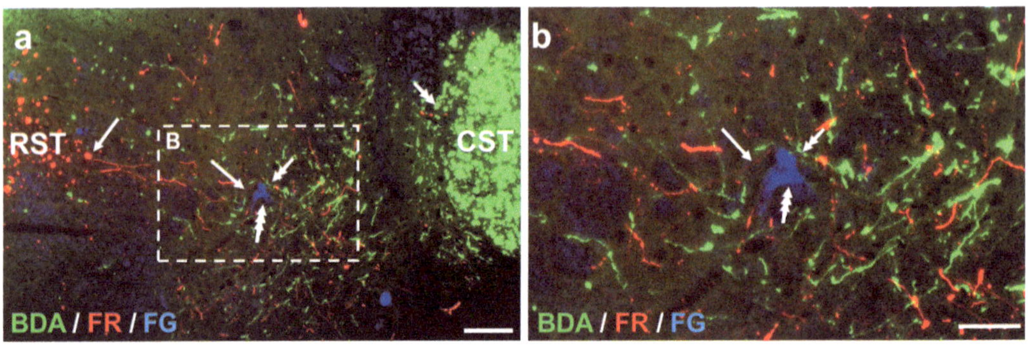

Fig. 12. (**a**) A representative image shows BDA anterogradely labeled CST (*green*), FR anterogradely labeled RST, and FG retrogradely labeled propriospinal neurons (*blue*) in a cross-section of the C2 spinal cord. Note that numerous CST (*double arrow*) and RST (*single arrow*) axons arborized and overlapped in the intermediate gray matter. These axons were in close association with a neuron (*triple arrow*) whose axon projected to the L2 spinal cord, evidenced by the retrograde labeling of FG in the cell body. (**b**) High magnification of boxed are in (**a**) clearly shows labeling of the RST (*red*, *arrow*), CST (*green*, *double arrow*), and a propriospinal neuron (*blue*, *triple arrow*), and their associations. Scale bar, 100 μm (**a**); 50 μm (**b**).

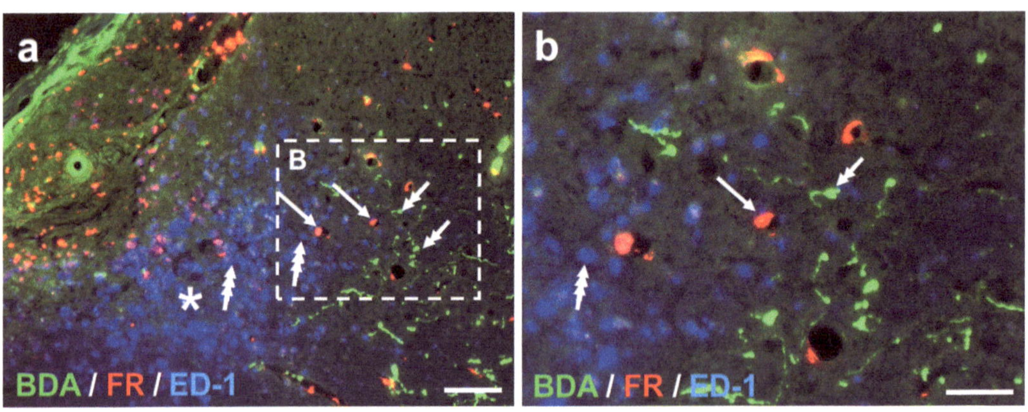

Fig. 13. (**a**) A representative image shows the sparing of FR-labeled RST axons (*red, arrow*), BDA-labeled CST axons (*green, double arrow*), and ED-1-labeled macrophages (*blue, triple arrow*) in a cross-section of the lesion site (asterisk). (**b**). High magnification of in (**a**) further confirms the labeling of FR-positive RST axons (*red, arrow*), BDA-positive CST axons (*green, double arrow*) in association with ED-1-positive macrophages (*blue, triple arrow*). Scale bar, 100 μm (**a**); 50 μm (**b**).

6. Incubate sections in biotinyl tyramide 1:75 diluted in amplication diluent (SAT700, NEN® Life Science Products, Boston, MA) for 10 min.
7. Wash sections with 0.01 M PBS for 5 min × 3.
8. Incubate sections in ExtrAvidin FITC (E2716, Sigma-Aldrich) or ExtrAvidin TRITC (E3011, Sigma-Aldrich) 1:500 in PBS-T + 0.1% NaN_3.
9. Wash sections with 0.01 M PBS for 5 min × 3.
10. Coverslip slides using a fluorescent mounting medium.

5. Quantification of Anterogradely Labeled Axons

The aim of using tract tracing study is not only to reveal desired tracts in the nervous system, but also to provide quantitative data on the changes of density and number after injuries or treatments. Several tracers, such as BDA and CTB, fill axon arbors completely, thus allowing serial reconstruction of axonal arborization. Appropriate software programs are available and they usually require a motorized stage of the microscope, interfaced to a personal computer to record the coordinates of the points which are drawn. The axons labeled are drawn either at the microscope (using a cross in the eyepiece), or at the camera lucida focused on the computer screen or directly on the screen (when the image is acquired by a video camera). Usually, these commercial software programs include appropriate programs for quantitative analysis and for figure editing. Axons, therefore, may be compared in the distance of their die back or regeneration, and in the number and density of remyelination or regeneration (18, 35, 36).

6. Application of Anterograde Tract Tracing in Spinal Cord Injury Research

Modern anterograde axonal tracing strategies can be combined with a variety of other techniques and can therefore be widely applied in the SCI research field. In particular, in vivo labeling with fluorescent tracers allows not only the definition of axonal sorting according to their anatomical connections, but also makes them readily accessible to a wide range of additional approaches. These approaches include (1) prelabeling of an axonal tract for its identification after an injury and/or treatment; (2) multiple labeling strategies using different fluorescent dyes in vivo or in fixed tissues to label different fiber tracts; (3) IHC technique to characterize specific neurotransmitters, cell types, and molecules; (4) retrograde tracing technique to visualize specific neuronal origins; and (5) electron microscopy (EM) for ultrastructural characterization of specific axons and synapses. The following are some application examples.

6.1. Single Labeling

6.1.1. Anterograde Tracing of Regenerating Axons with PHA-L

Phaseolus vulgaris leucoagglutinin was injected into the spinal cord at 6 mm rostral to the a Schwann cell (SC) bridge transplant at 30 days after a mid-thoracic spinal cord hemisection followed by transplantation of an SC-seeded guidance channel into the lesion gap and infusion of neurotrophins brain-derived neurotrophic factor (BDNF) and neurotrophin-3 (NT-3) into the distal host spinal cord. Ten days after injection, PHA-L labeled axons were visualized through and beyond the SC graft (Fig. 6).

6.1.2. Anterograde Tracing of Sensory Axons with CTB

Cholera toxin B subunit (1%) was directly injected into bilateral sciatic nerves of a normal adult rat and an adult rat that received dorsal spinal hemisection at C4. Three days after CTB injection, animals were perfused with 4% PFA. CTB anterogradely labeled ascending sensory axons were visualized in the dorsal column of the spinal cord by immunohistochemistry or immunofluorescence staining (Fig. 7).

6.1.3. Anterograde Tracing of the Corticospinal Tract and Rubrospinal Tract with BDA

BDA (MW 10,000) was injected into the motor cortex or red nucleus of normal adult rats. Two weeks after the injection, BDA was anterogradely transported in the CST or RST of the spinal cord (Figs. 2 and 3). BDA labeled axons can be visualized by immunohistochemistry staining (Fig. 8) or immunofluorescent staining in both cross and longitudinal sections (Fig. 9).

6.2. Double Labeling

6.2.1. Delivering Two Tracers into Two Different Brain Regions to Simultaneously Label CST and RST

BDA (10%) was injected into the motor cortex and FR (10%) was injected into the red nucleus, simultaneously, at 4 weeks after a C5 dorsal hemisection. Animals were perfused, spinal cords removed and fixed for analysis 2 weeks after the tracer injection (Fig. 10).

6.2.2. Anterograde Tracing Combined with other Immunofluorescence Staining

In this study, CST was anterogradely labeled with BDA and the section was immunofluorescently stained for BDA and GFAP, a marker for astrocytes, after SCI. BDA was injected into the motor cortex at 4 weeks after a C5 dorsal hemisection. The animal was perfused with 4% PFA and spinal cord was removed, sectioned, and immunostained with BDA and GFAP (Fig. 11).

6.3. Triple Labeling

6.3.1. Double Anterograde Tracings Combined with a Retrograde Tracing

In this study, double anterograde tracings with BDA and FR were combined with a single retrograde tracing of FG. BDA (10%) was injected into the motor cortex, FR (10%) was injected into the red nucleus, and FG (1%) was injected into the L2 spinal cord of a normal adult rat. The rat was perfused with 4% PFA and spinal cords were removed. Transverse sections of the C2 spinal cord segment were collected and stained for BDA immunostaining and were visualized under a fluorescent microscope (Fig. 12).

6.3.2. Double Anterograde Tracings Combined with an Immunofluorescence Staining

In this study, double anterograde tracings with BDA and FR were combined with an immunofluorescence staining of macrophages/monocytes using an anti-ED-1 antibody in adult rats that received unilateral spinal cord dorsal hemisection at C4. Rats were perfused with 4% PFA and spinal cords were removed. Transverse sections of the lesion site were collected and stained for BDA and then ED-1 immunostaining and were visualized under a fluorescent microscope (Fig. 13).

7. Limitations, Potential Problems and Solutions

1. Although tracers like PHA-L and BDA are excellent anterograde tracers, they also display some retrograde transport (37). The degree of retrograde transport varies from one neuronal system to another. In general, large injection sites result in a more extensive retrograde labeling than smaller injection sites.
2. It is difficult for carboyanine dyes to combine immunohistochemistry because of the need for detergents in most IHC procedures.
3. Although CST collateral branches in the gray matter can be visualized by BDA immunofluorescence staining of sections mounted on slides (Fig. 7), we found that immunohistochemistry using ABC approach can provide greater morphological details of BDA-labeled axons. Additionally, prolonging prepermeating time (to 1 h) and using free floating incubation can achieve more complete and detailed staining of these axonal branches (Fig. 6). However, care should be taken for the free-floating incubation of injured tissues since they are fragile particularly at the injury epicenter.
4. If fluorescent dyes are used for axonal tracing, it is better to transform the fluorescence to a permanent staining product (such as ABC staining) because photo-bleaching can make density quantification inaccuracy. Also it is usually difficult to have a good resolution of their image with the video camera (the sensitivity is usually low for single processes) or the camera lucida.
5. Table 1 is a summary of potential problems, possible reasons and solutions in the application of anterograde tracers modified from (33).

Table 1
Potential problems, possible reasons and solutions in the application of anterograde tracers modified from ref. (33)

Problem	Possible reason	Solution
Labeling not visible or very dim	Tracer conjugates are incompatible with filter sets on the microscope	1. Ensure all filter sets are compatible with the tracers 2. Ensure the UV light source is of good quality to provide enough light intensity for fluorescent labeling 3. Increase the exposure time if having problems capturing photos of the labeling
	Improper animal survival time	1. Determine proper survival time according to previous studies and the purpose of experiment 2. Adjust survival time according to the distance over which transport is required and do time-point experiments in some cases
	Improper injections	1. Ensure the animal is positioned correctly in the stereotaxic device to make sure the coordinates are reliable 2. Increase the injection pressure or the tip diameter if there is a complete absence of dye in the injection sites (or they are very small and faint) due to blood clog in the pipette 3. Slow the injection speed or prolong the needle holding time to avoid tracer leakage
	Improper storage of the tracer	1. Always avoid repeated freeze–thaw cycles of tracers 2. Always avoid the solution stored at 4°C for long time 3. If the solutions have to be stored at −20°C, ensure that they are gradually defrosted 4. Prepare fresh solutions immediately prior to surgeries. Check pH of the solution
	Improper concentrations	1. Increase the tracer concentration if the labeling pattern is very low
	Improper ABC/DAB procedure for visualizing	1. Set up a positive control 2. Check and repeat the procedure
High background fluorescence	Improper perfusion	1. Increase the duration of the initial saline perfusion 2. Ensure that the perfusion begins very quickly once the animal has been anesthetized before any clotting begins
	Improper concentrations	1. Decrease the tracer concentration
Fading of labeling	Photo-bleaching as a result of prolonged exposure	1. Ensure exposure to light at all stages of the procedure is minimized 2. Mount two adjacent series of sections to account for any possible fading: one for viewing the labeling patterns and the other for taking quality pictures
	Photo-bleaching due to prolonged storage	1. View samples and take pictures as soon as possible after staining 2. Coverslip sections with anti-fade medium as quickly as possible 3. Use ABC staining

References

1. Glees P, Meyer A, Meyer M (1946) Terminal degeneration in the frontal cortex of the rabbit following the interruption of afferent fibres. J Anat 80(Pt 2):101–106.3
2. Nauta WJ, Gygax PA (1951) Silver impregnation of degenerating axon terminals in the central nervous system: (1) technic. (2) Chemical notes. Stain Technol 26(1):5–11
3. Cowan WM, Gottlieb DI, Hendrickson AE, Price JL, Woolsey TA (1972) The autoradiographic demonstration of axonal connections in the central nervous system. Brain Res 37(1):21–51
4. Raju DV, Smith Y (2006) Anterograde axonal tract tracing. In: Crawley J, Gerfen C, Rogawski M et al (eds) Current protocols in neuroscience. John Wiley & Sons, Inc., Hoboken, NJ, pp 1.14.1–1.14.11
5. Thanos S, Bonhoeffer F (1983) Investigations on the development and topographic order of retinotectal axons: anterograde and retrograde staining of axons and perikarya with rhodamine in vivo. J Comp Neurol 219(4):420–430
6. Gerfen CR, Sawchenko PE (1984) An anterograde neuroanatomical tracing method that shows the detailed morphology of neurons, their axons and terminals: immunohistochemical localization of an axonally transported plant lectin, *Phaseolus vulgaris* leucoagglutinin (PHA-L). Brain Res 290(2):219–238
7. Glover JC, Petursdottir G, Jansen JK (1986) Fluorescent dextran-amines used as axonal tracers in the nervous system of the chicken embryo. J Neurosci Methods 18(3):243–254
8. Brandt HM, Apkarian AV (1992) Biotin-dextran: a sensitive anterograde tracer for neuroanatomic studies in rat and monkey. J Neurosci Methods 45(1–2):35–40
9. Veenman CL, Reiner A, Honig MG (1992) Biotinylated dextran amine as an anterograde tracer for single- and double-labeling studies. J Neurosci Methods 41(3):239–254
10. Rajakumar N, Elisevich K, Flumerfelt BA (1993) Biotinylated dextran: a versatile anterograde and retrograde neuronal tracer. Brain Res 607(1–2):47–53
11. Ruitenberg MJ, Levison DB, Lee SV, Verhaagen J, Harvey AR, Plant GW (2005) NT-3 expression from engineered olfactory ensheathing glia promotes spinal sparing and regeneration. Brain 128(Pt 4):839–853
12. Kuypers HG, Ugolini G (1990) Viruses as transneuronal tracers. Trends Neurosci 13(2):71–75
13. Dahlstrom A (1971) Axoplasmic transport (with particular respect to adrenergic neurons). Philos Trans R Soc Lond B Biol Sci 261(839):325–358
14. Jeffrey PL, Austin L (1973) Axoplasmic transport. Prog Neurobiol 2(3):207–255
15. Kobbert C, Apps R, Bechmann I, Lanciego JL, Mey J, Thanos S (2000) Current concepts in neuroanatomical tracing. Prog Neurobiol 62(4):327–351
16. Thanos S, Fischer D, Pavlidis M, Heiduschka P, Bodeutsch N (2000) Glioanatomy assessed by cell-cell interactions and phagocytotic labelling. J Neurosci Methods 103(1):39–50
17. Giloh H, Sedat JW (1982) Fluorescence microscopy: reduced photobleaching of rhodamine and fluorescein protein conjugates by n-propyl gallate. Science 217(4566):1252–1255
18. Liu Y, Wang X, Lu CC et al (2008) Repulsive Wnt signaling inhibits axon regeneration after CNS injury. J Neurosci 28(33):8376–8382
19. Deller T, Naumann T, Frotscher M (2000) Retrograde and anterograde tracing combined with transmitter identification and electron microscopy. J Neurosci Methods 103(1):117–126
20. Reiner A, Honig MG (2006) Dextran amines: versatile tools for anterograde and retrograde studies of nervous system connectivity. In: Zaborszky L, Wouterlood FG, Lanciego JL (eds) Neuroanatomical tract tracing 3. Springer, New York, pp 304–335
21. Fritzsch B, Sonntag R (1991) Sequential double labelling with different fluorescent dyes coupled to dextran amines as a tool to estimate the accuracy of tracer application and of regeneration. J Neurosci Methods 39(1):9–17
22. Nance DM, Burns J (1990) Fluorescent dextrans as sensitive anterograde neuroanatomical tracers: applications and pitfalls. Brain Res Bull 25(1):139–145
23. Gimlich RL, Braun J (1985) Improved fluorescent compounds for tracing cell lineage. Dev Biol 109(2):509–514
24. Schmued LC, Albertson C, Slikker W Jr (1997) Fluoro-Jade: a novel fluorochrome for the sensitive and reliable histochemical localization of neuronal degeneration. Brain Res 751(1):37–46
25. Schmued LC, Snavely LF (1993) Photocon version and electron microscopic localization of the fluorescent axon tracer fluoro-ruby (rhodamine-dextran-amine). J Histochem Cyto chem 41(5):777–782
26. Shibata-Iwasaki R, Dekimoto H, Katsuyama Y, Kikkawa S, Terashima T (2007) Anterograde labeling of the corticospinal tract in jimpy mutant mice by Di I injection into the motor cortex. Arch Histol Cytol 70(5):297–301

27. Holmqvist BI, Ostholm T, Ekstrom P (1992) DiI tracing in combination with immunocytochemistry for analysis of connectivities and chemoarchitectonics of specific neural systems in a teleost, the Atlantic salmon. J Neurosci Methods 42(1–2):45–63
28. Cucchiaro JB, Uhlrich DJ (1990) *Phaseolus vulgaris* leucoagglutinin (PHA-L): a neuroanatomical tracer for electron microscopic analysis of synaptic circuitry in the cat's dorsal lateral geniculate nucleus. J Electron Microsc Tech 15(4):352–368
29. Angelucci A, Clasca F, Sur M (1996) Anterograde axonal tracing with the subunit B of cholera toxin: a highly sensitive immunohistochemical protocol for revealing fine axonal morphology in adult and neonatal brains. J Neurosci Methods 65(1):101–112
30. Luppi PH, Sakai K, Salvert D, Fort P, Jouvet M (1987) Peptidergic hypothalamic afferents to the cat nucleus raphe pallidus as revealed by a double immunostaining technique using unconjugated cholera toxin as a retrograde tracer. Brain Res 402(2):339–345
31. Luppi PH, Fort P, Jouvet M (1990) Iontophoretic application of unconjugated cholera toxin B subunit (CTb) combined with immunohistochemistry of neurochemical substances: a method for transmitter identification of retrogradely labeled neurons. Brain Res 534(1–2):209–224
32. Ericson H, Blomqvist A (1988) Tracing of neuronal connections with cholera toxin subunit B: light and electron microscopic immunohistochemistry using monoclonal antibodies. J Neurosci Methods 24(3):225–235
33. Conte WL, Kamishina H, Reep RL (2009) Multiple neuroanatomical tract-tracing using fluorescent Alexa Fluor conjugates of cholera toxin subunit B in rats. Nat Protoc 4(8): 1157–1166
34. Holets VR, Hokfelt T, Ude J et al (1987) A comparative study of the immunohistochemical localization of a presumptive proctolin-like peptide, thyrotropin-releasing hormone and 5-hydroxytryptamine in the rat central nervous system. Brain Res 408(1–2):141–153
35. Bamber NI, Li H, Lu X, Oudega M, Aebischer P, Xu XM (2001) Neurotrophins BDNF and NT-3 promote axonal re-entry into the distal host spinal cord through Schwann cell-seeded mini-channels. Eur J Neurosci 13(2):257–268
36. Hsu JY, Xu XM (2005) Early profiles of axonal growth and astroglial response after spinal cord hemisection and implantation of Schwann cell-seeded guidance channels in adult rats. J Neurosci Res 82(4):472–483
37. Chen S, Aston-Jones G (1998) Axonal collateral-collateral transport of tract tracers in brain neurons: false anterograde labelling and useful tool. Neuroscience 82(4):1151–1163

Chapter 40

Assessments of Gliogenesis After Spinal Cord Injury

Drew L. Sellers and Philip J. Horner

Abstract

Spinal cord injury (SCI) results in a large proliferative response that serves to restore homeostatis and replenish cellular deficits. Postinjury repair and recovery can be interrogated as a function of the cell fates adopted by progenitors within the lesion. Administration of bromodeoxyuridine (BrdU) (by ip injection) and/or a recombinant retrovirus (by intraspinal injection) is used routinely to label progenitors in an injured spinal cord. In combination with immunofluorescence, confocal microscopy assesses the progeny of labeled-progenitors. Colocalization of phenotypic markers with BrdU or retroviral-reporters is used to determine the differentiation profile of progenitors to assess gliogenesis after SCI. This chapter introduces a brief history of labeling proliferative cells, required materials and tools to phenotype progeny, typical procedures, and variations.

Key words: Gliogenesis, Astrocytes, Oligodendrocytes, Microglia, Spinal cord injury, Cell proliferation, Mouse, Cell labeling, Retrovirus, BrdU, Promoter-specific reporter

1. Introduction

Spinal cord injury (SCI) sets in motion processes that react to restore homeostasis and cellular deficits (1–4). Previous research has demonstrated that the adult spinal cord retains a complement of stem and progenitor cells that maintain the potential to produce astrocytes, oligodendrocytes, and neurons when cultured in vitro (5, 6). However, progenitor cells are restricted to produce astrocytes and oligodendrocytes by the instructive niche that exists in the adult spinal cord (7). Therefore, neural stem cells (NSCs) and progenitor populations divide acutely after injury to primarily form astrocytes and myelinating oligodendrocytes (8–13). Numerous studies have examined how NSCs respond and differentiate as a consequence of therapeutic interventions intended to affect functional recovery after SCI. Common methods to mark dividing cells, such as bromodeoxyuridine (BrdU) or retroviral transfection, are used to characterize NSC proliferation after trauma.

Jun Chen et al. (eds.), *Animal Models of Acute Neurological Injuries II*, Springer Protocols Handbooks, DOI 10.1007/978-1-61779-782-8_40,

Consequently, coincident localization of BrdU or retroviral reporter immunofluorescence (IF) with phenotypic markers (or a transgenic reporter) by confocal microscopy is the standard methodology used to assess gliogenesis after SCI.

Originally, BrdU labeling was developed as an alternative approach to determine the proliferative index of tumors (14, 15) and was introduced to study cell proliferation in the developing nervous system by Nowakowski et al. (16) BrdU (a thymidine analog) is incorporated into the DNA of dividing cells during the S-phase of the cell cycle. As such, BrdU is used for birth dating and monitoring cell proliferation. Since BrdU crosses the blood–brain barrier (BBB), cycling cells in the spinal cord can be labeled by intracerebroventricular (icv), intraperitoneal (ip), intravenous (iv) injection, or orally ingestion (via water). While ip injection and oral administration are the most common modes of BrdU-administration (17), BrdU has a short half-life (approximately 10 min in human) (18). As such, establishing a precise cell birth-date is accomplished easily by a single ip injection at a known time-point, however, for sustained labeling, oral delivery may be preferred. As a result, BrdU is used systemically to label all dividing cells at specific time-points after injury, which allows newborn progenitors to be evaluated throughout the entire spinal cord. In spite of its ability to label a large population of dividing progenitors, the BrdU method carries with it an important caveat. Specifically, BrdU generically label all dividing cells, including NSCs, blood-borne progenitors, myofibroblasts, and vascular cells. Therefore, BrdU labeling alone cannot be used to parse discreet populations of dividing cells.

In contrast to the use of systemically delivered BrdU (which labels disparate progenitor populations), phenotypically distinct NSC populations can be identified by retroviral labeling. Retroviral technologies infect only dividing populations and take advantage of gene-specific promoters to control reporter-protein expression within transfected NSCs. Ectopic gene expression (i.e., reporter genes) by promoter-specific retroviruses has been used for decades to trace cell lineages (19–21). Due to the large proliferative responses at sites of SCI, replication-deficient recombinant retrovirus can be used to label mitotically active cells/progenitors. The ability to manipulate a retroviral-genome permits retroviruses to be engineered with a generic promoter (i.e., CMV), or a lineage specific-promoter to drive expression of a reporter protein (i.e., EGFP) to label alls proliferative cells or a subset of progenitors, respectively. Consequently, the lineage and fate-choice of a progenitor's progeny can be determined without dilution. Additionally, EGFP diffuses throughout the cytoplasm and permits a cellular morphology to be interrogated and co-labeled by phenotypic markers (discussed in more detail below). As a result, retroviruses can be engineered to permit NSCs to be investigated by subtype to determine how specific progenitor populations contribute to the gliogenesis after SCI.

The following sections briefly outline the materials to phenotype cells by immunofluorescence that proliferates in the spinal cord after traumatic injury: how to prepare, immunostain, and evaluate confocal micrographs to count, tally, and report gliogenesis within a spinal cord lesion.

2. Materials and Instruments

2.1. IF Solutions

To perform immunofluorescent stains, the following materials are needed:

1. PBS.
2. 10% Triton-X 100 stock (in water).
3. 10% Serum (serotype appropriate for secondary antibodies; i.e., if antibodies are derived from Donkey, Donkey Serum Albumin, Jackson Labs).
   In addition, to mount coverslips onto slides and attenuate photo-bleaching:
4. Gelvitol (DABCO antifade).

2.2. Spinal Cord Sectioning Instruments

To perform the sectioning, the following equipment is required:

1. Rat or mouse spinal cord slicer matrix, with 1.0 mm coronal slice intervals, and razor blades (Fig. 1).
2. Tissue blocking mold.
3. Tissue-Tek optimal cutting temperature (OCT) compound.
4. Isopentane bath.

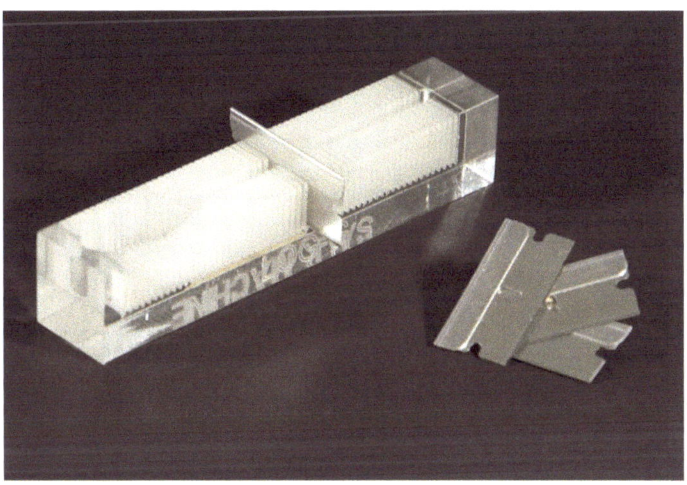

Fig. 1. A spinal cord matrix (*left*) and a razor blade (*right*).

5. Liquid nitrogen.
6. Cryostat.

2.3. Staining Equipment

1. Light-tight humidified chamber (slide-box with wetted paper in the bottom).
2. Orbital shaker.

2.4. Confocal Microscope and Processing Tools

1. Confocal microscope with Krypton-Argon laser (additional UV laser and Far Red diode/laser is optional to allow addition markers to be evaluated on a single slice).
2. Image software that allow *z*-stack images to be visualized and analyzed (ImageJ, Volocity, Microbrightfield, MetaMorh, and many others).
3. Fluorescent microscope with motorized stage, camera, and computer with stereology software (e.g., MicroBrightfield, Zeiss, MCID).

3. Procedures

3.1. Preparation of TTC Solution

1. Dissolve TTC in normal saline at a concentration of 2%.
2. Pour about 25 mL 2% TTC solution into a Petri dish or a small beaker.
3. Prewarm TTC solution to 37°C in water bath.

3.2. Spinal Cord Sectioning

1. At a specific time-point after injury, exsanguinate each animal by transcardiac perfusion with saline and then fix the tissue with 4% paraformaldehyde (PFA). Remove the entire spinal column and postfix (4% PFA) for 12–24 h at 4°C.
2. Carefully dissect spinal cord from the spinal column, and cyro-protect by equilibration in 30% sucrose (step wise from 10, 20, and 30% for 24 h, each, at 4°C).
3. Place spinal cord into the matrix-block, dorsal side up.
4. Align the spinal cord longitudinally within the matrix-block, dorsal side up.
5. Slice spinal cord with wet blades at intervals of 1 mm (generally 4–6 slices).
6. Lift spinal cord slices and blades (one by one) off the matrix, mind the orientation of each slice carefully and transfer to tissue blocking mold (rostral face down), repeat for each slice to form a 2×2 (2×3, *x*/*y*-plane) tissue slice-matrix: rostral faces down, in the *z*-plane; dorsal aspects up, in the *y*-plane.
7. Fill mold with OCT and deep freeze by floating the mold on liquid nitrogen-chilled isopentane. Carefully monitor the

proximity of the isopentane bath to the liquid nitrogen. One wants to ensure the coldest temperature possible without freezing the isopentane.

8. Once the OCT solidifies, transfer to −80°C for storage until cyrosectioning.
9. Cryosection the tissues, 10–20 µm sections, onto superfrost slides in a 1-in-6 series (i.e., section 1 onto slide 01, section 2 onto slide 02 on through to section 6 onto slide 06; repeat, where section 7 goes onto slide 01 until the capacity of the slides is reached. Move onto the next six slides and repeat until the tissue block is expired.) Slides are stored at −80°C until analysis.

3.3. Immunofluorescent Staining

1. Obtain a set of 1-in-6 slide (i.e., slide 01, 07, 13, etc.) for each animal to be analyzed. The frequency of 1-in-6 has been empirically found to be a good starting point for analyzing dividing cells in mouse and rat SCI in our lab. Nevertheless, the frequency of sampling may need to be optimized for the type of lesion or region to be analyzed.
2. Rinse OTC away in PBS (three rinses, 10 min each).
3. *For BrdU-labeled tissue (only)*: Incubate slide in prewarmed 2 N HCl for 30 min at 37°C. Rinse slides 2–3× in PBS before neutralizing in Borate buffer (10 mM, pH 8.3), 30 min at room temperature then rinse 3×, 10–15 min, with PBS.
4. Block tissue in PBS, 0.1% Triton-X 100, 2% donkey serum (ensure serum used is specific to antibodies used for secondary detection to minimize background).
5. Dilute primary antibodies, accordingly, in fresh blocking media.
6. Overlay primary antibody diluate onto tissue sections, in a humidified chamber and incubate at 4°C overnight with gentle shaking (orbital shaker).
7. Rinse slides 3–5 times in PBS (10–15 min each).
8. Dilute fluorescently conjugated secondary antibodies, accordingly, in blocking solution and incubate 1–2 h with gentle shaking (orbital shaker) at room temperature (protect from light to prevent photobleaching the secondary antibodies).
9. Rinse slides 3–5 times PBS (20–30 min each).
10. Mount coverslips with DABCO to attenuate photobleaching.

3.4. Confocal Microscopy to Interrogate Cell Phenotype

1. Image sections and acquire *z*-stack images (via an oil immersion, ×40 objective, NA >1.1) of spinal regions containing labeled progenitors.
2. Optimize laser settings to ensure optical fidelity and resolution of individual optical sections (planes 1 µm, preferable).

3. Repeat imaging at 3–5 randomly chosen locations within the targeted region for each tissue section within the complete series.
4. Analyze optical sections of each *z*-stack to determine which molecular markers colocalize onto each labeled-progenitor, maintain a tally for each marker, and analyze each section for all animals.

3.5. Determining the Differentiation Profile of Progenitors

1. Determine the "panel" of molecular markers (Table 1) required to analyze the cellular phenotypes of interest.
2. Repeat the immunofluorescent staining for each marker of the panel, on each animal.
3. Quantify the number of progenitors within each random field of view.
4. Identify and maintain an animal-tally for each progenitor that shows colocalization (Fig. 2) with specific molecular markers (a minimum acceptable sampling is 100 progenitors per animal per immunostain) (22).
5. The differentiation profile (for each phenotype, of each animal) is expressed as a percent of the whole population.

3.6. Presenting the Differentiation Profiles

The target area can be presented in any one of the three ways:

1. Lesion area of each slice.
2. Lesion ratio of each slice: lesion area divided by the area of contralateral hemisphere.
3. Total lesion volume: the sum of the lesion volumes of all slices.

Table 1
General antibody palette to characterize gliogenesis after SCI

Antibody	Target	Source (species)	Dilution
APC/CC1	Immature oligodendrocytes	Calbiochem (mouse)	1:200
Iba1	Microglia	Wako (rabbit)	1:500
GFAP	Astrocytes	Adv. immuno (guinea pig)	1:1,500
NG2	Progenitors cells	Chemicon (rabbit)	1:500
OX-42	Macrophages	Serotec (mouse)	1:250
s100B	Astrocytes	S Want (rabbit)	1:2,000
Vimentin	Reactive astrocytes	Chemicon (chicken)	1:1,000
BrdU	Proliferating cells	Novus biological (rat)	1:200
CSPG	Chondroitin proteoglycan (gliotic scar)	Sigma (mouse)	1:200

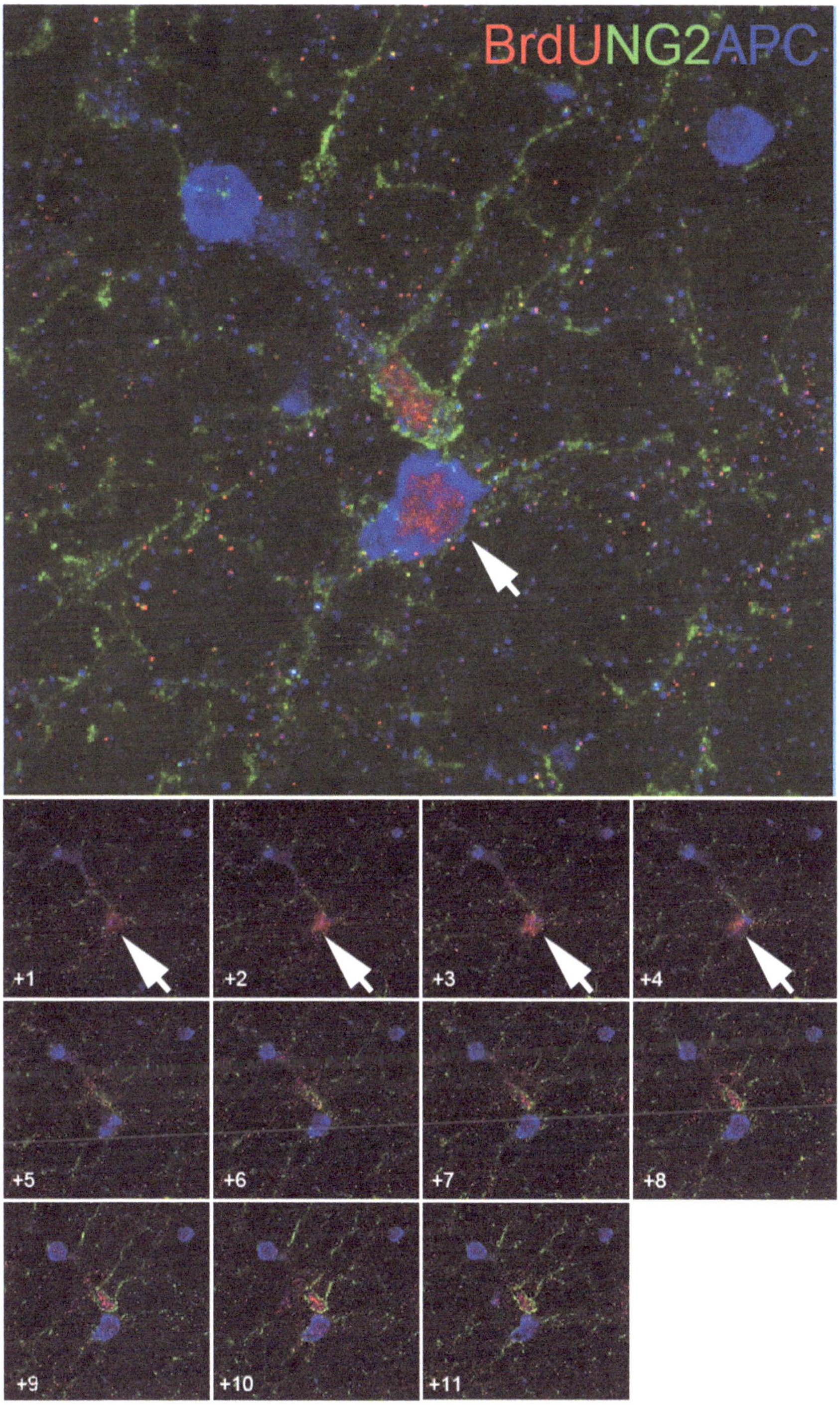

Fig. 2. Representative confocal image. A projected-stack of bromodeoxyuridine (BrdU)-labeled tissue stained for BrdU, NG2, and APC. Tissues section is imaged on a BioRad-Nikon TE3000 confocal microscope. The individual *z*-sections of the stack are displayed as 1 μm optical sections through an 11 μm *z*-series stack.

4. Labeling Mitotically Active Cells

4.1. Bromodeoxyuridine Labeling

BrdU can be used to label dividing cells via intraperitoneal injection (dose: 50 mg/kg). As such, progenitors can be labeled at single or multiple time points (after injury) in accordance with the injection regime.

4.2. Retroviral-Mediated Transduction to Label Dividing Progenitors

Ectopic gene expression by promoter-specific and generic retroviruses has been used for decades to trace cell lineages (19–21). Since retroviruses infect mitotically active cells only, expression of the reporter gene (i.e., EGFP) will mark a progenitor's cytoplasm. However, viral particle diffusion is limited; thus, intraspinal injections (1–3 μL) of high-titer promoter-specific reporter-virus ($\geq 1 \times 10^9$ infection particles per microliter) will label progenitors within a restricted zone (~1 mm around the injection site).

4.3. Unbiased Quantification of Dividing Cells

To determine the total number of cells per target region an unbiased sampling technique must be used. The effects of tissue swelling due to edema and shrinkage due to degeneration can lead to large changes in the target area and dramatically influence the predicated number of proliferating cells over time. Hence, the use of an unbiased or stereological method to quantify both target volume and proliferating cell number is recommended.

1. Using a stereology program to define the target area (sometimes referred to the template or reference area) to be analyzed (e.g., lesion, spared white or gray matter). This can be done with an automate system, such as StereoInvestigator by MicroBrightfield Inc. or by interrogating images, manually on a computer screen that have been collected using a grid system.
2. In thick sections (>20 μm), the use of optical fractionator is recommended at a magnification of ×40 or greater.

4.4. Reporting the Number of Dividing Cells and Lineages

1. Dividing cell density can be calculated by dividing the number of cells counted per area of the template region (Fig. 3).
2. Cell density per tracing area can also be extrapolated to the number of cells per cubic millimeter of tissue.
3. A labeling index is calculated by taking the total number of BrdU-immunoreactive cells per region and dividing it by the total number of DAPI-labeled nuclei from the corresponding template region.
4. The number of dividing cells of a particular phenotype can be estimated by taking the percent labeling of each lineage from section 3 and multiplying it by the BrdU or retroviral density from the Stereology.

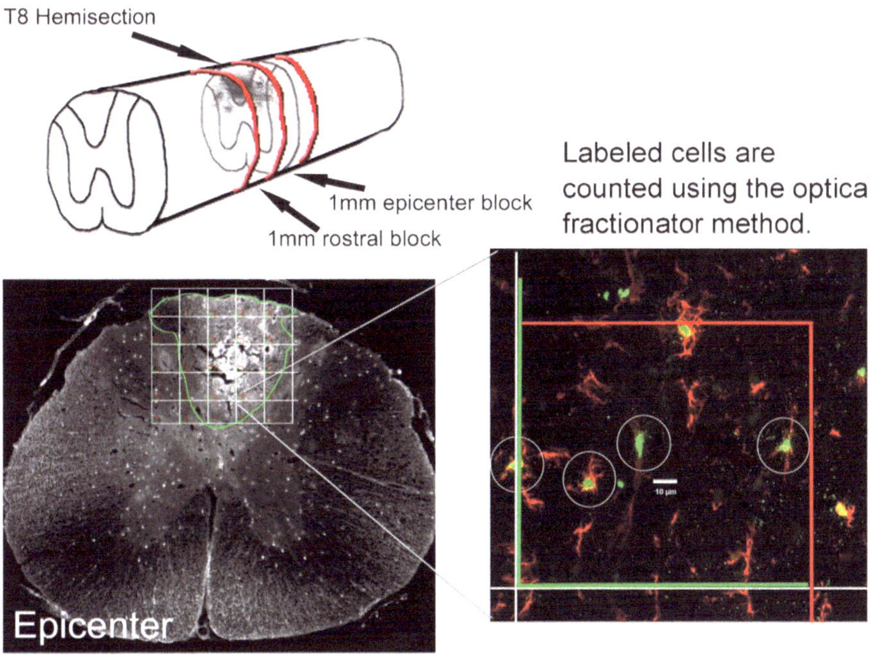

Fig. 3. Stereological analysis of BrdU or retroviral labeling. Shows a schematic representation of the lesion and tissue blocking procedure (described above). In this example, analysis is focused on the dorsal columns and a 1-in-6 series was taken from 1 mm coronal blocks at the epicenter and 1 mm rostral and caudal. A stereological analysis using the optical fractionator method is used to count individual dividing cells (*green* nuclei within each *white circle*) within the counting-frame (bound by *red* and *green* arms; cells intersected by the *green* arm are counted, whereas cells that fall along the red arm are not counted).

4.5. Technical Limitations to Consider When Deciding Between BrdU and Retroviral Labeling Paradigms

Despite the ease of administration and the potential to label a greater number of progenitors, BrdU-labeling possess significant analytical challenges when trying to colocalize BrdU-immunofluorescence with cytoplasmic and cell surface phenotype-markers. Since, BrdU is restricted to label DNA within the nucleus, the distribution of BrdU-immunofluorescence must be correlated via proximal localization with nonnuclear markers. As a result, the accuracy (by which phenotype is determined) can be skewed because one must resolve individual cytoplasmic/cell surface-markers to attribute localization with BrdU-staining via proximity. Additionally, BrdU-detection requires tissues to be treated with 2 N HCl to allow BrdU antibody-epitope recognition. Consequently, each IF stain must be optimized to preserve the cytoplasmic and cell surface marker-epitopes to permit phenotype analysis. Finally, BrdU is serially diluted each time a cell divides. In a highly proliferative population, signal dilution can occur and it may be difficult to assess whether cells have died leading to the loss of BrdU signal or if they have divided repeatedly and dilute the signal.

4.6. Technical Limitations for Retroviral Labeling Paradigms

Since efficacy is limited by dwell time, of the viral particles within a restricted zone (injection site), retroviruses require a high titer ($\geq 10^9$ particles per microliter). Additionally, retroviral administration is limited to 2–3 μL to avoid secondary damage to the tissue and requires a second surgery to administer. In conjunction, intraspinal injections should be preformed with a pulled glass-micropipette to minimize secondary gliosis that occurs with metallic needles. Retroviruses are also limited by the stringent regulation the promoter, which regulates expression of the reporter gene. Low promoter activity, or short-lived promoter activity, of a cell lineage can reduce the sensitivity of this system due to insufficient expression of the reporter. Therefore, combining retroviral reporters with transgenic mice can mitigate these sensitivity issues. For example, the use of a retroviral-delivered Cre recombinase, under a lineage-specific promoter, to a ROSA reporter mouse can be used as a promoter-trap to label cells that have divided permanently. This model system not only increases sensitivity, but also allows one to assess the final differentiation of a lineage by morphology.

References

1. Bunge RP, Puckett WR, Hiester ED (1997) Observations on the pathology of several types of human spinal cord injury, with emphasis on the astrocyte response to penetrating injuries. Adv Neurol 72:305
2. Reier PJ, Houle JD (1988) The glial scar: its bearing on axonal elongation and transplantation approaches to CNS repair. Adv Neurol 47:87
3. Fawcett JW, Asher RA (1999) The glial scar and central nervous system repair. Brain Res Bull 49:377
4. Fitch MT, Silver J (1997) Glial cell extracellular matrix: boundaries for axonal growth in development and regeneration. Cell Tissue Res 290: 379
5. Shihabuddin LS, Ray J, Gage FH (1997) FGF-2 is sufficient to isolate progenitors found in the adult mammalian spinal cord. Exp Neurol 148: 577
6. Weiss S et al (1996) Multipotent CNS stem cells are present in the adult mammalian spinal cord and ventricular neuroaxis. J Neurosci 16:7599
7. Shihabuddin LS, Horner PJ, Ray J, Gage FH (2000) Adult spinal cord stem cells generate neurons after transplantation in the adult dentate gyrus. J Neurosci 20:8727
8. Yamamoto S, Yamamoto N, Kitamura T, Nakamura K, Nakafuku M (2001) Proliferation of parenchymal neural progenitors in response to injury in the adult rat spinal cord. Exp Neurol 172:115
9. Frisen J, Johansson CB, Torok C, Risling M, Lendahl U (1995) Rapid, widespread, and longlasting induction of nestin contributes to the generation of glial scar tissue after CNS injury. J Cell Biol 131:453
10. McTigue DM, Wei P, Stokes BT (2001) Proliferation of NG2-positive cells and altered oligodendrocyte numbers in the contused rat spinal cord. J Neurosci 21:3392
11. Levine JM (1994) Increased expression of the NG2 chondroitin-sulfate proteoglycan after brain injury. J Neurosci 14:4716
12. Horky LL, Galimi F, Gage FH, Horner PJ (2006) Fate of endogenous stem/progenitor cells following spinal cord injury. J Comp Neurol 498:525
13. Sellers DL, Maris DO, Horner PJ (2009) Postinjury niches induce temporal shifts in progenitor fates to direct lesion repair after spinal cord injury. J Neurosci 29:6722
14. Hoshino T et al (1989) Prognostic implications of the bromodeoxyuridine labeling index of human gliomas. J Neurosurg 71:335
15. Struikmans H et al (1997) S-phase fraction, 5-bromo-2′-deoxy-uridine labelling index, duration of S-phase, potential doubling time, and DNA index in benign and malignant brain tumors. Radiat Oncol Investig 5:170
16. Nowakowski RS, Lewin SB, Miller MW (1989) Bromodeoxyuridine immunohistochemical determination of the lengths of the cell cycle and

the DNA-synthetic phase for an anatomically defined population. J Neurocytol 18:311

17. Kuhn HG, Dickinson-Anson H, Gage FH (1996) Neurogenesis in the dentate gyrus of the adult rat: age-related decrease of neuronal progenitor proliferation. J Neurosci 16:2027
18. Kriss JP, Maruyama Y, Tung LA, Bond SB, Revesz L (1963) The fate of 5-bromodeoxyuridine, 5-bromodeoxycytidine, and 5-iododeoxycytidine in man. Cancer Res 23:260
19. Brand AH, Perrimon N (1993) Targeted gene expression as a means of altering cell fates and generating dominant phenotypes. Development 118:401–415
20. Cepko CL et al (1998) Lineage analysis using retroviral vectors. Methods 14:393
21. Golden JA, Fields-Berry SC, Cepko CL (1995) Construction and characterization of a highly complex retroviral library for lineage analysis. Proc Natl Acad Sci U S A 92:5704–5708
22. Mouton PR (2002) Principles and practices of unbiased stereology: an introduction for bioscientists. The John Hopkins University Press, Baltimore, p 214

Chapter 41

Assessing Microvessels After Spinal Cord Injury

Theo Hagg, Richard L. Benton, Janelle M. Fassbender, and Scott R. Whittemore

Abstract

This chapter focuses on the assessment of spinal cord microvessels which are lost at the injury epicenter within 24 h after spinal cord contusion or compression in adult rats and mice. The penumbral blood vessels undergo angiogenesis during the first and second week possibly contributing to recovery. Rescue of blood vessels results in better tissue preservation and functional outcomes. We describe various markers, intravenous labeling techniques, quantitative histological measurements, permeability assays, collection of purified microvessels, genetic mouse models, and endothelial cell culture methods to assess vascular status after spinal cord injury.

Key words: Angiogenesis, Cell death, Degeneration, Endothelial cell, Histological, Intravenous, Mouse, Permeability, Rat, Spinal cord injury

1. Introduction

The capillaries of the central nervous system (CNS) are part of the neurovascular unit (NVU) which comprises endothelial cells (ECs), mural cells or pericytes, and astrocyte end feet (1). Together, tight junctions among the ECs, astrocytic end feet, and the basement membrane around the ECs form the blood–brain barrier (BBB) and the blood–spinal cord barrier (BSCB). Perhaps because ECs have a high glucose transport rate and a high density of mitochondria, they are much more vulnerable to traumatic injury and ischemia than their peripheral counterparts. For example, following contusive spinal cord injury (SCI), ECs start to degenerate within 30 min, lipid peroxidation of ECs occurs (2), and most microvessels are lost at the epicenter over the first postinjury day (3–5). The epicenter is also characterized by reduced microvascular perfusion

Jun Chen et al. (eds.), *Animal Models of Acute Neurological Injuries II*, Springer Protocols Handbooks, DOI 10.1007/978-1-61779-782-8_41, © Springer Science+Business Media, LLC 2012

and hemorrhage starting with small petechial bleeding soon after the injury. Surviving blood vessels in the penumbra are dysfunctional as noted by reduced glucose transporter-1, disruption of tight junctions, and pathological permeability (5–7). These vascular changes contribute to ischemia and edema, and the injured ECs mediate infiltration of detrimental leukocytes in the spinal cord (8). These pathological changes are thought to lead to progressive tissue loss at the injury epicenter in experimental animal models as well as in humans (9–11).

Vascular protection as a route to neuroprotection has been recognized as a viable therapeutic strategy. Treatment with nimodipine plus systemic pressors maintains blood flow and axonal conduction in acutely injured rats (12, 13). The steroid tirilazad, antioxidants, or selenium maintain blood flow and reduce permeability following SCI (14–16). The matrix metalloproteinase (MMP) inhibitor GM6001 reduces vascular dysfunction, white matter loss, and locomotor dysfunction in mice (17). Inhibition of transient receptor potential melastatin 4 (Trpm4) and K_{ATP}-sensitive potassium channel subunit SUR1, which are expressed by damaged ECs, reduces microvascular degeneration, tissue loss, and functional deficits (18, 19). Intravenous treatment with the angiopoietin-1 (Ang-1) ligand of the EC-selective Tie2 receptor in combination with an agonist peptide of the αvβ3 integrin is efficacious in rescuing microvessels and greatly improving functional outcomes (20).

Of interest to repair strategies is the finding that at 7–14 days following SCI in rodents, surviving blood vessels in the penumbra undergo angiogenesis, which is transitory in rats (3, 5). Such angiogenic microvessels can readily be detected by markers for proliferation and cell migration, such as MT1-MMP, as well as by a disintegrin and metalloproteinase family member 8 (ADAM8) (21). Our results suggest that the new vessels are beneficial (20), but this has yet to be determined. New microvessels might be involved in natural repair processes, including collateral perfusion and possibly promotion of sprouting (22) or being pathways for growing axons (3).

One problem with rodent models, particularly mice, is the small size of the spinal cord, which makes it difficult to measure vascular perfusion and function in real time with high resolution. Mice have the advantage of genetic models and hopefully existing technologies will be improved to suit their dimensions. Such techniques include noninvasive in vivo blood flow measurements using ultrasound, laser Doppler, laser speckle imaging, near-infrared spectroscopy, micro CT, PET scanning with EC-selective ligands, magnetic resonance imaging (MRI), real-time multiphoton microscopy, optical frequency domain imaging, and luminosity measurements with fluorescent probes (23–30). This chapter describes predominantly postmortem assessments.

2. Microvascular Markers

Table 1 presents the blood vessel markers that we have used in our histological analyses for studies into mechanisms and treatments of SCI. For additional markers, see ref. (24).

2.1. Tissue Fixation

Proper fixation is crucial to preserving the morphology of the antigens of interest. Attempting to detect a labile antigen after a harsh fixation increases background or decreases detection due to loss of antigenicity. Transcardial perfusion with 4% paraformaldehyde followed by immersion in fixative overnight results in a thorough fixation that adequately preserves tissue architecture. However, this is too harsh for the EC antigens, PECAM and ZO-1, and the majority of the inflammatory cell antigens. For these and similar labile antigens, transcardial perfusion with 0.1 M PBS and immediate freezing following fresh dissection of the cord preserve their antigenicity. The fresh frozen spinal cord can be sectioned on a cryostat without cryoprotection, and stored at −20°C. For antibody staining, slides should be warmed at 37°C for at least 30 min followed by fixation with cold methanol for 10 min, a protocol which is sufficient to fix tissue on the slide. Blocking nonspecific staining can be achieved by adding 10% serum (use the same species as the secondary antibody) in 0.1 M Tris-buffered saline (TBS) with 0.1% Triton X-100 and 0.5% bovine serum albumin (BSA). Incubate at room temperature for a minimum of 1 h, with further reduction of background achieved by increasing incubation times to 3 h at room temperature or overnight at 4°C. See ref. (31) for a description of appropriate negative controls.

After collection of fixed spinal cords, they are cryoprotected in 30% sucrose in 0.1 M phosphate-buffered saline (pH 7.4) overnight. We typically embed multiple spinal cords in TissueTek (Sakura Finetek, Torrance, CA) and freeze the block at −20°C for cutting on a cryostat. Alternatively, fresh frozen cords can be directly embedded in the freezing medium. To do this, 11-mm-long segments of spinal cords including the injury site are laid flat on the bottom of a plastic embedding container. The epicenters of all segments are aligned so that the same region is collected during subsequent cutting. It helps to mark the epicenter with a Sharpie before dissecting out the spinal cord from the bony spinal column. Each block should have at least one animal from each experimental group to ensure equal staining conditions. Next, several series of sections are cut, typically at 20 μm, and thaw mounted directly on charged glass microscope slides. For transverse (cross-sectional) sections, we collect every second section in a segment spanning from 5 mm rostral to 5 mm caudal from the visible epicenter. The slide containing the true injury epicenter is first determined for each animal using eriochrome cyanine staining for myelin as noted

Table 1
Blood vessel markers used in histological analyses for studies into mechanisms and treatments of SCI

Markers	Comments	Ab type/species	Company/catalog#	Dilution for FL	Injury	Angiogenesis
Endothelial cells						
Platelet endothelial cell adhesion molecule (PECAM-1) or CD31	General marker	m/r	BD Pharmingen 550274 clone MEC13.3	50–1,000	Down	Up
von Willebrand factor/factor VIII (vWF)	Clotting competency	p/rb	Dako A0082	200	Down	
Glucose transporter type 1 (Glut-1)	Metabolic potential	p/rb	Millipore AB1340	100	Down	
Occludin	Mature BBB	p/rb	Zymed clone Z-T22	50	Down	
ZO-1	Tight junction	p/rb	Zymed clone ZR1	50	Down	
Claudin-5	Tight junction	p/rb	Zymed clone Z43.JK	50	Down	
Rat endothelial cell antigen (RECA-1)	Rat-specific general marker	m/mo	Serotec MCA970 clone HIS52	25	Down	
Endothelial barrier antigen	Rat-specific mature BBB	m/mo IgM	Sternberger Monoclonals SMI71	100	Down	
Laminin-1 (LN)	Basement membrane	p/rb	Sigma L9393	100–300	Down	Up
Collagen IV	Basement membrane	p/rb	Biodesign International T40263R	100	Down	Up
ADAM8		p/rb	Millipore AB19017		Up	Up
α1β1 Integrin		m/mo	Millipore MAB2513		Up	Up
αv/β3 Integrin		p/rb	Millipore AB1930/AB1932	1,000		
α5β1 Integrin		m/r	Millipore MAB1984	100		
MMP14/MT1-MMP	Growth tips	p/rb	Millipore AB8104	200		Up
Injected BrdU	Proliferating cells	m/mo	Millipore MAB3510	30,000		Up
		p/sh	Biodesign	100		Up
Ki67	Proliferating cells	p/rb	Novocastra Laboratories NCL-Ki67p	500		Up
Apoptosis-inducing factor	Apoptosis	p/rb	Millipore AB16501	100	Up	
TUNEL	Apoptosis	ApopTag kit	Millipore S7160 or S7165		Up	

Pericytes/mural cells						
Desmin	Also smooth muscle cells	p/rb	Millipore AB907	50		
NG2	Activated pericytes (and other cells)	p/rb	Millipore AB5320	50	Up	Up
Astrocytes						
GFAP		m/mo	Millipore MAB3402	1,000	Up	
		p/rb	Dako Z0334	500	Up	

For additional markers, see ref. (24)

Ab antibody; *FL* fluorescence immunohistochemistry; *m* monoclonal; *mo* mouse; *p* polyclonal; *r* rat; *rb* rabbit; *sh* sheep. All antibodies are IgG unless otherwise indicated

by the maximum myelin loss (20). The true epicenter serves as a reference point for aligning the other histological stains and analyses. By collecting 5 series, up to 12 different markers (3 for each of the 4 remaining series) can be analyzed using regular RGB fluorophores and channels. If an intravenous fluorescent dye or peptide is used to label the blood vessels, one of the channels is lost, which means only two additional markers per series can be used.

2.2. Immunofluorescent Staining Procedure

After deciding the slide numbers for each animal that contains the epicenter, select three slides per 1 mm distance from the epicenter that are to be analyzed. The slides are warmed for 20 min on a slide warmer, the mounting media (e.g., Tissue Tek) is gently removed, and a ring of wax is applied around the sections with a PAP pen (Invitrogen, Carlsbad, CA). This ring enables the use of small amounts of solutions which are deposited into the "dam" with a pipetter. The slides are rinsed in 0.1 M TBS for 10 min. This can be done either by placing the slides in a jar or first removing the prior solution using vacuum. Nonspecific staining is blocked by incubating the slides for 1 h in 10% donkey serum (or the serum of the species in which the secondary antibody is made) in TBS containing 0.3% Triton X-100 (TBST), all at room temperature. Next, the slides are incubated overnight at 4°C with the primary antibody diluted in TBST containing 5% serum of the species in which the secondary antibodies are made. After washing three times for 10 min in TBS, sections are incubated for 1 h in TBST containing 5% serum and 1:500 (or other dilutions) of appropriate Fab′ fragment secondary antibodies conjugated to TRITC (red)-, FITC (green)-, or AMCA (blue) (Invitrogen, Carlsbad, CA; Alexa594, Alexa488, or Alexa405, respectively). The use of Fab′ fragment secondary antibodies greatly reduces nonspecific staining in fixed sections of injured spinal cord. If there is a potential exposure to UV light, the sections are shielded by a dark box during incubations. After another wash of three times for 10 min in TBS, the sections are coverslipped with antifade Gel/Mount aqueous mounting media (SouthernBiotech, Birmingham, AL).

3. Intravital Vascular Labeling

Perfused blood vessels are most relevant to tissue protection and function. Therefore, labeling the ECs with intravenous injections of reagents that bind to the luminal side of the blood vessels shows whether they are perfused. When this is done 15–30 min before transcardial perfusion for histological processing, it gives a snapshot of the vascularity in the spinal cord at the end of an experiment. Various methods have been used by other groups, e.g., intravenous injection of fluorescently (FITC, TRITC) conjugated gelatin, which fills the blood vessels, or dextrans which nonselectively bind to blood vessels whether or not they have ECs. One of the drawbacks of such reagents is that they fade under long UV exposure during analyses. The fluorescent dye DiI (Molecular Probes) is stable and has been infused transcardially during perfusion (32), but its poor solubility in aqueous solutions can lead to precipitation and less than optimal filling of all microvessels.

Another approach is the injection of reporter-conjugated lectins to bind luminal structures associated with blood vessels. Lectins are specific carbohydrate-binding proteins of nonimmune origin, which have been widely used to characterize the glycocalycal carbohydrate phenotype of blood vessels in situ (33–35). In the context of SCI, they have been used as histochemical reagents to identify changes in blood vessel reactivity following injury (36). The *Lycopersicon esculentum* (tomato) agglutinin (LEA) lectin exhibits specificity for *N*-acetyl-D-glucosamine β(1,4) *N*-acetyl-D-glucosamine oligomers and, when delivered intravascularly, identifies perfused microvessels in various tissues (37, 38). In contrast, the lectin IB4 specifically binds terminal α-galactosyl residues expressed by various cells including, but not limited to, ECs (34, 35, 39, 40) in normal mouse, rabbit, rat, and human tissues (41). Intravital application of IB4 has been used to specifically identify neovascular structures within tumor vascular networks (42). We have utilized the intravascular binding of these two lectins to recognize normal/nonreactive and angiogenic/dysfunctional microvessels following SCI in both mouse (5, 20, 21, 43, 44) and rat (45).

There are several advantages to employing this approach to assess microvascular plasticity in the injured spinal cord. First of all, the affinity of the lectin–glycocalyx complex is quite high, which results in stable labeling of perfused microvessels. Also, utilizing intravascular fluorophore-conjugated lectin delivery allows for the subsequent immunohistochemical processing of injured spinal tissues for either detection of various endothelial epitopes (see Figs. 1 and 2) or characterization of the lesion microenvironment. Lastly, and perhaps most importantly, the combinatorial use of intravital lectins to label perfused blood vessels with immunohistochemical approaches allows one to make a statement about the function of pathologically activated blood vessels. This is an important issue in light of previous findings of a significant lack of perfusion in "immunohistochemically intact" spinal capillary beds following injury (46).

4. Quantification of Vascularity

Quantitative histological data are essential for investigating vascular mechanisms and comparing the effectiveness of treatments following SCI. What follows are descriptions of the methods we regularly use. We have not yet applied unbiased stereological methods, but those

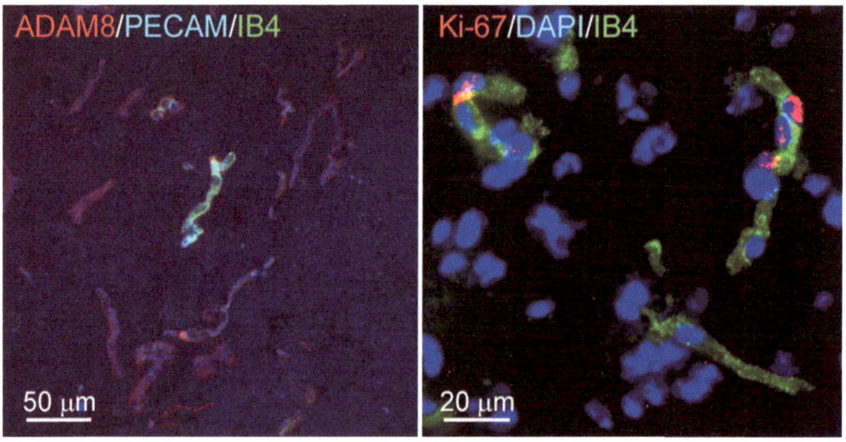

Fig. 1. Two spinal cord sections stained for two markers plus intravenous labeling with *Griffonia simplicifolia* (*Bandeiraea*) isolectin B4 (IB4) as examples of the methods we use to characterize spinal cord microvessels.

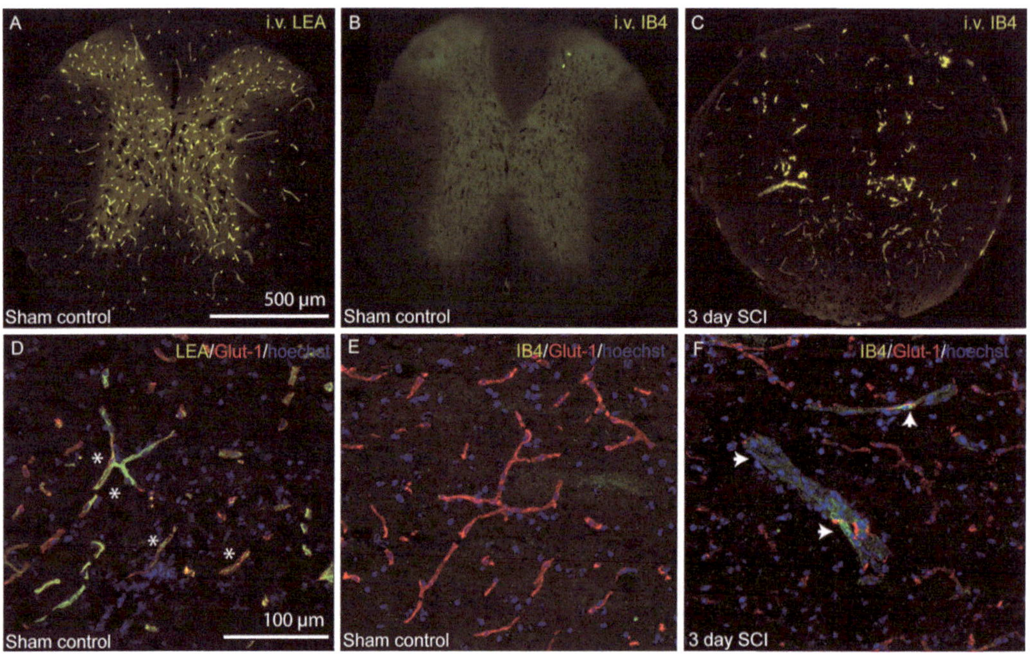

Fig. 2. Selective labeling of normal and angiogenic microvessels in the mouse spinal cord. Intravascular delivery of FITC-conjugated LEA lectin specifically binds the glycocalyx of intact blood vessels (**a**) also expressing the pan-CNS endothelial epitope glucose transporter, type I (Glut-1) (**d**; *asterisks*). By contrast, intravital infusion of the lectin IB4 does not bind normal spinal blood vessels (**b**, **e**). After traumatic SCI, reactive blood vessels in/around the injury site are transformed in their affinity for IB4 (**c**), with IB4-reactive vessels exhibiting altered metabolic potential as evidenced by a lack of Glut-1 immunoreactivity (**f**; *arrowheads*). Scale bars = 500 μm (**a**–**c**) and 100 μm (**d**–**f**).

would be very useful for estimating the total length of the microvasculature in a given volume with the orbital/spherical probe of stereology software. This is more difficult in sections thinner than 30–40 μm because the probe size becomes so small. One has to select a cutoff diameter to identify capillaries, but this might be a problem when dealing with pathological blood vessels which can be wider due to vasodilation. Other efforts are being developed to reconstruct the vascular bed in three dimensions at high resolution and those would be very useful for investigating regional differences and the relationships between neural cells and the vasculature (47).

The density of spinal cord microvessels can be determined by detecting EC or vascular markers within a region of interest (ROI). Following contusive SCI in rats, a cavity forms which helps with defining the injured area. In mice, the lost tissue is replaced by a fibroblast-rich stroma defined as the heterodomain (5, 7). This structure is characterized by a dysregulation of laminin deposition resulting in an area of dense extravascular laminin. In longitudinal sections, the heterodomain can be traced as an ROI to determine vascular density in the epicenter of the contusion (Fig. 3a, b). The heterodomain ROI can be used to define the lesion penumbra by expanding a 500-μm area surrounding the epicenter. This approach allows for analyses of perfusion status and vascular density in pathological tissue that is not destroyed by the primary injury event (43). For early time points post SCI before the heterodomain is formed (>3 days), we define the astrocyte-free lesion epicenter by staining with GFAP. If more specific analyses of the injured spinal cord are required, additional ROIs can be drawn, i.e., separating the gray and white matter (Fig. 3c). Within the ROI, image analysis

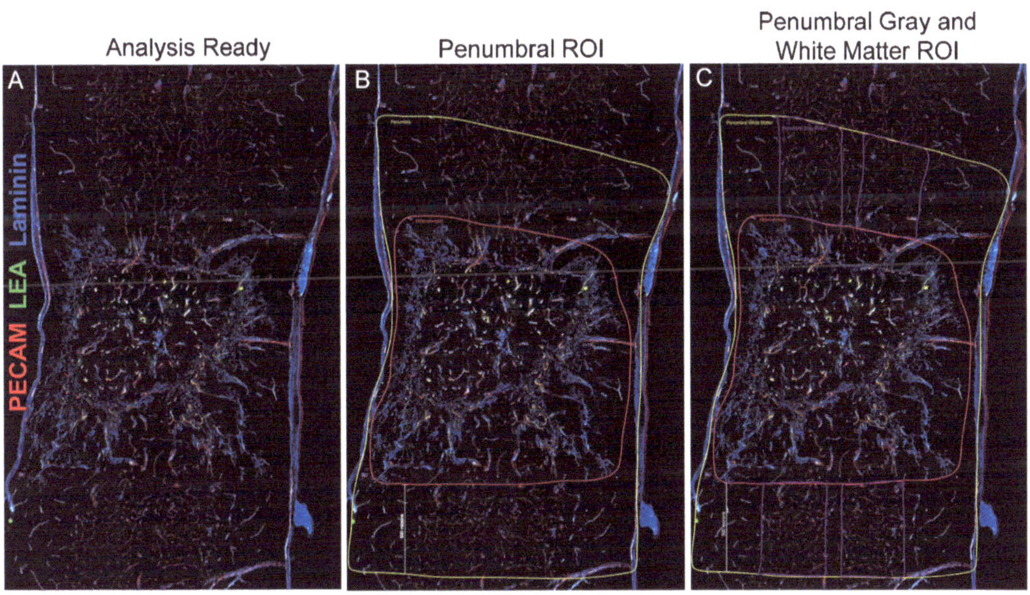

Fig. 3. Different regions in and around the injury epicenter can be identified and analyzed. (**a**) Longitudinal spinal cord section through the injury epicenter that is defined by laminin immunoreactivity, 7 days post-SCI. In (**b**) this epicenter is outlined in red and the penumbral region defined by extending 500 μm from the epicenter as indicated by the yellow line. In (**c**), the gray and white matter have been further identified by the purples lines with the two large medial boxes both rostral and caudal to the epicenter outlining the gray matter.

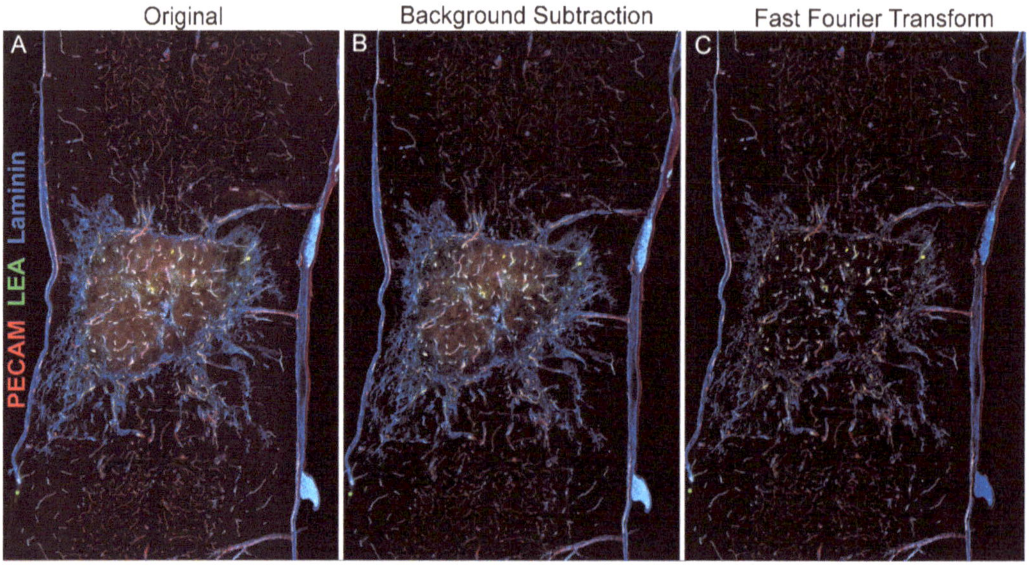

Fig. 4. Background subtraction followed by FFT of images facilitates quantitative analyses. (**a**) Raw, untransformed image of a longitudinal spinal cord section through the injury epicenter, 7 days post-SCI. In (**b**), the background has been subtracted and in (**c**) the data has been modified by fast Fourier transform to further reduce background staining, especially in the injury epicenter.

programs can be used to automatically detect pixels above a user-specified intensity to quantify density of the microvasculature within the penumbra.

If the tissue after staining has high background or nonspecific staining, fast Fourier transformation (FFT) can be used to decrease background without bias. The FFT computes the discrete frequency spectrum of a dataset. When performed on an image, the FFT algorithm generates a graphical output of the amplitude of this frequency-domain representation of the image's contents. This decomposition permits the removal of information contained at specific spatial frequencies. The data from frequencies representing the lowest signal-to-noise contents of the image are deleted, and an inverse FFT is performed to regenerate the original image devoid of the signal found at noisy frequencies. An example of a "raw" and "transformed" image is shown in Fig. 4. This image adjustment is done using commercially available software, and is applied to all images in any given study to ensure consistent comparison across all samples.

To get an estimate of the number of blood vessels at given distances from the epicenter, we have sampled a dorsal region in the midline with the bottom border on the central canal and an area in the ventrolateral region of the cord. When images are taken with a ×20 objective in mice, this primarily samples the penumbral region of the injury site, particularly when analyzing only spinal cord levels that contain the penumbra. For example, following a 50-kdyn

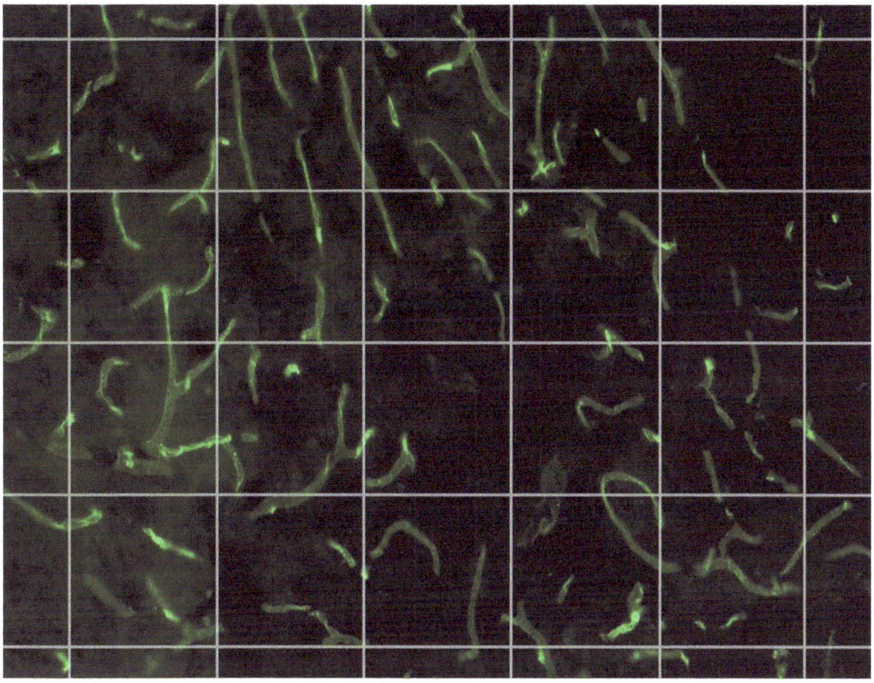

Fig. 5. Image of a region of a normal mouse spinal cord showing microvessels labeled with intravenous injection of LEA. A measure of the number of blood vessels can be made by manually counting the number of intersections with the vertical and horizontal lines of a 100-μm grid.

contusive injury with the IH impactor in mice, the penumbra can be found in sections through the epicenter and at 1 mm rostral and caudal from it. To assess the number of microvessels, we use LEA labeling and manually count the number of blood vessels intersecting horizontal and vertical lines of a 100-μm grid (Fig. 5). We have tried to automate this with software, such as ImageJ or NIH Image/SCION, but find that the human interpretation of what is a blood vessel and what is background is superior.

5. Vascular Function Assays

Blood flow is a global measure of blood vessel functionality and has been measured in rats following SCI and various treatments (48). However, this method does not measure tissue oxygen levels. Doppler analyses of intravenously injected microbubbles may also be useful but have not been tried in the SCI field. Other global methods that might at some point in time be useful are micro CT with intravenous contrast agents, luminometry with intravenous fluorescent labels, and PET scanning with EC-selective ligands. The small size of the spinal cord, particularly in mice, makes the distinction from surrounding tissue difficult. Micro CT could theoretically be used to document microvessels, but their small diameter currently precludes this. At the microscopic level, blood flow into and out of the spinal cord can be documented by measuring red blood cell movement using microscopy and water-immersion lenses placed onto the dura. To analyze blood vessels at greater depths, two-photon microscopy has been used (49).

BSCB breakdown and pathological permeability (leakiness) are among the earliest consequences of acute traumatic SCI. This can readily be documented by intravenous injection of reagents of different sizes that normally do not cross the BSCB. The localization of the leaky blood vessels can be shown by intravenous injection of Evans blue dye or horseradish peroxidase (HRP) followed by histology to detect the fluorescent Evans Blue or using DAB as a substrate for HRP (7, 50, 51). Dextrans of different sizes could also be used to probe the permeability of the BSCB. Dextrans have also been used to quantify permeability in the injured spinal cord (52). We and others have used a sensitive Luciferase assay of extracts from injured spinal cord segments (7, 20, 53). These reagents are also injected around half to an hour before analysis.

6. Rapid Isolation of Spinal Microvessels

A significant body of data has been collected focused on molecular responses to SCI. Most of these data have been generated using whole-tissue preparations, which are then subjected to various high-throughput biochemical analyses. While these approaches have been invaluable for discovering molecular mechanisms post injury, the injured spinal microenvironment is a complex mixture of cells and extracellular material. Thus, interpretation of the precise cell type responsible for these molecular signatures requires subsequent experiments to determine cellular localization. The cellular and molecular events attributed to vascular dysfunction remain poorly understood, partially due to the lack of methods to isolate pure populations of ECs from the injured spinal cord suitable for high-throughput molecular analyses. We have recently developed and utilized novel techniques to isolate highly enriched populations of microvascular fragments from injured mouse (44) and rat (45) spinal tissue.

While several protocols exist for isolation of CNS microvascular ECs from brain (54–56), only two reports had described primary isolation of microvessels from the porcine (57) and murine (58) spinal cord. These approaches yield a low purity of the microvascular EC preparation and it takes a long time to obtain the cells, both issues making them unsuitable for PCR-based molecular analyses of EC mRNA content (59). Our recently described technique (44) is unique in several ways. First of all, combining intravital

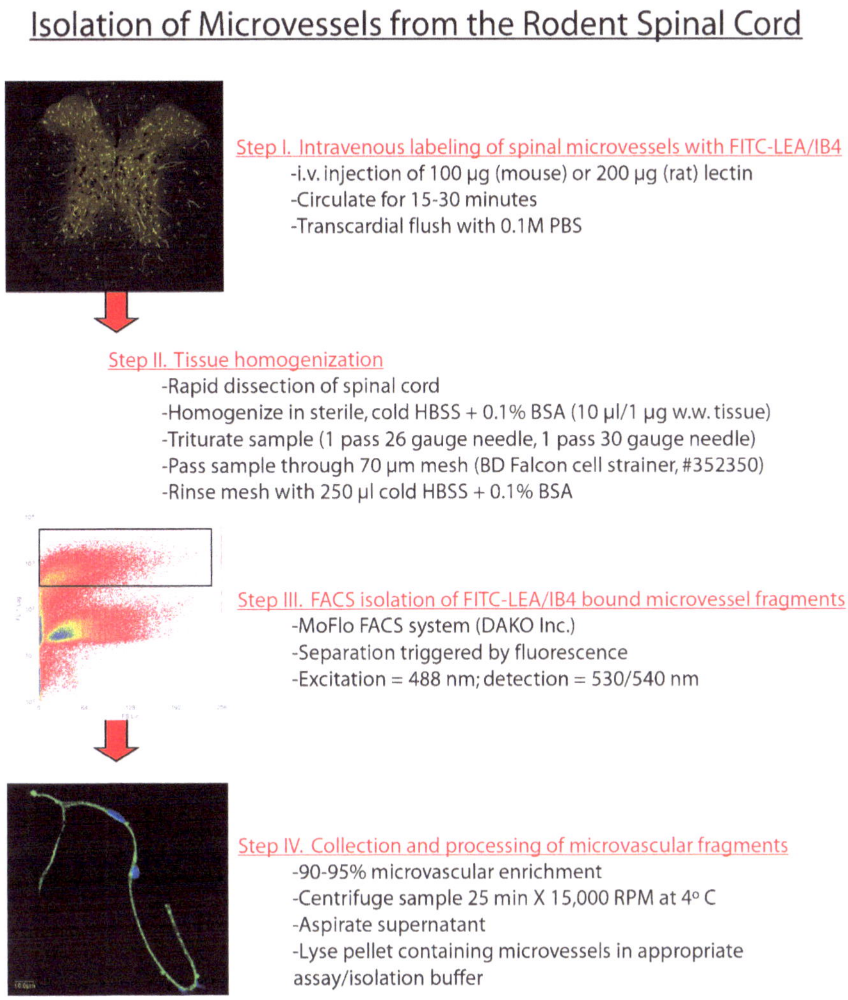

Fig. 6. Schematic overview of spinal microvascular isolation protocol.

labeling of spinal microvascular ECs using FITC-lectin allows for the isolation of analyzable amounts of enriched spinal EC material from relatively small quantities of spinal tissue (≈100 mg). In contrast, previously described protocols utilized several grams of CNS tissue as their biologic input (60, 61). This enhanced efficiency is facilitated by both the acute purity of the FACS-isolated spinal microvascular ECs as well as the implementation of qRT-PCR-based microarray approaches (44, 45). Further, this method results in very rapid (≈1.0–1.5 h) isolation of high-purity ECs from the spinal cord (Fig. 6). This is compared to variable lengths of time required for the more commonly applied gradient centrifugation isolation of CNS-derived ECs, with the more rapid isolation (≈1 h) associated with appreciable amounts of other cells in the preparation (60). Longer protocols (≈4–8 h) yield preparations of higher purity (54), but are not optimal for transcriptomic assessment of pathologic activation due to the degradation of target transcripts. Finally, this new preparation is amenable to a number of biochemical analyses, including ex vivo enzymology, immunoblotting, RT-PCR (44, 45), and mass spectroscopy (unpublished observation), greatly increasing the feasibility of experiments designed to directly address EC pathology and plasticity in the context of SCI.

7. Genetic Models

Genetic mouse models can be used as reporters for ECs which could help in the analyses or purification of ECs or inserting transgenes or conditionally deleting genes of interest selectively in ECs. The EC-specific receptor tyrosine kinase *Tie2* (also known as Tek) has been used to drive the expression of *lacZ* (62) (JAX Stock Number: 002856) and EGFP (63) (JAX Stock Number: 003658) in the mouse. The one caveat to using the Tie2-GFP mice is that EGFP expression is high in arterial and much weaker in venous vessels (J.B. Hoying, personal communication). Tie2-GFP transgenic rats have also been developed (64). The advantage of these models is that transgene expression is observed only in vasculature. One potential problem to control for is the fact that Tie2 can also be expressed by a subclass of proangiogenic monocytes (65) and hematopoietic stem cells (66). While there are no published reports using these animals to study SCI, the potential to use them to facilitate the understanding of vascular responses after SCI is readily evident. Double immunostaining for the respective reporter identifies molecules that can be directly linked to spinal cord ECs. An alternative mouse model is to use the universal reporter ROSA-Enhanced Yellow Fluorescent Protein (EYFP) mouse (67) (JAX Stock Number: 006148). These mice contain an EYFP gene inserted into the *Gt(ROSA)26Sor* locus. Expression of *EYFP* is blocked by an upstream *loxP*-flanked STOP sequence. When bred to mice with a *cre* recombinase gene under the control of a promoter of interest, the STOP sequence of the targeted gene is deleted in the tissue of interest and *EYFP* expression is observed. If these mice are crossed with mice expressing *cre* under the control of the EC-specific VE-cadherin promoter (68) (JAX Stock Number: 006137), all vasculature will specifically express EYFP. Figure 7 shows a three-dimensional brain microvascular culture that contains 80% wild-type and 20% ROSA-EYFPxVE-cadherin *cre* microvessels. Crossing these *cre* mice with *loxP* mice of any gene of interest could produce interesting models for investigating the role of those genes in EC behaviors. Given the fact that the vasculature is so important for embryonic development, use of doxycycline- or tamoxifen-inducible *cre* mice (69) or the injection of *cre*-expressing virus into the spinal cord is likely necessary.

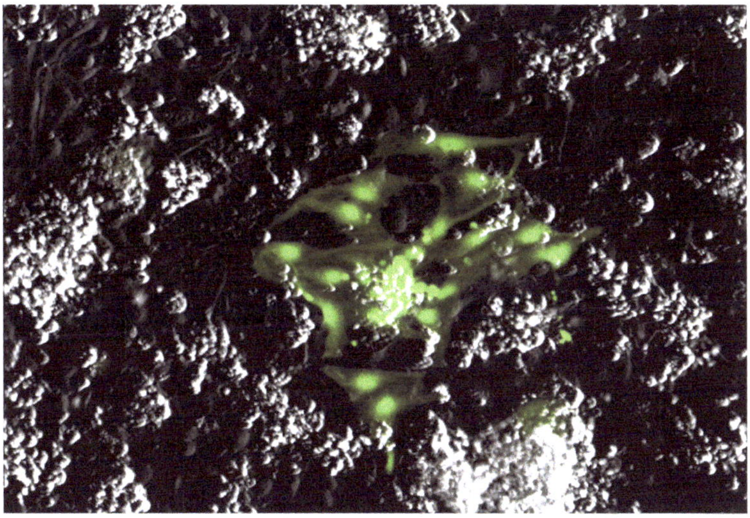

Fig. 7. Phase contrast and EYFP immunofluorescent photomicrograph of a three-dimensional brain microvascular fragment culture consisting of 80% wild-type and 20% EYFP-expressing ECs grown for 4 days. Courtesy of S.A. Myers.

8. Spinal Cord EC Cultures

There are no good primary cell culture models of spinal cord vasculature. However, much can be gained from their use, provided one understands their limitations. Those limitations are the following. (1) In vivo, spinal vasculature is three dimensional with multiple cell types (ECs, pericytes, smooth muscle cells, astrocytes, neurons) comprising the NVU (1). It is impossible to accurately reconstruct the NVU in a cell culture dish. (2) Isolation of ECs typically involves dissociation into microvascular fragments or single cells followed by some form of purification. The yield of microvessels/ECs from CNS tissue, especially spinal cord, is very low. This necessitates either very small experiments or subsequent passage and expansion. There is always a concern about how faithfully passaged cells reflect the properties of the primary isolate. Also, primary CNS ECs, especially from mouse spinal cord, senesce after a few passages. Thus, yield is again an issue. Nonetheless, methods have been developed for successful culture of passaged spinal cord ECs (58). Primary ECs from non-CNS tissues are readily grown in high yield (70), but again the question of their relevance to CNS ECs is of concern. For example, CNS ECs seem to be much more vulnerable to trauma than peripheral ones. If one wants to use a cell culture system that better approximates the in vivo microvascular environment, three-dimension collagen cultures of microvascular fragments that include EC tubes and mural cells have been developed (71). These cultures can also be implanted into the flank and will enosculate

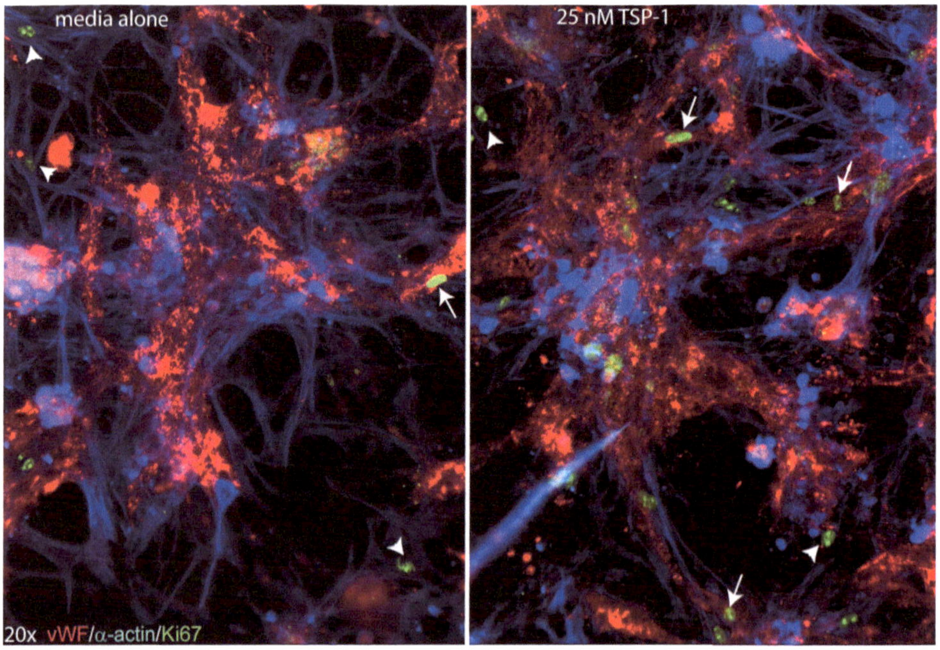

Fig. 8. Cortical microvascular fragments from adult mice (>6 weeks) were seeded in three-dimensional matrigel gels and grown in the absence or presence of TSP-1. After 8 days in culture, they were immunostained for von Willibrand factor (vWF, to identify ECs), α-actin (to identify smooth muscle cells), and the proliferation marker Ki67. Note the increase in proliferating cells that appear to be both ECs (*arrows*) and smooth muscle cells (*arrowheads*). Courtesy of S.A. Myers.

with host vasculature (72). We have adapted this system to brain microvascular fragments. These cultures have been used to examine the effects on CNS microvasculature of thrombospondin-1 (Fig. 8), a molecule identified to be acutely upregulated after SCI (45). Immortalized EC lines, such as bEnd3 (ATCC CRL-2299), have also been used to model ECs in vitro. The advantage of these transformed cell lines is that they are clonal and can be passaged indefinitely. We have used human aorta-derived ECs to test the biological activity of some of our reagents (20, 73). As discussed above, how well they and transformed cell lines reflect primary ECs is an unresolved question. For a detailed review of the concerns that must be considered in in vitro angiogenesis assays, details that are pertinent to all assays of EC function in cell culture, see ref. (74).

Microvessels with associated mural cells (vascular smooth muscle cells and/or pericytes) can be purified from mouse cortex and spinal cord (71). Fifteen mice (at least 6 weeks old) are required for a yield of approximately 200,000 microvessels. Mice are anesthetized and cortices extracted by detaching the head from the body, peeling back the skin, and carefully removing the skull from posterior to anterior using Rongeurs. The cerebellum is removed and the optic nerves detached to release brain into L15 with 2% FBS and 1% penicillin/streptomycin. White matter should be removed by separating the cortical hemispheres and scooping out the

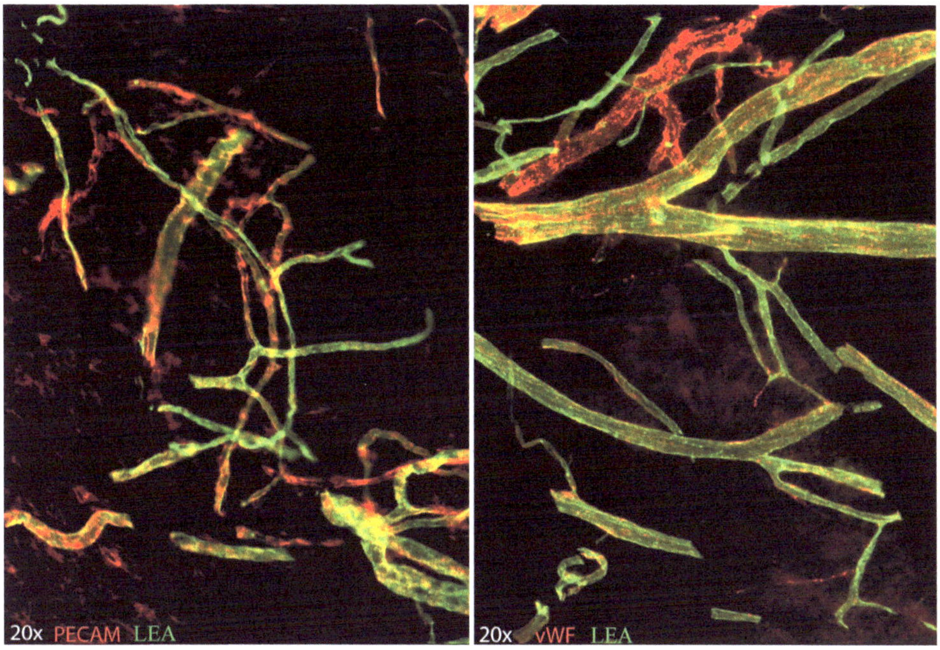

Fig. 9. Three-dimensional microvessel constructs can be implanted subdermally for ex vivo studies. Microvessel cultures similar to those from Fig. 8 were implanted under the skin for 30 days. Intravital labeling with intravenous LEA (*green*) shows that the construct inosculates with the host vasculature and retains ECs as shown by immunostaining for PECAM (*left*) or vWF (*right*). Courtesy of S.A. Myers.

subcortical structures (diencephalon and hippocampal formation). After collecting and removing white matter from all brains, the cortices are rinsed with 1% BSA in PBS and minced with long curved scissors. Tissue is incubated with agitation for 20 min at 37°C in 2.5 mL collagenase/mL of tissue and 0.1 mg/mL of DNase I. Digested tissue is then resuspended in 15% Dextran and centrifuged at 4,000×*g* at 4°C for 20 min. The supernatant is removed being careful not to disturb the delicate vascular pellet. The pellet is resuspended in 1% BSA in PBS and passed sequentially through pre-wetted 500 and 20-μm screens. The vascular fragments retained on the 20-μm screen are retrieved by soaking in 1% BSA in PBS and transferred to a 50-mL conical tube by washing the screen thoroughly. Under the microscope, count microvascular fragments and seed at 90,000–100,000 MV/mL in collagen-coated wells in DMEM/F12 with 20% platelet-derived serum, 100 μg/mL EC growth supplement, and heparin.

Alternatively, microvessels can be seeded in 4 mg/mL matrigel and solidified for three-dimensional assays in vitro. They can also be implanted subdermally for ex vivo studies of angiogenesis and vascular network formation (Fig. 9). The density of microvessels within the matrigel constructs is similar to plating in a dish and the isolation process is the same as stated above. Since the microvessels remain intact, there is not a need to eliminate smooth muscle cells as these cells contribute to angiogenesis and maturation of the newly formed microvasculature.

Acknowledgments

This work was supported by grants from the Kentucky Spinal Cord and Head Injury Research Trust and National Institutes of Health (NS045734 and RR015576) and by Norton Healthcare and the Commonwealth of Kentucky Challenge for Excellence (SRW, TH). We thank the many people who over the years have contributed to the development of these techniques, including Sheila Arnold, Darlene Burke, Charles Crawford, Jessica Darnall, Toros Dincman, Justin Geralds, Mark Gruenthal, Shu Han, Russ Howard, David Loy, Melissa Maddie, Ed Mahoney, Scott Myers, Christine Nunn, Rollie Reid, and Sheher Sun.

References

1. del Zoppo GJ (2010) The neurovascular unit in the setting of stroke. J Intern Med 267:156–171
2. Carrico KM, Vaishnav RA, Hall ED (2009) Temporal and spatial dynamics of peroxynitrite-induced oxidative damage after spinal cord contusion injury. J Neurotrauma 26(8):1369–1378
3. Loy DN, Crawford CH, Darnall JB, Burke DA, Onifer SM, Whittemore SR (2002) Temporal progression of angiogenesis and basal lamina deposition after contusive spinal cord injury in the adult rat. J Comp Neurol 445:308–324
4. Casella GT, Bunge MB, Wood PM (2006) Endothelial cell loss is not a major cause of neuronal and glial cell death following contusion injury of the spinal cord. Exp Neurol 202:8–20
5. Benton RL, Maddie MA, Minnillo DR, Hagg T, Whittemore SR (2008) *Griffonia simplicifolia* isolectin B4 identifies a specific subpopulation of angiogenic blood vessels following contusive spinal cord injury in the adult mouse. J Comp Neurol 507:1031–1052
6. Noble LJ, Wrathall JR (1989) Distribution and time course of protein extravasation in the rat spinal cord after contusive injury. Brain Res 482:57–66
7. Whetstone WD, Hsu JY, Eisenberg M, Werb Z, Noble-Haeusslein LJ (2003) Blood-spinal cord barrier after spinal cord injury: relation to revascularization and wound healing. J Neurosci Res 74:227–239
8. Donnelly DJ, Popovich PG (2008) Inflammation and its role in neuroprotection, axonal regeneration and functional recovery after spinal cord injury. Exp Neurol 209:378–388
9. Norenberg MD, Smith J, Marcillo A (2004) The pathology of human spinal cord injury: defining the problems. J Neurotrauma 21:429–440
10. Fleming JC, Norenberg MD, Ramsay DA, Dekaban GA, Marcillo AE, Saenz AD, Pasquale-Styles M, Dietrich WD, Weaver LC (2006) The cellular inflammatory response in human spinal cords after injury. Brain 129:3249–3269
11. Bramlett HM, Dietrich WD (2007) Progressive damage after brain and spinal cord injury: pathomechanisms and treatment strategies. Prog Brain Res 161:125–141
12. Fehlings MG, Tator CH, Linden RD (1989) The effect of nimodipine and dextran on axonal function and blood flow following experimental spinal cord injury. J Neurosurg 71:403–416
13. Guha A, Tator CH, Smith CR, Piper I (1989) Improvement in post-traumatic spinal cord blood flow with a combination of a calcium channel blocker and a vasopressor. J Trauma 29:1440–1447
14. Hall ED, Wolf DL (1986) A pharmacological analysis of the pathophysiological mechanisms of posttraumatic spinal cord ischemia. J Neurosurg 64:951–961
15. Hall ED (1988) Effects of the 21-aminosteroid U74006F on posttraumatic spinal cord ischemia in cats. J Neurosurg 68:462–465
16. Hall ED, McCall JM, Means ED (1994) Therapeutic potential of the lazaroids (21-amino-

steroids) in acute central nervous system trauma, ischemia and subarachnoid hemorrhage. Adv Pharmacol 28:221–268
17. Noble LJ, Donovan F, Igarashi T, Goussev S, Werb Z (2002) Matrix metalloproteinases limit functional recovery after spinal cord injury by modulation of early vascular events. J Neurosci 22:7526–7535
18. Simard JM, Tsymbalyuk O, Ivanov A, Ivanova S, Bhatta S, Geng Z, Woo SK, Gerzanich V (2007) Endothelial sulfonylurea receptor 1-regulated NC Ca-ATP channels mediate progressive hemorrhagic necrosis following spinal cord injury. J Clin Invest 117:2105–2113
19. Gerzanich V, Woo SK, Vennekens R, Tsymbalyuk O, Ivanova S, Ivanov A, Geng Z, Chen Z, Nilius B, Flockerzi V, Freichel M, Simard JM (2009) De novo expression of Trpm4 initiates secondary hemorrhage in spinal cord injury. Nat Med 15:185–191
20. Han S, Arnold SA, Sithu SD, Mahoney ET, Geralds JT, Tran P, Benton RL, Maddie MA, D'Souza SE, Whittemore SR, Hagg T (2010) Rescuing vasculature with intravenous angiopoietin-1 and alphavbeta3 integrin peptide is protective after spinal cord injury. Brain 133: 1026–1042
21. Mahoney ET, Benton RL, Maddie MA, Whittemore SR, Hagg T (2009) ADAM8 is selectively up-regulated in endothelial cells and is associated with angiogenesis after spinal cord injury in adult mice. J Comp Neurol 512:243–255
22. Dray C, Rougon G, Debarbieux F (2009) Quantitative analysis by in vivo imaging of the dynamics of vascular and axonal networks in injured mouse spinal cord. Proc Natl Acad Sci U S A 106:9459–9464
23. Zhao H, Webb RH, Ortel B (2002) Review of noninvasive methods for skin blood flow imaging in microcirculation. J Clin Eng 27:40–47
24. McDonald DM, Choyke PL (2003) Imaging of angiogenesis: from microscope to clinic. Nat Med 9:713–725
25. Mostany R, Portera-Cailliau C (2008) A method for 2-photon imaging of blood flow in the neocortex through a cranial window. J Vis Exp 25:678
26. van Laar PJ, van der Grond J, Hendrikse J (2008) Brain perfusion territory imaging: methods and clinical applications of selective arterial spin-labeling MR imaging. Radiology 246:354–364
27. Wuestenfeld JC, Herold J, Niese U, Kappert U, Schmeisser A, Strasser RH, Braun-Dullaeus RC (2008) Indocyanine green angiography: a new method to quantify collateral flow in mice. J Vasc Surg 48:1315–1321
28. Vakoc BJ, Lanning RM, Tyrrell JA, Padera TP, Bartlett LA, Stylianopoulos T, Munn LL, Tearney GJ, Fukumura D, Jain RK, Bouma BE (2009) Three-dimensional microscopy of the tumor microenvironment in vivo using optical frequency domain imaging. Nat Med 15:1219–1223
29. Boas DA, Dunn AK (2010) Laser speckle contrast imaging in biomedical optics. J Biomed Opt 15:011109
30. Fukumura D, Duda DG, Munn LL, Jain RK (2010) Tumor microvasculature and microenvironment: novel insights through intravital imaging in pre-clinical models. Microcirculation 17:206–225
31. Saper CB (2005) An open letter to our readers on the use of antibodies. J Comp Neurol 493:477–478
32. Li Y, Song Y, Zhao L, Gaidosh G, Laties AM, Wen R (2008) Direct labeling and visualization of blood vessels with lipophilic carbocyanine dye DiI. Nat Protoc 3:1703–1708
33. Balding P, Gold ER (1975) Observations on the reaction of en(a-)cells with sophora japonica haemagglutinin. Z Immunitatsforsch Exp Klin Immunol 145:156–165
34. Brabec RK, Peters BP, Bernstein IA, Gray RH, Goldstein IJ (1980) Differential lectin binding to cellular membranes in the epidermis of the newborn rat. Proc Natl Acad Sci U S A 77: 477–479
35. Franz S, Frey B, Sheriff A, Gaipl US, Beer A, Voll RE, Kalden JR, Herrmann M (2006) Lectins detect changes of the glycosylation status of plasma membrane constituents during late apoptosis. Cytometry A 69:230–239
36. Noble LJ, Mautes AE, Hall JJ (1996) Characterization of the microvascular glycocalyx in normal and injured spinal cord in the rat. J Comp Neurol 376:542–556
37. Thurston G, Baluk P, Hirata A, McDonald DM (1996) Permeability-related changes revealed at endothelial cell borders in inflamed venules by lectin binding. Am J Physiol 271:H2547–H2562
38. Jilani SM, Murphy TJ, Thai SN, Eichmann A, Alva JA, Iruela-Arispe ML (2003) Selective binding of lectins to embryonic chicken vasculature. J Histochem Cytochem 51:597–604
39. Peters BP, Goldstein IJ (1979) The use of fluorescein-conjugated *Bandeiraea simplicifolia* B4-isolectin as a histochemical reagent for the detection of alpha-D-galactopyranosyl groups. Their occurrence in basement membranes. Exp Cell Res 120:321–334
40. Laitinen L (1987) *Griffonia simplicifolia* lectins bind specifically to endothelial cells and some epithelial cells in mouse tissues. Histochem J 19:225–234

41. Hayes CE, Goldstein IJ (1974) An alpha-D-galactosyl-binding lectin from *Bandeiraea simplicifolia* seeds. Isolation by affinity chromatography and characterization. J Biol Chem 249:1904–1914
42. Niethammer AG, Xiang R, Becker JC, Wodrich H, Pertl U, Karsten G, Eliceiri BP, Reisfeld RA (2002) A DNA vaccine against VEGF receptor 2 prevents effective angiogenesis and inhibits tumor growth. Nat Med 8:1369–1375
43. Benton RL, Maddie MA, Gruenthal MJ, Hagg T, Whittemore SR (2009) Neutralizing endogenous VEGF following traumatic spinal cord injury modulates microvascular plasticity but not tissue sparing or functional recovery. Curr Neurovasc Res 6:124–131
44. Benton RL, Maddie MA, Worth CA, Mahoney ET, Hagg T, Whittemore SR (2008) Transcriptomic screening of microvascular endothelial cells implicates novel molecular regulators of vascular dysfunction after spinal cord injury. J Cereb Blood Flow Metab 28:1771–1785
45. Benton RL, Maddie MA, Dincman TA, Hagg T, Whittemore SR (2009) Transcriptional activation of endothelial cells by TGFbeta coincides with acute microvascular plasticity following focal spinal cord ischaemia/reperfusion injury. ASN Neuro 1:e00015
46. Koyanagi I, Tator CH, Lea PJ (1993) Three-dimensional analysis of the vascular system in the rat spinal cord with scanning electron microscopy of vascular corrosion casts. Part 2: acute spinal cord injury. Neurosurgery 33:285–291
47. Bjornsson CS, Lin G, Al-Kofahi Y, Narayanaswamy A, Smith KL, Shain W, Roysam B (2008) Associative image analysis: a method for automated quantification of 3D multi-parameter images of brain tissue. J Neurosci Methods 170:165–178
48. Guha A, Tator CH, Rochon J (1989) Spinal cord blood flow and systemic blood pressure after experimental spinal cord injury in rats. Stroke 20:372–377
49. Davalos D, Lee JK, Smith WB, Brinkman B, Ellisman MH, Zheng B, Akassoglou K (2008) Stable in vivo imaging of densely populated glia, axons and blood vessels in the mouse spinal cord using two-photon microscopy. J Neurosci Methods 169:1–7
50. Noble LJ, Maxwell DS (1983) Blood-spinal cord barrier response to transection. Exp Neurol 79:188–199
51. Benton RL, Whittemore SR (2003) VEGF165 therapy exacerbates secondary damage following spinal cord injury. Neurochem Res 28: 1693–1703
52. Qu ZX, Xu J, Perot PL Jr, Hogan EL (1991) A sensitive fluorometric method for measurement of vascular permeability in spinal cord injury. J Neurotrauma 8:149–156
53. Smith R, Brown EH, Shum-Siu A, Whelan A, Burke D, Benton RL, Magnuson DS (2009) Swim training initiated acutely after spinal cord injury is ineffective and induces extravasation in and around the epicenter. J Neurotrauma 26(7):1017–1027
54. Wu Z, Hofman FM, Zlokovic BV (2003) A simple method for isolation and characterization of mouse brain microvascular endothelial cells. J Neurosci Methods 130:53–63
55. Hemmings SJ, Storey KB (1999) Brain gamma-glutamyltranspeptidase: characteristics, development and thyroid hormone dependency of the enzyme in isolated microvessels and neuronal/glial cell plasma membranes. Mol Cell Biochem 202:119–130
56. Diglio CA, Grammas P, Giacomelli F, Wiener J (1982) Primary culture of rat cerebral microvascular endothelial cells. Isolation, growth, and characterization. Lab Invest 46:554–563
57. Joo F, Dux E, Szucs A (1982) Microvessels from the spinal cord: isolation procedure and characterization of the fraction. J Neurochem 39:263–266
58. Ge S, Pachter JS (2006) Isolation and culture of microvascular endothelial cells from murine spinal cord. J Neuroimmunol 177:209–214
59. Yousif S, Marie-Claire C, Roux F, Scherrmann JM, Decleves X (2007) Expression of drug transporters at the blood-brain barrier using an optimized isolated rat brain microvessel strategy. Brain Res 1134:1–11
60. Enerson BE, Drewes LR (2006) The rat blood-brain barrier transcriptome. J Cereb Blood Flow Metab 26:959–973
61. Shusta EV, Boado RJ, Mathern GW, Pardridge WM (2002) Vascular genomics of the human brain. J Cereb Blood Flow Metab 22:245–252
62. Schlaeger TM, Bartunkova S, Lawitts JA, Teichmann G, Risau W, Deutsch U, Sato TN (1997) Uniform vascular-endothelial-cell-specific gene expression in both embryonic and adult transgenic mice. Proc Natl Acad Sci U S A 94:3058–3063
63. Motoike T, Loughna S, Perens E, Roman BL, Liao W, Chau TC, Richardson CD, Kawate T, Kuno J, Weinstein BM, Stainier DY, Sato TN (2000) Universal GFP reporter for the study of vascular development. Genesis 28:75–81
64. Ohtsuki S, Kamiya N, Hori S, Terasaki T (2005) Vascular endothelium-selective gene induction by Tie2 promoter/enhancer in the brain and retina of a transgenic rat. Pharm Res 22:852–857
65. De PM, Venneri MA, Galli R, Sergi SL, Politi LS, Sampaolesi M, Naldini L (2005) Tie2 identifies

a hematopoietic lineage of proangiogenic monocytes required for tumor vessel formation and a mesenchymal population of pericyte progenitors. Cancer Cell 8:211–226

66. Arai F, Hirao A, Ohmura M, Sato H, Matsuoka S, Takubo K, Ito K, Koh GY, Suda T (2004) Tie2/angiopoietin-1 signaling regulates hematopoietic stem cell quiescence in the bone marrow niche. Cell 118:149–161

67. Srinivas S, Watanabe T, Lin CS, William CM, Tanabe Y, Jessell TM, Costantini F (2001) Cre reporter strains produced by targeted insertion of EYFP and ECFP into the ROSA26 locus. BMC Dev Biol 1:4

68. Alva JA, Zovein AC, Monvoisin A, Murphy T, Salazar A, Harvey NL, Carmeliet P, Iruela-Arispe ML (2006) VE-cadherin-Cre-recombinase transgenic mouse: a tool for lineage analysis and gene deletion in endothelial cells. Dev Dyn 235:759–767

69. Monvoisin A, Alva JA, Hofmann JJ, Zovein AC, Lane TF, Iruela-Arispe ML (2006) VE-cadherin-CreERT2 transgenic mouse: a model for inducible recombination in the endothelium. Dev Dyn 235:3413–3422

70. Bicknell R (ed) (1996) Endothelial cell culture. Cambridge University Press, New York

71. Hoying JB, Boswell CA, Williams SK (1996) Angiogenic potential of microvessel fragments established in three-dimensional collagen gels. In Vitro Cell Dev Biol Anim 32:409–419

72. Shepherd BR, Chen HY, Smith CM, Gruionu G, Williams SK, Hoying JB (2004) Rapid perfusion and network remodeling in a microvascular construct after implantation. Arterioscler Thromb Vasc Biol 24:898–904

73. Nakashima S, Arnold SA, Mahoney ET, Sithu SD, Zhang YP, D'Souza SE, Shields CB, Hagg T (2008) Small-molecule protein tyrosine phosphatase inhibition as a neuroprotective treatment after spinal cord injury in adult rats. J Neurosci 28:7293–7303

74. Staton CA, Reed MW, Brown NJ (2009) A critical analysis of current in vitro and in vivo angiogenesis assays. Int J Exp Pathol 90:195–221

Chapter 42

Physiological Assessment of Spinal Cord Injury

Lisa B.E. Shields, Yi Ping Zhang, and Christopher B. Shields

Abstract

The severity of injury and degree of recovery following experimental spinal cord injury (SCI) utilizing several electrophysiological tests (SSEP, MEP, H-reflex) are described in other chapters. This chapter summarizes tests of spinal cord physiology, including spinal cord blood flow, spinal cord field potential mapping, EMG, and PO_2 monitoring of the spinal cord. Physiological assessment of the spinal cord provides objective evaluation of nonelectrophysiological importance that is of value in assessing responses to novel therapies following experimental SCI.

Key words: Spinal cord injury, Spinal cord blood flow, Field potential mapping, EMG, Spinal cord temperature, Spinal cord oxygen

1. Introduction

Physiological assessment of an organ addresses its functional status with a specific set of tests that document whether it is working normally. For example, evaluation of the cardiovascular system is conducted by determining the blood pressure, EKG, angiography, and cardiac output. The spinal cord transmits ascending and descending signals between the brain and peripheral nervous system and organizes segmental reflexes. Proper metabolic support and suitable external and internal environments must be maintained for normal electrophysiological performance of the spinal cord. In addition to electro-conductive tests used to evaluate the functional integrity of the spinal cord, physiological assessment also includes the status of spinal cord blood flow (SCBF) and CSF circulation. Physiological assessment of the spinal cord covers a wide array of tests, and measurement of these functions provides information as to the severity of the spinal cord injury (SCI) and effectiveness of treatment.

Jun Chen et al. (eds.), *Animal Models of Acute Neurological Injuries II*, Springer Protocols Handbooks,
DOI 10.1007/978-1-61779-782-8_42,

Electrophysiological studies, such as motor-evoked potential (MEP), SSEP, and H-reflex are utilized in vivo and repeatedly to evaluate descending, ascending, and segmental function of the spinal cord, respectively. Injury to spinal cord axons, myelin, and synaptic connections causes structural damage leading to neurological dysfunction. Therapeutic intervention and/or inherent spinal cord plasticity may mitigate the disability. A detailed description of these electrophysiological tests (MEP, SSEP, H-reflex) is provided in this section. Other tests used to evaluate the status of the spinal cord are briefly described in this chapter, including SCBF, tissue oxygen levels, EMG, and field potential mapping.

2. Spinal Cord Blood Flow

Assessment of SCBF provides important information pertaining to blood circulation and tissue perfusion of the spinal cord. Blood flow to the spinal cord decreases as a result of poor systemic blood pressure or extrinsic tissue pressure against blood vessels. SCBF does not precisely reflect tissue oxygenation levels or the metabolic status of the spinal cord but is a valuable measurement which is often concordant with the extent of tissue damage and recovery. Several methods are utilized to measure SCBF in experimental studies (1) including:

2.1. Laser Doppler Flow

SCBF may be measured by determining the velocity of red blood cells in blood vessels. Laser Doppler flow (LDF) studies are based on the degree of erythrocyte motion using a monochromatic laser. The frequency of the Doppler shift emitted from red blood cells flowing intravascularly may be used to determine the velocity of red blood cells in spinal cord tissue. LDF is a noninvasive method that is designed to continuously measure SCBF. LDF has been successfully utilized in several animal models (2–7). This test is conducted after performing a laminectomy to expose dura over the spinal cord. A fiberoptic probe is placed on intact dura using a micromanipulator to attain its precise position. Compression of spinal cord tissue by the fiberoptic probe must be avoided to prevent distortion of the spinal cord and alteration of the SCBF by extrinsic pressure. Furthermore, respiratory movements causing attenuation of pressure of the probe against the target spinal cord should be avoided to prevent spurious recordings. LDF measurements of SCBF are simple and reliable for superficial recordings of blood flow. A major drawback is that LDF is unable to measure blood flow at the depths of spinal cord due to poor tissue penetration by the laser beam.

2.2. Hydrogen Clearance

Guha et al. have adapted the hydrogen clearance technique to measure changes in SCBF following SCI (8, 9). This technique is based on the assumption that the clearance time of saturated hydrogen from the spinal cord reflects SCBF. In a well-perfused spinal cord, tissue hydrogen is rapidly cleared by blood flowing through vessels due to rapid gas equilibration between blood and spinal cord. Hydrogen undergoes an electrochemical reaction in tissue where the current is measured by a polarized electrode. After exposure to hydrogen is stopped, focal tissue blood flow can be calculated from the decay curve of current recorded from the polarized electrode. Prolonged hydrogen clearance indicates low SCBF. This procedure is performed by a laminectomy followed by a durotomy. The platinum/iridium hydrogen clearance electrode is inserted into the area of the spinal cord being studied to measure current changes. Hydrogen (5%) is administered through the ventilator for 10 min to saturate spinal cord tissue. The SCBF can be estimated from the current decay curve during the 20 min desaturating phase. The natural-log plot of the clearance curve is used to compute the value of the SCBF (10). This technique is able to measure SCBF at various depths of the spinal cord depending on the electrode placement. This invasive procedure and the long duration required for baseline calibration renders the hydrogen clearance method unpopular.

During the hydrogen clearance technique, spinal fixation is necessary to avoid tissue damage induced by poor technique. Although several spinal stereotaxic frames are available (Tritech Research Inc., Los Angeles, CA or Harvard Apparatus, Holliston, MA), the most reliable rodent spine stabilizer called the LISA (Louisville Impactor System Apparatus) (Louisville, KY). This frame is recommended for use in several spinal cord procedures in which the target vertebra needs to be suspended to prevent interference caused by respiratory and other movements (Fig. 1).

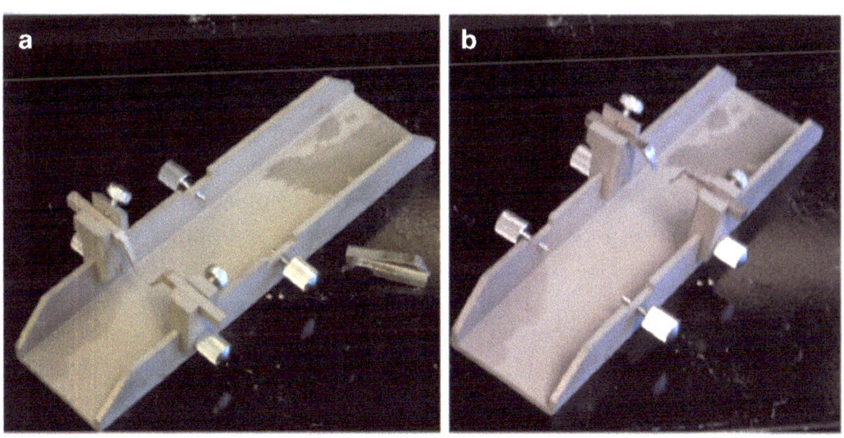

Fig. 1. These photographs illustrate the rat spine stabilizer, a device that prevents movement and enhances the precision of several procedures on the spinal cord. (**a**) Clamp position using long arms for cervical vertebrae fixation. (**b**) Short steel arms used for fixation of thoracic vertebrae.

2.3. Microspheres

Microsphere technology has been used to measure blood flow experimentally for many organs, including the spinal cord. Mechanisms inherent in this method are conceptually simple. Microspheres are injected into the left ventricle of the heart, and measurement of the number of microspheres trapped in the target tissue correlates well with that organ's blood flow. Several types of microspheres have been utilized, specifically, colored (auto-fluorescent) microbeads (11) and radiolabeled microbeads (12). After the microsphere suspension is injected into the left ventricle, they are distributed throughout the circulation. The spinal cord is removed postmortem, and the number of microspheres is measured to estimate SCBF. Under microscopic examination, the number of microspheres is counted per unit volume or weight of spinal cord. Microspheres may be injected at different times with each injection consisting of a different color or isotope. Each microsphere is used to measure the SCBF at each point in time. In SCBF measurement using color-labeled microspheres, the microspheres are counted either manually or automatically using flow cytometry. SCBF measurement using radiolabeled microspheres is calculated by the radioactivity within the target spinal cord using a Gamma Counter. The degree of radiation emitted from radiolabeled microspheres depends on the specific energy windows of the particular isotope. Use of this technique is a terminal study as the animal is killed to perform this test. A shortcoming of this method is that it may interfere with routine or specialized histological assessment of the spinal cord.

3. Spinal Cord Electro-physiological Assessment

The spinal cord serves as a pathway that conveys neural signals between the peripheral nervous system and brain. Compound action potential (CAP) recordings can be recorded on in vitro preparations. Isolated spinal cord white matter has been used as a model for several pathological conditions, allowing characterization of intercellular communication and synaptic physiology (13–16).

In vivo recordings of ascending and descending conduction using evoked potentials as well as segmental circuits measured by the H-reflex are discussed in the following section.

3.1. Intraspinal Field Potentials Mapping

Efferent connections, extending from the central to the peripheral nervous system, and afferent fibers are bundled together in a spatially organized fashion in the spinal cord. Predictable positive and negative field potentials can be elicited in different laminae of the lumbar cord after mesencephalic stimulation (17). Similarly, afferent somatosensory stimulation also evokes receptive fields of the spinal cord in a somatotopic order. Certain circuits are represented

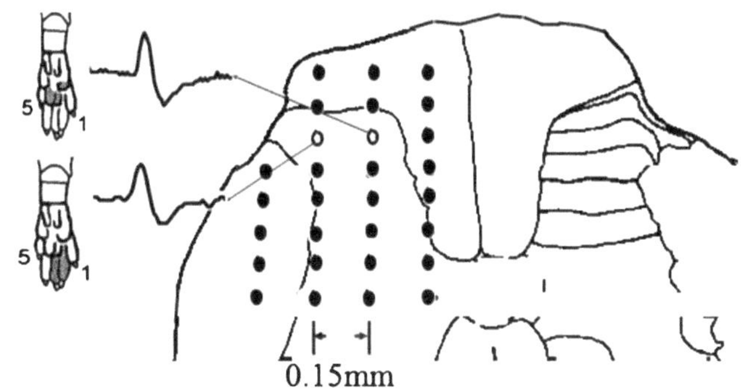

Fig. 2. Field potentials of the L5 dorsal spinal cord evoked by pinching the hindpaw of a mouse. The mouse was anesthetized using urethane and suspended on the mouse spine stabilizer. Note that the potential evoked from the second and third digits is 0.15 mm lateral to that evoked from the paw-pad.

in a point-to-point map representation. Intraspinal field potentials may be recorded extracellularly and intracellularly and demonstrate alterations in evoked potentials following efferent or afferent stimulation. Changes in the location, size, and modality of these receptive fields after pathological events may be used to analyze spinal cord function both anatomically and physiologically. Mapping changes of intraspinal field potentials demonstrate reorganization of circuits and the shifting nature of neural excitability. Alteration in receptive field properties reveals a gain or loss of electrophysiological function that represents mechanisms of plasticity. Field potentials may also be elicited in visceral or viscerosomatic stimulation, including evaluation of urogenital ascending neural circuits (18–21).

Field potential recording of sensory function on the lumbar enlargement of mice is evoked by stimulation of the ipsilateral hindpaw (Fig. 2). Animals are initially anesthetized with urethane. A T13-L1 laminectomy is performed to expose the lumbar enlargement. The dura is opened, and a 15 μm diameter tungsten electrode is stereotactically inserted into an array of recording sites in the spinal cord. At each site, the potential is recorded following a skin pinch of the ipsilateral foot using forceps. Different types of sensory stimulation (electric currency, brush strokes, skin pinch) activate different components of the spinal cord, each producing a different response pattern. Neuronal activation is recognized by its unique firing frequency amplified through a speaker. We recommend that field potential mapping is performed on a vibration-free table and that the spine is stabilized by the LISA to reduce movement induced by respiration which may induce motion artifact (Fig. 1).

3.2. Electromyography in Spinal Cord Research

Locomotion is the primary function altered following SCI, therefore, monitoring muscle activity is valuable after experimental SCI. MEPs, H-reflex, and the F-response are all examples of induced muscle activity. Free-run EMG is also a useful technique to measure electrophysiological activity of muscles in static and dynamic phases to demonstrate motor control. Resting EMG is recorded from neurologically intact animals and is quiescent. Following SCI, spontaneous resting muscle activity is increased and may be characterized by rapid firing and the behavior of spasticity. Progression of this state may be monitored by EMGs that express prolonged muscle contractions and increased excitability of muscles to noxious and non-noxious cutaneous stimulation (22, 23). Dynamic EMGs monitor muscle activity during locomotion. EMGs record the onset, duration, interval, and coordination of muscle activity and serve as reliable indicators of locomotor function which correlates well with patterns of movement (24, 25). Combined with kinematic recordings of gait, information from dynamic EMGs identifies and characterizes locomotor deficits following SCI.

We use a bandpass of 10–500 Hz when recording free-run EMGs. Small needle electrodes are inserted into target muscles in acute experiments. Chronic recordings may require intramuscular electrodes for single motor unit recordings or gross EMGs. Most chronic indwelling electrodes are made of 25–50 μm Teflon-coated platinum/iridium wires. For gross recordings of EMGs, two electrodes are implanted separately—one in the muscle belly and the other in the distal end of the muscle. For single-unit recordings, two wires are twisted tightly together. Chronic recording electrodes need to be tunneled subcutaneously from the recording point to the exterior. If intracranial electrodes are used, the electrodes are fixed to the skull using dental cement and tunneled to the point of exit from the neck. This decreases the risk of electrode damage caused by the animals (26–28).

4. Miscellaneous Assessments

Several other physiological tests of spinal cord function, such as pH, PCO_2, and PO_2 may be measured following SCI by intraparenchymal microelectrode recordings (29). pH, PCO_2, and PO_2 levels in the CSF may be used as a proxy for these tests in spinal cord tissue to avoid the trauma induced by electrode insertion into the spinal cord (30). Spinal cord oxygen levels may also be estimated by measuring the oxygen saturation of erythrocytes using oximetry. This technique is based on the difference in the spectrum of light absorbed by the deoxygenated and oxygenated hemoglobin in red blood cells. Oxygen saturation of the spinal cord may be monitored continuously and conveniently using oximetry (31). Temperature is another parameter of spinal cord function which is associated with autonomic function of thermoregulation and may be impaired following SCI (32). The ability to measure spinal cord temperature is vital in studying the role of hypothermia as a spinal cord protectant. Core temperature of the body is similar to spinal cord temperature, however, the actual cord temperature must be measured during treatment using focal hypothermia (33, 34). We prefer measuring the temperature from the cord surface to avoid damage induced by inserting intraparenchymal probes. Superficial temperature probes are easily manufactured using a 44 AWG microthermocouple.

SCI is a devastating injury and is frequently associated with a poor clinical prognosis. Progress focusing on the understanding of cellular signaling pathways is slowly revealing mechanisms underlying SCI. Current techniques in molecular biology and novel innovations in therapeutic methods are being developed to protect and restore the function of damaged circuits of the spinal cord which has renewed the hope of curing SCI. The effectiveness of any therapeutic strategy needs to be evaluated in vitro and in vivo to provide morphological and behavioral assessments of the proposed treatment. Both physiological and electrophysiological evaluations provide another layer to evaluate spinal cord function that is discussed in this section of the book. Quantitative analysis of physiological parameters is of value in determining the nature and severity of SCI as well as the degree of recovery in experimental SCI research.

With improvement of modern technology and rapid innovation in monitoring physiological parameters, physiological evaluation will become more important. Methods to assess electrophysiological function by MEPs, sensory evoked potentials (SSEP), and Hoffman's reflex (H-reflex) are described in the following chapters.

References

1. Kindt GW (1971) Autoregulation of spinal cord blood flow. Eur Neurol 6(1):19–23
2. Kurosawa M et al (2007) Contribution of supraspinal and spinal structures to the responses of dorsal spinal cord blood flow to innocuous cutaneous brushing in rats. Auton Neurosci 136(1–2):96–99
3. Lee JY et al (2008) Effect of electrical stimulation of the cervical spinal cord on blood flow following subarachnoid hemorrhage. J Neurosurg 109(6):1148–1154
4. Lindsberg PJ et al (1992) Laser-Doppler flowmetry in monitoring regulation of rapid microcirculatory changes in spinal cord. Am J Physiol 263(1 Pt 2):H285–H292
5. Malik Y et al (2009) Laser-Doppler measurements of spinal cord blood flow changes during hemilaminectomy in chondrodystrophic dogs with disk extrusion. Vet Surg 38(4):457–462
6. Simonovich M et al (2008) Real-time monitoring of mitochondrial NADH and microcirculatory blood flow in the spinal cord. Spine 33(23): 2495–2502
7. Toda H et al (2008) Responses of dorsal spinal cord blood flow to noxious mechanical stimulation of the skin in anesthetized rats. J Physiol Sci 58(4):263–270
8. Guha A et al (1989) Spinal cord blood flow and systemic blood pressure after experimental spinal cord injury in rats. Stroke 20(3):372–377
9. Guha A et al (1989) Improvement in post-traumatic spinal cord blood flow with a combination of a calcium channel blocker and a vasopressor. J Trauma 29(10):1440–1447
10. Macdonald AG et al (1993) Cerebral blood flow measurement using the hydrogen clearance technique: a microcomputer-based system for data collection and analysis. Med Biol Eng Comput 31(4):426–429
11. Etz CD et al (2009) Selective cerebral perfusion at 28 degrees C—is the spinal cord safe? Eur J Cardiothorac Surg 36(6):946–955

12. Sandor P et al (2003) Significance of endogenous opioids in the maintenance of cerebral and spinal vascular CO_2-sensitivity in deep hemorrhagic hypotension. Brain Res Bull 59(6):433–438
13. Imaizumi T et al (1998) Resistance to anoxic injury in the dorsal columns of adult rat spinal cord following demyelination. Brain Res 779(1–2):292–296
14. Pryor J, Shi R (2006) Electrophysiological changes in isolated spinal cord white matter in response to oxygen deprivation. Spinal Cord 44(11):653–661
15. Sinha K et al (2006) Functional changes in genetically dysmyelinated spinal cord axons of shiverer mice: role of juxtaparanodal Kv1 family K+ channels. J Neurophysiol 95(3): 1683–1695
16. Velumian AA et al (2010) Modular double sucrose gap apparatus for improved recording of compound action potentials from rat and mouse spinal cord white matter preparations. J Neurosci Methods 187(1):33–40
17. Noga BR et al (1995) Field potential mapping of neurons in the lumbar spinal cord activated following stimulation of the mesencephalic locomotor region. J Neurosci 15(3 Pt 2): 2203–2217
18. Hubscher CH (2006) Estradiol-associated variation in responses of rostral medullary neurons to somatovisceral stimulation. Exp Neurol 200(1):227–239
19. Hubscher CH (2006) Ascending spinal pathways from sexual organs: effects of chronic spinal lesions. Prog Brain Res 152:401–414
20. Kaddumi EG, Hubscher CH (2007) Urinary bladder irritation alters efficacy of vagal stimulation on rostral medullary neurons in chronic T8 spinalized rats. J Neurotrauma 24(7): 1219–1228
21. Kaddumi EG, Hubscher CH (2007) Changes in rat brainstem responsiveness to somatovisceral inputs following acute bladder irritation. Exp Neurol 203(2):349–357
22. Kitzman PH et al (2007) Gabapentin suppresses spasticity in the spinal cord-injured rat. Neuroscience 149(4):813–821
23. Kitzman PH (2009) Effectiveness of riluzole in suppressing spasticity in the spinal cord injured rat. Neurosci Lett 455(2):150–153
24. Gorska T et al (2009) Recovery of overground locomotion following partial spinal lesions of different extent in the rat. Behav Brain Res 196(2):286–296
25. Majczynski H et al (2007) Comparison of two methods for quantitative assessment of unrestrained locomotion in the rat. J Neurosci Methods 163(2):197–207
26. Carp JS et al (2001) Effects of chronic nerve cuff and intramuscular electrodes on rat triceps surae motor units. Neurosci Lett 312(1):1–4
27. Eken T (1998) Spontaneous electromyographic activity in adult rat soleus muscle. J Neurophysiol 80(1):365–376
28. Eken T et al (2008) Development of tonic firing behavior in rat soleus muscle. J Neurophysiol 99(4):1899–1905
29. Okada Y et al (1993) Depth profiles of pH and PO_2 in the isolated brain stem-spinal cord of the neonatal rat. Respir Physiol 93(3): 315–326
30. Ulus F et al (2009) Alterations in cerebrospinal fluid PO(2), PCO(2), and pH measurements during and after experimental thoracic aortic cross-clamping. Ann Vasc Surg 23(1): 122–127
31. Phillips JP et al (2006) An optical fiber photoplethysmographic system for central nervous system tissue. Conf Proc IEEE Eng Med Biol Soc 1:803–806
32. Kalincik T et al (2009) Local response to cold in rat tail after spinal cord transection. J Appl Physiol 106(6):1976–1985
33. Dimar JR et al (2000) The role of directly applied hypothermia in spinal cord injury. Spine 25(18):2294–2302
34. Lee JR et al (2007) Effects of intrathecal bupivacaine in conjunction with hypothermia on neuronal protection against transient spinal cord ischemia in rats. Acta Anaesthesiol Scand 51(1):60–67

Chapter 43

Electrophysiological Assessment of Spinal Cord Function on Rodents Using tcMMEP and SSEP

Yi Ping Zhang, Lisa B.E. Shields, and Christopher B. Shields

Abstract

Electrophysiological assays following experimental spinal cord injury objectively evaluate neurological function in the rodent. Major descending and ascending tracts can be monitored noninvasively using motor-evoked potentials and somatosensory-evoked potentials, respectively. This chapter summarizes the methods, procedures, and materials used to carry out these tests in our laboratory. Interpretation of results is also discussed.

Key words: Motor-evoked potentials, Transcranial magnetic stimulation-induced motor-evoked potentials, Somatosensory-evoked potentials, Electrophysiological assessment

1. Introduction

Electrophysiological assays have been developed to monitor the integrity of ascending and descending pathways of the spinal cord on humans for decades and have recently been modified for use in experimental animals. The value of motor-evoked potentials (MEPs) and somatosensory-evoked potentials (SSEPs) are recognized. Electrophysiological tests of spinal cord function have been studied in different animal models, but rodents are primarily used for neuroscience research due to their lower cost and easy maintenance as well as the availability of genetically modified mice. Transgenic and knockout mice mimic many human diseases that provide powerful tools to study specific genes that control pathological processes of spinal cord disease. Genetically modified mice are readily available due to recent advances in genetic engineering. However, electrophysiological assessment in these species has lagged due to the small size of rodents.

MEPs assess the integrity of the descending motor pathways in the spinal cord and are used clinically to evaluate alterations in

Jun Chen et al. (eds.), *Animal Models of Acute Neurological Injuries II*, Springer Protocols Handbooks,
DOI 10.1007/978-1-61779-782-8_43,

spinal cord function (1, 2). This technique is valuable when used on primates, cats, and rats (3–12). Recently, MEPs have been adapted for use in mice following spinal cord injury (SCI) (13–15). MEPs were initially elicited using high-voltage electrical stimulation to activate the motor cortex (16–18). Animals were anesthetized to minimize pain inherent in the use of electrical stimulation. However, anesthetic agents distort, delay, or eliminate the onset of MEP resulting in false positive responses that may be interpreted as pathological events (19–29). Thus, several common anesthetic agents, such as pentobarbital and halothane, are contraindicated when evaluating MEP in animal research (30). To obviate false positives caused by anesthetic or sedative agents on motor conductivity, painless magnetic stimulation was introduced to avoid discomfort induced by painful electrical stimulation. Nonsedated rodents readily tolerate magnetic stimulation when applied transcranially, over the spine, or peripheral nerves to induce MEP responses from the distal extremities that are simple, reliable, and reproducible (31). Magnetic stimulation-induced motor-evoked potentials (MMEPs) are also feasible on mice (13, 15).

SSEPs have been used (32) to assess the sensory component of spinal cord function intraoperatively (33). Absence of SSEP is an indicator of poor sensorimotor function both clinically and experimentally. SSEPs monitor only sensory function traversing the posterior columns of the spinal cord; however, it is commonly used as a proxy for both sensory and motor components. This test is performed by stimulating peripheral nerves of the forelimb and hindlimb and recording evoked responses from the cerebral cortex or cervical spinal cord. Experimentally, the SSEPs are recorded by an electrode anchored to the skull (34–39). The SSEP response (like MEP) is attenuated by anesthetic agents or sedative drugs (40–43). The feasibility and reliability of MEP and SSEP have been confirmed on nonanesthetized rodents (15). In this chapter, the materials and protocols to perform MEP and SSEP on nonanesthetized rodents are described, and the potential benefits and limitations in the laboratory are discussed.

2. Experimental Use of tcMMEP and SSEP

2.1. Pathways

In humans, SSEPs are utilized to assess proprioceptive function of the dorsal columns, and MEP test the integrity of the descending corticospinal tracts. However, in rodents the pathways which are measured by transcranial magnetic stimulation induced motor-evoked potentials (tcMMEPs) and SSEPs are different.

The major descending motor pathway in the rodent is the ventral funiculus of the spinal cord. This was confirmed by lesion studies performed in our laboratories. Graded laceration lesions in mice extending from the dorsal to the ventral surface of the spinal cord demonstrated that tcMMEP responses remain intact until >75% of the spinal cord diameter was transected (13) (Fig. 1).

The ventrolateral funiculus (VLF) pathway transmitting tcMMEP has been further verified by focal demyelination studies and graded contusion SCI (Fig. 2). All the data confirms that the VLF transmits tcMMEP signals.

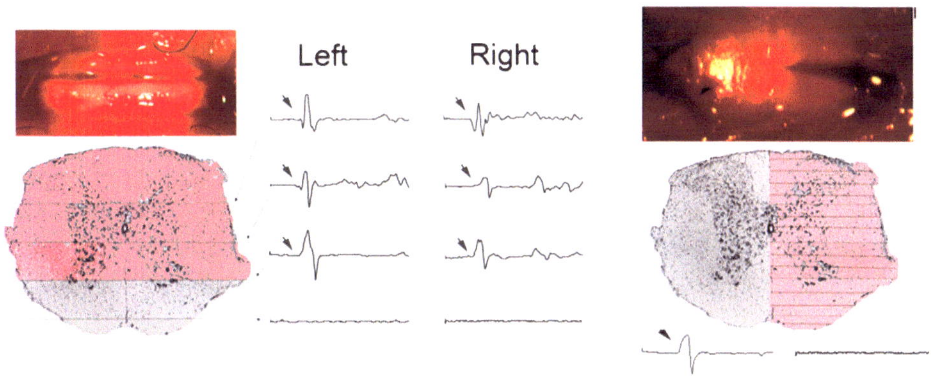

Fig. 1. Following bilateral (*left*) and unilateral (*right*) laceration lesions of the mouse at the T9 spinal cord level (using the LISA). The *top* pictures demonstrate lesions produced in the spinal cord. The *lower* diagrams indicate the area of laceration SCI (*pink*). Bilateral spinal cord lesions demonstrate that tcMMEP conduction occurs entirely within the ventral funiculus. MEP responses could be elicited until the lesion depth reached 1.4 mm (75% of the cord diameter). The unilateral lesion indicates that tcMMEP conduction is ipsilateral in the acute stage. The ipsilateral tcMMEP is partially restored during the chronic stage following SCI which may be due to activation of unknown connections across the midline or activation of other putative compensatory mechanisms.

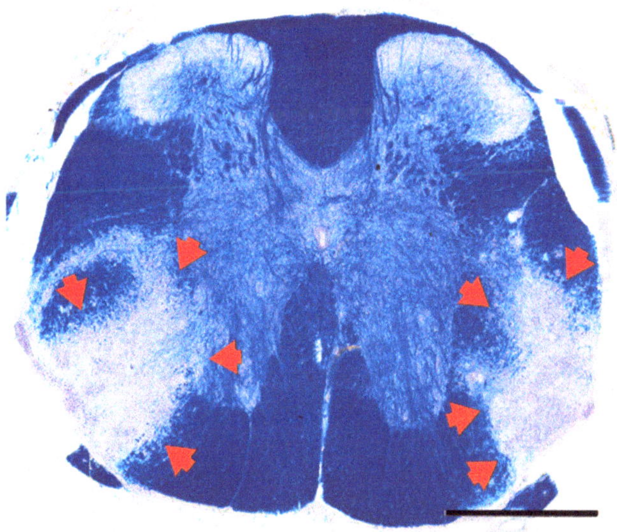

Fig. 2. Demyelination lesions of the ventrolateral funiculus (VLF). The *white area* (outlined by *red arrowheads*) was demyelinated using 0.1% lysolecithin (bar = 0.5 mm) (eriochrome cyanide stain). Following VLF demyelination, the tcMMEP responses could not be elicited.

TcMMEP conduction pathways are concentrated in the descending VLF of the rodent spinal cord; however, the ascending SSEP responses are more widely distributed. SSEP conduction is transmitted predominantly via the posterior columns in mammals, but at least three spinal tracts are involved in the cat and monkey, namely the posterior columns, the spinocervicothalamic tract, and the VLF (44–46). SSEP responses elicited from the rodent can vary depending on the characteristics and intensity of the stimulus. Ascending sensory connections to the cerebellum arise from the ventral and dorsal spinocerebellar tracts, and another ascending sensory pathway connects to the sensory motor gyrus of the cerebral cortex which is transmitted through the medial lemniscus in the dorsal column and the ventral spinothalamic pathway. Many ascending somatosensory inputs may make extensive synaptic contacts in the thalamus or the reticular formation before reaching the cerebral cortex. Thus, SSEP ascending projections to the brain are largely redundant. Recently, Muramatsu et al. obtained SSEP from the cerebellar hemisphere which exhibited negative peaks that were not affected by dorsal column lesions in rats (38). Electrophysiological signals recorded on the rodent brain may combine responses from the cerebellum and the cerebrum. Even with extensive laceration lesions in the dorsal funiculus, SSEP responses persist (Fig. 3). The ability to elicit cortical SSEP following a severe dorsal column lesion depends on the stimulus intensity and post-lesion interval.

Our studies show that SSEPs conduct more diffusely than tcMMEPs in the spinal cord. When the stimulus intensity is not controlled, the value of SSEP in evaluation of dorsal spinal cord function in the rodent is limited. The redundancy of SSEP following rat SCI is well recognized and is presumably due to multiple ascending connections and rapidly developing compensatory mechanisms. Following moderate 25 g/cm SCI by a weight drop device (and even after a 50 g cm SCI), there were no changes of SSEP (4, 15).

2.2. Applications of Electrophysiological Monitoring in Rodents

TcMMEP and SSEP have been used to assess different neurological disorders, such as SCI (47), brain/spinal cord ischemia (48–52), and demyelination diseases (53). In the laboratory, tcMMEP and SSEP objectively assess neurological function and the effectiveness of different therapeutic strategies that may reflect restoration of white matter function following an increase in the ischemic tolerance, a block in the apoptosis cascade, reestablishment of neural connectivity by tissue transplantation, or enhancement of axon regeneration or remyelination. Beside its usefulness in screening for effects of experimental therapies, tcMMEP have been used to test neural transmission on genetically modified mice. TcMMEP performed on null and LINGO-1 null mice produce no significant

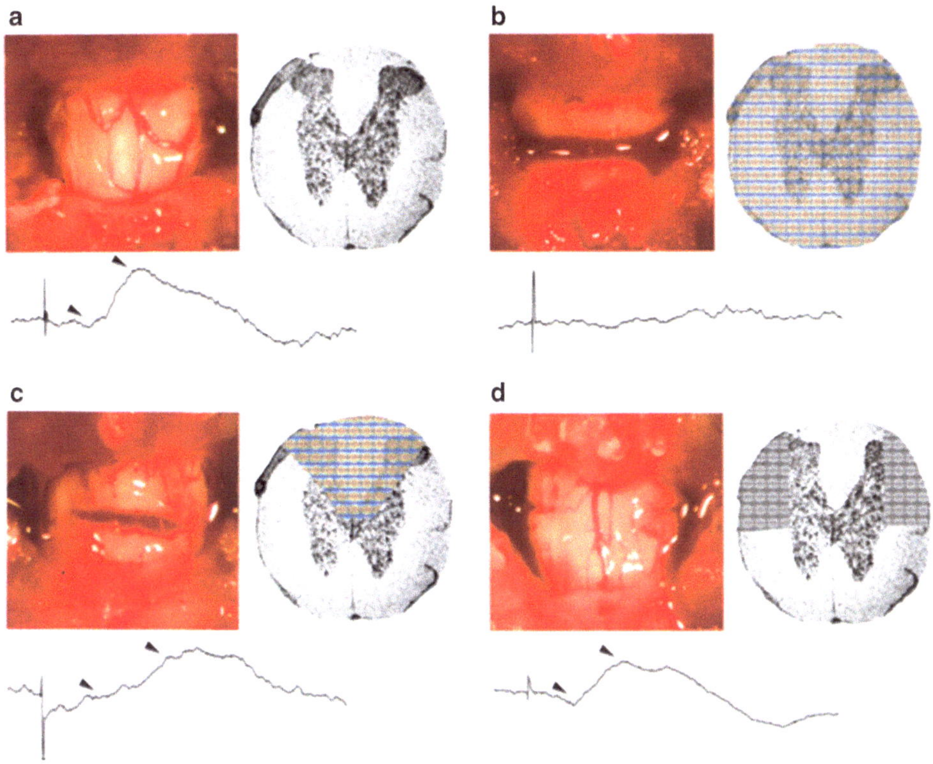

Fig. 3. The normal early SSEP response in rats is the P13/N18 complex (**a**). After complete transection of the spinal cord, SSEPs do not return (**b**). The dorsal column of the spinal cord contributes to SSEP conduction. Following dorsal column lacerations, SSEPs are distorted but still preserved (**c**). The SSEP conduction is not transmitted via the dorsolateral funiculus (DLF) as those responses are not altered after bilateral DLF lesions (**d**).

changes in latency and amplitude of responses, although these proteins are integrally related to oligodendrocyte myelin (54–56). Another study using NKX2.2(+/−)/NKX6.2(−/−) double knockout mice demonstrated a significant suppression of tcMMEP responses suggesting a role of homeobox genes in affecting nervous system development (4, 15). By these examples, tcMMEP and SSEP may be used as tools to characterize the electrophysiologic phenotype of genetically altered animals (47, 57).

3. Methods and Procedures

3.1. Equipment and Materials

3.1.1. tcMMEP Materials

1. *Evoked potential machine (EP)*: We use Cadwell systems (Quartum, Excel, Sierra and Cascade) (Cadwell Laboratories, Kennewick, WA) that are all capable of recording tcMMEP responses in rodents.

2. *Magnetic stimulator and coils*: Single source magnetic pulses applied to the skull generate tcMMEP in rodents. Stimulators designed for high frequency magnetic stimulation, such as Magstim Rapid (Magstim Company Ltd., Wales, UK) or MagPro (MagVenture, Farum, Denmark) are unsuitable to elicit tcMMEP and are used only for the treatment of patients suffering from depression or other psychiatric disorders. The Cadwell MES-10 and the Magstim 200 are capable of inducing an adequate magnetic pulse to create tcMMEP in rats. We prefer the Cadwell MES-10 stimulator on mice because the area of activation over the small brain area requires a stronger stimulus. The small magnetic coil (2.5 cm) conforms better to the rodent skull, but we do not recommend this coil as it does not produce adequate magnetic pulses to stimulate brain activation. The geometry and radius of the coil affects the induced electric field (58). Considering all options, we recommend that a 5 cm single coil without casing connected to the Cadwell MES-10 stimulator be used to induce optimal tcMMEP responses.
3. *Subdermal needle electrodes* are used for CMAP recordings from muscles as well as serving as the ground electrode.
4. *A wooden board* is used to restrain animals. The platform is 9″ × 3.5″ for rats and 4.5″ × 2″ for mice.
5. *Stockinet*: A thin and porous stockinet is used to restrain the rodent to the wooden platform.
6. *Thumb tacks*: 10–12.

3.1.2. SSEP Materials

1. *EP machine*: Any evoked potential machine such as that manufactured by Cadwell Laboratories (same as above).
2. *Epidural recording electrodes* were made from 0.2 mm silver wire and gold pins that are anchored to the skull.
3. *Subdermal needle electrodes* are used for grounding.
4. *Ring bipolar stimulating electrodes* are placed on the lower limb. That electrode delivers a painless low electrical current stimulus to the posterior tibial nerve (Fig. 4).
5. *Wooden board* (same as above).
6. *Stockinet* (same as above).
7. *Thumb tacks* (same as above).
8. *Conductive paste* is used to reduce impedance between skin and electrodes.

3.2. tcMMEP Recording

Nonanesthetized rodents are restrained on the wooden board by wrapping them with a gauze stockinet (15) beginning at the head, then progressing to the trunk and legs. Active needle electrodes are inserted into the target muscle bellies, and reference electrodes are inserted near the distal tendon of that muscle (Fig. 5).

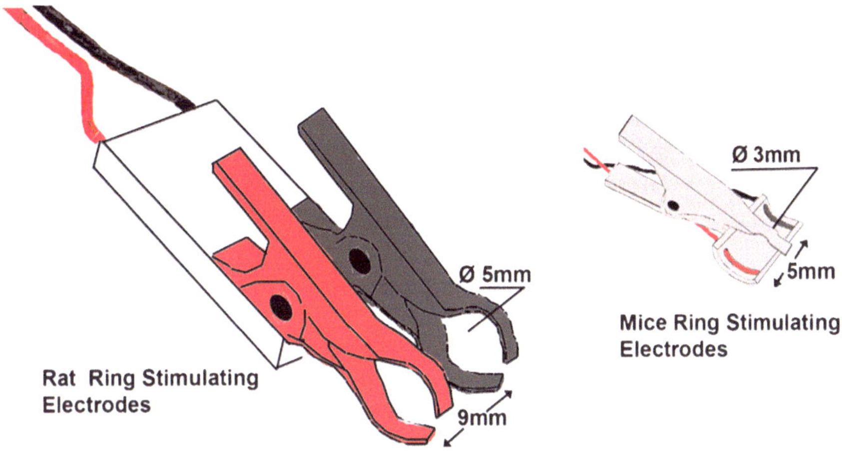

Fig. 4. A mild electrical current is used to stimulate the posterior tibial nerve transcutaneously. The specially designed ring stimulating electrode is clamped on the leg, and a thin layer of conductive paste is used to line the ring overlying the posterior tibial nerve.

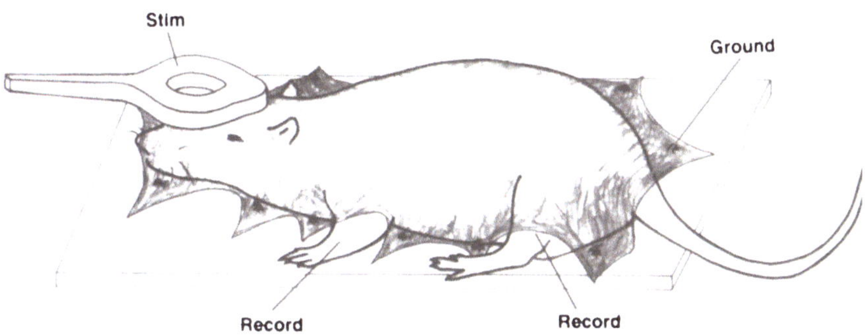

Fig. 5. Illustration of a restrained rodent undergoing a tcMMEP test. The head is stimulated through the stockinet using the magnetic coil. Limbs are exposed for the placement of recording electrodes. Ground electrodes are placed in the tail.

The common digital extensor and gastrocnemius muscles are used to monitor forelimb and hindlimb functions in the rat, respectively. Only hindlimb muscles are tested in the mouse with the ground electrode inserted at the base of the tail. Needle electrodes are secured with adhesive tape. TcMMEPs are elicited using a magnetic stimulator connected to a 5.0 cm doughnut-shaped coil placed over the cranium. The excitability of the brain of rats and mice is different due to the different brain sizes. In normal adult rats, the magnetic stimulation intensity threshold using the Cadwell MES-10 (maximal magnetic field of 2 T) is approximately 25% maximum output. In the mouse brain, threshold excitability is lower than in the rat, and maximum stimulation (100%) is required to elicit MEP responses using a 5 cm coil. In our experience, the Magstim can be used to obtain tcMMEP in the rat but not in the mouse.

Onset latencies are measured in milliseconds (ms). TcMMEP amplitudes are measured from the peak-to-trough of the compound muscle action potential (CAMP) and recorded in millivolts (mV). Since the tcMMEP has a large amplitude, it does not require averaging to reduce the influence of background noise. The onset latency in the normal rat is approximately 6 ms, and 5 ms for normal mice. The screen sweep is set at 25 ms to cover both the early and delayed responses. In the NKX6.2/NKX2.2 null animal, the tcMMEP onset latency was twice the duration of onset latency obtained from a wild type mouse (15).

The onset latency and amplitude of tcMMEP are also affected by the animal's age. If the animal is <5 weeks old, tcMMEPs have not matured (Fig. 6) and, therefore, reliable responses will not be obtained. This lack of tcMMEP development is directly related to the state of myelin maturation in the spinal cord (unpublished data).

3.3. SSEP Recording

SSEPs are recorded percutaneously by subcutaneous needle electrode placement into the scalp (59) or by the implantation of ball-shaped electrodes placed over the sensory cortex in the epidural space. Epidural electrodes diminish artifact arising from muscle contraction and greatly enhance the recording quality. Epidural electrodes must be placed under anesthesia. Three pairs of holes are created in the skull using a 1.2 mm drill, followed by passage of 0.2 mm diameter silver wires in the epidural space between the skull holes. Wires passing through the holes are twisted and anchored to the skull using dental cement. Active electrodes are placed 2 mm posterior to the bregma and 2 mm lateral to the sagittal suture with the reference wire located adjacent to the olfactory area. Pins are anchored to the skull using dental cement to prevent them from being dislodged by the forepaw during grooming (15). This allows the electrodes to remain in place over extended periods. In mice, a single hole is placed in the skull to fixate a silver ball epidurally. The 1 mm diameter silver ball electrodes are manufactured by melting the wire on a flame. If the silver wire comes in contact with skin, it can be insulated using a thin coat of nail polish. SSEP can be performed in the unanesthetized state 1 day after electrode placement. The animal is restrained as described above using a stockinet, and the grounding needle electrode is inserted into the animal's back. The intensity of the electric stimulus applied to the ring electrode needs to be well controlled. A current that causes a minimal flicker of the digit is adequate to obtain a response; however, a strong stimulus will cause pain and stress. The P13/N18 early SSEP complex is elicited with the stimulus set at 0.7 mA, 100 μs pulse width, and a frequency of 0.3 Hz. Twenty signals are averaged to distinguish the SSEP response from background EEG activity. Painless peripheral magnetic stimulation may induce SSEP responses; however, protective shielding surrounding the rat is

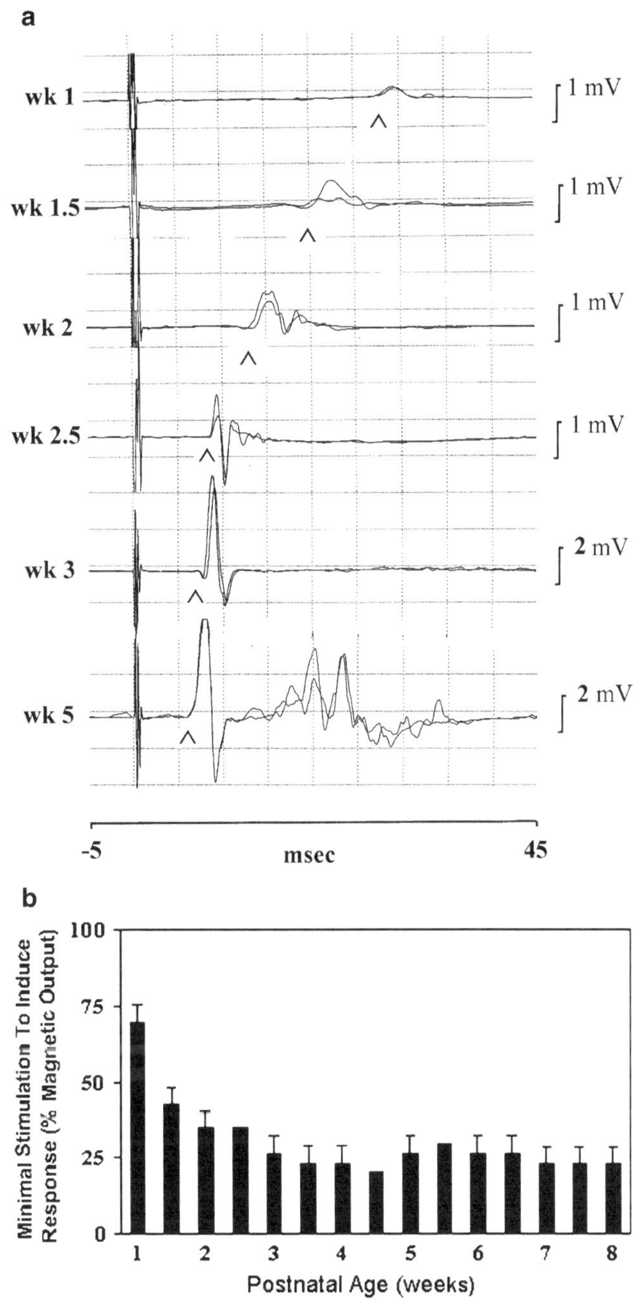

Fig. 6. (**a**) TcMMEP of immature rats display delayed onset latency and diminished amplitude that reach maturity at 5 weeks of age. (**b**) TcMMEP excitability increases rapidly within the first few weeks of life. For rats between 3 and 4 weeks old, the threshold stimulation to induce MEP responses is 25% of the maximal output.

necessary to perform this technique successfully. Without shielding, huge artifacts are elicited by magnetic stimulation that obscure small SSEP signals (15).

4. Discussion and Conclusion

Evoked potentials represent functional conductivity through the CNS and PNS. TcMMEP and SSEP represent conduction through motor and sensory pathways, respectively. However, electrophysiological findings do not closely correspond to the behavioral performance. For example, selective kainic acid (KA) injections into gray matter destroyed interneurons in the upper lumbar enlargement of the spinal cord. If α-motor neurons were partially preserved and animals were unable to move their hindlimbs, then the tcMMEP remained intact. This suggests that descending pathways are intact (including motoneurons), and the behavioral motor deficit is due to malfunction of the interneuron circuit responsible for coordinating locomotor functions (60, 61). Thus, in this scenario evoked potential assay and behavioral assessment are independent.

In rodents tcMMEP are transmitted through the VLF of the spinal cord. This test is exquisitely sensitive to loss of function following contusion SCI as well as following demyelination injury of the VLF. Following a precisely graded SCI contusion, the tcMMEP appearance corresponds with alterations in the BBB locomotor scale. Using these methods (BBB, tcMMEP), we can assess small functional differences in the magnitude of SCI with as little as 0.2 mm incremental change in tissue displacement (39). Thus, tcMMEP can be used to detect minimal differences in the severity of SCI, and tcMMEP may be used to monitor VLF function following SCI in that region (62). Following a demyelination injury created by the injection of 1% lysolecithin into the spinal cord, tcMMEP can provide a functional measure that corresponds to demyelination which closely reflects histological alterations (63).

We observed in rodents that following a unilateral SCI (Brown-Séquard syndrome), chronic signs do not persist as in humans due to the rapid return of motor function on the side of the lesion. Accordingly, tcMMEP responses in rodents also recover rapidly (unpublished data). This is an example of electrophysiological differences between species that reflect the difficulties in using rodents and other animals prior to initiating clinical trials.

SSEPs are conducted predominantly through the dorsal columns of the spinal cord (64, 65). In the adult rat, SSEPs are not permanently abolished following a selective injury to the dorsal columns (66) (Fig. 3) because multiple ascending SSEP projections may account for SSEP functional redundancy. SSEP may be useful as a method to monitor electrophysiological function following spinal cord ischemia or trauma, however, SSEPs are of little value in predicting the potential for recovery between the mild and moderate injury levels (4, 15). Mechanisms of cortical SSEP redundancy are unknown but may be due to multiple ascending connections or signal propagation in subcortical structures (67).

We observed that recovery of cortical SSEPs following injury is greater when cervical SSEPs are recorded.

Several methods to assess neurological behavior have been developed, including BBB, BMS, inclined plane, grid walking, footprint by Treadscan system, Von Frey Hair, and Hargreave's test. However, they are less valuable than the objective method of measuring evoked potentials in assessing conduction through the spinal cord. In unanesthetized rodents SSEP and tcMMEP, modified for use in the rodent, can reliably record ascending and descending conductive function of the spinal cord. These methods are valuable in SCI research as we search for metrics to measure minor changes following putative therapeutic modalities. However, conclusion that therapy is successful should be cautious due to the high spontaneous recovery rate in rodents.

References

1. Gonzalez AA et al (2009) Intraoperative neurophysiological monitoring during spine surgery: a review. Neurosurg Focus 27(4):E6
2. MacDonald DB (2002) Safety of intraoperative transcranial electrical stimulation motor evoked potential monitoring. J Clin Neurophysiol 19(5): 416–429
3. Adamson J et al (1989) Effects of selective spinal cord lesions on the spinal motor evoked potential (MEP) in the rat. Electroencephalogr Clin Neurophysiol 74(6):469–480
4. Baskin DS, Simpson RK Jr (1987) Corticomotor and somatosensory evoked potential evaluation of acute spinal cord injury in the rat. Neurosurgery 20(6):871–877
5. Dull ST et al (1990) Amplitude and latency characteristics of spinal cord motor evoked potentials in the rat. Electroencephalogr Clin Neurophysiol 77(1):68–76
6. Fehlings MG et al (1988) Motor and somatosensory evoked potentials recorded from the rat. Electroencephalogr Clin Neurophysiol 69(1):65–78
7. Fehlings MG et al (1991) The electrophysiological assessment of the pyramidal and nonpyramidal tracts of the spinal cord of rats. Electroencephalogr Clin Neurophysiol Suppl 43:287–296
8. Keller BP et al (1992) The effects of propofol anesthesia on transcortical electric evoked potentials in the rat. Neurosurgery 30(4): 557–560
9. Shiau JS et al (1992) The effect of graded spinal cord injury on the extrapyramidal and pyramidal motor evoked potentials of the rat. Neurosurgery 30(1):76–84
10. Wang CX et al (1993) Motor evoked potentials in a static load model of spinal cord injury in the rat. Neurosurgery 32(2):269–273
11. Zappulla RA et al (1988) Noncortical origins of the spinal motor evoked potential in rats. Neurosurgery 22(5):846–852
12. Zileli M, Schramm J (1989) [Spinal and muscular evoked response following single stimulation of the motor cortex of the rat]. EEG EMG Z Elektroenzephalogr Elektromyogr Verwandte Geb 20(2):106–111
13. Hill RL et al (2009) Anatomic and functional outcomes following a precise, graded, dorsal laceration spinal cord injury in C57BL/6 mice. J Neurotrauma 26(1):1–15
14. Lu M et al (2008) [Establishment and assessment of the mouse model for spinal cord injury]. Zhongguo Xiu Fu Chong Jian Wai Ke Za Zhi 22(8):933–938
15. Zhang YP et al (2007) Use of magnetic stimulation to elicit motor evoked potentials, somatosensory evoked potentials, and H-reflexes in non-sedated rodents. J Neurosci Methods 165(1):9–17
16. Marsden CD et al (1983) Direct electrical stimulation of corticospinal pathways through the intact scalp in human subjects. Adv Neurol 39:387–391
17. Merton PA, Morton HB (1980) Stimulation of the cerebral cortex in the intact human subject. Nature 285(5762):227
18. Merton PA et al (1982) Scope of a technique for electrical stimulation of human brain, spinal cord, and muscle. Lancet 2(8298):597–600
19. Fishback AS et al (1995) The effects of propofol on rat transcranial magnetic motor evoked potentials. Neurosurgery 37(5):969–974

20. Ghaly RF et al (1990) Effects of incremental ketamine hydrochloride doses on motor evoked potentials (MEPs) following transcranial magnetic stimulation: a primate study. J Neurosurg Anesthesiol 2(2):79–85
21. Hargreaves SJ, Watt JW (2005) Intravenous anaesthesia and repetitive transcranial magnetic stimulation monitoring in spinal column surgery. Br J Anaesth 94(1):70–73
22. Kawaguchi M et al (1996) Effect of isoflurane on motor-evoked potentials induced by direct electrical stimulation of the exposed motor cortex with single, double, and triple stimuli in rats. Anesthesiology 85(5):1176–1183
23. Kawaguchi M, Furuya H (2006) [Motor evoked potentials]. Masui 55(3):294–301
24. Pechstein U et al (1998) Isoflurane plus nitrous oxide versus propofol for recording of motor evoked potentials after high frequency repetitive electrical stimulation. Electroencephalogr Clin Neurophysiol 108(2):175–181
25. Scheufler KM et al (2005) The modifying effects of stimulation pattern and propofol plasma concentration on motor-evoked potentials. Anesth Analg 100(2):440–447
26. Sloan TB, Heyer EJ (2002) Anesthesia for intraoperative neurophysiologic monitoring of the spinal cord. J Clin Neurophysiol 19(5):430–443
27. Taniguchi M et al (1993) Effects of four intravenous anesthetic agents on motor evoked potentials elicited by magnetic transcranial stimulation. Neurosurgery 33(3):407–415
28. Toda Y (1992) [The effect of anesthetic agents on descending spinal cord evoked potential and the compound muscle action potentials elicited by stimulation at the cerebral motor cortex and the spinal cord]. Nippon Seikeigeka Gakkai Zasshi 66(4):279–290
29. Ubags LH et al (1998) Influence of isoflurane on myogenic motor evoked potentials to single and multiple transcranial stimuli during nitrous oxide/opioid anesthesia. Neurosurgery 43(1):90–94
30. Glassman SD et al (1993) Anesthetic effects on motor evoked potentials in dogs. Spine 18(8):1083–1089
31. Linden RD et al (1999) Magnetic motor evoked potential monitoring in the rat. J Neurosurg 91(2 Suppl):205–210
32. Allison T (1962) Recovery functions of somatosensory evoked responses in man. Electroencephalogr Clin Neurophysiol 14: 331–343
33. Chandanwale AS et al (2008) Intra-operative somatosensory-evoked potential monitoring. J Orthop Surg (Hong Kong) 16(3):277–280
34. Eftekharpour E et al (2005) Structural and functional alterations of spinal cord axons in adult Long Evans Shaker (LES) dysmyelinated rats. Exp Neurol 193(2):334–349
35. Fehlings MG et al (1989) The relationships among the severity of spinal cord injury, motor and somatosensory evoked potentials and spinal cord blood flow. Electroencephalogr Clin Neurophysiol 74(4):241–259
36. Hurlbert RJ et al (1996) A novel sensory evoked potential for selective monitoring of the ventral spinal cord: from bench to bedside. Electroencephalogr Clin Neurophysiol Suppl 46:221–231
37. Jou IM et al (2003) The effects of intrathecal tramadol on spinal somatosensory-evoked potentials and motor-evoked responses in rats. Anesth Analg 96(3):783–788
38. Muramatsu H et al (2009) Evoked potentials elicited on the cerebellar cortex by electrical stimulation of the rat spinocerebellar tract. Surg Neurol 72(4):395–400
39. Zhang YP et al (2008) Spinal cord contusion based on precise vertebral stabilization and tissue displacement measured by combined assessment to discriminate small functional differences. J Neurotrauma 25(10):1227–1240
40. Haghighi SS et al (1996) Suppression of spinal and cortical somatosensory evoked potentials by desflurane anesthesia. J Neurosurg Anesthesiol 8(2):148–153
41. Marcus MA et al (1997) Spinal somatosensory evoked potentials after epidural isoproterenol in awake sheep. Can J Anaesth 44(1): 85–89
42. Peterson DO et al (1986) Effects of halothane, enflurane, isoflurane, and nitrous oxide on somatosensory evoked potentials in humans. Anesthesiology 65(1):35–40
43. Solenkova AV et al (2000) [Epidural anesthesia in surgical interventions on the spine and spinal cord. II. Effects of epidural anesthesia on somatosensory evoked potentials]. Anesteziol Reanimatol 4:32–38
44. Eidelberg E, Woodbury CM (1972) Apparent redundancy in the somatosensory system in monkeys. Exp Neurol 37(3):573–581
45. Handwerker HO, Zimmermann M (1972) Cortical evoked responses upon selective stimulations of cutaneous group 3 fibers and the mediating spinal pathways. Brain Res 36(2): 437–440
46. Norrsell U, Wolpow ER (1966) An evoked potential study of different pathways from the hindlimb to the somatosensory areas in the cat. Acta Physiol Scand 66(1):19–33

47. Yamada K et al (2008) Modulation of the secondary injury process after spinal cord injury in Bach1-deficient mice by heme oxygenase-1. J Neurosurg Spine 9(6):611–620
48. Cizkova D et al (2007) Functional recovery in rats with ischemic paraplegia after spinal grafting of human spinal stem cells. Neuroscience 147(2):546–560
49. Fukuda S (2006) [Somatosensory evoked potential]. Masui 55(3):280–293
50. Janusz MT et al (1997) Experimental use of somatosensory evoked potential for intraoperative identification of spinal cord blood supply. J Invest Surg 10(4):195–203
51. Lips J et al (2002) The role of transcranial motor evoked potentials in predicting neurologic and histopathologic outcome after experimental spinal cord ischemia. Anesthesiology 97(1):183–191
52. Qayumi AK et al (1997) Animal model for investigation of spinal cord injury caused by aortic cross-clamping. J Invest Surg 10(1–2): 47–52
53. Iuliano BA et al (1994) Motor and somatosensory evoked potentials in mice infected with Theiler's murine encephalomyelitis virus. J Neurol Sci 123(1–2):186–194
54. Mi S et al (2005) LINGO-1 negatively regulates myelination by oligodendrocytes. Nat Neurosci 8(6):745–751
55. Mi S et al (2008) LINGO-1 and its role in CNS repair. Int J Biochem Cell Biol 40(10): 1971–1978
56. Nie DY et al (2006) Oligodendrocytes regulate formation of nodes of Ranvier via the recognition molecule OMgp. Neuron Glia Biol 2:151–164
57. Abe Y et al (2003) Decreased neural damage after spinal cord injury in tPA-deficient mice. J Neurotrauma 20(1):43–57
58. Cohen LG et al (1990) Effects of coil design on delivery of focal magnetic stimulation. Technical considerations. Electroencephalogr Clin Neurophysiol 75(4):350–357
59. Claus D, Weitbrecht W, Neundorfer B (1985) Pentobarbital: the influence on somatosensory conduction in the rat. In: Schramm J, Jones SJ (eds) Spinal cord monitoring. Springer, Berlin, pp 90–94
60. Hadi B et al (2000) Lasting paraplegia caused by loss of lumbar spinal cord interneurons in rats: no direct correlation with motor neuron loss. J Neurosurg 93(2 Suppl):266–275
61. Magnuson DS et al (1999) Comparing deficits following excitotoxic and contusion injuries in the thoracic and lumbar spinal cord of the adult rat. Exp Neurol 156(1):191–204
62. Cao Q et al (2005) Functional recovery in traumatic spinal cord injury after transplantation of multineurotrophin-expressing glial-restricted precursor cells. J Neurosci 25(30): 6947–6957
63. Loy DN et al (2002) Functional redundancy of ventral spinal locomotor pathways. J Neurosci 22(1):315–323
64. Cusick JF et al (1979) Spinal cord evaluation by cortical evoked responses. Arch Neurol 36(3):140–143
65. Powers SK et al (1982) Spinal cord pathways mediating somatosensory evoked potentials. J Neurosurg 57(4):472–482
66. Onifer SM et al (2007) Loss and spontaneous recovery of forelimb evoked potentials in both the adult rat cuneate nucleus and somatosensory cortex following contusive cervical spinal cord injury. Exp Neurol 207(2):238–247
67. Kimura T et al (2008) Impulse propagation along thalamocortical fibers can be detected magnetically outside the human brain. J Neurosci 28(47):12535–12538

Chapter 44

Operant Conditioning of Spinal Cord Reflexes in Rats

Xiang Yang Chen and Jonathan R. Wolpaw

Abstract

The H-reflex, the electrical analog of the spinal stretch reflex (SSR) is mediated largely by a wholly spinal, primarily two-neuron pathway. Because this pathway is influenced by descending pathways from the brain, these spinal reflexes can be operantly conditioned. Motivated by a paradigm in which reward depends on reflex size, monkeys, humans, rats, and mice can gradually increase or decrease the SSR or the H-reflex. The reflex change develops gradually over weeks and involves neuronal and synaptic plasticity at multiple sites in the spinal cord and the brain. This simple learning phenomenon is important both as a basic science model and as the basis for a new approach to restoring useful function after spinal cord injury or in other disorders. As a basic science model, it provides access to the locations, mechanisms, and behavioral impact of the complex plasticity associated with acquisition and maintenance of even the simplest motor skills, and it offers a unique opportunity to study the role of corticospinal tract activity in health and disease. As a new therapeutic approach to spinal cord injury or other neuromuscular disorders, it offers the opportunity to target each person's specific functional deficits and to produce pathway-specific changes designed to ameliorate these deficits and restore more effective motor function. This article describes the methodology of H-reflex conditioning in the laboratory rat. The conditioning protocol described here is comparable in its key features to the reflex conditioning protocols that have proved effective in monkeys, mice, and humans.

Key words: H-reflex, Activity-dependent plasticity, Spinal cord injury, Rehabilitation

1. Introduction

Spinal cord reflexes are simple behaviors produced by CNS pathways within the spinal cord. The spinal stretch reflex (SSR), also called the knee jerk or M1, is the initial response to sudden muscle stretch (1–5). It is the simplest spinal cord reflex, and is produced largely by a two-neuron monosynaptic pathway consisting of the Ia afferent neuron from the muscle spindle, its synapse on the alpha motoneuron, and the motoneuron itself. Sudden muscle stretch excites the

Jun Chen et al. (eds.), *Animal Models of Acute Neurological Injuries II*, Springer Protocols Handbooks,
DOI 10.1007/978-1-61779-782-8_44, © Springer Science+Business Media, LLC 2012

Ia afferent, which in turn excites motoneurons innervating the same muscle and its synergists, thereby causing muscle contraction that opposes the sudden stretch.

The H-reflex is the electrical analog of the SSR. It is similar to the SSR, except that the H-reflex is elicited by direct electrical stimulation of the Ia afferent fiber, thereby bypassing the muscle spindle. Because the reflex arc is influenced by descending pathways from the brain (3, 4), the SSR or the H-reflex can be operantly conditioned. Motivated by a paradigm in which reward depends on reflex size, monkeys, humans, rats, and mice can gradually increase or decrease the SSR or the H-reflex (6–11). This conditioning develops gradually and involves plasticity in the spinal cord itself, since conditioned change remains even after all descending activity is abolished (12).

The conditioning-induced change in reflex size occurs over days and weeks of exposure to the conditioning protocol, and the larger (up-conditioning mode) or smaller (down-conditioning mode) reflex that results is a simple motor skill (i.e., "an adaptive behavior acquired throughout practice" (13)). Its acquisition is associated with neuronal and synaptic plasticity at multiple sites in the spinal cord and the brain (14). Recent studies indicate that reflex conditioning affects locomotor function in normal animals and can improve walking in spinal cord-injured rats (13, 15). It should also be noted that comparable conditioning protocols can be used to change other simple spinal cord reflexes (16).

Reflex conditioning is important both as a basic science model and as the basis for a new approach to restoring useful function after spinal cord injury or in other disorders. As a basic science model, it provides access to the locations, mechanisms, and behavioral impact of the complex plasticity associated with acquisition and maintenance of even the simplest motor skills (see refs. (17–19) for review). Furthermore, because the corticospinal tract is the only major spinal cord tract essential for H-reflex conditioning, it offers a unique opportunity to study the role of corticospinal tract activity in health and disease. As a new therapeutic approach, spinal reflex conditioning offers the opportunity to target each person's specific functional deficits and to produce pathway-specific changes designed to ameliorate these deficits and restore more effective motor function. It can also be used to evaluate corticospinal tract function after injury and to assess the effects of therapeutic interventions (15, 20, 21). In addition, reflex conditioning could be an important new approach to restoring motor function in people with partial spinal cord injuries and other chronic neuromuscular disorders. Reflex conditioning protocols should be especially useful when techniques for achieving significant regeneration become available and methods for guiding spinal cord plasticity so as to produce functionally appropriate spinal reflex function become essential (22).

2. Materials

2.1. Animals

Adult Sprague Dawley rats of either sex, weighing 250–400 g.

2.2. Chronic Stimulating and Recording Electrodes

Nerve stimulating cuff: A silicon rubber cuff (Silastic laboratory tubing, Dow Corning Co.) containing a pair of stainless steel multi-stranded fine-wire (Part # AS631, Cooner Wire, 9186 Independence Ave., Chatsworth, CA 91331) electrodes. For the rat posterior tibial nerve (which innervates the triceps surae muscles), the cuff measurements are 3.18 mm OD, 1.98 mm ID, and 3–4 mm length. The wires are inserted in the cuff so that each circles the nerve and they are 1.5 mm apart.

EMG recording electrode pairs: Two stainless steel multi-stranded fine-wire (Part # AS631, Cooner Wire) twisted together with the last 0.5 cm of each stripped of insulation and 3–5 mm separating the bare ends. A #20 hypodermic needle is used to insert the pair into the muscle belly longitudinally.

2.3. Instruments

A computer-controlled stimulating and recording system, including computer, an EMG amplifier, a nerve-cuff stimulating unit, a commutator, an oscilloscope, and a pellet dispenser (for food reward). The computer uses the custom software package Elizan (Wadsworth Center) that performs the functions described below.

3. Methods

3.1. Implantation of Chronic Stimulating and Recording Electrodes

Each rat is implanted under general anesthesia (ketamine HCl (80 mg/kg) and xylazine (10 mg/kg) (intraperitoneal)) with the EMG recording electrodes in a muscle and the nerve-stimulating cuff on its nerve. Here, we use the right soleus muscle as an example. To record soleus EMG activity, a pair of fine-wire electrodes with the final 0.5 cm stripped is placed in the right soleus muscle. To elicit the right soleus H-reflex, the silicon rubber cuff containing the stimulating electrodes is placed around the right posterior tibial nerve just proximal to the soleus branch. The Teflon-coated wires from the nerve cuff and the muscle pass subcutaneously to a small connector plug (MS363 6-pin pedestal, Plastics One Inc.) mounted on the skull with stainless steel screws and dental cement. The rat is then placed under a heat lamp and given an analgesic (Demerol, 0.2 mg, i.m.). Once awake, it receives a second dose of analgesic and is returned to its cage.

3.2. Animal Lifestyle

Figure 1a (left) shows the soleus H-conditioning protocol in the rat. Data collection begins 20 days after the implantation surgery and lasts for 70 days. Throughout this period, the animal lives in a

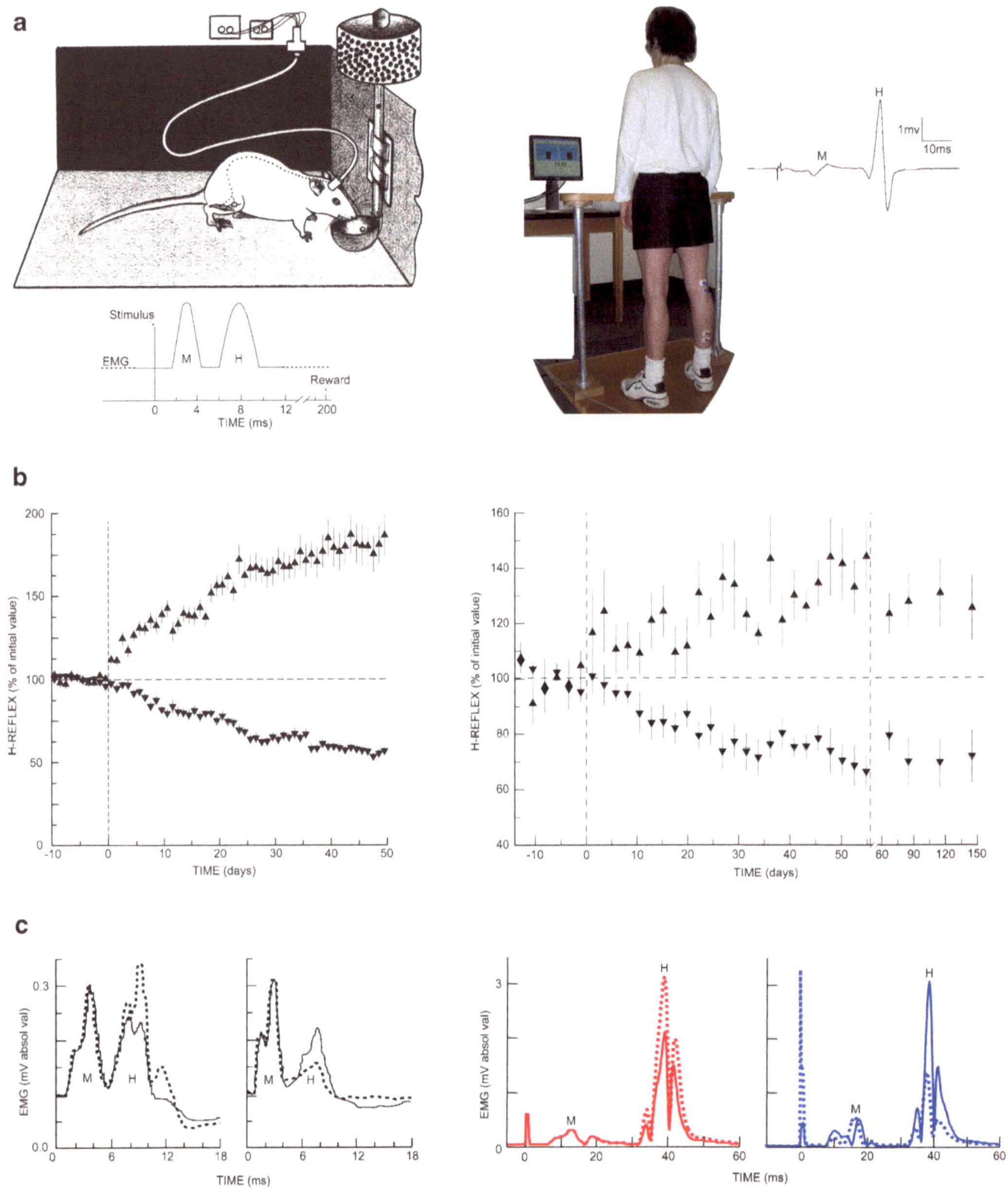

Fig. 1. The H-reflex operant conditioning protocol and its results in rats and humans. (**a**) The H-reflex conditioning protocol. *Left*: Soleus muscle activity is monitored 24 h/day in a rat with chronically implanted EMG electrodes and a posterior tibial nerve cuff. The implant wires travel subcutaneously to a head-mounted connector and then through a flexible cable and a commutator to an amplifier and a stimulator. The rat can move freely about the cage. Whenever the absolute (i.e., rectified) value of soleus EMG remains in a specified range for a randomly varying 2.3–2.7-s period, a nerve cuff stimulus elicits a threshold M response (i.e., a direct muscle response (23)) and an H-reflex. The trace shows a single trial (with EMG displayed as absolute value). A rat averages 2,000–6,000 trials per day. *Right*: Soleus activity is monitored in a person with EMG recording electrodes over the muscle and with nerve-stimulating electrodes on the popliteal fossa over the tibial nerve. The person maintains a standard erect posture facing a video screen that displays the current level (absolute value) of soleus EMG in relation to a defined range. Whenever the absolute value of soleus EMG stays in this range for several seconds, tibial nerve stimulation elicits a threshold M response and an H-reflex. The trace shows a single trial (with EMG displayed in traditional fashion as actual (i.e., unrectified) value). A person completes three 225-trial sessions per week

standard rat cage with a 40-cm flexible cable (6TCS cable, Plastics One Inc.) attached to the skull plug. The cable, which allows the animal to move freely about the cage, carries the wires from the electrodes to a commutator (SL6C commutator, Plastics One Inc.) above the cage and from there to an EMG amplifier (gain = 1,000, bandwidth 100–1,000 Hz) and a stimulus isolation unit. All animals have free access to water and food, except that during H-reflex conditioning they receive food mainly by performing the task described below. Animal well-being is carefully checked several times each day, and body weight is measured weekly. Laboratory lights are dimmed from 2100 (9 pm) to 0600 (6 am) each day.

3.3. H-Reflex Measurement and Conditioning

The computer system monitors the ongoing soleus EMG continuously, 24 h/day, 7 days/week. Whenever the absolute value (equivalent to the full-wave rectified value) of background (i.e., ongoing) EMG remains within a defined range for a randomly varying 2.3–2.7-s period, the computer initiates a trial. In each trial, it stores the digitized EMG (absolute value digitized at 5,000 Hz) for the most recent 50 ms (i.e., the background EMG interval), delivers a stimulus pulse (typically, 0.5 ms) to the nerve cuff, and stores the EMG for another 100 ms. The nerve cuff stimulus is initially set to produce a small M response (also called "M wave") (i.e., it is set just above M-response threshold), and then is automatically adjusted by the computer after each trial to maintain the amplitude

Fig. 1. (continued) (modified from ref. (11)). (**b**) The course of H-reflex changes. *Left*: For the first 10 days (from day -10 to 0), the rat is exposed to the control mode, in which no reward is given and the H-reflex is simply measured to determine its initial size. For the next 50 days, it is exposed to the up-conditioning or down-conditioning mode, in which a food-pellet reward is given whenever the H-reflex is above (HRup mode) or below (HRdown mode) a criterion value. The background EMG and the M response stay constant throughout. Successful conditioning (defined as a change of at least 20% in the correct direction) occurs in 75–80% of the rats (the others remain within 20% of their control value). The graphs show average (±SEM) daily H-reflex sizes for 55 successful HRup rats (*up-pointing triangle*) and 72 successful HRdown rats (*down-pointing triangle*). In both groups, mode-appropriate change in H-reflex size develops gradually over the 50 days. If a rat is switched from the up mode to the down mode (or vice versa), the H-reflex change reverses in the same gradual fashion (updated from refs. (9, 14)). *Right*: Each person completes three sessions per week. For the first six sessions (from day −14 to 0), the person is exposed to the control mode, in which the H-reflex is simply measured to determine its initial size. For the next 24 sessions (days 0–56), the person is exposed to the HRup- or HRdown-conditioning mode, in which for each trial the video screen gives immediate feedback indicating whether the H-reflex is above (HRup mode) or below (HRdown mode) a criterion value. After completing these 24 conditioning sessions, people return for four follow-up sessions over the next 3 months. The background EMG and the M response stay constant throughout. Successful conditioning occurs in about 80% of the people. The graphs show average (±SEM) daily H-reflex sizes for six successful HRup people (*up-pointing triangle*) and eight successful HRdown people (*down-pointing triangle*). In both groups, mode-appropriate change in H-reflex size develops gradually over the 24 conditioning sessions. In the follow-up sessions, the H-reflex increase in the HRup group is smaller but still evident, and the H-reflex decrease in the HRdown group persists unchanged (from ref. (11)). (**c**) Examples of H-reflex change. *Left*: Average post-stimulus EMG (absolute value) for representative days from an HRup rat (*left*) and an HRdown rat (*right*) under the control mode (*solid*) and near the end of HRup or HRdown conditioning (*dashed*). The H-reflex is larger after up conditioning and smaller after down conditioning. In contrast, background EMG (shown here by EMG at time zero) and M responses are not changed (from refs. (9, 14)). *Right*: Average post-stimulus EMG (absolute value) for representative sessions from an HRup person (*left*) and an HRdown person (*right*) under the control mode (*solid*) and near the end of HRup or HRdown conditioning (*dashed*). The H-reflex is larger after up conditioning and smaller after down conditioning. Background EMG and M responses do not change (stimulus artifacts are present at 0 ms) (from ref. (11)).

of the EMG in the M-response interval (typically, 2.0–4.5 ms after stimulation) unchanged throughout the days and weeks of data collection. The threshold M response reflects stimulation of a few large motoneuron axons (23). Thus, by maintaining a constant M response over days and weeks, the protocol ensures that the effective stimulus strength remains stable despite any changes that might occur (e.g., tissue growth around the nerve cuff).

In the control mode, the computer simply digitizes and stores the absolute value of soleus EMG for 100 ms following the stimulus. Under the up-conditioning (HRup) or down-conditioning (HRdown) mode, the computer gives a reward (i.e., a 20-mg food pellet) 200 ms after nerve stimulation whenever EMG amplitude in the H-reflex interval (typically, 6.0–10.0 ms after stimulation) is above (HRup mode) or below (HRdown mode) a criterion value. In the course of its normal activity, the animal usually satisfies the background EMG requirement, and thus receives nerve cuff stimulation 2,000–6,000 times per day.

For each rat, data are collected first under the control mode for 20 days and then under the HRup mode or HRdown mode for 50 more days. Under the HRup or HRdown mode, the reward criterion value is initially set on the basis of the control-mode data, and subsequently adjusted as needed each day, so that the rat receives an adequate amount of food (e.g., about 1,000 reward pellets per day for a 500-g rat). Thus, in the HRup mode, a typical rat is rewarded for the largest 1,000 H-reflexes of each day, while in the HRdown mode a typical rat is rewarded for the smallest 1,000 H-reflexes of each day. All rats continue to gain weight throughout data collection.

3.4. Data Analysis

Background EMG amplitude is calculated as average absolute value of EMG during the 50 ms prior to nerve stimulation. M-response size is calculated as the average absolute value of EMG in the M-response interval (typically, 2.0–4.5 ms after the tibial nerve stimulation) minus average background EMG amplitude. H-reflex size is calculated as the average absolute value of EMG in the H-reflex interval (typically, 6.0–10.0 ms after the nerve stimulation) minus average background EMG amplitude, and is usually expressed in units of background EMG amplitude. In all rats, background EMG and M-response size remain stable throughout data collection. To determine the final effect on H-reflex size of HRup or HRdown conditioning, average H-reflex size for the final 10 days (i.e., days 41–50) of conditioning is calculated as percent of initial H-reflex size (i.e., average for final ten control-mode days). An H-reflex size change of at least 20% in the correct direction qualifies as successful conditioning (9, 24). Overall, conditioning is successful in 75–80% of animals. In the other 20–25%, H-reflex size remains within 20% of its control value.

Figure 1b (left) shows average H-reflex size for 127 successful rats for the final 10 days under the control mode and for 50 days under the up- or down-conditioning mode. The H-reflex does not change under the control mode. It gradually increases during up-conditioning or gradually decreases during down-conditioning.

Figure 1c (left) shows examples of post-stimulus EMG from an up-conditioned rat (left) and a down-conditioned rat (right) before conditioning (solid lines) and at the end of conditioning (dashed lines). In these rectified (i.e., absolute value) and averaged traces, the H-reflex is increased at the end of up conditioning and is decreased after down conditioning with respect to the control-mode H-reflex without any change in the background EMG level or in the M response.

The protocols and courses of H-reflex conditioning are comparable in humans (Fig. 1 (right)), monkeys, mice, and rats (see refs. (7, 9–11)). It should be noted that, while animals typically perform several thousand trials per day, humans perform only 225 trials three times per week and still show nearly comparable change in H-reflex size. In all four species, reflex size gradually increases during up conditioning and gradually decreases during down conditioning. The similarities in the time course and extent of reflex conditioning suggest that the phenomenon is comparable in these different species. (These striking interspecies similarities also suggest that reflex conditioning in animals could be successful with considerably fewer trials than the thousands per day provided by the current protocols.).

4. Notes

It is important to maintain stable background EMG level and M-response size over the prolonged course of H-reflex conditioning. This typically requires daily or twice-daily monitoring of their values for the past 12–24 h, and appropriate adjustments as needed in the background EMG range and M-response target value used by the protocol. In general, daily average background EMG should be maintained within 10% of its average value for the initial control period (i.e., the last ten control-mode days before imposition of the HRup or HRdown mode), and daily average M-response size should be maintained within 25% of its average value for the initial control period. For the final ten conditioning days, which are generally used to assess the success of conditioning, more severe limits (i.e., ±5% for background EMG and ±10% for M response) are preferable.

The H-reflex conditioning protocol can be combined with anatomical and/or additional behavioral studies. For example, a retrograde tracer, such as CTB-Fluor 488, can be injected into the soleus muscles to label its motoneurons, and other labeling procedures can be used concurrently to label specific populations of synaptic terminals on the motoneurons (e.g., refs. (25–27)). In addition, assessments of muscle behavior and limb kinematics

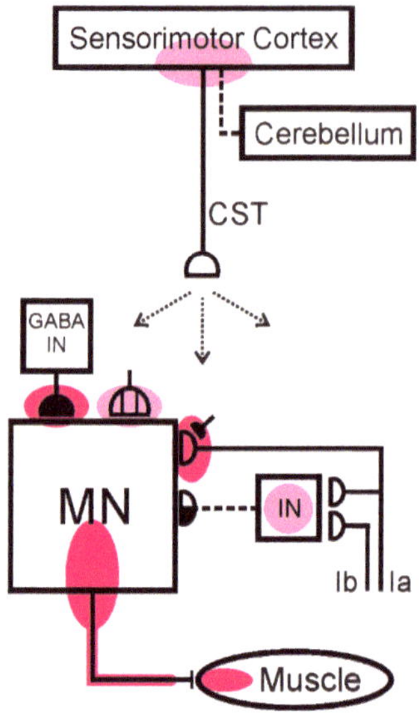

Fig. 2. Summary of the current understanding of spinal and supraspinal plasticity associated with H-reflex conditioning. The *shaded ovals* indicate the sites of definite (*red*) or probable (*pink*) spinal or supraspinal plasticity associated with H-reflex conditioning. *MN* motoneuron; *CST* main corticospinal tract; *IN* spinal interneuron; *GABA IN* GABAergic interneuron. Open synaptic terminals are excitatory, *solid ones* are inhibitory, half-open ones could be either, and the subdivided one is a cluster of C terminals. Dashed pathways imply the possibility of intervening spinal interneurons. The monosynaptic and probably oligosynaptic H-reflex pathway from Ia and Ib inputs to the motoneuron is shown. The sites of definite or probable plasticity include the motoneuron membrane (firing threshold and axonal conduction velocity); motor unit properties; GABAergic terminals and C terminals on the motoneuron; the Ia afferent synaptic connection; interneurons and their terminals conveying oligosynaptic group I inhibition or excitation to the motoneuron; and sensorimotor cortex. The essential roles of the corticospinal tract (originating in sensorimotor cortex) and cerebellar output to cortex are indicated. The spinal cord plasticity that is directly responsible for H-reflex conditioning appears to be induced and maintained by cortical plasticity that itself depends for its long-term survival on the cerebellum (modified from ref. (14)).

during treadmill locomotion can assess the wider functional effects of H-reflex conditioning (e.g., refs. (13, 15)).

By placing stimulating cuffs on several nerves and EMG electrodes in multiple muscles, it is possible to monitor and even condition the reflexes of multiple muscles simultaneously. Furthermore, the same protocol can be used to change other spinal reflexes, such as reciprocal inhibition (16).

Figure 2 summarizes the current understanding of the complex spinal and supraspinal plasticity associated with H-reflex conditioning.

Acknowledgments

Work in the authors' laboratories is supported by NIH grants NS22189, HD36020, NS061823, and HD032571 and by the New York State Spinal Cord Research Fund.

References

1. Matthews PBC (1972) Mammalian muscle receptors and their central actions. Williams and Wilkins, Baltimore, pp 319–409
2. Lee RG, Tatton WG (1975) Motor responses to sudden limb displacement in primate with specific CNS lesions and in human patients with motor system disorders. Can J Neurol Sci 2:285–293
3. Burke RE, Rudomin P (1977) Spinal neurons and synapses. In: Brooks VB (ed) Handbook of physiology. Section I: The nervous system. Volume I: cellular biology of neurons, Part II. Williams and Wilkins Company, Baltimore, pp 877–944
4. Baldissera F, Hultborn H, Illert M (1981) Integration in spinal neuronal systems. In: Brooks VB (ed) Handbook of physiology. Section I: The nervous system. Volume II: motor control, Part I. Williams and Wilkins Company, Baltimore, pp 509–595
5. Henneman E, Mendell LM (1981) Functional organization of motoneuron pool and inputs. In: Brooks VB (ed) Handbook of physiology. Section I: the nervous system. Volume II: motor control, Part I. Williams and Wilkins Company, Baltimore, pp 423–507
6. Wolpaw JR, Braitman DJ, Seegal RF (1983) Adaptive plasticity in the primate spinal stretch reflex: initial development. J Neurophysiol 50:1296–1311
7. Wolpaw JR (1987) Operant conditioning of primate spinal reflexes: the H-reflex. J Neurophysiol 57:443–458
8. Evatt ML, Wolf SL, Segal RL (1989) Modification of human spinal stretch reflexes: preliminary studies. Neurosci Lett 105:350–355
9. Chen XY, Wolpaw JR (1995) Operant conditioning of H-reflex in freely moving rats. J Neurophysiol 73:411–415
10. Carp JS, Tennissen AM, Chen XY, Wolpaw JR (2006) H-reflex operant conditioning in mice. J Neurophysiol 96:1718–1727
11. Thompson AK, Chen XY, Wolpaw JR (2009) Acquisition of a simple skill: task-dependent adaptation plus long-term change in the human soleus H-reflex. J Neurosci 29:5784–5792
12. Wolpaw JR, Lee CL (1989) Memory traces in primate spinal cord produced by operant conditioning of H-reflex. J Neurophysiol 61:563–572
13. Chen Y, Chen XY, Jakeman LB et al (2005) The interaction of a new motor skill and an old one: H-reflex conditioning and locomotion in rats. J Neurosci 25:6898–6906
14. Wolpaw JR, Chen XY (2009) Operant conditioning of reflexes. In: Squire LR (ed) Encyclopedia of neuroscience, vol 7. Academic, Oxford, pp 225–233
15. Chen Y, Chen XY, Jakeman LB et al (2006) Operant conditioning of H-reflex improves locomotion after spinal cord injury in rats. J Neurosci 26:12537–12543
16. Chen XY, Chen Y, Chen L et al (2006) Operant conditioning of reciprocal inhibition in rat soleus muscle. J Neurophysiol 96:2144–2150
17. Wolpaw JR, Tennissen AM (2001) Activity-dependent spinal cord plasticity in health and disease. Annu Rev Neurosci 24:807–843
18. Wolpaw JR (2006) The education and re-education of the spinal cord. Prog Brain Res 157:261–280
19. Wolpaw JR (2010) What can the spinal cord teach us about learning and memory? Neuroscientist 16(5):532–549
20. Chen XY, Chen Y, Chen L, Tennissen AM, Wolpaw JR (2006) Corticospinal tract transection permanently abolishes H-reflex down-conditioning in rats. J Neurotrauma 23:1743–1750
21. Chen XY, Wolpaw JR (2002) Probable corticospinal tract control of spinal cord plasticity in rats. J Neurophysiol 87:645–652
22. Wolpaw JR (2006) The education and re-education of the spinal cord. Prog Brain Res 157:261–280
23. Brown WF (1984) The physiological and technical basis of electromyography. Butterworths, Boston
24. Wolpaw JR, Herchenroder PA, Carp JS (1993) Operant conditioning of the primate H-reflex: factors affecting the magnitude of change. Exp Brain Res 97:31–39
25. Wang Y, Pillai S, Wolpaw JR et al (2006) Motor learning changes GABAergic terminals on spinal motoneurons in normal rats. Eur J Neurosci 23:141–150
26. Pillai S, Wang Y, Wolpaw JR, Chen XY (2008) Effects of H-reflex up-conditioning on GABAergic terminals on rat soleus motoneurons. Eur J Neurosci 28:668–674
27. Wang Y, Pillai S, Wolpaw JR, Chen XY (2009) H-reflex down-conditioning greatly increases the number of identifiable GABAergic interneurons in rat ventral horn. Neurosci Lett 452:124–129

Chapter 45

Cellular and Molecular Biological Assessments of Inflammation and Autoimmunity After Spinal Cord Injury

Yi Zhang, Daniel P. Ankeny, and Phillip G. Popovich

Abstract

Inflammation and autoimmune responses after spinal cord injury (SCI) are complex processes involving sequential cellular and molecular changes in cells of the innate and adaptive immune systems. In preclinical animal models of SCI and human SCI, immune responses have been implicated in the progression of both tissue injury and repair. Routine characterization of these changes includes analyses of intraspinal cytokines and chemokines, activation of glia, and the influx and activation of myeloid and lymphoid cells. In this chapter, a cursory overview is provided of cellular and molecular inflammation after SCI; however, the primary goal is to describe the most common and reliable qualitative and quantitative methods used to assess these changes. Through the use of these techniques, it may be possible to achieve a better understanding of how inflammation and autoimmunity affect outcome after SCI.

Key words: Spinal cord injury, Inflammation, Autoimmunity, Cytokines, Chemokines, Lymphocytes, Microglia, Astrocytes, Macrophages, Neutrophils, Dendritic cells, Biological assessments

1. Introduction

Mechanical trauma to the spinal cord causes compression and shearing of neurons/axons, glia, and vascular cells leading to irreversible necrotic cell death (1). Additional progressive destruction of spared tissue surrounding the lesion site is known as secondary injury. Secondary injury is a constellation of cellular, biochemical, and physiological cascades, including hypoxia/ischemia, free radical formation/lipid peroxidation, excitotoxicity, enhanced proteolytic activity, inflammation, and autoimmunity that develop days to months after the primary mechanical injury (2).

Inflammation is a coordinated physiological response to tissue injury mediated by cells and mediators of the immune system. Normally, the onset of inflammation is dominated by activation of innate immunity (e.g., neutrophils, complement, monocytes/

Jun Chen et al. (eds.), *Animal Models of Acute Neurological Injuries II*, Springer Protocols Handbooks, DOI 10.1007/978-1-61779-782-8_45,

macrophages) with subsequent activation of adaptive immune cells that defend the host against viruses, bacteria, and other microorganisms. This response is also elicited in response to tissue injury and is an integral component of restoring tissue homeostasis.

The adaptive immune system comprises T and B lymphocytes and antibodies. T and B cells possess specialized receptors that convey exquisite sensitivity to a nearly infinite barrage of pathogenic stimuli. Ligation of these receptors by the appropriate antigen(s) triggers cellular proliferation and differentiation into effector cells whose primary function is to eliminate or neutralize the antigen(s). Most lymphocytic responses are directed toward protein antigens found on bacteria or viruses. However, lymphocytes also can recognize and react to proteins, carbohydrates, lipids, and nucleic acids found in the host. These self-directed reactions are termed autoimmune responses and when they cause tissue pathology, manifest as autoimmune disease.

Inflammation and immune system activation are important contributors to secondary injury and also to the repair of injured tissues after spinal cord injury (SCI) (3, 4). These complex responses begin within minutes after SCI spreading throughout the damaged spinal cord over a period of days and weeks. The cellular and molecular immune responses induced by SCI are characterized by: (1) enhanced intraspinal synthesis and release of pro-inflammatory cytokines, chemokines, and other vasoactive substances (e.g., histamine, complement proteins) by neurons, microglia, astrocytes, endothelia, and infiltrating leukocytes (5–8); (2) activation and proliferation of resident spinal microglia and astrocytes (7, 9); (3) influx of circulating innate immune cells, including neutrophils and monocytes/macrophages from the circulation (10, 11); and (4) systemic activation and intraspinal accumulation of lymphocytes, including those reactive with self-antigens (i.e., autoimmune cells) (12–15).

2. General Aspects and Working Protocols

2.1. Release of Pro-inflammatory Cytokines and Chemokines

Cytokines are category of intercellular signaling molecules that coordinate developmental, physiological, and pathological processes throughout the body. Cytokines influence cell function by binding to specific cell-surface receptors which in turn regulate expression of genes encoding transcription factors, cytokines, additional surface receptors, or intracellular signaling cascades that provide feedback regulation of cytokine synthesis (16, 17). A variety of pathways including those that are downstream of ligand-gated ion channels, G-protein-coupled receptors, and tyrosine kinase receptors participate in cytokine-induced cascades.

Chemokines are a family of low molecular weight chemotactic cytokines that are classified into groups based on the presence and position of conserved cysteine residues (18). Chemokines are synthesized at sites of inflammation where they establish a concentration gradient and activate endothelia culminating in the recruitment of immune effector cells to the affected area (19). Rigorous analysis of intraspinal chemokine and cytokine expression can provide insight into the nature of a developing or ongoing inflammatory or autoimmune reaction after SCI and/or the tissue response to a specific treatment.

In general, a range of pro-inflammatory cytokines and chemokines are produced in the injured spinal cord, typically increasing within 15 min of SCI then returning toward baseline expression within 24–48 h (6, 7, 20). TGF-β1, a presumed anti-inflammatory cytokine, reaches maximal expression ~7 days postinjury (21, 22). A secondary increase in a subset of cytokines (e.g., IL-1β and TNF-α) also has been described in SCI mice (8). SCI-induced chemokine expression also is transient and follows a time course similar to that of most cytokines although prolonged expression of a subset of lymphocyte-specific chemokines has been described in injured mouse spinal cord (13, 23). In some instances, in situ hybridization is used to characterize the spatial distribution of chemokine/cytokine gene expression at the cellular level (8, 22); however, changes in gene expression (e.g., PCR) or protein synthesis [e.g., enzyme-linked immunosorbent assay (ELISA), Western blotting, and immunohistochemistry] are quantified using mRNA or protein obtained from injured tissue homogenates (5, 6, 20, 23, 24). In acute or chronic human SCI subjects, injury-associated changes in intraspinal cytokines and chemokines can be measured in CSF samples using ELISA (24, 25).

2.1.1. mRNA Analysis of Chemokines, Cytokines, and Other Inflammatory Factors in Spinal Cord

Quantitative PCR (Q-PCR) is a powerful and commonly used technique for detecting relative or absolute changes in gene expression between experimental and control samples (26–28). For detecting changes in cytokines, chemokines, or immune modulatory neurotrophins, mRNA is isolated from homogenates of injured and uninjured spinal cords then is copied to cDNA for use in Q-PCR reaction. A more focused analysis of gene expression can be accomplished by isolating mRNA from individual cells using laser capture microdissection (LCM) (29, 30). Before an analysis can be completed, gene-specific primer pairs must be synthesized. These will vary as a function of the experimental question. A number of commercial providers are available for custom primer design and synthesis. Below is a list of common inflammatory genes that are important for different aspects of postinjury inflammation and autoimmunity.

- Pro-inflammatory cytokines and chemokines: IL-1β, IL-6, IL-11, IL-12, TNF-α, lymphotoxins (LTs), IFN-γ, T-lymphocyte chemoattractant CXCL10, gamma-interferon-inducible protein (IP-10), monocyte chemoattractant protein (MCP-1), and macrophage inflammatory protein-1 alpha/beta (MIP-1α/β).

- Neurotrophins and immune-regulatory cytokines: neurotrophin (NT)-3, BDNF, TGF-β, IL-4, IL-5, IL-10, and IL-13.
- Adhesion molecules: intracellular and vascular adhesion molecules-1 (ICAM-1 and VCAM-1).
- Lipid inflammatory mediators: cyclooxygenase 2 (COX-2) and prostaglandins (PGs).

Procedures for RNA Isolation and Quantitative Real-Time PCR

- *Isolating RNA*: Total RNA is purified from a segment of spinal cord centered on the impact site (or similar length of uninjured tissue) using Trizol (Invitrogen, Carlsbad, CA) followed by RNeasy (Qiagen, Valencia, CA) binding. Total RNA is quantified by spectrophotometry. Many commercially available kits are available for RNA isolation (e.g., Qiagen's RNeasy Lipid Mini Kit).
- *Preparing cDNA*: cDNA is prepared from RNA by reverse transcription (e.g., SuperScript II First-Strand Synthesis System with random primers; Invitrogen). Using this kit, reverse transcription involves the following procedures:
  1. Add the following components to a nuclease-free 0.2 ml PCR tube:
     (a) Oligo(dt) 12–18 (500 μl/ml): 1 μl.
     (b) Total RNA: 10 μl.
     (c) 10 mM dNTPmix: 1 μl.
  2. Mix gently and spin down.
  3. Heat mixture to 65°C for 5 min and quick chill on ice. Collect the contents of the tube by brief centrifugation and add:
     (a) 5× first-strand buffer: 4 μl.
     (b) 0.1 M DTT: 2 μl.
     (c) RNaseOUT: 1 μl.
  4. Mix contents of the tube gently then incubate at 42°C for 2 min.
  5. Add 1 μl of SuperScript II RT (Invitrogen Cat: 18064-014) and mix by pipetting gently up and down.
  6. Incubate at 42°C for 50 min.
  7. Inactivate the reaction by heating at 70°C for 15 min.
  8. Store cDNA at −20°C until used for PCR.
- *Performing real-time PCR*: To perform real-time PCR, the user needs cDNA (~10 ng), primer pairs, and some form of

fluorescent detection kit to reveal amplification of the PCR product. A comparison of the advantages and disadvantages of the SYBR®-Green and TaqMan®-based detection techniques can be found at (http://www.appliedbiosystems.com/absite/us/en/home/applications-technologies/real-time-pcr/taqman-and-sybr-green-chemistries.html). All cytokine primers and probes to be used with Applied Biosystems PCR machines (described below) are designed using Primer Express software (Applied Biosystems). GAPDH or other appropriate housekeeping genes are used as a stable endogenous control (i.e., "housekeeping" genes). The following protocol uses SYBR®-Green detection. SYBR®-Green kits are available from many companies (e.g., Applied Biosystems, Stratagene, and Qiagen). The PCR procedure consists of the following:

1. Add the following components in the well of the real-time plate:
   (a) SYBR®-Green Master mix: 10 μl.
   (b) RNase-DNase free water: 5 μl.
   (c) Primer (forward 5 μM): 2 μl.
   (d) Primer (reverse 5 μM): 2 μl.
   (e) cDNA: 1 μl.
2. Centrifuge the plate at 400 × *g* for 1 min and set the plate into real-time PCR machine.

- An increase in SYBR®-Green fluorescence reveals accumulation of PCR product. Past data from our lab has used an ABI 7300 Fast Real-Time PCR system (Applied Biosystems); however, other systems are available and each comes with different software to perform PCR and analyze data. Refer to the user's guide for the machine you are using.
- Standard curves are generated for each gene using a control cDNA dilution series. Melting point analyses are performed for each reaction to confirm single amplified products.

2.1.2. Enzyme-Linked Immunosorbent Assay Analysis of Cytokines

Cytokines in plasma or tissue samples, or intracellular cytokines released from splenocytes, lymph node cells, or purified glia can be measured using commercially available solid-phase ELISA kits (R&D Systems, Minneapolis, MN) or self-coated plates using the sandwich technique.

Procedures

1. Coat high-quality plastic 96-well round bottom plate (e.g., Dynatech or Falcon with 40 μl/well of capture antibody (e.g., goat anti-mouse IL-X). Capture antibody should be diluted to a final concentration of 2–10 μg/ml in 1× phosphate buffered saline (PBS)/0.02% sodium azide.
2. Incubate plate at 37°C for 2 h or overnight at 4°C.

3. Wash plate ×4 with a minimum of 150 μl/wash 1× PBS.
4. Block plate with 150 μl/well of PBS + 1% bovine serum albumin (BSA). Incubate plate at 37°C for a minimum of 1 h (or 4°C for 18 h).
5. Wash plate ×6 with PBS/0.05% Tween-20.
6. Add 30 μl/well of standards or test samples at appropriate dilution. Incubate at 37°C for 2 h (or overnight at 4°C). To prepare standard solutions, dilute purified mouse cytokine stocks solution using 1:2 serial dilutions such that the final concentrations range from 2 ng/ml to 4 pg/ml.
7. Wash plate ×6 with PBS/0.05% Tween-20.
8. Add 40 μl/well biotin-conjugated anti-mouse antibody specific for the cytokine (or other protein) of interest. Dilute antibody to a final concentration of 1 μg/ml in 1× PBS + 1% BSA. Incubate plate at 37°C for 2 h (or overnight at 4°C).
9. Wash plate ×6 with PBS/0.05% Tween-20 and then ×4 with PBS.
10. Dissolve 1 mg of *p*-nitrophenyl phosphate per 1 ml DEM solution for a final concentration of 1 mg/ml. Add 70 μl/well to visualize reaction, and then measure optical density (OD) at a wavelength of 405 nm.

2.2. Assays to Define Lymphocyte Phenotype and Function

T lymphocytes (T cells) are not normally found in the uninjured spinal cord but they increase in number after SCI (31). Intraspinal T cell infiltration occurs in parallel with the activation of microglia and the influx of monocytes and B lymphocytes, predominantly at the injury epicenter and nearby penumbra (15, 32, 33). Once activated, presumably via interactions with antigen and antigen presenting cells (e.g., microglia, B cells), T cells produce cytokines and chemokines that initiate and modulate intracellular signal transduction of CNS and non-CNS cells (13, 31). Both beneficial and deleterious roles have been attributed to activated T cells after SCI (14, 34).

The principal function of B cells is to make antibodies (immunoglobulins, Ig) which are secreted proteins that bind specifically to their target antigen. The formation of immune complexes (i.e., antibodies bound to antigen) leads to activation of the complement system, Fc-receptor bearing cells (e.g., microglia/macrophages) or antibody-dependent cellular cytotoxicity. The net effect of this coordinated response is neutralization and/or elimination of the antigen. After SCI, a long-lasting B cell response with pathogenic potential is elicited in the periphery and in the CNS. This response consists of enhanced B lymphopoiesis in bone marrow and spleen, prolonged synthesis of CNS-reactive autoantibodies and infiltration/accumulation of B cells and antibodies at the site of injury (12, 15). Myelin-reactive T cells, antibodies and pro-inflammatory chemokines, and cytokines are elevated in the circulation and cerebrospinal fluid (CSF) after SCI in humans suggesting that lymphocyte activation may be clinically significant (24, 35).

Indices of lymphocyte activation (including autoimmune responses) after SCI can be inferred by analyzing injury-induced changes in the time-course and composition of T and B lymphocyte subsets and their effector molecules (e.g., cytokines and antibodies). These changes may vary as a function of time postinjury and whether they occur in the spinal cord or peripheral lymphoid tissues (e.g., spleen, lymph node).

2.2.1. Immunohistochemical Detection of Intraspinal Lymphocytes

Antibodies against select cell-surface antigens can be used to distinguish T cells (i.e., CD3) or T cell subsets (CD4 and CD8) from B cells (i.e., CD45R/B220, CD19, and CD79). Lymphocyte activation can be evaluated using antibodies against additional cell-surface antigens, often requiring double or triple staining for multiple antigens (e.g., CD3 plus CD80 or CD86). These antibodies can be incubated on paraformaldehyde-fixed, frozen spinal cord sections for subsequent microscopic analyses. The staining protocol described below is also applicable for detecting and quantifying changes in intraspinal microglia/macrophages (see Sect. 2.3.1) (15, 32, 33, 36).

Materials

- 5GFTP+.
  Normal Goat Serum (Vector, S-1000): 500 μl.
  1% Fish Gelatin (Sigma, G7765) in 0.1 M PBS: 1,000 μl.
  10% BSA (Fisher, BP1605) in 0.1 M PBS: 1,000 μl.
  1% Triton X-100 (Sigma, X-100) in distilled water: 2,000 μl.
  0.1 M PBS, pH 7.4: 5,500 μl.
- Antibodies (available from multiple commercial sources) and specificities:
  - Anti-CD4 or CD8: labels T-helper cells (Th) or cytotoxic T cells (Tc), respectively.
  - Anti-CD45R/B220: labels mature B lymphocytes.
  - Secondary antibodies: use species-specific biotinylated or fluorophore-conjugated antibodies that recognize the primary antibody.

Procedures

1. Mice or rats are anesthetized then perfused transcardially with cold PBS followed by 4% paraformaldehyde/PBS.
2. Spinal cords are removed from the spinal canal, postfixed for 24 h, cryoprotected in 30% sucrose/PBS solution then embedded and frozen in OCT medium (Sakura Finetek, Torrance, CA).

3. Using a cryostat, serial sections of spinal cord are cut (10 μm) through the injury site and then are thaw mounted onto SuperFrost Plus slides (Fisher Scientific, Pittsburgh, PA). Slides can be stored at −20 to −80°C.
4. To perform immunohistochemistry, take slides from the freezer and air-dry at room temperature (RT) for 1–2 h. To minimize antibody use, circumscribe the sections with a hydrophobic PAP pen.
5. To minimize tissues lifting off of the slide, dry the sections for an additional 1–2 h on a slide warmer heated to 37°C.
6. Incubate the slide with 5GFTP+ for 60 min in a humid chamber placed on a rotary shaker set to 60 rpm.
7. Aspirate the 5GFTP+ and incubate the slide with primary antibody (diluted in 5GFTP+) overnight at 4°C in a humid chamber. Primary and secondary antibody dilutions should be empirically determined for individual tissue samples.
8. The next day, aspirate primary antibody off the slide then place the slide in 0.1 M PBS in a Coplin jar. Rinse ×2 for 5 min at room temperature on a rotary shaker (60 rpm).
9. Shake then aspirate 0.1 M PBS off the slide using suction. Incubate the slide with a fluorochrome (or biotin)-conjugated secondary antibody (diluted in 5GFTP+) for 60 min in a humid chamber on a rotary shaker (60 rpm). If staining with fluorescent antibodies, all incubations should be performed in the dark.
10. Aspirate secondary antibody off the slide using suction then place in a Coplin jar with 0.1 M PBS. Rinse ×2 for 2 min at 60 rpm.
11. Coverslip using mounting medium. If using fluorescent antibodies, use a medium that does not have autofluorescent properties (e.g., ImmuMount). Store labeled slide in the dark at 4°C. If using biotinylated second antibodies, positive cells can be visualized using a streptavidin-conjugated horseradish peroxidase (or similar enzyme detection) system (e.g., ABC Elite kit from Vector Laboratories, Burlingame, CA) and diaminobenzidine (DAB) as a substrate.
12. Total numbers of intraspinal T or B cells are estimated by manually counting CD4+(Th)/CD8+(Tc) or CD45R/B220+ profiles using light or fluorescence microscopy in equidistant sections spanning the length of the injured tissue.

2.2.2. Quantitative Analysis of Lymphocytes in Blood or Peripheral Lymphoid Tissue Using Flow Cytometry

Flow cytometry allows rapid, quantitative analysis of cell surface and cytoplasmic proteins. It is an antibody-based application but offers select advantages over immunohistochemistry, most notably the ability to quantify multiple cell parameters simultaneously. This is based on laser-mediated excitation of a wide range of fluorophores and sophisticated fluorescence detection hardware that can simultaneously read the emission spectra of various fluorophores. Also, the relative magnitude and intensity of this fluorescence can be measured on mixed cell populations that can be distinguished from one another based on cell size or granularity. A practical application might in include an experiment that is designed to quantify SCI-induced changes in the ratio of different T cell subsets (e.g., CD4 vs. CD8 vs. Th17 vs. Tregs), B cells and macrophages or dendritic cells (DCs) using a mixed splenocyte preparation. Using a single cell preparation from one spleen, each of these distinct lymphocyte subsets could be labeled and detected using different fluorophore-conjugated antibodies. The downside of flow cytometry is that since the tissue must be homogenized to prepare cell suspensions, it is not possible to evaluate potential cell–cell interactions or to localize specific cell subsets to discrete anatomical locations in the tissue of interest. Immunohistochemistry with light or fluorescent microscopy is better for these latter types of analyses.

Procedures for Preparing Lymphocytes from Lymphoid Tissue or Blood for Analysis by Flow Cytometry

- *Preparation of splenocytes or lymph node cells*: After excision, prepare single cell suspension from individual spleens or lymph nodes by expressing them through 40-μm nylon mesh strainers. Wash strainers ×3 with 5 ml RPMI + 5% FBS (Atlas Biologicals, Ft. Collins, CO). Lyse red blood cells with 0.8% ammonia chloride and then wash splenocytes with RPMI + 5% FBS. Resuspend cells in RPMI 1640 medium containing 10% FBS, 0.01 M HEPES, 0.1 nM nonessential amino acids, 1 mM sodium pyruvate, 5×10^{-5} M β-mercaptoethanol (Sigma), 0.01% penicillin–streptomycin and 2 nM L-glutamine.
- *Preparation of cells from blood*: Isolate blood (e.g., submandibular bleed, venous puncture) then immediately add diluted heparin (1:10 heparin in saline/PBS). Heparanized capillary tubes also can be used. Spin tubes for 8 min at 250 × *g* then remove supernatant. Lyse red blood cells with 0.8% ammonia chloride. Spin tubes for 5 min at 250 × *g* then remove supernatant. Resuspend cells in PBS.

Antibody Labeling for Flow Cytometry

- Wash and resuspend cells in PBS then, to minimize nonspecific binding by primary antibodies, incubate for 15 min at 4°C in 0.25 μl Fc-block (anti-mouse-CD16/CD32).
- Label cells with APC, FITC, or PE-conjugated antibody specific for the antigen(s) of interest [e.g., B cells (CD45R/B220, CD19, and IgG), T cells (CD4 and CD8), or DCs (CD11c)] then incubate for 60 min at 4°C in dark.
- Wash cells ×3 in PBS and then resuspend at a final concentration of 5×10^{4}–1×10^{6} cells (in ~50 μl) cells/ml in PBS for analysis. There are different commercial sources for cytometers

(e.g., FACS Calibur, Becton-Dickinson, Sunnyvale, CA) and the procedure for analysis will vary by system. Users should consult their core facility manager or the user manual for operational and data analysis procedures specific for their system.

2.2.3. ELISA Analysis

ELISA Analysis of Circulating Nonspecific Total (i.e., IgM + IgG + IgA + IgE) or Isotype Specific-Immunoglobulins (e.g., IgM, IgG1, or IgE Isotypes)

Procedures

1. Collect blood by submandibular bleeds or other method. Keep the blood at 4°C for 1.5–2 hr to clot, and then centrifuge at 2,000 × *g* for 20 min. The supernatant (serum) is stored at −80°C until used in ELISA.
2. Coat Dynatech ELISA 96-well round bottom plate with 2 μg/ml anti-total IgM/IgG antibodies (Southern Biotechnology Associates, Birmingham, AL, USA), or appropriate concentration of anti-specific IgG/IgM antibody diluted in PBS/0.02% sodium azide.
3. Incubate for 2 h at 37°C or overnight at 4°C.
4. Wash plate ×4 with 1× PBS.
5. Block plate with 150 μl/well of PBS + 1% BSA. Incubate plate at 37°C for a minimum of 1 h (or 4°C for 18 h).
6. Wash plate ×6 with PBS/0.05% Tween-20.
7. Add 20 μl/well of Standards and Test samples in complete RPMI or Hanks + 5% FCS. Incubate at 37°C for 2 h (or overnight at 4°C). To prepare standard solutions, dilute purified immunoglobulin stock solution for IgM, IgG using 1:2 serial dilutions such that the final concentrations range from 1 μg/ml to 0.5 ng/ml.
8. Wash plate ×6 with PBS/0.05% Tween-20.
9. Add 40 μl/well of detecting alkaline phosphatase antibody (stock = 1 mg/ml) diluted 1:1,000 in 1× PBS + 1% BSA. Incubate plate at 37°C for 2 h (or overnight at 4°C).
10. Wash plate ×6 times with PBS/0.05% Tween-20 and then four times with PBS.
11. Dissolve 1 mg of *p*-nitrophenyl phosphate per 1 ml DEM solution for a final concentration of 1 mg/ml. Add 70 μl/well to visualize bound antibodies, and then measure OD (optical density) at 405 nm.

ELISA Analysis of Antigen-Specific Antibodies

The presence and concentration of antibodies with discrete antigen specificity can be assayed using a modification of the ELISA procedures presented in Sect. 2.2.3.1. However, instead of coating plates with a capture antibody, plates are coated with the specific target antigen of interest (e.g., a myelin or neuronal antigen), then serum or purified antibody samples are incubated and antibodies that bind to the target antigen are detected as outlined above. Triplicate or quadruplicate wells are commonly used for each sample to be assayed for the presence of antigen-specific antibodies.

Procedures

1. Coat plates with target antigen (e.g., purified myelin basic protein) at 10–100 μg/ml diluted in 1× PBS/0.02% sodium azide. Antigen concentrations will vary as a function of the expected prevalence of a given antibody.
2. Incubate for 2 h at 37°C or overnight at 4°C.
3. Wash plate ×6 with PBS/0.05% Tween-20 to remove unbound target antigen.
4. Block nonspecific binding using 150 μl/well of PBS + 1%BSA. Incubate plate at 37°C for a minimum of 1 h (or 4°C for 18 h).
5. Wash plate ×4 with PBS/0.05% Tween-20.
6. Apply appropriate volume of diluted serum/purified antibody samples to plate and incubate for 2 h or overnight as described above. Required volumes and concentrations are dependent on the prevalence of the specific antibody in the samples. Typical volumes and dilutions of serum are 10–50 μl of serum that is undiluted or used at 1:1,000+ dilutions.
7. Wash plate ×6 with PBS/0.05% Tween-20.
8. Add 40 μl/well of alkaline phosphatase antibody (stock = 1 mg/ml) diluted 1:1,000 in 1× PBS + 1% BSA. Incubate plate at 37°C for 2 h (or overnight at 4°C). Choice of detection antibody depends on whether the investigator is interested in total specific Ig levels (i.e., using anti-mouse IgM + IgG + IgA + IgE to detect all anti-MBP antibodies) or only specific isotypes of a given antibody (e.g., anti-mouse IgG1 antibodies that bind MBP).
9. Wash plate ×6 with PBS/0.05% Tween-20 and then ×4 with PBS.
10. Dissolve 1 mg of *p*-nitrophenyl phosphate per 1 ml DEM solution for a final concentration of 1 mg/ml. Add 70 μl/well to visualize bound antibodies and then measure OD at 405 nm.

2.2.4. Assessing Lymphocyte Proliferation

The proliferative capacity of a given lymphocyte population is indicative of the ability of that cell population to respond to a specific stimulus. If one is interested in knowing whether immunological responsiveness to a potential autoantigen (e.g., a myelin protein) has increased (or decreased) as a result of SCI, testing the ability of that antigen to elicit proliferation in purified lymphocytes is a useful assay. This type of assay is typically performed using purified cell populations in combination with stimulation parameters that are specific for the cell type of interest (e.g., incubation with activating anti-CD3 or anti-Ig antibodies for T cells and B cells, respectively). Lymphocytes derived from blood, spleen, or lymph node are cultured at ~1×10^5/100 μl in 96-well plate and then incubated with the autoantigen of interest or polyclonal activators (mitogen stimulation), including phorbol myristate acetate (PMA; 250 ng/ml) plus ionomycin (250 ng/ml) for 48 h at 37°C under 5% CO_2 (12, 37). Polyclonal activation of T or B lymphocytes can also be achieved using concanavalin A (conA) or lipopolysaccharide (LPS), respectively (38). The magnitude of proliferation can be measured using ELISA-based or radioisotope detection methods (12, 37–39).

2.3. Activation and Proliferation of Resident Glia (Microglia and Astrocytes) and Infiltration of Monocyte-Derived Macrophages

Microglia comprise ~10% of glia and are distributed ubiquitously throughout the CNS (40). Microglia respond rapidly to pathological changes in the CNS by altering their morphology, increasing de novo expression of cell surface molecules and by synthesizing or releasing chemokines, cytokines, and other soluble factors, including neurotrophins, enzymes, or reactive oxygen/nitrogen intermediates (7, 41). Activated microglia have been implicated in posttraumatic secondary injury and also in orchestrating repair of the injured spinal cord. As the resident antigen presenting cells of the CNS, microglia are also believed to regulating autoimmune reactions in the injured CNS (42, 43).

Microglia, like monocyte-derived macrophages, derive from myeloid precursor cells and can therefore be identified in tissue sections and via flow cytometry using antibodies that bind to antigens that define cells of the myeloid lineage. Accordingly, at sites of pathology, it is difficult to distinguish between macrophages that were derived from microglia or those derived from blood monocytes. Only relative differences in the mean fluorescence intensity of CD45 can be used to distinguish between these macrophage subsets (44, 45). In the nonpathological CNS, resident parenchymal/resting microglia express low levels of CD45 while fully differentiated macrophages (including perivascular microglia) are designated at $CD45^{high}$. In pathological tissue specimens, infiltrating monocyte-derived macrophages also are $CD45^{high}$ and activated microglia upregulate CD45 expression, some becoming $CD45^{high}$-expressing cells while other express low or intermediate levels of CD45. A flow cytometer is needed to quantify these relative expression patterns of CD45. Other commonly used phenotypic markers that specifically distinguish microglia and macrophages from other cells within the spinal cord include complement component 3b receptor (iC3bR/CD11b), CD68, the intracellular calcium binding protein Iba-1 and the F4/80 antigen. Species-specific antibody clones recognize some of these and other antigens. For example, OX42 labels rat CD11b whereas the M1/70 clone labels mouse CD11b. Several lectins, including tomato lectin and the *Griffonia* lectins identify microglia and macrophages irrespective of species. These antibodies can be used in flow cytometric assays or immunohistochemistry to detect microglia/macrophages (46).

Neutrophils circulate in the blood stream and are among the first leukocytes to respond to pathogenic or traumatic inflammatory stimuli (11, 47, 48). Activated neutrophils release a cadre of pontentially injurious inflammatory mediators, including neutrophil elastase and oxygen-free radicals (49). Accordingly, therapeutic agents that inhibit neutrophil activation have been shown to alleviate neurological impairment in rat models of SCI (50). Recent data suggest that acute neutrophil infiltration is crucial for triggering delayed healing cascades in the injured spinal cord. Indeed, depletion of neutrophils using anti-Ly6G/Gr-1 antibodies exacerbated tissue injury and functional recovery in SCI mice (10). The same Ly6G/Gr-1 antibodies can be used to label neutrophils for immunohistochemistry or flow cytometry (33, 51). Thus, it is important to understand the kinetics, distribution, and magnitude of change of infiltrating neutrophils in response to SCI.

Dendritic cells are professional antigen presenting cells that express lymphocyte co-stimulatory molecules, migrate to lymphoid organs, and secrete cytokines to initiate immune responses (52). They can propagate or reduce autoimmune reactions by either activating or tolerizing T lymphocytes to self (auto) antigens body (self-antigens). A variety of complex factors, including antigen concentration, cytokine milieu, activation state of DCs and type of antigen will determine the outcome of the T cell:DC interaction. DCs can be distinguished from myeloid progenitor-derived microglia and macrophages using a pan-surface marker CD11c using either flow cytometry or immunohistochemistry (see protocols described above and below).

Astrocytes are a major subtype of glia in the CNS and are a prominent component of the cellular response to SCI (9). Normally, astrocytes perform many functions in the CNS, including biochemical and structural support of endothelia [which form the blood–brain barrier (BBB)], provision of nutrients and water across the BBB and maintenance of ionic homeostasis (9, 53). In response to injury or disease in the brain or spinal cord, astrocytes remove excess glutamate and K^+ ions from the extracellular space, provide neurons with metabolic substrates, perform antioxidant functions, and synthesize/release growth factors and cytokines important for regulating inflammation and autoimmune responses (54, 55). In response to SCI and other CNS insults, astrocytes undergo morphological and functional changes that are associated with clearly defined changes in gene expression (9, 56). These changes are associating with diverse effects on immunological functions after SCI. For example, selective deletion of reactive astrocytes after SCI exacerbates inflammation and postinjury pathology (9). Conversely, inhibition of NF-κB signaling in astrocytes attenuates post-SCI neuroinflammation and confers neuroprotection in a mouse contusion model of SCI (57). Therefore, evaluation of astrocytic phenotype and function is an important consideration in the understanding of posttraumatic neuroinflammation.

2.3.1. Immunohistochemical Detection of Resident Immune-Effector Glia (Microglia and Astrocytes) and Infiltrating Leukocytes in Injured Spinal Cord

Materials and Applications

- 5GFTP+(see Sect. 2.2.1).
- Antibodies (available from multiple commercial sources) and specificities:
  - Anti-CD11b (Mac-1 alpham chain), ED-1 antibodies: activated microglia and/or monocyte-derived macrophages.
  - Anti-GFAP: astrocytes.
  - CD11c Ab: Dendritic cells.
  - Anti-Ly6G/Gr-1: neutrophils; enzymatic detection of myeloperoxidase (MPO) is also commonly used for detecting neutrophils in spinal cord (11, 58).
  - Secondary antibodies: use species-specific biotinylated or fluorophore-conjugated antibodies that recognize the primary antibody.

Procedures (see Sect. 2.2.1 Peroxidase or Immunohistochemical Labeling Procedures)

- The magnitude of intraspinal microglia/astrocytes activation and infiltration of macrophages, as well as cell distribution, can be detected with immunoreactivity for specific antibodies in tissue sections then quantified using manual profile counting, computer-assisted counting with unbiased sampling techniques (stereology) and/or digital image analysis (36, 59, 60).

2.3.2. Flow Cytometric Analysis of Intraspinal Microglia/Macrophages

As described above (Sect. 2.3), a wide range of fluorescently conjugated antibodies are available for detecting microglia and macrophages and using flow cytometry, it is possible to detect and quantify relative differences in the expression of these markers (e.g., CD45 and CD11b).

Materials and Solutions

- Autoclaved fire-polished pipettes with cotton filters (~5 per animal).
- *Solutions*
  - Kreb's Ringer's buffer (KRB)—500 ml at 10×.

    500 ml UltraPure H_2O.

    35.35 g NaCl.

    1.8 g KCl.

    0.83 g KH_2PO_4.

    12.85 g D-Glucose.

    10.7 g $NaHCO_3$.

    Filter sterilize and store at 4°C.
  - KRB with heparin: 1 L.

    100 ml KRB 10× KRB.

    900 ml UltraPure H_2O.

    2 ml Heparin at 1,000 U/ml (final at 2 U/ml).

- KRB/BSA—500 ml.

  50 ml 10× KRB.

  0.45 g BSA fraction V.

  0.45 ml 0.15 M $MgSO_4$.

  Bring final volume to 500 ml with tissue culture grade H_2O.

  Filter sterilize then store at 4°C.

- KRB/trituration (KRB/TRIT)—100 ml.

  99 ml KRB/BSA.

  8 mg DNase I, type II.

  52 mg Trypsin inhibitor.

  1 ml 0.15 M $MgSO_4$.

  Mix and filter sterilize, aliquot at 10 ml and store at −20°C.

- KRB enzyme inhibiting solution (KRB/EIS)—100 ml.

  16 ml KRB/TRIT.

  84 ml KRB/BSA.

  Aliquot at 10 ml, and store at −20°C.

- Isotonic Percoll.

  10 ml 1.5 M NaCl—add to Percoll slowly with swirling.

  90 ml stock Percoll.

  Filter sterilize.

- 70% Percoll in KRB/BSA.

  70 ml Isotonic Percoll.

  3 ml 10× KRB/BSA.

  27 ml UltraPure H_2O.

- 35% Percoll in KRB/BSA.

  35 ml Isotonic Percoll.

  6.5 ml 10× KRB/BSA.

  58.5 ml UltraPure H_2O.

  200 μl Phenol Red solution.

- Flow buffer.

  1 ml FBS.

  100 μl 1,000× (100 mM) sodium azide stock (final at 0.1%).

  Bring volume up to 100 ml with tissue culture grade PBS and filter sterilize.

- DNase I, type II solution.

  Dilute to 0.15 mg/ml (or 20,000 U/ml) in Neurobasal-A media.

  Filter sterilize and store at −20°C.

Procedures of Isolating Microglia/Macrophages from Spinal Cord for Analysis by Flow Cytometry

1. Pipette 3 ml KRB/BSA solution into 60 mm Petri dishes (one per animal) and keep on ice.
2. Perfuse animal with ice-cold KRB with heparin for 10 min, remove spinal cord, and place in dish on ice. Mince spinal cord with sterile razor blade then transfer tissue and solution from the dish to a labeled 15 ml polystyrene tube on ice. Add 2 ml KRB/BSA to dish and transfer wash to sample tube (total 5 ml).
3. Add 56 μl of 2.5% Trypsin to each tube and triturate ×5 with 5 ml pipette. Incubate 30 min at 37°C in shaker-incubator at 175 rpm.
4. Add 5 ml KRB/EIS solution plus 1 tube DNase I, type II (3,000 total units) per sample tube. Mix and incubate for 5 min at RT then triturate ×10 with 10 ml pipette.
5. Centrifuge at 400 × *g* for 5 min then discard supernatant.
6. Add ~2 ml (~1 full Pasteur pipette) KRB/TRIT to each sample tube and triturate ×10 with fire-polished pipette. Add ~2 ml KRB/TRIT to each sample tube and wait for tissue to settle.
7. Remove top 2 ml from sample tube and transfer to new 15 ml polystyrene tube. Triturate remaining 2 ml in sample tube ×10 with fire-polished pipette. Add ~2 ml KRB/BSA and wait for tissue to settle.
8. Remove top 2 ml from sample tube and transfer to new 15 ml sample tube. Triturate remaining 2 ml in sample tube ×10 with fire-polished pipette. Remove final 2 ml from sample tube and transfer to new 15 ml sample tube.
9. Centrifuge at 400 × *g* for 5 min. Discard supernatant with Pasteur pipette. Resuspend pellet in ~2 ml 70% Percoll in KRB/BSA.
10. Triturate ×5 with fire-polished pipette. Add ~2 ml 70% Percoll in KRB/BSA to sample tube. Add ~4 ml KRB/BSA to new polystyrene tube. Underlay KRB/BSA with ~4 ml 35% Percoll in KRB/BSA. Underlay existing two layers with the ~4 ml of 70% Percoll in KRB/BSA containing cells/tissue.
11. Centrifuge at 1,250 × *g* with slow break and slow acceleration for 20 min. Carefully aspirate myelin debris layer (upper interface). Carefully collect microglial layer (lower interface) and transfer to 12 × 75 polystyrene flow tube (not more than 1.5 ml total).
12. Dilute as much as possible with cold HBSS and centrifuge at 750 × *g* for 7 min then carefully discard supernatant using Pasteur pipette.
13. Wash pellet with 4 ml HBSS then centrifuge at 750 × *g* for 7 min. Discard supernatant using Pasteur pipette, then add

400 μl flow buffer. Vortex and transfer cell solution to new 500 μl tubes.

14. Centrifuge at 380 × *g* for 5 min. Discard supernatant and resuspend with 500 μl flow buffer.
15. Take equal volumes from sample tubes and pool accordingly to create tubes for isotype controls, cells alone and setup tubes.
16. Centrifuge at 380 × *g* for 5 min. Discard supernatant and resuspend with 2.5 μl of 1:10 diluted Fc block. Incubate on ice for 15 min.
17. Stain with antibodies according to specific protocol (most antibodies can be used undiluted at 1–3 μl and 30 min incubations in dark on ice).
18. For final resuspension, add 250 μl flow buffer (if only extracellular staining was performed and flow cytometry will be performed within 2 h).
    (a) For extracellular staining and longer than 2 h, use 1% PFA in PBS.
    (b) For intracellular staining, use stabilization buffer.
19. Transfer contents to flow tubes and run for flow cytometric analysis.

References

1. Balentine JD (1978) Pathology of experimental spinal cord trauma. I. The necrotic lesion as a function of vascular injury. Lab Invest 39:236–253
2. Jones TB, McDaniel EE, Popovich PG (2005) Inflammatory-mediated injury and repair in the traumatically injured spinal cord. Curr Pharm Des 11:1223–1236
3. Donnelly DJ, Popovich PG (2008) Inflammation and its role in neuroprotection, axonal regeneration and functional recovery after spinal cord injury. Exp Neurol 209:378–388
4. Popovich PG, Longbrake EE (2008) Can the immune system be harnessed to repair the cns? Nat Rev Neurosci 9:481–493
5. Klusman I, Schwab ME (1997) Effects of pro-inflammatory cytokines in experimental spinal cord injury. Brain Res 762:173–184
6. Lee YL, Shih K, Bao P, Ghirnikar RS, Eng LF (2000) Cytokine chemokine expression in contused rat spinal cord. Neurochem Int 36: 417–425
7. Rice T, Larsen J, Rivest S, Yong VW (2007) Characterization of the early neuroinflammation after spinal cord injury in mice. J Neuropathol Exp Neurol 66:184–195
8. Pineau I, Lacroix S (2007) Proinflammatory cytokine synthesis in the injured mouse spinal cord: multiphasic expression pattern and identification of the cell types involved. J Comp Neurol 500:267–285
9. Faulkner JR, Herrmann JE, Woo MJ, Tansey KE, Doan NB, Sofroniew MV (2004) Reactive astrocytes protect tissue and preserve function after spinal cord injury. J Neurosci 24: 2143–2155
10. Stirling DP, Liu S, Kubes P, Yong VW (2009) Depletion of ly6g/gr-1 leukocytes after spinal cord injury in mice alters wound healing and worsens neurological outcome. J Neurosci 29:753–764
11. Carlson SL, Parrish ME, Springer JE, Doty K, Dossett L (1998) Acute inflammatory response in spinal cord following impact injury. Exp Neurol 151:77–88
12. Ankeny DP, Lucin KM, Sanders VM, McGaughy VM, Popovich PG (2006) Spinal cord injury triggers systemic autoimmunity: evidence for chronic b lymphocyte activation and lupus-like autoantibody synthesis. J Neurochem 99:1073–1087
13. Jones TB, Hart RP, Popovich PG (2005) Molecular control of physiological and

pathological T-cell recruitment after mouse spinal cord injury. J Neurosci 25:6576–6583
14. Jones TB, Basso DM, Sodhi A, Pan JZ, Hart RP, MacCallum RC, Lee S, Whitacre CC, Popovich PG (2002) Pathological cns autoimmune disease triggered by traumatic spinal cord injury: implications for autoimmune vaccine therapy. J Neurosci 22:2690–2700
15. Ankeny DP, Guan Z, Popovich PG (2009) B cells produce pathogenic antibodies and impair recovery after spinal cord injury in mice. J Clin Invest 119:2990–2999
16. Rothwell NJ (1999) Annual review prize lecture cytokines – killers in the brain? J Physiol 514(Pt 1):3–17
17. Haddad JJ (2002) Cytokines and related receptor-mediated signaling pathways. Biochem Biophys Res Commun 297:700–713
18. Rollins BJ (1997) Chemokines. Blood 90:909–928
19. Luster AD (1998) Chemokines–chemotactic cytokines that mediate inflammation. N Engl J Med 338:436–445
20. Nesic O, Xu GY, McAdoo D, High KW, Hulsebosch C, Perez-Pol R (2001) IL-1 receptor antagonist prevents apoptosis and caspase-3 activation after spinal cord injury. J Neurotrauma 18:947–956
21. Semple-Rowland SL, Mahatme A, Popovich PG, Green DA, Hassler G Jr, Stokes BT, Streit WJ (1995) Analysis of TGF-beta 1 gene expression in contused rat spinal cord using quantitative RT-PCR. J Neurotrauma 12:1003–1014
22. McTigue DM, Popovich PG, Morgan TE, Stokes BT (2000) Localization of transforming growth factor-beta1 and receptor mrna after experimental spinal cord injury. Exp Neurol 163:220–230
23. McTigue DM, Tani M, Krivacic K, Chernosky A, Kelner GS, Maciejewski D, Maki R, Ransohoff RM, Stokes BT (1998) Selective chemokine mrna accumulation in the rat spinal cord after contusion injury. J Neurosci Res 53:368–376
24. Hayes KC, Hull TC, Delaney GA, Potter PJ, Sequeira KA, Campbell K, Popovich PG (2002) Elevated serum titers of proinflammatory cytokines and cns autoantibodies in patients with chronic spinal cord injury. J Neurotrauma 19:753–761
25. Kwon BK, Tetzlaff W, Grauer JN, Beiner J, Vaccaro AR (2004) Pathophysiology and pharmacologic treatment of acute spinal cord injury. Spine J 4:451–464
26. Gibson UE, Heid CA, Williams PM (1996) A novel method for real time quantitative RT-PCR. Genome Res 6:995–1001
27. Heid CA, Stevens J, Livak KJ, Williams PM (1996) Real time quantitative PCR. Genome Res 6:986–994
28. Wong ML, Medrano JF (2005) Real-time PCR for mRNA quantitation. Biotechniques 39: 75–85
29. Longbrake EE, Lai W, Ankeny DP, Popovich PG (2007) Characterization and modeling of monocyte-derived macrophages after spinal cord injury. J Neurochem 102:1083–1094
30. Kigerl KA, Lai W, Rivest S, Hart RP, Satoskar AR, Popovich PG (2007) Toll-like receptor (tlr)-2 and tlr-4 regulate inflammation, gliosis, and myelin sparing after spinal cord injury. J Neurochem 102:37–50
31. Popovich PG (2000) Immunological regulation of neuronal degeneration and regeneration in the injured spinal cord. Prog Brain Res 128:43–58
32. Sroga JM, Jones TB, Kigerl KA, McGaughy VM, Popovich PG (2003) Rats and mice exhibit distinct inflammatory reactions after spinal cord injury. J Comp Neurol 462:223–240
33. Kigerl KA, McGaughy VM, Popovich PG (2006) Comparative analysis of lesion development and intraspinal inflammation in four strains of mice following spinal contusion injury. J Comp Neurol 494:578–594
34. Schwartz M (2000) Beneficial autoimmune t cells and posttraumatic neuroprotection. Ann N Y Acad Sci 917:341–347
35. Kil K, Zang YC, Yang D, Markowski J, Fuoco GS, Vendetti GC, Rivera VM, Zhang JZ (1999) T cell responses to myelin basic protein in patients with spinal cord injury and multiple sclerosis. J Neuroimmunol 98:201–207
36. Popovich PG, Wei P, Stokes BT (1997) Cellular inflammatory response after spinal cord injury in Sprague-Dawley and Lewis rats. J Comp Neurol 377:443–464
37. Popovich PG, Stuckman S, Gienapp IE, Whitacre CC (2001) Alterations in immune cell phenotype and function after experimental spinal cord injury. J Neurotrauma 18:957–966
38. Mosmann T (1983) Rapid colorimetric assay for cellular growth and survival: application to proliferation and cytotoxicity assays. J Immunol Methods 65:55–63
39. Ahmed SA, Gogal RM Jr, Walsh JE (1994) A new rapid and simple non-radioactive assay to monitor and determine the proliferation of lymphocytes: an alternative to [3 h] thymidine incorporation assay. J Immunol Methods 170: 211–224
40. Kreutzberg GW (1996) Microglia: a sensor for pathological events in the CNS. Trends Neurosci 19:312–318

41. Watanabe T, Yamamoto T, Abe Y, Saito N, Kumagai T, Kayama H (1999) Differential activation of microglia after experimental spinal cord injury. J Neurotrauma 16:255–265
42. Shrikant P, Benveniste EN (1996) The central nervous system as an immunocompetent organ: role of glial cells in antigen presentation. J Immunol 157:1819–1822
43. Aloisi F (2001) Immune function of microglia. Glia 36:165–179
44. Ford AL, Goodsall AL, Hickey WF, Sedgwick JD (1995) Normal adult ramified microglia separated from other central nervous system macrophages by flow cytometric sorting. Phenotypic differences defined and direct ex vivo antigen presentation to myelin basic protein-reactive cd4+ T cells compared. J Immunol 154:4309–4321
45. Sedgwick JD, Schwender S, Imrich H, Dorries R, Butcher GW, ter Meulen V (1991) Isolation and direct characterization of resident microglial cells from the normal and inflamed central nervous system. Proc Natl Acad Sci USA 88:7438–7442
46. Beck KD, Nguyen HX, Galvan MD, Salazar DL, Woodruff TM, Anderson AJ. Quantitative analysis of cellular inflammation after traumatic spinal cord injury: evidence for a multiphasic inflammatory response in the acute to chronic environment. Brain133:433–447
47. Nathan C (2006) Neutrophils and immunity: challenges and opportunities. Nat Rev Immunol 6:173–182
48. Hausmann ON (2003) Post-traumatic inflammation following spinal cord injury. Spinal Cord 41:369–378
49. Taoka Y, Okajima K, Uchiba M, Murakami K, Kushimoto S, Johno M, Naruo M, Okabe H, Takatsuki K (1997) Role of neutrophils in spinal cord injury in the rat. Neuroscience 79:1177–1182
50. Taoka Y, Okajima K (1998) Spinal cord injury in the rat. Prog Neurobiol 56:341–358
51. Tjoa T, Strausbaugh HJ, Maida N, Dazin PF, Rosen SD, Noble-Haeusslein LJ (2003) The use of flow cytometry to assess neutrophil infiltration in the injured murine spinal cord. J Neurosci Methods 129:49–59
52. Banchereau J, Steinman RM (1998) Dendritic cells and the control of immunity. Nature 392:245–252
53. Bush TG, Puvanachandra N, Horner CH, Polito A, Ostenfeld T, Svendsen CN, Mucke L, Johnson MH, Sofroniew MV (1999) Leukocyte infiltration, neuronal degeneration, and neurite outgrowth after ablation of scar-forming, reactive astrocytes in adult transgenic mice. Neuron 23:297–308
54. Sofroniew MV, Vinters HV. Astrocytes: biology and pathology. Acta Neuropathol 119: 7–35
55. Peuchen S, Bolanos JP, Heales SJ, Almeida A, Duchen MR, Clark JB (1997) Interrelationships between astrocyte function, oxidative stress and antioxidant status within the central nervous system. Prog Neurobiol 52:261–281
56. Carson MJ, Thrash JC, Walter B (2006) The cellular response in neuroinflammation: the role of leukocytes, microglia and astrocytes in neuronal death and survival. Clin Neurosci Res 6:237–245
57. Brambilla R, Bracchi-Ricard V, Hu WH, Frydel B, Bramwell A, Karmally S, Green EJ, Bethea JR (2005) Inhibition of astroglial nuclear factor kappab reduces inflammation and improves functional recovery after spinal cord injury. J Exp Med 202:145–156
58. Schultke E, Griebel RW, Juurlink BH. Quercetin attenuates inflammatory processes after spinal cord injury in an animal model. Spinal Cord 48:857–861
59. Donnelly DJ, Gensel JC, Ankeny DP, van Rooijen N, Popovich PG (2009) An efficient and reproducible method for quantifying macrophages in different experimental models of central nervous system pathology. J Neurosci Methods 181:36–44
60. Popovich PG, Yu JY, Whitacre CC (1997) Spinal cord neuropathology in rat experimental autoimmune encephalomyelitis: modulation by oral administration of myelin basic protein. J Neuropathol Exp Neurol 56:1323–1338

Chapter 46

Neurobehavioral Assessments of Spinal Cord Injury

Jed S. Shumsky and John D. Houlé

Abstract

Experimental spinal cord injuries can be categorized as being mild, moderate, or severe, based upon the extent of tissue damage and the severity of functional deficit incurred. The level of injury and unilateral or bilateral nature of an injury also greatly impact upon whether deficits involve single or multiple limbs. Having these features in mind, the investigator is directed towards performing the appropriate neurobehavioral tests to assess the functional capabilities of injured animals before and after therapeutic interventions have been applied. An ideal test should use quantitative methods that can provide some insight into the physiological mechanisms involved in functional recovery.

Key words: Somatosensory testing, Behavioral assessment, Locomotion, Spinal cord injury, Functional recovery

1. Introduction

For a comprehensive review of behavioral testing in experimental animals after spinal cord injury (SCI), we refer the reader to these excellent reviews (1–3). Our goal here is to describe our approach to selecting an appropriate battery of functional assessments for different SCI models. Behavioral methods can be used to assess the degree of deficit and functional significance of interventions. Experienced personnel are needed to ensure the standardized testing and analysis required to provide the most reproducible results and to facilitate the comparison of the effects of different treatments. Test selection continues to be the subject of considerable debate. Several factors influence these choices, such as (1) individual investigator preference, (2) availability of appropriate testing equipment, (3) test reliability, and (4) test sensitivity. While not always possible, an ideal test should be rapid, reliable, and sensitive to the location and extent of the injury, the degree to which the intervention

Jun Chen et al. (eds.), *Animal Models of Acute Neurological Injuries II*, Springer Protocols Handbooks,
DOI 10.1007/978-1-61779-782-8_46, © Springer Science+Business Media, LLC 2012

improves the integrity of the injured region, and the degree to which intervention affects plasticity beneficially. Most spinal cord injury (SCI) models involve a partial lesion and tests should assess not only the degree of damage to the intended structure, but also residual impairment. This is also necessary because restoration of function is usually incomplete. A number of convenient sensorimotor tests have been developed that are not altered by repeated testing, do not require extensive training, aversive motivation, or food deprivation, and appear to be very useful for assessing established models of SCI. Food and water deprivation procedures are required for motivation to complete some tests, but must be used carefully as these manipulations may have profound effects on functional outcome (see ref. (4) for review).

It is usually necessary to test a range of behaviors in each animal to more fully describe the behavioral effects resulting from lesions of different sizes and locations and interventions with different characteristics. Preferably, a test uses quantitative methods that can provide some insight into physiological mechanisms and that are reproducible in other laboratories. Order of testing can matter, as testing on one apparatus can tire an animal or interfere with another testing procedure. Timing of testing remains another major issue. Typically, it is necessary to establish baseline performance on a behavioral test prior to injury. Even when a wealth of baseline data exists, habituation to the behavioral apparatus prior to injury is often beneficial in reducing the variability of behavioral performance after injury. Testing within 3–7 days of injury is useful to establish the degree of deficit from which to gauge recovery, and tests should be performed at the same time of day within a study. Usually, the recovery phase is monitored until the animal reaches plateau performance (usually, 4–8 weeks post injury depending upon the lesion model). This can pose issues for experimental design that can often be addressed in a small pilot study to determine the specific parameters for the selected injury model and its neurobehavioral assessments.

2. Neurobehavioral Assessments for Midthoracic Transection Injuries

A midthoracic transection injury in adult animals produces very severe deficits of the hind limbs, where animals are completely paralyzed below the injury and experience no somatosensation. Reflexive behaviors, however, are intact and are the usual targets for behavioral assessment. The BBB score (see below) can identify improved hindlimb movement. In addition, treadmill locomotion, with the trunk supported, can elicit reflex stepping movements, which can be measured kinematically. This test provides information about motor function even in animals that cannot independently support their own weight. Urodynamic parameters also show substantial deficits with this injury model. The Tail Flick test can also be used to measure the level of spinal hyperreflexia. If interventions are successful, then one may be able to employ tests for higher functioning animals as described below.

2.1. BBB Testing

The BBB test (5, 6) is a widely used and replicable method of evaluating loss and recovery of overground hindlimb motor function after spinal injury (see Chap. 47). Although originally developed for incomplete injuries, the BBB test can be used after spinal transection. Rats are placed in an enclosure and scored by two individuals according to the BBB scale at weekly intervals. It is not particularly sensitive for transection injuries as the range of typical scores is quite restricted to locomotor-like movements at the lower end of the scale. We have successfully used pharmacologic challenge with serotonin receptor agonists (7, 8) to reveal differences in the expression of locomotor-like movements of the hind limbs among treatment groups.

2.2. Treadmill Locomotion

We have used treadmill locomotion to demonstrate transplant-mediated recovery after complete spinal transection in neonates (9, 10) and adults (11). Prior to injury, rats are trained to walk on a treadmill at a speed of 8 m/min (13 cm/s) in 5–15-min sessions, 5 days/week over a period of several weeks, until they reach a plateau of performance. Performance outcomes can vary with treadmill speed; therefore, postinjury testing can be done at slower speeds that, depending upon the experimental design, may be increased to further challenge the animal. Treadmill locomotion is videotaped once a week. On the testing days, the percentage of weight-supported hindlimb stepping during the second minute of locomotion is measured. A *weight-supported step* is one in which the hindquarters are elevated above the surface of the treadmill and where from foot placement to foot placement only the foot touches the treadmill surface. After SCI, *nonweight-supported step* is one in which hindlimb stepping movements are made, but either the hindquarters or belly are in contact with the surface of the treadmill. Rats that cannot support their hindquarters propel themselves forward using their forelimbs. In our experience, the performance of rats injured as neonates shows the following distribution: those that develop <20% weight-supported stepping, those that develop 20–60% weight-supported stepping, and those that develop >60% weight-supported stepping (9). Treadmill locomotion testing has the advantage that kinematic analyses from the videotapes can be more sensitive to changes in motor performance than are measures of weight-supported locomotion. Kinematics of the hindlimb movements can be obtained during treadmill locomotion even if the steps are not weight supporting, as in rats spinalized as adults, if the rat's trunk is supported by a sling.

2.3. Locomotor Function in Cats

Training of stepping on the treadmill begins 1 week after injury for 15 min/day, 5 days/week. In the first few weeks, manual stimulation of the perineum, base of the tail, and/or abdominal regions is necessary to induce or maintain locomotion in the hind limbs, and the limbs are assisted in their stepping. In addition, it may be necessary for the hindquarters to be lifted and supported by the experimenter. The amount of weight support needed will decrease with time and the cats are expected to be able to achieve some weight support by 14–24 days after the start of locomotor training. Cats are considered to have reached full weight support when they can walk for 5–10 min without experimenter's support (12, 13). Testing begins at the lower treadmill speed (0.2 m/s) and progresses in increments of 0.2 m/s, up to 1.0 m/s. Maximum speed is defined as the fastest speed at which 10 consecutive steps can be executed (and 10 consecutive steps can be executed at all the lower speeds). Locomotor performance is routinely videotaped and analyzed. Stick figures representing each limb segment's orientation as a function of the step cycle are constructed and evaluated.

The extent of locomotor recovery can be assessed using quantitative measures extracted from the kinematics measurements. Maximum speed of stepping is a good general indicator of the locomotor recovery following spinal transection (14). Another measure of hindlimb motor recovery is to evaluate the stride length both pre and post spinalization (12) (using the same treadmill speed) and to compare the percentage of recovery. The statistical test for stride length is Student's paired *t*-test. A comparison of mean maximum velocity and average stride length between groups of cats is done using Student's unpaired *t*-test. A group size of five animals allows one to determine differences in mean maximum speed of about 0.27 m/s (for our power analysis, we use the standard error of de Leon, et al. (13) to estimate the variance in maximum speed achievable after training, and a significance level of 0.05). Average stride lengths between the two groups should prove a more powerful test due to the larger sample size for this variable.

Quadrupedal locomotion requires communication between the hindlimb spinal pattern generators and the forelimbs. The interlimb-phase interval analysis is used to indicate the consistency of the interval between the end of forelimb stance and the end of the ipsilateral swing. The interlimb swing phase analysis provides information on the synchrony of the swing phases. The swing phases are considered coordinated if the swing phase of any of the other limbs overlaps the swing phase of the hindlimb under analysis (right hindlimb) for 40% or more of that limb's swing phase (15). Interlimb coordination is compared under the same conditions for each animal, i.e., treadmill speed of about 0.4 m/s. Averages of the phase interval and phase synchrony for a number of step cycles can be constructed and compared. In addition, ipsi- and contralateral limb coordinations are examined once bilateral kinematic measurements are recorded.

2.4. Thermal Sensitivity/Tail Flick and Von Frey Tests

Preoperative and postoperative thresholds to thermal and mechanical stimuli are measured and compared. Lowered thresholds, i.e., exaggerated responses to light touch (tactile allodynia) and temperature (thermal allodynia) following SCI may indicate the development of central pain (16). Responses to thermal stimuli are evaluated using a hot plate test (17). Rats are placed on a platform and a radiant heat source (25–29°C) is positioned beneath a paw or the tail. If the paw is not withdrawn, the thermal stimulus is removed after 30 s. The withdrawal latency is measured in four trials, averaged, and the ratio between the hemisected and intact side for both the forepaw and hindpaw is calculated (18). We find that rats hemisected as adults develop decreased paw removal latencies to a mild thermal stimulus, whereas rats given transplants of fibroblasts genetically modified to produce BDNF or NT-3 show significantly longer latencies. Calibrated Von Frey filaments test response to mechanical stimuli. Rats standing on a screen receive stimulation by applying filaments (0.5–50 g intensity) in an ascending sequence to the glabrous skin on the paws until the filament bends, for a total of six trials, and then the threshold response (50% flexion) for paw flexion is calculated (19). Alternately, an up–down method may be employed (20).

2.5. Urodynamics

2.5.1. Voiding Volume

Bladders are manually expressed before placing the animals in the metabolic cages (Nalgene Metabolic Cage, Nalge Co., New York, USA) for 24 h to evaluate voiding behavior. The voided urine is collected on an electronic scale (FORT250, World Precision Instruments, Sarasota, FL, USA) connected to a microcomputer for recording micturition frequency and volume (21, 22). Data are recorded and stored using data acquisition software (WINDAQ, DATAQ Instruments, Akron, USA). Frequency and volume of micturition episodes are measured preoperatively and then weekly after SCI. Typically, thoracic-injured animals reach a new plateau of voiding volume that is 1.5–2 times preoperative values because of bladder hypertrophy.

2.5.2. Cystometry

In terminal experiments, rats are anesthetized under isoflurane inhalation and a polyethylene catheter is implanted into the bladder. Following 2% isoflurane inhalation, the bladder is exposed by a midline lower abdominal incision and a polyethylene catheter (PE-60, Clay-Adams, New Jersey, USA) implanted into the bladder through the dome, as described previously (21–23). The catheter is tunneled subcutaneously and exited through the skin at the back of the rat. Animals are removed from gas anesthesia, placed in a restraining cage (KN-326, Natsume, Tokyo, Japan), and allowed to recover for 1–2 h. The bladder catheter is connected to a pressure transducer (BLPR, World Precision Instruments) and a microinjection pump (STC-523, Terumo, Tokyo, Japan). Room-temperature saline is infused at a rate of 0.1 ml/min, and intravesical pressure recorded to compare urodynamic parameters in each group of animals. The maximal voiding pressure, postvoid residual urine, bladder capacity, and the frequency of DHR per micturition episode are compared in each group. DHR is defined as rhythmic intravesical pressure increases greater than 5 mmHg from baseline without a release of fluid from urethra. Saline voided from the urethral meatus is collected to determine the voided volume. The remaining intravesical contents are collected by withdrawing the residual saline from the bladder through the intravesical catheter and then expressing the bladder manually by applying pressure on the abdominal wall. Bladder capacity is calculated as the voided volume plus residual volume.

3. Neurobehavioral Assessments for Midthoracic Contusion Injuries

This type of injury produces substantial-to-intermediate hindlimb deficits that can be evaluated using CatWalk, Exploratory Rearing, and Grid tests in addition to those listed above (BBB, treadmill, urodynamics, thermal sensitivity, and Von Frey tests). Depending upon the injury severity, deficits can range from almost complete paralysis and loss of somatosensation below the injury to subtle deficits of hindlimb coordination or paw placements, with possible allodynia.

3.1. CatWalk Gait Analysis

By using this apparatus for locomotor gait analysis (24–26), we are able to measure and interpret the dynamics of locomotion, such as interlimb coordination, limb base of support, stride length, and swing duration (see Chap. 49). The Catwalk apparatus has an enclosed glass walkway. Light from a fluorescent tube enters the long edge of the floor and is entirely internally reflected. When animals cross the glass floor, light escapes from the floor and illuminates where the paws make contact. Animals are filmed from underneath while traversing the walkway. By using the Catwalk program, paw prints are labeled and gait parameters are measured or computed. The following gait parameters are assessed: (1) regularity index, which grades the degree of interlimb coordination. In intact (fully coordinated) animals, the regularity index is 100%, but may appear slightly lower due to collection of incomplete step sequences. (2) Base of support, which measures the distance between either the front or hind limbs. (3) Stride length, which measures the distance between consecutive steps of an individual paw during locomotion. (4) Swing duration, which measures the time between swing phases for each limb. (5) Phase dispersion, which is a more quantitative measure of interlimb coordination.

3.2. Exploratory Rearing Test

When placed in a Plexiglas cylinder (17.8 cm diameter × 35.5 cm height), animals spontaneously rear and contact the walls with their forepaws. Numbers and types of rears are analyzed during the 3-min recorded observation period in the cylinder (16). Different types of rears of increasing heights are observed: (1) crouch—animals lift both forepaws off the ground but do not extend the trunk; (2) tripod—animals lift both forepaws off the ground and support themselves with the tail; (3) rear—animals rear and extend the trunk with the hind limbs flexed; and 4) stand—animals rear and extend the trunk with the hind limbs extended. Occasionally, an animal progresses through several types of rears before touching down, in which case it is counted as a single rear of the highest type. Data are expressed as the number of rears per minute and percentage of functional rears (number of rears and stands/total rears).

3.3. Grid Test

Similar to locomotion across a horizontal ladder, rats walk on an elevated grid (36 cm L × 38 cm W × 30 cm H, with 3 × 2 cm openings) for 2 min. Total number of steps by each forepaw and hindpaw and the number of misses made by each forepaw and hindpaw are recorded separately for both right and left limbs. This task examines both sensory and motor functions. To avoid missteps, rats palpate the wire to place paws properly on the wire. Injured rats are deficient in this task. Neonatal spinal rats receiving fetal transplants are almost as impaired as spinal rats on both horizontal ladders (9) and grids. Transplants of fetal tissue into incomplete lesions in neonates (27) or transplants of fibroblasts modified to secrete NT-3 into adults (28) decrease errors in the grid test. Improvement in this test may indicate improved sensorimotor integration in spinal animals with transplants.

3.4. Swimming

By providing partial weight support in a buoyant medium, swimming tests, such as the Louisville Swim Scale (29), can be used to assess motor function in moderate-to-severely injured animals (see Chap. 51).

4. Neurobehavioral Assessments for Cervical Injuries

This category would include full and partial hemisections, overhemisections, and contusions. In most cases, cervical lesions affect hindlimb and forelimb locomotion and thus produce deficits in the BBB, CatWalk, and Grid tests (as above) as well as Cylinder, Grip, Reaching, Rope, Staircase Reaching, Swimming, and Tapered Beam tests (see Chaps. 48 and 50). Forelimb function is more affected by a cervical injury, and thus the Cylinder and Reaching tests are particularly sensitive. Disruption of ascending sensory pathways leads to deficits in Patch Removal, Somatosensory Discrimination, Tactile Placing, Thermal Sensitivity, and Von Frey tests. Often, cervical injuries are performed unilaterally because bilateral deficits reduce the ability of animals to reach food and water. Assessment of the uninjured side can still be useful to detect unexpected deficits and compensatory behaviors.

Unilateral damage to the cervical spinal cord causes persistent deficits in somatosensory function and use of the ipsilateral forelimbs can be easily quantified (30) using simple, reliable, and sensitive behavioral tests. These tests have been useful in studies of recovery of function following central nervous system injury, as well as for assessing efficacy of behavioral or pharmacological interventions (30). Frequently, recovery of function is defined as reinstatement of some preinjury ability or abilities, which were lost or diminished following damage. An additional approach is to measure recovery of function in terms of the reinstatement of symmetrical tactile sensation or limb use, which is analogous to handedness, in that plasticity is not altered unless specifically manipulated by limb-specific practice or other interventions.

4.1. Cylinder (Limb Preference) Test

This test is a measure of spontaneous forelimb usage (16, 31) and can be used in conjunction with the exploratory rearing test. Baseline behavior (handedness) is measured before a unilateral injury. The rats are placed in a clear Plexiglas cylinder (20 cm in diameter and 30 cm high) for 3 min. The testing sessions are videotaped and forelimb usage can be scored blindly at a later date. The following behaviors are scored: (a) independent use of the left or right forelimb for contacting the wall of the cylinder during a rear to initiate a weight-shifting movement or to regain the center of gravity while moving laterally in a vertical posture along the wall; (b) simultaneous, or near-simultaneous, use of both the left and right forelimb to contact the wall of the cylinder during a rear and for lateral movements along the wall. Each behavior is expressed in terms of (1) the percent use of the contralateral (nonimpaired) forelimb relative to the total number of ipsilateral, contralateral, and simultaneous (both) limb use observations; (2) percent use of the ipsilateral (impaired) forelimb relative to the total number of ipsilateral, contralateral, and simultaneous (both) limb use observations; and (3) percent simultaneous (both) limb use relative to the total number of ipsilateral, contralateral, or simultaneous (both) limb use observations.

4.2. Grip Force Assessment

The animals undergo grip force assessment of the involved and uninvolved forelimb at baseline and during the recovery period (32–34). The grip force test is a test of spontaneous behavior and does not require training. Because the rubrospinal tract has been observed to make direct projections onto the motoneurons of muscles that control the wrist and digits of the rat, grip force assessment is potentially a sensitive measurement of recovery of input from the red nucleus. The animal's forepaw is placed on a wire grid (1 mm wire diameter), which is attached to a force transducer (sampling rate of 1,000 Hz). Then, the rat is pulled in a horizontal plane parallel to the grid until the animal's grip is broken. The grip force of each animal is recorded as the averaged peak force from three trials performed on two separate days.

4.3. Patch Removal (Somatosensory Asymmetry) Test

Somatosensory asymmetry is assessed using this bilateral tactile stimulation test (35–37), though it is time consuming to administer and score and therefore is used selectively. Animals are first tested to indicate the presence of a somatosensory asymmetry. The animal is removed from the home cage and adhesive patches (Avery adhesive backed labels, 113 mm^2) are placed on the distal–radial aspect of both forelimbs. After being returned to the home cage, rats contact and remove the stimuli one at a time using their teeth. The order and latency of stimulus contact and removal are recorded for each of the four trials. To increase the reliability of somatosensory asymmetry testing, rats are tested in the home cage and familiarized with the adhesive patches preoperatively and before testing; rats that do not respond to and remove the patches are not used.

The order of contact determines whether animals show a bias for the stimulus on the forelimb either unaffected or affected by the injury. If the animal shows either a 25 or 75% preference for removing the stimulus from the unaffected forelimb first, a fifth trial will be given. All animals with a bias (removal of the stimulus from the unaffected limb first on more than 70% of the trials) are then tested to determine the magnitude of the somatosensory asymmetry. In the magnitude of asymmetry phase of the test, the size of the affected limb (A) stimulus is progressively increased and the size of the unaffected limb (U) stimulus is simultaneously decreased by an equal amount (14.1 mm^2). Thus, a sufficient increase in the A/U ratio causes a reversal of the original bias if the rat begins to respond to the stimulus placed on the affected limb first. The A/U ratio necessary to reverse the initial bias is roughly proportional to the degree of damage to sensory systems (36, 38). This test measures cutaneous sensitivity and sensorimotor integration. A similar test has been shown to identify recovery of function mediated by transplants into neonatal hemisections (27).

4.4. Reaching

This test requires substantial preoperative training and is also used selectively. Initially, animals are habituated to a Plexiglas reaching chamber (45 cm deep × 40 cm tall × 12.5 cm wide) that has a 1-cm-wide slit cut out and centered on the 12.5-cm-wide front wall. On the outside of the front wall, a 3.2-cm-deep × 12.5-cm-wide shelf is mounted so that the top surface is 4 cm above the chamber floor. Two food pellet wells are drilled out and centered to each edge of the slit on the front wall and placed 1.5 cm away from the outside front wall. Following habituation to the reaching chamber, animals are operantly trained to shuttle to the back of the cage and then to come forward and reach with the right forelimb through the slit in order to grasp and consume a food pellet that is placed in the contralateral food well. Shuttling after a reach creates a natural separation between reaches and encourages consistency in how the animal orients and initiates itself to perform the next reach (39). The animal is considered to be successfully trained when it reaches with the right forelimb, grasps the food pellet with greater than 50% success in pellet retrieval, shuttles to the back of the chamber, and returns to the front for its next attempt.

4.4.1. Qualitative Components

The reach-to-grasp task is scored using a movement rating scale (40–42), which assesses ten different components of a reach. Each component is given a rating of 0 if the movement appears normal, 1 if the movement appears somewhat abnormal but recognizable, or 2 if the movement is absent or compensated by moving other parts of its body. The ratings from each component of the reach are summed to get a deficit score from each trial (0 = no deficit; and 20 = maximum deficit). A total of five reaches from each animal are scored from video replay and averaged. The ten components of the reach are the following: (1) *limb lift*—the forelimb is lifted from the floor by the upper arm and the digits are brought to the body's midline; (2) *digits close*—as the forelimb is lifted, the digits are semiflexed and the paw is supinated so that the palm faces the midline of the body; (3) *aim*—the elbow is adducted by the upper arm so that the forearm and digits are aligned with the body's midline and the paw is held just under the mouth; (4) *limb advance*—the head is lifted and the limb is advanced above and beyond the pellet; (5) *digits open*—the digits are extended and opened as the limb is advanced; (6) *pronation*—the elbow is abducted by the upper arm and the paw is pronated over the food; (7) *grasp*—as the palmar pads touch the food, the digits close around it, and closure can occur before or during paw withdrawal; (8) *supination I*—as the paw is withdrawn, it is dorsiflexed and supinated 90° by a movement around the wrist and by adduction of the elbow; (9) *supination II*—as the rat sits back on its haunches, the paw is supinated by a further 90° and is ventroflexed so that the food is brought up to its mouth; and (10) *release*—the digits are opened and the food is transferred to the mouth.

4.4.2. Quantitative Assessment

The reach-to-grasp task can be performed by counting the number of successful grasps over the number of reaching attempts. It can also be assessed kinematically in the reaching chamber described above using a synchronized, high-speed (500 frames/s), 2-camera digital recording system (Redlake). The two cameras are positioned to get a lateral view of the right side and a frontal view. A ventral view is recorded from the frontal camera angle via a mirror placed at 45° beneath the Plexiglas floor of the reaching chamber. The animals' right forelimb is shaved and ink marks placed over the shoulder, elbow, and wrist joints and the tips of the digits for kinematic analysis using WINanalyze tracking software (Mikromak).

Quantitative measures that can be assessed during the reach-to-grasp task are the following: (1) percent successful reaches; (2) peak wrist velocity; (3) number of velocity reversals; (4) deviation of wrist trajectory from a straight line; (5) amount of paw pronation; (6) amount of supination during *supination I* phase; and (7) spread of digits during *digits open* and *pronation* phases. A successful reach is when the rat makes contact with the food pellet, grasps it to remove it from the pellet well on the platform, and eats the pellet from its paws. A reach during which the animal "rakes" the pellet from the well and does not grasp the pellet or if the animal drops the pellet once it has retrieved it from the platform does not count as a successful reach. The percent of successful reaches is calculated as (no. of successful reaches/no. of unsuccessful reaches) × 100.

Both qualitative and quantitative assessments of the reach-to-grasp task are performed during the prelesion phase after the animals are trained on the task. Post injury, qualitative reaching assessment is performed weekly, while quantitative reaching is assessed only at early and late designated time points because of technical demands. These tests can detect differences in function associated with rubrospinal or corticospinal damage (43) and, therefore, should be useful in detecting recovery associated with the repair of specific pathways. Damage to the corticospinal tract reduces success more than damage to the rubrospinal system; both injuries impair paw rotation and grasping, and red nucleus lesions eliminate the arpeggio paw movement. Neural transplantation into overhemisection sites in neonates (27) or corticospinal tract lesions in adults (44) reestablished targeted reaching. Trained reaching movements are amenable to EMG (45) and kinematic analyses (42, 43) which can help in understanding the altered strategies used by animals with impairments and animals with interventions that promote recovery.

4.5. Rope Test

Animals learn to traverse a rope 4.4 cm in diameter and 83 cm in length within a few trials and uninjured rats cross the rope quickly with few errors. Rats are videotaped for five consecutive trials. Both objective and ranking scoring methods are used. Slips of one limb and falls from the rope are counted (falls are scored as two slips). The total number of slips and falls constitute the Total Error/Step. For ranking, two to three observers view videotapes of rats traversing the rope and rate the behavior on a 4-point system: 0—normal, 1—good, 2—moderate, and 3—poor. Each rat is evaluated according to limb placement, weight support, and posture. An animal with a score of 9 has the maximum deficit. Each rat is observed three times and the final deficit score is the mean of the three independent observations. More detailed studies of gait pattern can be made, in which step sequences are examined to reveal alternative strategies for crossing the rope. Since rats may use altered strategies to walk on the rope, kinematic measurements from videotapes are used to supplement error scores. Horizontal rope walking tests both postural adjustment and hindlimb locomotion. Hemisected animals that walk well on narrow beams and have high BBB scores are deficient in this test; hemisected animals with grafts of BDNF fibroblasts show significantly better recovery (31).

4.6. Somatosensory (Tactile) Discrimination

Animals are trained to reach for two horizontally placed stimuli in a chamber (46, 47) similar to that described for the single-pellet reach-to-grasp task. The chambers are identical in structure, except that for the tactile discrimination task a Plexiglas frame is placed on top of the pellet shelf. This frame contains holes (about 1.5 mm in diameter) on the vertical wall of the frame contralateral to the preferred paw for reaching. In one hole, a 2.5-cm-long (1.5 mm diameter) drill bit is attached to a short piece of pasta by a nylon cable tie. The other hole contains a normal piece of pasta. Because the same force would break off each stimulus item, the rats learn to use tactile discrimination to choose the piece of pasta for consumption. The animals are given 50 trials/day for 8 days and the location of the pasta and metal drill bit is randomly alternated in location. Correct responses and errors are scored as follows: (1) a trial is scored as correct when the animal grasps the metal drill bit stimulus, but does not apply sufficient force to break it and subsequently chooses and breaks the pasta stimulus and (2) an error is scored when the rat breaks off the metal drill bit stimulus. The percentage of correct responses is determined by dividing the number of correct trials by the number of correct plus error trials and multiplying by 100.

4.7. Staircase-Reaching (Pellet Retrieval from Multiple Levels) Test

We have used the staircase-reaching test to examine reach performance at multiple levels of shoulder flexion (42, 48). The staircase-reaching test offers several advantages as follows: (1) the shelf of the single-pellet reaching apparatus is 4.0 cm above the floor and, therefore, requires good shoulder muscle control to aim and contact the pellet. The staircase-reaching test is able to test reaching function independent of shoulder control, which may be impaired after dorsolateral funiculotomy because it can disrupt part of the ventral gray matter at the C3/C4 spinal cord level that contains the motoneuron pool that controls the shoulder musculature. (2) The single pellet test allows the possibility for the rat to "rake" the pellet through the slot for retrieval rather than grasp it. The staircase test requires the animal to make a coordinated grasp and lift the pellet to its mouth with no possibility for "raking." The animals are trained on the staircase-reaching test for 2 weeks prior to the lesion surgery (2–15-min sessions/day). Reaching performance is measured at baseline, 1 and 8 weeks post injury. During these testing periods, the animals have 2–15-min testing sessions per day for 5 days. The number of pellets retrieved is counted, and the average of the three best (most pellets retrieved) sessions is recorded for each animal.

4.8. Swimming

This task permits the study of both forelimb and hindlimb function in the absence of weight bearing or postural demands. Normal rats swim in a straight line with both forelimbs adducted beneath the chin (planning). The body is propelled forward by alternating hindlimb movements. Deficits in planning by the impaired forelimb occur after C3/C4 partial hemisections but are partially reversed by BDNF fibroblast transplants (31). Non-alternating hindlimb movements, a more primitive pattern, persist after C3/C4 hemisections, but alternating patterns are partially restored after BDNF fibroblast transplants. Kinematic analysis of patterns from videotapes permits more detailed description of the deficits and recovery of locomotor function in the absence of weight bearing. While this test is also useful for studying hindlimb locomotor patterns, it is somewhat stressful to the rats and, therefore, is not used routinely.

4.9. Tactile Placing

Forelimb placing asymmetry behaviors are scored using the vibrissae-elicited forelimb placing test (49). Animals are held by their torsos allowing forelimbs to hang free. While holding the animal, the experimenter applies gentle up and down movements in space prior to place testing, which facilitates muscle relaxation and eliminates struggling movements. Independent testing of each forelimb is induced by brushing the respective vibrissae on the edge of a table top once per trial for ten trials. Intact animals place the forelimb of both sides quickly onto the countertop. Animals with unilateral damage show varying degrees of impaired limb placing ability while still placing the unimpaired limb reliably. A correct response is scored each time the rat places its forelimb on the edge of the tabletop in response to the vibrissae stimulation. Typically, animals place successfully in 100% of the trials with the unimpaired limb. Tactile placing may be used to assess the degree of control by the corticospinal tract.

4.10. Tapered (Ledged) Beam Test

This test examines locomotor function across a challenging terrain and the ledge reduces postural compensatory adjustments, thereby highlighting locomotor effects (49, 50). It was developed as a more sensitive measure of forelimb and hindlimb function compared with walking across a rope or narrow beam. Another advantage of this procedure is the automatic compensation provided by stepping on the supporting ledge beneath the tapered beam. Dr. Schallert initially developed the procedure for models of ischemic injury and we have adapted it for use with SCI. Motor function after SCI is sufficiently different from ischemic injury such that validation of a novel scoring system is required. We have created a scoring method that is sensitive to deficits and recovery following SCI and have added this procedure to our battery of behavioral tests.

The beam is raised and reinforced onto a wooden platform by wooden rods. The animals' home cage is positioned at the narrow end of the beam, which provides an incentive for traversing the beam. There are five trials for each animal. The beam is divided into five sections and the middle three quadrants are assigned point values. This test focuses on hindlimb ambulatory movement to determine if there is a full slip or a half slip. A full slip takes place when the animal's foot touches the Plexiglas ledge under the beam. A half slip occurs when the animal's foot slips below the beam but does not touch the transparent ledge. The total number of slips for each hind limb is determined and the total weighted points for each trial are added together to give a total value for each limb.

5. Rehabilitation Training (Exercise)

As reviewed by Edgerton et al. (51), there are fundamental reasons to work towards the development of activity-based rehabilitation approaches, including the observation that anatomical and physiological plasticity continues for a prolonged postinjury period and rewiring of spinal circuitry may benefit from the rhythmic sensory input related to specific task retraining. We have been developing methods of rehabilitation training using passive rotation of hind limbs on a motorized bicycle (52) and motorized exercise wheels (53). Treadmill training, step and stance training, and swimming have all been used by others as rehabilitative interventions (20, 54–57).

5.1. Cycling

We have a custom-made apparatus, consisting of a unit of four parallel mounted bicycles and entrained by a single electric motor with variable speed control. It is fitted with adapters to accommodate rats of different sizes. Rats are suspended above a pedal set by an adjustable sling. The hind limbs are attached to the pedals with 3 M micropore Surgery tape (3 M, St Paul, MN). Training begins at 5 days post transection and continues for the duration of the experiment (with a several-day break after any additional surgery is performed). All exercised animals are trained for two 30-min sessions with 10-min rest in between, 5 days a week at 45 rpm. Passive rotation of rat hind limbs on a motorized bicycle can reduce the extent of muscle atrophy (52), restore the rate-dependent depression response of the H reflex (58), and improve membrane properties of spinal motoneurons below the level of a transection injury (59). We have observed that animals that possess some degree of hindlimb function, either because of incomplete injury or as a result of a successful intervention, tend to resist placement on the bicycles and must be monitored carefully to avoid the risk of hindlimb injury. In this case, treadmill training may be considered as an alternative depending upon the experimental design.

5.2. Activity Wheels

Highly functional animals with a unilateral cervical injury can be trained using the rat forced exercise walking wheel system (Model 80806, Lafayette Instruments, Lafayette, IN). The apparatus consists of six wheels (13.38″ diameter and a width of 4.9″), rotating on a motorized belt assembly. Animals are trained on the wheel for 1 week for a period of 10 min at 14.0 m/min prior to surgery. Injured animals are exercised five times a week for 20 min/day beginning at 2.5 m/min, with increases in speed commensurately with the increase in walking capacity of each animal (53). We also used this system to validate a qualitative wheel walking assessment that examines weight support, the ability to grip an individual rung and successful steps, where placement on a rung is maintained.

6. Conclusion

Multiple sensorimotor tasks are available to evaluate the extent of functional deficits following SCI in experimental animals, and we have touched on a few of the more useful ones in general use at this time. Testing is often conducted early after injury and thereafter on a weekly basis to detect and chart spontaneous recovery of function as well as to provide a measure of recovery after therapeutic intervention. Animals with injury at different levels and of different regions (partial vs. complete) of the spinal cord will present with variable deficits and it is critical that appropriate behavioral assessments are chosen to reflect changes in sensory and/or motor outcome after SCI. Specifics of animal handling, test sequence, time of day, and postoperative care can influence behavioral outcomes, and researchers should strive to maintain consistency in all aspects of the study. Many labs are involved in the evaluation of general and task-specific training after SCI and advances in this area will expand our understanding of the role of physical rehabilitation in recovery of function after SCI.

References

1. Metz GA, Merkler D, Dietz V, Schwab ME, Fouad K (2000) Efficient testing of motor function in spinal cord injured rats. Brain Res 883:165–177
2. Basso DM (2004) Behavioral testing after spinal cord injury: congruities, complexities, and controversies. J Neurotrauma 21:395–404
3. Sedý J, Urdzíková L, Jendelová P, Syková E (2008) Methods for behavioral testing of spinal cord injured rats. Neurosci Biobehav Rev 32:550–580
4. Schallert T, Tillerson JL (1999) Intervention strategies for degeneration of dopamine neurons in Parkinsonism: optimizing behavioral assessment of outcome. In: Emerich DF (ed) Innovative models of CNS diseases. Humana Press, Clifton NJ
5. Basso DM, Beattie ME, Bresnahan JC (1995) A sensitive and reliable locomotor rating scale for open field testing in rats. J Neurotrauma 12:1–21
6. Basso DM, Beattie ME, Bresnahan JC (1996) Graded histological and locomotor outcomes after spinal cord contusion using the NYU weight-drop device versus transection. Exp Neurol 139:244–256

7. Nothias JM, Mitsui T, Shumsky JS, Fischer I, Antonacci MD, Murray M (2005) Combined effects of neurotrophin secreting transplants, exercise, and serotonergic drug challenge improve function in spinal rats. Neurorehab Neural Repair 19:296–312
8. Kao T, Shumsky JS, Jacob-Vadakot S, Himes BT, Murray M, Moxon KA (2006) Role of the 5-HT2C receptor for improving weight supported stepping in adult rats spinalized as neonates. Brain Res 1112:159–168
9. Miya D, Tessler A, Giszter S, Mori F, Murray M (1997) Fetal transplants alter the development of function after spinal cord transection in newborn rats. J Neurosci 17:4856–4872
10. Kim D, Murray M, Simansky KJ (2001) The serotonergic 5-HT(2C) agonist m-chlorophenylpiperazine increases weight-supported locomotion without development of tolerance in rats with spinal transections. Exp Neurol 169:496–500
11. Stackhouse SK, Shibayama M, Bowes M, Murray M (1997) Fetal tissue transplants improve hindlimb function in adult spinal rats. Soc Neurosci Abstr 23:906
12. Belanger M, Drew T, Provencher J, Rossignol S (1996) A comparison of treadmill locomotion in adult cats before and after spinal transection. J Neurophysiol 76:471–491
13. de Leon RD, Tamaki H, Hodgson JA, Roy RR, Edgerton VR (1999) Hindlimb locomotor and postural training modulates glycinergic inhibition in the spinal cord of the adult spinal cat. J Neurophysiol 82:359–369
14. Brustein E, Rossignol S (1998) Recovery of locomotion after ventral and ventrolateral spinal lesions in the cat. I. Deficits and adaptive mechanisms. J Neurophysiol 80:1245–1267
15. Howland DR, Bregman BS, Tessler A, Goldberger ME (1995) Transplants enhance locomotion in neonatal kittens whose spinal cords are transected: a behavioral and anatomical study. Exp Neurol 135:123–145
16. Shumsky JS, Tobias CA, Tumolo M, Long WD, Giszter SF, Murray M (2003) Delayed transplantation of fibroblasts genetically modified to secrete BDNF and NT-3 into a spinal cord injury site is associated with limited recovery of function. Exp Neurol 184:114–130
17. Hargreaves K, Dubner R, Brown F, Flores C, Joris J (1988) A new and sensitive method for measuring thermal nociception in cutaneous hyperalgesia. Pain 32:77–88
18. Christensen MD, Everhart AW, Pickelman JT, Hulsebosch CE (1996) Mechanical and thermal allodynia in chronic central pain following spinal cord injury. Pain 68:97–107
19. Yezierski RP, Liu S, Ruenes GL, Kajander KJ, Brewer KL (1998) Excitotoxic spinal cord injury: behavioral and morphological characteristics of a central pain model. Pain 75:141–155
20. Hutchinson KJ, Gómez-Pinilla F, Crowe MJ, Ying Z, Basso DM (2004) Three exercise paradigms differentially improve sensory recovery after spinal cord contusion in rats. Brain 127:1403–1414
21. Mitsui T, Kakizaki H, Shibata T, Tanaka H, Matsuoka I, Koyanagi T (2003) Transplanted immortalized neural stem cells into injured spinal cord promote recovery of voiding function in rats. J Urology 170:1421–1425
22. Mitsui T, Fischer I, Shumsky JS, Murray M (2005) Transplants of fibroblasts expressing BDNF and NT-3 promote recovery of bladder function following spinal contusion injury in rats. Exp Neurol 194:410–431
23. Yoshiyama M, Nezu FM, Yokoyama O, De Groat WC, Chancellor MB (1999) Changes in micturition after spinal cord injury in conscious rats. Urology 54:923–933
24. Cao Y, Shumsky JS, Sabol MA, Kushner RA, Tessler A, Strittmatter S, Hamers FPT, Lee DHS, Rabacchi SA, Murray M (2008) Nogo-66 receptor antagonist peptide (NEP1-40) administration promotes functional recovery and axonal growth after lateral funiculus injury in the adult rat. Neurorehab Neural Repair 22:262–278
25. Hamers FP, Lankhorst AJ, van Laar TJ, Veldhuis WB, Gispen WH (2001) Automated quantitative gait analysis during overground locomotion in the rat: its application to spinal cord contusion and transection injuries. J Neurotrauma 18:187–201
26. Koopmans GC, Deumens R, Honig WM, Hamers FP, Steinbusch HW, Joosten EA (2005) The assessment of locomotor function in spinal cord injured rats: the importance of objective analysis of coordination. J Neurotrauma 22: 214–225
27. Diener PS, Bregman BS (1998) Fetal spinal cord transplants support the development of target reaching and coordinated postural adjustments after neonatal cervical spinal cord injury. J Neurosci 18:763–778
28. Grill R, Murai K, Blesch A, Gage FH, Tuszynski MH (1997) Cellular delivery of neurotrophin-3 promotes corticospinal axonal growth and partial recovery after spinal cord injury. J Neurosci 17:5560–5572
29. Smith RR, Burke DA, Baldini AD, Shum-Siu A, Baltzley R, Bunger M, Magnuson DS (2006) The Louisville Swim Scale: a novel assessment of

hindlimb function following spinal cord injury in adult rats. J Neurotrauma 23:1654–1670
30. Liu Y, Kim D, Himes BT, Chow S, Murray M, Tessler A, Fischer I (1999) Transplants of fibroblasts genetically modified to express BDNF promote regeneration of adult rat rubrospinal axons and recovery of forelimb function. J Neuroscience 19:4370–4387
31. Kim D, Schallert T, Liu Y, Browarak T, Nayeri N, Tessler A, Fischer I, Murray M (2001) Transplantation of genetically modified fibroblasts expressing BDNF in adult rats with a subtotal hemisection improves specific motor and sensory functions. Neurorehab Neural Repair 15:141–150
32. Cabe PA, Tilson HA, Mitchell CL, Dennis R (1978) A simple recording grip strength device. Pharmacol Biochem Behav 8:101–102
33. Meyer OA, Tilson HA, Byrd WC, Riley MT (1979) A method for the routine assessment of fore- and hindlimb grip strength of rats and mice. Neurobehav Toxicol 1:233–236
34. Anderson KD, Gunawan A, Steward O (2007) Spinal pathways involved in the control of forelimb motor function in rats. Exp Neurol 206:318–331
35. Schallert T, Upchurch M, Lobaugh N, Farrar SB, Spirduso WW, Gilliam P, Vaughn D, Wilcox RE (1982) Tactile extinction: distinguishing between sensorimotor and motor asymmetries in rats with unilateral nigrostriatal damage. Pharmacol Biochem Behav 16:455–462
36. Schallert T, Upchurch M, Wilcox RE, Vaughn DM (1983) Posture-independent sensorimotor analysis of inter-hemispheric receptor asymmetries in neostriatum. Pharmacol Biochem Behav 18:753–759
37. Schallert T, Hernandez TD, Barth TM (1986) Recovery of function after brain damage: severe and chronic disruption by diazepam. Brain Res 379:104–111
38. Barth TM, Jones TA, Schallert T (1990) Functional subdivisions of the rat somatic sensorimotor cortex. Behav Brain Res 39:73–95
39. Whishaw IQ, Pellis SM, Gorny BP (1992) Medial frontal cortex lesions impair the aiming component of rat reaching. Behav Brain Res 50:93–104
40. Whishaw IQ, Pellis SM, Gorny B, Kollo B, Tetzlaff W (1993) Proximal and distal impairments in rat forelimb use in reaching follow unilateral pyramidal tract lesions. Behav Brain Res 56:59–76
41. McKenna JE, Whishaw IQ (1999) Complete compensation in skilled reaching success with associated impairments in limb synergies, after dorsal column lesion in the rat. J Neurosci 19:1885–1894
42. Stackhouse SK, Murray M, Shumsky JS (2008) The effect of cervical dorsolateral funiculotemy on reach-to-grasp function in the rat. J Neurotrauma 25:1039–1047
43. Whishaw IQ, Gorny B, Sarna J (1998) Paw and limb use in skilled and spontaneous reaching after pyramidal tract, red nucleus and combined lesions in the rat. Behav Brain Res 93: 167–183
44. Li Y, Field PM, Raisman G (1997) Repair of adult rat corticospinal tract by transplants of olfactory ensheathing cells. Science 277:2000–2002
45. Hyland BI, Jordan VMB (1997) Muscle activity during forelimb reaching movements in rats. Behav Brain Res 85:175–186
46. Ballermann M, Tompkins G, Whishaw IQ (2000) Skilled forelimb reaching for pasta guided by tactile input in the rat as measured by accuracy, spatial adjustments, and force. Behav Brain Res 109:49–57
47. Ballermann M, McKenna J, Whishaw IQ (2001) A grasp-related deficit in tactile discrimination following dorsal column lesion in the rat. Brain Res Bull 54:237–242
48. Montoya CP, Campbell-Hope LJ, Pemberton KD, Dunnett SB (1991) The "staircase test": a measure of independent forelimb reaching and grasping abilities in rats. J Neurosci 36:219–228
49. Schallert T, Woodlee MT, Fleming SM (2002) Disentangling multiple types of recovery from brain injury. In: Krieglstein J, Klump S (eds) Pharmacology of cerebral ischemia. Stuttgart, Medpharm Scientific Publishers, pp 201–216
50. Schallert T, Woodlee MT (2005) Orienting and placing. In: Whishaw IQ, Kolb B (eds) The behavior of the laboratory rat. New York, Oxford University Press, pp 129–140
51. Edgerton VR, Kim SJ, Ichiyama RM, Gerasimenko YP, Roy RR (2006) Rehabilitative therapies after spinal cord injury. J Neurotrauma 23:560–570
52. Houle JD, Morris K, Skinner RD, Garcia-Rill E, Peterson CA (1999) Effects of fetal spinal cord tissue transplants and cycling exercise on the soleus muscle in spinalized rats. Muscle Nerve 22:846–856
53. Sandrow-Feinberg HR, Izzi J, Shumsky JS, Zhukareva V, Houle JD (2009) Forced exercise as a rehabilitation strategy after unilateral cervical spinal cord contusion injury. J Neuro trauma 26:721–731
54. de Leon RD, Hodgson JA, Roy RR, Edgerton VR (1999) Retention of hindlimb stepping ability in adult spinal cats after the cessation of step training. J Neurophysiol 81:85–94
55. Edgerton VR, de Leon RD, Harkema SJ (2001) Topical review: retraining the injured spinal cord. J Physiol (Lond) 533:15–22

56. Magnuson DS, Smith RR, Brown EH, Enzmann G, Angeli C, Quesada PM, Burke D (2009) Swimming as a model of task-specific locomotor retraining after spinal cord injury in the rat. Neurorehab Neural Repair 23:535–545
57. Smith RR, Shum-Siu A, Baltzley R, Bunger M, Baldini R, Burke DA, Magnuson DS (2006) Effects of swimming on functional recovery after incomplete spinal cord injury in rats. J Neurotrauma 23:908–919
58. Skinner RD, Houle JD, Reese NB, Berry CL, Garcia-Rill E (1996) Effects of exercise and fetal spinal cord implants on the H-reflex in chronically spinalized adult rats. Brain Res 729:127–131
59. Beaumont E, Houle JD, Peterson CA, Gardiner PF (2004) Passive exercise and fetal spinal cord transplant both help to restore motoneuronal properties after spinal cord transection in rats. Muscle Nerve 29:234–242

Chapter 47

Basso, Beattie, and Bresnahan Scale Locomotor Assessment Following Spinal Cord Injury and Its Utility as a Criterion for Other Assessments

Darlene A. Burke and David S.K. Magnuson

Abstract

A wide variety of approaches have been used to help restore locomotor function after spinal cord injury (SCI). In the rat model, these range from biological techniques such as cell transplantation and axon regeneration to retraining methods such as activity-based rehabilitation. Correspondingly, there are many types of behavioral assessments employed to assess functional status and to help determine the best treatment approaches following (SCI). One assessment of locomotor function following SCI in the rat is the Basso, Beattie, and Bresnahan (BBB) scale, which is a categorical scale that subjectively documents limb movements and walking characteristics in an open-field environment. A categorical scale such as this that measures "global" attributes of locomotor ability lends itself well for use as a criterion to determine if an animal is functionally able to participate in other, more objective assessments. The BBB is a useful indicator of the basic overground locomotion of the animal and whether or not the animal would be capable of performing more difficult motor-related assessment tests that require specific abilities (agility, strength, and paw placing), such as the sensorimotor tests of horizontal ladder rung walking or treadmill-based gait analysis. Increasing the number and scope of the assessments performed in addition to the BBB provides a broader view of the functional recovery of animals following a variety of spinal cord injuries/severities and increased sensitivity to functional improvements. The optimal process for increasing the number and types of assessments is the use of dedicated personnel with long-term experience to ensure consistency and reliability in assessment scoring and repeatability. These factors are vital for comparisons within an individual laboratory's ongoing research and for establishing standardization comparable among outside laboratories.

Key words: Behavior assessments, Spinal cord injury, Rat, Mouse

1. Introduction

Spinal cord injury (SCI) research in the rat aiming to improve functionality has been diverse, ranging from invasive treatments for remyelination (1, 2) or cell transplantation spurring axonal

Jun Chen et al. (eds.), *Animal Models of Acute Neurological Injuries II*, Springer Protocols Handbooks,
DOI 10.1007/978-1-61779-782-8_47, © Springer Science+Business Media, LLC 2012

reconnection (3) to rehabilitation training or activity-based regimens (4–6). Accordingly, a variety of functional assessment tools are needed to determine the efficacy of treatments affecting a variety of mechanisms (7). The Basso, Beattie, and Bresnahan (BBB) Scale of locomotion was developed to assess limb movements and walking characteristics in rats with mid-thoracic level SCI (8). More severe injuries resulting in paralysis are characterized in the BBB scale by assessing the degree of hip, knee, and ankle limb movements in the absence of standing or stepping. Once weight support is achieved, walking parameters are assessed based on factors such as coordination, paw position, and trunk stability.

This type of scale that categorizes detailed locomotor characteristics is very useful as a predictor of how and whether a rat has the ability to perform other types of behavioral assessments. The BBB offers the opportunity for a quick, efficient, and detailed assessment of locomotor function that can be performed easily prior to other assessments, such as gait analysis (9), horizontal ladder (10), or grid walking (11). Furthermore, using performance on the BBB scale to determine applicability of other behavior tests eliminates animals that cannot perform the task to a minimum standard. When animals lacking sufficient ability to perform a task are included in an assessment, scores may be skewed either too high or too low. This compromises the data and possibly causes another undesired result: high variability in the scores. Thus, the benefits of establishing BBB criteria for other assessments include lower variability and improved consistency in the resulting data.

2. Materials

- BBB: A round, plastic wading pool with a slightly textured surface, 100 cm in diameter, 17 cm in height; Alternate: a round metal feeding trough, approximately 150 cm in diameter, 35 cm in height (see Fig. 1).
- Horizontal Ladder Beam, Dual Species (Rats and Mice) (Columbus Instruments, Columbus, OH, www.colinst.com) (see Fig. 2).
- Alternate to Horizontal Ladder Beam: Grid Walking Apparatus: plastic mesh (0.5 mm thickness) with diamond shaped openings (4.5 cm × 5 cm) that is held taut across a square frame (approx. 188 cm^2) and raised to eye level for viewing; timer; counter (see Fig. 3).
- Single Lane Gait Analysis Treadmill (see Beare et al., this volume).

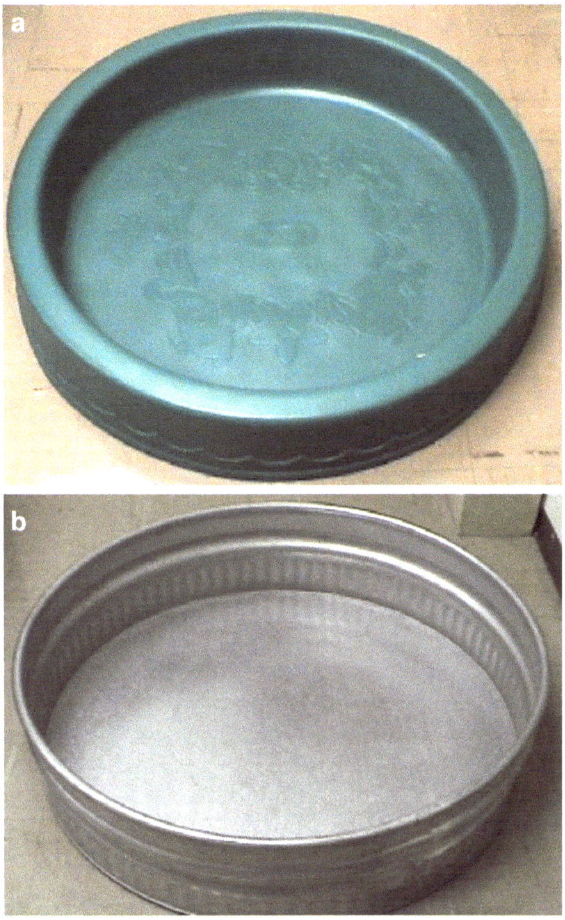

Fig. 1. A plastic wading pool (*top*) is used to assess rats with the Basso, Beattie, and Bresnahan locomotor scale. An alternative to the smaller plastic pool is the use of a metal trough (*bottom*). This offers a larger testing surface, both taller and wider, and is used to assess the much faster walking mice with the BMS locomotor scale.

2.1. Multiple Assessment Tests with Dedicated Core Personnel

One way to facilitate the effort to conduct more assessment tests in SCI research is through the use of dedicated personnel that can perform a wide variety of assessments. This produces members with long-term experience in a wide range of injury models and ensures consistency and reliability in assessment scoring. Other advantages include the ability to increase sample sizes and, using a greater number of assessments, incorporating a wider diversity of assessments.

3. Procedures

3.1. Behavioral Tests

3.1.1. Basso, Beattie, and Bresnahan

The BBB locomotor rating scale has been well documented and is widely used in the field of SCI research (7, 8). Training from the developers at their facility is provided (The OSU Spinal Cord

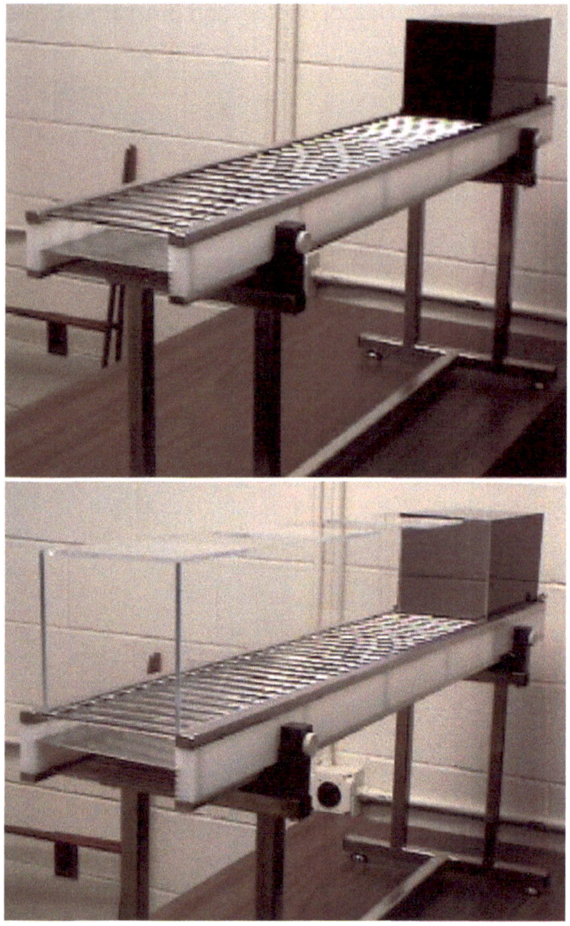

Fig. 2. The ladder beam (*top*) is a sensorimotor assessment test in which rats must walk across the rungs without a limb slipping through between the rungs. A plexiglass cover (*bottom*) can be made to enclose the rungs making it easier to train the rats to cross the length of the ladder through the tunnel. In this model (Columbus Instruments) mice can be assessed by turning the beam over and testing on the reverse side where the rungs are closer together.

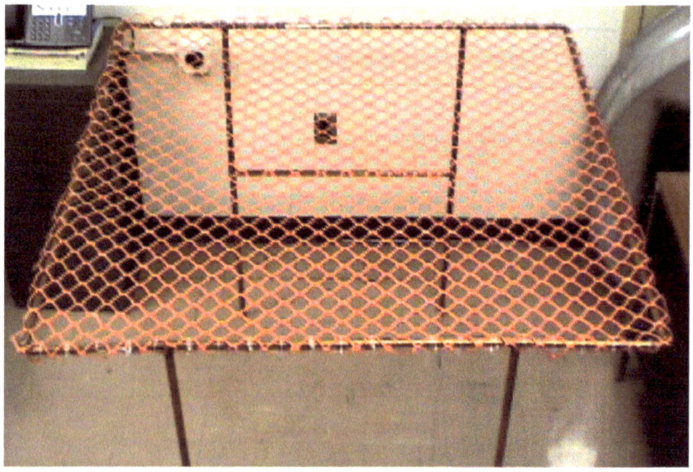

Fig. 3. The grid test surface is comprised of a plastic mesh attached to a square frame and raised in height for viewing the hindlimb footfalls. Rats are scored according to the number of footfall errors they make when their limb falls through the holes as they walk across the surface.

Injury Training Program: http://www.ninds.nih.gov/funding/areas/repair_and_plasticity/sci_training.htm).

- Briefly, each rat is placed in the wading pool for 4 min. The frequency of limb movements and walking characteristics are scored by two trained raters, blinded to the experimental grouping of the rats (see Table 1). BBBs typically are performed weekly, but are necessary at least at preinjury baseline and at the study terminal time point post-injury/treatment.

3.1.2. Horizontal Ladder Rung Walking Test: Hindlimb and/or Forelimb

The ladder test is a sensorimotor test that assesses the rat's ability to walk across the rungs of a horizontal ladder without a limb falling through.

- Each animal is trained and tested individually. Rats are trained to walk from the end of the ladder toward the box for a total of 3–5 crossings without hesitating or changing direction. They can be trained to walk only toward the box or in either direction.
- Definition/criteria of an acceptable crossing:
  - Walks across the rungs with little or no hesitation.
  - No dragging of limbs across the rungs at any time.
  - At least three high quality crossings.
  - Both raters must agree on judgments about the crossings.
- Automated counter: Using three persons to conduct the test in this manner will yield the most accurate results. In addition to the two raters, one is needed to monitor the counter since not all crossings are complete and the counter must continually be reset. The two raters watch and agree on the acceptability of each crossing. The third person records the number of errors (see Table 2). Both forelimb and hindlimb footfalls are tallied by the counter. Two disadvantages of the automated counter are its inability to discriminate forelimb from hindlimb footfalls and when/if two paws hit the plate simultaneously or a single paw hits the plate multiple times after a single misstep.
- Manual counting: On each crossing the errors are counted by two raters. A footfall error is defined as a misstep where at least three-fourths of the limb extends between two rungs. Another error for that limb cannot be counted until the limb is moved to the next rung. After each crossing the number of errors observed by the two raters is discussed. Scorers should agree within 1–2 errors and, if 2, the middle (average) value is used. If the error count of the two raters is off by more than 2, the trial should not be used and another crossing should be made (see Table 2). The inter-rater agreement should equal approximately 95%. Two advantages of manually counting footfall errors are that missteps (forelimb and/or hindlimbs) can be

Table 1
The Basso, Beattie, and Bresnahan (BBB) scale ranges from 0 to 21, assessing hindlimb movements and walking characteristics (8)

BBB locomotor rating scale

I. Early stage of recovery (hindlimb joint movements)
0. No observable hindlimb (HL) movement
1. Slight movement of one or two joints, usually the hip and/or knee
2. Extensive movement of one joint or extensive movement of one joint and slight movement of one other joint
3. Extensive movement of two joints
4. Slight movement of all three joints of the HL (hip, knee, and ankle)
5. Slight movement of two joints and extensive movement of the third
6. Extensive movement of two joints and slight movement of the third
7. Extensive movement of all three joints of the HL

II. Intermediate stage of recovery (coordination in stepping ability)
8. Sweeping with no weight support or plantar placement of the paw with no weight support
9. Plantar placement of the paw with weight support in stance only (i.e., when stationary) or occasional, frequent, or consistent weight supported dorsal stepping and no plantar stepping
10. Occasional weight supported plantar steps, no FL-HL (for ellimb-hindlimb) coordination
11. Frequent to consistent weight supported plantar steps and no FL-HL coordination
12. Frequent to consistent weight supported plantar steps and occasional FL-HL coordination
13. Consistent weight supported plantar steps and frequent FL-HL coordination

III. Late stage of recovery (details, refinement of locomotion)
14. Consistent weight supported plantar steps, consistent FL-HL coordination, and predominant paw positions during locomotion are rotated or frequent plantar stepping, consistent FL-HL coordination, and occasional dorsal stepping
15. Consistent FL-HL coordination and no toe clearance or occasional toe clearance during forward limb advancement. Predominant paw position is parallel to the body at initial contact
16. Consistent FL-HL coordination during gait and toe clearance occurs frequently during forward limb advancement. Predominant paw position is parallel at initial contact and rotated at lift off
17. Consistent FL-HL coordination during gait and toe clearance occurs frequently during forward limb advancement. Predominant paw position is parallel at initial contact and at lift off
18. Consistent FL-HL coordination during gait and toe clearance occurs consistently during forward limb advancement. Predominant paw position is parallel at initial contact and rotated at lift off
19. Consistent FL-HL coordination during gait and toe clearance occurs consistently during forward limb advancement. Predominant paw position is parallel at initial contact and lift off Tail is down part or all of the time
20. Consistent coordinated gait; consistent toe clearance. Predominant paw position is parallel at initial contact and lift off, but there is trunk instability. The tail is consistently up
21. Coordinated gait, consistent toe clearance, predominant paw position is parallel throughout stance, consistent trunk stability, and tail is consistently up

Table 2
The scoring sheet for the ladder test includes recording the number of errors observed in five crossings, their quality (high or low), a comment line, and experimental information including the plate level position below the rungs (three out of five crossings are averaged)

LADDER TESTING FORM

Experiment ______________ Raters ______________

Date ______________ ______________

Test Time (post-injury) ______________ Plate Level: 1 2 3 4 5 6 7 (Circle)

CROSSING

[Animals' walking quality: hesitation, slow, fast]

ID	1 [H L]	2 [H L]	3 [H L]	4 [H L]	5 [H L]	3 Best
	1 [H L]	2 [H L]	3 [H L]	4 [H L]	5 [H L]	
	1 [H L]	2 [H L]	3 [H L]	4 [H L]	5 [H L]	
	1 [H L]	2 [H L]	3 [H L]	4 [H L]	5 [H L]	
	1 [H L]	2 [H L]	3 [H L]	4 [H L]	5 [H L]	
	1 [H L]	2 [H L]	3 [H L]	4 [H L]	5 [H L]	

counted more accurately compared to the automated counter and either two or three persons can perform the assessment. There is a minor disadvantage in that two persons require training and experience to yield agreement and accurate results, but this is largely accomplished at baseline.

- The score is the average of the errors on the three crossings with the highest quality (out of 5).
- The plate below the rungs should not be too close to the rungs to ensure that the animals will walk on the rungs and not the plate surface. For young adult female Sprague-Dawley, 190–250 g, placing the plate at level 5 works well.
- Optional: Use a treat (e.g., low sugar cereal) as incentive to walk across the rungs to the box.
- Because quicker walking can lead to more errors, additional measures such as the number of steps and the amount of time to complete each crossing can be recorded and used to normalize the data.

3.1.3. Grid Walking

The grid test is another sensorimotor assessment, similar to the ladder rung test (11). It is conducted across an "open field" which is a less comfortable setting for most rats compared to the ladder apparatus that forms a narrow runway. It is slightly more difficult than horizontal ladder walking for some rats. Two raters count the number of footfalls of the rat as they walk across the grid surface.

- Rats are assessed for footfalls during 2 min of *actual* walking time.
- The definition of a footfall error is the same as for the ladder rung test, defined as a misstep where at least three-fourths of the limb falls between two rungs. Additional errors are not scored when repeated attempts by an animal fail to move the paw out of the hole between the grids. The observed missteps are discussed and agreed upon by the raters as they occur.

3.2. BBB Criterion for Behavioral Tests

3.2.1. Horizontal Ladder Beam

- When used as a criterion for the horizontal ladder rung walking test rats must have consistent weight support and frequent to consistent plantar stepping on both hindlimbs (BBB = 11). It is necessary for rats to have consistent weight supported walking because rats without consistent weight support will drag their limbs over the rungs rather than step, artificially lowering the count of footfall errors. In contrast, when the hindpaw of these rats falls through the rungs, they may make repeated unsuccessful attempts to raise the paw to move to the next rung, thus repeatedly hitting the counting platform below the rungs and increasing the count of footfall errors, as they continue down the ladder. A high frequency of dorsal stepping also disqualifies rats from testing on the horizontal ladder.

Without the ability to plantar place the paw on the rung, they are unable to perform the task because the paw automatically slips off the rung at each step.

- Prior training of rats is not usually necessary. Baseline assessments serve this function, familiarizing the rats with the procedure. A treat may be used as reinforcement, if desired. A second baseline test is performed on rats whose error rate is too high in the first exposure to the test (e.g., greater than an average of three errors in the three highest quality crossings out of five).

3.2.2. Criterion for Grid Walking

- The BBB criterion score for the grid test is identical to that of the horizontal ladder rung test. Rats need consistent weight support and frequent to consistent plantar stepping on both hindlimbs.

3.2.3. Criterion for Automated Gait Analysis

- The BBB walking criterion for treadmill testing (see Beare et al., this volume) is lower than horizontal ladder testing. Rats with at least frequent weight supported stepping can walk on the treadmill. Dorsal stepping is not a disqualification.

3.3. BBB Subscale

The BBB subscale score is calculated from the information on the BBB scoring sheet and is used to describe a subset of the characteristics of late phase recovery. After consistent weight support, coordination is recorded (occasional, frequent, and consistent) and because some aspects of walking are evident before consistent coordination, the subscore can be used to represent the emergence of the finer details of locomotion. These include toe clearance, paw position, trunk instability, and tail position (12). Animals with consistent weight supported walking can be scored in the following ways: BBB = 11: consistent plantar stepping and no dorsal stepping; BBB = 11: frequent plantar stepping and occasional dorsal stepping; BBB = 10: occasional plantar stepping and frequent dorsal stepping (FAQ Regarding the Basso, Beattie, Bresnahan Locomotor Rating Scale, revised version, 1994, personal communication). Note BBB scores of 10 or 11 can be obtained with occasional or frequent weight supported plantar stepping, respectively, but neither would receive a subscore. Rats with consistent weight supported dorsal stepping (100% dorsal, no plantar, BBB = 9) are less common but would receive a subscore since they have consistent weight support.

- The frequency of occurrence of the four characteristics is assigned a value for both the left and right sides individually which are then summed to produce the subscore. Maximum total score equals 13 (see Table 3, top).
  - Toe clearance, each left and right side: none = 0, occasional = 1, frequent = 2, and consistent = 3.

Table 3
Subscores of the Basso, Beattie, and Bresnahan (BBB) (*top*) and Basso Mouse Scale (BMS) (*bottom*) are tabulated from the information below and can be inset onto the corner of the BBB and BMS scoring sheets

BBB SUBSCORE

SUBSCORE: BBB=Consistent Wt. Support		
Toe Clearan. Lt & Rt	0=None 1=Occas. 2=Freq. 3=Consis.	L ___ R ___
Paw Rotation Lt & Rt	0=Rotated 1=1Rotated 2=Parallel	L ___ R ___
Trunk Instabil.	0=Yes 1=No	___
Tail Up	0=No 1= Yes & No 2=Yes	___
Max=13	Total	___

BMS SUBSCORE

SUBSCORE Tally	
Plantar Stepping	Score
Freq = 0; Consistent = 1	L ___ R ___
Coordination	
None = 0; Some = 1; Most = 2	___
Paw Position	
Rotated thru out = 0 Parallel and rotated = 1 Parallel thru out = 2	L ___ R ___
Trunk	
Severe = 0; Mild = 1; Normal = 2	___
Tail	
Down, Up&Down = 0; Up = 1	___
TOTAL SUBSCORE	☐

  - Paw rotation, each left and right side: rotated at both the initial contact and lift off = 0; either the initial contact or lift off is parallel = 1; and both the initial contact and lift off are parallel = 2.
  - Trunk instability: yes = 0 and no = 1.
  - Tail position: always down = 0, up and down = 1, and consistently up = 2.
- The major advantage to measuring the subscore is its ability in some cases to detect significant differences between experimental groups when the overall BBB does not. A disadvantage is the large variability that sometimes results when rats within the same experimental group have large differences in their rate of recovery.

3.4. Rat Strain: Fischer-344

Extensive handling and gentling of these rats for 1–2 weeks prior to injury is the most productive way to familiarize and train Fischer-344 (F-344) rats on these behavioral tasks to obtain the most reliable assessment scores.

- BBB: The locomotor abilities of F-344 rats are different from Sprague-Dawley, Long-Evans, and Wistar (13). Following mid-thoracic mild-to-moderate impactor contusion injury F-344 rats tend to recover toe clearance relatively quickly compared to other strains, such as Sprague-Dawley.
- Horizontal Ladder: Prior training of F-344 is necessary for this assessment. Pretraining to walk across the rungs into the box located at the opposite end should be conducted at least twice a week for approximately 2 weeks until a continuous, steady pace of walking by the rat is achieved. Treats in the box help to increase their motivation to cross.
- Single Lane Gait Analysis Treadmill: F-344 rats walk more slowly on the treadmill compared with the Sprague-Dawley rats and the speed should be reduced. Even with prolonged experience in walking on the treadmill they tend to remain uncomfortable during testing.

Fischer-344 rats are different from several other strains of rats. In addition to BBB, behavioral outcome differences compared to other rat strains have been reported, including paw withdrawal from mechanical stimuli and heat (14); they also have been found to differ in morphology, rearing limb support, horizontal ladder, and kinematics measurements (13). Thus, caution must be used in interpretation of outcomes using F-344 rats and comparisons with outcomes of experiments using other rat strains.

3.5. Mice Behavior Tests

3.5.1. Basso Mouse Scale

The Basso Mouse Scale (BMS) of locomotion assesses the frequency of limb movements and walking characteristics (15, 16). Mice are scored individually by two trained raters, blinded to the experimental grouping of the rats. On-site training workshops are conducted by the developers of the scale (The OSU Spinal Cord Injury Training Program: http://www.ninds.nih.gov/funding/areas/repair_and_plasticity/sci_training.htm).

- Mice are observed in a large wading pool such as the metal trough used for BBB testing (see Sect. 2, BBB and Fig. 1) for 4 min by two trained, experienced raters, blinded to the experimental group of the mice. Limb movements, frequency of stepping, and walking characteristics are recorded and scored.

3.5.2. Horizontal Ladder Beam: Forelimb and Hindlimb

The procedures for the horizontal ladder test using mice are nearly identical to the training and testing of rats. The only differences are as follows:

- The ladder rungs are closer together.
- The automatic counter must be used because of the very quick walking speed of the mice.
- The plate below the rungs must be moved closer to the rungs due to the smaller size of the mice (e.g., plate levels 1 or 2).

3.5.3. Automated Gait Analysis

Details of this technique are described elsewhere (see Beare et al., this volume).

3.6. BMS Criterion for Behavior Tests

3.6.1. Criterion for Horizontal Ladder Beam

- Mice must have frequent weight supported plantar stepping and a score of at least 3, 4 to qualify for testing on the horizontal ladder rung walking test. At least one side must have plantar stepping, a score of 4, because if dorsal stepping, a score of 3, it is more difficult to step on the rungs and they usually slip off. Some mice with a score of 3, 4 are able to keep themselves stable enough to perform the task while others are unable.
- Mice exhibiting limb dragging across the rungs rather than stepping are disqualified because they do not consistently support their weight while walking over the rungs. A low frequency of dorsal stepping does not exclude testing.
- Prior training of most strains of mice is not necessary. Baseline assessments familiarize them with the procedure. A second baseline test should be performed on any groups of mice with a high number of errors in their first exposure to the test (e.g., greater than an average of three errors in the three highest quality crossings out of five).

3.6.2. Criterion for Automated Gait Analysis

- The BMS criteria score for treadmill walking is the ability of weight supported stepping (see Beare et al., this volume). Dorsal stepping is not a disqualification.

3.7. BMS Subscale

The BMS subscore is calculated from the BMS test scoring sheet in the same manner as the BBB subscore is determined from the BBB test (15, 16) (see Table 3, bottom). Mice must be capable of at least frequent stepping and have a score of 5 on both the left and right sides to calculate a subscore. The BMS subscale score is calculated on both the left and right sides from the following characteristics (maximum total = 11):

- Plantar stepping, each left and right side: frequent = 0 and consistent = 1.
- Coordination: none = 0, some = 1, and most = 2.
- Paw rotation, each left and right side: rotated at both the initial contact and lift off = 0; either the initial contact or lift off is parallel = 1, and both the initial contact and lift off are parallel = 2.
- Trunk instability: severe = 0, mild = 1, and normal = 2.
- Tail position: down or up and down = 0 and consistently up = 1.

4. Summary

There are a wide variety of behavioral assessment tests available for use in animal studies following SCI (see Table 4). Depending upon the injury level, severity, treatment parameters, and goals of the experiment, locomotor, sensorimotor, and sensory functioning can be measured in the hindlimb and, for some tests, the forelimb as

Table 4
There are many behavioral assessment tests available for testing rats following spinal cord injury

		Injury Severity				
Type	Assessment	Mild	Moderate	Severe	Hindlimb	Forelimb
Locomotor	BBB	X	X	X	X	---
Locomotor	BBB Subscore	X	X	---	X	---
Locomotor	Treadmill Kinematics	X	X	X	X	X
Locomotor	Shallow water walking	X	X	X	X	X
Locomotor	LSS	X	X	X	X	P
Locomotor	LSS Subscore	X	X	X	X	P
Locomotor	Swim Kinematics	X	X	X	X	P
Sensorimotor	Ladder Beam	X	X	---	X	X
Sensorimotor	Grid Walking	X	X	---	X	X
Sensory	Tail Flick	X	X	---	---	---
Sensory	von Frey	X	X	X	X	X
Sensory	Hargreaves	X	X	X	X	X

P = possible

An "X" shows the capability of each test to measure function according to injury severity. Some tests measure the hindlimbs alone and others can assess both the hindlimbs and forelimbs well. The ability to maximize the number and variety of assessments performed per study can be accomplished by using a small group of Core members trained in administering these assessment tests. This practice greatly expands the scope of experimental studies and the positive results from particular types of assessments help to focus the direction of future studies.

References

1. Talbot JF, Loy DN, Liu Y, Qui MS, Bunge MB, Rao MS, Whittemore SR (2005) Endogenous Nkx2.2^{+}/Olig2^{+} oligodendrocyte precursor cells fail to remyelinate the demyelinated adult rat spinal cord in the absence of astrocytes. Exp Neurol 192:11–24
2. Cao Q, He Q, Wang Y, Cheng X, Howard RM, Zhang Y, DeVries WH, Shield CB, Magnuson DS, Xu XM, Kim DH, Whittemore SR (2010) Transplantation of ciliary neurotrophic factor-expressing adult oligodendrocyte precursor cells promotes remyelination and functional recovery after spinal cord injury. J Neurosci 30: 2989–3001
3. Xu XM, Onifer SM (2009) Transplantation-mediated strategies to promote axonal regeneration following spinal cord injury. Respir Physiol Neurobiol 169:171–182
4. Smith RR, Burke DA, Baldini AD, Shum-Siu A, Baltzley R, Bunger M, Magnuson DSK (2006) The Louisville swim scale: a novel assessment of hindlimb function following spinal cord injury in adult rats. J Neurotrauma 23(11):1654–1670
5. Edgerton VR, de Leon RD, Harkema SJ, Hodgson JA, London N, Reinkensmeyer DJ, Roy RR, Talmadge RJ, Tillakaratne NJ, Timoszyk W, Tobin A (2006) Retraining the injured spinal cord. J Physiol 533:15–22
6. Engesser-Cesar C, Ichiyama RM, Nefas AL, Hill MA, Edgerton VR, Cotman CW, Anderson AJ (2007) Wheel running following spinal cord injury improves locomotor recovery and stimulates serotonergic fiber growth. Eur J Neurosci 25:1931–1939
7. Muir GD, Webb AA (2000) Mini-review: assessment of behavioural recovery following spinal cord injury in rats. Eur J Neurosci 12:3079–3086
8. Basso DM, Beattie MS, Bresnahan JC (1995) A sensitive and reliable locomotor rating scale for open field testing in rats. J Neurotrauma 12:1–21
9. Beare JE, Morehouse JR, DeVries WH, Enzmann GU, Burke DA, Magnuson DSK, Whittemore SR (2009) Gait analysis in normal and spinal contused mice using the TreadScan® system. J Neurotrauma 26:2045–2056
10. Metz GA, Whishaw IQ (2002) Cortical and subcortical lesions impair skilled walking in the ladder rung walking test: a new task to evaluate fore- and hindlimb stepping, placing, and co-ordination. J Neurosci Methods 115:169–179
11. Kunkel-Bagden E, Bregman BS (1990) Spinal cord transplants enhance the recovery of locomotor function after spinal cord injury at birth. Exp Brain Res 81:25–34
12. Basso DM (2004) Behavioral testing after spinal cord injury: congruities, complexities, and controversies. J Neurotrauma 21:395–404
13. Webb AA, Gowribai K, Muir GD (2003) Fischer (F-344) rats have different morphology, sensorimotor and locomotor abilities compared to Lewis, Long-Evans, Sprague-Dawley and Wistar rats. Behav Brain Res 144:143–156
14. Mills CD, Hains BC, Johnson KM, Hulsebosch CE (2001) Strain and model differences in behavioral outcomes after spinal cord injury in rat. J Neurotrauma 18(8):743–756
15. Basso DM and Fisher LC (2003) The Basso mouse scale for locomotion (BMS) is a more sensitive indicator of recovery than the BBB scale in mice with spinal cord injury. J Rehabil Res Dev 40(6), Suppl. 3, 58 (abstract)
16. Basso DM, Fisher LC, Anderson AJ, Jakeman LB, McTigue DM, Popovich PG (2006) Basso Mouse Scale for locomotion detects differences in recovery after spinal cord injury in five common mouse strains. J Neurotrauma 23(5):635–659

Chapter 48

High-Speed Video Recording Used for the Analysis of Hand Shaping in Overground Walking, Cylinder Exploration, Skilled Reaching, and Rung Walking in Rats

Ian Q. Whishaw and Lori-Ann Sacrey

Abstract

Hand shaping in the rat (*Rattus norvegicus*) is adapted to many functions, including walking, climbing, exploration, and skilled manipulation. An understanding of hand movement can provide insights into the neural control of movement and the impairments that result from brain injury. Here, hand movements are described in a number of widely used tests: overground walking, horizontal or inclined ladder rung walking, exploring a vertical wall of a cylinder, and skilled reaching for food. High-speed (250–1,000 frames/s) video recording with frame-by-frame behavioral and kinematic analyses reveals the details of hand shaping and its sensory control. Three hand gestures are common to all tasks: (1) release—the hand is pushed off and lifted from a substrate; (2) collection—the digits are closed and flexed through limb transport; and (3) manipulation—the hand and digits are shaped to contact a target. Manipulation features specialized digit use and varied sensory control (olfaction, vibrissae and sinus hairs, and tactile senses) in the different tasks. Conserved release and collection vs. the variability of manipulation reveal the multisensory skilled action used in control of the hand.

Key words: Comparisons of limb movements in the rodent, Ladder rung climbing, Ladder rung walking, Limb movements in the cylinder test, Evolution of forelimb movement, Evolution of skilled movement, Elements of movement, Hand shaping in laboratory tasks, Rat limb movements, Sensory control of hand movements, Skilled reaching, Walking movements

1. Introduction

Hand movements are used for many behaviors, including walking, climbing, and exploration. They are affected by many neurological diseases, including stroke (1–3), Huntington's disease (4), Parkinson's disease (5–8), and spinal cord injury (9–11). At present, the complexity and the sensory control of the movements are underappreciated, but they can be subject to analysis through the use of high-speed video filming methods that are now inexpensive.

Jun Chen et al. (eds.), *Animal Models of Acute Neurological Injuries II*, Springer Protocols Handbooks, DOI 10.1007/978-1-61779-782-8_48,

One way of investigating the range of hand movements used by the rat is to examine their use in standardized laboratory tests. The present paper describes hand movements in Long–Evans rats in four tasks that have been widely used for neural investigations in the rat: overground locomotion (9–11), rung walking (12), cylinder exploration (13, 14), and skilled reaching (15–22). In each of the tasks, the movements of the rats are video recorded using a high-speed camera that captures movements up to 1,000 frames/s. The video record is analyzed frame by frame biometrically (Peak Motus) and descriptively to reveal the component features of movements and their sensory control.

2. Methods

2.1. Feeding and Food Restriction

Prior to and during training, rats are gradually food restricted to 90–95% of their body weight by feeding them 20 g once daily (Purina rat chow). The week prior to training, each rat receives 20–45-mg dustless precision banana-flavored pellets (product #F0021; Bioserve, Frenchtown, NJ, USA) 1 h prior to their daily feeding. These banana pellets serve as reward to motivate behavior in the behavioral tasks. Once training begins, only rat chow is given as food in the home cages until the end of the study.

2.2. Overground Walking

The overground walking test is conducted in a straight alley that is a 45-cm-long box that was 6 cm wide with walls that were 19 cm high, measured from the height of the floor. The rats are placed at one end of the alley and walked spontaneously to a refuge (home cage) that was placed at its end.

2.3. Ladder Rung Walking

The horizontal ladder rung walking test apparatus consists of side walls made of clear Plexiglas and metal rungs (3 mm diameter) that create a floor with a minimum distance of 1 cm between rungs (12). The side walls are 1 m long and 19 cm high measured from the height of the rungs. The ladder is elevated 30 cm above the ground with a neutral start cage and a refuge (home cage) at the end. Because animals were habituated during training, the elevation of the apparatus was unlikely to cause anxiety. The width of the alley is adjusted to the size of the animal so that it is about 1 cm wider than an animal, thus preventing the animal from turning around. The rats are tested with the ladder placed horizontally or inclined to 45°.

2.4. Cylinder Exploration

The cylinder (13, 14) is high enough (30 cm) so that rats cannot reach the top edge by rearing. The diameter of the cylinder (20 cm) allowed rats to freely turn while scanning the horizontal surface.

The cylinder is sufficiently heavy that it does not move when the animal supports its weight against the wall. The cylinder is placed on a glass table, beneath which is an inclined mirror. Animals are video recorded from a ventral perspective through the mirror or through the wall of the cylinder.

2.5. Skilled Reaching

The single pellet-reaching box (17, 22) is made of clear Plexiglas with the dimensions 45 cm long × 14 cm wide × 35 cm high. At the center of the front wall, there is a vertical slot 1 cm wide extending from 2 cm above the floor to a height of 15 cm. On the outside of the wall, in front of the slot, there is a 2-cm-deep shelf mounted 3 cm above the floor. Two indentations on the surface of the shelf are located 2 cm from the inside of the wall in order to hold the food targets. The indentations are aligned with the edges of the slot. This location prevents a rat from lapping the food with its tongue and, because the hand pronates medially to grasp, requires the use of the contralateral hand for retrieval.

The first week of training consists of daily 10-min sessions for each rat. The objective is to introduce the rat to the reaching cage with pellets on the shelf and to have the rat retrieve the pellet by hand or tongue. Once a rat successfully retrieves pellets, pellets are moved further away on the shelf to encourage reaching with the hand. After a rat demonstrates a preference for one hand by making the majority of reaching attempts with it, individual pellets are placed into the indentation contralateral to that hand. During the second week, rats continue to receive daily 10-min training sessions, consisting of discrete trials with intertrial intervals during which they are shaped to leave the slot, walk to the rear wall of the cage, and turn and approach the slot again for the next pellet. Additionally, by withholding food on semi-random trials, rats are shaped to sniff the shelf for a pellet and to reach only if a pellet is present. Thus, each rat eventually learns to orient to the food pellet with its nose, transport its limb through the slot, grasp the food pellet, and retract its hand through the slot to release the food into its mouth. After 2–3 weeks, each rat is given 20 daily trials until group success levels exceed 50%.

2.6. Video Recording

The high-speed video camera was a Photron (Fastcam-ultima APX, Rev. 1.08; High Speed Imaging, http://www.hsi.ca) that filmed up to 1,000 frames/s with the assistance of a cold light source (23). Many low-cost high-speed cameras are commercially available, however. High-speed settings of 500–1,000 frames/s are used for data analysis, while slower settings are used to construct pictorial images. Images are captured from any perspective. To freeze the movements without compromising the quality of the image, the shutter speed is set at 1/1,000 s. The camera can be set in "end trigger mode" to capture representative reaches that can be saved or discarded on editing. Representative reaches are captured using Final Cut Express HD (V.3.5; http://www.apple.com) for subsequent analysis.

2.7. Kinematic Analysis

Representative trials are captured with Final Cut Express HD (V.3.5; http://www.apple.com) and analyzed by the frame-by-frame motion measurement software "Peak Motus" v. 8.3.0 2-D digitizing system (Peak Performance Technologies, Inc., Centennial, CO). The data are acquired via a manual mode, digitizing the moving points by cursor. Different parts of the body are digitized in order to form *x* and *y* coordinates for movements.

3. Behaviors

3.1. Hand Shaping

For all of the tasks, forelimb movement consisted three arm, hand, and digit actions: release, collection, and manipulation. Release is the movement associated with lifting the limb from the substrate. Collection describes the shaping of the hand as it is transported to its next target. Manipulation describes the movement of the hand and digits as they contact the target.

3.2. Overground Walking

The movements of release, collection, and manipulation are illustrated for overground locomotion in Fig. 1. Release is the action in which the hand pushes off from the substrate and is then carried forward by movement of the limb. In the first portion of the forward movement, the digits apply force to assist movement, but once they lose contact with the substrate they close, are initially passive, and are carried by the limb. Collection is an action in which the digits flex until the hand makes a fist and it occurs as the limb is lifted beneath the body. Maximal digit flexion is associated with a partial supination of the hand. Manipulation is an action in which the hand is positioned to again contact the floor. In doing so, the digits are extended and opened, and the hand is pronated so that digits 5 then 4, 3, and finally 2 sequentially contact the floor. At the same time, the palm contacts the surface laterally to medially.

Because the rat is moving forward and the hand is being carried by the limb in the swing phase of the step, and the digits are flexing and then extending, the velocities of the rat's body, the forelimb, and the digits all have a different profile. These are illustrated as resultant velocities of the eye, nose, wrist, and tip of digit 3 in Fig. 2. The eye maintains a relatively constant vertical position during walking and so its velocity represents the forward movement of the body. The nose is lowered toward the substrate and then raised during walking and so its resultant velocity changes across a stepping cycle. The velocity of the wrist accelerates as the hand is lifted from the substrate and then slows until the hand again contacts the substrate. The velocity of the digits, as represented by the tip of digit 3, is biphasic. Velocity is slow and lags behind the wrist during release, and then accelerates and decelerates to collection, which occurs at the point of maximal velocity of the wrist. A second peak of velocity occurs as the digits extend

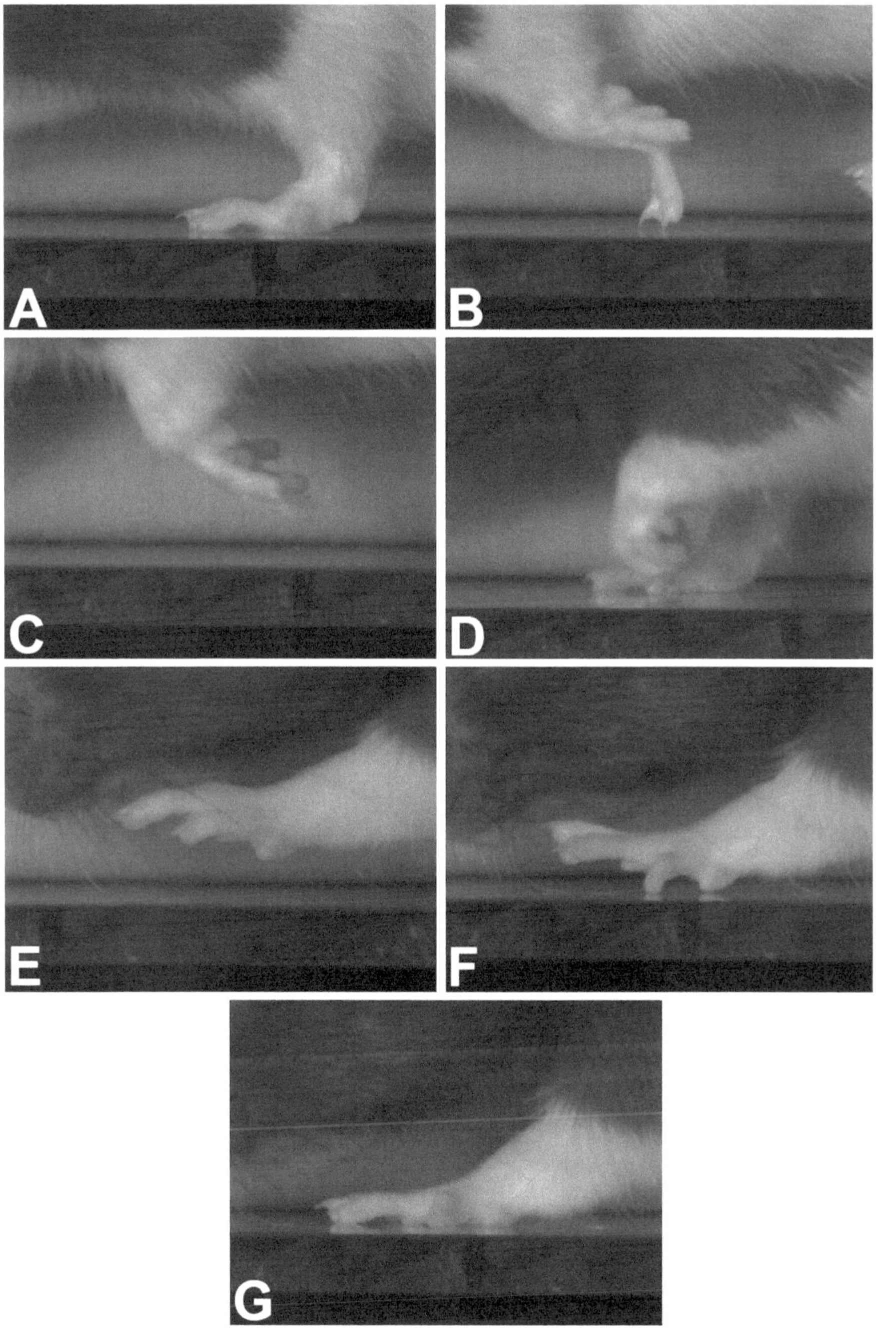

Fig. 1. Hand shaping in the swing phase of a forelimb step in overground walking. Release: The hand is pushed off from the surface first from the palm and then from digits 2 to 3 (**a**, **b**). Collection: The digits are flexed and closed as they are carried forward by the forelimb (**c**, **d**). Manipulation: The digits extend, open, and pronate to contract the floor (**e**–**g**).

during the manipulatory phase of the action in preparation for contact with the floor surface. Manipulation is associated with the slowing of the hand relative to the floor so that when it contacts the floor it is not moving relative to the floor.

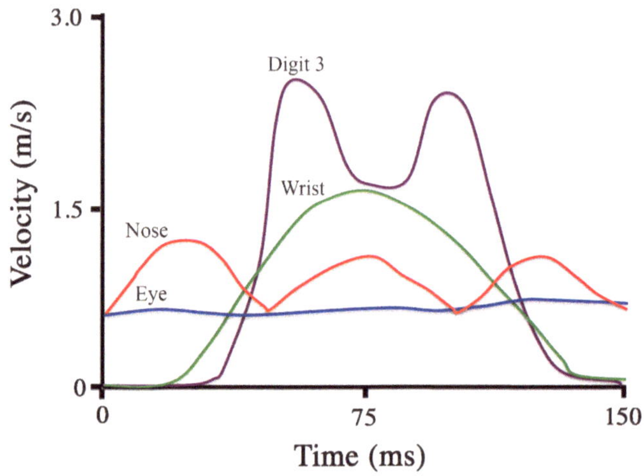

Fig. 2. Resultant velocity of the eye, nose, wrist, and the tip of digit 3 during a single step made by a rat walking on a flat surface. The eye holds a relative constant vertical position and so represents the forward speed of the head and trunk. The nose of the rat bobs with the step cycle of each diagonal limb pair so that it almost contacts the floor surface and then lifts, giving a rhythmic velocity profile. The velocity representation from the wrist begins with the paw on the surface of the floor and ends with the paw again placed on the floor. The first velocity peak of the digits represents digit flexion and the second peak represents digit extension. Collection is represented by the dip in velocity of digit 3 in the midpoint of the swing phase of the step cycle.

3.3. Ladder Rung Walking

The movements of release, collection, and manipulation associated with walking on the rungs of a horizontal ladder are illustrated in Fig. 3a. Similar movements used in climbing the rungs of an inclined ladder are illustrated in Fig. 4b. For release, the rat pushes off from the rung with the palm of its hand and then the digits close and are carried forward by the movement of the limb. For collection, the digits flex to make a fist and flexion is associated with a slight supination of the hand. For manipulation, the digits extend and the palm of the hand is placed on the rung while the digits remain extended. For the inclined ladder condition, the movements are very similar, except for manipulation in which the digits flex to grasp the rung at the completion of manipulation. It is noteworthy that in initial training the way that a hand was placed on a rung is variable, with the digits sometimes contacting the rung first, often sliding from the rung with contact, and sometimes completely missing the rung. Once the rats are well practiced on the task, however, contact is made with the surface of the palm.

3.4. Cylinder Exploration

The movements of release, collection, and manipulation associated with exploration of the cylinder wall are illustrated in Fig. 4. Because exploration of the cylinder wall is associated with the movement of rearing, it is less obvious that the rats pushed off from the surface of the floor with the hand and digits during the movement of release. As the digits lose contact with the floor

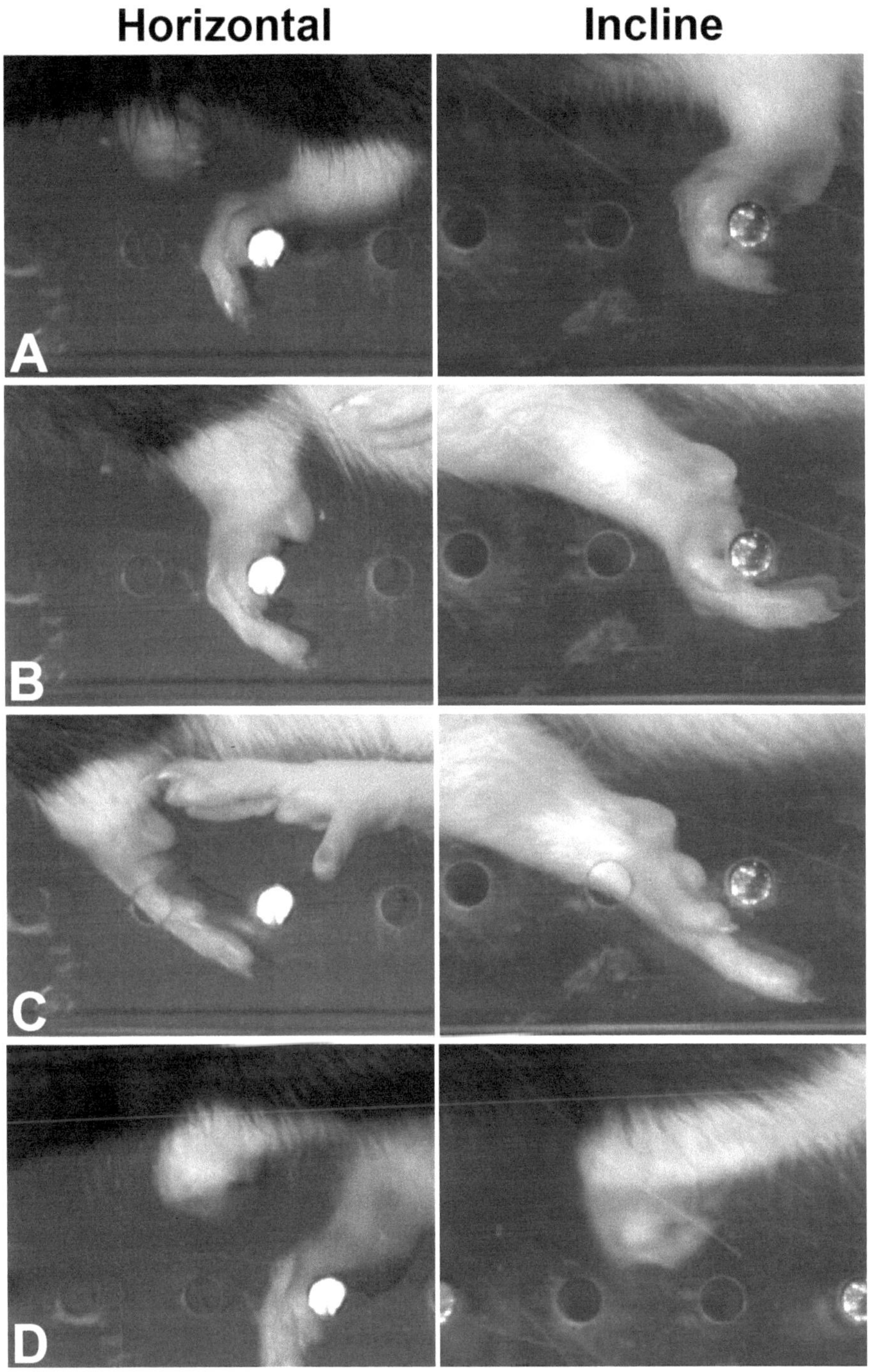

Fig. 3. Hand shaping in the swing phase of a forelimb step in horizontal (*left*) and inclined (*right*) rung walking. Release: The hand is pushed off from the surface (**a**, **b**). Collection: The digits are flexed and closed as they are carried forward by the forelimb (**c**, **d**). Manipulation: The digits extend and open, but only the palm contacts the rung (**e**–**g**).

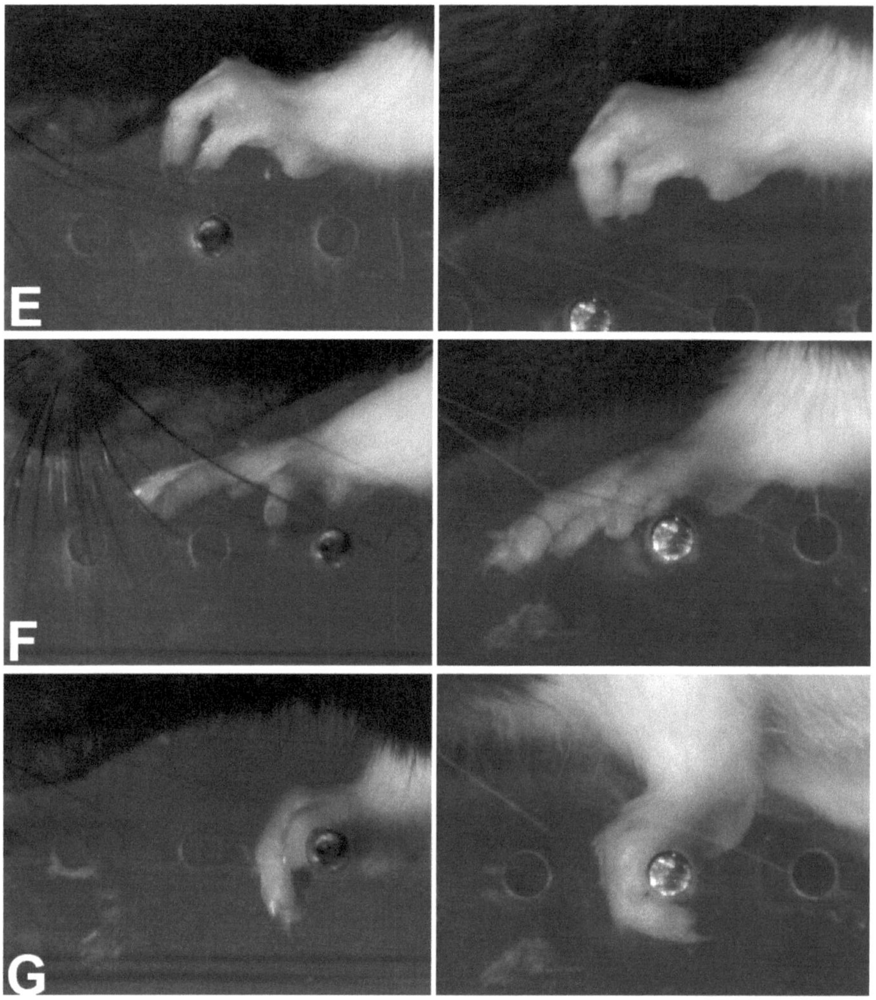

Fig. 3. (continued)

surface, they close. As the limb is lifted, the digits flex to make a fist for collection. Collection is associated with a slight supination of the hand. As the hand is moved toward the cylinder wall, the digits extend and then partially open as they contact the wall digit tips first. A rat might support itself with its digit tips against the wall or press the hand surface fully against the wall of the cylinder. Wall exploration varies as the speed of each rear is different, and in addition a rat might hold a limb in the release position or in the collected position for some time before reaching out to contact the wall. If a rat adjusts its posture once it contacts the wall, it might withdraw the hand into a collected position and then reach forward to make another wall contact. It might also move the hand to a new position without withdrawing it to the collected position. Withdrawing the limb into a collected position is usually done with a large movement while smaller movements are used when the limb is simply repositioned.

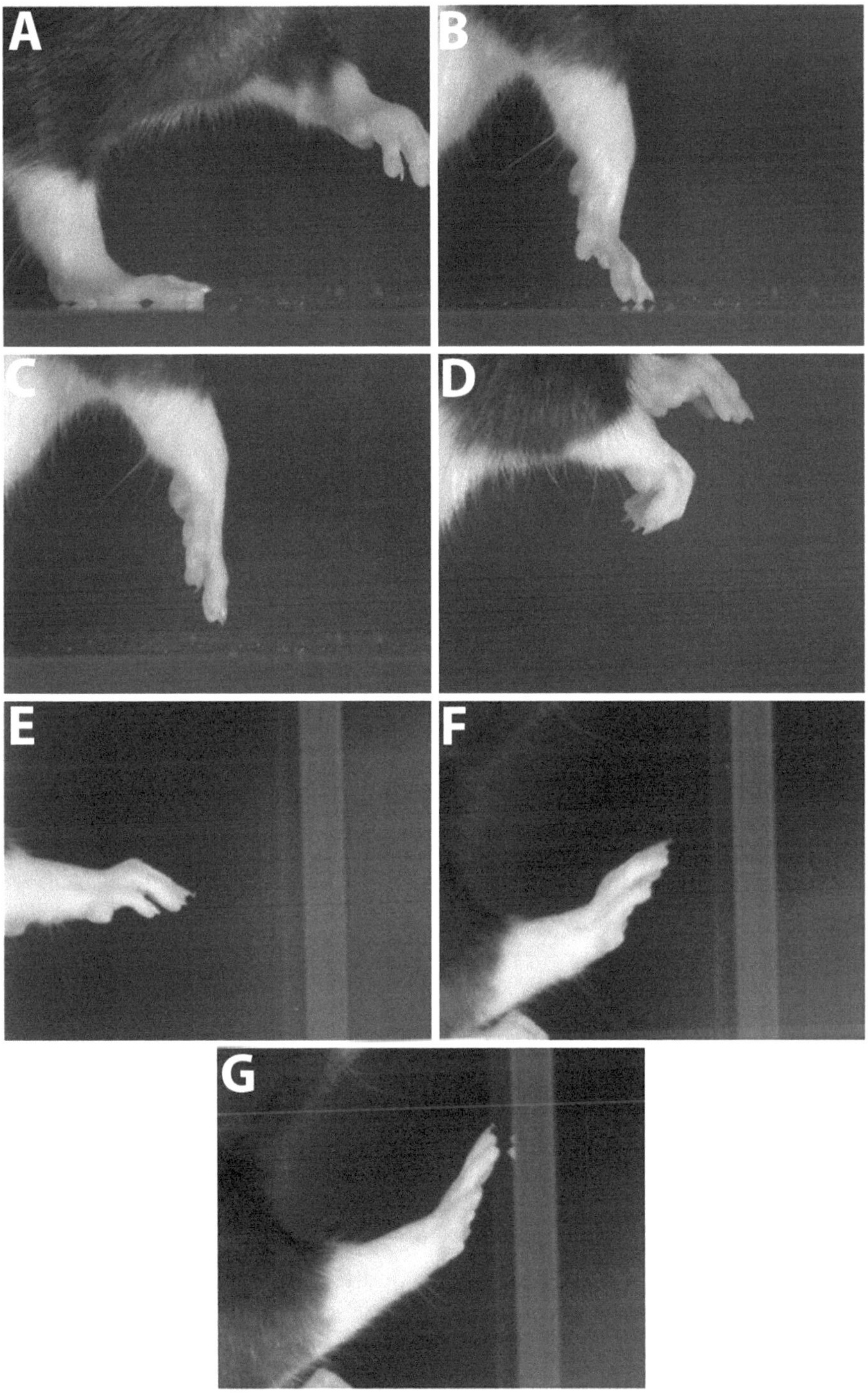

Fig. 4. Hand shaping in the swing phase of a forelimb step in cylinder exploration. Release: The forelimb is pushed off from the floor (**a**, **b**). Collection: The digits close and flex as the forelimb is lifted (**c**, **d**). Manipulation: The digits extend and open to contact the wall of the cylinder (**e**–**g**).

3.5. Skilled Reaching

The movements of release, collection, and manipulation associated with skilled reaching are illustrated in Fig. 5. Release occurs just as the rat is coming to a stop, or is stopped, at the front of the box at which the shelf containing the food is located. For release, the hand is lifted from the floor and the digits are closed. For collection, the hand is brought to the midpoint of the body, and the digits are flexed to make a fist. As the digits are collected, the hand is supinated so that the palm of the hand is in a vertical position. Sometimes, the movement from release to manipulation is continuous. At other times, the movement of the hand is arrested, and remains relatively still in the collected position, before being advanced to the slot. Manipulation begins with the movement of the hand to the slot and extension is almost complete as the hand enters the slot. As the hand is lowered to the shelf, it is pronated in an arpeggio movement in which digits 5 through 2 are sequentially lowered to the shelf. With pronation, the digits are opened. As the hand contacts the food pellet, the digits close to grasp the food, and the hand remains in place until the grasp is complete.

3.6. Task Comparisons

A comparison of the duration of the three postures in the different tasks is obtained by measuring the durations of release (from the last frame of hand contact with the substrate until complete digit flexion), collection, and manipulation (from the first frame of digit extension to the point of hand contact with the substrate). Figure 6 represents the averages of three forelimb cycles obtained from each of the six rats. For overground walking and rung walking, the duration of the three hand postures was quite similar with the shortest period of time spent in collection. For exploration of the cylinder wall, the duration of limb movement was much longer overall than it was for locomotion, but again collections were maintained for the shortest period of time. In skilled reaching, however, duration of collection was distinctively longer.

Whereas the movements of release and collection are very similar in the different tasks, manipulation varies with each task. For overground walking, the digits contacted the surface of the floor in an arpeggio movement (digits 5 through 2) and the entire ventral digit and hand surface eventually pressed against the floor to support the animal's weight. For contacting the rungs of the ladder, there is little use of the digits, and they are extended and partially opened, and contact is made with the palm of the hand. For horizontal rung walking, the digits do not flex to grasp the rung, but they do flex to grasp for inclined rung walking. When exploring the walls of the cylinder, the digits extend and partially open for contact with the wall, but contact is always made with the digit tips. The palm of the hand is only pressed against the wall after digit tip contact is made. The manipulatory movement of skilled

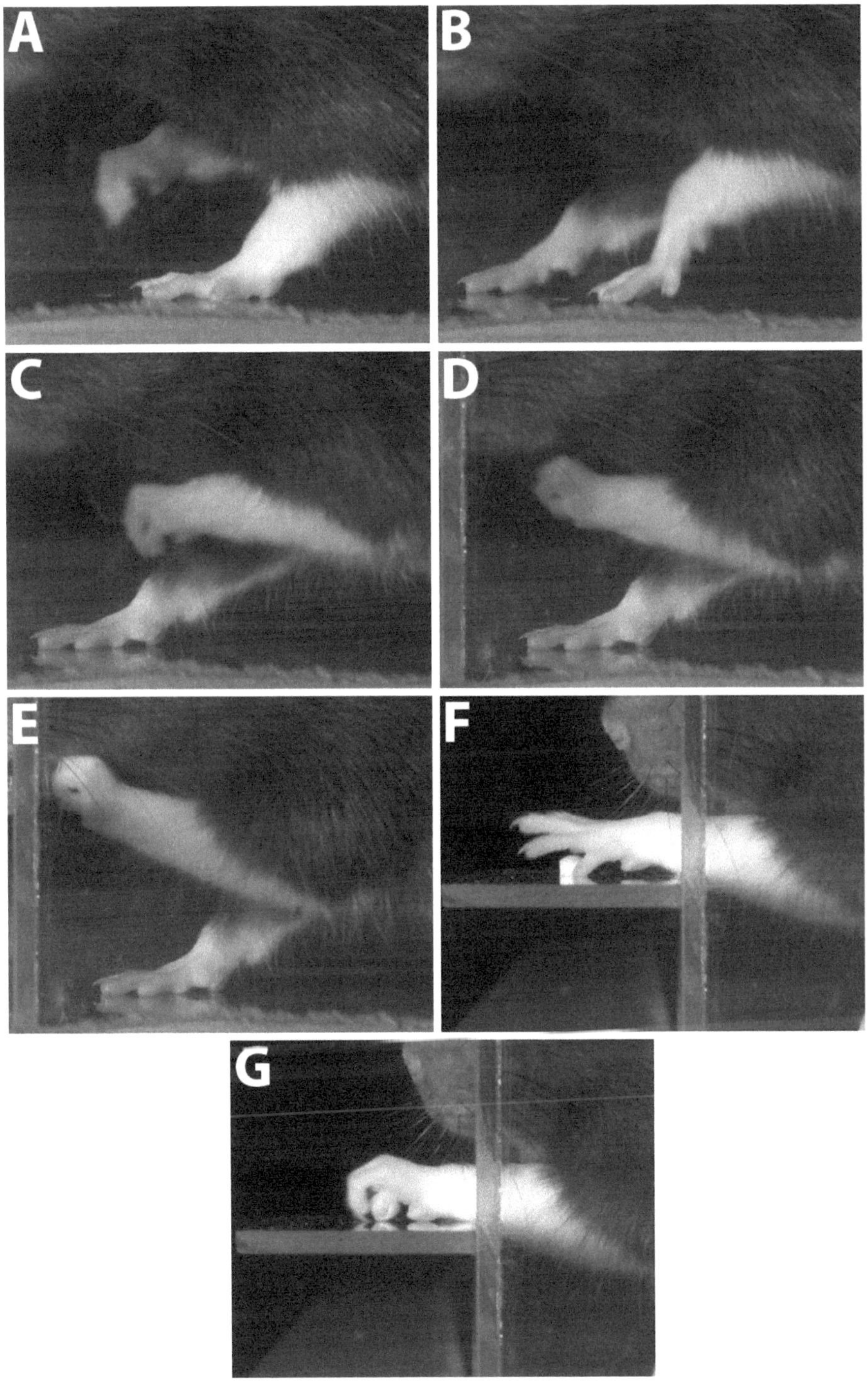

Fig. 5. Hand shaping in the advance phase of a forelimb in skilled reaching. Release: The hand is pushed off from the surface (**a**, **b**). Collection: The digits are flexed and closed as they are carried forward by the forelimb (**c**, **d**). Manipulation: The digits are extended and opened as the paw is pronated over the food for grasping (**e**–**g**).

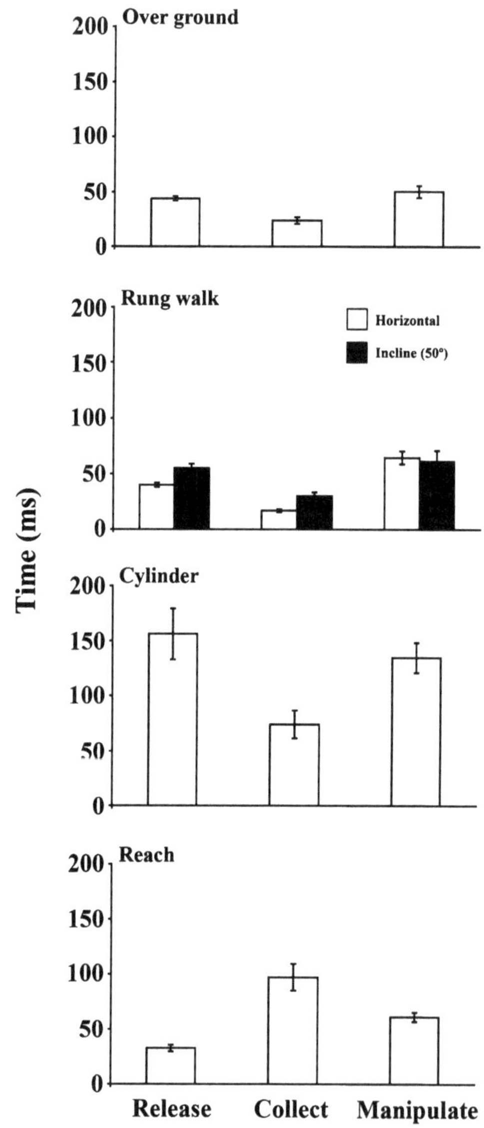

Fig. 6. Duration (mean and standard error) of hand shaping movements of release, collection, and manipulation in overground walking, rung walking, cylinder exploration, and skilled reaching.

reaching involves a very clear extension of the digits prior to their opening. Digit opening only occurs with the arpeggio movement (placement of digits 5 through 2 on the shelf). At the completion of the arpeggio movement, the digits are fully closed and flexed to grasp the food. Thus, use of the digits varies by task in the timing of opening and extending, in contact with the target, and in their action on the target.

4. Digit Use

There are a number of specializations in the way different digits are used. These are summarized for each of the tasks and are illustrated in Fig. 7.

4.1. Overground Walking

When contacting the surface at the end of the swing phase of a step, digit 5 fully opened and extended to be the first digit to contact the floor surface and so likely played a role in guiding the movement of the other digits and hand as they pronated onto the floor (Fig. 7a). When pushing off from the substrate in overground walking, digits 3 and 4 appeared to be principally involved in applying force as they were the last digits to have contact with the floor surface (Fig. 7b).

4.2. Ladder Rung Walking

For ladder rung walking, the digits were used in different ways depending upon whether the ladder is horizontal or inclined. With the ladder in a horizontal position, hand contact with the rung is made with the palm of the hand and the digits were not used for either contacting the rung or grasping (Fig. 7c). On some contacts with the rung, digit 5 does contact the rung first but then moves over the rung. It is not clear whether this contact is intentional or accidental. With the ladder in an inclined position, all of the digits flex and close on the rung. Nevertheless, digits 5 and 2 appeared to take a firmer grip on the rung than did digits 3 and 4 (Fig. 7d). For both horizontal and vertical rung walking, digit 1 is positioned directly over the rung with its tip against the rung, and so may be involved in balancing and grasping. Its small size prevented an examination of how it contributes to rung walking. Although these patterns of rung contact were typical of well-trained rats, it is observed that in preliminary training the rats often stepped on the rungs with their digits.

4.3. Cylinder Exploration

In exploring the wall of the cylinder, the digits and hand are used in a number of ways. The wall might be contacted only by tips of the longest digits, digits 3 and 4 (Fig. 7e), or by the tips of digits 2 through 5 (Fig. 7f). When fully supporting the weight of all of the digits, the palm might be pressed against the wall. During any sequence of wall exploration, a rat might use one or a number of hand contact postures as it moves across the surface of the wall.

4.4. Skilled Reaching

In skilled reaching, digit 5 is the first digit to contact the shelf floor as the hand is pronated over the food (Fig. 7g). Digits 2–5 all play a role in grasping as all flex and close after the hand contacts the food pellet (Fig. 7h). Depending upon the reach and the rat, the food is grasped most frequently with a whole hand grasp using digits 2 through 5, grasped between digits 2 and 3, between

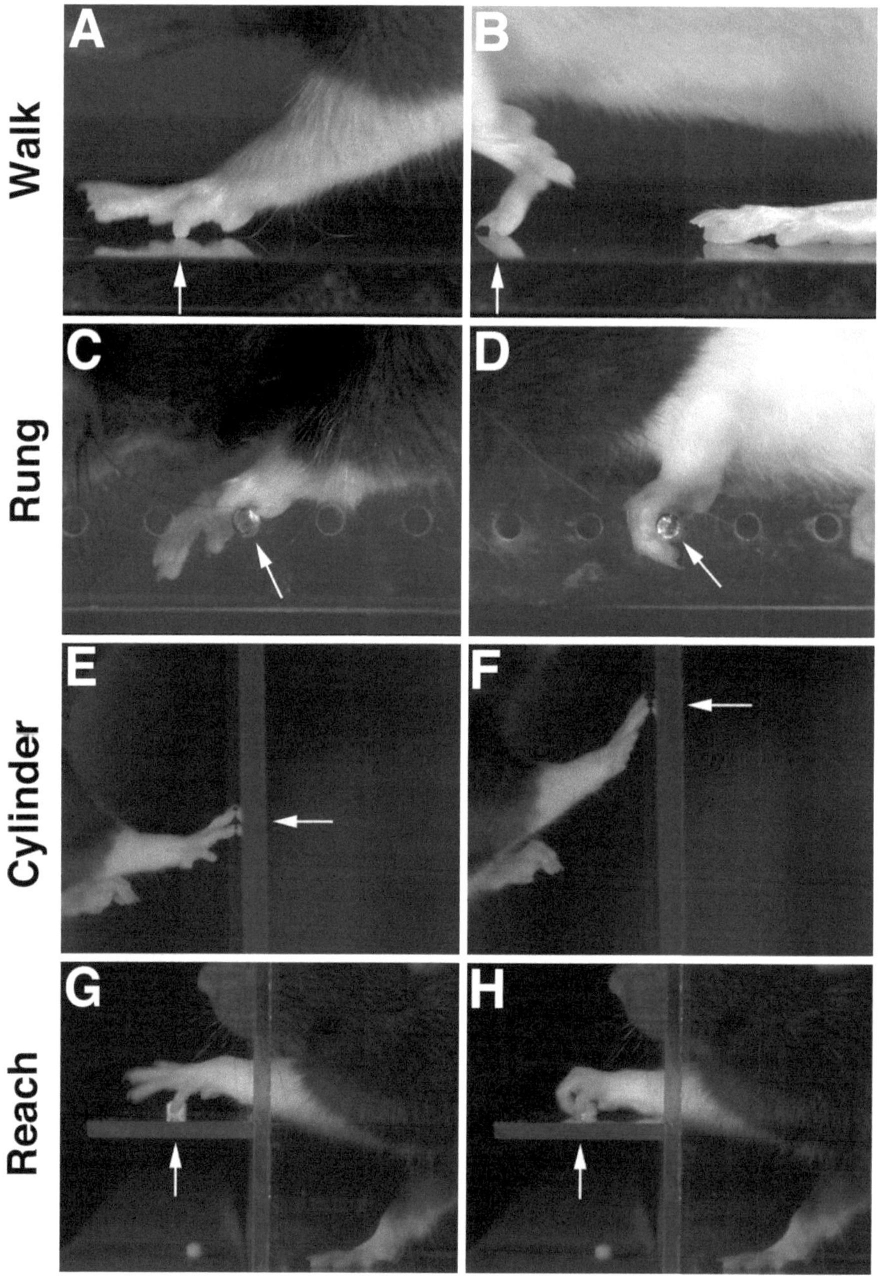

Fig. 7. Independent digit contributions to manipulation in different behaviors. Overground walking: (**a**) Digits 3 and 4 push the hand off from the floor surface; (**b**) digit 5 opens and extents to make the first contact with the floor surface. Ladder rung walking: (**c**) Only digit 1 contacts the rung on a horizontal ladder; (**d**) all digits contribute to grasping on an inclined ladder. Cylinder exploration: (**e**) The tips of digits 2 and 3 contact the wall surface; (**f**) digits 2–5 contact the wall surface. Skilled reaching: (**g**) Digit 5 extends and opens to contact the shelf surface to initiate pronation; (**h**) digits 2–5 flex and close to grasp the food pellet.

digits 3 and 4, or between digits 4 and 5. The actual grasp used on any reach appears to be opportunistic and depends upon where the hand is placed on the shelf.

5. Sensory Control

Sensory guidance is different for the different tasks and includes contact with the vibrissae, sniffing, and hand and digit contact. Illustrations of some of the sources of sensory guidance used in the different tasks are given in Fig. 8.

5.1. Overground Walking

For overground walking, the rhythmical vertical movement of the snout suggested that information concerning the substrate is obtained from the vibrissae as they whisked the substrate (Fig. 8a). Often, the rats place a forepaw at a floor location at which they had whisked. Additionally, the snout might be lifted from the surface and held up without obvious surface whisking. Because the vibrissae are very long, information concerning the substrate is obtained both during surface whisking and nonwhisking. Information that guides placement of the hand on the ground also comes from the sinus hairs on the ulnar surface of the wrist (Fig. 8b). As the hand approaches the floor surface, the sinus hairs contact the surface and continue to bend until the hand is placed on the surface. Because the sinus hairs are of different sizes, they provide information concerning the hand's movement relative to the surface of the floor and its distance from the floor. As the hand approached the floor, digit 5 became fully opened and extended so that it is the first digit to contact the surface. It is possible that digit 5 provides information that guides pronation of the hand onto the floor surface (Fig. 8b).

5.2. Ladder Rung Walking

Information from the vibrissae on the ventral surface of the snout and/or from sniffing provides information to guide placement of the hand on each rung. The vibrissae contact the rung well before the head reaches a rung (Fig. 8c). On about half of all steps, the rats lower their chin surface onto a rung so that the mental vibrissae on their ventral chin surface bend. For other steps, the rats additionally lower the snout to a rung and seemingly sniff the rung before placing a hand on that rung (Fig. 8d). The sinus hairs on the wrist provide information about the location of the rungs because, as the hand approaches a rung, they project into the space between the rungs.

5.3. Cylinder Exploration

Scanning of the vertical cylinder with the snout is complex, with the rats making whisking movements at the surface while sniffing and scanning (Fig. 8e). Sometimes, the rats appeared to place a hand at a location that they had sniffed and whisked, but more

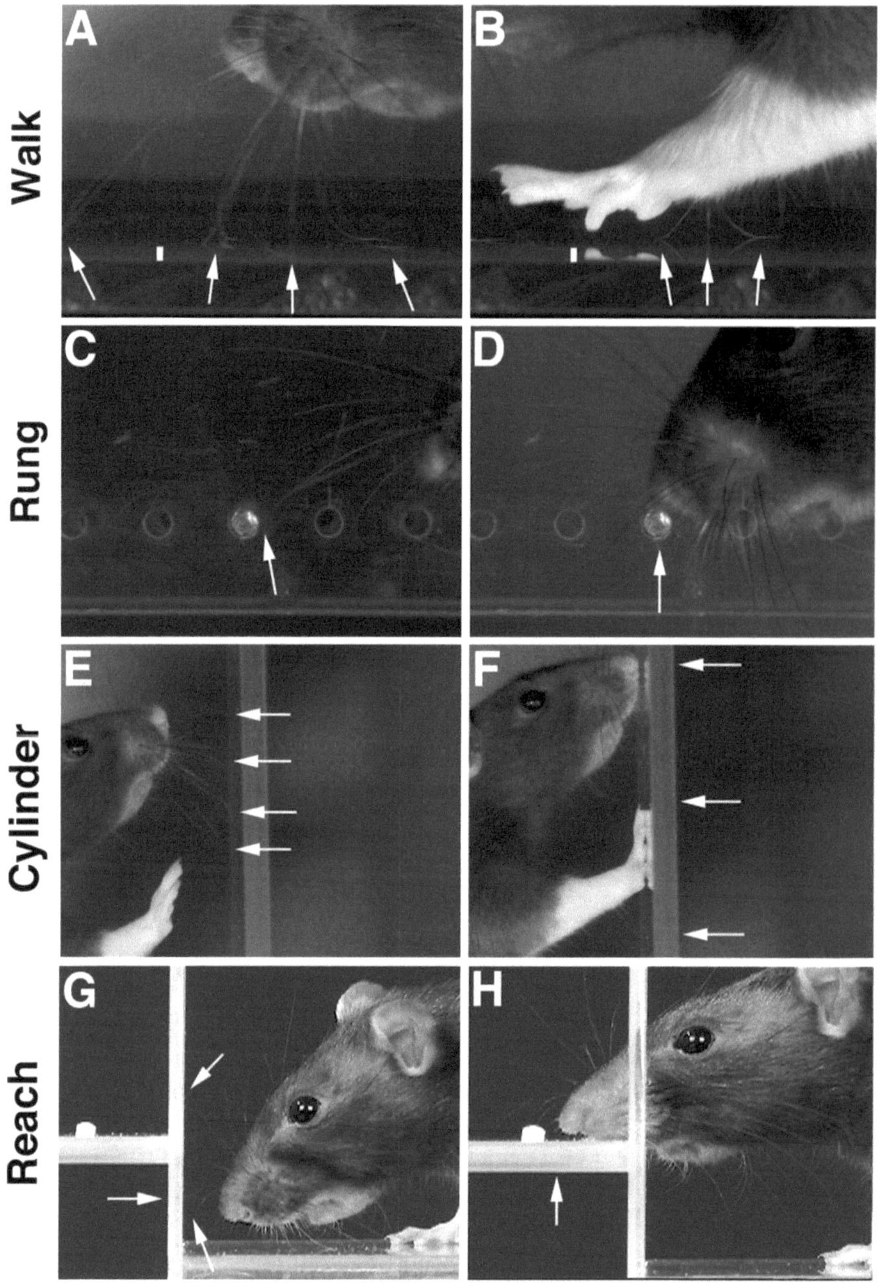

Fig. 8. Sensory contributions to manipulation in different behaviors. Overground walking: (**a**) The nose is lowered to the floor and the vibrissae contact the floor (*arrows*) at a location that the rat places its hand; (**b**) sinus hairs on the wrist bend as the hand approaches the floor. Note also that digit 5 makes the first contact with the floor. Ladder rung walking: (**c**) Vibrissae contact a rung in advance of the nose reaching the rung; (**d**) the rat sniffs the rung on which the hand steps. Cylinder exploration: (**e**) The mystacial vibrissae whisk the wall surface in advance of the hand contacting the wall; (**f**) the nose contacts the wall surface. Skilled reaching: (**g**) Vibrissae contact the slot through which the rat reaches; (**h**) the rat sniffs to locate the food pellet.

often there is no relation between whisking and hand placement. Nevertheless, vibrissae contact helps place the hands on the surface of the wall. The rats also sniff the wall, and often place a hand at a location at which they had sniffed (Fig. 8f). Information about the wall also comes from the digit tips as the rats often use them to touch and balance on the wall surface while at other times information comes from the digits surface as they contact the wall.

5.4. Skilled Reaching

It is likely that the rats located the slot using the vibrissae (Fig. 8g). It has been previously shown that rats use olfaction to locate food pellets in the skilled reaching task, and this result is confirmed by high-speed video records that show that the rats sniff the food pellets prior to grasping (Fig. 8h). Once the rats reached into the slot, it is likely that the sinus hairs on the wrist contributed to locating the shelf.

6. Discussion

There is richness in the movements and the sensory control of the hands of the rats that can be revealed using high-speed video recording. Although there are many tasks in which rats use their hands and digits, it is likely that the range of tasks used here provides a representative sample of hand movements. Because most hand movements are completed in less than a third of a second, high-speed video recording with a high shutter speed provides a reliable method of analysis. The methods allow a detailed visualization of not only the arm and hand, but also the digits. A video record provides a rich description and still allows movements of interest to be digitized for biometric analysis. Inspection of the video record can also identify the sensory control of forelimb use. Whereas it is likely that vision would contribute to the performance of similar tasks in primates, rat skilled reaching depends upon olfaction (21) and tactile information from the vibrissae, sinus hairs, and digits (24).

There are three phases in hand use, release, collection, and manipulation. These terms are not intended to be limiting because it is likely that the movements could be further subdivided. Rather, the terms represent a heuristic for comparing commonalities in forelimb movements across different behaviors. In release, the hand and digits are pushed off from a substrate and are then carried by the movement of the forelimb. In collection, the digits are closed and flexed to make something of a fist. In manipulation, the digits are extended and opened for contacting a surface or grasping. The relative durations of these actions were not fixed across tasks. Collection comprises a similar briefer portion of limb transport during overground locomotion, rung walking, and wall contact in the cylinder and occurs as the hand is supinated and is in motion relative to the body. In contrast, collection comprises the longest portion of limb movement in skilled reaching and the collected limb is also held in an arrested position relative to the body before the hand is advanced through the aperture leading to the target food item.

There are specializations in the way that the digits are used in release and manipulation movements. For overground locomotion, skilled reaching, and rearing in the cylinder task, digits 3 and 4 are the last digits in contact with the floor and they exert force to push the limb from the floor. Because these are the longest digits, it is not surprising that they play a role in force application. It is likely that these digits have a long evolutionary history in aiding stepping because it is these digits that become specialized as hooves in Artiodactyla (third and fourth digits) and Perissodactyla (third digit). During surface contact in overground locomotion and in skilled reaching, digit 5 is the first digit to contact the surface and this contact initiates the arpeggio movement by which the hand is pronated to contact the surface. Lateral-to-medial pronation of the hands and feet is a conserved movement in many animals (16). For horizontal rung walking, once the rats are well practiced, the digits do not play a preeminent role. The rats contacted and supported themselves on the rung with the palm of the hand. Digit 5 may contribute to grasping because it is positioned with its tip directly above the rung. When climbing an inclined ladder, digits 2 through 5 curl to grasp the rung with most of the grasp coming from the more laterally placed digits. Wall exploration in the cylinder task features variable contact with the tips of digits 2–5. For skilled reaching, as has been described, the rats use a number of grasping movements, including grips that feature mainly digits 4 and 5 or grasps that feature digits 2 and 3 (16). It is likely that if still other tasks were examined other patterns of hand and digit use would be revealed. For example, in previous studies of food handling, it is reported that rats, when taking food from their mouth for holding, shape the digits in proportion to food size (22).

The flexibility of manipulatory movements in different behaviors suggests that movements of manipulation require a greater neural commitment than do the movements of release and collection. It is these movements that have a greater cortical representation and appear most affected by neurological insults. It is unknown the extent to which cortical and subcortical systems, including the spinal cord circuits, mediate manipulation, but it is reported that overground locomotion appears less dependent upon corticospinal projections than does skilled reaching (10, 11).

That different digits specialize in different features of hand movement asserts that specialization in digit use is present in animals that are ancestral to primates. This specialization likely provides the substrate for the evolution of more flexible digit control in primates. The specialization of digit use implies that the neural control of different digits in the rat may also be somewhat different. Although neurobiological studies of the rat have been directed toward examining whether rat digits can be moved independently, it may be more revealing to consider their differential use in different tasks.

A distinguishing feature of different manipulatory movements is sensory control. In all of the tasks, the vibrissae (24–27) contribute to hand placement as they are positioned to palpate surfaces well before the hand makes contact with them. Additionally, sinus hairs on the wrist bend as the hand is placed on a surface and so are positioned to contribute to hand placing. Digit 5 plays a special role in hand placing during overground walking and skilled reaching because for these movements it is the first digit that touches the surface (the floor for walking or the shelf for skilled reaching). In the cylinder task, the digit tips of digits 4 and 5 also selectively contact the wall during rearing. Here, the digit tips are used for support and they play a role in exploring the wall surface. Olfaction and nose contact also contribute to rung walking. Although it could not be determined whether the rats actually sniffed a rung before stepping on it, they did clearly place the snout close to the rung on which they then stepped. Considered together, the sensory control of the hand in the rat is not only task dependent, but also complex in its dependence on different senses.

Acknowledgments

Supported by Natural Sciences and Engineering Research Council of Canada.

References

1. Alaverdashvili M, Leblond H, Rossignol S, Whishaw IQ (2008) Cineradiographic (video X-ray) analysis of skilled reaching in a single pellet reaching task provides insight into relative contribution of body, head, oral, and forelimb movement in rats. Behav Brain Res 192:232–247
2. Gharbawie OA, Williams PT, Kolb B, Whishaw IQ (2008) Transient middle cerebral artery occlusion disrupts the forelimb movement representations of the rat motor cortex. Eur J Neurosci 28:951–963
3. Durukan A, Strbian D, Tatlisumak T (2008) Rodent models of ischemic stroke: a useful tool for stroke drug developments. Curr Pharm Des 14:359–370
4. Fricker-Gates RA, Smith R, Muhith J, Dunnett SB (2003) The role of pretraining on skilled forelimb use in an animal model of Huntington's disease. Cell Transplant 12:257–264
5. Döbrössy MD, Dunnett SB (2006) The effects of lateralized training on spontaneous forelimb preference, lesion deficits, and graft-mediated functional recovery after unilateral striatal lesions in rats. Exp Neurol 199:373–383
6. Barnéoud P, Parmentier S, Mazadier M, Miquet JM, Boireau A, Dubédat P, Blanchard JC (1995) Effects of complete and partial lesions of the dopaminergic mesotelencephalic system on skilled forelimb use in the rat. Neuroscience 67:837–848
7. Whishaw IQ, O'Connor RB, Dunnett SB (1986) The contributions of motor cortex, nigrostriatal dopamine and caudate-putamen to skilled forelimb use in the rat. Brain 109:805–843

8. Falkenstein G, Rosenthal C, Reum T, Morgenstern R, Döbrössy M, Nikkhah G (2009) Pattern of long-term sensorimotor recovery following intrastriatal and accumbens DA micrografts in a rat model of Parkinson's disease. J Comp Neurol 515:41–55
9. Muir G (2005) Locomotion. In: Whishaw IQ, Kolb B (eds) The laboratory rat. Oxford University Press, New York, pp 150–161
10. Muir GD, Whishaw IQ (1999) Complete locomotor recovery following corticospinal tract lesions: measurement of ground reaction forces during overground locomotion in rats. Behav Brain Res 103:45–53
11. Kanagal SG, Muir GD (2009) Task-dependent compensation after pyramidal tract and dorsolateral spinal lesions in rats. Exp Neurol 216: 193–206
12. Metz GA, Whishaw IQ (2002) Cortical and subcortical lesions impair skilled walking in the ladder rung walking test: a new task to evaluate fore- and hindlimb stepping, placing, and coordination. J Neurosci Methods 115:169–179
13. Hua Y, Schallert T, Keep RF, Wu J, Hoff JT, Xi G (2002) Behavioural tests after intracerebral hemorrhage in the rat. Stroke 33:2478–2484
14. Schallert T, Woodlee MT (2005) Orienting and placing. In: Whishaw IQ, Kolb B (eds) The laboratory rat. Oxford University Press, New York, pp 129–140
15. Whishaw IQ, Coles BL (1996) Varieties of paw and digit movement during spontaneous food handling in rats: postures, bimanual coordination, preferences, and the effect of forelimb cortex lesions. Behav Brain Res 77:135–148
16. Whishaw IQ, Gorny B (1994) Arpeggio and fractionated digit movements used in prehension by rats. Behav Brain Res 60:15–24
17. Whishaw IQ, Pellis SM (1990) The structure of skilled forelimb reaching in the rat: a proximally driven movement with a single distal rotary component. Behav Brain Res 41:49–59
18. Whishaw IQ, Pellis SM, Gorny BP (1992) Skilled reaching in rats and humans: evidence for parallel development or homology. Behav Brain Res 47:59–70
19. Whishaw IQ, Pellis SM, Gorny B, Kolb B, Tetzlaff W (1993) Proximal and distal impairments in rat forelimb use in reaching follow unilateral pyramidal tract lesions. Behav Brain Res 56:59–76
20. Whishaw IQ, Sarna JR, Pellis SM (1998) Rodent-typical and species-specific limb use in eating: evidence for specialized paw use from a comparative analysis of ten species. Behav Brain Res 96:79–91
21. Whishaw IQ, Tomie J (1989) Olfaction directs skilled forelimb reaching in the rat. Behav Brain Res 32:11–21
22. Whishaw IQ (2005) Prehension. In: Whishaw IQ, Kolb B (eds) The behaviour of the laboratory rat: a handbook with tests. Oxford University Press, Oxford, pp 162–170
23. Sacrey LA, Alaverdashvili M, Whishaw IQ (2009) Similar hand shaping in reaching-for-food (skilled reaching) in rats and humans provides evidence of homology in release, collection, and manipulation movements. Behav Brain Res 204:153–161
24. Ballermann M, Tompkins G, Whishaw IQ (2000) Skilled forelimb reaching for pasta guided by tactile input in the rat as measured by accuracy, spatial adjustments, and force. Behav Brain Res 109:49–57
25. Dyck RH (2005) Vibrissae. In: Whishaw IQ, Kolb B (eds) The behavior of the laboratory rat. Oxford University Press, New York, pp 81–89
26. Jenkison EW, Glickstein M (2000) Whiskers, barrels, and cortical efferent pathways in gap crossing by rats. J Neurophysiol 84: 1781–1789
27. Green EC (1955) The anatomy of the rat. Hafner Publishing Company, New York

Chapter 49

Automated Gait Analysis Following Spinal Cord Injury

Jason E. Beare, Johnny R. Morehouse, David S.K. Magnuson, and Scott R. Whittemore

Abstract

Successful animal models of spinal cord injury (SCI) are dependent on outcome measures sensitive to minute functional differences following injury. Computer-assisted gait analyses provide an objective measurement of numerous gait characteristics, including but not limited to velocity, stance time, swing time, stride length, hindlimb base of support (track width), and toe spread. Using the TreadScan® software system (CleverSys, Inc., Reston, VA), animals can be filmed ventrally as they traverse a clear treadmill-driven belt at a controlled speed, and gait characteristics are automatically calculated by the software. This chapter details the training and use of this sensitive software system as well as the type and scope of data experimenters can expect when analyzing SCI rodents.

Key words: Behavioral assessments, Gait analyses, Locomotor function, Spinal cord injury

1. Introduction

Rodent models of spinal cord injury (SCI) allow researchers to study not only the locomotor deficits associated with SCI, but also the potential benefits of therapeutic agents. Reliable assessment measures of SCI must be objective, informative, and reproducible. Current tools for assessment include a variety of measures from open-field scales, such as the Basso, Beattie, and Bresnahan (BBB) (1) scale for rats and the Basso Mouse Scale (BMS) (2) for mice, to more task-specific measures of function, such as the beam (3) and ladder rung task (4–6).

Computer-assisted footprint analyses, such as the CatWalk system, allow for an objective measurement of gait characteristics following SCI. Velocity, stance time, swing time, and stride length are all measurable as the animal traverses a clear stationary walkway (7–9), although experimenters are unable to control for speed and must calculate all gait characteristics manually. One advantage to

Jun Chen et al. (eds.), *Animal Models of Acute Neurological Injuries II*, Springer Protocols Handbooks, DOI 10.1007/978-1-61779-782-8_49,

the CatWalk system, however, is the fact that gait parameters are measured as the animals walk through space; on TreadScan®, animals are stationary, as the movement is driven by the motorized belt.

The TreadScan® software system provides an automated method for obtaining similar data to the CatWalk system, but with faster throughput and the ability to control for speed. With little to no training, rats and mice are capable of walking on the clear, motor-driven treadmill belt, while the software records and later analyzes myriad gait parameters, resulting in fast, easily interpretable data. The software has proven sensitive to differences between sham animals and injured animals, and has found gait differences between injury severity (mild vs. moderate contusion SCI) (10). The limitation of the TreadScan® system is that animals must be capable of at least occasional plantar stepping in order to be tested on the treadmill, thereby limiting the severity of SCI to a mild-to-moderate injury. However, some gait characteristics show marked deficits following SCI that do not recover up to 10 weeks following injury, indicating that these parameters may offer insightful outcome measures of improvement following treatment with a therapeutic agent (10) (discussed in more detail in Sect. 3.2).

The following sections discuss the materials and equipment required with the TreadScan® system, the method for training and optimizing the software for top performance, and the organization and analyses of the obtained data for fastest throughput and ease of interpretation.

2. Materials and Instruments

1. Single Lane Gait Analysis Treadmill (Columbus Instruments, Columbus, OH) (Fig. 1).
2. TreadScan® software system (CleverSys Inc., Reston, VA), preferably installed on a desktop computer with a 1-TB hard drive and 4 GB of RAM.
3. High-speed digital video camera mounted below the clear treadmill belt. Should record at least 100 frames/s. Most setups include a horizontally mounted camera and an angled mirror to record the ventral view of the animal as it walks.

3. Procedures

3.1. Setup and Software Training

The efficacy of the TreadScan® system is highly dependent on successful training of the software to properly identify foot placement, calibration distances, etc. Insufficient training may compromise the

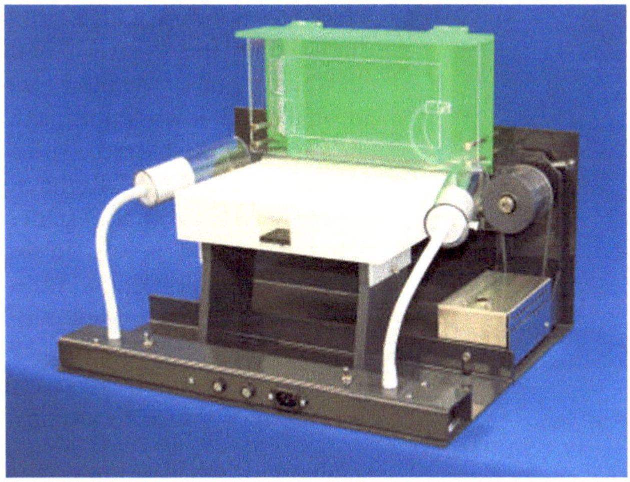

Fig. 1. Gait Analysis Treadmill: Exer Gait (Columbus Instruments, Columbus, OH). Treadmill speeds can be adjusted from 0 to 50 cm/s. Adjustable external lighting allows for optimal color contrast between the feet of the animal and the green background. Running lane shown is for the mouse; a separate larger running lane (Exer Gait XL) is available for rats.

integrity of the data. The following sections outline the optimal training procedures and potential pitfalls to avoid during the training process.

1. Set up and install equipment and software according to respective manuals. If treadmill includes external light source, position lighting to maximize color differentiation of the paw from the background. This is more difficult with lighter colored strains; less light directed into the treadmill chamber and/or a dark cover over the green chamber lid may enhance color contrast and allow for better paw identification by the software. Recording software also includes brightness and gain functions that can maximize contrast. For example, for Sprague Dawley rats, it is often necessary to lower the brightness and increase the gain in order to make the feet stand out as more red tinted against the white body.
    - Proper lighting conditions are crucial for optimal software performance, so extra time spent perfecting the lighting is rewarded with improved functionality.
2. Calibrate the treadmill area by placing a small ruler or object of known size on the treadmill belt and capturing a picture using the recording software included with TreadScan®. Save the picture and open it in TreadScan®. To begin calibrating, click the "Calib" option, and then "Enter Calib Mode." First, the corners of the walking lane (or where the animal is expected to appear throughout the entire walking session) need to be selected. Right click on the top left corner of the region, and

select "Region Top Left Point." Repeat for all four corners of the walking lane. Next, calibrate horizontal distances by right clicking on one of the marks on the ruler or on the left edge of an object of known length, and then select "Mark X1." Right click another mark on the ruler or the right edge of the object and select "Mark X2." Right click anywhere on the screen and select "Set Horizontal Distance," and then enter the distance between the two points (in cm). Repeat for vertical (*Y*-axis) distances by right clicking and selecting "Mark Y1" and "Mark Y2." When all distances are selected and calibrated, right click anywhere on the screen and select "Exit Calibration." Click "Calib," and then "Save Calib" to save the calibration file. This file can be reused for future recordings.

- Note: If the treadmill chamber or the camera moves or is altered in any way, a new calibration file should be created.

3. Record an animal on the treadmill using the recording software for use as a TreadScan® training video. For mice, adjust the Shutter option in the recording software to record at 100 frames/s; for rats, record at 60 frames/s in order to capture enough step cycles.

   - Note: It is important to change the frames/s setting in TreadScan® to match the frames/s of the recorded video.

4. Remove the animal and capture a background image. In TreadScan®, load the video, calibration file, and background file. Use the foot model training function of the software to carefully trace a small portion of each paw while it is in contact with the treadmill belt (Fig. 2). Use extreme caution not to include any pixels outside of the paw, as the software uses color differentiation to identify each paw placement. Take one sample per frame for an entire stance cycle (initial contact to liftoff) for each paw. More samples than this may negatively impact the software training. Foot models can be altered manually in the red, green, or blue color spectrum to maximize paw identification and minimize extraneous background. Click the "Foot Model (Wide)" tab for each foot to display the color histograms during analysis. If the foot boxes are too large, resulting in misidentification of paws, the histograms may be narrowed by clicking and dragging the numbers for each color. Conversely, if the foot boxes are too small, resulting in paws not being identified at all, the histograms may be widened to allow for more colors to be identified as part of the foot.

   - Manual adaptation of the histograms is a trial-and-error process. A successful foot model has nice sharp bell curves for each of the red, green, and blue color spectra. However, even the best foot model may require minor changes to histograms for optimal performance.

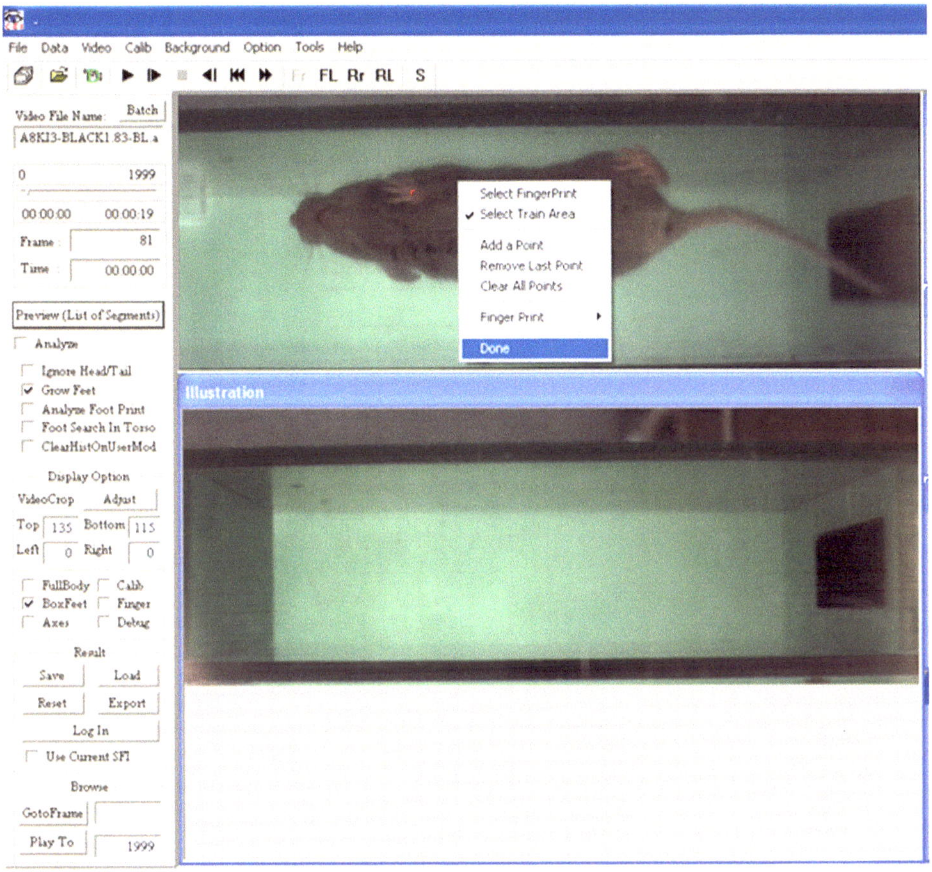

Fig. 2. Manual identification of the paw during software training. Note the small selection of the middle of the paw (*red box*). This insures that no extraneous pixels are included in the sample. Also note that, in this setup, the image on the screen is a mirror image of the animal. Thus, the paw being sampled is the front right (Fr) paw. A later version of TreadScan® (Version 3) offers an option to turn off the mirror image.

5. It is important to remember that the video file is a mirror image when selecting the paws (Fig. 2). In Version 3 of the software, the mirror image is an option that may be turned off by deselecting a check box.

6. TreadScan® software includes an initial settings file that can be altered for each individual rodent strain as necessary. Reference the TreadScan® manual for descriptions of all settings parameters. Foot-specific parameters are the most important for effective software training, especially Stance Combine Dist., Stride Combine Limit, and Grow Feet Expansion settings (Fig. 3). The software can be trained for any number of strains of mouse or rat; simply save a foot model and settings file with optimal paw identification for each strain.

   - Note: New foot models may be necessary if lighting conditions change drastically from one study to the next.

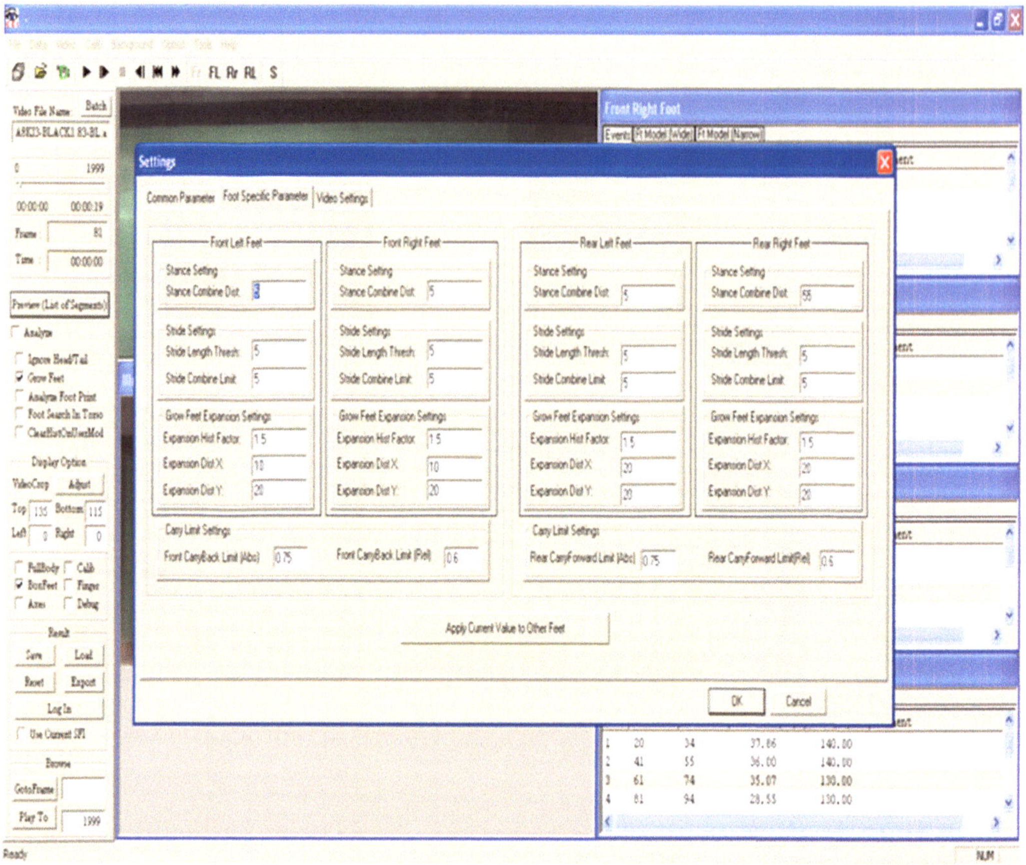

Fig. 3. Settings: If the software is not identifying the feet optimally, the settings file may be altered to correct the sensitivity. For example, if multiple step cycles are incorrectly combined into one step cycle, the Stance Combine Dist. may be lowered. If the nose or tail is incorrectly identified as a paw, the Grow Feet Expansion Settings may be altered. Some trial and error is necessary when modifying the settings file; consult the CleverSys manual for further instruction.

It is important to verify that the existing foot models and settings files are properly identifying the feet at the beginning of each new study; if not, create new files.

3.2. Treadmill Training and Recording

1. Mice perform optimally with minimal treadmill training. Although ungentled mice and rats can perform on the treadmill, gentling the animals prior to treadmill introduction greatly reduces spontaneous non-locomotor behaviors, such as exploration or "freezing up" during the task. After a few gentling sessions, baseline recording may be performed on the first introduction to treadmill, if so desired. More than minimal training on the treadmill may lead to adverse training effects, including refusal to walk and/or changes in baseline gait characteristics (8). It is recommended that exposure to the treadmill is kept to a minimum, i.e., animals should only be placed

on the treadmill for a single baseline recording and no more than weekly recording following SCI.

2. Place animal in treadmill chamber for 20 s to acclimate. Turn on the treadmill, slowly at first, and then ramp up speed until animal is consistently walking in place (or walking at the preferred speed of the experiment; see below). Record walking for 2,000 frames (equivalent to 100 frames/s for 20 s for mice, 60 frames/s for 33 s for rats), pausing the recording if the animal is not walking consistently (i.e., runs to the front and rides back, explores the chamber, etc.). Twelve or more consecutive step cycles are optimal, but three or more time frames containing a minimum of four consecutive step cycles may be substituted if the animal proves uncooperative. After recording, stop the treadmill, remove the animal, and save the recorded file.
   - Save the recording as studyname-animal#-timepoint.avi. This allows for easy identification of the video file.
   - The software allows the experimenter to convert large .avi files to smaller .mpg files; however, it is recommended that all analyses are performed on the original .avi file. Convert to .mpg files only after analyses are complete to save disc space.
3. After the first animal is recorded for each treadmill session, capture a background image specific to that particular day. Changes in ambient lighting, positioning of the external lighting, time of day, etc. can cause minute changes in the background, so it is important to capture a new background image for each treadmill session. Additionally, check the video of the first animal in TreadScan® to ensure that the existing foot model and settings files correctly identify the paws before moving on to further animal recordings.
4. For an uncooperative animal, it may prove useful to modify the animal's behavior by turning the treadmill on at a low speed and keeping it on until the animal walks properly to the front of the chamber. When the animal correctly performs the task, turn the treadmill off as a reward. This negative reinforcement often extinguishes the exploratory, non-locomotor behaviors within a few minutes.
   - From time to time, an animal will simply not perform the treadmill task. In this case, the animal may be removed from the treadmill and set aside for a few minutes while other animals are recorded. Come back to the uncooperative animal again later and try recording. If the animal still will not walk, a Styrofoam bumper may be added to the back of the treadmill lane as an aversive stimulus to induce walking. If the bumper is required, be sure to cover it with

green tape to match the color of the background, and also capture a separate background image to use for analysis of this file.

It is important to note that recording procedures depend on the goal of the study. Previous treadmill studies have shown that some gait parameters, such as hindlimb swing time (10, 11) and track width (10), are unaffected by treadmill speed but do exhibit marked deficits that are consistent up to 10 weeks following SCI (Fig. 4). Thus, these gait characteristics may be useful in testing the efficacy of a proposed treatment. In this case, animals may be allowed to walk at their own best speed; that is, each individual animal can walk at whatever speed that allows for best performance on the treadmill. If, however, the experimenter is interested in speed-dependent measures such as hindlimb stance time or stride length, each animal must be walked at the same speed to allow for comparison.

The speed of the treadmill also depends on the goals of the study. Uninjured mice often struggle to keep up with treadmill speeds above 20–25 cm/s, so 15 cm/s is often used as the optimum walking speed in the mouse (10, 11). Similarly, optimal walking speeds for the uninjured rat are within the range of 25–30 cm/s. Injured animals rarely achieve walking speeds equal to sham or baseline animals, and thus treadmill speeds should be adjusted accordingly (5–10 cm/s for moderately injured mice; 10–15 cm/s for rats).

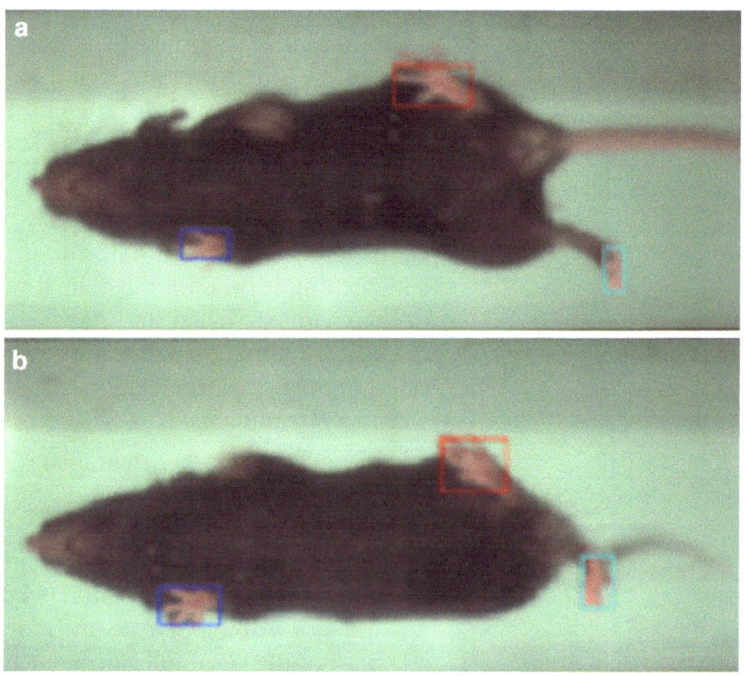

Fig. 4. (**a**) Sham animal (BMS = 9) walking at 10 cm/s 6 weeks post injury. (**b**) SCI animal (BMS = 6) walking at 10 cm/s 6 weeks post injury. Note the narrower toe spread of the right hindlimb (*red box*) and the narrower hindlimb base of support following injury.

3.3. Analysis

1. Once the software has been properly calibrated and trained, open the file of interest, appropriate background, foot model, output, and settings files. Uncheck the "Analyze" box, click "Preview," and watch the file carefully in an attempt to find the best 10–15 step cycles. Attempt to choose step cycles in which the animal is consistently stepping with minimal lateral or longitudinal movement. If there are not 10–15 consecutive step cycles, choose multiple smaller time windows with no less than three consecutive step cycles each. Enter the appropriate starting and stopping frames for the time(s) of interest in the "Preview (List of Segments)" box, click "Add," and then click "Close" to close the preview window.
2. Return to the beginning of the video file, and check the "Analyze" box. Click "Play" to start analysis. If toe spread data is desired, manually advance the frames and identify the toes of the rear paws by selecting the appropriate paw (i.e., RR for rear right), then right click on the video screen, and click "Select FingerPrint" (Fig. 5). Starting at the top of the video screen, identify each toe and the base of the foot, then right click, and select "Done." Identify five to eight toe spread data points for each rear paw, one data point per step cycle, and then allow the software to analyze the remainder of the selected frames. It is important to collect toe spread data from the same point in the step cycle every time, i.e., initial contact.
3. During analysis, insure that the software is correctly identifying each paw. Correct any misidentifications manually using the foot model tool discussed in Sect. 3.1. Select the foot that is being misidentified (i.e., RR for rear right), trace a small portion of that paw using the foot model function, and click "Done." The foot box should shift to correctly identify the paw. Manually advance a few frames to make sure that the paw is now correctly identified throughout the step cycle. If not, the foot model and/or the settings file may require minor editing.
4. Upon completion of the software analysis, manually check that each individual step cycle is identified correctly by holding the Shift key and double clicking each individual step for each paw. Criteria for a correctly analyzed step include proper identification of initial contact and liftoff, and the animal should be walking consistently. Steps that include any of the following should be excluded from the analysis:
   - Initial contact or liftoff is misidentified.
   - The animal is not plantar stepping.
   - A twitch or spasm is identified as a step.
   - The animal is not actively walking, i.e., riding back on the treadmill.

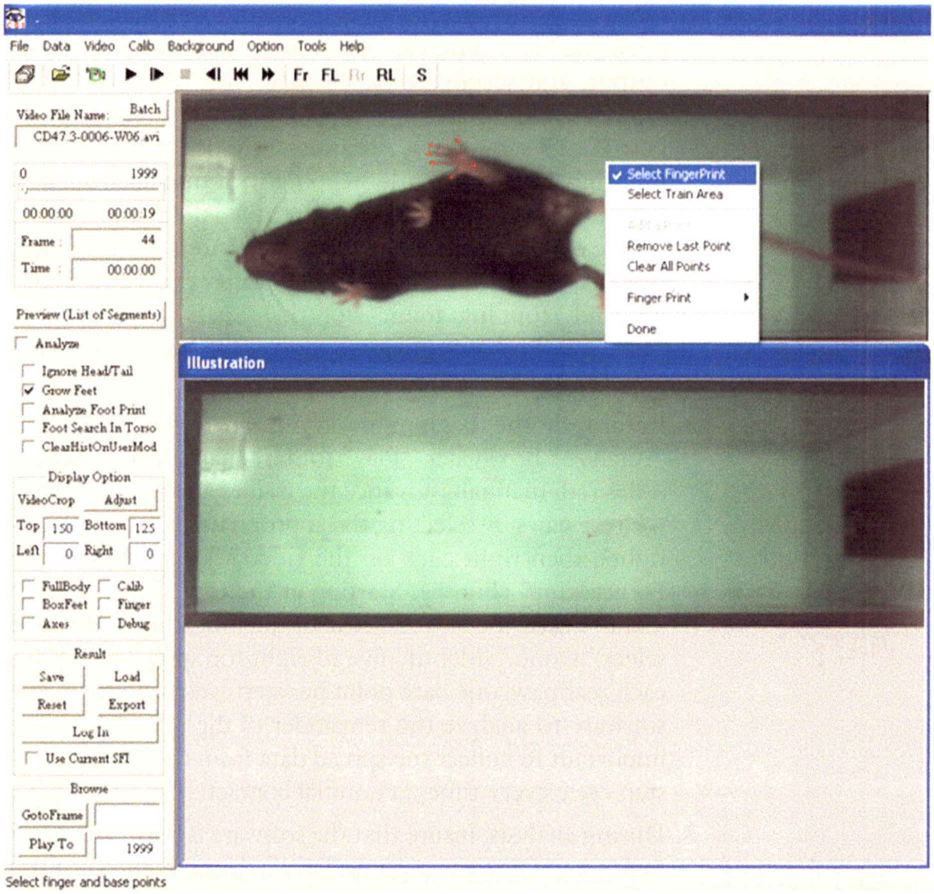

Fig. 5. Manually identifying the toes for toe spread analysis. *Red dots* indicate proper identification of toes and base of foot. Toe spread data points should always be taken from the same point in the stance cycle (i.e., at the initial contact). Select the foot of interest (in this example, rear right), right click the video screen, and click "Select FingerPrint." Starting at the top of the screen, left click each of the toes from top to bottom, and then select the base of the foot. Right click and select "Done" to finish sampling. Collect five to eight data points for each hind paw for each animal.

- Multiple steps are combined into a single step.
- Other anatomical markers are misidentified as paws, i.e., nose, genitals, base of tail, etc.

5. When all steps have been checked for accuracy and any incorrect steps deleted, save the file. Click "Export" to send the data to a Microsoft Excel file for further analysis. Save the Excel file as discussed in Sect. 3.2: studyname-animal#-timepoint.xls.
   - Example: CD47-123-w2.xls for animal #123 in a CD47 study 2 weeks post injury.
6. Naming the files in this manner allows the experimenter to rearrange the data for statistical analysis quickly and effortlessly using the University of Louisville's downloadable TreadScan Add-in discussed below.

3.4. Data Management: TreadMerge Add-In

The TreadMerge Add-in is designed to streamline the data analysis process following TreadScan® video analysis, especially for older versions of the software. TreadMerge merges all individual animal files into one Excel document, and then arranges the data automatically into easily manageable tabs. TreadMerge is a simple, easy-to-use pop-up menu incorporated into Excel. Go to http://louisville.edu/kscirc/about/cobre-cores/core-d-electrophysiology-and-behavioral-assessment.html to download the free add-in and install it by carefully following the steps in the downloadable TreadMerge Installation Manual. There are downloadable add-ins for Microsoft Excel 1997–2003 and for Excel 2007. Version 3 of TreadScan® includes a Group Export function that works similarly to TreadMerge. Please review the TreadScan® manual for information on this function.

1. After all animals have been run for a particular treadmill session, save all exported Excel files together in a single folder. Be sure to follow the proper naming protocol previously discussed (studyname-animal#-timepoint.xls). Only save files that should be merged in this folder; extraneous files prohibit TreadMerge from executing properly.
2. In Microsoft Excel 2007, click the Data tab and select TreadMerge Add-in (Fig. 6). On the TreadMerge Form, browse for and select the folder to be arranged, and click "TreadArrange." This rearranges the data to be merged into a single tab. The TreadMerge Form automatically closes upon completion.
3. Once the data have been arranged, individual gait parameters, such as stance time, swing time, track width, etc., may be merged into a new Excel workbook. Open the TreadMerge Form again, select the gait parameters of interest by checking the individual boxes (or select "All Data" to merge all gait parameters), and click "Merge" (Fig. 6). This combines the data of interest from all animals into a new workbook.
4. Finally, to organize all of the newly merged data into easily manageable tabs, reopen TreadMerge and click "Data Arrange." Save the new file. All of the selected gait parameters for all animals are given individual tabs, one per gait parameter, and the data is arranged for easy statistical analysis.

3.5. Optional Manual Assessments: Regularity Index and Plantar Stepping Index

1. While the TreadScan® software contains a built-in measure of coordination, it is not as well established as other measures of interlimb coordination. However, the ventral view recorded by TreadScan® allows for two quick assessments which can further distinguish behaviorally similar animals based on coordination.
2. Regularity index (RI)—measure of interlimb coordination during uninterrupted locomotion (8). The RI is primarily used

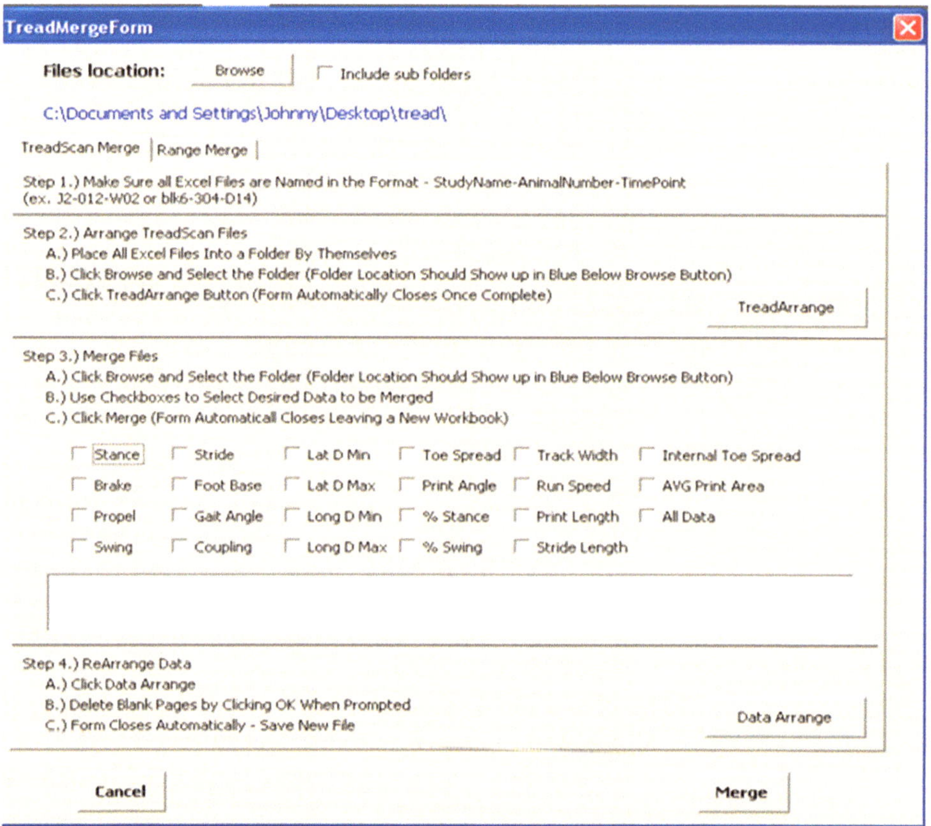

Fig. 6. TreadMerge Form in Microsoft Excel 2007. By selecting the folder of interest and clicking "TreadArrange," all exported TreadScan® data is rearranged into a single Excel tab. Clicking "Merge" then combines the data for all animals into a single Excel file. Finally, click "Data Arrange" to organize the data into a more manageable, easier-to-navigate format for data analysis.

to distinguish behaviorally high-scoring animals based on coordination.

3. To calculate RI, first assign each paw a number. For example, front right = 1, front left = 2, rear right = 3, and rear left = 4. Using the same time frames identified for TreadScan® analysis, make note of the limb sequence during stepping. This can actually be done while TreadScan® is analyzing. Write down the number for each footfall during the step cycles of interest, in order of placement. Make note of any dorsal steps by underlining the corresponding footfall.
   - Example: Front right–rear left–front left–dorsal rear right would be denoted 1423.
4. Once all footfalls are recorded, go back and make a "slash" mark (/) after every fourth footfall to delineate each four footfall step cycle. A normal step sequence pattern (NSSP) occurs when each of the four paws is included exactly once in a given

step cycle. Any step cycle that includes multiple footfalls of any one paw (which logically leads to no step for at least one other paw in that step cycle) or any otherwise coordinated step cycle that includes a dorsal step is considered uncoordinated.

- Example: 1423 is a coordinated step cycle. 1421 is uncoordinated. 142$\underline{3}$ is uncoordinated.

$$\text{RI} = \frac{\text{NSSP} \times 4}{\text{PP}} \times 100(\%),$$

where NSSP is the number of normal step sequence patterns (or coordinated step cycles) and PP is the total number of paw placements analyzed (6).

- Example: 1423/1324/1432/1243/1423/1423/1423. There are 7 total step cycles (and 28 total paw placements), and each one is coordinated. $\text{RI} = ((7 \times 4) / 28) \times 100(\%) = 100\%$ coordinated.
- Example: 1423/1321/1432/1232/1423/142$\underline{3}$/1423. Again, there are 7 total step cycles and 28 total paw placements; however, only 4 of the step cycles are coordinated. Two step cycles included multiple footfalls for the same paw, while one step cycle includes a dorsal step. Thus, $\text{RI} = ((4 \times 4) / 28) \times 100(\%) = 57.14\%$ coordinated.

5. Plantar Stepping Index (PSI)—percentage of plantar hindlimb steps relative to forelimb steps. PSI is primarily used to distinguish animals that have lower behavioral scores due to low coordination.
6. To calculate PSI, use the same method as the RI to assign a number to each paw and record the footfalls for the step cycles of interest.

$$\text{PSI} = \frac{\text{PHL}}{\text{PFL}} \times 100(\%),$$

where PHL is the total number of plantar hindlimb steps and PFL is the total number of plantar forelimb step (12).

- Example: 1423/1324/1432/1243/1423/1423/1423.w There are 14 plantar hindlimb steps and 14 plantar forelimb steps, so $\text{PSI} = (14 / 14) \times 100(\%) = 100\%$ hindlimb:forelimb interaction.
- Example: 1423/1321/1432/1232/1423/142$\underline{3}$/1423. There are 11 plantar hindlimb steps and 16 plantar forelimb steps, so $\text{PSI} = (11 / 16) \times 100(\%) = 68.75\%$ hindlimb:forelimb interaction.

Again, these manual calculations are not necessary for proper TreadScan® analyses. Much like task-specific behavioral measures such as the beam and ladder rung tasks, they may provide additional insight into differences between animals. It is recommended that the automatic gait parameter analyses of the TreadScan® system are used in conjunction with an open-field assessment (BMS) and any task-specific measures of interest, allowing the experimenter to obtain a wealth of information on locomotor abilities following SCI.

References

1. Basso DM, Beattie MS, Bresnahan JC (1995) A sensitive and reliable locomotor rating scale for open field testing in rats. J Neurotrauma 12:1–21
2. Basso DM, Fisher LC, Anderson AJ, Jakeman LB, McTigue DM, Popovich PG (2006) Basso Mouse Scale for locomotion detects differences in recovery after spinal cord injury in five common mouse strains. J Neurotrauma 23:635–659
3. Farooque M, Suo Z, Arnold PM, Wulser MJ, Chou CT, Vancura RW, Fowler S, Festoff BW (2006) Gender-related differences in recovery of locomotor function after spinal cord injury in mice. Spinal Cord 44:182–187
4. Apostolova I, Irintchev A, Schachner M (2006) Tenascin-R restricts posttraumatic remodeling of motoneuron innervation and functional recovery after spinal cord injury in adult mice. J Neurosci 26:7849–7859
5. Farr TD, Liu L, Colwell KL, Whishaw IQ, Metz GA (2006) Bilateral alteration in stepping pattern after unilateral motor cortex injury: a new test strategy for analysis of skilled limb movements in neurological mouse models. J Neurosci Methods 153:104–113
6. Hsu JY, McKeon R, Goussev S, Werb Z, Lee JU, Trivedi A, Noble-Haeusslein LJ (2006) Matrix metalloproteinase-2 facilitates wound healing events that promote functional recovery after spinal cord injury. J Neurosci 26: 9841–9850
7. Clarke KA, Still J (1999) Gait analysis in the mouse. Physiol Behav 66:723–729
8. Hamers FP, Lankhorst AJ, van Laar TJ, Veldhuis WB, Gispen WH (2001) Automated quantitative gait analysis during overground locomotion in the rat: its application to spinal cord contusion and transection injuries. J Neurotrauma 18:187–201
9. Hamers FP, Koopmans GC, Joosten EA (2006) CatWalk-assisted gait analysis in the assessment of spinal cord injury. J Neurotrauma 23: 537–548
10. Beare JE, Morehouse JR, DeVries WH, Enzmann GU, Burke DA, Magnuson DS, Whittemore SR (2009) Gait analysis in normal and spinal contused mice using the TreadScan system. J Neurotrauma 26:2045–2056
11. Leblond H, L'Esperance M, Orsal D, Rossignol S (2003) Treadmill locomotion in the intact and spinal mouse. J Neurosci 23:11411–11419
12. Kuerzi J, Brown EH, Shum-Siu A, Siu A, Burke D, Morehouse J, Smith RR, Magnuson DSK (2010) Task-specificity vs ceiling effect: step-training in shallow water after spinal cord injury. Exp Neurol 224(1):178–187

Chapter 50

Forelimb Functional Assessments of Rats and Mice After Cervical Spinal Cord Injury

Stephen M. Onifer

Abstract

Traumatic spinal cord injury (SCI) frequently occurs to the cervical segments. Regaining arm and hand function is the highest priority for persons with tetraplegia after cervical SCI. Given the tremendous impact that therapeutic changes of hand and arm function would have on the quality of their lives and since veterinary care leading to chronic cervical SCI survival became realizable, experimental research in nonhuman primates, canines, felines, rats, and mice is being conducted. The experiments aim to determine the extent and anatomical–physiological mechanisms of spontaneous hand and arm (also referred to as forepaw and forearm or forelimb) function recovery, and the lack thereof, as well as to develop treatments to improve it. The purposes of this chapter are (1) to report on forelimb functional assessment tests that have been developed and adapted for experiments using various strains of adult rats and mice with cervical SCI and (2) to provide general protocols of some frequently used ones assessing general usage, tactile discrimination, and skilled movements to guide future investigations. These are the Limb-Use Asymmetry or Cylinder Test, Grip Strength Test, Tactile Stimulation Test, and the Staircase Test.

Key words: Cervical spinal cord injury, Rat, Mouse, Assessment, Tetraplegia, Arm, Hand, Forelimb, Forepaw, Manual dexterity

1. Introduction

Arm and hand dysfunctions occur to humans, both adults and children, after a variety of traumas to and diseases of their central and peripheral nervous systems, including cervical spinal cord, brachial plexus, peripheral nerve, and brain injuries as well as stroke. Traumatic spinal cord injury (SCI) frequently occurs to the cervical segments (1). The resulting complete or incomplete tetraplegia, as well as respiratory complications (for example, see ref. 2), are devastating and lead to enormous personal and financial hardships. A recent survey revealed that regaining arm and hand function was the highest priority for persons with tetraplegia after cervical SCI (3).

Jun Chen et al. (eds.), *Animal Models of Acute Neurological Injuries II*, Springer Protocols Handbooks,
DOI 10.1007/978-1-61779-782-8_50,

Given the tremendous impact that therapeutic changes of hand and arm function would have on the quality of their lives and since veterinary care leading to chronic cervical SCI survival became realizable, experimental research in nonhuman primates, canines, felines, rats, and mice is being conducted. The experiments aim to determine the extent and anatomical–physiological mechanisms of spontaneous hand and arm (also referred to as forepaw and forearm or forelimb) function recovery, and the lack thereof, as well as to develop treatments to improve it. The purposes of this chapter are (1) to report on forelimb functional assessment tests that have been developed and adapted for experiments using various strains of adult rats and mice with cervical SCI and (2) to provide general protocols of frequently used ones assessing general usage, tactile discrimination, and skilled movements to guide future investigations. One important anatomical consideration for these and translational studies is that the majority of descending motor corticospinal tract axons in the rat and mouse spinal cord are not located in the dorsolateral funiculi. Instead, they are found below the dorsal columns in the dorsal funiculi (for adult rat review, see ref. 4). The dorsal columns of rodents and other mammals similarly contain ascending sensory primary afferents from the upper extremities and the rest of the lower body (for review, see ref. 5). Both descending and ascending axon pathways are found throughout the dorsolateral funiculi and the rest of the white matter (for adult rat review, see ref. 4). Forelimb motor neurons are distributed throughout the adult rat mid-cervical to upper thoracic segments (6).

The four primary types of traumatic injury occurring to the adult human cervical spinal cord are (1) the frequent contusion evolving to cavity formation, (2) massive compression, (3) laceration, and (4) solid cord (7–9). A literature search using PubMed (10) and "cervical spinal cord injury," "rats," or "mouse" revealed that a variety of skilled and unskilled forelimb functional assessment tests have been utilized for adult rats, primarily, and mice, injured with models of these traumatic injury types. The tests quantified and/or scored aspects of reaching, grasping, retrieval, exploratory, locomotor, and somatosensory behaviors. Those that assessed forelimb function of cervical spinal cord-injured animals while they were not or were locomoting are shown in Tables 1 and 2, respectively. The literature search results in these Tables for rats are from 2005 to 2010 since a very thorough review of forelimb functional assessment tests for rats with cervical SCI was published in 2005 (11). Some of the tests in Table 2 include new scales that have been developed to assess the forelimbs of cervical spinal cord-injured rats. Details about using tests in Table 2 as well as additional tests to evaluate the hind limbs of locomoting rats and mice following traumatic SCI can be found within other chapters of this section. Importantly, treatments of experimental cervical SCI have been evaluated with some of these forelimb functional assessment tests.

Table 1
Forelimb functional assessments after cervical SCI of non-locomoting adult rats and mice since 2004, each in chronological order

	Contusion	Compression	Laceration	Other
Grip strength test	C5, ESCID, bilateral (29); C5, ESCID, bilateral (39)[a]; C4, IHI, unilateral (49)[a,b]; C5–8, IHI, bilateral (46)[b]; C5, ESCID, bilateral (41)[a,b]; C4, IHI, unilateral (50)[a,b]; C5, IHI, bilateral (44)[b,c]	C4, sharpened forceps, bilateral, dorsal funiculus (51); C4, sharpened forceps, bilateral, dorsal funiculus (40)[a]	C5, C6, C7, T1, Moria microknife, unilateral, lateral hemisection (42)[b,c]; C5, Moria microknife, unilateral, lateral hemisection (43)[b]; C5, Moria microknife, unilateral, medial stab, lateral stab (45)[b]; C5, Moria microknife, unilateral lateral hemisection (47)[b,c]; C4, Moria microknife, unilateral, lateral hemisection (48)[b]	
Grooming test	C5, NYU/MASCIS Impactor, unilateral (30)[b]; C5, IHI, unilateral (61)[a,b]		C4, microscissors, unilateral, lateral hemisection (58)[b]	
Inclined plane	C5, ESCID, bilateral (29); C4–5, Blunt Force Impactor, unilateral (21)[a]; C5, ESCID, bilateral (41)[a,b]	C7–T1, extradural clip compression, bilateral (85)[a]; C7–T1, extradural clip compression, bilateral (86)[a,b]		

(continued)

Table 1 (continued)

	Contusion	Compression	Laceration	Other
Limb-use asymmetry or cylinder test	C5, ESCID, bilateral (29)[b]; C5, NYU/MASCIS Impactor, unilateral (30)[b]; C4–5, Blunt Force Impactor, unilateral (21)[a]	C4/5, sharpened forceps, unilateral, dorsolateral funiculus (35)[a]; C4, sharpened forceps, unilateral, lateral funiculus (23)[a,b]	C5/6, iridectomy scissors, overhemisection (28)[a,b]; C3, modified needle, bilateral, dorsolateral funiculus (32)[b]; C4/5, iridectomy scissors, overhemisection (34)[a]	C3–4, microscalpel and aspiration, unilateral, subtotal hemisection (24)[a,b]; C4, aspiration, unilateral, dorsolateral funiculus (25)[a]; C3–4, microscissors, aspiration, unilateral, partial lateral hemisection (26)[a,b]; C3, aspiration, unilateral, lateral hemisection (31)[a,b]; C4, iridectomy scissors and aspiration, overhemisection (36)[a]; C3–4, microscissors, aspiration, unilateral, partial lateral hemisection (27)[a,b]; C4, aspiration, unilateral, lateral hemisection (22)[a,b]; C3/4, microscalpel, microscissors, aspiration, unilateral, dorsolateral funiculus, lateral funiculus, dorsal horn (33)[b]
Mechanical threshold, hyperalgesia/allodynia/Von Frey hairs[d]	C5–8, IHI, bilateral (46)[b]	C4/5, sharpened forceps, unilateral, dorsolateral funiculus (35)[a,b]; C4/5, sharpened forceps, bilateral, dorsal funiculus (87)[a,b]	C5, Moria microknife, unilateral, lateral hemisection (43)[b]; C4–5, wire knife and scalpel blade, unilateral, dorsal funiculus (75)[a,b]	
Mouse pectoral hair loss scale	C5, IHI, bilateral (44)[b,c]			
Placing response	C5, ESCID, bilateral (29)[b]; C5, ESCID, bilateral (41)[a,b]	C4, sharpened forceps, bilateral, dorsal funiculus (51); C4, sharpened forceps, bilateral, dorsal funiculus (40)[a]		

Reaching tests		C4/5, sharpened forceps, unilateral, dorsolateral funiculus (35)[a,b]; C4, sharpened forceps, bilateral, dorsal funiculus (40)[a]	C5, Moria microknife, unilateral, lateral hemisection (43)[b]; C4–5, wire knife and scalpel blade, unilateral, dorsal funiculus (75)[a,b]; C4, microknife, unilateral, dorsal funiculus (76)[a,b]; C5/6, iridectomy scissors, overhemisection (28)[a,b]; C5, Moria microknife, unilateral, medial stab, lateral stab (45)[b]; C3, microblade, unilateral, dorsolateral quadrant (16)[a,b]; C2, modified 25G needle, bilateral, dorsal funiculus, dorsal columns (90)[b]; C3, modified needle, bilateral, dorsolateral funiculus (32)[b]; C2/3, modified 25G needle, bilateral, dorsolateral funiculus and dorsal columns or dorsal funiculus (91)[b]; C4/5, iridectomy scissors, overhemisection (34)[a,b]; C3, modified 25G needle and fine forceps, bilateral, dorsolateral funiculus, dorsolateral funiculus and pyramidal tract (92)[b]; C3, microblade, unilateral, dorsolateral quadrant (17)[a,b]	C1–2, radio frequency, dorsal corticospinal tract (88)[a,b]; C2/3, $sPLA_2$, unilateral, dorsolateral funiculus (68)[b]; C3/4, microscalpel, microscissors, aspiration, unilateral, dorsolateral funiculus, lateral funiculus, dorsal horn (33)[b]; C1/2, heat, bilateral, dorsal funiculus and gray matter (89)[a,b]
Staircase test		C4, sharpened forceps, bilateral, dorsal funiculus (51); C4, sharpened forceps, bilateral, dorsal funiculus (40)[a,b]		C3/4, microscalpel, microscissors, aspiration, unilateral, dorsolateral funiculus, lateral funiculus, dorsal horn (38)[b]

(continued)

Table 1 (continued)

	Contusion	Compression	Laceration	Other
Tactile stimulation test[d]	C5, ESCID, bilateral (39)[a]	C4, forceps, bilateral, dorsal funiculus (55)[a,b]	C3, microwire knife, bilateral, dorsal funiculus (56)[a,b]; C4, Vibraknife™, bilateral, dorsal column, dorsal funiculus, dorsolateral funiculus, dorsal hemisection (53)[b]; C4, microscissors, unilateral, lateral hemisection (57)[a,b]; C4, microscissors, unilateral, lateral hemisection (58)[b]	C4, iridectomy scissors and aspiration, overhemisection (36)[a]
Thermal sensitivity, threshold, hyperalgesia[d]		C4/5, sharpened forceps, unilateral, dorsolateral funiculus (35)[a,b]; C4/5, sharpened forceps, bilateral, dorsal funiculus (87)[a,b]	C4–5, wire knife and scalpel blade, unilateral, dorsal funiculus (75)[a,b]	C3–4, microscalpel and aspiration, unilateral, subtotal hemisection (24)[a]; C4, aspiration, unilateral, lateral hemisection (22)[a,b]
Weight support forelimb hanging	C5, ESCID, bilateral (29); C5, ESCID, bilateral (41)[a,b]			

ESCID Electromagnetic SCI device; *IHI* Infinite Horizon Impactor; *NYU/MASCIS* New York University/Multicenter Animal SCI Study

[a] Pre-injury acclimation, training, evaluation

[b] Treatment examined

[c] Mice

[d] Sensory assessment

Table 2
Forelimb functional assessments after cervical SCI of locomoting adult rats since 2005

Task	Contusion	Compression	Laceration	Other
Footfalls, footslips, misplacements—grid runway	C5, ESCID, bilateral (29)[b]; C5, ESCID, bilateral (41)[a,b]	C4, forceps, bilateral, dorsal funiculus (55)[a,b]	C4/5, iridectomy scissors, overhemisection (34)[a,b]	C4, iridectomy scissors and aspiration, overhemisection (36)[a]
Footfalls, footslips, misplacements—horizontal ladder	C5, NYU/MASCIS Impactor, unilateral (30)[b]	C4, forceps, bilateral, dorsal funiculus (55)[a,b]; C4, sharpened forceps, unilateral, lateral funiculus (23)[a,b]; C4, sharpened forceps, bilateral, dorsal funiculus (40)[a]	C4–5, wire knife and scalpel blade, unilateral, dorsal funiculus (75)[a,b]; C3, microblade, unilateral, dorsolateral quadrant (16)[a,b]; C2, modified 25G needle, bilateral, dorsal funiculus, dorsal columns (90)[b]; C3, modified needle, bilateral, dorsolateral funiculus (32)[b]; C2, modified 25G needle, bilateral, dorsal funiculus (93)[b]; C2/3, modified 25G needle, bilateral, dorsolateral funiculus and dorsal columns or dorsal funiculus (91)[b]; C3, modified 25G needle and fine forceps, bilateral, dorsolateral funiculus, dorsolateral funiculus and pyramidal tract (92)[b]; C3, micro-blade, unilateral, dorsolateral quadrant (17)[a,b]; C4, microscissors, unilateral, lateral hemisection (57); C4, microscissors, unilateral, lateral hemisection (58)[b]	
Footfalls, footslips, misplacements—horizontal rope			C5, Moria microknife, unilateral, lateral hemisection (43)[b]; C3, microwire knife, bilateral, dorsal funiculus (56)[a,b]	C3–4, microscissors, aspiration, unilateral, partial lateral hemisection (26)[a,b]; C3, aspiration, lateral hemisection (31)[a,b]

(continued)

Table 2 (continued)

Task	Contusion	Compression	Laceration	Other
Footfalls, footslips, misplacements—inclined ladder			C4, microscissors, unilateral, lateral hemisection (57); C4, microscissors, unilateral, lateral hemisection (58)[b]	
Footfalls, footslips, misplacements—open-field grid	C4, IHI, unilateral (49)[a,b]; C4, IHI, unilateral (50)[a,b]; C5, IHI, unilateral (94)[a,b]; C5, IHI, unilateral (61)[a,b]; C5, IHI, unilateral (95)[a,b]		C4, Vibraknife™, bilateral, dorsal hemisection (53)[b]	C3–4, microscalpel and aspiration, unilateral, subtotal hemisection (24)[a]; C5, aspiration, unilateral (94)[a,b]
Footprints—beam, runway, walkway	C5, ESCID, bilateral (29); C5, ESCID, bilateral (39)[a]; C5, ESCID, bilateral (41)[a,b]	C4, forceps, bilateral, dorsal funiculus (55)[a,b]; C4, sharpened forceps, bilateral, dorsal funiculus (51)[a]; C4, sharpened forceps, unilateral, lateral funiculus (23)[a,b]	C4–5, wire knife and scalpel blade, unilateral, dorsal funiculus (75)[a,b]	C2/3, $sPLA_2$, unilateral, dorsolateral funiculus (68)[b]
Forelimb locomotor assessment scale—alley	C5–8, IHI, bilateral (96)[b]			
Forelimb locomotor scale—open field	C4, IHI, unilateral (49)[a,b]; C4, IHI, unilateral (50)[a,b]; C5, IHI, unilateral (94)[a,b]; C5, IHI, unilateral (61)[a,b]; C5, IHI, unilateral (95)[a,b]			C4, aspiration, unilateral, lateral hemisection (22)[a,b]; C5, aspiration, unilateral (94)[a,b]
Forelimb movement measures—walkway	C5, IHI, bilateral (97)[a,b]			
Forelimb swing—open field				C3, aspiration, lateral hemisection (31)[a,b]

Gait	C7, weight drop, bilateral (98)[a]; C7, weight drop, bilateral (99)			
Gait—Catwalk	C5, NYU/MASCIS Impactor, unilateral (30)[b]		C4/5, iridectomy scissors, overhemisection (34)[a,b]	C4, aspiration, unilateral, lateral hemisection (22)[a,b]
Gait—SIMI motion analysis system			C4, microscissors, lateral hemisection (100)[a,b]	
Ground reaction forces and gait—runway			C2, modified 25G needle, bilateral, dorsal funiculus, dorsal columns (90)[b]; C3, modified 25G needle, bilateral, dorsolateral funiculus (32)[b]; C2, modified 25G needle, bilateral, dorsal funiculus (93)[b]; C2/3, modified 25G needle, bilateral, dorsolateral funiculus and dorsal columns or dorsal funiculus (91)[b]; C3, modified 25G needle and fine forceps, bilateral, dorsolateral funiculus, dorsolateral funiculus, and pyramidal tract (92)[b]	
Locomotor rating scale—open field			C4, microscissors, unilateral, lateral hemisection (101)[b]; C4, microscissors, unilateral, lateral hemisection (57)[b]; C4, microscissors, unilateral, lateral hemisection (58)[b]	
Wheel walking	C4, IHI, unilateral (50)[a,b]			

ESCID Electromagnetic SCI device; *IHI* Infinite Horizon Impactor; *NYU/MASCIS* New York University/Multicenter Animal SCI Study
[a]Pre-injury acclimation, training, evaluation
[b]Treatment examined

They also have been used for studying adult rats and mice with experimental brachial plexus and forelimb peripheral nerves injuries (for reviews, see refs. 12–15).

2. Forelimb Functional Assessment Tests Materials and Methods

Choosing of which developed forelimb functional assessment test(s) to use should take into consideration whether it is appropriate for the hypothesis being tested, valid, reliable, sensitive, convenient, and can provide mechanistic information about functional recovery, or the lack thereof, as indicated by the literature about it. Apparatuses and other equipment are commercially available; however, some of these may need to be made. To successfully assess behaviors of laboratory rats and mice, they should be housed in the facility after arrival from the vendor for at least 1 week before being used. This allows the investigator time to do a physical examination and to reduce stress by familiarizing the animals to being handled around their trunks or nape of the neck rather than being picked up by their tails, especially after surgery. Exposure to the testing situation as well as establishing limb preference and training, if necessary, prior to collection of baseline data and/or post-surgery data is essential. For testing, animals should be transported to a quiet testing room at the same time each day and allowed to acclimate for at least 15 min before blinded individuals begin the procedures. Lastly, daily testing post surgery and when it begins needs to be carefully considered as this has been seen to improve test-specific forelimb use recovery compared to less frequent testing and impair recovery of another behavior (16, 17).

2.1. Limb-Use Asymmetry or Cylinder Test

Adult rats and mice naturally support themselves while rearing to explore and when landing using their forelimbs and palmar surfaces of their forepaws with extended digits. This test, which has a number of names, is based on their spontaneous exploratory behavior while in cylinders or home cage. While in a cylinder, rats and mice predominantly use both forelimbs simultaneously for support during rearing and landing (18–20). Each forelimb is used less and equivalently. Various types and levels of cervical SCIs in adult rats, including both unilateral and bilateral third–sixth cervical segments (C3–6) contusion, dorsolateral and lateral funiculotomies, lateral hemisection, and overhemisection, have been seen to acutely and chronically disrupt this pattern (for review, see ref. 11) (Table 1). Moreover, the disrupted pattern was found to significantly recover following various treatments (for review, see ref. 11), including blocking SUR1-regulated $NC_{Ca\text{-}ATP}$ channels (21), Nogo receptor antagonism (22), dietary restriction (23), as well as engraftments of adult human bone marrow stromal cells (24), alginate encapsulated brain-derived neurotrophic factor (BDNF)-producing adult rat skin fibroblasts (25), adult human olfactory neuroepithelial neurosphere-forming cells (26, 27), and embryonic rat spinal cord combined with neurotrophin-3 (NT-3) (28). To date, there are no reports of this test being used to assess mice with cervical SCI. However, untrained adult mice when placed in a cylinder (8 cm diameter × 30 cm height) had significant decreases of ipsilesional and both forepaws' use for support as well as a significant increase of contralesional forepaw use acutely and chronically after unilateral pyramidotomy (20).

The general procedure for this test is as follows.

A. First weeks

1. Naive animals should be handled in advance of being assessed. This can be done by gently removing each animal from its home cage with a hand around its trunk in random order to avoid order bias. Place animal on a flat surface.
2. Gently grasp animal around its trunk and return to its home cage.
3. Acclimating the animals to the testing apparatus appears to be optional. This can be done by placing each animal into the testing apparatus and leaving there until at least 20 exploratory movements along the wall and at least 10 landing movements have occurred (19). Frequently placing animals in the testing apparatus does not influence test outcome (18).

B. Baseline and post-surgery data

1. One trial at baseline (22, 23, 26, 27, 29–33) as well as at least once weekly (22, 23, 25, 27, 30, 31, 33, 35) or one time (24, 26, 28, 29, 36) post surgery have been reported for this test. Multiple trials during a test session, but with no intertrial interval, also have been reported (36).
2. For a trial, place a mirror in the same position behind a transparent, open-topped, plexiglass or glass cylinder (diameter × height: 8 × 30 cm (20); 17.5 × 26.5 cm (28, 34, 36); 17.8×35.5 cm (22, 24, 31, 33); 19×20 cm (21); 20 × 25 cm (29); 20 × 30 cm (25–27, 35); 20 × 46 cm (30); 40 × 46 cm (32)). The mirror allows forelimb movements to be seen when the animal's back is to the observer (19). The testing chamber should be high enough so that the animal cannot reach its top edge during rearing and wide enough that the animal has at least a 2-cm space between the wall and both its snout and base of its tail when not rearing (19). The home cage could be used, especially at the beginning of the dark cycle for less active animals (19). It is not known whether static or randomly placed environmental cues are necessary.

3. In random order to avoid order bias, gently remove animal from its home cage by a hand around its trunk. Place it in testing chamber for at least 20 exploratory movements along the wall and at least 10 landing movements (19, 23, 29), at least 10 contacts (28), 3 min (21, 22, 24, 25, 28, 33, 34, 36), 4 min (31), or 5 min (20, 26, 27, 30, 32, 35). Videotape animal, for example, with a Panasonic SVHS camera at a shutter speed of 1/1,000 s (28) so that the mirror can be seen or from a ventral perspective through the testing apparatus floor (32).
4. Gently grasp animal around its trunk and return to its home cage.
5. Clean testing apparatus with paper towel and then begin procedure with another randomly selected animal.
6. Frame by frame of the video recording, for example, with a Mitsubishi SVHS VCR (28), count the number of times each forelimb was used individually, that is, when 5 s or more passed without the other forelimb making the initial and total contact (23, 30), and the number of times both forelimbs were used simultaneously to contact the wall during vertical exploration and floor during landing (19, 20, 32, 34). Counts are not done while one forepaw is anchored and the other is making contact with different parts of the testing apparatus (19, 30). Counts for both forelimbs are given during lateral exploration when both paws change position (19, 30). Each of the independent and both forelimb counts can be expressed as a percentage of total limb movements (19, 20, 24–27, 29, 33). Should the cervical SCI be unilateral, results from the ipsilesional or contralateral forelimbs during individual and both contacts can be added together (21–25, 28, 33, 35, 36). A forelimb-use asymmetry score can be obtained by subtracting the percentage of contacts by the ipsilesional forelimb from those by the contralesional forelimb (19). The order of the forelimb contacts, forepaw surfaces used, and whether the digits were extended during the initial and total contact also could be recorded. Since hindlimb dysfunction could affect the performance and interpretation of the results (31), the number of rears could be recorded (21, 24, 28, 34).

2.2. Grip Strength Test

This test is based on adult rats and mice naturally grasping an object near them with their forepaws and then releasing it when being pulled or lifted by their tails (37). Using a lab-made apparatus consisting of a triangular bar connected to a dynamometer (37), we first reported that bilateral C7 contusion injury significantly reduced the rat's forepaw grip strength recorded when both forepaws released their grip on the bar as they were gently being pulled away from it by their tail (38). Other types and levels of cervical SCIs in rats and mice have since been seen to acutely and chronically reduce forepaw grip strength, including both unilateral and bilateral C4–T1 contusion, dorsal funiculotomy, medial and lateral stabs, and lateral hemisection (for review, see ref. 11) (Table 1). Significant forepaw grip strength recovery has been found following various treatments, including anti-ASC neutralizing antibodies (39), chondroitinase ABC combined with enriched housing rehabilitation (40), and hypothermia (41).

The general procedure for this test is as follows.

A. First weeks

1. In random order to avoid order bias, gently remove animal from its home cage by a hand around its trunk or nape of the neck. Handle so that it gets used to having its forelimbs manipulated (42–44).
2. Place on a lab-made (37, 38) or commercially available (designed by TSE-Systems, distributed by SciPro, Inc. (42–48); San Diego Instruments (29, 39, 41, 49, 50); Linton Instruments (40, 51)) Grip Strength Meter daily for 5 min (42).
3. Gently grasp animal and return to its home cage. Clean test apparatus with paper towel and then begin the procedure with another randomly selected animal.
4. Once the animals appear calm when handled and on the apparatus, begin daily training for 3–5 weeks until reproducible grip strengths are obtained (38, 42, 43). With hind limbs elevated, allow the animal to grasp bar or mesh of the apparatuses or bars of the lateralized apparatus (40, 51) with all four digits of both limbs. Gently pull backwards in a straight line by the trunk or tail until the grip is released. Should assessment of one forelimb be required, alternate taping (for example with Micropore™ surgical tape (44)) or restrain one limb and allow animal to grasp bar or mesh with free limb (42, 43).

B. Baseline and post-surgery data

1. Baseline data for this test have been acquired during 4–6 consecutive daily trials, 5 days per week, for 5 weeks (38); 4 trials per forepaw each day, 3–5 days per week, for 2 weeks (42, 43, 47); 15–25 trials per forepaw per day separated by 2 s for 8 days (44, 48); or 3 or 6 successive trials on 1 day (29, 50, 51).
2. Post-surgery data have been acquired during 4–6 consecutive daily trials, 5 days per week (38); 4 trials per forepaw each day, 3–4 days per week (42, 43, 47); 15–25 trials per forepaw

per day (44, 48); 6 trials on 1 day per week (50); or 3 successive trials on 1 or 2 days (29, 39–41, 51).

3. We observed that bilateral contusion injury at the adult rat C7 spinal cord resulted in an impairment of spontaneous forepaw digits' extension (38). However, the digits were partially extended when performing this test and during other goal-directed behaviors. To avoid confusing a grasp with forepaw digit contracture and/or spasticity, it is critical for grip strength to be measured during baseline and post-surgery testing, The rats and mice need to be seen grasping the bar or mesh by extending and flexing their digits as well as extending their digits during release (42–44).
4. With hind limbs elevated, allow the animal to grasp bar(s) or mesh of the apparatuses. Gently pull backwards in a straight line by the trunk or tail until the grip is released. Record force.

2.3. Tactile Stimulation Test

This test is based on adult rats and mice quickly sensing and attending to (sensory) and then removing (motor) with their forepaws or teeth adhesive items placed on their forelimbs and forepaws, respectively (for review, see refs. 11, 20, 52) (Table 1). From an investigation of sensory dysfunction after cervical SCI, we reported that the time to attend to adhesive stickers on rat forepaws significantly increased 1 week after bilateral lesion of their C4 dorsal columns and dorsal funiculi, but not dorsolateral funiculi (53). Significantly increased sense times for up to 6 weeks after bilateral C4 dorsal funiculi lesions also were seen (54, 55). Persistent, significantly increased removal times of adhesive stickers with their teeth have been observed after bilateral C3 and C4 dorsal funiculi lesions (54–56) and C4 lateral hemisection, which also damages gray matter (19, 57, 58). Significantly recovered sense and removal times have been reported following retinoic acid receptor β2 delivery to corticospinal neurons' (55) and locomotor training on various surfaces (57). To date, there are no reports of this test being used to assess mice with cervical SCI. However, trained adult mice had significant increases of ipsilesional forepaw sense and removal times acutely and chronically after unilateral pyramidotomy (20).

The general procedure for this test is as follows.

A. Week 1

1. In random order to avoid order bias, gently remove animal from its home cage by a hand around its trunk or by the nape of the neck and place on a flat surface.
2. Gently grasp animal with a towel around its trunk or by a hand at the nape of the neck and lift it vertically.
3. Another individual should handle and wipe with a gauze pad the forelimb or forepaw similarly to as when the adhesive sticker(s) is applied.

4. Gently place animal onto the bottom of a clean test cage, without bedding, and leave there for the maximum duration of the trial, for example 1–5 min (20, 53, 56–58).
5. Gently grasp animal and return to its home cage.
6. Clean test cage with paper towel and then begin the procedure with another randomly selected animal.
7. Repeat the procedure two to five more times with each animal when at least 5–10 min have elapsed.

B. Week 2 and beyond for baseline and post-surgery data

Of note, we saw that C4 dorsal hemisection significantly increased attention times for 2 weeks (53). They were not different than those of sham rats at 3 and 4 weeks. Interestingly, attention times of trained rats assessed only at the fourth week after dorsal hemisection were significantly increased. This contrasts with persistent deficits in sense times seen after bilateral C4 dorsal funiculi lesions (54, 55). These results suggest that a plasticity or training effect may have occurred with repeated assessments after dorsal hemisection.

1. Once weekly (20, 53–58) in random order to avoid order bias, gently remove an animal from its home cage by a hand around its trunk and place on a flat surface.
2. Gently grasp the animal with a towel around its trunk and lift it vertically.
3. Another individual gently wipes the forelimb or forepaw and applies an adhesive sticker to the radial or ulnar aspects of the wrist (19, 52) or palm (20, 53–55, 57, 58). Adhesive stickers that have been used include those from AVERY® (13-mm diameter dot, 19 × 14-mm rectangle (52); 11.3- and 14.1-mm square (19); 6-mm diameter dot (53)) or tape (0.3 × 1-in. rectangle (54); 0.1 × 1-in. rectangle (56); 10 × 5-mm rectangle (20); 5-mm square (57, 58)). For the outcome measures, good adhesion of the sticker to the surface that it is applied to is critical to reduce variability of the data. Placement of more than one sticker should be done in a random order.
4. Gently place animal onto the bottom of a clean test cage, without bedding, as quickly as possible after applying stickers.
5. Using two clocks, two individuals should record sense, attention, or contact times and removal time of each sticker. Sense, attention, or contact times begin at placement into the cage and end when the sticker is sensed as indicated by paw shake (54), when the forepaw is rotated dorsally exposing the sticker to the mouth (20, 53) or contacted with the mouth (19, 52).

6. Once the desired time has been achieved or the maximum duration of the trial has elapsed, gently grasp animal with a towel around its trunk. Gently remove remaining stickers and return animal to its home cage.
7. Clean test cage with paper towel and then begin the procedure with another randomly selected animal.
8. Repeat the procedure two to five more times (53, 56–58) with each animal when at least 5–10 min have elapsed.
9. Depending upon whether the cervical SCI was unilateral or bilateral, averaged times for each or both limbs can be compared (52–56). Additionally, differences in the asymmetry index (57, 58) and order or magnitude of somatosensory asymmetry using varying sizes of stickers (19) can be evaluated.

   Adult rats also quickly respond to adhesive items placed on their head (36, 39, 52, 59). Similar to the Grooming Test (60) used for rats with unilateral contusive (30, 61) and lateral hemisection (58) cervical SCI, scale scores have been applied to the grooming movements of the forelimbs.

2.4. Staircase Test

Similar to humans, adult rats can perform skilled reaching, grasping, and retrieval movements for objects, especially food (62). The first reports of the effects of cervical SCI on adult rat performance of these movements were by Schrimsher and Reier (59, 63). They used a lab-made testing apparatus that required food-restricted and trained rats to reach horizontally for a food pellet while supporting themselves with the other limbs. They found that bilateral contusion at C4–5 markedly impaired reach for 4 months so that there was no pellet retrieval (59). They then saw that while pellets were retrieved, bilateral dorsolateral and ventrolateral funiculotomies significantly hindered digit flexion during grasp up to 1 month and reach for 1 week, respectively (63). Bilateral dorsal funiculotomies did not significantly affect reach, grasp, or pellet retrieval; however, targeting and digit dysfunctions were observed. We next used the Staircase Test apparatus (64) to evaluate the effects of bilateral contusion at C7 on adult rat performance of these movements (38). We found that bilateral contusion injury at C7 significantly impaired reach and pellet retrieval for 1 month by rats that were trained and food restricted during training only.

An advantage of the Staircase Test apparatus is that tetraplegic animals can lay on a platform rather than having to support themselves with their dysfunctional limbs. Moreover, they reach vertically rather than horizontally for pellets that are on both sides of the platform, thus avoiding the effects of shoulder dysfunction and limb preference. The apparatus is designed so that only the forelimb ipsilateral to the pellets can be used. Therefore, ipsilesional effects of unilateral cervical SCI can be examined. The Staircase Test apparatus has subsequently been used to show a significant decrease in pellet retrieval with the ipsilesional forelimb by trained and food-restricted rats for 2 months after unilateral C3–4 dorsolateral funiculotomy (33). Moreover, chronic pellet retrieval deficits of trained rats caused by bilateral C4 dorsal funiculotomy were found to be significantly attenuated following chondroitinase ABC treatment alone (51) and combined with food reach, grasp, and retrieval rehabilitation (40). Interestingly, they were worse following chondroitinase ABC treatment and enriched housing (40). To date, there are no reports of this test being used to assess mice with cervical SCI. However, trained and food-restricted adult mice had a significant decrease in pellet retrieval with the contralesional forelimb for 3 weeks after unilateral cortical injury during the Staircase Test in a modified testing apparatus (65). No reach or pellet retrieval deficit of the mouse ipsilesional forelimb was observed for 1 month following unilateral pyramidotomy (20).

The general procedure for this test is as follows.

A. First weeks

1. Various forms of food limitation have been used as part of training and testing for this test. For adult rats, these include daily food restriction of 12–15 g (14, 33) or to 80–85% (38) and 90% (33) of the ad libitum body weight as determined from sham or naive littermates included in the experiment (38). For adult mice, fasting for 20 h with ad libitum feeding on nontraining/test days has been used (65). At the same time of food limitation, pellets to be used for testing (rats: 45 and 190 mg purified rodent, P. J. Noyes Company Inc. (38); 45 mg Noyes dust-free sucrose reward, Research Diet, Inc. (14); 45 mg chocolate-flavored dustless precision purified, Bio-Serv, Inc. (33); mice: 20 mg sucrose, banana, chocolate, grain, Bio-Serv, Inc., P. J. Noyes Company Inc. (65)) should be placed onto the bedding of the home cage to familiarize the animal with them. Individual housing is recommended to avoid competition for food. It is important that free feeding should occur after daily training/testing. Lastly, food limitation's effects on performance of other tests that are being utilized at the same time should be taken into consideration (66).
2. At the same time that food limitation begins, gently remove animal from its home cage by a hand around its trunk or by the nape of its neck in random order to avoid order bias. Place it into a lab-made (33, 38, 64) or commercially available (Lafayette Instrument (65); Campden Instruments (14)) Staircase Test apparatus daily for 15 min (38).

3. Gently grasp animal and return to its home cage. Clean test apparatus with paper towel and then begin the procedure with another randomly selected animal.
4. Begin training once the animals appear calm when handled and in the apparatus as well as have reached the appropriate body weight and are eating pellets left in their cage. Training for rats has been with two 45-mg or one 190-mg pellets per well of each stair for 13 weeks, 5 days per week, 1 trial per day, and for 15 and then 5 min per trial when pellet retrieval occurred (38) and with two or three 45-mg pellets per well for 2 weeks, 15 min per trial (14, 33). Training for mice has been with two 20-mg pellets per well for 15 days, 15 min per trial (65) and with two 20-mg pellets per well for 3 days and then one pellet per well of the lower stairs for 2 weeks (20). Criteria for training are essential because the post-surgery results of those animals not meeting them are affected. They include being able to retrieve or clear pellets from all wells and the floor of the apparatus (38), at least eight pellets retrieved and at least five steps reached (14), and all of the pellets on one staircase retrieved (33). Importantly, strain of rat studied and well depth also have been seen to play a role in pellet retrieval training and performance (14, 67).

B. Baseline and post-surgery data

1. Baseline data for this test have been acquired during 5 days per week, one trial per day, and 5 min per trial (38), 15 or 30 min (14, 20, 40, 51), and 1 week, two trials per day, and 15 min per trial (33). Once trained, Baird et al. (65) evaluated test session length, food limitation, and reward diet for adult male and female mice. They found over a period of 2 weeks that 30- and 60-min test sessions with banana, grain, and chocolate-flavored pellets when preceded by a 20-h fast resulted in the most reaches and pellets retrieved by all mice.
2. Post-surgery data have been acquired during 4 weeks, 5 days per week, one trial per day, and 5 min per trial (38) or twice, one or two trials per day, and 15 min per trial (33, 40, 51). Out of concern for drastic post-cervical SCI decreases of body weight, we performed testing on a free-feeding schedule (38). Importantly, this did not alter the performance of sham rats compared to their baseline data. Free-feeding schedules immediately prior to and after surgery also have been used.
3. It is essential that animals are watched during post-surgery testing since they may attempt to retrieve pellets with their tongues.

4. A number of methods have been developed to quantify reach, grasp, and retrieval movements. The number of pellets remaining in the wells and on the floor of the apparatus has been counted. The total number of pellets retrieved and eaten was obtained from this and the number of pellets in the wells at the start of the test (14, 33, 38, 40, 51, 65). We also developed an index of reaching ability called the well clearance score (38). For each well, the number of pellets cleared from each well, that is, knocked down onto a lower step or floor, grasped and dropped, or grasped and eaten, was multiplied by the distance between the top of the platform and the bottom of the well. Another measure of reach has been the furthest well from which pellets were displaced (14, 40, 51, 65). Time spent attempting to retrieve pellets also has been determined (40).
5. Videotaping of the Staircase Test has not been reported. However, analyses of the video recordings could reveal information about the kinematics of the reach, grasp, and retrieval movements. This has been done with other skilled reach, grasp, and retrieval tests.

We have modified the Staircase Test apparatus so that adult rats reached horizontally for single pellets in tubes on both sides of the platform that they lay upon (68). We also fasted the rats overnight on alternate days during training and utilized ad libitum feeding on nontraining days. Using these modifications and 45 mg chocolate-flavored dustless precision purified pellets (Bio-Serv, Inc.), we found that a pellet retrieval deficit of the ipsilesional forelimb was correlated with the extent of C2–3 dorsolateral funiculus demyelination and remyelination (68). Other apparatuses and tests have been used to assess adult rat skilled reaching, grasping, and retrieval after cervical SCI (for review, see ref. 11) (Table 1). Since 2005, they have included Directed Forepaw (69), Single and Multiple Pellets (70–76), Pasta (77–79), Noodles (80), and Skilled Target (81) Reaching Tests with Components and Arpeggio Rating Scales (71–73, 75, 81, 82). All of these required that the cervical spinal cord-injured rats support themselves with their dysfunctional limbs during performance. As with the Staircase Test, the strain of rat used for these tests may have an effect on the results (83, 84). Significant recovery of reaching, grasping, and retrieval movements while performing these tests has been observed following various treatments, including administration of high dose of the ROCK inhibitor Y27632 (75), engraftment of embryonic rat spinal cord combined with BDNF or NT-3 (28), task-specific rehabilitation (16, 17), enriched housing and task-specific rehabilitation (34), as well as food reach, grasp, and retrieval rehabilitation alone and combined with chondroitinase ABC administration plus chondroitinase ABC alone (40). Interestingly, those movements were worse following BDNF administration to the red nucleus alone and combined with engraftment of olfactory ensheathing cells (35) as well as chondroitinase ABC treatment and enriched housing (40).

References

1. https://www.nscisc.uab.edu/public_content/facts_figures_2009.aspx
2. Special Issue (2009) Spinal cord injury—neuroplasticity and recovery of respiratory function. In: Sieck GC, Mantilla CB (eds) Respiratory Physiol Neurobiol, 169:83–226
3. Anderson KD (2004) Targeting recovery: priorities of the spinal cord-injured population. J Neurotrauma 21:1371–1383
4. Tracey D (2004) Ascending and descending pathways in the spinal cord. In: Paxinos G (ed) The Rat Nervous System, 3rd edn. Elsevier Academic Press, San Diego, pp 149–164
5. Kaas JH, Qi HX, Burish MJ, Gharbawie OA, Onifer SM, Massey JM (2008) Cortical and subcortical plasticity in the brains of humans, primates, and rats after damage to sensory afferents in the dorsal columns of the spinal cord. Exp Neurol 209:407–416
6. McKennna JE, Prusky GT, Whishaw IQ (2000) Cervical motoneuron topography reflects the proximodistal organization of muscles and movements of the rat forelimb: a retrograde carbocyanine dye analysis. J Comp Neurol 419:286–296
7. Bunge RP, Puckett WR, Becerra JL, Marcillo A, Quencer RM (1993) Observations on the pathology of human spinal cord injury. A review and classification of 22 new cases with details from a case of chronic cord compression with extensive focal demyelination. Adv Neurol 59:75–89
8. Bunge RP, Puckett WR, Hiester ED (1997) Observations on the pathology of several types of human spinal cord injury, with emphasis on the astrocyte response to penetrating injuries. Adv Neurol 72:305–315
9. Norenberg MD, Smith J, Marcillo A (2004) The pathology of human spinal cord injury: defining the problems. J Neurotrauma 21:429–440
10. http://www.ncbi.nlm.nih.gov/pubmed, 2009
11. Webb AA, Muir GD (2005) Sensorimotor behaviour following incomplete cervical spinal cord injury in the rat. Behav Brain Res 165:147–159
12. Nichols CM, Myckatyn TM, Rickman SR, Fox IK, Hadlock T, Mackinnon SE (2005) Choosing the correct functional assay: a comprehensive assessment of functional tests in the rat. Behav Brain Res 163:143–158
13. Tos P, Ronchi G, Nicolino S, Audisio C, Raimondo S, Fornaro M, Battiston B, Graziani A, Perroteau I, Geuna S (2008) Employment of the mouse median nerve model for the experimental assessment of peripheral nerve regeneration. J Neurosci Methods 169:119–127
14. Galtrey CM, Fawcett JW (2009) Characterization of tests of functional recovery after median and ulnar nerve injury and repair in the rat forelimb. J Peripher Nerv Syst 12: 11–27
15. Ibrahim AG, Kirkwood PA, Raisman G, Li Y (2009) Restoration of hand function in a rat model of repair of brachial plexus injury. Brain 132:1268–1276
16. Girgis J, Merrett D, Kirkland S, Metz GA, Verge V, Fouad K (2007) Reaching training in rats with spinal cord injury promotes plasticity and task specific recovery. Brain 130:2993–3003
17. Krajacic A, Ghosh M, Puentes R, Pearse DD, Fouad K (2009) Advantages of delaying the onset of rehabilitative reaching training in rats with incomplete spinal cord injury. Eur J Neurosci 29:641–651
18. Liu Y, Kim D, Himes BT, Chow SY, Schallert T, Murray M, Tessler A, Fischer I (1999) Transplants of fibroblasts genetically modified to express BDNF promote regeneration of adult rat rubrospinal axons and recovery of forelimb function. J Neurosci 19:4370–4387
19. Schallert T, Fleming SM, Leasure JL, Tillerson JL, Bland ST (2000) CNS plasticity and assessment of forelimb sensorimotor outcome in unilateral rat models of stroke, cortical ablation, parkinsonism and spinal cord injury. Neuropharmacology 39:777–787
20. Starkey ML, Barritt AW, Yip PK, Davies M, Hamers FP, McMahon SB, Bradbury EJ (2005) Assessing behavioural function following a pyramidotomy lesion of the corticospinal tract in adult mice. Exp Neurol 195:524–539
21. Simard JM, Tsymbalyuk O, Ivanov A, Ivanova S, Bhatta S, Geng Z, Woo SK, Gerzanich V (2007) Endothelial sulfonylurea receptor 1-regulated NC Ca-ATP channels mediate

progressive hemorrhagic necrosis following spinal cord injury. J Clin Invest 117:2105–2113
22. Cao Y, Shumsky JS, Sabol MA, Kushner RA, Strittmatter S, Hamers FP, Lee DH, Rabacchi SA, Murray M (2008) Nogo-66 receptor antagonist peptide (NEP1-40) administration promotes functional recovery and axonal growth after lateral funiculus injury in the adult rat. Neurorehabil Neural Repair 22:262–278
23. Plunet WT, Streijger F, Lam CK, Lee JH, Liu J, Tetzlaff W (2008) Dietary restriction started after spinal cord injury improves functional recovery. Exp Neurol 213:28–35
24. Neuhuber B, Timothy Himes B, Shumsky JS, Gallo G, Fischer I (2005) Axon growth and recovery of function supported by human bone marrow stromal cells in the injured spinal cord exhibit donor variations. Brain Res 1035:73–85
25. Tobias CA, Han SS, Shumsky JS, Kim D, Tumolo M, Dhoot NO, Wheatley MA, Fischer I, Tessler A, Murray M (2005) Alginate encapsulated BDNF-producing fibroblast grafts permit recovery of function after spinal cord injury in the absence of immune suppression. J Neurotrauma 22:138–156
26. Xiao M, Klueber KM, Lu C, Guo Z, Marshall CT, Wang H, Roisen FJ (2005) Human adult olfactory neural progenitors rescue axotomized rodent rubrospinal neurons and promote functional recovery. Exp Neurol 194:12–30
27. Xiao M, Klueber KM, Guo Z, Lu C, Wang H, Roisen FJ (2007) Human adult olfactory neural progenitors promote axotomized rubrospinal tract axonal reinnervation and locomotor recovery. Neurobiol Dis 26:363–374
28. Lynskey JV, Sandhu FA, Dai HN, McAtee M, Slotkin JR, MacArthur L, Bregman BS (2006) Delayed intervention with transplants and neurotrophic factors supports recovery of forelimb function after cervical spinal cord injury in adult rats. J Neurotrauma 23: 617–634
29. Pearse DD, Lo TP Jr, Cho KS, Lynch MP, Garg MS, Marcillo AE, Sanchez AR, Cruz Y, Dietrich WD (2005) Histopathological and behavioral characterization of a novel cervical spinal cord displacement contusion injury in the rat. J Neurotrauma 22:680–702
30. Gensel JC, Tovar CA, Hamers FP, Deibert RJ, Beattie MS, Bresnahan JC (2006) Behavioral and histological characterization of unilateral cervical spinal cord contusion injury in rats. J Neurotrauma 23:36–54
31. Houle JD, Tom VJ, Mayes D, Wagoner G, Phillips N, Silver J (2006) Combining an autologous peripheral nervous system "bridge" and matrix modification by chondroitinase allows robust, functional regeneration beyond a hemisection lesion of the adult rat spinal cord. J Neurosci 26:7405–7415
32. Muir GD, Webb AA, Kanagal S, Taylor L (2007) Dorsolateral cervical spinal injury differentially affects forelimb and hindlimb action in rats. Eur J Neurosci 25:1501–1510
33. Stackhouse SK, Murray M, Shumsky JS (2008) Effect of cervical dorsolateral funiculotomy on reach-to-grasp function in the rat. J Neurotrauma 25:1039–1047
34. Dai H, MacArthur L, McAtee M, Hockenbury N, Tidwell JL, McHugh B, Mansfield K, Finn T, Hamers FP, Bregman BS (2009) Activity-based therapies to promote forelimb use after a cervical spinal cord injury. J Neurotrauma 26:1719–1732
35. Bretzner F, Liu J, Currie E, Roskams AJ, Tetzlaff W (2008) Undesired effects of a combinatorial treatment for spinal cord injury—transplantation of olfactory ensheathing cells and BDNF infusion to the red nucleus. Eur J Neurosci 28:1795–1807
36. Kim BG, Dai HN, Lynskey JV, McAtee M, Bregman BS (2006) Degradation of chondroitin sulfate proteoglycans potentiates transplant-mediated axonal remodeling and functional recovery after spinal cord injury in adult rats. J Comp Neurol 497:182–198
37. Meyer OA, Tilson HA, Byrd WC, Riley MT (1979) A method for the routine assessment of fore- and hindlimb grip strength of rats and mice. Neurobehav Toxicol 1:233–236
38. Onifer SM, Rodriquez JF, Santiago DI, Benitez JC, Kim DT, Brunschwig J-PR, Pacheco JT, Perrone JV, Llorente O, Hesse DH, Martinez-Arizala A (1997) Cervical spinal cord injury in the adult rat: assessment of forelimb dysfunction. Restor Neurol Neurosci 11:211–223
39. de Rivero Vaccari JP, Lotocki G, Marcillo AE, Dietrich WD, Keane RW (2008) A molecular platform in neurons regulates inflammation after spinal cord injury. J Neurosci 28: 3404–3414
40. García-Alías G, Barkhuysen S, Buckle M, Fawcett JW (2009) Chondroitinase ABC treatment opens a window of opportunity for task-specific rehabilitation. Nat Neurosci 12:1145–1151
41. Lo TP Jr, Cho KS, Garg MS, Lynch MP, Marcillo AE, Koivisto DL, Stagg M, Abril RM, Patel S, Dietrich WD, Pearse DD (2009) Systemic hypothermia improves histological and functional outcome after cervical spinal

cord contusion in rats. J Comp Neurol 514:433–448
42. Anderson KD, Abdul M, Steward O (2004) Quantitative assessment of deficits and recovery of forelimb motor function after cervical spinal cord injury in mice. Exp Neurol 190:184–191
43. Anderson KD, Gunawan A, Steward O (2005) Quantitative assessment of forelimb motor function after cervical spinal cord injury in rats: relationship to the corticospinal tract. Exp Neurol 194:161–174
44. Aguilar RM, Steward O (2010) A bilateral cervical contusion injury model in mice: assessment of gripping strength as a measure of forelimb motor function. Exp Neurol 221:38–53
45. Anderson KD, Gunawan A, Steward O (2007) Spinal pathways involved in the control of forelimb motor function in rats. Exp Neurol 206:318–331
46. Anderson KD, Sharp KG, Steward O (2009) Bilateral cervical contusion spinal cord injury in rats. Exp Neurol 220:9–22
47. Blanco JE, Anderson KD, Steward O (2007) Recovery of forepaw gripping ability and reorganization of cortical motor control following cervical spinal cord injuries in mice. Exp Neurol 203:333–348
48. Strong MK, Blanco JE, Anderson KD, Lewandowski G, Steward O (2009) An investigation of the cortical control of forepaw gripping after cervical hemisection injuries in rats. Exp Neurol 217:96–107
49. Sandrow HR, Shumsky JS, Amin A, Houle JD (2008) Aspiration of a cervical spinal contusion injury in preparation for delayed peripheral nerve grafting does not impair forelimb behavior or axon regeneration. Exp Neurol 210:489–500
50. Sandrow-Feinberg HR, Izzi J, Shumsky JS, Zhukareva V, Houle JD (2009) Forced exercise as a rehabilitation strategy after unilateral cervical spinal cord contusion injury. J Neurotrauma 26:721–731
51. García-Alías G, Lin R, Akrimi SF, Story D, Bradbury EJ, Fawcett JW (2008) Therapeutic time window for the application of chondroitinase ABC after spinal cord injury. Exp Neurol 210:331–338
52. Schallert T, Upchurch M, Lobaugh N, Farrar SB, Spirduso WW, Gilliam P, Vaughn D, Wilcox RE (1982) Tactile extinction: distinguishing between sensorimotor and motor asymmetries in rats with unilateral nigrostriatal damage. Pharmacol Biochem Behav 16:455–462
53. Onifer SM, Zhang YP, Burke DA, Brooks DL, Decker JA, McClure NJ, Floyd AR, Hall J, Proffitt BL, Shields CB, Magnuson DS (2005) Adult rat forelimb dysfunction after dorsal cervical spinal cord injury. Exp Neurol 192:25–38
54. Bradbury EJ, Moon LD, Popat RJ, King VR, Bennett GS, Patel PN, Fawcett JW, McMahon SB (2002) Chondroitinase ABC promotes functional recovery after spinal cord injury. Nature 416:636–640
55. Yip PK, Wong LF, Pattinson D, Battaglia A, Grist J, Bradbury EJ, Maden M, McMahon SB, Mazarakis ND (2006) Lentiviral vector expressing retinoic acid receptor beta2 promotes recovery of function after corticospinal tract injury in the adult rat spinal cord. Hum Mol Genet 15:3107–3118
56. Lu P, Jones LL, Tuszynski MH (2005) BDNF-expressing marrow stromal cells support extensive axonal growth at sites of spinal cord injury. Exp Neurol 191:344–360
57. Martinez M, Brezun JM, Zennou-Azogui Y, Baril N, Xerri C (2009) Sensorimotor training promotes functional recovery and somatosensory cortical map reactivation following cervical spinal cord injury. Eur J Neurosci 30:2356–2367
58. Martinez M, Delcour M, Russier M, Zennou-Azogui Y, Xerri C, Coq JO, Brezun JM (2010) Differential tactile and motor recovery and cortical map alteration after C4-C5 spinal hemisection. Exp Neurol 221:186–197
59. Schrimsher GW, Reier PJ (1992) Forelimb motor performance following cervical spinal cord contusion injury in the rat. Exp Neurol 117:287–298
60. Bertelli JA, Mira JC (1993) Behavioral evaluating methods in the objective clinical assessment of motor function after experimental brachial plexus reconstruction in the rat. J Neurosci Methods 46:203–208
61. Tom VJ, Sandrow-Feinberg HR, Miller K, Santi L, Connors T, Lemay MA, Houlé JD (2009) Combining peripheral nerve grafts and chondroitinase promotes functional axonal regeneration in the chronically injured spinal cord. J Neurosci 29:14881–14890
62. Whishaw IQ, Pellis SM, Gorny BP (1992) Skilled reaching in rats and humans: evidence for parallel development or homology. Behav Brain Res 47:59–70
63. Schrimsher GW, Reier PJ (1993) Forelimb motor performance following dorsal column, dorsolateral funiculi, or ventrolateral funiculi lesions of the cervical spinal cord in the rat. Exp Neurol 120:264–276
64. Montoya CP, Campbell-Hope LJ, Pemberton KD, Dunnett SB (1991) The "staircase test": a measure of independent forelimb reaching

and grasping abilities in rats. J Neurosci Methods 36:219–228
65. Baird AL, Meldrum A, Dunnett SB (2001) The staircase test of skilled reaching in mice. Brain Res Bull 54:243–250
66. Maurissen JP, Marable BR, Andrus AK, Stebbins KE (2003) Factors affecting grip strength testing. Neurotoxicol Teratol 25: 543–553
67. Nikkhah G, Rosenthal C, Hedrich HJ, Samii M (1998) Differences in acquisition and full performance in skilled forelimb use as measured by the 'staircase test' in five rat strains. Behav Brain Res 92:85–95
68. Titsworth WL, Onifer SM, Liu NK, Xu XM (2007) Focal phospholipases A2 group III injections induce cervical white matter injury and functional deficits with delayed recovery concomitant with Schwann cell remyelination. Exp Neurol 207:150–162
69. Whishaw IQ, O'Connor WT, Dunnett SB (1986) The contributions of motor cortex, nigrostriatal dopamine and caudate-putamen to skilled forelimb use in the rat. Brain 109:805–843
70. Whishaw IQ, Pellis SM (1990) The structure of skilled forelimb reaching in the rat: a proximally driven movement with a single distal rotatory component. Behav Brain Res 41:49–59
71. Whishaw IQ, Pellis SM, Gorny B, Kolb B, Tetzlaff W (1993) Proximal and distal impairments in rat forelimb use in reaching follow unilateral pyramidal tract lesions. Behav Brain Res 56:59–76
72. Whishaw IQ, Gorny B, Sarna J (1998) Paw and limb use in skilled and spontaneous reaching after pyramidal tract, red nucleus and combined lesions in the rat: behavioral and anatomical dissociations. Behav Brain Res 93:167–183
73. Metz GA, Whishaw IQ (2000) Skilled reaching an action pattern: stability in rat (*Rattus norvegicus*) grasping movements as a function of changing food pellet size. Behav Brain Res 116:111–122
74. Whishaw IQ, Metz GA (2002) Absence of impairments or recovery mediated by the uncrossed pyramidal tract in the rat versus enduring deficits produced by the crossed pyramidal tract. Behav Brain Res 134:323–336
75. Chan CC, Khodarahmi K, Liu J, Sutherland D, Oschipok LW, Steeves JD, Tetzlaff W (2005) Dose-dependent beneficial and detrimental effects of ROCK inhibitor Y27632 on axonal sprouting and functional recovery after rat spinal cord injury. Exp Neurol 196:352–364
76. Ruitenberg MJ, Levison DB, Lee SV, Verhaagen J, Harvey AR, Plant GW (2005) NT-3 expression from engineered olfactory ensheathing glia promotes spinal sparing and regeneration. Brain 128:839–853
77. Ballermann M, Tompkins G, Whishaw IQ (2000) Skilled forelimb reaching for pasta guided by tactile input in the rat as measured by accuracy, spatial adjustments, and force. Behav Brain Res 109:49–57
78. Ballermann M, McKenna J, Whishaw IQ (2001) A grasp-related deficit in tactile discrimination following dorsal column lesion in the rat. Brain Res Bull 54:237–242
79. Ballermann M, Metz GA, McKenna JE, Klassen F, Whishaw IQ (2001) The pasta matrix reaching task: a simple test for measuring skilled reaching distance, direction, and dexterity in rats. J Neurosci Methods 106:39–45
80. Yamamoto M, Raisman G, Li Y (2009) Loss of directed fore-limb reaching after destruction of spinal grey matter. Brain Res 1265:47–52
81. Diener PS, Bregman BS (1998) Fetal spinal cord transplants support the development of target reaching and coordinated postural adjustments after neonatal cervical spinal cord injury. J Neurosci 18:763–778
82. McKenna JE, Whishaw IQ (1999) Complete compensation in skilled reaching success with associated impairments in limb synergies, after dorsal column lesion in the rat. J Neurosci 19:1885–1894
83. VandenBerg PM, Hogg TM, Kleim JA, Whishaw IQ (2002) Long-Evans rats have a larger cortical topographic representation of movement than Fischer 344 rats: a microstimulation study of motor cortex in naïve and skilled reaching-trained rats. Brain Res Bull 59:197–203
84. Whishaw IQ, Gorny B, Foroud A, Kleim JA (2003) Long-Evans and Sprague-Dawley rats have similar skilled reaching success and limb representations in motor cortex but different movements: some cautionary insights into the selection of rat strains for neurobiological motor research. Behav Brain Res 145:221–232
85. Ackery A, Robins S, Fehlings MG (2006) Inhibition of Fas-mediated apoptosis through administration of soluble Fas receptor improves functional outcome and reduces posttraumatic axonal degeneration after acute spinal cord injury. J Neurotrauma 23:604–616
86. Baptiste DC, Austin JW, Zhao W, Nahirny A, Sugita S, Fehlings MG (2009) Systemic polyethylene glycol promotes neurological recovery

and tissue sparing in rats after cervical spinal cord injury. J Neuropathol Exp Neurol 68: 661–676
87. Andrews MR, Czvitkovich S, Dassie E, Vogelaar CF, Faissner A, Blits B, Gage FH, Ffrench-Constant C, Fawcett JW (2009) Alpha9 integrin promotes neurite outgrowth on tenascin-C and enhances sensory axon regeneration. J Neurosci 29:5546–5557
88. Keyvan-Fouladi N, Raisman G, Li Y (2005) Delayed repair of corticospinal tract lesions as an assay for the effectiveness of transplantation of Schwann cells. Glia 51:306–311
89. Yamamoto M, Raisman G, Li D, Li Y (2009) Transplanted olfactory mucosal cells restore paw reaching function without regeneration of severed corticospinal tract fibres across the lesion. Brain Res 1303:26–31
90. Kanagal SG, Muir GD (2007) Bilateral dorsal funicular lesions alter sensorimotor behaviour in rats. Exp Neurol 205:513–524
91. Kanagal SG, Muir GD (2008) Effects of combined dorsolateral and dorsal funicular lesions on sensorimotor behaviour in rats. Exp Neurol 214:229–239
92. Kanagal SG, Muir GD (2009) Task-dependent compensation after pyramidal tract and dorsolateral spinal lesions in rats. Exp Neurol 216:193–206
93. Kanagal SG, Muir GD (2008) The differential effects of cervical and thoracic dorsal funiculus lesions in rats. Behav Brain Res 187:379–386
94. Tom VJ, Kadakia R, Santi L, Houlé JD (2009) Administration of chondroitinase ABC rostral or caudal to a spinal cord injury site promotes anatomical but not functional plasticity. J Neurotrauma 26:2323–2333
95. Sandrow-Feinberg HR, Zhukareva V, Santi L, Miller K, Shumsky JS, Baker DP, Houle JD (2010) PEGylated interferon-beta modulates the acute inflammatory response and recovery when combined with forced exercise following cervical spinal contusion injury. Exp Neurol 223(2):439–451
96. Anderson KD, Sharp KG, Hofstadter M, Irvine KA, Murray M, Steward O (2009) Forelimb locomotor assessment scale (FLAS): novel assessment of forelimb dysfunction after cervical spinal cord injury. Exp Neurol 220:23–33
97. Sharp J, Frame J, Siegenthaler M, Nistor G, Keirstead HS (2010) Human embryonic stem cell-derived oligodendrocyte progenitor cell transplants improve recovery after cervical spinal cord injury. Stem Cells 28:152–163
98. Collazos-Castro JE, Muñetón-Gómez VC, Nieto-Sampedro M (2005) Olfactory glia transplantation into cervical spinal cord contusion injuries. J Neurosurg Spine 3:308–317
99. Collazos-Castro JE, Soto VM, Gutiérrez-Dávila M, Nieto-Sampedro M (2005) Motoneuron loss associated with chronic locomotion impairments after spinal cord contusion in the rat. J Neurotrauma 22: 544–558
100. Ying Z, Roy RR, Zhong H, Zdunowski S, Edgerton VR, Gomez-Pinilla F (2008) BDNF-exercise interactions in the recovery of symmetrical stepping after a cervical hemisection in rats. Neuroscience 155:1070–1078
101. Martinez M, Brezun JM, Bonnier L, Xerri C (2009) A new rating scale for open-field evaluation of behavioral recovery after cervical spinal cord injury in rats. J Neurotrauma 26:1043–1053

Chapter 51

Swimming as an Assessment of Hindlimb Function in Animals with Traumatic Spinal Cord Injury

Edward H. Brown, Johnny R. Morehouse, and David S.K. Magnuson

Abstract

In the field of spinal cord injury, there is a need for accurate and versatile behavioral assessments. Locomotor rating scales (BBB and BMS; see Burke and Magnuson, Chap. 47) form the foundation of behavioral assessments, but combining them with objective techniques provides a more comprehensive view for the characterization of animal models and postinjury deficits and functional recovery. The swimming scale used in our training model is the Louisville Swim Scale (LSS). It has proven to be a sensitive and robust assessment of hindlimb function in adult rats with T9 contusions injuries ranging from mild to moderately severe and sensitively detects improvements brought about by swim training. We have developed an objective measure to complement the LSS based on two-dimensional (2-D) kinematics. The 2-D kinematics was found to be highly sensitive, interpretable, and inexpensive. The recent addition of an automated Excel add-in has decreased analysis time, improved reliability and significantly enhanced throughput. Taken together, the LSS and 2-D kinematics provide excellent tools to aid in distinguishing between different injury severities and functional improvements following an experimental therapy.

Key words: Swimming, Kinematics, LSS, SCI, Plasticity, Rehabilitation

1. Introduction and Background

1.1. Clinical Locomotor Rehabilitation

The most widely accepted rehabilitation strategy for patients with spinal cord injury (SCI) is body weight-supported treadmill training (BWSTT) (1, 2). The patients are placed into an upright posture with their body weight being supported by a harness while therapists assist the legs during stepping on the treadmill. The combination of treadmill belt propulsion and afferent feedback provided by the therapists helps to initiate the stepping pattern which in most patients improves with training. Patients with incomplete injuries have shown significant improvements in treadmill stepping after weeks or months of training (3–5).

Jun Chen et al. (eds.), *Animal Models of Acute Neurological Injuries II*, Springer Protocols Handbooks, DOI 10.1007/978-1-61779-782-8_51, © Springer Science+Business Media, LLC 2012

1.2. Assessments for Humans After SCI

Two of the most common assessments for the patients after SCI are the ASIA scale and stepping kinematics. These two criteria represent a functional scale and an objective measure of function after injury. These tools are used to provide a prognosis and chart the progress of patients during the recovery phase. Additional measures used to assess function in patients with SCI are the functional independence measure (FIM), and the 10-m walking test (6). These assessments allow clinicians to grade patients based on sensorimotor function, self care behaviors, communication, psychological adjustment, cognition, and ambulatory ability. Used in tandem with these measures (ASIA, FIM, and 10-m walking test), kinematic analysis provides information about the trajectory of the leg, excursion of the joints, and the temporal parameters associated with stepping (i.e., stance vs. swing). These two valuable assessment classifications allow researchers to identify functional deficits, characterize the BWSTT rehabilitation model, and provide a method to document functional improvements.

1.3. Animal Models That Led to BWSTT

The BWSTT approach arose from work using the fully spinally transected cat model (7–9). In this model, adult cats received complete low-thoracic transections and were retrained to walk with their hindlimbs on a treadmill while their forelimbs are positioned on a platform. The experimenters assisted the cats by holding the tail for balance and provided manual stimulation to the base of the tail to encourage stepping. Implications from the "spinal cat" work suggests that the isolated lumbar spinal cord contains central pattern generating (CPG) circuitry that can be retrained with repeated exposure to the proper pattern of afferent feedback from the hindlimbs (10). These studies have provided understanding of the link between retraining and the plasticity of neurons in the spinal cord. In addition, this animal model provided evidence that the CPG can be accessed and retrained in the absence of supra-spinal input. This is a very important concept in the field of rehabilitation with far reaching clinical implications.

1.4. Swimming as a Model of Task-Specific Locomotor Retraining and Modes of Assessment

In order to improve patient recovery, it is vital for scientists to develop appropriate animal models and create innovative translational assessment tools. Central to our work is a swim training model used to investigate the various elements involved in retraining following injury in the adult rat. In addition to characterizing this model, we have developed assessments that provide detailed behavioral data that are accurate and comprehensive. Recently, we have shown that rats with moderately severe T9 contusions can be retrained to swim starting at 2 weeks postinjury (11, 12). We have also demonstrated that supplemental cutaneous feedback provided to the hindpaw, in the form of inverted centrifuge tubes suspended from the bottom of the swimming tank, can enhance hindlimb activity during the retraining process. This model is based on the original work by Muir and Steeves (13), who reported similar results in chicks. The swim-training model allows even severely injured animals to achieve the benefits of training without the requirement of weight support. The buoyancy provided by the water allows the animal to achieve full range of motion and a higher frequency of stroke cycles compared to treadmill or overground stepping. Another strength of the model is that the limb movement is generated by the inherent ability of the spinal cord, without manual assistance or external stimulation. Consistent with other animal models, there are some common assessment tools and methods of data analysis in our protocol. The first is a subjective behavioral scale, the LSS (11, 12), modeled after the well-known BBB open field locomotor scale (14). The LSS allows experimenters to assign animals a numerical score that corresponds to their overall swimming ability. This subjective measure supplies information about robust aspects of swimming behavior, such as hindlimb movement, alternation, and body position. However, the objective swimming kinematics technique, developed in our laboratory, is used to describe less obvious aspects of swimming behavior like limb position and joint angle excursions. Our kinematic approach is sensitive enough to detect differences between animals with roughly similar LSS scores. We use the objective and subjective assessments together to get a detailed understanding of swimming performance and persistent deficits. We strive to continually improve our kinematic assessments and analysis techniques to assist in the translation and interpretation of results following the application of experimental therapies.

2. Assessments for Swimming Behavior

2.1. The Louisville Swimming Scale

The LSS is a novel swimming assessment developed in our laboratory (11, 12) in order to accurately observe improvements in swimming behavior. The LSS is an 18-point scale based on common behaviors while swimming, including hindlimb movement, forelimb dependency, hindlimb alternation, lateral trunk instability, and body angle. Two observers simultaneously assess an animal in the swimming tank and produce a score that represents the given behavior. This tool has proven useful and accurate in assessing swimming behavior and the LSS score correlates significantly with the BBB score in otherwise untrained or untreated animals (11, 12, 14).

2.1.1. Scoring Categories

The LSS involves three ranges of scores: 0–5, 6–11, and 12–17. The ranges describe poor, intermediate, and good swimming behavior. The individual behaviors are scored based on frequency

Louisville Swimming Scale (LSS)

HL Movement	HL Alternating	FL Dependency	Trunk Instability	Body Angle
0 N	0 N/S	4 N	4 N or S(Mi)	2 N/mi
1 S	1 O	3 S	3 O(Mi) or S(Mo)	1 Mo
2 O	2 F	2 O	2 F/C(Mi) or O(Mo)	0 S
3 F	3 C	1 F	1 F/C(Mo) or O(Se)	
4 C		0 C	0 F/C(Se)	

Score Categories

N-None
O-Occasional
F-Frequent
C-Consistent

Instability Tilt scale

N-None	0°
Mi-Mild	1° to 20°
Mo-Moderate	21° to 60°
S-Severe	61° and up

Body Angle

N/Mi	0° to 10°
MO	11° to 45°
S	46° or greater

HL Movement

None	0
Seldom	1%-5%
Occasional	6%-50%
Frequent	51%-95%
Consistent	96%-100%

12 and up
Good Swimmer
Consistent HL movement. Little to no FL dependency.
Mostly alternating HL coordination. Usually no notable TI.

5 to 11
Intermediate Swimmer
Usually frequent to consistent HL movement.
Tend to depend on FL frequently or consistently.
Some synchronized HL coordination may be noted.
TI is evident.

0 to 4
Poor Swimmer
Occasional to no HL movement.
Consistent FL dependency.
Frequent to consistent trunk instability.

Some notes on scoring+
FL dependency: The use of one forearm occasionally to steer (usually along the edge of tank) should not be noted as FL dependency
Trunk Instability: Should only be noted while swimming; (Uninjured rats exhibit occasional TI when floating.)

Fig. 1. LSS score sheet.

of occurrence during the 4-min testing period. Behaviors, such as forelimb dependency, hindlimb movement, and hindlimb alternation, are classified as None (0%), Seldom (1–5%), Occasional (6–50%), Frequent (51–95%), or Consistent (96–100%). Trunk instability, defined as lateral twisting of the trunk, is classified as none (0°), mild (1–20°), moderate (21–60°), or severe ($x > 61°$). Body angle, formed between the dorsum of the back and the surface of the water, is classified as mild (0–10°), moderate (11–45°), or severe ($x > 46°$). Each percentage and degree range is associated with its own numerical score. The individual scores are summed to generate the final LSS score, which is agreed upon by the two observers. For details on the LSS, reference the score sheet in Fig. 1 and (11, 12).

2.2. Swimming Kinematics

Although the kinematics of stepping and swimming in rats has been described (15–18), kinematic assessments are very labor intensive, rely on expensive hardware and software packages, and can be difficult to interpret (19). The swimming kinematics setup that we developed addresses many of the aforementioned issues. Working in tandem with our core personnel, we have been able to streamline the setup process, minimize the equipment cost, and develop a simple analysis technique that is accurate and easy to interpret. The 2-D swimming kinematic technique has become a primary outcome measure due to its sensitivity and fit to our injury model.

2.2.1. Hindlimb Components, Point Placement, and Angles Measured

The 2-D swimming kinematics describes the angular changes in the hindlimb joints of the rat. The three joints commonly measured are the hip, knee, and ankle. The angles used to measure the excursion (range of motion) of each joint are formed by identifying specific anatomical landmarks. The skin overlying each landmark is made visible with a tattoo or permanent Sharpie marker. Infrared cameras and reflective markers cannot be used in the water due to changes in water temperature and light absorption and refraction. For our purposes, the knee will not be individually measured due to the absence of a stable anatomical landmark (i.e., skin slippage). However, our technique still incorporates the excursion of all three joints by dividing the hindlimb into a proximal and distal component (Fig. 2). This simplified division of the hindlimb into two components allows us to measure the cumulative movement of all three joints with two angles instead of three. The anatomical landmarks used to form the two angles are the iliac crest (denoted "I"-*palpable notch on the ilium*), the greater trochanter (denoted "H"-*hip*), the lateral malleolus (denoted "A"-*ankle*), and the metatarsophalangeal joint (denoted "T"-*toe*). The IHA angle (proximal component) is a three-point angle formed between the iliac crest, hip, and ankle marks. It should be noted that the IHA angle involves movement of the hip and knee joint. The HAT angle (distal component) is a three point angle formed between the hip, ankle, and toe marks. It should be noted that the HAT angle involves the movement of the knee and ankle (see Fig. 2 for clarification).

2.2.2. Elliptical Fourier Analysis and Toe Velocity

During swimming in uninjured animals, the IHA and HAT angles are nearly 180° out of phase. When these angles are plotted against each other (an angle–angle plot), the plot has the shape of an ellipsoid. The area inside the ellipsoid describes the out-of-phase excursion of both angles on a stroke-by-stroke basis. On the angle–angle plot, the HAT angle is labeled on the *x*-axis and the IHA angle is labeled on the *y*-axis. The ellipsoid shape is formed due to a slightly larger excursion of the HAT vs. the IHA in uninjured animals. If both angles had identical excursions and were perfectly out

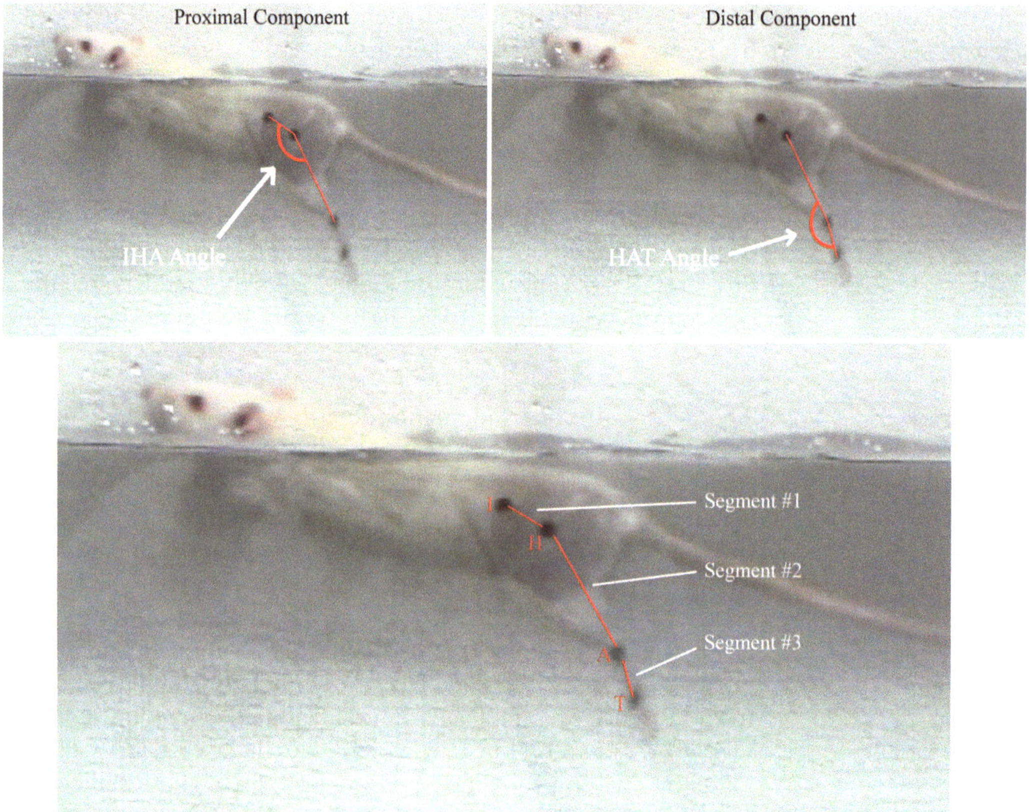

Fig. 2. Limb components.

of phase, the resulting shape would be a circle. In addition to area calculations, the centroid of the ellipse represents the median angle throughout each stroke, supplying information about the relative position of the hindlimb (flexed vs. extended). The centroid represents an *x*–*y* coordinate identifying the center of the ellipsoid (see Fig. 3). Changes in limb movement or swimming pattern will be detected as alterations in ellipsoid area and centroid position. A smaller ellipsoid area would signify smaller HAT and IHA excursions. A larger centroid *x*- or *y*-value would translate to a more extended limb position during the stroke cycle and conversely a smaller centroid *x*- or *y*-value would translate to a more flexed limb position during the stroke cycle (see Fig. 4). Thus, a quantitative description of hindlimb movement during swimming can be made using the area of the ellipsoid, the mean peak and trough (extension and flexion) of the HAT and IHA angles, and the centroid of the angle–angle plot (19). We also have stick figure analysis at our disposal for graphic depictions of the limb during swimming (refer to Fig. 3). The ellipsoids are analyzed using an Elliptical Fourier Analysis as described by Ferson et al. (20) using the freeware program EFAW (http://life.bio.sunysb.edu/morph/). The EFAW program provides Excel output files with the ellipsoid areas and

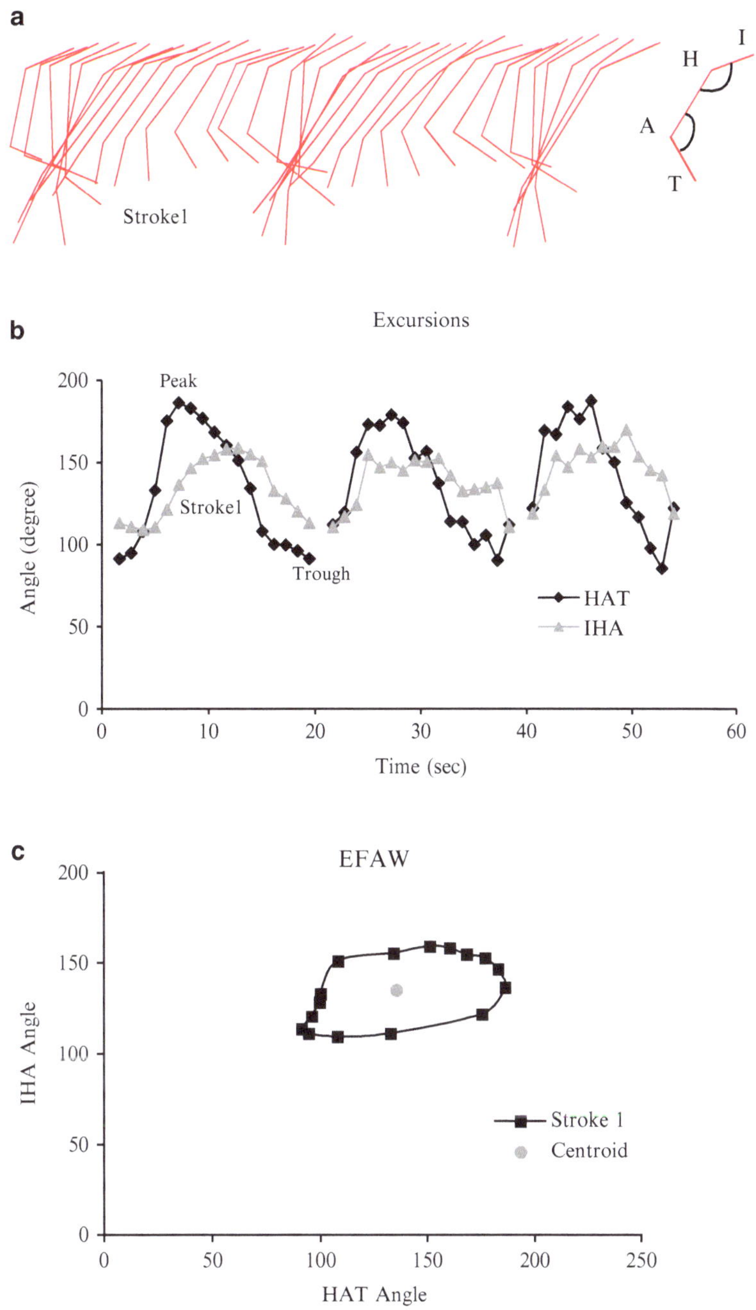

Fig. 3. (**a**) Shown are stick figures of the right hindlimb through three complete stroke cycles. The limb is represented as three segments (I-H, H-A and A-T) and two angles (I-H-A, and H-A-T). (**b**) Shown are angle excursions (y) over time (x) for the three stroke cycles shown in **a**. (**c**) Shown is the angle-angle plot (I-H-A on the x-axis vs H-A-T on the y-axis) for stroke 1 from **a**, above. The centroid for that stroke, calculated using EFAW is also shown.

centroids for each stroke cycle. To complement the EFAW calculations, the software package MaxMate (from Innovision Systems, Inc.) can be used to calculate the mean toe velocity (relative to the hip)

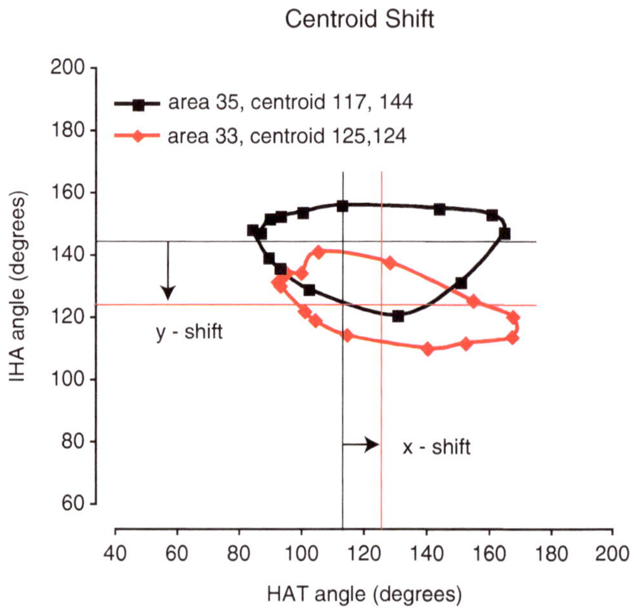

Fig. 4. Area and centroid shifts.

for each rat during swimming. The toe velocity is obtained by analyzing the movement of the toe mark relative to the hip mark over time and provides an objective measure of the force being generated during the stroke (extension) and recovery (flexion) phases of the swim cycle. MaxMate creates a velocity vs. time graph with accompanying velocity data from the digitized ASCII file. This data is useful when comparing animals with similar limb trajectories (see Fig. 5).

3. Setup Protocol

3.1. Tank Specs

The tank used for swimming is 60-in. long, 6.5-in. wide, and 12-in. deep. The tank was constructed out of 3/8th-in. thick Plexiglas glued together and sealed using silicon caulk. One end of the tank has a ramp covered with a neoprene pad, which permits the rats to comfortably exit the water.

3.2. Tattoos and Point Placement

Prior to filming, the rat's hindquarter is shaved and green tattoos (Aramis micro-tattoo punch, Braintree, MA) are placed in the skin overlying the iliac crest (I) and greater trochanter (H). The tattoos allow for consistent and accurate point placement throughout the duration of the study. Black Sharpie pens are used to reinforce the

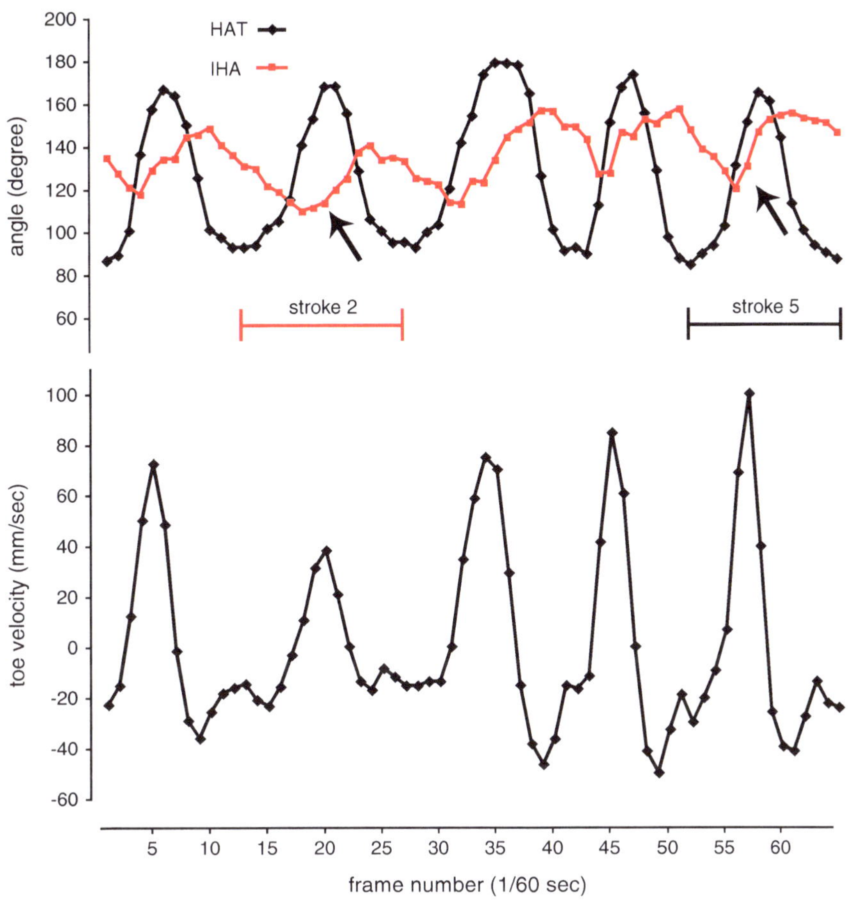

Fig. 5. Toe velocity.

tattoo marks in addition to being placed on the lateral malleolus (ankle—A) and metatarsophalangeal joint (toe—T).

3.3. Video Recording and File Types

The animals are filmed using a Basler 602f high-resolution digital camera (Basler Vision Technologies, Exton, PA) running at 60 Hz connected to an active PC via fire wire. The tripod stand, holding the camera, is positioned 18-in. from the side of the tank. The far side of the tank is covered with a white towel creating maximum image contrast between the white fur of the rat and the black Sharpie marks. Next, the experimenter manually checks for proper focus and lighting criteria. The first step is to adjust the iris and focus on the lens of the camera. Fine tuning of the lighting is done by adjusting the brightness and gain options on the video acquisition software. We use the video software DVR Explorer (Advanced Digital Vision, Natick, MA), but almost any digital video system should suffice. The focus and lighting settings are rechecked once an animal is placed into the tank. The digital video is acquired in AVI format. Each AVI file is opened in the software package MaxTraq (Innovision Systems, Columbiaville, MI) and the black Sharpie marks are identified and digitized using an image tracking function, checked for mistakes, and corrected manually. Upon completion, the digitized points are used to create the HAT and IHA angles via the 3-point angle tool (Fig. 2). After digitization is complete, the MaxTraq file is saved as a custom ASCII file (make sure the "frame number" and "include angle" boxes are checked). Note: Since the HAT angle is always labeled on the x-axis, while in MaxTraq create the HAT angle first (i.e., Angle#1). This will place the HAT angle data in the first column (x-value column) of the spreadsheet when the ASCII file is opened in Excel. The processed files are further analyzed in Excel and MaxMate (the plug-in for Excel developed by Innovision systems).

4. Analysis Technique for Swimming Kinematics

4.1. Introduction to the Automated Add-In

From this point on, the analysis can be done by using an "add-in" developed by one of the co-authors, Johnny Morehouse. This add-in is designed to speed up the post-video data analysis of swimming kinematics. The main functions included are calculations of the average peak HAT/IHA angles, average trough HAT/IHA angles, Elliptical Fourier Analysis (EFAW) and toe velocity. The add-in also merges multiple Excel files into one spreadsheet. This is all performed though a simple and easy-to-use pop-up menu incorporated into Excel. In the following section, several tabs from the add-in will be identified and described. Each tab has one designated task, highlighted in the tab title, with defined step-by-step instructions demonstrating how to complete each part of the analysis. The tabs are positioned in sequential order from left to right (see screen shots, Figs. 6–9). In other words, the method starts with the first tab on the left (Peaks/Troughs tab) and moves to the right (Excursions tab, EFAW tab, Toe Velocity tab) until the desired outcomes are completed.

4.2. 2-D Swimming Kinematics (Automated Analysis)

Additional information and add-in downloads can be accessed at: http://louisville.edu/kscirc/about/cobre-cores/core-d-electrophysiology-and-behavioral-assessment.html (21).

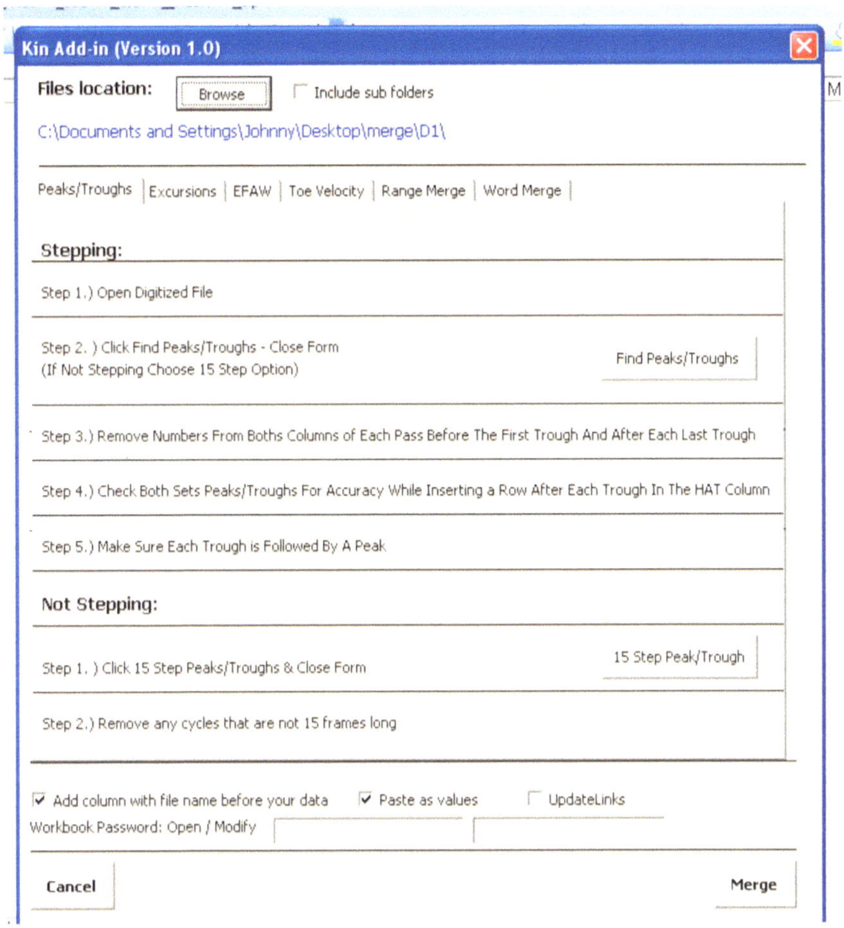

Fig. 6. The first page of the add-in is designated for peaks and troughs (P/T). The first tab of the automated add-in is used to find the peak and trough of the HAT and IHA angles. Locating the peak and trough HAT/IHA angles allows the analyst to separate the entire digitized video into separate swimming strokes. This step is necessary for calculating angle excursions, ellipsoid areas, and centroids on a stroke-by-stroke basis.

5. Final Points, Problems, and Solutions

5.1. Strengths of Swimming Kinematics

- Swimming kinematics represents an objective measure of swimming behavior. This technique can be consistently replicated with a minimal amount of equipment cost and technician training.
- The analysis technique is highly sensitive to minute changes in behavior that can be observed and described by the area, centroid, and toe velocity.

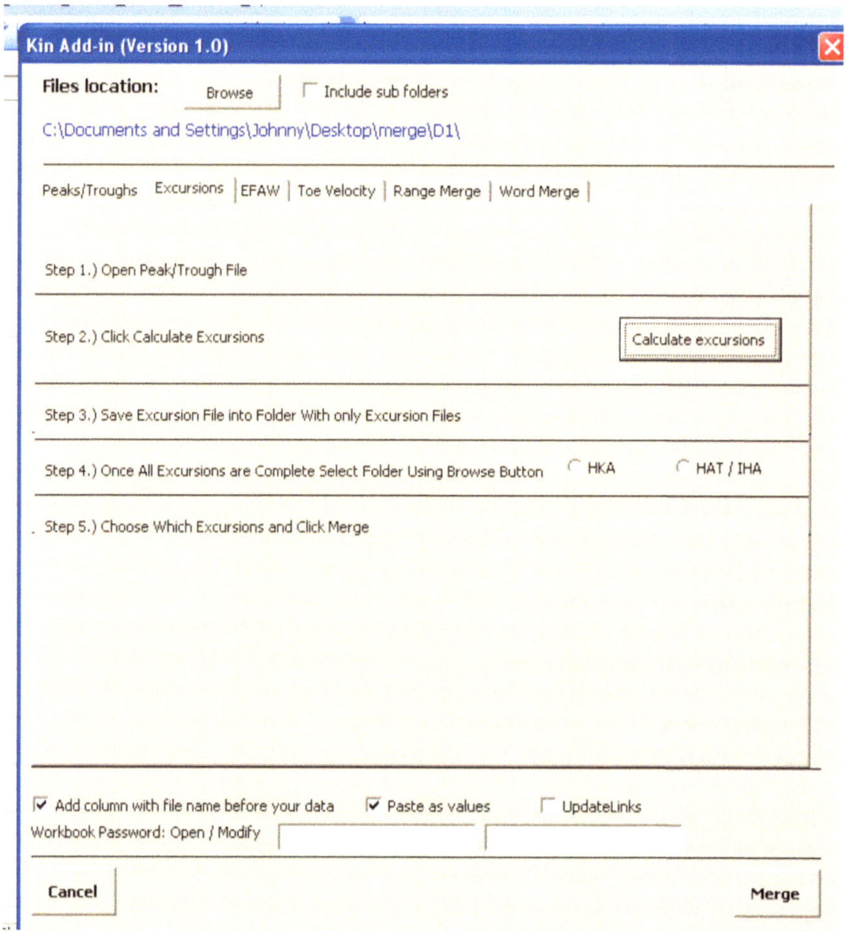

Fig. 7. The next page of the add-in is designated for calculating the mean peak, trough, and excursion of the HAT/IHA. After completion of the previous step, the "Excursions" tab is used to calculate the mean peak, trough, and excursion of the HAT/IHA angles. This step automatically calculates the means and merges the data into a separate Excel worksheet. The peak values represent the average maximum extension of each angle during swimming for each animal. The trough values represent the maximum flexion of each angle during swimming for each animal. The excursion is calculated by subtracting the trough values from the peak values for each individual stroke. The excursion represents the range of motion for the given angle.

- The data analysis phase is relatively high throughput, simple to implement, and comprehensive.

5.2. Problems and Solutions

- Accurate execution of this method is dependent upon a few criteria. Before filming, each animal must be cleanly shaved and receive fresh Sharpie marks.
- The lighting in the room and the lighting settings on the recording software must be checked and adjusted for quality video analysis.

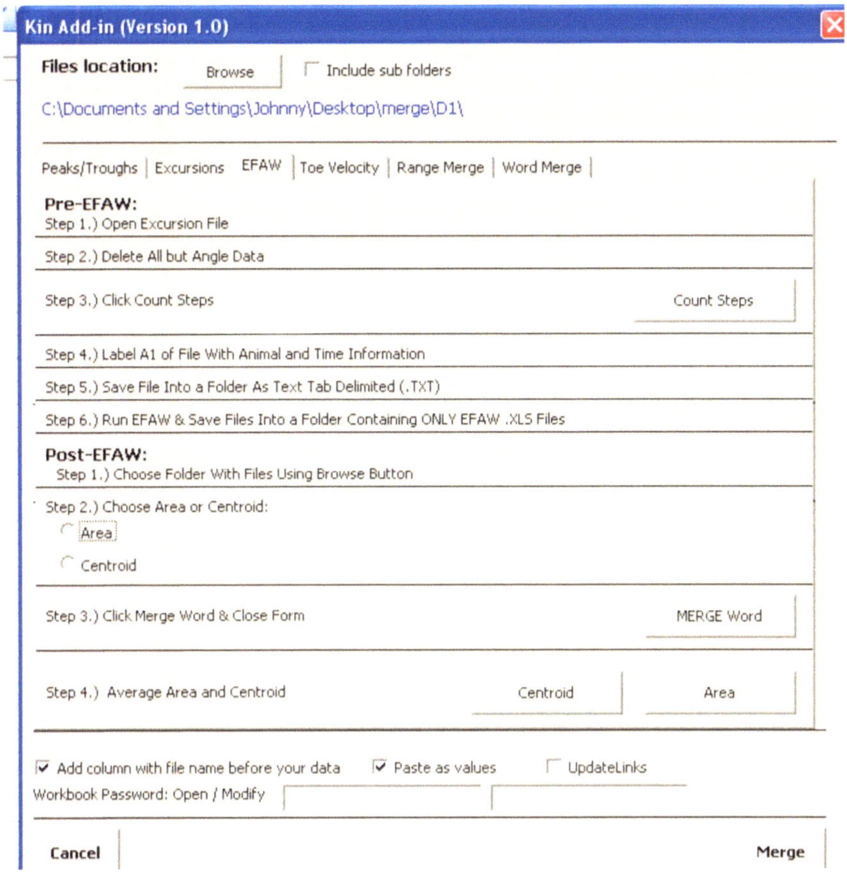

Fig. 8. The next page of the add-in is designated for EFAW. The step is designated for the Elliptical Fourier Analysis. This tab involves a set of steps to be completed before and after the data is run processed using the EFAW program. The EFAW program is a separate program that is not automatically linked to this add-in. The Pre-EFAW section sets up the data in a suitable fashion to run through the EFAW program. Next, the data contained within and Excel output file generated by the EFAW is used to automatically calculate the desired means and the output is merged into one Excel worksheet.

- In some situations, an animal may exhibit severe trunk instability that causes twisting of the trunk during swimming. If this occurs, the points on the body may not be in view of the camera. Should this occur, the animal should be given ample time to rest between passes and should be given several opportunities to perform a suitable pass. If after many attempts the animal is unable to control its trunk, that animal cannot be included for the current time point.
- During filming, it is critical that the area of interest (AOI) of the camera is large enough to capture the entire body of the animal and at least three consecutive stroke cycles.
- When performing the post-video analysis (Add-in), each step must be carefully checked before moving to the next step.

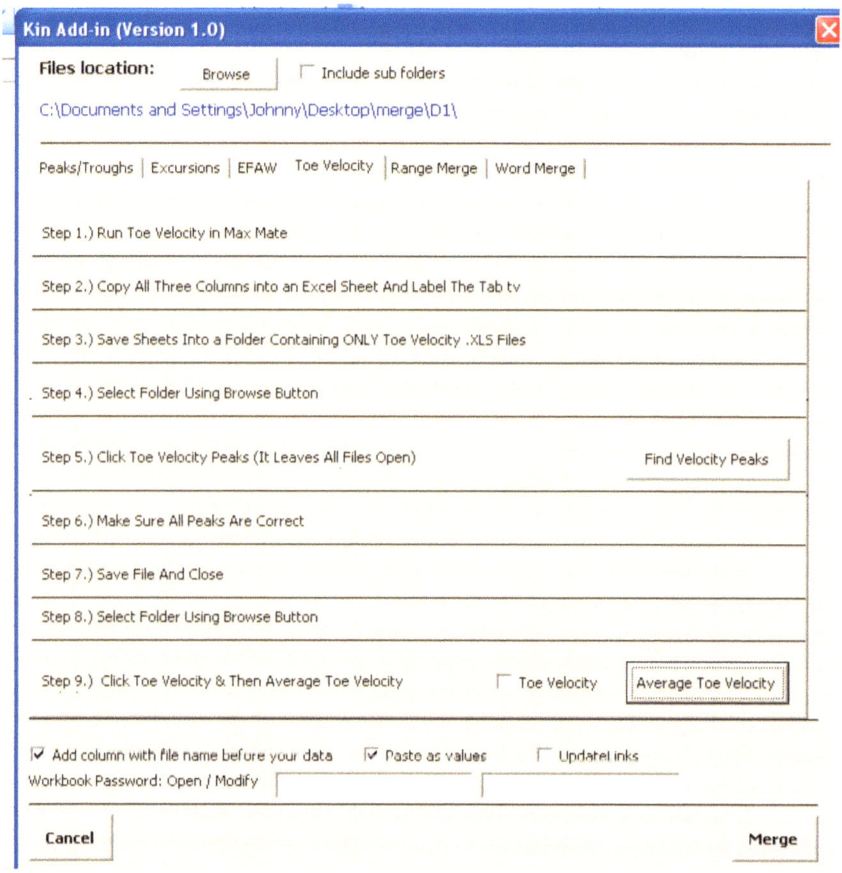

Fig. 9. The next page of the add-in is designated for toe velocity. The last tab is used for calculating the toe velocity during swimming. The toe velocity is calculated relative to the hip. For our purposes, the peak velocity is the only value recorded. The goal is to observe the maximum velocity generated by the animal during swimming. The toe velocity data has proven valuable in separating animals with similar hindlimb trajectories. The peak velocities are determined automatically, the mean is calculated and the data is merged into a separate Excel worksheet.

References

1. Wernig A, Muller S (1992) Laufband locomotion with body weight support improved walking in persons with severe spinal cord injuries. Paraplegia 30(4):229–238
2. Wernig A, Muller S, Nanassy A, Cagol E (1995) Laufband therapy based on 'rules of spinal locomotion' is effective in spinal cord injured persons. Eur J Neurosci 7(4):823–829
3. Dobkin B, Apple D, Barbeau H, Basso M, Behrman A, Deforge D, Ditunno J, Dudley G, Elashoff R, Fugate L, Harkema S, Saulino M, Scott M, Spinal Cord Injury Locomotor Group (2006) Weight-supported treadmill vs overground training for walking after acute incomplete SCI. Neurology 66(4):484–493
4. Behrman A, Lawless-Dixon A, Davis S, Bowden M, Nair P, Phadke C, Hannold E, Plummer P, Harkema S (2005) Locomotor training progression and outcomes after incomplete spinal cord injury. Phys Ther 85(12):1356–1371
5. Harkema SJ (2001) Neural plasticity after human spinal cord injury: application of locomotor training to the rehabilitation of walking. Neuroscientist 7(5):455–468 (review)
6. Ditunno J (2010) Outcome measures: evolution in clinical trials of neurological/functional

recovery in spinal cord injury. Spinal Cord 48(9):674–684

7. Barbeau H, Rossignol S (1987) Recovery of locomotion after chronic spinalization in the adult cat. Brain Res 412(1):84–85
8. Berlanger M, Drew T, Provencher J, Rossignol S (1996) A comparison of treadmill locomotion in adult cats before and after spinal transection. J Neurophysiol 76(1):471–491
9. Rossignol S, Chau C, Brustein E, Belanger M, Barbeau H, Drew T (1996) Locomotor capacities after complete and partial lesions of the spinal cord. Acta Neurobiol Exp (Wars) 56(1):449–463
10. Barriere G, Simmers J, Combes D (2008) Multiple mechanisms for integrating proprioceptive inputs that converge on the same motor pattern-generating network. J Neurosci 28(35): 8810–8820
11. Smith R, Burke D, Baldini A, Shum-Siu A, Baltzley R, Bunger M, Magnuson D (2006) The Louisville Swim Scale: a novel assessment of hindlimb function following spinal cord injury in adult rats. J Neurotrauma 23(11):1654–1670
12. Smith R, Shum-Siu A, Baltzley R, Bunger M, Baldini A, Burke D, Magnuson D (2006) Effects of swimming on functional recovery after incomplete spinal cord injury in rats. J Neurotrauma 23(6):908–919
13. Muir G, Steeves J (1995) Phasic cutaneous input facilitates locomotor recovery after incomplete spinal injury in the chick. J Neurophysiol 74(1):358–368
14. Basso D, Beattie M, Bresnahan J (1996) Graded histological and locomotor outcomes after spinal cord contusion injury using the NYU weight-drop device versus transection. Exp Neurol 139(2):244–256
15. Gruner J, Altman J (1980) Swimming in the rat: analysis of locomotor performance in comparison to stepping. Exp Brain Res 40(4): 374–382
16. Gruner J, Altman J, Spivak N (1980) Effects of arrested cerebellar development on locomotion in the rat. Cinematographic and electromyographic analysis. Exp Brain Res 40(4): 361–373
17. Thota A, Carlson S, Jung R (2001) Recovery of locomotor function after treadmill training of incomplete spinal cord injured rats. Biomed Sci Instrum 37:63–67
18. Hamers F, Lankhorst A, van Laar T, Veldhuis W, Gispen W (2001) Automated quantitative gait analysis during overground locomotion in the rat: its application to spinal cord contusion and transection injuries. J Neurotrauma 18(2):187–201
19. Magnuson D, Smith R, Brown E, Enzmann G, Angeli C, Quesada P, Burke D (2009) Swimming as a model of task-specific locomotor retraining after spinal cord injury in the rat. Neurorehabil Neural Repair 23(6):535–545
20. Ferson S, Rohlf F, Koehn R (1995) Measuring shape variation of two-dimensional outlines. Syst Zool 34:59–68. Freeware program EFAW http://life.bio.sunysb.edu/morph/
21. Kinematic Analysis ADD-IN. http://louisville.edu/kscirc/about/cobre-cores/core-d-electrophysiology-and-behavioral-assessment.html

Chapter 52

Neuroimaging Assessment of Spinal Cord Injury in Rodents

Mehmet Bilgen

Abstract

In vivo preclinical imaging of spinal cord injury (SCI) in rodent models is sought after for obtaining clinically relevant neuropatholocial information in translational research. The ability to noninvasively monitor the spatial and temporal evolution of injury response allows understanding and characterizing the injury model, identifying the surrogate markers of the lesion, and also objectively evaluating the rationale and efficacy of any potential therapy. Magnetic resonance neuroimaging of injured rodent SC is challenging, but yet extensive studies have demonstrated that it is also feasible and plays an indispensable role in longitudinal studies. It is possible to acquire high quality images from both rat and mouse with special considerations and hardware developments. This chapter is organized to summarize the current state of art in instrumentation in our laboratory, methods and procedures, and materials used to successfully perform magnetic resonance microscopy and to interpret the results obtained with different imaging modalities.

Key words: Preclinical, Translational, Magnetic resonance imaging, Spinal cord injury, Contrast agent, Gadolinium, Manganese, Contrast enhancement, Dynamic contrast-enhanced MRI, Manganese-enhanced MRI, Magnetization transfer imaging, Diffusion tensor imaging, Fiber tractography

1. Introduction

Rodent models have assumed a prominent role in preclinical spinal cord injury (SCI) research (1). Main goal of the ongoing experiments is to understand the variations in the spatiotemporal course of the injury response in different strains of normal or genetically engineered rodents or in those subjected to a specific intervention. The findings from these explorations are expected to move the field forward by producing potential new targets for translational therapeutic strategies aimed at enhancing the neuronal protection, regeneration, and functional recovery following SCI in human patients (2).

Jun Chen et al. (eds.), *Animal Models of Acute Neurological Injuries II*, Springer Protocols Handbooks, DOI 10.1007/978-1-61779-782-8_52,

Magnetic resonance imaging (MRI) has found extensive applications in pharmaceutical research and drug discovery with animal models (3). MRI is also the preferred radiological modality for in vivo investigation of SCI (4). It allows comprehensive and noninvasive assessment of the injury characteristics. The same animal can be scanned repetitively over a period of time. Multiple images from different views can be acquired to evaluate the neuropathology in its natural progression. This capability allows each animal acts as its own historical control in a longitudinal study (5). Such approach reduces the variability between the animals and cuts down the number of animals considerably, and hence improves the statistical reliability of the acquired data.

MRI in Fig. 1 helps visualizing the pathology of an injured mouse SC in its natural setting. The vertebral dislocation seen about two vertebral levels rostral to the injury is likely occurred during stabilizing the animal for performing the injury. SC compression by the dislocated vertebrae introduces additional neurological deficits. Behavioral test score from this animal would suggest significant deficit compared to those obtained from a mouse that did not experience the dislocation, but just the injury alone. If histological tools were the only option for analyzing the SC tissue, the unanticipated dislocation would possibly be missed, making the interpretation of the outlying behavioral score from this animal difficult. Confirming the discrepancy in the model early on by MRI visualization clearly justifies eliminating this mouse from the study and discarding the associated data. This example demonstrates the

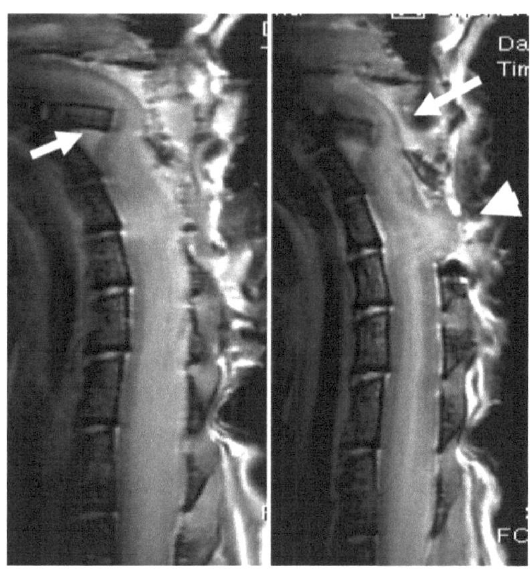

Fig. 1. Injured mouse spinal cord (SC). *Arrow head* points to the injury. *Arrow* points to the vertebral dislocation which introduces compression on the SC and causes neurological deficits difficult to interpret and account for in studies with traditional postmortem histological analysis.

vital role of MRI in validating the consistency of the SCI model. MRI has also merits and offers a guidance in the selection of a proper injury model (mild, moderate, or severe) to examine the efficacy of a potential therapeutic strategy aimed at improving the outcome from SCI (6).

In the past, numerous studies have used MRI modalities to comprehensively investigate SCI and contributed to our understanding of the lesion development and injury properties in rodent models. The existing MRI modalities provided valuable multiparametric information on the anatomy, structure, vasculature, and function of the SC under normal and pathological conditions. In the following, we describe the necessary infrastructure and basic instrumentation to successfully perform high quality MR scans longitudinally on injured rodent SCs, and also discuss the implementation of high resolution imaging protocols and other practical issues.

2. Materials and Instrumentation

Rodent models of SCI involve wild-type, transgenic or knock-out mutant models of rat or mouse, both male and female genders and about 3 months old. The weights of the animals are about 25 g for mouse and 250 g for rat.

2.1. Injury Device

An injury device is needed to produce a SCI model that mimics special aspects of real injuries induced in human by an accident or a fall. Numerous devices have been developed to damage the SC in various animals with a reproducible protocol under a computer controlled environment and quantifiable neurofunctional outcome (i.e., motor deficits). The mechanisms of producing injuries in these models are primarily based on tissue deformation applied to an exposed SC by using weight-drop techniques, electromagnetic devices, pneumatic impact devices, and others based on aneurysm clips or an inflatable balloon (ref. (7) and references therein). The biomechanical parameters that define the mechanical deformation of the SC tissue also determine the severity of the induced injury. Changing the parameters produces graded variations in morphology, neuropathology and biochemistry of the injured cord tissue and causes differences in neurobehavioral patterns during recovery. Some of the existing devices were developed for a specific species, i.e., either for mouse or for rat. Some devices are capable of accommodating different size animals and therefore can be used for both mouse and rat provided that appropriate bit of either circular or rectangular in shape and size is installed for contusing the SC.

Quality control of the consistency of the induced injuries is assured by different approaches. One approach tests the reproducibility of the prescribed shaft motion under different setting and loading conditions. This can be accomplished by impacting a soft paraffin or putty block with a smooth top surface as a media. The penetration depth of the dent produced by the impact on the surface is inspected under magnification and compared with those obtained using different tips and injury parameter values. Simultaneously, the velocity or motion of the injury tip is sampled and incorporated into the comparison for further confirming the repeatability of the injury process (7).

2.2. Scanner

Neuroimaging of rodents requires using high magnetic field MR scanner to produce high resolution images from the underlying injured SC. The images depict the spatial distribution of hydrogen atom in the medium under investigation. There are two commercial companies Bruker Inc. (Billerica, MA) and Varian Inc. (Palo Alto, CA) who manufacture and market high magnetic field MR scanners that can be used for imaging rodent SC. The bore dimensions of the scanner vary to accommodate a shim coil and different size gradient coils. Placing a sample in the scanner causes a slight perturbation in the magnetic field that would otherwise be uniform. Shim coil is used to make the field uniform at and nearby the isocenter of the magnet. The strengths of the gradient coils in the current commercial scanners can be as high as 1,000 mT/m for in vivo mouse imaging and 400 mT/m for rat imaging. Each scanner is supplied with a variety of volume or surface coils to image rat or mouse. The scanner is also equipped with a sophisticated electronic console accommodating a host computer, radio frequency transmitter, and receiver operating at different carrier frequency for different spin system, associated radio frequency (rf) amplifiers, demodulators, timer and data acquisition boards, shim coil current supply, gradient amplifiers with pre-emphasis circuitry. Meters are included for monitoring the coolant levels and temperature of the gradient coil. For setting up the parameters in the imaging sequences and postprocessing or analysis of the acquired data, Bruker Inc. provides Paravision and Varian Inc. uses VNMRJ softwares for interfacing the electronic console to the main control computer.

One may ask “why do we need a high magnetic field scanner for imaging the rodent spinal cord?” This question may simply be answered by the basic physical principles of the nuclear magnetic resonance. Number of excitable spins in a material exposed to an external magnetic field increases linearly with the field strength. This means that higher magnetic field leads to stronger signal. Corollary to this relationship is that more excitable spins in a given voxel produces stronger signal associated with that voxel. Electronic noise or ambient noise level in the electronic circuitry of the scanner remains unaffected by the magnetic field strength. Consequences of the signal increasing with magnetic field but noise remaining unchanged are twofolds. The first consequence is the improvement in signal-to-noise ratio when the dimension of the voxel is kept unchanged at different magnetic fields. The second one is that increase in signal allows reducing the voxel dimension while maintaining the signal-to-noise ratio at a constant level. This allows increasing spatial resolution without compromising the image quality at higher magnetic fields. The smaller details of the underlying medium are better visualized in a high resolution image.

Another question that often rises is that "what is the connection between the gradient strength and spatial image resolution?" A mathematical transformation is used to reconstruct the acquired data in spatial domain or to spatially localize the distribution of the hydrogen atoms. To satisfy this transformation formulation, the uniform magnetic field of the scanner is linearly perturbed along the x, y, and z directions by using gradient coil. When the bandwidth of the scanner's electronic circuitry is fixed, increasing the gradient strength allows reducing the voxel volume, thereby increasing the spatial resolution.

Radio frequency coil is also an important component of the operating scanner. The coil is used to transmit and receive the signals. Volume coil accommodate the whole or part of the sample and provide uniform excitation and detection. Surface coil is used for imaging a specific region of the sample. In some applications, both volume and surface coils are used in combination to limit the effects of nonuniform excitation of the surface coil when used alone. Coil dimension defines the magnitude of ambient noise injected from the environment. Smaller dimension coils are more desirable for reducing the electronic noise. The payoff of using small coil is realized as decreased voxel volume and hence again increased spatial resolution. In addition to the volume and surface coils, a special coil can be used for imaging the injured spinal cord at substantially higher qualities. This coil is described next.

2.3. Implantable Radio Frequency Coil

For acquiring high quality MRI data from injured SCs of rodent models, a special rf coil system was developed for applications in both rat and mouse (8–10). The system basically consists of two coils that act together as the primary and secondary windings of an electrical transformer. The combined set is used for both transmission of rf pulses and reception of the MR signal. An important feature of these coils is that the secondary coil is surgically implanted at the level of the injury for imaging. This coil configuration offers compact footprint of the rf excitation/signal detection and allows small field-of-view (FOV) selection due to the small dimension of the imaging coil. These features are all beneficial in terms of improving signal-to-noise ratio and producing high-resolution images from the tissue of interest only. This arrangement also provides a practical solution for spatially localized imaging and enhances the quality of the MR data acquired in longitudinal SCI studies. Signal transmission and reception with the implanted coil is achieved via mutual-inductive coupling with a larger coil placed external to the body. Strengthening the mutual coupling results in inductive overcoupling and increases the functionality of the implantable coil (11). As needed, one can modify the frequency response characteristics of the implantable coil with fixed electrical elements via an external coil with tuning and matching capacity. This ability is highly desirable since it allows reliable and repeatable MR data acquisition in studying the progression of the neuropathology in injured SC over time on the same animal.

For operations at 2, 7, and 9.4 T, which corresponds to the tuning frequencies of 85, 300, and 400 MHz, implantable coils for imaging SC in rat or mouse can be fabricated large enough from 18 or 22 G soft bare copper wires (CDA alloy 102, MWS Wire Industries, Westlate Village, CA). The wire pieces are shaped to form a "rooftop" shape to provide proper rf excitation on the SC and soldered to ceramic chip capacitor(s) (American Technical Ceramics, Huntington Station, NY) to establish a resonance circuit. Depending on the configuration and magnetic field strength, the coil is constructed from a single capacitor of 75 pF (12) or two or more distributed capacitors (8.2, 12, or 16 pF) (8, 9). The coil is coated with biologically inert silicone elastomer (Factor II, Inc., Lakeside, AZ) for electrical isolation from its surrounding tissue. The coil is inductively coupled to an external pickup coil—a simple rectangular-shaped wire loop or a volume coil. The coil combination is then tuned at the desired frequency and impedance matched for proper operation.

2.4. Monitoring and Gating System

It is critical to use a monitoring and gating system while scanning live animal. The system should be MR compatible meaning its parts must be nonmagnetic. There are several commercial companies that manufacture such systems. The system is assembled with various leads which are properly connected to or positioned on the animal to sense various physiological signs, such as electrocardiogram (ECG), temperature, respiratory, and/or blood flow in tail. The unit displays these measurable signals in the computer screen and also provides a gating interface to the scanner for timing the acquisition. Combining the respiratory and ECG signals to gate the scanner allows acquiring high quality SC images free from artifacts due to breathing or heart motion.

It is important to maintain the body temperature of the scanned animal at about 37°C. This requires circulating warm air in the bore of the scanner or placing the animal in a cradle with capability to circulate warm water. Some monitoring systems also include components to allow such functionalities.

2.5. Exogenous Contrast Agents

In MRI, hydrogen is imaged as it interacts with its surrounding, which produces differences in relaxation times in a tissue-dependent fashion. Endogenous contrast depicted in an MR image may represent variations in anatomy, structure, function, or metabolic activity. To improve the visibility and hence diagnosis of certain aspects of the underlying tissues, additional contrast can be introduced by using injectable agents manufactured with paramagnetic ions. When present, these exogenous agents alter the longitudinal and transverse relaxation times T1 and T2 of the tissue and result in a higher or lower signal levels depending on the mechanism of its action. Currently, gadolinium (Gd), iron oxide, and manganese (Mn)-based MR contrast agents are available in various chemical formulations for clinical or experimental applications.

The same contrast agents are also used in experimental SCI research. The utility of gadolinium-based agent in investigating injured SC was demonstrated in quantitatively mapping the changes in blood–spinal cord barrier (BSCB) permeability (13–16). Manganese ion was used for functional imaging (17) and visualizing fiber connectivity in corticospinal tract (CST) (18, 19). Iron oxide particles were used for tracking stem cells implanted at the lesion site (20).

3. Methods

3.1. Anesthesia

The animals can be sedated under the influence of different anesthetic agents for surgical procedures described below. During the scan, moving the animal in and out of the scanner for extending the anesthesia effect may result in losing the spatial reference and therefore special care needs to be exercised not to do so. Delivering isoflurane gas anesthesia to the animal through a nose mask or intubation helps keeping the animal sleep throughout the scan without any need to take the animal out of the scanner provided that the animal's well-being is assessed using a monitoring system. For MRI procedures, either rat or mouse is subjected to an initial anesthesia using an induction chamber (5% isoflurane) and later maintained with a mixture (1.5% isoflurane, 30% oxygen and air), but adjusted slightly to keep the animal alive.

3.2. Surgical Procedures and Inducing Contusion-Type SCI

Spinal cord of rat or mouse is injured in a sterile surgical environment. Back hair of the animal is shaved off in a rectangular area from the vertebral levels C7 to L2. An incision centering over the spinous processes of vertebral body is made from the levels T4 to L1. A freer elevator is used to bluntly dissect the paraspinal musculature to expose the posterior elements of the spine. Then, a rongeur is used to perform a laminectomy at the thoracic level of interest to expose the SC while leaving the dura intact.

The laminectomized animal is moved over the support plate of the injury device, described above. The exposed posterior surface of the SC is positioned under the injury tip. The animal's vertebral body is stabilized by two hemostatic forceps by attaching the vertebral projections at the rostral and caudal sides of the laminectomy. Injury was produced by using an injury bit with its tip properly selected for rat or mouse applications. For injuring mouse SC, a 1-mm diameter injury bit with rounded tip parameters; 0.75 m/s for the velocity of the bit motion during the impact, 0.5 mm for the depth of tissue deformation on the dorsal surface of the cord, and 85 ms for the duration of the compression (21, 22). For injuring rat SC, a rectangular (1 mm × 2 mm) injury tip and the following parameters are used: deformation depth = 0.5 mm for producing mild injury and 1.5 mm for producing severe injury, impact velocity = 1.5 m/s, and duration = 80 ms (7).

3.3. Coil Implantation

The procedure for coil implantation is described in refs. (9, 10). The coil is positioned to center the injured section of the SC. The spinous processes below the coil edges are removed in order to stabilize the coil and to support the coil to lie flat against the rib cage. The coil is temporarily held in place with two sutures through the overlying muscles, and fine tuning is achieved by slightly bending the coil. Once proper tuning is achieved, the fascia of the overlying paraspinal muscles are sutured at the midline to hold the coil in place, and the skin is closed with a water tight seal. The animals were then placed in a heated cage to maintain the body temperature while recovering.

3.4. MRI Scan Procedures and Parameters

The injured animal is anesthetized by using isoflurane anesthesia delivered via a nose mask or intubation, as described in Sect. 3.1. The sedated animal is placed supine such that the implanted coil lies against the external coil mounted on a Plexiglas cradle (9). If inductive-overcoupling configuration is preferred, the animal is placed in a volume coil (10, 11). Once the coil system is tuned and matched, the cradle is inserted into the scanner and tuning and matching is rechecked again for any drift. Scout images are then acquired rapidly to confirm if the lesion is positioned at the isocenter of the magnet.

MR allows multimodal imaging. In each modality, data acquisition is governed by a specific imaging sequence that is associated with certain set of parameters. Spin echo and gradient echo are two fundamental sequences and others can be derived from these two by means of temporally manipulating the rf excitation/reception and gradient coil activities. Typical acquisition parameters are repetition time (TR), echo time (TE), field-of-view (FOV), acquisition matrix, slice thickness, and number of experiments or averaging (NEX). FOV and acquisition matrix together define the spatial dimension of each pixel. Combining this information with slice thickness describes the voxel size.

In the following, we will cover the flowing modalities for:

(a) Anatomical imaging.
(b) Magnetization transfer imaging (MTI).
(c) Angiography.
(d) Dynamic gadalonium-enhanced MRI.
(e) Manganese-enhanced MRI (MEI).
(f) Diffusion tensor imaging (DTI).

3.4.1. Quantitative Characterization of the MR Properties of Rodent SC

The intrinsic MR properties, such as proton density and longitudinal (T1) and transverse (T2) relaxation times, determine the mechanism that is responsible for the endogenous contrast seen between gray and white matters on the anatomical images of SC. The proton density remains independent but the relaxation times change with the magnetic field strength. These properties were spatially mapped and the measured values at 7 and 9.4 T were reported in refs. (8, 23). The knowledge of these parameters at the field strength of interest is essential for optimizing the imaging protocols in normal SC, establishing sensitivity and specificity of the MR parameters to the pathological changes in the injured SCs, and also describing the signal enhancement features achieved with the use of contrast agents. The basic MR properties of SC should be taken into account when planning scans at high magnetic fields and the imaging parameters should be manipulated accordingly to improve the image quality. Since, the paramagnetic contrast agents, including Gd and Mn, shorten both relaxation times, careful attention must be paid to optimize the acquisition in both pre- and postcontrast conditions.

3.4.2. Anatomical Magnetic Resonance Imaging

Conventional anatomical images acquired with optimized protocols exhibit superb soft tissue contrast, enabling in vivo visualization of the location and extent of the neuropathology with exquisite details, as shown in Fig. 2. For axial spin echo imaging

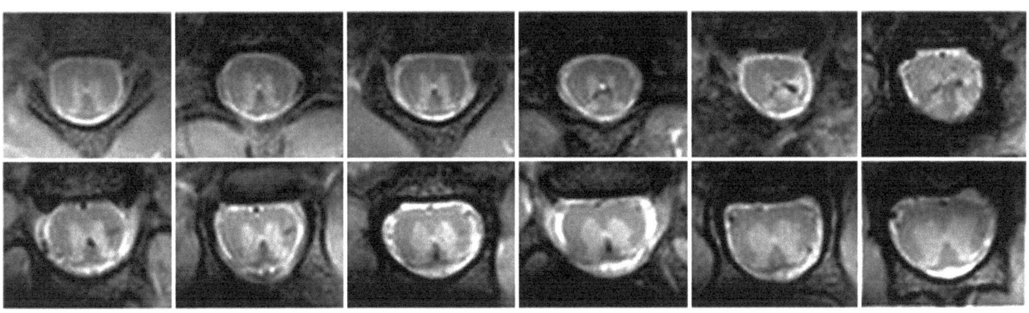

Fig. 2. Serial in vivo spin-echo images of an injured rat spinal cord acquired on postinjury day 28 with implantable coil.

of rat SC, the acquisition parameters are $T_R/T_E = 2{,}500/12$ ms, FOV = 20 mm, image matrix = 128 × 128, slice thickness = 1 mm, and NEX = 2. Similar parameter values can also be used for mouse imaging.

3.4.3. Magnetization Transfer Imaging

Magnetization transfer provides additional degree of contrast in MRI and addresses the need for noninvasively detecting demyelination or myelin loss occurring as a consequence of SCI (24). MT depends on the existence of two separate pools of water in tissues: free and bound water. The "free water" is highly mobile and MR visible. MRI reflects the distribution of this water pool. The bound water is generally not visible. However, it is possible to indirectly detect the bound water using MT technique where an rf pulse with a narrow bandwidth with a few kHz frequency offset from the resonance frequency (off resonance pulse) is applied to saturate the protons associated with the bound water pool. Magnetization from the bound water pool is transferred to the free water pool that results in the attenuation of free water MR signal. The degree of attenuation is proportional to the concentration of the bound water molecules. Because of the frequency offset, rf pulse does not directly affect the free water signal. Since the majority of bound water is associated with myelin, the degree of MT, generally expressed as the magnetization transfer ratio, can be used as a measure of the myelin integrity in injured SC (25, 26). Axial MT images can be acquired using spin echo sequence with and without the application of the MT pulse of duration = 1.5 s and offset frequency = 5 kHz using the other parameters as in Sect. 3.4.2 (9).

3.4.4. Magnetic Resonance Angiography

Magnetic resonance angiography (MRA) offers means to image vascular system noninvasively. Vascular anatomy of the rodent spine and SC exhibits close similarity to that of human (27, 28). Vascular organization in spine consists of a complex network of arteries and veins, respectively supplying and draining blood to and from the cord at multiple directions and levels. Angiograms are acquired using the time of flight technique without using exogenous contrast agents. Typical imaging parameters are $T_R/T_E = 45/4$ ms and flip angle (FA) = 45°, and applied over a volume of 40 × 20 × 20 mm^3, where the former represents the readout direction along the cord. The raw data is sampled at a rate of 256 × 128 × 128. A selective background suppression centered on the second phase encode direction is applied over a slab thickness of 16 mm. The angiogram of an injured rat SC acquired with these parameters on postinjury day 3 is depicted in Fig. 3. The vasculature is seen absent at the injury site. Lack of vasculature has important implications in recovery from SCI considering the necessity of blood supply for the survival and repair of the injured neuronal tissue.

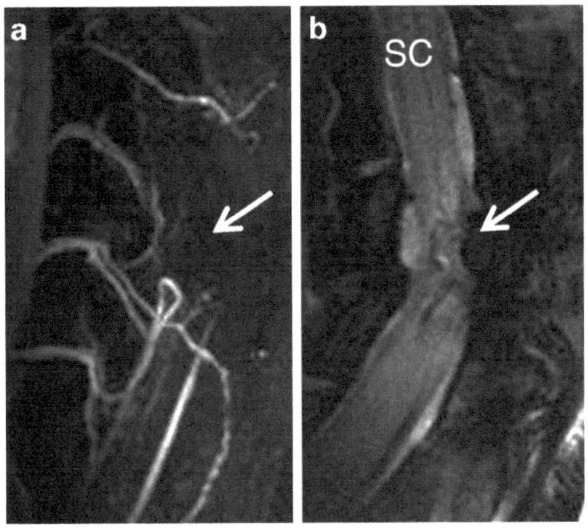

Fig. 3. (**a**) Sagittal angiogram of an injured spinal cord (SC) on postinjury day 3 and (**b**) anatomical T2-weighted reference image. *Arrows* point to the injury. Please note the lack of vascular filling at the injury site (*left arrow*).

3.4.5. Dynamic-Contrast Enhanced Magnetic Resonance Imaging

Regional damage to the neurovasculature is one of the earliest pathological events following trauma to the SC. The integrity of the blood spinal cord barrier (BSCB) can be noninvasively probed using dynamic contrast-enhanced magnetic resonance imaging (DCE-MRI) (13, 15). Gadopentetate dimeglumine (Gd, Magnevist, Berlix, Wayne, NJ) is one of the complexes used clinically as a contrast agent. This relatively large molecule does not cross the intact BSCB and is confined to the vasculature in normal SC. If the barrier is compromised, the contrast agent escapes the capillary and this region appears bright on the T1-weighted image. The BSCB leakiness can be quantified by estimating the barrier permeability by acquiring images repetitively following the administration of Gd and using appropriate pharmacokinetic modeling (14, 16). After this quantification, the spatial and temporal variation of the BSCB permeability in the injured cord can be mapped. The leakiness of the BSCB can also be exploited to detect if any pharmacological treatment is capable of closing or preventing the vessels from being broken further. Another important feature of the Gd-based contrast agents is that they are extracted by the kidneys within 3–4 h of injection. This allows DCE-MRI scans performed daily.

During the scan, T1-weighted image is first acquired using spin-echo sequence with ($T_E/T_R = 13/1{,}000$ ms) and establishes the precontrast data set. The imaging planes and other parameter values are the same as those in Sect. 3.4.2. Then, bolus Gd is administrated intravenously at a dose of 0.1 mmol/kg (<5 s) through the catheterized jugular vein. Immediately after, the post-contrast T1-weighted images are acquired repetitively for up to 1 h. This temporal data is processed to quantitatively assess the BSCB permeability changes using transfer rate constant from

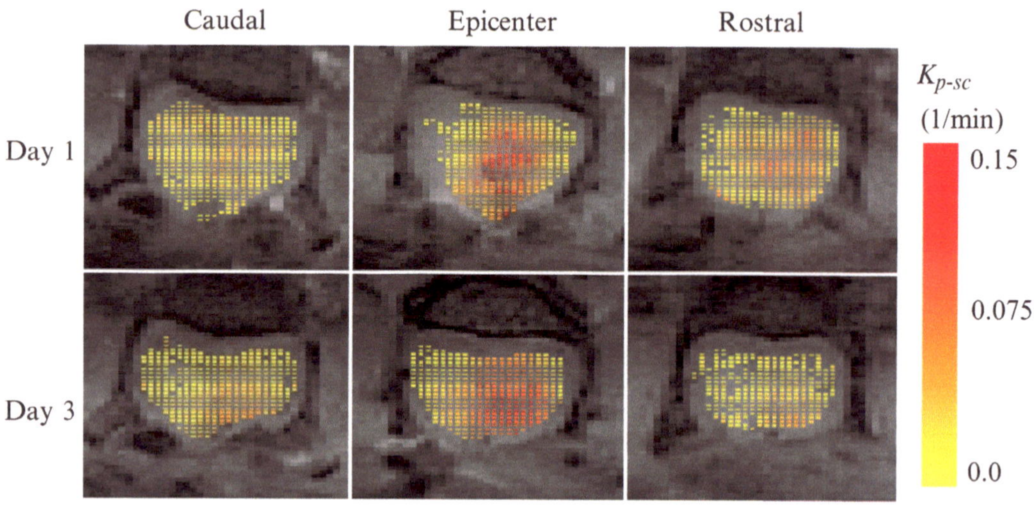

Fig. 4. Color coded blood–spinal cord barrier permeability ($K_{p\text{-}sc}$)-maps at the epicenter of SCI in mouse. The backgrounds are the T1-weighted precontrast images. The color from *yellow* to *red* indicates linearly increasing compromise in barrier function. The *areas with color* towards *red* indicate regions with more vascular damage causing the barriers become more permeable to blood constituents. Modified from ref. (16).

plasma-to-spinal cord in a pharmacokinetic model (16). Figure 4 shows quantitative distribution of BSCB permeability. The permeability estimates are overlaid as a new layer with color coding on the precontrast T1-weighted images in the background. In the maps, color coding towards red means increased BSCB dysfunction. The red zones represent regions with greater leakage of the contrast agent. These current acquisition protocols and procedures provide measurable parameters that can serve as sensitive and specific in vivo neurovascular biomarkers for comprehensively evaluating the vascular remodeling, and may offer prognostic value for functional recovery from SCI.

3.4.6. Manganese-Enhanced MRI

Manganese (Mn) is an essential trace metal required for the development and function of the brain. It has an ionic radius similar to that of calcium-ion, and can enter cells through calcium pathways, such as voltage-gated calcium channels and is confined to the intracellular compartment (29). Mn remains compartmentalized after exogenous administration, and can be used for delineating targeted tissue elements (30). In addition, divalent manganese ion (Mn^{++}) is paramagnetic, affecting the longitudinal relaxation rate of surrounding water protons, and therefore acts as an excellent MRI contrast agent (31, 32). The broad range of the applications of MEI in brain and cardiac studies has been described in a special issue of NMR in Biomedicine journal (33). By exploiting the paramagnetic nature of Mn, MEI can be used as an alternative method for anterograde neuronal tract tracing and functional imaging in injured SC (17). Mn enhancement in specific neuronal pathways is known to be produced by natural stimuli, such as olfactory stimulation or voluntary actions such as singing in song birds. Based on these, it has been hypothesized that increased motor activity associated with electrical stimulation of the motor cortex might amplify Mn contrast uptake and transport in CST, and thereby lead to better delineation of this tract with respect to the background and the neighboring tissue on Mn-enhanced images. Electrical stimulation of motor cortex was shown to be highly effective in labeling the CST not only in normal cords, but also in injured cords below the site of the lesion (19, 34). Since, one of the most important consequences of SCI is damage to the neural network, MEI can be used to trace the CST in live rodents and to detect fiber connectivity in injured SC (18).

Mn delivery involves a setup consisting of a micropositioner and Hamilton syringe, both attached to a stereotaxic frame, to slowly administer Mn into the target tissue using a pulled glass pipette. The animal is anesthetized by intramuscular delivery of ketamine hydrochloride to leave the excitability of the cortical motor system intact. The head of the anesthetized animal is fixed in a stereotaxic frame (Kopf Instruments, Tujunga, CA) and craniotomy is performed to open two bilateral burr holes with 1.0 mm diameter for accessing the motor cortex. An additional hole is opened to secure a titanium screw serving as a reference electrode for electrical stimulation.

Using 1 M $MnCl_2$ in a 1 μl Hamilton syringe with a tapered, graduated micropipette tip, a total of 0.2 μl of this solution is focally delivered by a direct stereotaxical injection at 0.5 mm below the surface of the cortex through the center of each burr hole. The solution is injected slowly over a period of 5 min to prevent backflow. Following the Mn delivery, the cortex is electrically stimulated with a 1 mm diameter stainless steel electrode and the screw serving as the reference. The stimulus involves trains of biphasic pulses (each phase 0.2 ms, negative first) applied at 100 Hz for an "on" period of 5 s followed by an "off" period of 5 s. The stimulus current is adjusted until an evoked visible motor response is produced in the forelimb, hindlimb or tail, and applied for 90 min. If the strength of the evoked movement declined noticeably, frequency and current values adjusted to maintain a constant motor response. These procedures were then repeated for the opposite cortex through the second burr hole. After the stimulation, the electrodes are removed, and the skin was sutured tightly. The animal is then left to recover in its cage.

After 24 h of recovery, animal is anesthetized with isoflurane for MEI scan. T_1-weighted volumetric images covering the brain and SC at the cervical and thoracic levels are acquired using a 3D gradient echo sequence ($T_R/T_E = 45/4$ ms and FA = 45°). The data is sampled on a matrix = 128 × 128 × 64 ranging over a volume of 70 × 32 × 32 mm^3, and processed and interpolated to

256 × 256 × 128 pixels for the final display. Maximum intensity projections are generated to delineate the Mn enhancement in the CST over a desired thickness in axial or sagittal planes. Inversion recovery spin echo (IR-SE) acquisition provides richer contrast enhancement. In sagittal view, IR-SE image can optimally be acquired using $T_R/T_E/T_I = 2{,}000/12/550$ ms, FOV = 70 × 32 mm^2, image matrix = 256 × 128, slice thickness = 2 mm, and NEX = 4. The corresponding parameters for the axial image are $T_R/T_E/T_I = 2{,}000/17/550$ ms, FOV = 22 × 22 mm^2, image matrix = 128 × 128, slice thickness = 2 mm, and NEX = 4.

If high resolution data from excised cord is desired, the multislice spin echo images are acquired in axial planes using $T_R/T_E = 2{,}500/12$ ms, FOV = 10 × 10 mm^2, image matrix = 128 × 128, slice thickness = 2 mm, and NEX = 2. The corresponding parameters for axial IR-MEI are $T_R/T_E/T_I = 2{,}000/15/550$ ms, FOV = 10 × 10 mm^2, image matrix = 128 × 128, slice thickness = 2 mm, and NEX = 4. Figure 5 shows images acquired with these protocols from an injured rat. The Mn-labeling of the CST at sections above and below the injury together with the partial Mn-labeling seen at the epicenter suggests the presence of connected corticospinal fibers across the injury (19).

3.4.7. Diffusion Tensor Imaging

DTI provides additional image contrast by exploiting the perpetual and random motion of water molecules in biological tissues (35). In the absence of boundaries, water diffusion is isotropic and the unrestricted motion is characterized by a single parameter, generally referred to as apparent diffusion coefficient. In organized or semi-organized structures, such as axons of the white matter in SC, however, water molecules diffuses in preferential directions, i.e., faster along the axons as compared to the transverse direction. This directionality is referred to as anisotropy and contributed by axonal structure and myelin sheeting. The degree of anisotropy reflects axonal integrity that is not evident on conventional MRI. Anisotropic diffusion can be visualized by an ellipsoid that is characterized theoretically by a second-order tensor. This tensor describes a second moment (covariance) in conditional probability distribution of finding a water molecule at a new position that is different from its initial position after some time. In this view, the ellipsoid essentially represents the isosurface of a multivariate normal probability distribution for the anisotropic water diffusivity. The DT is defined by a symmetric and positive definite matrix with elements satisfying specific conditions of reciprocity requirement and physical principle of reversibility (36, 37). For a given voxel, each element of DT is estimated from diffusion-weighted magnetic resonance images using relationship between the measured signal magnitudes and the strength and orientation of the applied diffusion gradients. A typical DTI scan involves an initial baseline image acquired without applying diffusion gradients,

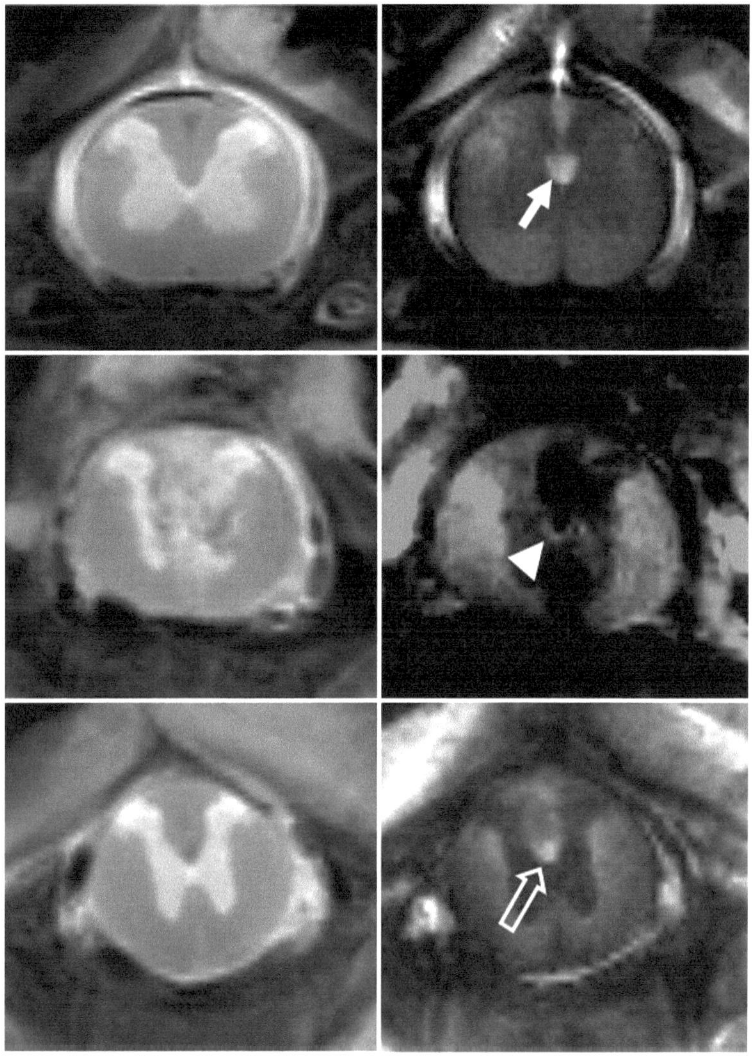

Fig. 5. *Left column*: anatomical axial spine echo images from three slice locations (top-rostral to injury, middle-injury epicenter and bottom-caudal to injury). *Right column*: the corresponding MEIs. *Arrow* points to the Mn-enhanced contrast in corticospinal tract. The partial signal enhancement, depicted by the *arrowhead*, is likely to represent a portion of the tract that is populated with intact fibers. By continuously projecting through the injury site, these intact fibers transport Mn from rostral to caudal sections, as indicated by the labeling in the tract below the injury (*open arrow*). Modified from ref. (19).

which is followed by the acquisition of at least six diffusion-weighted images with diffusion sensitizing gradients applied in six independent directions (110,101,011,-110,-101,0-11). Early acquisition protocol was developed using spin-echo sequence modified to have diffusion gradient pulses placed symmetric around the 180° rf pulse. Later, echo planar imaging (EPI) was implemented to reduce the total acquisition time. In DTI of mouse SC, diffusion weighting can be achieved using diffusion gradient strength = 80 mT/m, width (δ) = 6.5 ms, and separation (Δ) = 11 ms to produce a *b*-value of 342 s/mm^2 (22). The other imaging parameters can be set to T_R/T_E = 2,000/26 ms, FOV = 12 mm × 8 mm, acquisition matrix = 64 × 64, slice thickness = 2 mm, and NEX = 1. When multishot, multislice, blipped EPI is employed for the rat SC, the corresponding parameters are set to number of shots = 8, T_R/T_E = 2,000/40.2 ms, NEX = 4, FOV = 25 mm × 25 mm, image matrix = 128 × 128, and slice thickness = 1 mm while δ = 12 ms and Δ = 18.9 ms (38). With these settings, when the diffusion gradient strength is 200 mT/m, the *b*-value can attain as high as 6,200 s/mm^2.

DTI contains a large amount of multivariate data. Visualization of this information in an orientation-independent manner is a challenging problem. Various indices are commonly used as a way of reducing the DT data into simple scalar quantities that could be displayed as an image in a straightforward manner. This reduction is centered on the three eigenvalues (λ_1, λ_2, and λ_3) of the DT. The eigenvalues quantify the mean diffusion distances along the three principal directions as specified by the corresponding orthogonal eigenvectors. Each biologic tissue has a unique diffusivity pattern, and any sign of deviation from this pattern indicates abnormality that can be used for detecting diseases, including neuro-degenerative processes in the central nervous system. The principal diffusivity, i.e., the largest eigenvalue (λ_1), is aligned with the polar axis of the diffusion ellipsoid or the corresponding eigenvector. For describing the diffusion in neuronal fibers of SC, $\lambda_1 = \lambda_{||}$ is denoted as the longitudinal diffusivity of water molecules along the axonal fibers, and $\lambda_\perp = (\lambda_2 + \lambda_3)/2$ is an expression of the transverse diffusivity of the molecules moving perpendicular to the direction of the fibers. In injured SC, reduction in $\lambda_{||}$ indicates compromise in axonal integrity, and increase in $\lambda_\perp$ suggests axonal swelling or myelin loss (39, 40). One of the commonly used scalar invariant quantities to characterize diffusion in tissues is the fractional anisotropy (FA). The FA index was computed by algebraically combining the eigenvalues. FA receives a value closer to 0 for fully isotropic diffusion as in CSF, and to 1 for high degree of anisotropy as in white matter. Scalar quantities to represent the complete tensorial information are limited, and techniques were developed for visualizing the anisotropic diffusion in 3D using Mohr Diagram method (36, 37).

As an extension of DTI, fiber tractography was developed to provide quantitative information to visualize the architectural features and to study the connectivity and continuity of neural pathways in coherently organized nervous system in vivo (41). Based on specific acquisition schemes, the principal eigenvector of DT represents local fiber orientation allowing for the reconstruction of fiber bundles and tracts. This technique allows improving our understanding of the fiber reorganization and spatial temporal dynamics of the fiber connectivity during the course of postinjury in SC.

4. Notes

In longitudinal studies, the frequency response characteristics of the implantable coil with fixed electrical elements drifts with time due to the changes in the tissue environment surrounding the coil at the implantation site (11). Implementation of the inductive overcoupling principle provides a compensation mechanism for this drift. When used, flexibility can be achieved in externally tuning the resonant frequency and matching the impedance of the implanted coil remotely by using a volume coil. Such ability allows reliable and repeatable acquisition of MR data, which is highly desirable when studying the progression of the neuropathology in injured SC over time on the same animal.

Although the inductively overcoupled coil system is capable of providing high quality images, it also poses a limitation because of the physics involved in the MR data acquisition over small FOV. The images acquired inherently contain wraparound signals from the tissue outside the selected FOV because of excitation by the volume coil. Due to wraparound, separating signals between the regions of interest and background noise becomes difficult, even after adjusting the contrast and brightness on the display. Strong intensity artifacts can be caused by the motion of heart, but other motionless regions appear significantly low in intensity. Nevertheless, employing respiratory and ECG gating minimizes these artifacts.

Evolution of neuroinflammatory responses in injured rat and mouse SC exhibits distinctly different spatiotemporal courses. Multiparametric MRI modalities provide vast amount of information from the injury site but comprehensive and robust interpretation of the underlying morphology and neuropathology is still lacking (5, 6, 21, 22, 28). Further correlative studies are required to correctly relate what is observed on the images to the readings from postmortem histology with different stains and immunohistochemistry. Efforts are also needed for discovering special biomarkers and defining their specificity and sensitivity in identifying the different pathological features of the injury.

In principle, both MEI and DTI can provide data to detect the axonal fibers. Comparing the techniques, each has its own advantages and disadvantages. The MEI is simple and straightforward to implement, but is invasive since it needs Mn injection. Besides, there is always an issue of toxicity with Mn. On the other hand, the DTI is completely noninvasive and can be applied repetitively. But, it is susceptible to image artifacts especially when EPI is employed for the acquisition and also computationally more intensive.

However, the use of both techniques is justified to explore if MEI can play an in vivo gold standard role for testing the accuracy of the DTI-based measurements (19).

Concerning the use of Mn, we note that this agent is not extracted from the body as rapidly as Gd. This limits Mn injection only once for the tract tracing purposes once its toxicity is taken into account. Since both Gd and Mn are T1-shortening MR contrast agents, their simultaneous use in the same subject would produce intensity enhancement that may be difficult to specify its origin, i.e., whether the observed enhancement is from Mn or Gd. For this reason, the experimental protocol should carefully be crafted to address this specificity issue when both agents are involved in the study.

The selection of the SCI model has important translational implications (6). Severe injuries exhibit characteristics that match well with those seen in clinics, but also possess extensive neuronal tissue damage or loss, which makes the quantitative characterization of the lesion size from the end-point MRI data difficult (42). Current preclinical efforts are focused on developing drugs for improving neuroprotection by slowing down the neurodegeneration processes and promoting neuronal recovery. The mild- or moderate injury models may be a good candidate to test the efficacy of these drugs in the pipeline. If the goal is to understand the role of specific genes in the repair and recovery of the SC from an injury, then these models may also be preferable as test system to characterize the injury response in rat or mouse. Again, the selection of the SCI model is critical in the sense that it should have the capacity to yield quantifiable lesion dimensions when evaluated using MRI. Mild- or moderate SCI models seem to be a good candidate for this purpose and we suggest their use for SCI drug screening purposes in preclinical translational research (43).

References

1. Akhtar AZ, Pippin JJ, Sandusky CB (2008) Animal models in spinal cord injury: a review. Rev Neurosci 19:47–60
2. Blesch A, Tuszynski MH (2009) Spinal cord injury: plasticity, regeneration and the challenge of translational drug development. Trends Neurosci 32:41–47
3. Beckmann N, Mueggler T, Allegrini PR, Laurent D, Rudin M (2001) From anatomy to the target: contributions of magnetic resonance imaging to preclinical pharmaceutical research. Anat Rec 265:85–100
4. Demaerel P (2006) Magnetic resonance imaging of spinal cord trauma: a pictorial essay. Neuroradiology 48:223–232
5. Bilgen M, Rumboldt Z (2008) Neuronal and vascular biomarkers in syringomyelia: investigations using longitudinal MRI. Biomark Med 2:113–124
6. Chou PC, Tatar I, Bilgen M (2008) Multimodal MRI evaluation of acute mild-contusive injury in mouse spinal cord. Neuroanatomy 7:83–92
7. Bilgen M (2005) A new device for experimental modeling of central nervous system injuries. Neurorehabil Neural Repair 19:219–226
8. Bilgen M, Elshafiey I, Narayana PA (2001) In vivo magnetic resonance microscopy of rat spinal cord at 7 T using implantable RF coils. Magn Reson Med 46:1250–1253
9. Bilgen M (2004) Simple, low-cost multipurpose RF coil for MR microscopy at 9.4 T. Magn Reson Med 52:937–940
10. Bilgen M (2007) Magnetic resonance microscopy of spinal cord injury in mouse using a

miniaturized implantable RF coil. J Neurosci Methods 159:93–97

11. Bilgen M (2006) Inductively-overcoupled coil design for high resolution magnetic resonance imaging. Biomed Eng Online 5:3
12. Bilgen M, Abbe R, Liu SJ, Narayana PA (2000) Spatial and temporal evolution of hemorrhage in the hyperacute phase of experimental spinal cord injury: in vivo magnetic resonance imaging. Magn Reson Med 43:594–600
13. Bilgen M, Abbe R, Narayana PA (2001) Dynamic contrast-enhanced MRI of experimental spinal cord injury: in vivo serial studies. Magn Reson Med 45:614–622
14. Bilgen M, Narayana PA (2001) A pharmacokinetic model for quantitative evaluation of spinal cord injury with dynamic contrast-enhanced magnetic resonance imaging. Magn Reson Med 46:1099–1106
15. Bilgen M, Dogan B, Narayana P (2002) In vivo assessment of blood-spinal cord barrier permeability: serial dynamic contrast enhanced MRI of spinal cord injury. Magn Reson Imaging 20:337
16. Tatar I, Chou PC, Desouki MM, El Sayed H, Bilgen M (2009) Evaluating regional blood spinal cord barrier dysfunction following spinal cord injury using longitudinal dynamic contrast-enhanced MRI. BMC Med Imaging 9:10
17. Bilgen M, Dancause N, Al-Hafez B, He YY, Malone TM (2005) Manganese-enhanced MRI of rat spinal cord injury. Magn Reson Imaging 23:829–832
18. Bilgen M, Peng W, Al-Hafez B, Dancause N, He YY, Cheney P (2006) Electrical stimulation of cortex improves corticospinal tract tracing in rat spinal cord using manganese-enhanced MRI. J Neurosci Methods 156:17–22
19. Bilgen M (2006) Imaging corticospinal tract connectivity in injured rat spinal cord using manganese-enhanced MRI. BMC Med Imaging 6:15
20. Jendelova P, Herynek V, Urdzikova L, Glogarova K, Kroupova J, Andersson B, Bryja V, Burian M, Hajek M, Sykova E (2004) Magnetic resonance tracking of transplanted bone marrow and embryonic stem cells labeled by iron oxide nanoparticles in rat brain and spinal cord. J Neurosci Res 76:232–243
21. Bilgen M, Al-Hafez B, Alrefae T, He YY, Smirnova IV, Aldur MM, Festoff BW (2007) Longitudinal magnetic resonance imaging of spinal cord injury in mouse: changes in signal patterns associated with the inflammatory response. Magn Reson Imaging 25:657–664
22. Bilgen M, Al-Hafez B, Berman NE, Festoff BW (2005) Magnetic resonance imaging of mouse spinal cord. Magn Reson Med 54:1226–1231
23. Bilgen M, Al-Hafez B, Malone TM, Smirnova IV (2005) Ex vivo magnetic resonance imaging of rat spinal cord at 9.4 T. Magn Reson Imaging 23:601–605
24. Wolff SD, Balaban RS (1994) Magnetization transfer imaging: practical aspects and clinical applications. Radiology 192:593–599
25. Gareau PJ, Weaver LC, Dekaban GA (2001) In vivo magnetization transfer measurements of experimental spinal cord injury in the rat. Magn Reson Med 45:159–163
26. McGowan JC, Berman JI, Ford JC, Lavi E, Hackney DB (2000) Characterization of experimental spinal cord injury with magnetization transfer ratio histograms. J Magn Reson Imaging 12:247–254
27. Bilgen M, Al-Hafez B, He YY, Brooks WM (2005) Magnetic resonance angiography of rat spinal cord at 9.4 T: a feasibility study. Magn Reson Med 53:1459–1461
28. Bilgen M, Al-Hafez B (2006) Comparison of spinal vasculature in mouse and rat: investigations using MR angiography. Neuroanatomy 5:12–18
29. Aschner M, Vrana KE, Zheng W (1999) Manganese uptake and distribution in the central nervous system (CNS). Neurotoxicology 20:173–180
30. Sloot WN, Gramsbergen JB (1994) Axonal transport of manganese and its relevance to selective neurotoxicity in the rat basal ganglia. Brain Res 657:124–132
31. Kang YS, Gore JC (1984) Studies of tissue NMR relaxation enhancement by manganese. Dose and time dependences. Invest Radiol 19:399–407
32. Mendonca-Dias MH, Gaggelli E, Lauterbur PC (1983) Paramagnetic contrast agents in nuclear magnetic resonance medical imaging. Semin Nucl Med 13:364–376
33. Koretsky AP, Silva AC (2004) Manganese-enhanced magnetic resonance imaging (MEMRI). NMR Biomed 17:527–531
34. Huang ZL, He YY, Bilgen M (2007) Corticospinal tract tracing in mouse using manganese-enhanced contrast. J Neurol Sci 10:38–44; http://www.jns.dergisi.org
35. Le Bihan D, Mangin JF, Poupon C, Clark CA, Pappata S, Molko N, Chabriat H (2001) Diffusion tensor imaging: concepts and applications. J Magn Reson Imaging 13:534–546
36. Bilgen M, Elshafiey I, Narayana PA (2002) Mohr diagram representation of anisotropic diffusion tensor in MRI. Magn Reson Med 47:823–827
37. Bilgen M, Narayana PA (2003) Mohr diagram interpretation of anisotropic diffusion indices in MRI. Magn Reson Imaging 21:567–572

38. Elshafiey I, Bilgen M, He R, Narayana PA (2002) In vivo diffusion tensor imaging of rat spinal cord at 7 T. Magn Reson Imaging 20:243–247
39. Gaviria M, Bonny JM, Haton H, Jean B, Teigell M, Renou JP, Privat A (2006) Time course of acute phase in mouse spinal cord injury monitored by ex vivo quantitative MRI. Neurobiol Dis 22:694–701
40. Kim JH, Budde MD, Liang HF, Klein RS, Russell JH, Cross AH, Song SK (2006) Detecting axon damage in spinal cord from a mouse model of multiple sclerosis. Neurobiol Dis 21:626–632
41. Mori S, Kaufmann WE, Davatzikos C, Stieltjes B, Amodei L, Fredericksen K, Pearlson GD, Melhem ER, Solaiyappan M, Raymond GV, Moser HW, van Zijl PC (2002) Imaging cortical association tracts in the human brain using diffusion-tensor-based axonal tracking. Magn Reson Med 47:215–223
42. Chou P, Gunsoy ML, Desouki MM, Shunmugavel A, Khan M, Sayed HE, Nguyen S, Singh I, Bilgen M (2009) Preclinical screening of a potential drug treatment for spinal cord injury using magnetic resonance imaging. In: 17th Annual scientific meeting and exhibition of ISMRM. Honolulu, HI, 20–25 April 20–25 2009
43. Chou P, Shunmugavel A, Sayed HE, Desouki MM, Nguyen SA, Khan M, Singh I, Bilgen M (2011) Preclinical use of longitudinal magnetic resonance imaging for screening the efficacy of S-nitrosoglutathione in treating spinal cord injury. J Magn Reson Imag 33:1301–1311

INDEX

A

Jun Chen et al. (eds.), *Animal Models of Acute Neurological Injuries II*, Springer Protocols Handbooks,
DOI 10.1007/978-1-61779-782-8, © Springer Science+Business Media, LLC 2012

D

F

M

N

Q

R

T

X

MIX
Papier aus verantwortungsvollen Quellen
Paper from responsible sources
FSC® C105338

If you have any concerns about our products,
you can contact us on
ProductSafety@springernature.com

In case Publisher is established outside the EU,
the EU authorized representative is:
Springer Nature Customer Service Center GmbH
Europaplatz 3, 69115 Heidelberg, Germany

Printed by Libri Plureos GmbH
in Hamburg, Germany